D0081130

FINITE MATHEMATICS

Also Available from McGraw-Hill

SCHAUM'S OUTLINE SERIES IN MATHEMATICS & STATISTICS

Most outlines include basic theory, definitions and hundreds of example problems solved in step-by-step detail, and supplementary problems with answers.

Related Titles on the Current List Include:

Advanced Calculus
Advanced Mathematics for Engineers & Scientists
Analytic Geometry
Basic Mathematics for Electricity & Electronics
Basic Mathematics with Applications to Science & Technology
Beginning Calculus
Boolean Algebra & Switching Circuits
Calculus
Calculus for Business, Economics, & the Social Sciences
College Algebra
College Mathematics
Combinatorics
Complex Variables
Descriptive Geometry
Differential Equations
Differential Geometry
Discrete Mathematics
Elementary Algebra
Essential Computer Mathematics
Finite Differences & Difference
  Equations
Finite Mathematics
Fourier Analysis
General Topology
Geometry
Group Theory
Laplace Transforms
Linear Algebra
Mathematical Handbook of Formulas & Tables
Mathematical Methods for Business & Economics
Mathematics for Nurses
Matrix Operations
Modern Abstract Algebra
Numerical Analysis
Partial Differential Equations
Probability
Probability & Statistics
Real Variables
Review of Elementary Mathematics
Set Theory & Related Topics
Statistics
Technical Mathematics
Tensor Calculus
Trigonometry
Vector Analysis

SCHAUM'S SOLVED PROBLEMS SERIES

Each title in this series is a complete and expert source of solved problems with solutions worked out in step-by-step detail.

Titles on the Current List Include:

3000 Solved Problems in Calculus
2500 Solved Problems in College Algebra and Trigonometry
2500 Solved Problems in Differential Equations
2000 Solved Problems in Discrete Mathematics
3000 Solved Problems in Linear Algebra
2000 Solved Problems in Numerical Analysis
3000 Solved Problems in Precalculus

Available at most college bookstores, or for a complete list of titles and prices, write to: Schaum Division
McGraw-Hill, Inc.
1221 Avenue of the Americas
New York, NY 10020

FINITE MATHEMATICS

FOURTH EDITION

Daniel P. Maki
Professor of Mathematics
Indiana University

Maynard Thompson
Professor of Mathematics
Indiana University

McGraw-Hill, Inc.
New York St. Louis San Francisco Auckland Bogotá Caracas Lisbon
London Madrid Mexico City Milan Montreal New Delhi
San Juan Singapore Sydney Tokyo Toronto

FINITE MATHEMATICS

Copyright © 1996, 1989, 1983, 1978 by McGraw-Hill, Inc. All rights reserved. Printed in the United States of America. Except as permitted under the United States Copyright Act of 1976, no part of this publication may be reproduced or distributed in any form or by any means, or stored in a data base or retrieval system, without the prior written permission of the publisher.

This book is printed on acid-free paper.

6 7 8 9 10 DOC/DOC 0 9 8 7 6 5 4 3 2 1

ISBN 0-07-039763-5

This book was set in Times Roman by American Composition & Graphics, Inc.
The editors were Michael Johnson and Jack Maisel;
the production supervisor was Richard A. Ausburn.
The cover was designed by Circa '86.
The photo editor was Anne Manning.
R. R. Donnelley & Sons Company was printer and binder.

Part-Opening Photo Credits
Part 1: Fred Lyon/Photo Researchers
Parts 2 and 3: Comstock

Library of Congress Cataloging-in-Publication Data

Maki, Daniel P.
Finite mathematics / Daniel P. Maki, Maynard Thompson.—4th ed.
p. cm.
Includes index.
ISBN 0-07-039763-5
1. Mathematics I. Thompson, Maynard. II. Title.
QA39.2.M333 1996
510—dc20 94-48013

ABOUT THE AUTHORS

Daniel P. Maki received his B.S. in mathematics from Michigan Technological University and his M.S. and Ph.D. in mathematics from the University of Michigan. He has been on the faculty at Indiana University since 1966 and has also been a Fulbright Research Fellow at the University of Helsinki, and visiting Professor at the University of Michigan and Claremont Graduate School. Professor Maki served as a member of the Board of Governors of the Mathematical Association of America. His research interests include moment problems, orthogonal polynomials, signal processing, and algorithms for speech recognition by computer.

Maynard Thompson received his B.A. from DePauw University and his M.S. and Ph.D. from the University of Wisconsin. He joined the Indiana University faculty in 1962, and has held visiting positions at the University of Maryland, Georgia Tech, and at General Motors Research Laboratories. Professor Thompson chaired the CUPM Committee on Applied Mathematics, served as an MAA Visiting Lecturer, is a member of the MAA Consultant's Board, a member of the Consultant's Board for the Sloan Foundation program on The New Liberal Arts, a member of the Advisory Board for the Mathematical Competition in Modeling, and a member of the SIAM Education Committee. His primary research interests are in the applications of mathematics to problems arising in biology and medicine.

Contents

Preface

Probability and linear mathematics, the core of the traditional course in finite mathematics, provide some of the most basic and widely used mathematical tools in business and the social and life sciences. These core topics and their applications are presented in Parts I and II of this text; additional applications and related topics are introduced in Part III. Throughout the book there is an emphasis on ideas and techniques useful in solving problems.

You learn to solve problems by working problems. Therefore, we provide many exercises for you to use in developing and testing your problem-solving ability. Some are very easy, most are similar in difficulty to the discussions and the worked examples in the text, and a few are fairly tough. To solve problems, you must know where to start and how to proceed. We use discussions and examples to introduce and illustrate ideas and techniques to aid you in acquiring these skills. Some of the examples are straightforward computations, while others show you how to solve problems by combining several ideas and techniques. Examples are also used to illustrate the important method of breaking a problem into simpler problems, solving them one at a time, and then putting the results together to solve the original problem. Since we cannot provide examples of every type of problem you may encounter, we identify fundamental principles that should be helpful in unfamiliar situations.

Chapters 1 through 4 cover basic concepts in probability, and it is common for this material to constitute about one-half of a one-semester finite mathematics course. The other half of such a course is usually devoted to linear equations, matrices, and linear programming, and these topics are covered in

Chapters 5, 6, and 7. There are many applications in Chapters 1 through 7; these chapters can easily form a complete course, although we recommend that at least one chapter from Part III also be included. Chapter 8 is devoted to a topic that combines probability and matrices and serves as a useful mathematical model in many applications.

The fourth edition of *Finite Mathematics* contains improvements based on classroom teaching experience at Indiana University, both ours and that of our colleagues, and the suggestions of reviewers. The fourth edition includes:

- An expanded number of exercises, now over 1700
- Discussions of mathematical modeling, especially as related to linear systems and linear programming
- A chapter on logic for finite mathematics
- An increased emphasis on problem formulation in the chapter on systems of equations
- An expanded discussion of the geometry associated with systems of linear equations in more than two variables and with linear programming
- Material on linear programming organized into a chapter on the modeling and formulation of general linear programming problems, and the solution of problems in two variables, and another chapter on the simplex algorithm and duality
- Increased attention to the development of problem-solving strategies, especially in counting and probability
- A balance in the emphasis on ideas and techniques

The material on the geometry of linear equations and linear inequalities in three dimensions, which provides additional background for understanding some of the fundamental results of linear programming, can be omitted without loss of continuity.

In addition to revision within chapters, there has been a reorganization of chapters, including a shift of the chapter on the simplex algorithm and duality from Part 2 to Part 3. These organizational changes are in response to the increased accessibility of mathematical software and the frequencies with which certain material is included in finite mathematics courses.

The dependency relations between chapters are shown in the following diagram. Parts 1 and 2 are each sequentially dependent, but independent of each other. Chapter 8 requires familiarity with several topics from earlier chapters.

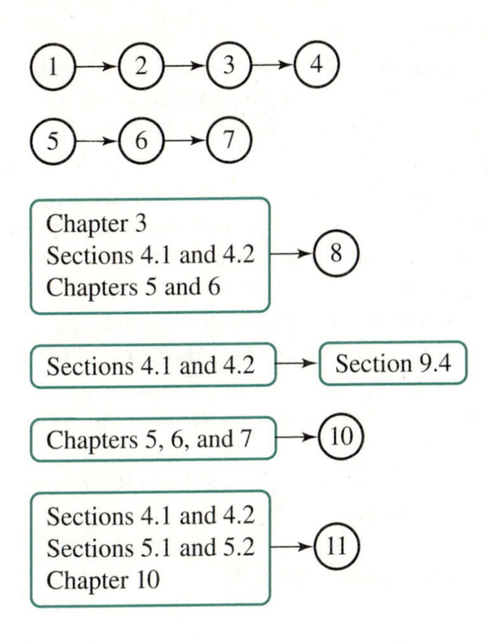

If an example or exercise is not strictly mathematical, then the setting in which the problem arises (biology, business, psychology, etc.) is noted in the margin next to the example or exercise. There is an index of applications on the inside covers.

Exercises that require the use of a calculator or computer are identified by an asterisk.

Each chapter is followed by a set of review exercises. These exercises are of varying degrees of difficulty (not necessarily graded from easy to hard) and are related to the topics of the chapter, although not always to a specific section.

Appendices A and B contain tables necessary to solve some of the problems in Chapters 3 and 4, and Appendix C contains the answers to odd-numbered exercises.

Supplements

A number of supplements have been prepared for both student and instructor:

The STUDENT'S SOLUTIONS MANUAL is available through the bookstore. It contains detailed solutions to the odd-numbered exercises in the text.

The INSTRUCTOR'S MANUAL contains a chapter-by-chapter overview, sample course outlines, case studies, and detailed solutions to the even-numbered exercises in the text. Also included are transparency masters which present important concepts from the text.

The PROFESSOR'S ASSISTANT is a computerized test generator that allows the instructor to create tests using questions generated from a standard testbank. This testing system enables the instructor to choose questions either manually or randomly by section, question type, difficulty level, and other criteria. This system is available for IBM, IBM-compatible, and Macintosh computers.

The PRINT TEST BANK is a printed and bound copy of the questions found in the standard test bank.

For further information about these supplements, please contact your local McGraw-Hill sales representative.

Acknowledgments

It is a pleasure to acknowledge the support and advice we received from our colleagues: Professors David Hoff, Andrew Lenard, Joseph Stampfli, and William Wheeler; the following exceptionally helpful reviewers: Barbara Barone, Hunter College; Craig J. Beham, University of Kentucky; Jerry Bloomburg, Essex Community College; Barbara Bohannon, Hofstra University; Russell Euler, Northwest Missouri State University; Martin E. Flashman, Humboldt State University; Curtis Olson, University of South Dakota; Jerry F. Reed, Mississippi State University; Cynthia Siegel, University of Missouri–St. Louis; and Jeffrey Watt, Purdue University; and the McGraw-Hill editors.

Daniel P. Maki

Maynard Thompson

FINITE MATHEMATICS

PART

1

Probability Models

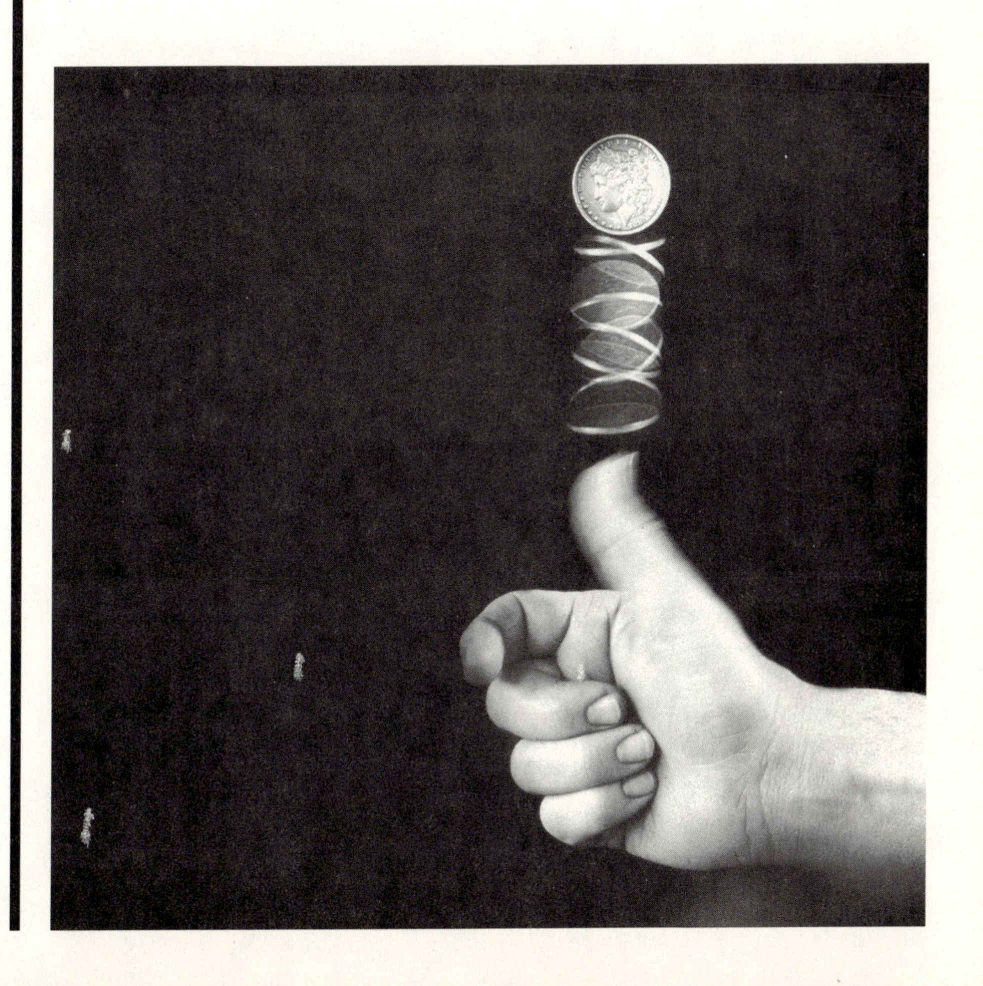

CHAPTER 1

Sets, Partitions, and Tree Diagrams

1.0 THE SETTING AND OVERVIEW

We develop a common notation and terminology for sets and set operations which will be helpful in explaining and understanding probability. The discussion begins with sets, ways of combining sets, and a connection between set operations and certain logical operations. A useful technique of representing sets with diagrams is developed. We introduce a special type of set needed in our work on probability, the set of outcomes of an experiment. We develop three key methods for counting the elements in particular kinds of sets. These methods are partitions, tree diagrams, and the multiplication principle, concepts that will be applied and extended in Chapters 2, 3, and 4.

1.1 REVIEW OF SETS AND SET OPERATIONS

The students in a finite mathematics class form a set. So do the workers in an office, the books on a shelf, and the courses taken by a student this semester. A *set* is a collection whose members are specified by a list or a rule. The items in the collection are the *elements* of the set. To use a rule to specify a set, the rule must make it possible to determine precisely which things are in the set and which are not. When a set is specified by a list, the usual practice is to list the elements, each one exactly once, between a pair of braces, Thus, the set S of names of states beginning with the letter A may be denoted by

$$S = \{\text{Alabama, Alaska, Arizona, Arkansas}\}$$

When a rule is used to specify a set, the usual practice is to write the rule after a symbol denoting a general element of the set followed by a colon and to include all the information in braces. For instance, to specify the set S with a rule, we can write

$$S = \{x\colon\ x \text{ is the name of a state beginning with the letter A}\}$$

In this expression the symbols "$S = \{x\colon \ldots\}$" are read "S is the set of all x such that . . . ."

Whether a list or a rule is used to specify a set, it is important to remember that either something belongs to a set or it does not. It cannot partly belong to a set, and it cannot belong to a set several times.

Example 1.1 The set I of even positive integers less than 15 can be represented in two ways:

$$I = \{n\colon\ n \text{ is an even positive integer less than } 15\}$$

and

$$I = \{2, 4, 6, 8, 10, 12, 14\}$$

■

In some cases a list of the elements provides the simplest and most useful representation of a set. In other cases a rule is preferable. For sets with large numbers of elements, the use of a rule is often the only practical way to define the set.

To indicate that x is an element of a set X, we write $x \in X$. To indicate that x is *not* an element of X, we write $x \notin X$. Thus in Example 1.1 we have $8 \in I$, but $5 \notin I$.

A set A is a *subset* of a set B, written $A \subset B$, if every element in A is also in B. If $A \subset B$ and $B \subset A$, then A and B have exactly the same elements, and we say that A and B are *equal*. We write $A = B$.

To illustrate the notation, let $F = \{4, 8, 12\}$ and $G = \{4, 5, 6\}$, and let I be the set of Example 1.1. Then $F \subset I$ since every element of F is also an element of I. The set G is not a subset of I since not all elements of G are in I. Indeed, $4 \in I$, $6 \in I$, but $5 \notin I$. In such a case it is sometimes convenient to write $G \not\subset I$.

It is often helpful to describe sets in terms of other sets. For instance, if

$$A = \{2, 4, 8, 12\}$$

$$B = \{2, 6, 10, 12, 14\}$$

and

$$I = \{2, 4, 6, 8, 10, 12, 14\}$$

then I is the set of all elements which are in A or in B or in both. Thus, we can view I as the set which results from combining A and B in a specific way. The operation which combines sets in this way is known as the *union*.

> Let A and B be sets. The set $A \cup B$, called the *union* of A and B, consists of all elements which are in A or in B or in both.
>
> $$A \cup B = \{x\colon\ x \in A \text{ or } x \in B\}$$

Note that in expressions such as "$x \in A$ or $x \in B$" the "or" is inclusive; that is, the condition is fulfilled if at least one of $x \in A$ or $x \in B$ holds. This includes the possibility that both hold.

To continue, let

$$A = \{2, 4, 8, 12\} \qquad B = \{2, 6, 10, 12, 19\} \qquad J = \{2, 12\}$$

Then J is the set of elements which are in both A and B. The operation which combines sets in this way is known as the *intersection*.

> Let A and B be sets. The set $A \cap B$, called the *intersection* of A and B, consists of all elements which are in both A and B.
>
> $$A \cap B = \{x\colon\ x \in A \text{ and } x \in B\}$$

Example 1.2 Let sets S, E, C, and M be defined as follows:

$$\begin{aligned} S &= \{\text{CT, MA, MD, CA, CO, MI, MN}\} \\ E &= \{\text{CT, MA, MD}\} \\ C &= \{\text{CA, CO, CT}\} \\ M &= \{\text{MA, MD, MI, MN}\} \end{aligned}$$

The elements of these sets are actually the standard abbreviations for names of states. The meaning is unimportant, however, and the elements can be viewed simply as symbols. Since each element of E is also an element of S, set E is a subset of S and we write $E \subset S$. Likewise, $C \subset S$ and $M \subset S$. However, since CT $\in E$ but CT $\notin M$, set E is not a subset of M, written $E \not\subset M$. Sets E and C have only the element CT in common, but sets E and M share the elements MA and MD. Therefore,

$$E \cap C = \{\text{CT}\} \qquad \text{and} \qquad E \cap M = \{\text{MA, MD}\}$$

Combining pairs of sets with the union operation, we have

$$E \cup C = \{\text{CA, CO, CT, MA, MD}\}$$
$$E \cup M = \{\text{CT, MA, MD, MI, MN}\}$$
and $$C \cup M = \{\text{CT, MA, MD, CA, CO, MI, MN}\} = S$$ ■

Note that the element CT which appears in both E and C appears only once in $E \cup C$. As we noted earlier, when a set is specified by a list, each element in the set must appear in the list exactly once. Also, if a set is specified by a list, then the order in which the elements appear in the list does not matter. Thus {CT, MA, MD} is the same as {MA, MD, CT} or {MD, CT, MA}.

In Example 1.2 the sets $C = \{\text{CA, CO, CT}\}$ and $M = \{\text{MA, MD, MI, MN}\}$ have no elements in common. That is, the set $C \cap M$ has no elements. Likewise, the set

$$S = \{x\colon\ x \text{ is the name of a state beginning with the letter B}\}$$

has no elements.

The set which contains no elements is known as the *empty set*, and it is denoted by $\varnothing$. By convention the empty set is considered to be a subset of every set.

Since the empty set has no elements, we see that for every set A,

$$A \cap \varnothing = \varnothing \qquad \text{and} \qquad A \cup \varnothing = A$$

Two sets A and B are *disjoint* if $A \cap B = \varnothing$.

The sets $C = \{\text{CA, CO, CT}\}$ and $M = \{\text{MA, MD, MI, MN}\}$ are disjoint since $C \cap M = \varnothing$.

The definitions of union and intersection were formulated for two sets. With the use of parentheses they can be used in expressions which involve more than two sets. For instance, if we have three sets A, B, and C, then the set of all elements which are in A and in B and in C can be written as either $A \cap (B \cap C)$ or $(A \cap B) \cap C$. In this case the parentheses do not matter, and thus we can write $A \cap B \cap C$ for short. Likewise the set of all elements in A or in B or in C (or in more than one of these sets) can be unambiguously denoted by $A \cup B \cup C$. However, when an expression involves both union and intersection operations, it is generally necessary to use parentheses in writing the expression. The operations within the parentheses are to be carried out first.

Example 1.3 Let $A = \{a, b, c\}$, $B = \{a, c, e\}$, and $C = \{a, d\}$. Then

$$\begin{aligned} A \cap B \cap C &= \{a\} \\ A \cup B \cup C &= \{a, b, c, d, e\} \\ (A \cap B) \cup C &= \{a, c\} \cup \{a, d\} = \{a, c, d\} \\ A \cap (B \cup C) &= \{a, b, c\} \cap \{a, c, d, e\} = \{a, c\} \end{aligned}$$

Here we have $(A \cap B) \cup C \neq A \cap (B \cup C)$. The parentheses are clearly crucial to the meaning of the expressions. ■

We have defined the intersection of two sets A and B (the set of elements which are in A *and* B) and the union of A and B (the set of elements which are in A *or* B or both). Thus we have operations on sets which associate naturally with "and" and "or." We turn next to an operation on sets which is the natural associate of the word "not."

A set U is said to be a *universal set* for a problem if all sets being considered in the problem are subsets of U. Given a universal set U, the *complement* of a subset A of U is the set of all elements in U which are not in A. The complement of A is written A'.

$$A' = \{x,\ x \in U \text{ and } x \notin A\}$$

Notice that if A and B are subsets of U, then the set of elements in A which are not in B can be written as $A \cap B'$. Such sets arise frequently in applications.

Example 1.4 Let $U = \{\text{CA, CO, CT, IL, IN}\}$, $X = \{\text{CA, CT, IL}\}$, $Y = \{\text{CO, CT, IN}\}$, and $Z = \{\text{CO, IN}\}$. Then

$$\begin{array}{lll} X' = \{\text{CO, IN}\} = Z, & Y' = \{\text{CA, IL}\} & Z' = \{\text{CA, CT, IL}\} = X \\ Y \cap Z' = \{\text{CT}\} & X \cap Z' = \{\text{CA, CT, IL}\} = X & Z \cap Y' = \varnothing \end{array}$$ ■

In addition to taking unions, intersections, and complements, there are other useful ways of building new sets from given ones. For example, suppose that a sociologist has enough money to conduct one survey. The survey can be conducted either by mail (M) or by phone (P) in one of three cities: Atlanta (A), Boston (B), or Cincinnati (C). Thus the choice for the sociologist can be viewed as selecting a method (M or P) and a city (A, B, or C). Each possible survey can be denoted by an *ordered pair* of elements, one from the set $\{M, P\}$ and one from the set $\{A, B, C\}$. Thus selecting a survey is clearly the same as selecting an element from the set

$$\{(M, A), (M, B), (M, C), (P, A), (P, B), (P, C)\}$$

The *cartesian product* of sets A and B, denoted by $A \times B$, is the set of all ordered pairs (a, b) where $a \in A$ and $b \in B$.

$$A \times B = \{(a, b): a \in A, b \in B\}$$

Example 1.5 Let

$$A = \{a, c, e\} \quad B = \{b, d, e\} \quad C = \{b, d\}$$

Then

$$A \times C = \{(a, b), (a, d), (c, b), (c, d), (e, b), (e, d)\}$$
$$B \times C = \{(b, b), (b, d), (d, b), (d, d), (e, b), (e\ d)\}$$
$$C \times C = \{(b, b), (b, d), (d, b), (d, d)\}$$

■

Since $A \times C$ in Example 1.5 is a set, the order of the elements within the braces is *not* important. In particular, we could also write

$$A \times C = \{(a, b), (c, b), (e, b), (a, d), (c, d), (e, d)\}$$

The order of the symbols within the parentheses *is* important: The element (x, y) is different from (y, x) for x different from y. Our next example illustrates this.

Example 1.6 *Sports*

A division of a football league consists of four teams: the Aardvarks (A), Bisons (B), Coyotes (C), and Dingos (D). Each game can be represented as an ordered pair of teams in which the first entry denotes the home team. With this notation the set of all possible games is a subset of the cartesian product of the set $L = \{A, B, C, D\}$ with itself, $L \times L$. Note that the cartesian product $L \times L$ contains elements such as (A, A), which are not legitimate games. In fact, the set of all possible games is

$$G = \{(A, B), (A, C), (A, D), (B, A), (B, C), (B, D), (C, A), (C, B), (C, D), (D, A), (D, B), (D, C)\}$$

and the set of all games involving the Aardvarks, Bisons, and Coyotes is

$$H = \{(A, B), (A, C), (B, A), (B, C), (C, A), (C, B)\}$$

■

Example 1.6 illustrates the importance of order in the construction of an ordered pair. For instance, the game between A and B with A as the home team is denoted by (A, B), while (B, A) denotes the game between the same teams with B as the home team. Thus, the ordered pair (A, B) is not the same as the ordered pair (B, A). However, as we noted earlier, the order of elements in the list specifying the set is unimportant, and set H can also be represented, e.g., as

$$H = \{(A, B), (B, A), (A, C), (C, A), (B, C), (C, B)\}$$

Exercises for Section 1.1

1. Let $R = \{a, b\}$, $S = \{a, c, f\}$, and $T = \{a, b, c, d, e\}$. Decide whether each of the following assertions is correct.
   (*a*) $S \subset T$ (*b*) $R \subset S$ (*c*) $b \in R \cap T$
2. With R, S, and T defined as in Exercise 1, decide whether each of the following assertions is correct.
   (*a*) $R \subset T$ (*b*) $(R \cup S) \subset T$ (*c*) $c \in S \cap T$
3. With R, S, and T defined as in Exercise 1, decide whether each of the following assertions is correct.
   (*a*) $R \subset (S \cap T)$ (*b*) $R \subset (S \cup T)$
4. With R, S, and T defined as in Exercise 1, find $(R \cup S) \cap T$.
5. Let $A = \{p, q, r\}$. Find all nonempty subsets B and C of A such that $B \cap C = \varnothing$ and $B \cup C = A$.
6. Let $U = \{1, 2, 3, 4, 5\}$ be a universal set with subsets $A = \{2, 4\}$ and $B = \{1, 5\}$. List the elements in each of the following sets.
   (*a*) A' (*b*) $A' \cap B'$ (*c*) $A \cup B$ (*d*) $(A \cup B)'$
7. The sets M, A, B, and C are defined as follows:

   M = {Minnesota, Michigan, Montana, Massachusetts}
   A = {Alabama, Arkansas, Michigan}
   B = {Montana, Michigan}
   C = {Alabama, Arkansas}

   Decide which of the following subset relationships are correct.
   (*a*) $B \subset M$ (*b*) $B \subset C$ (*c*) $C \subset A$
   (*d*) $C \subset B$ (*e*) $C \subset M$ (*f*) $A \subset (B \cup C)$
8. Let M, A, B, and C be as in Exercise 7, and set $U = M \cup A \cup B \cup C$. Decide which of the following equalities are correct.
   (*a*) $M \cap A = A$ (*b*) $A \cup C = A$
   (*c*) $B \cap C' = B$ (*d*) $B \cap C = B \cap M'$
9. Let $U = \{a, b, c, d, 4, 6\}$, $X = \{a, c, d, 6\}$, and $Y = \{a, b\}$.
   (*a*) Find X'. (*b*) Find Y'. (*c*) Find $X \cap Y'$.
10. Let $U = \{x, y, z, 1, 2, 3\}$, $A = \{y, z, 2\}$, $B = \{y, 1, 2\}$, and $C = \{x, 3\}$. List the elements in each of the following sets.
    (*a*) $A \cup B$ (*b*) $B \cap C$ (*c*) A'
    (*d*) $(A \cup B) \cap (B \cup C)$ (*e*) $(B \cap A') \cap C'$

11. Let U, A, B, and C be defined by

$$U = \{a, b, c, 1, 2, 3\}$$
$$A = \{a, b, c\} \qquad B = \{a, 2, 3\} \qquad C = \{1, 2, 3\}$$

List the elements in each of the following sets.
(a) $A \cup B$ (b) $B \cap C$ (c) $(A \cup B) \cap (B \cup C)$
(d) A' (e) $A \cap B'$ (f) $A \cup C'$

12. Let sets R, S, and T be defined as follows:

$R = \{x$: x lives in the residence halls at Gigantic State University$\}$
$S = \{x$: x is a student at Gigantic State University$\}$
$T = \{x$: x is an employee of Gigantic State University$\}$

Describe in words each of the following sets.
(a) $S \cap R$ (b) $S \cup T$ (c) $(S \cap T) \cup R$

13. Let sets A, B, C, and D be defined by

$A = \{x$: x owns a GM car$\}$
$B = \{x$: x works for GM$\}$
$C = \{x$: x is the president of GM$\}$
$D = \{x$: x owns stock in GM$\}$

Describe in words each of the following sets.
(a) $A \cap B$ (b) $B \cap A'$ (c) $(A \cup B) \cap D$ (d) $C \cap A$

14. Let X and Y be sets with $a \in X$ and $b \in Y$. Is it *always* true (yes or no) that $\{a, b\} \subset X \cup Y$? That $\{a, b\} \subset X \cap Y$?

15. Let A and B be subsets of a universal set U. Is it *always* true that
(a) $B \cap A' \subset A$ (b) $A \cap B \subset A \cup B$
(c) $A' \cap B' \subset (A \cap B)'$ (d) $A' \cup B' \subset (A \cup B)'$

16. Let A, B, and C be subsets of U. Is it *always* true that
(a) $A \subset A \cap (B \cup C)$ (b) $A' \subset (A \cup B \cup C)'$
(c) $A \cap B \subset A \cup (B \cap C)$ (d) $A \cup B \subset A \cup (B \cap C)$

17. Let A, B, C, and D be subsets of U with $A \subset B$ and $C \subset D$. Is it *always* true that
(a) $A \cap C \subset B \cap D$ (b) $A' \cap C' \subset B' \cap C'$

18. Let $U = \{w, x, y, z\}$. Find examples of subsets A and B of U which satisfy the stated condition.
(a) $A \cup B = A$ (b) $A \cap B = A$
(c) $A \cap B' = A$ (d) $A \cap B' = B \cap A'$

19. Let $U = \{1, 2, 3, 4, x, y\}$ be a universal set with subsets $X = \{1, 2, 3, x, y\}$, $Y = \{2, 4, y\}$, and $Z = \{2, x\}$. Use intersections, unions, and complements to express each of the following sets in terms of X, Y, and Z.

$$A = \{2, y\} \qquad B = \{1, 3, y\} \qquad C = \{2, 4, x, y\}$$

20. With X, Y, Z, and U as in Exercise 19, use intersections, unions, and complements to express the set $\{x\}$ in terms of X, Y, and Z.
21. List all subsets of the following sets.
    (*a*) $\{x\}$ (*b*) $\{x, y\}$ (*c*) $\{x, y, z\}$
22. Counting the empty set and the set itself, how many subsets does each of the following sets contain?
    (*a*) $\{x\}$ (*b*) $\{x, y\}$ (*c*) $\{x, y, z\}$ (*d*) $\{w, x, y, z\}$
    Is there a pattern? If so, what is the pattern? How many subsets does a set with seven elements contain?
23. Let $U = \{a, b, c, 2, 4, 6\}$ be a universal set with subsets X, Y, and Z. Suppose that $X \cup Y = \{b, c, 2, 4, 6\}$, $X \cap Y = \{b, 2, 4\}$, $Y' \cap Z' = \{a, c\}$, and $Z' = \{a, c, 2\}$. Find sets X, Y, and Z which satisfy these conditions.
24. Let $A = \{(1, 2), (2, 3), (3, 4)\}$ and $B = \{(1, 2), (3, 2), (3, 3)\}$. List the elements in $A \cap B$ and $A \cup B$.
25. Let $A = \{a, b, c\}$ and $B = \{a, b, d\}$.
    (*a*) List the elements in $A \times B$.
    (*b*) List the elements in $(A \times B) \cap (B \times A)$.
26. With A and B as in Exercise 25, find $(A \times A) \cap (B \times B)$.
27. Suppose $A \times B = \{(a, 1), (b, 1), (a, 2), (b, 2), (a, 3), (b, 3)\}$. Find A and B.
28. Let $A = \{u, x, z\}$, $B = \{x, y, z\}$ and $U = (A \times B) \cup (B \times A)$.
    (*a*) List the elements in U.
    (*b*) List the elements in $(A \times B)'$.
29. Let $U = \{-2, -1, 0, 1, 2\}$ and $S = \{-1, 0, 1\}$. Also, let $A = \{(x, y): x \in S, y = x^2\}$ and $B = \{(x, y): x \in U, y = x^2\}$. Is it true that
    (*a*) $A \subset B$ (*b*) $A \subset S \times S$
    (*c*) $B \subset S \times S$ (*d*) $B \subset U \times U$
30. Let A, B, S, and U be the sets given in Exercise 29. Find $A \times A$ and $B \times B$. Show that $A \times A \subset B \times B$. Is it always true that if $A \subset B$, then $A \times A \subset B \times B$? Why or why not?

1.2 VENN DIAGRAMS AND PARTITIONS

In working with sets and the relations between sets, it is often helpful to represent them with diagrams or pictures. A *Venn diagram* serves this purpose. In a Venn diagram, a universal set U and its subsets are pictured by using geometric shapes. By convention the set U is usually represented by a rectangle, and the subsets of U are usually circles inside the rectangle. For example, subset A of U is shown as the shaded region in Figure 1.1*a*, subset B is shown in Figure 1.1*b*, and subset A' is shown in Figure 1.1*c*.

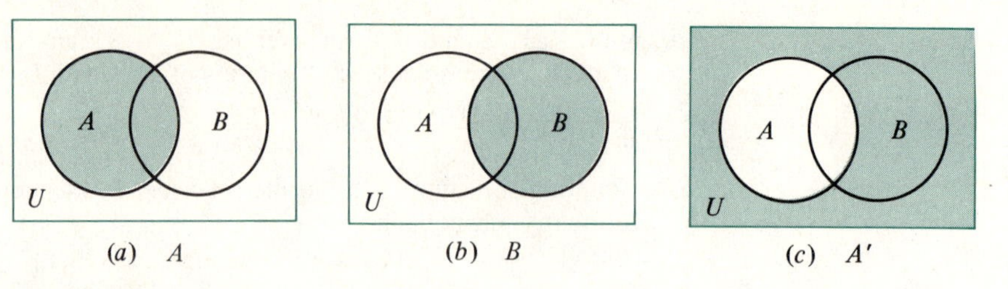

(*a*) A (*b*) B (*c*) A'

FIGURE 1.1

Subsets of U obtained by taking unions, intersections, and complements of A or B or both can also be represented by Venn diagrams. For instance, $A \cup B$ is illustrated in Figure 1.2*a*, and $A \cap B$ is illustrated in Figure 1.2*c*.

Two useful set equalities are known as *deMorgan's laws.*

For any subsets A and B of a universal set U

$$(A \cup B)' = A' \cap B'$$
$$(A \cap B)' = A' \cup B'$$

These equalities are illustrated in Figure 1.2. First, Figure 1.2*a* and *b* illustrate that $(A \cup B)' = A' \cap B'$. This relation can be read "The complement of a union is the intersection of the complements," and it follows from the definitions of union, intersection, and complement. Likewise, Figure 1.2*c* and *d* illustrate that $(A \cap B)' = A' \cup B'$, which can be read "The complement of an intersection is the union of the complements."

FIGURE 1.2

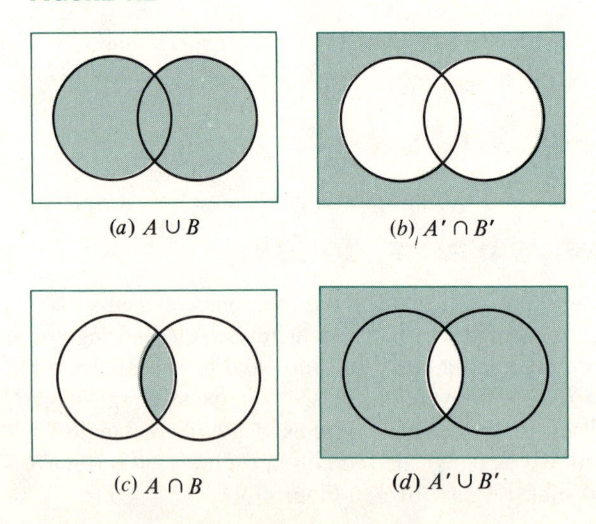

(*a*) $A \cup B$ (*b*) $A' \cap B'$

(*c*) $A \cap B$ (*d*) $A' \cup B'$

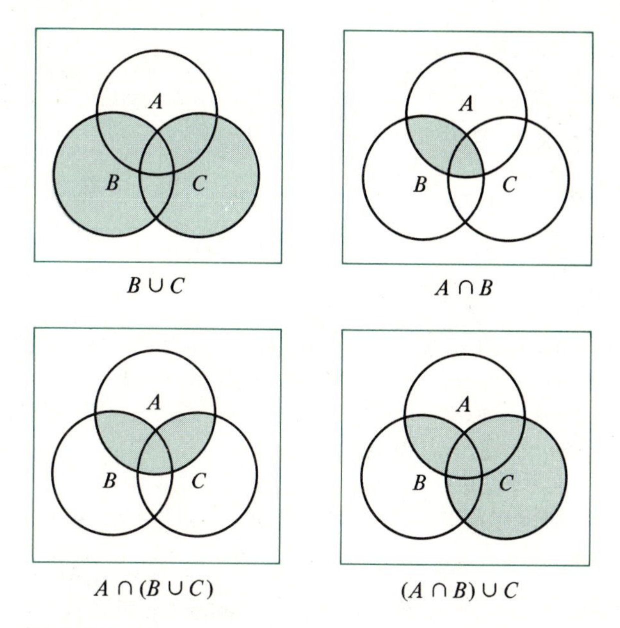

FIGURE 1.3

Other useful relations can be illustrated with Venn diagrams and verified by using the definitions. Among these relations are the following distributive laws:

$$A \cap (B \cup C) = (A \cap B) \cup (A \cap C)$$
$$A \cup (B \cap C) = (A \cup B) \cap (A \cup C)$$

These relations hold for any three sets A, B, and C. As we mentioned earlier, the parentheses are essential, and the expressions would be ambiguous without them.

Example 1.7 Figure 1.3 illustrates that in general the sets $A \cap (B \cup C)$ and $(A \cap B) \cup C$ are different. ■

FIGURE 1.4

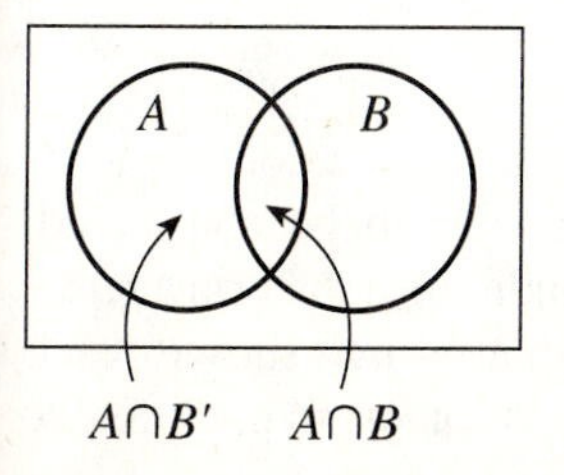

Venn diagrams provide us with a geometric way to represent the decomposition of a set into subsets. For example, in Figure 1.4 we illustrate how the set A can be decomposed into subsets $A \cap B$ (the football-shaped region) and $A \cap B'$ (the crescent-shaped region). The notion of decomposition of a set into subsets is extremely useful in the study of probability, and we consider the situation in greater detail.

Example 1.8 Let $X = \{2, 4, 5, 12, 15\}$.

Problem Find subsets X_1 and X_2 of X such that X_1 contains all elements in X which are less than 10 and X_2 contains all elements in X which are greater than 10.

Solution

$$X_1 = \{2, 4, 5\} \qquad \text{and} \qquad X_2 = \{12, 15\}$$

Note that X_1 and X_2 are such that $X_1 \cap X_2 = \varnothing$ and $X_1 \cup X_2 = X$. ■

Example 1.9 Let $X = \{\text{Austria, Brazil, Chile, Denmark, Egypt, France}\}$.

Geography

Problem Find subsets A, E, and S of U such that

$$A = \{x\colon\ x \text{ is a country in Africa}\}$$
$$E = \{x\colon\ x \text{ is a country in Europe}\}$$
$$S = \{x\colon\ x \text{ is a country in South America}\}$$

and find the union of these three sets and the intersection of each pair of them.

Solution The sets $A = \{\text{Egypt}\}$, $E = \{\text{Austria, Denmark, France}\}$, and $S = \{\text{Brazil, Chile}\}$ satisfy the conditions. Sets A, E, and S have the following properties:

$$A \cup E \cup S = X \qquad A \cap E = \varnothing \qquad A \cap S = \varnothing \qquad E \cap S = \varnothing \quad (1.1)$$

■

In Example 1.9 sets A and E are disjoint. Likewise sets A and S are disjoint, and sets E and S are disjoint. Relationships like this occur frequently enough for us to use a special expression to describe them.

The sets in a collection are said to be *pairwise disjoint* if every pair of sets in the collection is disjoint.

Also in Example 1.9, the result of each classification assigns a country to A, E, or S; the union $A \cup E \cup S$ is the set of all countries to be considered: $A \cup E \cup S = X$. Subsets A, E, and S of X which satisfy condition (1.1) form a *partition* of X. In general, a partition is the result of cutting up a set into subsets; each subset contains some elements of the set, and no two subsets can overlap. Formally,

A *partition* of a set X is a collection of nonempty subsets of X which are pairwise disjoint and whose union is the entire set X.

In the use of sets in finite mathematics, one of our main concerns is to count the number of elements in certain sets. This is the primary topic of the next section, and it will recur frequently in our work on probability. Partitions are especially useful in helping us count the number of elements in a set. To show how, we need some notation.

Let A be a set with a finite number of elements. The number of elements in A is denoted by $n(A)$.

For instance, if X, A, E, and S are the sets of Example 1.9, then $n(X) = 6$, $n(A) = 1$, $n(E) = 3$, and $n(S) = 2$. In this case we have

$$6 = n(X) = n(A) + n(E) + n(S) = 1 + 3 + 2$$

In fact, the definitions of a partition and the number of elements in a set lead to the following useful principle:

Partition Principle

If a set X is partitioned into subsets $X_1, X_2, \ldots, X_k$, then

$$n(X) = n(X_1) + n(X_2) + \cdots + n(X_k) \qquad (1.2)$$

If each of the subsets $X_1, X_2, \ldots, X_k$ has the same number of elements, then Equation (1.2) can be simplified to

$$n(X) = kn(X_1) \qquad (1.3)$$

This version of (1.2) will be useful when X is a cartesian product.

Example 1.10 *Course selection*

A student is to plan a schedule which consists of one science course and one humanities course. The student can choose a science course in astronomy (A), biology (B), or chemistry (C) and a humanities course in history (H), philosophy (P), religion (R), or theater (T).

Problem Determine the number of possible schedules.

Solution A schedule is a science course (A, B, or C) *and* a humanities course (H, P, R, or T). Thus, a class schedule can be represented as an ordered pair in which the first entry is a science course and the second entry is a humanities course. The set of class schedules S is a cartesian product of the set of science courses $X = \{A, B, C\}$ and the set of humanities courses $Y = \{H, P, R, T\}$. This cartesian product can be arranged in the array

$$\begin{array}{cccc} (A, H) & (A, P) & (A, R) & (A, T) \\ (B, H) & (B, P) & (B, R) & (B, T) \\ (C, H) & (C, P) & (C, R) & (C, T) \end{array}$$

There are three rows in the array, one corresponding to each element in the set $\{A, B, C\}$; and there are four columns in the array, one corresponding to each element in the set $\{H, P, R, T\}$. We can view the set S as partitioned into three subsets, one corresponding to each row of the array. Each row of the array contains four elements, and we conclude from (1.3) that

$$n(S) = n(\{A, B, C\}) \cdot n(\{H, P, R, T\}) = 3 \cdot 4 = 12$$ ■

The technique used in Example 1.10 for counting the number of elements in a set which can be represented as the cartesian product of two sets is perfectly general. We have the following rule.

If A and B are sets, then

$$n(A \times B) = n(A) \cdot n(B) \tag{1.4}$$

As we shall see in our next example, it is often necessary to consider cartesian products of more than two sets. The definition is similar to the definition of the cartesian product of two sets. For instance, the cartesian product $E \times F \times G$ of three sets is the set of all ordered triples (e, f, g), with $e \in E$, $f \in F$, and $g \in G$.

Example 1.11

Course selection

Suppose that the student of Example 1.10 also plans to take one language course, either French (F) or German (G). Then the possible class schedules can be represented by ordered triples (x, y, z) where $x \in \{A, B, C\}$, $y \in \{H, P, R, T\}$, and $z \in \{F, G\}$. That is,

$$S = \{A, B, C\} \times \{H, P, R, T\} \times \{F, G\}$$

The number of class schedules available to the student is $n(S)$. Hence, as in Example 1.10, $n(S)$ can be obtained by taking the product of the numbers of elements in each of the sets in the cartesian product:

$$n(S) = n(\{A, B, C\}) \cdot n(\{H, P, R, T\}) \cdot n(\{F, G\}) = 3 \cdot 4 \cdot 2 = 24$$ ■

The result illustrated in Example 1.11 for three sets can be extended to an arbitrary number of sets. The corresponding general result is as follows:

If $X_1, X_2, \ldots, X_k$ are sets, then

$$n(X_1 \times X_2 \times \cdots \times X_k) = n(X_1) \cdot n(X_2) \cdots n(X_k) \tag{1.5}$$

Exercises for Section 1.2

1. Let A, B, and C be subsets of a set U. Draw a Venn diagram to illustrate each of the following sets. In each case, shade the area corresponding to the designated set.
   (*a*) $A' \cap B$ (*b*) $A \cup B'$
   (*c*) $A \cup B \cup C$ (*d*) $(A \cup B) \cap C'$
2. Repeat Exercise 1 for the following sets.
   (*a*) $(A \cap B') \cup B$ (*b*) $A \cap B \cap C'$
   (*c*) $A' \cup (B \cap C)$ (*d*) $(A \cup B)' \cup (A \cap B)$
3. In each case determine which of points v, w, x, y, and z in Figure 1.5 are contained in the specified set.
   (*a*) $A \cup C$ (*b*) $A \cap B$ (*c*) $A \cup B$ (*d*) $B \cap C$
4. In each case determine which of points v, w, x, y, and z in Figure 1.5 are contained in the specified set.
   (*a*) $A \cap C$ (*b*) $A \cap B \cap C$
   (*c*) $A \cup B \cup C$ (*d*) $A \cap (B \cup C)$
5. Using Figure 1.5, decide which of the following statements are true and which are false.
   (*a*) $z \in A \cup C'$ (*b*) $y \in B \cup (A \cap C')$
   (*c*) $y \in (B \cup C) \cap A'$ (*d*) $v \in (B \cup C) \cap (A \cup C)$
6. Describe the shaded areas in each Venn diagram of Figure 1.6 by using the set operations of union, intersection, and complement and sets A, B, and C.
7. Repeat Exercise 6, using Figure 1.7.
8. Let A, B, C, and D be subsets of a universal set U.
   (*a*) Suppose $D \cap (B \cup C) = \varnothing$, and draw a Venn diagram to illustrate these sets.
   (*b*) Suppose $A \cap C = \varnothing$ and $D \cap B = \varnothing$, and draw a Venn diagram to illustrate these sets.

FIGURE 1.5

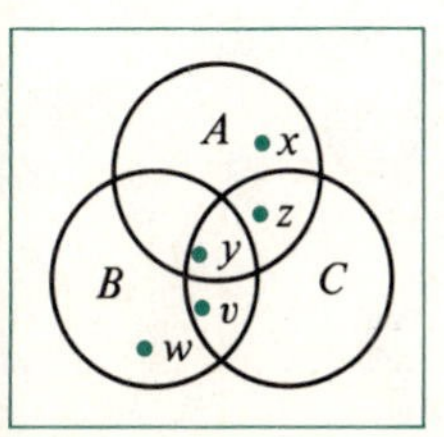

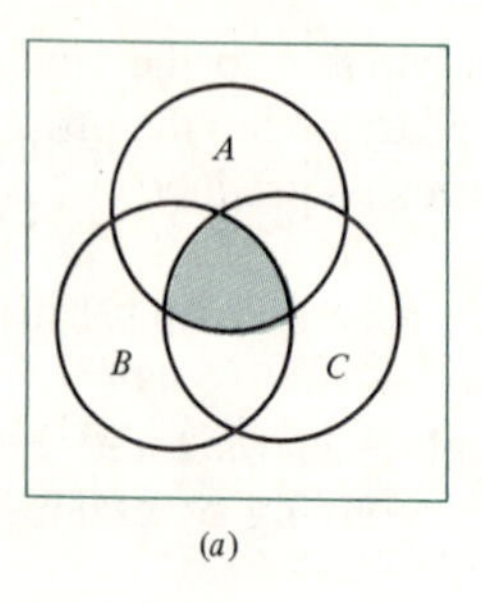

(a)

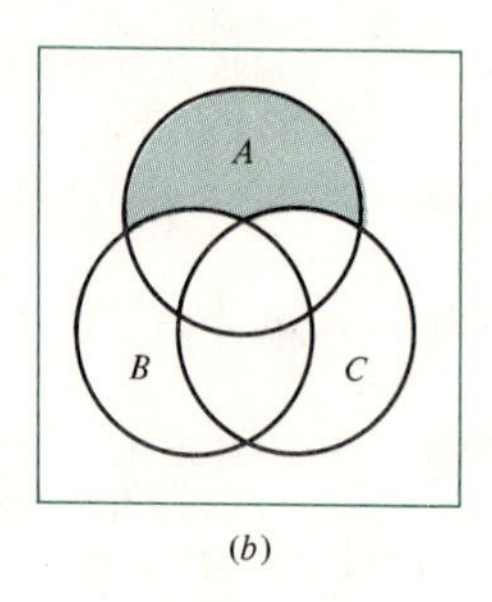

(b)

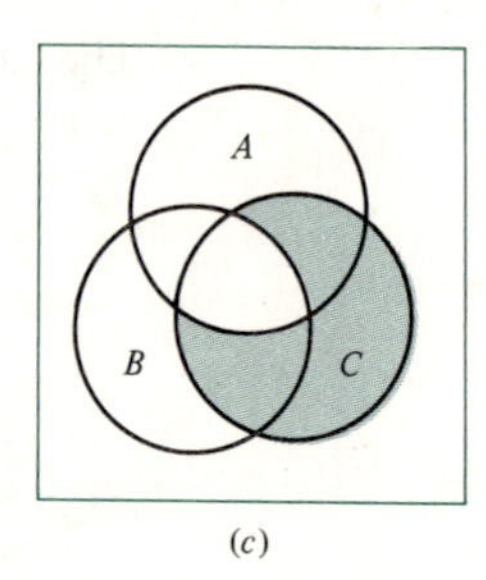

(c)

FIGURE 1.6

9. Determine which (if any) of the following set relations are true for *all* sets A and B. (*Hint*: Use Venn diagrams.)
   (*a*) $A' \cap B' = (A \cap B)'$ (*b*) $(A \cap B') \subset A'$
   (*c*) $A' \cap B' \subset (A \cup B)'$ (*d*) $(A \cap B)' \subset A'$
10. In each of the following expressions, determine which (if any) of the set relations $=$, $\subset$, or $\supset$ yields a true statement for all sets A and B when inserted in the blank.
    (*a*) $(A \cup B)' \cup (A \cap B')$ ________ B'
    (*b*) $(A \cup B)' \cup (A \cap B')$ ________ A'
11. Let U be a universal set with disjoint subsets A and B; $n(U) = 60$, $n(A) = 25$, and $n(B) = 30$. Find $n((A \cup B)')$.
12. Let U be a universal set with disjoint subsets A and B; $n(U) = 55$, $n(A) = 25$, and $n(B) = 10$. Find $n(A' \cup B)$.
13. Let U be a universal set with disjoint subsets A and B; $n(A) = 25$, $n(A') = 40$, and $n(B') = 30$. Find $n(A \cup B)$.
14. Let A and B be subsets of U. With the aid of the Venn diagram of Figure 1.1, find four disjoint sets whose union is U. Express these sets in terms of A and B and the proper symbols.

FIGURE 1.7

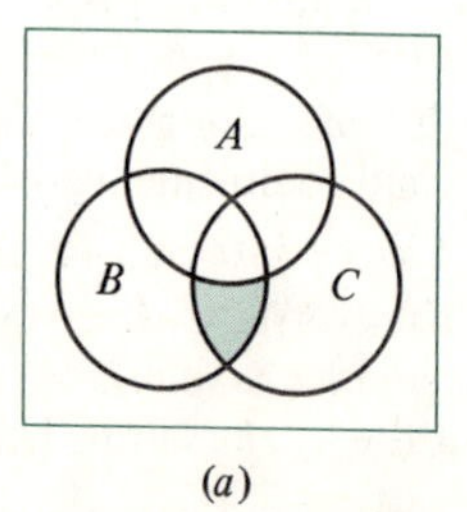

(a)

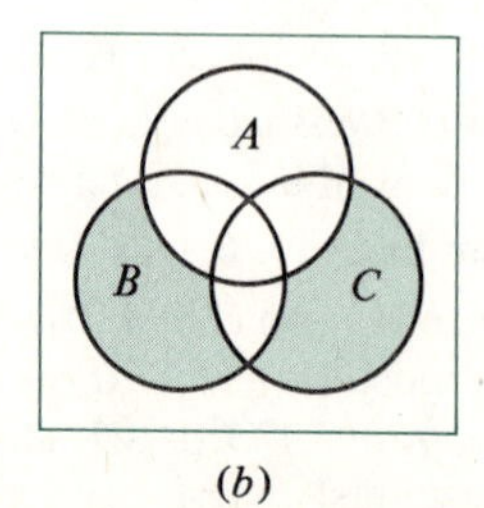

(b)

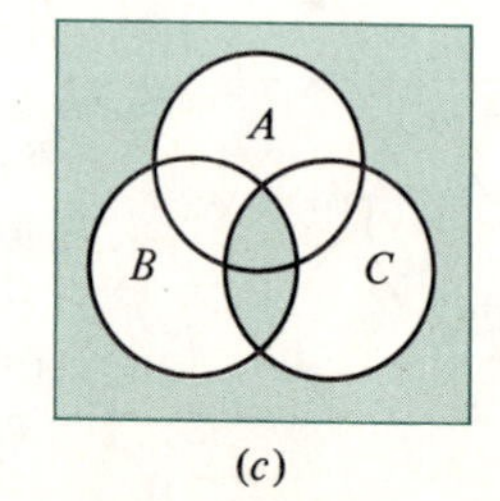

(c)

15. Let X, A, B, and C be defined by

$$X = \{a, b, c, 1, 2, 3\}$$
$$A = \{a, b, c\} \qquad B = \{a, 2, 3\} \qquad C = \{1, 2, 3\}$$

Which of the following pairs of subsets form a partition of X?
(*a*) A and B (*b*) A and C
(*c*) B and C (*d*) $(A \cup B)$ and $(C \cap B')$

16. Let $U = \{a, b, c, d, e\}$. Find a partition of U which contains exactly three sets. Find one which contains exactly four sets.

17. Let $A = \{1, 2, 3\}$ and $B = \{v, w\}$. By listing the elements in each of the three sets, show that $A \times \{v\}$ and $A \times \{w\}$ provide a partition of $A \times B$.

18. A set X can be partitioned into three subsets E, F, and $G \times H$. If $n(E) = 3$, $n(F) = 6$, $n(G) = 2$, and $n(H) = 5$, find $n(X)$.

19. Let $A = \{1, 7, 3, a, b\}$ and $B = \{3, c\}$. Determine
(*a*) $n(A \times B)$ (*b*) $n(B \times B \times B)$

20. If $n(A \times B) = 10$, $n(A \times C) = 15$, and $n(B \times C) = 6$, find $n(A)$.

Data surveys

21. A sociologist has a project which involves the collection of data. She is interested in data which can be obtained by mail, by phone, or in person in any of the cities of Atlanta, Boston, Chicago, Denver, or Elmira. She has funds for one project, that is, to collect data in one way from one city. Determine the number of possible ways to carry out the project.

Data surveys

22. Suppose that the sociologist of Exercise 21 can collect data in any one of 3 ways, in any one of 5 cities, and in any one of 4 months. Determine the number of possible ways to carry out the project.

23. Let A, B, and C be distinct subsets of a universal set U with $A \subset B \subset C$. Also suppose $n(A) = 3$) and $n(C) = 7$. In how many different ways can you select B such that $n(B) = 4$? Repeat this exercise with $n(B) = 5$.

24. Suppose $n(A) = 5$, $n(B) = 10$, and $n(C) = 20$. Which of the following sets has more elements, $A \times B \times C$ or $B \times B \times B$?

Travel

25. A traveler plans to fly from Chicago to London either directly or through New York. There are 4 airlines which fly directly from Chicago to London, 5 which fly from Chicago to New York, and 6 which fly from New York to London. A flight is distinguished by an airline and a route. How many flight options does the traveler have?

26. Suppose a set X can be partitioned into two sets Y and Z and that $Y = A \times A \times B$ and $Z = C \times D$. If $n(A) = 3$, $n(B) = 2$, $n(C) = 5$, and $n(D) = 8$, determine $n(X)$.

27. A set X is partitioned into subsets X_1, X_2, and X_3. The number of elements in X_1 is twice the number in X_2, and the number in X_3 is 5 times the number in X_2. If $n(X) = 40$, find $n(X_1)$, $n(X_2)$, and $n(X_3)$.

28. A set X with $n(X) = 45$ is partitioned into three subsets X_1, X_2, and X_3. If $n(X_2) = 2n(X_1)$ and $n(X_3) = 3n(X_2)$, find the number of elements in subset X_1.

29. A set X with $n(X) = 60$ is partitioned into subsets $X_1, \ldots, X_6$. If $n(X_1) = n(X_2) = n(X_3)$, $n(X_4) = n(X_5) = n(X_6)$, and $n(X_1) = 4n(X_4)$, find $n(X_1)$.

30. Let A_1 and A_2 be a partition of A, and let B_1 and B_2 be a partition of B. Is it true that $A_1 \times B_1$, $A_1 \times B_2$, $A_2 \times B_1$, and $A_2 \times B_2$ form a partition of $A \times B$? Why or why not?

1.3 SIZES OF SETS

We have seen that "the whole is equal to the sum of the parts" when we are dealing with subsets which form a partition of a set. This is summarized in the partition principle, formula (1.2). What if the sets of interest do not form a partition of another set? For instance, what if they are not disjoint? In such cases Venn diagrams and the partition principle are still useful when applied appropriately. We begin by analyzing a specific example in some detail. Our goal is both a technique and a very useful formula.

Example 1.12 A set U with nondisjoint subsets A and B has the following:

$$n(U) = 10 \qquad n(A) = 7 \qquad n(B) = 6 \qquad n(A \cap B) = 4$$

Problem Find $n(A \cup B)$.

Solution We use a Venn diagram with universal set U and subsets A and B. Inside subset $A \cap B$ we insert the number 4 to indicate that $n(A \cap B) = 4$. At this stage we have the diagram shown in Figure 1.8*a*. Next, since A has 7 elements and since 4 of them are in $A \cap B$, the portion of A not in $A \cap B$, that is, $A \cap B'$, must contain $7 - 4 = 3$ elements. We insert a 3 in the set $A \cap B'$ to indicate this. Likewise, since $n(B) = 6$, there must be 2 elements in $A' \cap B$. The information $n(A \cap B) = 4$, $n(A \cap B') = 3$, and $n(A' \cap B) = 2$ is shown in Figure 1.8*b*. Using this information, we see from the partition principle [formula (1.2)] that $n(A \cup B) = 3 + 4 + 2 = 9$. ■

It is helpful to examine this example more closely. Since $n(A \cup B) = 9$ while $n(A) + n(B) = 6 + 7 = 13$, it is clear that in general we cannot obtain $n(A \cup B)$ simply by adding $n(A)$ and $n(B)$. In fact, by examining Figure 1.8 we see that adding $n(A)$ and $n(B)$ actually counts the elements in $A \cap B$ twice. It follows that to find $n(A \cup B)$, we must subtract $n(A \cap B)$ from $n(A) + n(B)$. In this way each element in $A \cup B$ will be counted exactly once. In Example 1.12 we have

FIGURE 1.8

(*a*)

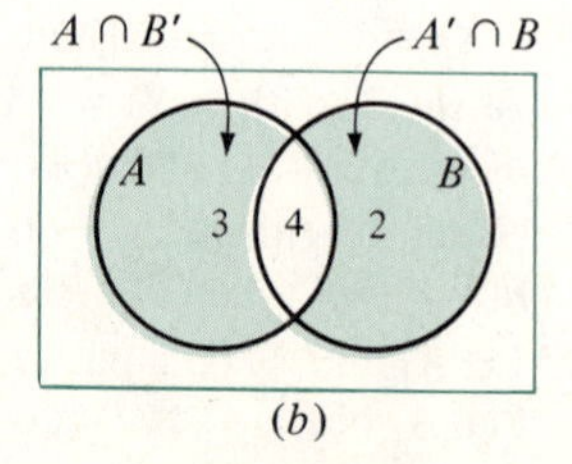

(*b*)

$$n(A \cup B) = (3 + 4) + (4 + 2) - 4 = 9 = n(A) + n(B) - n(A \cap B)$$

Our argument holds for any two sets A and B. We have the useful formula

$$n(A \cup B) = n(A) + n(B) - n(A \cap B) \tag{1.6}$$

Example 1.13

Entertainment

The operator of a radio station surveyed 100 listeners by asking them two questions:

(*a*) Do you prefer to have a weather update once an hour or more frequently?

(*b*) Do you prefer to have commercial breaks three times each hour (and more commercials at each break) or commercials spread evenly throughout the hour?

The results of the survey are that 60 people prefer to have a weather update once an hour, 45 people prefer to have commercials three times each hour, and 25 people prefer both an hourly weather update and commercials three times each hour.

Problem How many people preferred both more frequent weather updates and commercials spread evenly throughout the hour?

Solution We use a Venn diagram with a universal set U consisting of the 100 individuals who were surveyed. We also let W denote the subset of 60 people who prefer hourly weather updates, and we let C denote the subset of 45 people who prefer to have commercials three times each hour. From the data of the problem we know that $n(W \cap C) = 25$. Since $n(W) = 60$ and $n(C) = 45$, there must be $60 - 25 = 35$ individuals in $W \cap C'$ and $45 - 25 = 20$ individuals in $W' \cap C$. Thus we have the diagram and numbers shown in Figure 1.9.

From Figure 1.9 we see that $n(W \cup C) = 35 + 25 + 20 = 80$ and $n((W \cup C)') = 100 - 80 = 20$. We are interested in the individuals who are in W' (prefer more frequent weather updates) and who are also in C' (prefer commercials spread throughout the hour), and therefore are in $W' \cap C'$. Since $W' \cap C' = (W \cup C)'$, the answer to the problem is

$$n(W' \cap C') = n((W \cup C)') = 20$$

■

FIGURE 1.9

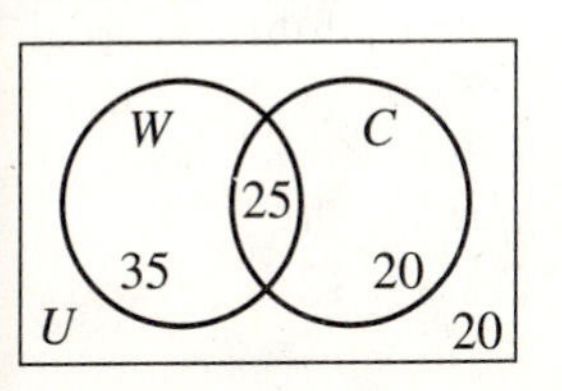

In Example 1.13 we used the fact that the number of elements in $(W \cup C)'$ is the difference between the number of elements in $W \cup C$ and the number of elements in the universal set U. This result holds for any subset A of U, and we have the formula

$$n(A') = n(U) - n(A) \tag{1.7}$$

The question raised in Example 1.13 can also be answered directly (without using a Venn diagram) by applying formulas (1.6) and (1.7). Proceeding in this way, we have

$$\begin{aligned} n(W \cup C) &= n(W) + n(C) - n(W \cap C) \\ &= 60 + 45 - 25 \\ &= 80 \end{aligned}$$

and therefore,

$$\begin{aligned} n[(W \cup C)'] &= n(U) - n(W \cup C) \\ &= 100 - 80 \\ &= 20 \end{aligned}$$

Nevertheless, *in general, it is best to draw the Venn diagram.* The diagram is a useful aid in organizing the information of the problem, and it helps us to spot mistakes which may result from using the formulas incorrectly. Also, the logic followed in using a Venn diagram is the same when there are three or more subsets of interest as when there are only two. This is illustrated in the following example.

Example 1.14 *Health care*

The Flabnomore Exerciser Company requires each of its employees to pass a yearly physical examination. The results of the most recent examination of 50 employees were that 30 employees were overweight, 25 had high blood pressure, and 20 had a high cholesterol count. Moreover, 15 of the overweight employees also had high blood pressure, and 10 of those with a high cholesterol count were also overweight. Of the 25 with high blood pressure, there were 12 who also had a high cholesterol count. Finally, there were 5 employees who had all three of these undesirable conditions. When the reports reached the desk of the president, Jox Chinup, he asked, "Don't we have any completely healthy employees around here?"

Problem Answer his question and find the exact number of employees free of all these symptoms.

Solution Let U be the set of all employees, with O, B, and C the sets of employees who are overweight, who have high blood pressure, and who have a high cholesterol count, respectively. The information gathered in the tests can then be summarized as follows:

$$\begin{array}{llll} n(U) = 50 & n(O) = 30 & n(B) = 25 & n(C) = 20 \\ n(O \cap B) = 15 & n(O \cap C) = 10 & n(B \cap C) = 12 & \\ n(O \cap B \cap C) = 5 & & & \end{array}$$

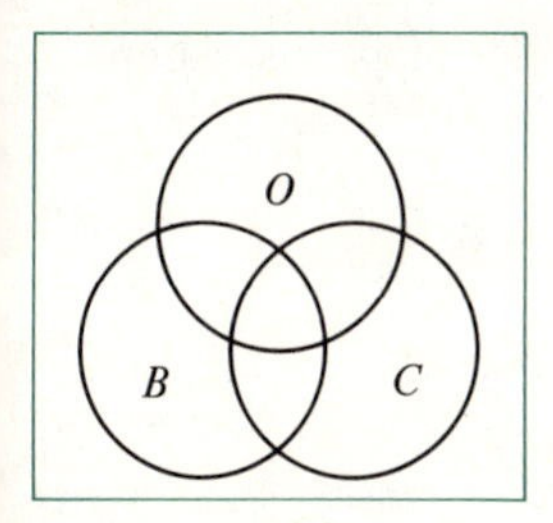

FIGURE 1.10

This problem is concerned with three subsets of U, and an appropriate Venn diagram is shown in Figure 1.10. Let us begin with the intersection $O \cap B \cap C$, which is known to have 5 elements, and place a 5 in that portion of the diagram, as in Figure 1.11*a*. Since $O \cap B$ has 15 elements, it follows that the portion of $O \cap B$ which is not in $O \cap B \cap C$ contains 10 elements. Likewise, since $O \cap C$ contains 10 elements, the part of $O \cap C$ not in $O \cap B \cap C$ contains 5 elements, and since $B \cap C$ contains 12 elements, the part of $B \cap C$ not in $O \cap B \cap C$ contains 7 elements. This information is shown on Figure 1.11*a*.

Next, since the numbers of elements in O, B, and C are known, we can use the information determined thus far to compute the number of elements in O but not in B or C, and so forth. This information is shown on Figure 1.11*b*. Note that each number on the Venn diagram gives the number of employees in the set corresponding to that particular portion of the Venn diagram in which the number is printed. Thus there are 25 employees in set B but only 3 in the part of B not in $O \cup C$, and consequently there is a 3 printed in the portion of the Venn diagram corresponding to $B \cap (O \cup C)'$. The seven regions bounded by arcs in Figure 1.11*b* represent a partition of $O \cup B \cup C$, and the number of elements in each of them is known. Using Equation (1.2), we have

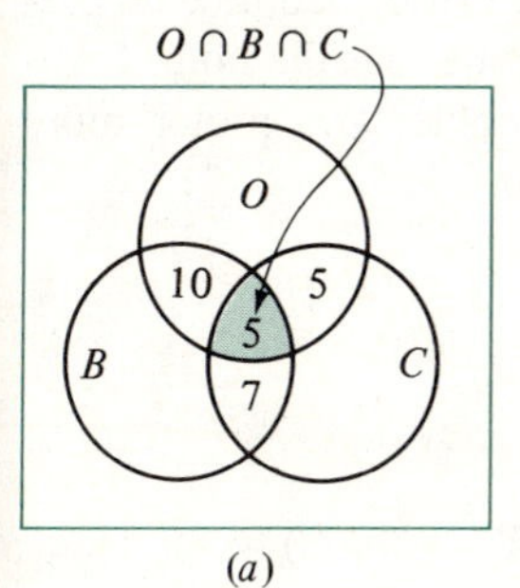

(*a*)

$$n(O \cup B \cup C) = 3 + 10 + 5 + 7 + 10 + 5 + 3 = 43$$

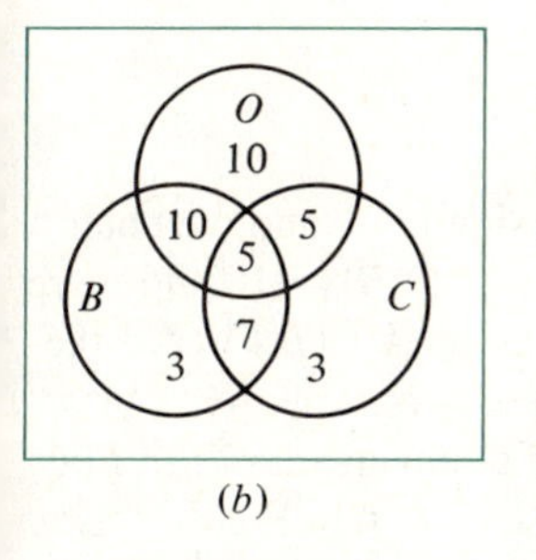

(*b*)

FIGURE 1.11

Therefore, since there are 50 elements in U, there are $50 - 43 = 7$ elements in $(O \cup B \cup C)'$. That is, 7 employees do not have high blood pressure or high cholesterol, nor are they overweight. ■

In some situations similar to that of Example 1.14, it initially appears that we do not have enough information to solve the problem. However, by introducing unknowns, or variables, to represent quantities we do not know, we can sometimes achieve our goal.

Example 1.15 *Entertainment*

In the setting described in Example 1.13, suppose that 100 people are surveyed, that 15 people prefer neither hourly weather updates nor commercials three times an hour, and that 20 people prefer both.

Problem How many people prefer exactly one of the two?

Solution Let E denote the set of people who prefer exactly one of the two: hourly weather updates or commercials three times an hour. Our goal is to determine the unknown quantity $x = n(E)$. Using the notation of Example 1.13, namely, W denotes the subset of people who prefer hourly weather updates and C denotes the subset of people who prefer commercials three times an hour, we can write the given information as $n((W \cup C)') = 15$ and $n(W \cap C) = 20$. The sets $(W \cup C)'$, E, and $W \cap C$ form a partition of the set of people sampled. Therefore, from (1.2) we have

$$n((W \cup C)') + n(E) + n(W \cap C) = 100$$

Since $n((W \cup C)') = 15$ and $n(W \cap C) = 20$, it follows that

$$x = n(E) = 100 - 15 - 20 = 65$$

■

In Example 1.15, it is impossible to determine the number of people who prefer hourly weather updates and commercials spread throughout the hour. Likewise, it is impossible to determine the number of people who prefer more frequent weather updates and commercials three times an hour.

Example 1.16 We are given sets A, B, and C with

$$n(A \cup B \cup C) = 85 \qquad n(A) = 50 \qquad n(B) = 40 \qquad n(C) = 35$$
$$n(A \cap B) = 18 \qquad n(B \cap C) = 12 \qquad n(A \cap B \cap C) = 5$$

Problem Find $n(A \cap C)$.

Solution We begin, as usual, "to work from the inside out," and we enter 5 in the set $A \cap B \cap C$, as shown in Figure 1.12*a*. Since $n(A \cap B) = 18$ and there are 5 elements in $A \cap B \cap C$, there must be 13 elements in $A \cap B \cap C'$. Likewise, there must be 7 elements in $A' \cap B \cap C$. However, from the given information, there is no direct way to calculate the number of elements in $A \cap B' \cap C$, and we denote this number by x. This information is also shown in Figure 1.12*a*. Next, using the partition of B shown in the figure and the numbers of elements in three of the four subsets (which we know), we conclude from the partition principle that there are 15 elements in $B \cap (A \cup C)'$. A similar argument can be

FIGURE 1.12

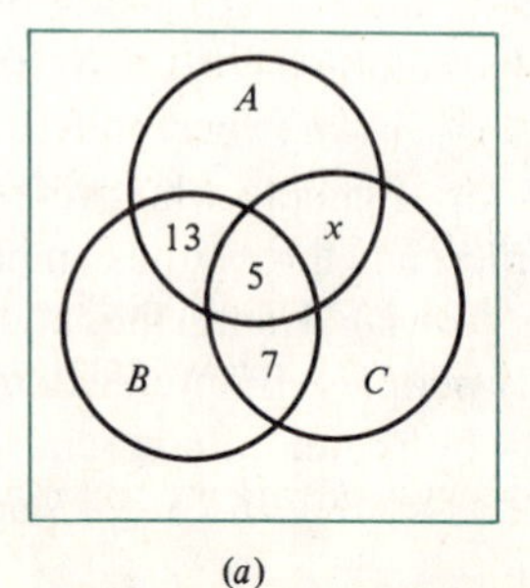

(*a*)

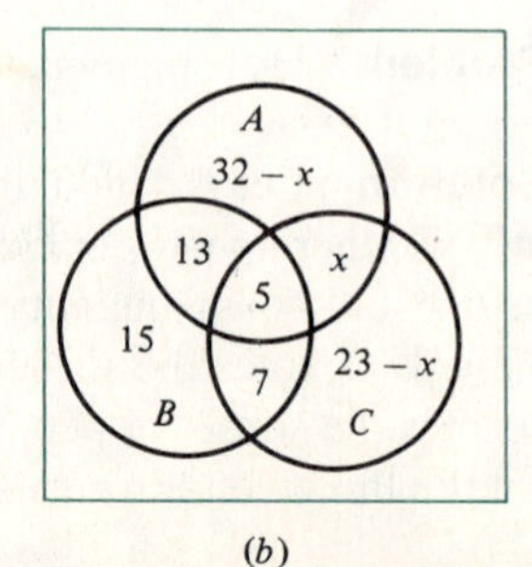

(*b*)

used with set A. Indeed, there is a partition of the set A into four sets, and we have the numbers of elements in three of these sets as 13, 5, and x. Since $n(A) = 50$, we must have $n(A \cap (B \cup C)') = 50 - 13 - 5 - x = 32 - x$. Using a similar argument and set C, we find $n(C \cap (A \cup B)') = 23 - x$. These results are shown in Figure 1.12b. Finally, all the sets shown in Figure 1.12b give a partition of $A \cup B \cup C$, and consequently

$$\begin{aligned} n(A \cup B \cup C) = 85 &= 5 + 7 + 13 + x + 15 + (23 - x) + (32 - x) \\ &= 95 - x. \end{aligned}$$

Thus $85 = 95 - x$, and consequently $x = n(A \cap B' \cap C) = 10$. Finally, since $n(A \cap B \cap C) = 5$, we have $n(A \cap C) = 10 + 5 = 15$. ■

Exercises for Section 1.3

1. Find $n(A \cup B)$ given $n(A) = 25$, $n(B) = 15$, and $n(A \cap B) = 8$.
2. Find $n(A \cap B)$ given $n(A) = 45$, $n(B) = 25$, and $n(A \cup B) = 58$.
3. Let A and B be subsets of U with $n(U) = 100$, $n(A) = 60$, $n(B') = 30$, and $n(A \cup B) = 75$. Find $n(A \cap B)$.
4. Let A and B be subsets of U with $n(U) = 120$, $n(A') = 55$, $n(B') = 30$, and $n(A \cup B) = 105$. Find $n(A \cap B)$.
5. Let A and B be subsets of U with $n(U) = 150$, $n(B') = 70$, $n(A \cap B') = 30$, and $n(A \cap B) = 75$. Find $n(A \cup B)$.
6. Let A and B be disjoint subsets of U with $n(U) = 45$, $n(A) = 12$, and $n(B') = 22$. Find $n(A' \cap B)$.

Voter preferences

7. A state legislature is considering increases in the gas tax and additional spending for highways, and 50 likely voters are asked their views. Of these voters, 25 favor additional spending for highways, 10 favor a gas tax increase, and 8 favor both. How many of these voters favor neither a gas tax increase nor additional spending for highways?

Product performance

8. An item is said to be defective if it has a major defect or a minor defect or both. In a batch of 25 defective items, 20 have major defects and 14 have minor defects. How many items in the batch have both major and minor defects?

Course enrollments

9. There are 30 students in a biology lab section. Of these, 18 are also taking mathematics, 8 are taking economics, and 3 are taking both mathematics and economics. How many are taking neither mathematics nor economics?

Food preferences

10. Fifty people are interviewed about their food preferences. Twenty of them like Greek food, 32 like Italian food, and 12 like neither Greek nor Italian food. How many like Greek but not Italian food?

Course enrollments

11. At the local high school, 34 students take a course in mathematics, 26 take a course in psychology, and 12 take both. How many students take exactly one of these two courses?

Data surveys

12. A survey of 150 college students results in the following data:

100 read the student paper published by their school.
20 read the local city paper.
85 who read the student paper *do not* read the city paper.

(*a*) How many of the students surveyed read the city paper but not the student paper?
(*b*) How many read *at least* one of the papers?

Product performance 13. A quality control analyst for Wonder Widget Company is reviewing the performance of 15 different types of widgets. He finds that in comparison with the previous year, 12 have improved reliability and 8 have improved durability. Only 2 types have improved neither reliability nor durability. How many types of widgets have improved both reliability and durability?

Product performance 14. An automobile tested by a national highway traffic safety commission was found to have 20 production defects. Of these, 11 were classified as major defects and 8 were design defects; 4 were neither major defects nor design defects. How many of the design defects were major?

Careers 15. There were 100 premedical students who were not admitted to medical school. Asked whether they would be interested in careers as medical technicians or registered nurses, 54 of these students expressed an interest in medical technology, 32 in nursing, and 23 in both. How many students were interested in neither of these careers?

Investments 16. An account representative for a securities firm has 400 customers. She recommends stocks but not bonds to 120, bonds but not stocks to 80, and neither stocks nor bonds to 95. To how many of her customers does she recommend both stocks and bonds?

17. Let A and B be subsets of U, with $n(U) = 50$. If $n(A) = 24$, $n(B') = 30$, and there are 8 elements of U in A which are not in B, find the number in B which are not in A.

18. Let A and B be subsets of U with $n(U) = 50$, $n(A' \cap B') = 20$, and $n(A \cap B) = 6$. Find the number of elements which are in A or in B but not in both.

Sports 19. During December, 84 people purchase new cross-country skis from a sporting goods store. Of these people, 22 already own downhill skis, 15 own cross-country skis, and 5 own both. How many people do not own skis of either type?

Course enrollments 20. In a mathematics class with 250 students, 100 are also taking accounting, 150 are taking economics, and 200 are in an English composition class. Of those students in the mathematics class who are also taking economics, 25 are *not* taking either accounting or English, 75 are taking both accounting and English, and 25 are taking English but not accounting.
(*a*) How many of the mathematics students are taking accounting and economics but not English?
(*b*) How many are taking mathematics and accounting but neither English nor economics?

Business 21. Of the 200 vehicles in a corporation motor pool, there are 90 with antilock brakes, 70 with compact disk (CD) players, and 40 with both. How many vehicles have exactly one of these features?

Business 22. In the situation described in Exercise 21, suppose that of the 200 vehicles there are 140 sedans and 60 minivans. Among the minivans there are 30 with antilock brakes, 40 with CD players, and 20 with both. How many minivans have antilock brakes but not a CD player?

Business 23. An accounting firm has partners who are specialists in specific areas. The areas of specialization and the number of partners with each specialty are shown in Table 1.1. If every partner is a specialist in at least one area, how many partners are there?

TABLE 1.1

Specialization	Number
Auditing	11
Consulting	9
Tax	12
Auditing and consulting	5
Auditing and tax	8
Consulting and tax	7
All three	3

Data surveys 24. A survey of 500 families provided the following data:

63 families subscribed to *The Wall Street Journal.*
41 families subscribed to *Rolling Stone.*
37 of the families who subscribed to *Rolling Stone* did not subscribe to *The Wall Street Journal.*

(*a*) How many families subscribed to both?
(*b*) How many families subscribed to neither?

Marketing 25. A market analyst at Healthful Drug Corporation is analyzing the results of a market survey on a new product, Acheaway pain reliever. Each individual surveyed was asked to respond (positively, neutral, or negatively) to the effectiveness of the drug, the side effects (if any), and its cost. There are 150 completed surveys. Of those surveyed, 60 responded positively to effectiveness, 50 responded positively to side effects, and 40 responded positively to cost. Also, 20 responded positively to both effectiveness and side effects, 15 to side effects and cost, 10 to cost and effectiveness, and 37 to none of the items. Find the number that responded positively to all three.

Product performance 26. Eighty-two individuals have complained to the Consumer Protection Agency about the 1993 Joltmobile. The information contained in the letters of complaint is summarized below.

25 complained about steering.
23 complained about comfort.
22 complained about visibility.
11 complained about steering and comfort.
7 complained about steering and visibility.
5 complained about all three.
33 complained about none of the three.

(*a*) How many people complained about comfort and visibility but not about steering?

(*b*) How many complained about *exactly* one of the three items: steering, comfort, and visibility?

27. Sets A, B, and C are subsets of a universal set U. Suppose $n(U) = 80$, $n(A) = 15$, $n(A \cap B) = 2$, $n(C) = 30$, $n((A \cup B)') = 35$, and $(A \cup B) \cap C = \varnothing$. Find:
(*a*) $n(C')$ (*b*) $n(A \cap C)$ (*c*) $n(A' \cap B)$

Entertainment

28. One hundred people who attend a movie preview are asked about the plot and the acting. Of these, 40 people liked the plot, 20 liked the plot and the acting, and 30 liked neither. How many people liked the acting?

Recreation

29. Each of the students in the Outdoor Club at Gigantic State University like at least one of the activities of hiking, camping, and canoeing. Of these students, 90 like either hiking or camping or both, 60 like canoeing, and 30 like all three. What can be said about the number of students who like canoeing and exactly one of hiking or camping?

Business

30. A corporation employs 95 people in the areas of sales, research, and administration. Some of the employees can function in more than one area; indeed, 10 can function in any of the three areas, 30 can function in sales and administration, 20 can function in sales and research, and 15 can function in administration and research. There are twice as many people in sales as in research and the same number in sales and in administration.

(*a*) How many can function in exactly one area?
(*b*) How many of these employees can function in sales?
(*c*) How many can function only in sales?

(There really is enough information to answer these questions.)

Data surveys

31. Data, including geographic location, city size, and marital status, on 200 recent graduates of Gigantic State University are collected by the Alumni Association. The results are as follows:

108 live in the west.
86 live in a large city.
68 are married.
41 live in the west in a large city.
23 are married and live in a large city.
19 are married and live in the west.
12 are married and live in a large city in the west.

How many are unmarried, do not live in a large city, and do not live in the west?

Data surveys

32. The Transportation and Parking Committee at Gigantic State University collects data from 100 students on how they commute to campus. The following data are obtained:

8 drive a car at least part of the time.
20 use the bus at least part of the time.

48 ride a bicycle at least part of the time.
38 do none of these.
No student who ever drives a car also uses the bus.

How many students who ride a bicycle also drive a car or use the bus?

1.4 SETS OF OUTCOMES AND TREES

Certain types of sets are of special interest in the study of probability. The elements of these sets represent outcomes of experiments. The term "experiment" is used here in a much more general sense than in the physical sciences. We refer to activities such as flipping coins, rolling dice, drawing cards, interviewing people, and testing things as examples of experiments. Outcomes describe the consequences of an experiment. For instance, in an experiment consisting of flipping an ordinary coin once and noting which side lands on top, the outcome is either a head or a tail; the set of outcomes is $S = \{\text{head, tail}\}$. It is helpful to introduce the terminology here but to defer a more detailed discussion of the concepts to Chapter 2.

> A *sample space* is a set consisting of all possible outcomes of an experiment.

Example 1.17 *Marketing*

An experiment consists of making two telephone calls, one after the other, and noting in order whether each is completed, i.e, whether someone answers the phone.

Problem Describe the set of outcomes of this experiment, and form a sample space.

Solution Either each call is completed, or it is not. We use C to denote the result of a call which is connected and N to denote the result that it is not connected. The result of making two calls in succession is a pair of letters, the results of the two calls in order. For instance, CN denotes the outcome that the first call is completed and the second is not. With this notation the possible outcomes of the two calls are the pairs CC, CN, NC, and NN. It is important to note that the outcome CN (the first call is completed but the second is not) is different from the outcome NC (the first call is not completed but the second is). The set of outcomes of the experiment can be represented by S:

$$S = \{CC, CN, NC, NN\}$$

■

Notice that each outcome of the experiment of Example 1.17 is an ordered pair, and we could have described the sample space as a cartesian product. However, in many cases it is not possible to describe the set of outcomes of an experiment as a cartesian product, and it is useful to have an approach which is applicable to all experiments.

Since the sample space is a set of outcomes of an experiment, the nature of the elements in the sample space depends on exactly how the experiment is conducted and how the outcomes are described. For instance, when we made the telephone calls of Example 1.17, suppose that we were interested only in the number of completed calls. In this case the sample space could be represented by the set

$$T = \{0, 1, 2\}$$

Here the outcome corresponding to both calls being completed is denoted by 2, the outcome of exactly one call being completed is denoted by 1, and the outcome of neither call being completed is denoted by 0. Not only are the sets S and T different, but even $n(S) \neq n(T)$. That is, the two sample spaces do not even have the same number of elements.

Example 1.18 An experiment consists of flipping a coin 3 times and recording the result of each flip (head or tail) in order.

Problem Describe the outcomes of this experiment, and form a sample space.

Solution Each time the coin is flipped, it comes up either heads (denoted H) or tails (denoted T). The experiment consists of *three* flips of the coin, so an outcome must tell what happened on each flip. If all three flips result in heads, it is natural to denote the outcome by HHH. Similarly, if the first two flips result in heads and the third flip results in a tail, then we denote the outcome by HHT. In this way we can represent any outcome as a sequence of three letters, each letter an H or a T. We must take care, however, that in listing the outcomes we do not omit any of them or include any one outcome more than once. One way to construct a list which includes each outcome exactly once is to use a diagram to represent the result of each flip as it occurs. The diagram shown in Figure 1.13 provides a systematic way of identifying all the possible outcomes of the experiment. We see that the sample space is

$$S = \{HHH, HHT, HTH, HTT, THH, THT, TTH, TTT\}$$ ■

Diagrams such as the one in Figure 1.13 are called *tree diagrams*. They are very useful in representing the outcomes of experiments which take place in

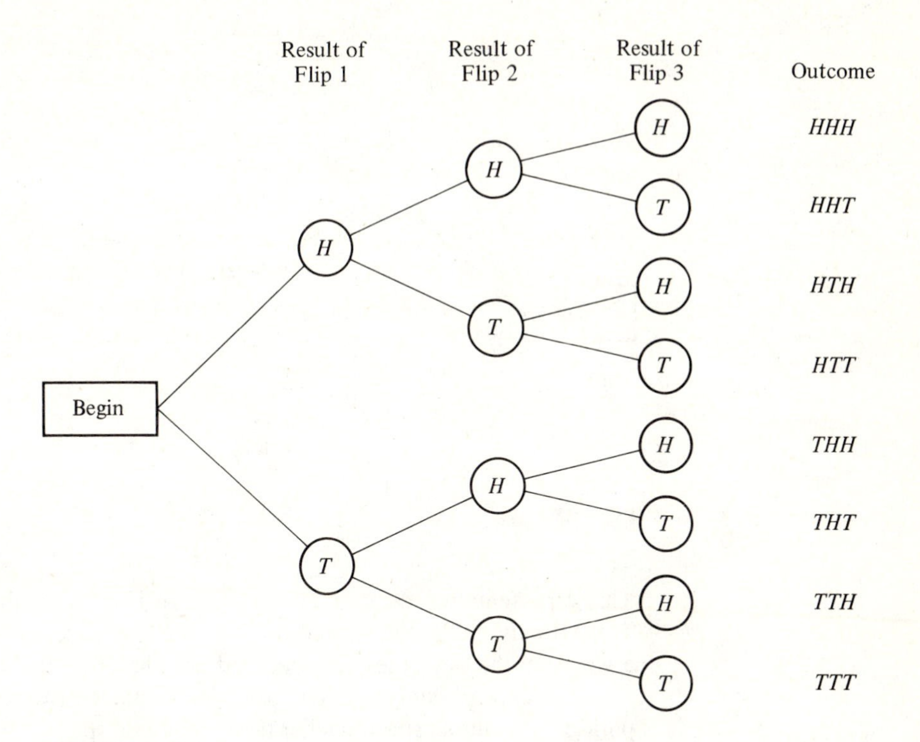

FIGURE 1.13

steps or stages. In a tree diagram, the results of each stage are represented by using the forks and branches of the tree. Each path through the tree (from left to right in our diagrams) represents the results of all stages of the experiment and therefore is one outcome of the experiment. For example, the diagram in Figure 1.14 represents an experiment with two stages. At the first stage there are two possible results, marked a and b. If result a occurs, then there are three possible results for the second stage, marked R, W, and G. If b occurs at the first stage, then there are two possible results for the second stage, marked R and W. It is clear from Figure 1.14 that the experiment has five outcomes. The sample space for the experiment is

$$S = \{aR, aW, aG, bR, bW\}$$

In the experiment with the tree diagram in Figure 1.14, a typical outcome such as aR could have been denoted by (a, R) or by a-R or in any of a number of other ways. Thus the sample space for the experiment could have been denoted by

$$S = \{(a, R), (a, W), (a, G), (b, R), (b, W)\}$$

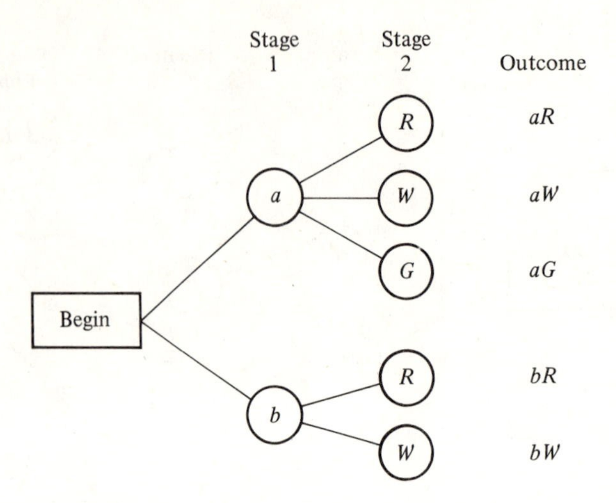

FIGURE 1.14

This representation of S and the one given just above differ only in notation. This will always be the case for us. That is, experiments will be defined in such a way that the outcomes are specified and therefore the sample space is determined except possibly for notation. This is the justification of our use of the phrase "the sample space" rather than "a sample space."

Example 1.19 A two-stage experiment consists of first selecting a bowl from the two shown in Figure 1.15 and noting its label (a or b) and then selecting a ball from that bowl and noting its color (R is red, W is white, and G is green).

Problem Describe the outcomes of this experiment, and find the sample space.

Solution The experiment has two stages: Select a bowl; then select a ball. Consequently, each outcome must include the result of stage 1 (the bowl) and the result of stage 2 (the color of the ball). Clearly, the tree diagram of Figure 1.14 depicts the experiment, and therefore the sample space is

$$S = \{aR, aW, aG, bR, bW\}$$

FIGURE 1.15

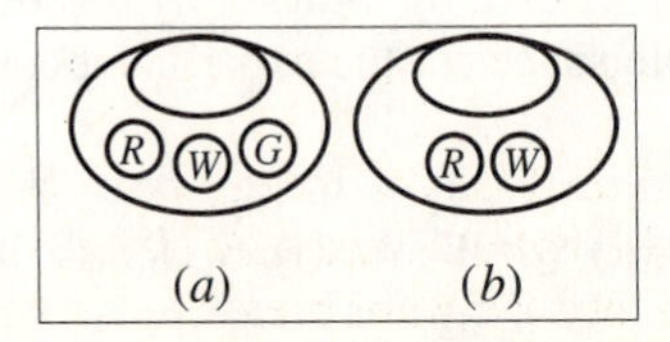

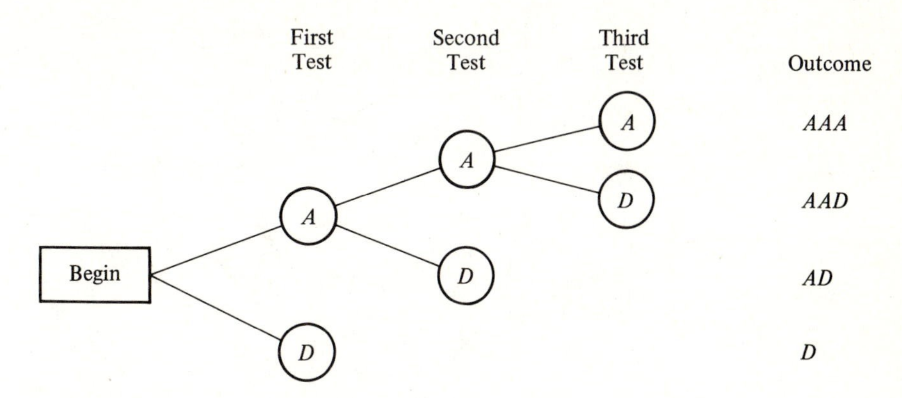

FIGURE 1.16

Example 1.20

Product testing

Videocassette recorder (VCR) tapes produced on an assembly line are either acceptable or defective. An experiment consists of checking tapes one after another until either a defective tape has been found or three tapes have been checked. Once a tape has been checked, it is set aside and not checked again.

Problem Draw a tree diagram to represent the experiment, and find the sample space.

Solution The first tape checked is either acceptable or defective. If it is defective, the testing process ends. If it is acceptable, then a second tape is checked and the process is repeated. Since at most three tapes are to be checked, the tree diagram is as shown in Figure 1.16. Given the notation that A represents the result that an acceptable tape is selected and D represents the selection of a defective tape, the sample space for the experiment is

$$S = \{D, AD, AAD, AAA\}$$ ■

Many of the problems which we shall meet in our study of probability do not require us to obtain the sample space S of an experiment. Instead they require only the *number* of outcomes $n(S)$. Of course, one way of determining the number of outcomes is to specify the sample space by a list and then count the elements in the list. However, for certain experiments there is a simple formula which can be used to obtain $n(S)$ without obtaining S. These experiments have a symmetric tree diagram. The next three examples illustrate that type of experiment and the formula.

Example 1.21

Food service

The local delicatessen has three types of meat—beef (B), ham (H), and turkey (T)—and two types of bread—light (L) and dark (D). An experiment consists of selecting a type of meat and then a type of bread.

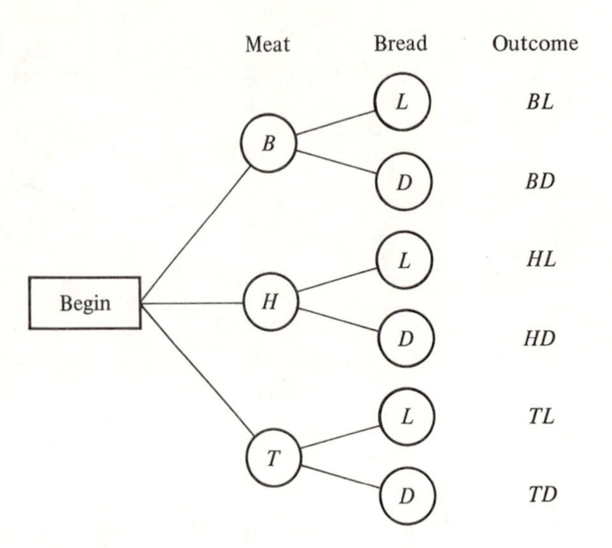

FIGURE 1.17

Problem Draw a tree diagram for this experiment, and find the sample space.

Solution The tree diagram is shown in Figure 1.17. The sample space is

$$S = \{BL, BD, HL, HD, TL, TD\}$$

■

It is useful to analyze this example in greater detail. Note that the tree diagram has the property that at the second stage (selecting bread) the number of results (2) is independent of the result at the first stage. That is, if the meat selected is beef, then there are 2 possible results for the second stage, and the same result holds if the meat is ham or turkey. Since there are 3 possible results at the first stage (B, H, T), and since for each of the results at the first stage there are 2 possible results at the second stage (L and D), it follows that there are $3 \times 2 = 6$ possible outcomes for the experiment: $n(S) = 6$.

Example 1.21 is a simple one in that the tree is easy to draw and it is clear that there are 6 possible outcomes for the experiment. The example could be made more complicated by increasing the number of possible results at each stage. However, if the structure of the experiment remains the same, it is still a simple matter to determine the number of possible outcomes of the experiment without drawing the tree diagram. For instance, suppose that at the first stage of a two-stage experiment any of 6 possible results can occur. Also, suppose that at the second stage there are 4 possible results for each of the results of the first stage. Then the associated tree diagram would have 6 branches at the first stage and each of these 6 branches would split into 4 branches at the second stage (see Fig-

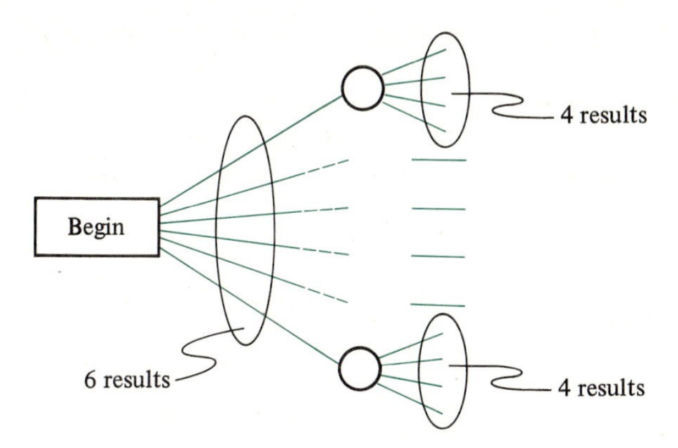

FIGURE 1.18

ure 1.18). The complete tree diagram represents $6 \times 4 = 24$ outcomes for the experiment.

In the general case of a two-stage experiment like that of Example 1.21, suppose that the first stage has n_1 results and the second stage has n_2 results (for each of the results at the first stage). Then the experiment has $n_1 \times n_2$ possible outcomes, which is the number of elements in the cartesian product of the sets of results for the first and second stages. Indeed, an outcome of a multistage experiment of this type is an ordered set of results of the component stages, i.e., an element of the cartesian product of the sets of results of the individual stages. Counting the number of outcomes is the same as counting the number of elements in a cartesian product. The general result can be formulated as the following principle.

Multiplication Principle

Consider a multistage experiment consisting of k stages. Suppose that the first stage has n_1 possible results, the second stage has n_2 possible results regardless of the result of stage 1, the third stage has n_3 results regardless of the results of the first two stages, . . . , and the kth stage has n_k results regardless of the results of the earlier stages. Then there are

$$n_1 \times n_2 \times n_3 \times \cdots \times n_k$$

elements in the sample space of the experiment.

Example 1.22 An experiment consists of flipping a coin twice, noting which face comes up each time, and then drawing a card from an ordinary deck and noting its suit.

Problem Find the number of elements in the sample space of this experiment.

Solution This experiment has three stages: flip a coin, flip a coin again, draw a card. The first stage has two results, so $n_1 = 2$. Likewise, $n_2 = 2$. There are four suits in an ordinary deck (spades, hearts, diamonds, and clubs), so $n_3 = 4$. It follows from the multiplication principle that there are $2 \times 2 \times 4 = 16$ elements in the sample space. ■

Example 1.23

Product testing

A doctor and a pharmaceutical manufacturer plan to test a new pain-relieving drug. The tests can be performed on either men or women, on people in any of four age groups, and in any of five geographical areas.

Problem Find the number of plans for the test from which the doctor must choose.

Solution A plan for the test, i.e., a choice of women or men, an age group, and a geographical location, is equivalent to a path in the tree diagram of the experiment. The multiplication principle can be applied in this case, and we conclude that there are $2 \times 4 \times 5 = 40$ outcomes and, consequently, 40 paths in the tree diagram. The doctor must choose among 40 plans. ■

Example 1.24

License plates

A state has used all the license plate numbers available under the current plate design, and alternatives are being considered. One of the proposed designs consists of a number from 1 through 99 followed by three letters selected from the standard alphabet with I and O removed (to avoid confusion with the integers 1 and 0). Thus, there are 24 choices for each letter.

Problem (*a*) How many different plates are available with the proposed design?

(*b*) How many plates do not include the letter Q?

(*c*) How many plates are there in which the first of the 3 letters is not a Q?

Solution (*a*) The proposed design is a number followed by 3 letters. The number is to be selected from the integers 1 through 99, for 99 choices. Each letter is to be selected from among the 26 letters with I and O removed, for 24 choices. Thus, by the multiplication principle, there are $99 \times 24 \times 24 \times 24 = 1{,}368{,}576$ different plates available under the proposed design.

(*b*) If the plate does not contain the letter Q, then there are only 23 possible choices for each of the letters, and there are $99 \times 23 \times 23 \times 23 = 1{,}204{,}533$ different plates available.

(*c*) If the first of the three letters is not a Q, then there are 23 choices for that letter. There are 24 choices for each of the other two letters. Therefore, there are $99 \times 23 \times 24 \times 24 = 1{,}311{,}552$ different plates available in which the first letter is not a Q. ■

Exercises for Section 1.4

Note: The result of rolling a die is the number of dots on the side which lands facing up.

Educational testing

1. An experiment consists of answering four true-false questions with either a T or an F. Find the sample space.

Educational testing

2. A multiple-choice test has 10 questions, and there are 5 choices for the answer to each question. An answer sheet has 1 answer for each question. How many different answer sheets are possible?
3. An experiment consists of drawing a card from an ordinary deck of cards and noting its suit, replacing that card and drawing another, again noting the suit. Describe the outcomes, and find the sample space of the experiment.
4. An experiment consists of drawing 4 cards, one after another, from an ordinary deck and noting the number of times a spade is selected. Find the sample space of this experiment.
5. An experiment consists of rolling a die twice and noting the sum of the results. Find the sample space of the experiment.

Marketing

6. An experiment consists of making three telephone calls, noting each time whether the phone is answered, there is a busy signal, or the phone rings but is unanswered. Describe the outcomes for this experiment and find the number of elements in the sample space.
7. An experiment consists of rolling a die twice, each time recording a 0 if the result is an odd number and the number if the result is an even number. Describe the outcomes of this experiment, and find the sample space.
8. Let $X = \{A, B, C\}$ and $Y = \{1, 2, 3, 4, 5\}$. A product code consists of 2 different symbols selected from set X followed by 2 not necessarily distinct symbols selected from the set Y. How many different product codes are there?

Biology

9. A wildlife biologist plans an experiment to determine the effect of a growth inhibitor on freshwater algae. There are 3 different ways in which the tests can be performed; the biologist can choose any one of 4 different lakes, and the tests can be performed during any single month from March through September, inclusive. Find the number of plans for the experiment from which the biologist must choose.

10. An experiment consists of flipping a coin and noting which side faces up and then rolling a die twice and noting the result of each roll. Find the number of elements in the sample space of this experiment.

Transportation 11. A shipper has 3 routes from New York to Chicago, 4 routes from Chicago to Denver, and 3 routes from Denver to Los Angeles. In how many ways can merchandise be shipped from New York to Los Angeles by using these routes?

Marketing 12. A businessman identifies 3 markets in which a new product can be test-marketed. In each market he has his choice of newspaper, TV, or direct-mail advertising and either regular or discount pricing. A marketing plan consists of a market, an advertising method, and a pricing policy. How many marketing plans are there?

Sales 13. A sales representative makes 5 calls to potential customers. Each call has 3 possible results: She reaches the customer in person, the phone is answered by someone other than the customer and she leaves a message, and the phone is unanswered. A phone record is a list of the results of each call. How many possible phone records are there?

Opinion survey 14. In a public opinion survey, each of 200 people ranks inflation (I), unemployment (U), and taxes (T) in order of economic importance. Responses are coded so that a response of UTI, for example, means that the responder believes that unemployment is of first importance, taxes second, and inflation third. Consider the survey of one person as an experiment, and find the sample space for this experiment.

Psychology 15. Of 4 rats in a cage in a psychology laboratory, 3 are untrained and 1 is trained. A rat is removed from the cage, and it is noted whether or not it is trained, and then it is put into another cage. Two more rats are removed and treated in the same way. Draw a tree diagram for this experiment, and find the sample space.

Psychology 16. An experiment is just as described in Exercise 15 except that each rat is replaced in the original cage before another is selected. Thus there are 4 rats available for each selection. Draw a tree diagram for this experiment. Find the number of elements in the sample space.

Product codes 17. A product code is formed from the symbols Q, R, T, V, and Z. The code consists of four symbols arranged one after another, for example, QQRT.
(*a*) How many different product codes are there?
(*b*) How many different product codes do not contain Z?
(*c*) How many different product codes contain exactly one T?

Travel 18. There are direct flights from Indianapolis to Denver, flights which stop at St. Louis, and flights with a stop at Kansas City. Each of these 3 types of flights can be taken either during the day or at night.
(*a*) An experiment consists of selecting a type of flight and then a time. Draw the tree diagram for this experiment.
(*b*) Another experiment consists of selecting a time and then the type of flight. Draw the tree diagram for this experiment.

Travel 19. In the situation described in Exercise 18, suppose that the only flights which stop at St. Louis are night flights. Repeat parts (*a*) and (*b*) of Exercise 18 in this situation.

Course selection 20. A student can take either mathematics or accounting in large or small classes during the day or at night and economics in small classes during the day or at night. An experiment consists of selecting a course, then a time, then a class size. Draw a tree diagram for this experiment, and find the sample space.

Sales 21. At each call on a customer, a sales representative either makes a sale or does not. She plans to make calls, one after another, until she makes 2 sales or a total of 5 calls. Her log for the day consists of a list of the calls, with sale or no sale noted for each. How many possible logs are there? (*Hint*: Draw a tree diagram.)

Recreation 22. A student eats in a restaurant on Friday, Saturday, and Sunday each week. Each evening she selects from a Chinese restaurant, a German restaurant, a Greek restaurant, an Italian restaurant and Hamburger Heaven, and she is willing to eat at a restaurant more than once. In how many ways can she arrange her restaurant visits in a specific week?

Recreation 23. Suppose that in the situation described in Exercise 22 the Greek and German restaurants are closed on Sunday evenings. In how many ways can the student now arrange her restaurant visits?

Recreation 24. A student is planning a ski trip. He can go to Colorado or New England. There are 3 possible ski areas for Colorado and 2 times that he can go to each ski area. There are 4 ski areas in New England and 2 times for 2 of the areas but only 1 time for the other 2 areas. A plan involves a location, a ski area, and a time. How many possible plans are there?

Sociology 25. A sociologist can conduct a study in a rural area or in a suburb. If it is conducted in a rural area, it can be done by mail or by telephone, and either a short or a long questionnaire can be used. If it is conducted in a suburb, then it can be done by mail, telephone, or personal interview, and either a short or a long questionnaire can be used. How many different plans for study are there?

26. A box contains 1 red, 1 white, and 2 green balls. An experiment consists of drawing balls in succession without replacement and noting the color of each until a red ball is drawn.
(*a*) Draw a tree diagram for this experiment.
(*b*) How many outcomes in the sample space?

Sales 27. A telephone sales representative makes a sale, gets a tentative commitment, or fails to make a sale on each call. Today she plans to make telephone calls, one after another, until she makes 3 sales, gets 2 tentative commitments, or fails to make a sale. Her log for the day consists of a list of the calls, with sale, tentative commitment, or no sale noted for each. How many possible logs are there?

Food service 28. Suppose that you have \$50 to spend on meals and that a meal in an expensive restaurant costs \$20 and a meal at a moderately priced restaurant costs \$10. An experiment consists of deciding on a sequence of meals (expensive or moderate) whose total cost is exactly \$50. Draw a tree diagram for this experiment, and find the sample space.

29. An experiment consists of flipping a coin and noting whether the result is a head or a tail. If it is a head, then the coin is flipped three more times and the number of heads is noted. If the result of the first flip is a tail, then the coin is flipped once more and the result, head or tail, is noted. How many outcomes are there in the sample space of this experiment?

30. A trial of an experiment consists of first flipping a coin until either a head comes up or there are three tails and then rolling a die until either a 6 comes up or the sum of the number of dots on all rolls is at least 4. How many outcomes in the sample space of this experiment?

IMPORTANT TERMS AND CONCEPTS

You should be able to describe, define, or give examples and use each of the following:

Set
Element
Subset
Set equality
Intersection
Union
Empty set
Disjoint sets
Partition
Partition principle
Universal set
Complement
Cartesian product
Venn diagram
Number of elements in a set
Experiment
Sample space
Tree diagram
Multiplication principle

REVIEW EXERCISES

1. Let $A = \{1, 2, 3, u, v, x, z\}$, $B = \{2, 4, x, y\}$, and $C = \{2, 3, x, z\}$.
   (*a*) Find $A \cap B$. (*b*) Find $A \cup B$. (*c*) Find $(A \cap B) \cup C$.
2. With sets A, B, and C defined as in Exercise 1, decide whether each of the following statements is true.
   (*a*) $B \subset A$ (*b*) $y \in (A \cap C)$ (*c*) $C \subset (A \cup B)$
3. Let $U = \{2, 4, 6, \ldots, 26\}$, and let A, B, and C be subsets of U defined as follows:

$$A = \{x: \ x \text{ is a multiple of } 3\}$$
$$B = \{x: \ x < 15\}$$
$$C = \{4, 6, 8, 18\}$$

   Find the elements in each of the following sets.
   (*a*) $A \cap B$ (*b*) $(A \cup B) \cap C$ (*c*) $A \cup (B \cap C)$
   (*d*) C' (*e*) $A \cap B'$ (*f*) $B \cap (A \cup C)'$
4. With sets U, A, B, and C defined as in Exercise 3, find the following numbers.
   (*a*) $n(B)$ (*b*) $n(A \cup C)$ (*c*) $n(B' \times C)$
5. Let $A = \{1, x, y\}$ and $B = \{1, y, z\}$. Find $(A \times B) \cap (B \times A)$.
6. Let A, B, and C be subsets of U. Use Venn diagrams to illustrate the following sets.
   (*a*) $A' \cup B$ (*b*) $B' \cap C$ (*c*) $(A \cup B') \cap C$
7. *Geography* Let M = {Maine, Michigan, Minnesota, Montana}, and let S be the subset of $M \times M$ defined by

$$S = \{(x, y): x \text{ is a state to the west of } y\}$$

   (*a*) List the elements of S.
   (*b*) Use $M \times M$ as a universal set, and determine S'.
8. Suppose A and B are subsets of U, $n(U) = 60$, $n(A' \cap B') = 12$, and $n(A' \cap B) = n(A \cap B') = 17$. Find $n(A \cap B)$. Find $n(A \cup B)$.

Sports

9. In a group of 75 students, 43 said they enjoyed tennis and 24 said they enjoyed swimming but not tennis. Which of the following can be determined:
   (*a*) The number of students who enjoy both sports
   (*b*) The number of students who enjoy neither sport

10. In how many different ways can a grade of A, B, C, D, or F be assigned to each of 20 students?

Course selection

11. In the core curriculum of a university, a student must select one of 4 physical science courses, one of 5 humanities courses, and one of 6 social science courses. In how many different ways can a student make these selections?

12. A universal set U with $n(U) = 60$ contains subsets A, B, and C with $n(A) = 18$, $n(A \cap B) = 8$, $n(C) = 12$, $n((A \cup B)') = 25$, and $(A \cup B) \cap C = \varnothing$.
   (*a*) Find $n(C')$. (*b*) Find $n(A \cap C)$. (*c*) Find $n(A' \cap B)$.

Music preferences

13. One hundred first-year students are asked about their music preferences. Their preferences are as follows: 65 like rock, 35 like country, and 10 like both rock and country. How many like rock but not country music?

Education

14. Consider the following data concerning 160 English majors:

    76 take French.
    85 take German.
    33 take French and German.
    35 take German and Russian.
    32 take French as their only foreign language.
    15 take French, Russian, and German.

    Every student takes at least one foreign language.
    (*a*) How many take Russian?
    (*b*) How many take French and Russian but not German?

Data surveys

15. The following information was gathered from a group of 200 undergraduates:

    39 own an imported car.
    32 own a domestically manufactured car.
    98 own a bicycle.
    69 own neither a car nor a bicycle.

    (*a*) If no student owns two cars, how many own both a bicycle and a car?
    (*b*) If you also know that 21 owners of imported cars do not own bicycles, how many owners of domestically manufactured cars do own bicycles?

Food preferences

16. Tina, Debbie, and Harriet decide to order a pizza. Tina is to select the crust—thin, thick, or deep dish; Debbie picks the size—small, medium, large, or jumbo; and Harriet picks one topping—sausage, pepperoni, green pepper, mushrooms, or black olives. How many different pizzas could be ordered?

17. A box contains 3 red, 1 white, and 2 green balls. An experiment consists of drawing balls in succession without replacement and noting the color of each ball selected until either a red ball or 2 green balls are drawn. Draw a tree diagram for this experiment, and list the elements in the sample space.

18. A box contains 1 red, 1 white, and 2 green balls. An experiment consists of drawing balls in succession without replacement and noting the color of each

ball selected until both a red and a white ball have been selected. Draw a tree diagram for this experiment, and list the elements in the sample space.

19. A box contains 1 red, 1 white, and 2 green balls. An experiment consists of drawing a ball and noting its color. If it is red, it is set aside: otherwise, it is replaced. A second ball is drawn, and its color is noted. Draw a tree diagram for this experiment, and list the elements in the sample space.

Product performance

20. The owner of an Italian restaurant is training a new cook, and she monitors each batch of sauce. A batch of sauce is classified as acceptable (A), having an appearance defect (P), or having a taste defect (T). Testing stops, and the cook is fired, as soon as there are two batches which have taste defects or one batch with an appearance defect. Testing stops and the cook is retained after three acceptable batches have been tested. Draw a tree diagram for this situation.

Data surveys

21. A survey is to be made of purchasing agents on their views of the economy. Each interview is to be made by phone or in person. If by phone, either a short-form or a long-form questionnaire can be used. If the interview is in person, then either a short or long questionnaire or a questionnaire consisting of open-ended questions can be used. A survey is defined by a method of interview and a type of questionnaire. How many possible surveys are there? What are they?

Sports

22. A workout for a jogger is determined by a route, a pace, and an exercise program. If the jogger has 4 routes, 3 paces, and 5 exercise programs, how many different workouts are there?

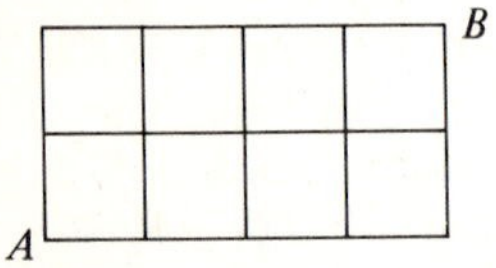

A B

FIGURE 1.19

23. How many different numbers between 10 and 500 can be formed with the digits in the set $\{1, 2, 4, 6, 8\}$ if no digit is repeated in any number?

24. How many different numbers between 10 and 900 can be formed with the digits in the set $\{1, 2, 3, 4, 6, 8, 9\}$ if no digit is repeated in any number? How many such numbers can be formed if digits can be repeated?

Construction

25. A utility company must lay lines from the point labeled A to the point labeled B on the map shown in Figure 1.19. The company is constrained to lay lines only in streets (shown by line segments), and it is interested in keeping its lines as short as possible. How many paths must be considered if all blocks are the same length? (*Hint*: Label all vertices and use a tree diagram.)

Psychology

26. A psychologist can conduct an experiment with a group of females, a group of males, or a mixed group. The experiment can be conducted in either 4 or 5 sessions, and the data can be recorded by hand or automatically. A design for the experiment consists of the specification of a group, a number of sessions, and a method for recording data. How many designs are there?

Manufacturing

27. A chair manufacturer has 4 types of upholstery, 2 types of wood, and 4 designs to choose from. If 2 of the designs allow any choice of upholstery and wood and the other 2 designs allow only a choice of upholstery, how many different chairs can he make? (*Hint*: Use a tree diagram in which the first choice is the design.

Marketing

28. A market analyst for Cleanup Inc. is planning the next market survey. Products which have been recommended by management for survey are laundry soap, bleach, shampoo, and toothpaste. A survey can be conducted in New England, the Gulf states, or the west coast. Each survey can be conducted by mail or in the supermarkets, and either a short or a long form can be used. Suppose that a survey consists of a set of products (one or more), an area, a method, and a form. Determine the number of possible surveys.

Task scheduling

29. A part-time secretary works 20 hours each week typing reports and market surveys. A report requires 5 hours of typing; a market survey requires 10 hours. Use a tree diagram to find the ways in which the work for a week can be organized so that exactly 20 hours are used and there is no partly completed job at the end of the week.

Business

30. A publisher produces oversized, normal-size, and pocket-size books that are in hardcover and/or paperback form. Oversized books are produced only in hardcover, pocket books are produced only in paperback, and normal-size books are produced in both hardcover and paperback. Hardcover books come in both standard and deluxe editions. An experiment consists of noting the size of a book, the type of cover, and (if applicable) whether it is a standard or deluxe edition. Draw a tree diagram for this experiment.

31. An experiment consists of flipping a coin one time after another and noting the result of each flip. The experiment stops when either the same side comes up twice in a row or the coin has been flipped 4 times. How many outcomes are there for this experiment?

Product codes

32. A product carries an identification number which consists of a letter, a single-digit number, and a four-digit number, for example, B-3-1018. The letter corresponds to the inspector, the single-digit number to the place of manufacture, and the four-digit number to the date (month and day) of manufacture. If there are 8 plants with 15 inspectors for each plant and if each month has 20 working days, how many possible identification numbers are there?

Food service

33. Suppose that you have \$50 to spend on meals and that a meal in an expensive restaurant costs \$15 and a meal at a moderately priced restaurant costs \$10. An experiment consists of deciding on a sequence of meals (expensive or moderate) whose total cost is exactly \$50. Draw a tree diagram for this experiment, and find the sample space.

Product performance

34. An engine inspector at an auto plant must inspect engines each hour in the following way: Each engine produced after the start of the hour is checked until either 5 good engines are found or 2 bad engines are found, whichever occurs first. View this inspection process as an experiment in which each outcome is a list (in order) of the engines tested with each engine labeled good or bad. How many elements are there in this sample space?

35. Let U be the set of integers 1 through 20. Find the largest subset X of U for which the sets X_1 and X_2 form a partition of X where

$$X_1 = \{n\colon\ n \in U \text{ and } n \text{ is a multiple of } 2\}$$
$$X_2 = \{n\colon\ n \in U \text{ and } n \text{ is a multiple of } 3\}$$

CHAPTER 2

Probabilities, Counting, and Equally Likely Outcomes

2.0 THE SETTING AND OVERVIEW

Each of us encounters situations involving uncertainty in our classes, in the workplace, and in everyday life. In this chapter we begin a study of probability—a mathematical tool which helps us understand and deal with uncertainty. Our study uses concepts from set theory and develops the idea of the probability of an event (a set of outcomes). To compute probabilities (numbers assigned to events), it is frequently necessary to count the number of elements in a set. Accordingly, we develop tools (counting principles) to help us count in situations where straightforward enumeration is inefficient. In some cases, we will be able to count the elements in a set by directly applying one of our tools; in other cases, two or more tools must be used together. We conclude the chapter by using these counting tools to compute probabilities in some special situations.

2.1 PROBABILITIES, EVENTS, AND EQUALLY LIKELY OUTCOMES

Probabilities are numbers assigned to sets of outcomes of an experiment; the probability of a set of outcomes represents the likelihood that one of the outcomes in the set will occur on a performance of the experiment. Probabilities arise in various ways. In some situations an experiment can be repeated any number of times under essentially identical conditions. In such circumstances an estimate of the probability (the likelihood) of the occurrence of a specific outcome of the experiment is given by the ratio of the number of times that outcome occurs to the number of times the experiment is repeated. That is, if a specific outcome occurs 62 times in 100 repetitions of an experiment, then the ratio 62/100 provides an estimate of the probability of that outcome. We call this method of assigning probabilities to outcomes the *relative frequency method*. Of course, a different number of repetitions or even the same number of repetitions another time may yield a different estimate of the probability. Assessing the relationship between different estimates and the number assigned as the probability of an outcome requires a deeper study of statistics than is possible in this book. Accordingly, if probabilities are assigned according to the relative frequency method, only one estimate will be given.

In many situations it is not practical to repeat an experiment hundreds of times, and the relative frequency method cannot be used. For some of these situations we will be able to use a method which is based on properties of the experiment. For instance, an ordinary coin has two faces which have different designs but are otherwise identical. It is reasonable to assume that if the coin is flipped hundreds (or thousands) of times, each face will come up about the same number of times. Likewise, it is reasonable to assume that if the coin is flipped only once, then it is just as likely to come up heads as to come up tails. Therefore, if we wish to assign probabilities to reflect these likelihoods, then we must assign the same probability to heads as to tails. Since there are two possible outcomes and we expect each to occur one-half of the time, we assign probability $\frac{1}{2}$ to each outcome. Similarly, if we have three balls colored blue, green, and red and otherwise identical and if a blindfolded person selects one ball arbitrarily, then it is reasonable to assume that each ball will be selected one-third of the time. We assign probability $\frac{1}{3}$ to the selection of the blue ball, $\frac{1}{3}$ to the selection of the green ball, and $\frac{1}{3}$ to the selection of the red ball to reflect this assumption. The method in which we assign probabilities by making assumptions and reasoning about the situation is called the *deductive method*.[1]

Using the deductive method, we can assign probabilities in situations not involving repeated experiments. For instance, even though a particular coin may not have been flipped repeatedly, we can assign probabilities by analyzing the situation and making assumptions. Such assumptions correspond to idealizations of real coins and may be used independently of any actual experiment. Thus in an example, we might consider a biased coin (as contrasted with a fair coin) for which the probability of a tail is $\frac{2}{3}$. This coin is an idealization of an actual coin which is somehow weighted so that if it were flipped many times, then tails would come up two-thirds of the time and heads one-third of the time.

[1]There is a third method of assigning probabilities to outcomes known as the *subjective*, or *measure of belief*, method. This method corresponds to the use of the term "probability" in the sentence: The probability of a flight to Mars with astronauts by the year 2010 is $\frac{1}{100}$. In this case the assigned probability is neither a sample frequency nor a probability deduced in the sense described above. Instead it is a subjective evaluation by the speaker of the likelihood that an event will take place. This is a common use of the term "probability," but we shall not pursue it in this book.

The term "experiment" has been used in this book several times (see also Section 1.4), and it is time for us to make it somewhat more precise. For us, activities such as testing VCR tapes for defects, flipping coins and noting heads or tails, and drawing balls and noting their colors are examples of experiments. An important part of the definition of an experiment is a clear specification of the outcomes of the experiment. For example, flipping a coin 3 times in succession and noting the result (heads or tails) of each flip define a different experiment than flipping a coin 3 times and noting the number of heads. In the first, a typical outcome might be recorded as HTH, and in the second as 2. A *performance* of an experiment is the carrying out of the activity which defines the experiment once. Note that in the experiment of flipping a coin 3 times, a single flip of the coin is not a performance of the experiment. Each performance of an experiment gives an outcome, and the set of all possible outcomes of an experiment is the sample space of the experiment (Section 1.4).

In addition to sample spaces, we are frequently interested in subsets of sample spaces. For instance, if we flip a coin 3 times, we might be interested in outcomes which include both heads and tails.

An *event* is a subset of a sample space of an experiment.

Example 2.1 A coin is flipped 3 times in succession, and the result is noted after each flip. The sample space (determined in Example 1.18) is

$$S = \{HHH, HHT, HTH, HTT, THH, THT, TTH, TTT\}$$

Problem Find the events (*a*) all heads, (*b*) exactly 2 heads, and (*c*) at least 2 heads.

Solution (*a*) $\{HHH\}$
(*b*) $\{HHT, HTH, THH\}$
(*c*) $\{HHT, HTH, THH, HHH\}$ ■

Note that the event "at least 2 heads" can be partitioned into the subevents "exactly 2 heads" and "exactly 3 heads." Thus

$$\{HHT, HTH, THH, HHH\} = \{HHT, HTH, THH\} \cup \{HHH\}$$

Partitioning events in this way is often a useful technique for solving certain types of counting and probability problems.

Example 2.2 *Sales*

A telephone sales representative makes successive calls to potential customers, and the result of each call, either a sale (S) or no sale (N), is noted. Calls are made until either 2 no sale calls are made or a total of 4 calls is made.

Problem Find the event E that exactly one sale is made and the event F that exactly one call results in no sale.

Solution A tree diagram for the situation is shown in Figure 2.1. The set listed under the column labeled "outcome" is the sample space for the problem. Inspection of the sample space shows that events E and F are

$$E = \{NSN, SNN\}$$
$$F = \{NSSS, SNSS, SSNS, SSSN\}$$

■

FIGURE 2.1

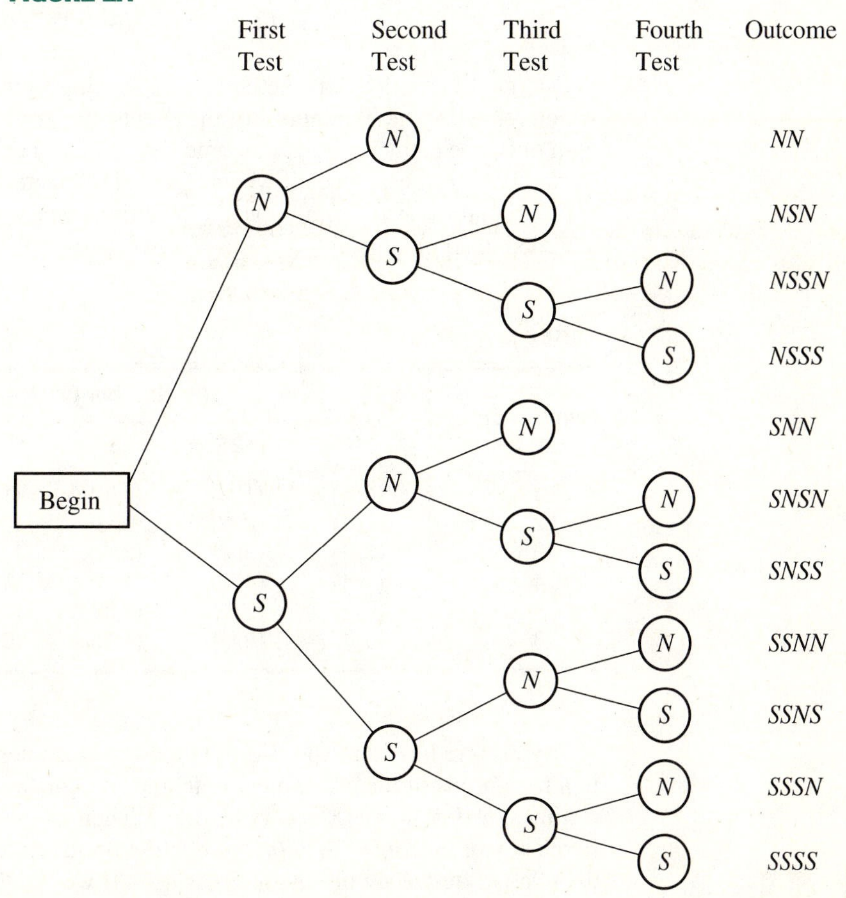

Example 2.3 An experiment consists of rolling a red die and a green die and noting the result of each roll.

Problem Find the sample space of the experiment, and identify the outcomes in events E and F, where

$$E = \{\text{the red die shows a } 3\}$$
$$F = \{\text{the sum of the results for the two dice is } 5\}$$

Solution We represent an outcome by an ordered pair (m, n), where m is the number of dots on the red die and n is the number of dots on the green die. With this notation the sample space can be represented as in the array in Table 2.1. Events E and F are

$$E = \{(3, 1), (3, 2), (3, 3), (3, 4), (3, 5), (3, 6)\}$$
$$F = \{(1, 4), (2, 3), (3, 2), (4, 1)\}$$

■

Our goal is to develop methods of assigning numbers to events, numbers which represent the likelihoods of the events occurring. An outcome is an event with only one element—that outcome. We begin by considering the assignment of numbers to these special events. Later, in Chapter 3, we study methods of assigning numbers to certain events in settings where we do not have numbers assigned to all outcomes.

TABLE 2.1

Number on Red Die	Number on Green Die					
	1	2	3	4	5	6
1	(1, 1)	(1, 2)	(1, 3)	(1, 4)	(1, 5)	(1, 6)
2	(2, 1)	(2, 2)	(2, 3)	(2, 4)	(2, 5)	(2, 6)
3	(3, 1)	(3, 2)	(3, 3)	(3, 4)	(3, 5)	(3, 6)
4	(4, 1)	(4, 2)	(4, 3)	(4, 4)	(4, 5)	(4, 6)
5	(5, 1)	(5, 2)	(5, 3)	(5, 4)	(5, 5)	(5, 6)
6	(6, 1)	(6, 2)	(6, 3)	(6, 4)	(6, 5)	(6, 6)

Suppose we have an experiment with three outcomes denoted 1, 2, and 3; and in n repetitions of the experiment outcome 1 occurs n_1 times, outcome 2 occurs n_2 times, and outcome 3 occurs n_3 times. Then $n_1 + n_2 + n_3 = n$. The relative frequency of outcome 1 is n_1/n, the relative frequency of outcome 2 is n_2/n, and the relative frequency of outcome 3 is n_3/n. If we use these frequencies as measures of likelihood, then we see that each is between 0 and 1 and that the sum of the likelihoods of all outcomes is 1.

Suppose that an experiment has a sample space S with n outcomes $\mathcal{O}_1, \mathcal{O}_2, \ldots, \mathcal{O}_n$. We have an *assignment of probabilities* to these outcomes if with each outcome $\mathcal{O}_i \in S$ there is associated a number w_i such that

$$0 \leq w_i \leq 1 \qquad \text{for } i = 1, 2, \ldots, n \tag{2.1}$$

$$w_1 + w_2 + \cdots + w_n = 1 \tag{2.2}$$

The probability of an outcome is also called the *weight* of that outcome. It is the probability associated with an outcome that we use as a measure of the likelihood of the event consisting of that outcome.

Different probability problems contain different information and ask different questions. In some cases the information provided enables us to determine the probabilities of all outcomes. This is the primary situation considered in this chapter.

Example 2.4 *Service times*

The equipment on a customer telephone service line records the length of incoming calls and classifies the calls according to length: a call of length less than 1 minute, 1 minute or more but less than 3 minutes, 3 minutes or more but less than 5 minutes, and 5 minutes or more. Denote the outcomes by $\mathcal{O}_1$, $\mathcal{O}_2$, $\mathcal{O}_3$, and $\mathcal{O}_4$, respectively. The choice $w_1 = \frac{1}{4}$, $w_2 = \frac{1}{4}$, $w_3 = \frac{1}{4}$, and $w_4 = \frac{1}{4}$ gives an assignment of probabilities. Another option is $w_1 = .15$, $w_2 = .25$, $w_3 = .40$, and $w_4 = .20$, and this choice also gives an assignment of probabilities. Although both choices are acceptable as assignments of probabilities, we need more information about the situation to know which choice is more appropriate. ■

Example 2.5 *Ecology*

A white-tailed deer population in a state park has grown beyond the number that can be supported by the available food supply, and the deer appear to be undernourished. The staff of the Department of Natural Resources investigates the situation by trapping several deer and checking them. The results are that undernourished deer outnumber those which appear adequately nourished by 5 to 1, and extremely undernourished deer outnumber those adequately nourished by 4 to 1.

Problem View the trapping and evaluating of a deer as an experiment with the outcomes "adequately nourished," "undernourished," and "extremely undernourished." Find an assignment of probabilities which reflects the data.

Solution Let $\mathcal{O}_1$, $\mathcal{O}_2$, and $\mathcal{O}_3$ denote the outcomes "adequately nourished," "undernourished," and "extremely undernourished," respectively, and let w_1, w_2, and w_3 be the associated probabilities. Since we know that undernourished deer outnumber those which appear adequately nourished by 5 to 1, we have $w_2 = 5w_1$; and since extremely undernourished deer outnumber those adequately nourished by 4 to 1, we have $w_3 = 4w_1$. Using these relations and Equation (2.2), we have

$$w_1 + w_2 + w_3 = w_1 + 5w_1 + 4w_1 = 10w_1 = 1$$

We conclude that

$$w_1 = .1 \qquad w_2 = 5w_1 = .5 \qquad w_3 = 4w_1 = .4$$

■

Up to this point we have assigned probabilities to outcomes only. We are also interested in the likelihoods of events, and we extend our use of the term "probability" from outcomes to events, i.e., sets of outcomes.

Suppose that we are given an experiment with a sample space S and an assignment of probabilities to outcomes. For any event $E \subset S$, the probability of E, denoted $\Pr[E]$, is defined to be the sum of the probabilities of the outcomes in E.

There are, of course, events consisting of a single outcome. By our definition, the probability of such an event is the same as the probability assigned to the outcome. Thus, for any outcome $\mathcal{O}$, the probability w associated with this outcome is equal to the probability of the event $\{\mathcal{O}\}$: $w = \Pr[\{\mathcal{O}\}]$. For notational convenience we use $\Pr[\mathcal{O}]$ to denote this probability.

Example 2.6 In the situation described in Example 2.4, suppose that the following assignment of weights is used:

Service times

$$w_1 = .15 \qquad w_2 = .25 \qquad w_3 = .40 \qquad w_4 = .20$$

Let E be the event that a call of length *at least* 1 minute is received.

Problem Find $\Pr[E]$.

Solution The event E is $\{\mathcal{O}_2, \mathcal{O}_3, \mathcal{O}_4\}$, and using the definition of the probability of an event,

$$\Pr[E] = w_2 + w_3 + w_4 = .25 + .40 + .20 = .85$$ ■

One of the conditions for an assignment of weights is that the sum of the weights of all outcomes in the sample space S be 1. Let E be an event in S, and let E' be the complement of E with respect to S as a universal set. Clearly those outcomes which are not in E must be in E', and consequently

$$\Pr[E] + \Pr[E'] = w_1 + w_2 + \cdots + w_n = 1$$

This relation, summarized below, is a useful tool in solving problems. We shall use it extensively in our future work on probability.

If E is any event in a sample space S and E' is the complementary event, then

$$\Pr[E] + \Pr[E'] = 1 \tag{2.3}$$

There are many situations in which *one* of $\Pr[E]$ or $\Pr[E']$ is easy to compute while the other is extremely difficult. The use of Equation (2.3) sometimes provides the only workable method of obtaining a certain probability.

Equally Likely Outcomes

In many situations in which the deductive method is used to assign probabilities, it can be argued that the probabilities of all outcomes should be the same. In such a case, since the sum of the probabilities of all outcomes is 1, it follows that the probability assigned to each outcome should be $1/n$, where n is the number of possible outcomes. For instance, rolling a die once and noting the result is an experiment with 6 outcomes, and if the die is fair, then each outcome should be assigned probability $\frac{1}{6}$. Also, flipping a coin and noting which face comes up is an experiment with 2 outcomes, and if the coin is fair, then each outcome should be assigned probability $\frac{1}{2}$. Experiments like this are said to be experiments with *equally likely outcomes*.

In experiments with equally likely outcomes, the probability assigned to each outcome is the same number, and this number is determined by finding the total number of outcomes n and taking $1/n$. Our first step is to find the total number of outcomes $n = n(S)$, the number of elements in the sample space.

Example 2.7 A bowl contains 3 balls: 1 blue, 1 white, and 1 green. An experiment consists of selecting a ball, noting its color, and then flipping a coin and noting the result, heads or tails.

Problem Find the number of outcomes. If the outcomes are equally likely, what weight should be assigned to each?

Solution There are 3 results for the color of the ball selected, and there are 2 results for the flip of the coin. Therefore, by the multiplication principle there are $3 \times 2 = 6$ outcomes for the experiment. Each outcome should be assigned probability $\frac{1}{6}$. ■

Example 2.8 An experiment consists of rolling a red die and a green die and noting the result of each roll.

Problem Assuming that the dice are fair and that all outcomes are equally likely, find the probability that should be assigned to each outcome.

Solution There are 6 numbers which can come up on the red die and 6 numbers which can come up on the green die. Therefore, $n(S) = 6 \times 6 = 36$; the 36 outcomes are shown in Table 2.1. Each outcome should be assigned probability $\frac{1}{36}$. ■

We have defined the probability of an event to be the sum of the probabilities of the outcomes in that event. If all outcomes are equally likely, then the probability of each outcome is $1/n(S)$. Since event E contains $n(E)$ outcomes, we have the following fundamental formula:

If an experiment with sample space S has equally likely outcomes, then for any event E the probability of E is given by

$$\Pr[E] = \frac{n(E)}{n(S)} \tag{2.4}$$

where $n(E)$ and $n(S)$ denote the number of elements in E and S, respectively.

It is important to remember that *Equation (2.4) is to be used only for experiments with equally likely outcomes*. For instance, it cannot be used to determine the probability of event E of Example 2.6 (a call of length at least 1 minute is received).

Example 2.9 An experiment consists of rolling a red die and a green die and noting the result of each roll.

Games of chance

Problem Find the probability that the sum is 8.

Solution Let S be the sample space, and let E be the event that the sum is 8. Then $n(S) = 36$ by Example 2.8. Using the notation of Example 2.3 [i.e., an outcome is denoted by (m, n) where m is the number of dots on the red die and n is the number of dots on the green die], we have

$$E = \{(6, 2), (5, 3), (4, 4), (3, 5), (2, 6)\}$$

and consequently $n(E) = 5$. Applying Equation (2.4), we have

$$\Pr[E] = \frac{n(E)}{n(S)} = \frac{5}{36}$$ ■

The assignment of probabilities based on an assumption of equally likely outcomes provides an example of a *probability measure*. It is known as the *equiprobable measure*. Probability measures in general will be defined and studied in Chapters 3 and 4. The equiprobable measure is frequently adopted in applications, and it gives an assignment of probabilities which corresponds closely to an intuitive idea of randomness. Indeed, when we use a phrase like "a card selected at random from an ordinary deck" without further comment regarding the likelihood of outcomes, it is understood that we assume all outcomes, i.e., all selections, to be equally likely. That is, for us the phrase "selected at random" indicates that an equiprobable measure is to be used. Also, we often use the term "fair" to refer to a die in which each face is equally likely to land up when the die is rolled or to a coin in which heads and tails are equally likely when the coin is flipped.

In Examples 2.8 and 2.9, the solution of the problem involves counting the number of elements in certain sets. Many sets S are too large or too complicated for us to list and count the elements. For these sets we need to use special counting techniques and formulas (such as the partition principle and the multiplication principle) to obtain $n(S)$. More of these techniques and formulas are developed in the next two sections. We then conclude the chapter by using these counting techniques and the formula $\Pr[E] = n(E)/n(S)$ to solve a variety of probability problems.

In some of the exercises for this section (Exercise 29, for instance) and elsewhere in the book, the phrase "without replacement" is used to describe experiments in which the item selected at one stage or step is not available to be selected again. If an item can be selected repeatedly, we use the phrase "with replacement."

Exercises for Section 2.1

Sales

1. A telephone sales representative makes successive calls to potential customers, and the result of each call is recorded as sale (S) or no sale (N). Calls are continued until either two successive sales are made, two successive calls result in no sale, or a total of 3 calls is made. Find the sample space for this situation.

Sales

2. In the situation described in Exercise 1, suppose calls are continued until either 2 successive sales are made, 2 successive calls result in no sale, or a total of 5 calls is made. Find the sample space for this situation.

Sales

3. In the situation described in Example 2.2, find the events

$$E = \{\text{at least 3 sales are made}\}$$
$$F = \{\text{the second call results in a sale}\}$$

4. An experiment consists of rolling a red die and a green die, noting the number on the red die and whether the green die is even (E) or odd (O). Thus, a typical outcome might be denoted $3E$, meaning that the red die shows a 3 and the green die is even. Find the sample space for this experiment.

5. In Example 2.3 find the events

$$E = \{\text{exactly 1 die has a result of 3}\}$$
$$F = \{\text{at least 1 die has a result of 3}\}$$
$$G = \{\text{the sum of the results is an odd number}\}$$
$$H = \{\text{the sum of the results is at most 7}\}$$

6. An experiment consists of rolling a die and noting whether the number of dots on top is even or odd. If the number is odd, then a coin is flipped once and the side landing up is recorded. If the number on the die is even, then a coin is flipped twice and the side landing up is recorded for each flip.
   (*a*) What is the sample space of this experiment? Be sure to define your notation.
   (*b*) What is event E that at least one head appears?
   (*c*) What is event F that no tails appear?
   (*d*) What is the relation between events E and F?

Economics

7. An economist reviews historical data on economic recoveries after recessions. On the basis of the data, she classifies each recovery as weak (W), strong (S), or exceptionally strong (E). Also she measures the increase in inflation during each recovery and classifies each recovery as having low (L), moderate (M), or high (H) inflation.
   (*a*) Find the sample space for the experiment of reviewing one recovery and noting its strength and inflation level.
   (*b*) Find the event that the recovery is strong.
   (*c*) Find the event that the recovery is strong or has low inflation.
   (*d*) Find the event that the recovery is strong and has low inflation.

Product performance

8. Automobile exhaust systems are to be given emission and vibration checks. The result of an emission check is satisfactory (S) or unsatisfactory (U), and the result of a vibration check is high vibration (H), low vibration (L), or no vibration (N).
   (*a*) Find a sample space for the experiment of checking one automobile.
   (*b*) Find the event that the emission check is satisfactory.

Opinion survey

9. A public opinion survey asks each of 200 individuals to rank inflation (I), unemployment (U), and taxes (T) in order of economic importance. Responses are coded so that, e.g., a response of UTI means that the responder believes that unemployment is of first importance, taxes second, and inflation third.
(*a*) If the survey is viewed as an experiment, what is the sample space?
(*b*) What is the event that inflation is of first importance?
(*c*) What is the event that taxes are more important than inflation?

Biology

10. A marine biologist monitors the changes in amounts of nutrients in a particular area of the ocean. The data are recorded as increased (I), unchanged (U), or decreased (D) on each of three consecutive visits to the area.
(*a*) What is the event that the amounts never decrease?
(*b*) What is the event that there are at least two increases?

Psychology

11. A psychology experiment consists of presenting a stimulus to a subject and noting the response. At each presentation of the stimulus, the subject is evaluated as making a strong association (S), a weak association (W), or no association (N). If the subject makes a strong association on the first presentation, the experiment ends. Otherwise it continues and a second stimulus is presented. The experiment ends after the second stimulus if the subject makes a strong association; otherwise it continues and a third stimulus is presented. The experiment ends after the response to the third stimulus (if one is presented) is recorded. An outcome is a sequence of responses to stimuli.
(*a*) Find the event that there is a strong association.
(*b*) Find the event that there is at least one weak association.
(*c*) Find the event that there is exactly one weak association.

Biology

12. A botanist plants 100 seeds in trays and monitors the results over a season. For each seed, he notes whether it germinates (G) or does not (N). If it germinates, he notes whether the plant survives (S) or dies (D), and if it survives, whether it flowers (F) or has only foliage (O). Finally, if it flowers, he notes the color of the flowers: pink (P), lavender (L), or white (W).
(*a*) Find the sample space for the experiment of monitoring a single seed over the season.
(*b*) Find the event that the seed germinates and survives.
(*c*) Find the event, i.e., the subset of the sample space, including those outcomes in which the seed germinates but not including the outcome where the plant has pink flowers.

13. An experiment with five outcomes has $w_1 = .2$, $w_2 = w_3$, $w_4 = .35$, and $w_5 = .15$. Find w_2.

14. A die is weighted so that outcome 1 is twice as likely as each of outcomes 2, 3, 4, 5, and 6, and these outcomes are equally likely. What probabilities (weights) should be assigned to the outcomes to reflect this information?

15. A die is weighted so that outcomes 1 and 2 are equally likely; outcome 1 is three times as likely as each of outcomes 3, 4, 5, and 6; and these outcomes are equally likely. What probabilities (weights) should be assigned to the outcomes to reflect this information?

16. A die is weighted so that the odd numbers are 3 times as likely to come up as the even numbers. All the even numbers are equally likely, and all the odd numbers are equally likely. What probabilities $w_1, w_2, w_3, w_4, w_5, w_6$ should be assigned to outcomes 1, 2, 3, 4, 5, 6, respectively?

17. Suppose an experiment has three possible outcomes $\mathcal{O}_1$, $\mathcal{O}_2$, and $\mathcal{O}_3$. Also suppose that $\mathcal{O}_1$ occurs with frequency $1/a$, outcome $\mathcal{O}_2$ with frequency $2/a$, and $\mathcal{O}_3$ with frequency $3/a$. What value must be assigned to a so that these frequencies can be used for an assignment of probabilities to the outcomes?
18. Consider a die constructed so that outcomes of 1, 2, 3, or 4 dots are equally likely, 5 is twice as likely as 1, and 6 is twice as likely as 5. What probabilities should be assigned to the outcomes to reflect this?
19. Consider a coin weighted so that when the coin is flipped, the result is heads 3 times as often as tails. What probabilities should be assigned to the outcomes "heads" and "tails" to reflect this weighting?
20. A die is weighted so that the odd numbers are equally likely, the even numbers are equally likely, and the odd numbers are k times as likely as the even numbers. If $\Pr[2] = \frac{1}{12}$, what is k?
21. Suppose an experiment has sample space $S = \{\mathcal{O}_1, \mathcal{O}_2, \mathcal{O}_3, \mathcal{O}_4, \mathcal{O}_5\}$. Also, let outcome $\mathcal{O}_i$ be assigned probability w_i, where $w_1 = w_2 = \frac{1}{5}$, $w_3 = \frac{2}{5}$, and $w_4 = w_5 = \frac{1}{10}$. Find $\Pr[E]$ and $\Pr[F']$ where $E = \{\mathcal{O}_2, \mathcal{O}_4\}$ and $F = \{\mathcal{O}_1, \mathcal{O}_4\}$.

Business

22. Of the 200 employees at Everest Securities, Inc., 65 are in retail sales, 45 are in institutional sales, 25 are in research, and the remainder are in administration. An experiment consists of selecting an employee at random and noting whether the employee is in sales. What probabilities should be assigned to the outcomes on the basis of the data?

Education

23. In a finite mathematics class with 300 students, 10 percent withdraw, 15 percent receive a grade of A, 20 percent receive B, 35 percent get C, 15 percent get D, and 5 percent receive F. Assign probabilities to the events "pass the course" and "withdraw or fail the course."

Biology

24. Two hundred duck hatchlings are weighed, and it is found that 38 are underweight, 12 are extremely underweight, and the remainder are normal weight. Consider an experiment consisting of weighing a single hatchling.
   (*a*) What are the possible outcomes of this experiment?
   (*b*) What probabilities should be assigned to these outcomes on the basis of the data?

Lotteries

25. A lottery is designed so that on average it pays $10 on 1 play out of 100, $1000 on 1 play out of 5000, and $10,000 on 1 play out of 100,000. All other plays have no return. Assign probabilities to the four possible outcomes of buying a lottery ticket.

Forestry

26. Consider an experiment which consists of planting 10,000 pine tree seeds. Suppose that it is known that about 90 percent of these seeds will germinate and of those that germinate 90 percent will live and grow into seedlings. Define the outcomes of the experiment of planting a single one of these seeds to be:
   (*a*) The seed does not germinate.
   (*b*) The seed germinates but dies.
   (*c*) The seed germinates and grows into a seedling.
   What probabilities should be assigned to each of these outcomes?
27. A product code consists of one number selected from {1, 2, 3, 5} followed by two letters, not necessarily distinct, selected from the set {Q, T, Z}. For example, 3QT is such a code. An experiment consists of selecting a code at random.

(*a*) How many codes are possible?
(*b*) What probability should be assigned to each code?
(*c*) Find the probability of the event that the code contains the number 3.

28. For the experiment described in Exercise 27:
(*a*) Find the probability of the event that the code contains exactly one Q.
(*b*) Find the probability of the event that the code contains at least one Q.

29. There are balls numbered 1 through 5 in a box. Two balls are selected at random in succession without replacement, and the number on each ball is noted.
(*a*) How many outcomes does this experiment have?
(*b*) What probability should be assigned to each?
(*c*) What probability should be assigned to the event that at least one ball has an odd number?

30. In the experiment described in Exercise 29, find the probability of the event that the sum of the numbers is less than 7.

31. A fair coin is flipped 3 times, and the result (heads or tails) is noted after each toss. How many outcomes does this experiment have? What probability should be assigned to each?

32. In the experiment of Exercise 31, find the probability of the event that there are at least two heads.

33. A box contains 3 balls: 1 red, 1 blue, and 1 yellow. The balls are drawn at random one after the other until all have been removed from the box. The color of each ball is noted as it is drawn. How many outcomes are there for this experiment?

34. The outcomes of the experiment described in Exercise 33 are equally likely.
(*a*) Find the probability that the red ball is selected first.
(*b*) Find the probability that the red ball is selected before the blue ball.

35. An experiment similar to that of Exercise 29 is conducted, except the first ball selected is replaced before the second ball is selected. Repeat Exercise 29 for this situation.

2.2 COUNTING ARRANGEMENTS: PERMUTATIONS

Many experiments which take place in steps or stages have outcomes, each of which can be viewed as a list of the results of successive steps in the experiment: an ordered list or arrangement. The number of elements in the set of outcomes of such an experiment can often be determined by using the multiplication principle. In fact, a special case of the multiplication principle, the *permutation principle*, is especially useful. Before stating the latter principle, we illustrate the type of situation to which it applies with two examples. Notice that in both examples, to apply the multiplication principle, we *create* the setting of an experiment; the experiment is not explicitly given in the statement of the problem.

Example 2.10

Committees

A program committee consists of 5 members: Audrey (A), Bill (B), Connie (C), David (D), and Emita (E). The committee is planning the next event, and one member is to handle publicity and another is to reserve a room for the program.

Problem Find the number of different ways of assigning pairs of committee members to the two tasks of publicity and room reservations. Two assignments are different if either task is handled by different people; in particular, if two people switch tasks, the result is a new assignment.

Solution Think of assigning people to the tasks as a two-stage experiment: the assignment of a person to handle publicity followed by the assignment of a person to reserve a room. The first stage, assigning someone to publicity, has 5 possible results. After the person handling publicity has been identified, and regardless of who has been given that assignment, there remain 4 people from which to select someone to reserve a room. Thus, by the multiplication principle, the number of outcomes of the experiment—the number of different ways to assign 2 people selected from the 5 available to 2 tasks—is $5 \times 4 = 20$. ■

Let us examine Example 2.10 more closely. We can think of assigning 2 people to 2 tasks as making a list of 2 names: The first person is to handle publicity, and the second is to reserve a room. If we use letters instead of names, then each list can be represented as an ordered pair of letters, (A, B), (D, E), etc. The list (A, B) means that Audrey handles publicity and Bill reserves a room. Clearly the list (A, B) is different from the list (B, A); in the latter, B handles publicity and A reserves a room. The multiplication principle can be applied to the formation of lists in exactly the way it was used in Example 2.10. We use the term "permutation" to denote an ordered list.

A *permutation* is an ordered list of elements selected from a set. Two permutations are *different* unless they consist of exactly the same elements in exactly the same order.

Recall that either an element occurs in a set or it does not. An element cannot belong to a set "several times." Consequently, the elements in a permutation are necessarily distinct—they have been selected from a set without replacement.

Example 2.11

How many permutations of 3 letters can be formed with letters selected from a set of 6 distinct letters?

Solution The first letter (first in the list) can be selected in 6 ways. After the first has been selected, the second can be selected in 5 ways. After the first 2 letters have been selected, there are 4 letters remaining and, consequently, the third letter can be selected in 4 ways. Using the multiplication principle, we conclude that the number of arrangements of 3 letters selected from 6 is

$$6 \times 5 \times 4 = 120$$ ■

As a variation of Example 2.10, suppose that the number of tasks in that example were expanded to include a third task, say, arranging for refreshments. Approaching this new problem in the same way as in Example 2.10, we ask for the number of different ways of making a list of 3 names: the first person handles publicity, the second reserves a room, and the third arranges for refreshments. By the multiplication principle, the number of different ways to assign 3 people selected from 5 people to do 3 tasks is $5 \times 4 \times 3 = 60$.

Another variation of Example 2.10 is illustrated by supposing that the committee consists of 6 members rather than 5. In this case 2 people selected from the 6 available can be assigned to the 2 tasks of Example 2.10 in $6 \times 5 = 30$ different ways.

Example 2.10, and its variations, and Example 2.11 illustrate a very common application of the multiplication principle. In such applications there is a set of objects (committee members in Example 2.10, letters in Example 2.11), and the goal is to count the number of ways a subset of objects can be selected from the entire set and arranged in order. When used in this way, the multiplication principle has a special name.

Permutation Principle

The number of permutations (ordered lists) of r objects selected from a set of n distinct objects is

$$n \times (n-1) \times \cdots \times (n-r+1)$$

Example 2.12 A city is divided into 3 districts which are to be surveyed for opinions on municipal services. There are 5 experienced survey takers available to do the job.

Opinion surveys

Problem How many different ways can surveyors be assigned to districts, exactly one surveyor to each district?

Solution There are 5 people available, and we must assign a person to each of 3 districts. Using the permutation principle with $n = 5$ and $r = 3$, we conclude that the assignment can be made in

$$5 \times 4 \times 3 = 60$$

ways. (Note that since $n - r + 1 = 5 - 3 + 1 = 3$, the smallest number which appears in the product is 3.) ■

Example 2.13 A map showing the 4 states of Arizona, Colorado, New Mexico, and Utah is to be colored so that each state is a different color. There are 8 colors available.

Geography

Problem In how many different ways can the map be colored? We consider two colorings to be different unless they have identical colors in each of the 4 states.

Solution We apply the permutation principle to conclude that the map can be colored in

$$8 \times 7 \times 6 \times 5 = 1680 \text{ different ways}$$ ■

It is convenient to introduce the notation $n!$ (read "n factorial") to denote $n \times (n-1) \times \cdots \times 2 \times 1$. Thus $5! = 5 \times 4 \times 3 \times 2 \times 1 = 120$, and $10! = 10 \times 9 \times 8 \times \cdots \times 2 \times 1 = 3{,}628{,}800$. When this notation is used, the expression $n \times (n-1) \times \cdots \times (n-r+1)$, which occurs in the permutation principle, can be written

$$\begin{aligned} n \times (n-1) \times \cdots \times (n-r+1) \\ &= \frac{n \times (n-1) \times \cdots \times (n-r+1) \times (n-r) \times \cdots \times 2 \times 1}{(n-r) \times \cdots \times 2 \times 1} \\ &= \frac{n!}{(n-r)!} \end{aligned}$$

We also define $0!$ to be 1. Therefore, when $r = n$, $(n-r)! = 0! = 1$.

Finally, it will be helpful to have a symbol for the expression $n!/(n-r)!$ which occurs frequently in applications. We introduce the symbol $P(n, r)$ by setting

$$P(n, r) = \frac{n!}{(n-r)!} \tag{2.5}$$

We refer to $P(n, r)$ as the *number of permutations of n objects taken r at a time*. With this notation the permutation principle can be restated as follows:

The number of permutations of n distinct objects taken r at a time is

$$P(n, r) = \frac{n!}{(n-r)!}$$

Remember that in permutations *order is important*. Thus $ABCD$ and $ACBD$ are different permutations of the 4 letters A, B, C, D. In fact, the total number of permutations of the 4 letters is

$$P(4, 4) = \frac{4!}{(4-4)!}$$
$$= \frac{4!}{0!} = \frac{4!}{1} = 4 \times 3 \times 2 \times 1 = 24$$

Example 2.14

Financial services

A financial services company has regional offices in Atlanta, Baltimore, Chicago, Denver, Rochester, Seattle, and Tampa. A consultant is to visit 4 different offices.

Problem (*a*) A schedule is a set of 4 different offices to visit and an order in which to visit them. How many schedules are there? (*b*) How many schedules include visits to Chicago or Denver or both?

Solution (*a*) There are 7 cities, and by the permutation principle there are $P(7, 4) = 7 \times 6 \times 5 \times 4 = 840$ schedules.

(*b*) We answer this question by finding the number of schedules which *do not* include visits to either Chicago or Denver, and we subtract this number from the answer to part *a*. Excluding Chicago and Denver, there are 5 cities. Therefore, there are $P(5, 4)$ schedules which do not include visits to either Chicago or Denver. We conclude that there are

$$P(7, 4) - P(5, 4) = (7 \times 6 \times 5 \times 4) - (5 \times 4 \times 3 \times 2) = 840 - 120 = 720$$

schedules which include either Chicago or Denver or both. ■

Our solution of Example 2.14*b* (there are other ways to solve this problem as well) illustrates a method based on solving two or more subproblems and then combining the solutions of the subproblems to solve the original problem. In this example we used the permutation principle twice, each time to solve a subproblem. First we used it to count all schedules, and then we used it to count the schedules which included neither Chicago nor Denver. Finally, the answer to (*b*) is the difference of these two numbers. This is a typical example of a multistep problem.

Example 2.15

Product codes

A product code consists of 3 different symbols. There are 5 different symbols available. We assume that the order in which the symbols appear in a product code is important and that two product codes are different unless they have exactly the same symbols in the same order.

Problem Find the number of distinct product codes which can be formed from 5 symbols.

Solution It is useful to think of a product code as a permutation of 3 symbols selected from 5 symbols. Adopting this point of view, we conclude from the permutation principle that the number of product codes is

$$P(5, 3) = \frac{5!}{(5-3)!} = \frac{5!}{2!} = 5 \times 4 \times 3 = 60$$

■

Example 2.16

Theater casting

A director of a community theater is conducting auditions for a play which has 6 distinct roles: 4 for females and 2 for males. There are 7 females trying out for the female roles and 8 males trying out for the male roles.

Problem How many possible casts are there? A cast is an assignment of people to roles, and two casts are different unless exactly the same people are assigned to the same roles.

Solution This problem cannot be solved by direct application of the permutation principle. Instead we must use both the permutation principle and the multiplication principle. Since we assume that all roles are distinguishable, there are $7 \times 6 \times 5 \times 4 = 840$ ways of selecting 4 females to play the female roles and $8 \times 7 = 56$ ways of selecting 2 males to play the male roles.

Consider the task of selecting a cast as a two-stage experiment: Select the female players, and then select the male players. The first stage has 840 results. For each of these results, there are 56 possible results of the second stage. Consequently, there are $840 \times 56 = 47{,}040$ outcomes of the experiment. That is, there are 47,040 possible casts (probably a good deal more than the director even cares to contemplate, let alone try out). ■

In the counting problems studied so far, we have considered only distinct objects. However, the techniques can be used in settings where not all the objects are distinct. We illustrate the method in a simple setting.

Example 2.17

How many different 4-letter "words" can be formed with the 4 letters in the word "book"?

Solution To use the techniques of this section, we begin by assuming that the two o letters are tagged in some way, with subscripts, for instance, to distinguish them. In such a case there are 4 distinct letters, and the number of permutations is

$$P(4, 4) = 24$$

If we imagine a list of these permutations, then each of the following appears: bo_1ko_2 and bo_2ko_1. Now suppose the subscripts are erased. Then the word "boko" appears *twice* in the list. Indeed, each permutation appears twice, and consequently there are $24 \div 2 = 12$ different words which can be formed with the letters of the word "book." ■

Example 2.18 How many different 8-letter words can be formed by using the 8 letters in the word "notebook"?

Solution As in Example 2.17, we begin by assuming the three o letters are tagged in some way (for example, $no_1tebo_2o_3k$). In such a case there would be 8 different "letters" and there would be $P(8, 8) = 8! = 40{,}320$ different words. However, a list of these words would include items such as $no_1tebo_2ko_3$, $no_2tebo_3ko_1$, and $no_3tebo_1ko_2$; and after the subscripts are removed, these will all be the same word. In fact, since the three o letters can be permuted in $P(3, 3) = 3! = 6$ ways, the list of 40,320 words can be divided into groups of 6 words, all of which are the same after the subscripts are removed. Thus the answer to the problem is the number of these groups. That is,

$$\frac{P(8, 8)}{P(3, 3)} = \frac{40{,}320}{6} = 6720$$ ■

The ideas introduced in Examples 2.17 and 2.18 can be extended and used in situations where there is more than one repeated letter. For instance, suppose we consider the situation described in Example 2.17 with the word "college". That is, how many different 7-letter words can be formed with the letters in the word "college"? Here we have a situation in which each of two letters, the letters e and l, appears twice. If, as in Example 2.17, we imagine that each of the two letters e is distinguished by a subscript, and each of the two letters l is distinguished similarly, then there are 7 distinct letters. These 7 distinct letters can be arranged in $P(7, 7)$ ways. Now, when we remove the subscripts on the letters e, then each of the resulting words appears twice [since $P(2, 2) = 2! = 2$], and after doing so we see there are $\frac{P(7, 7)}{2}$ distinct words. Finally, when we remove the subscripts on the letters l, then each of the resulting words appears twice, and consequently after doing so we see there are $\frac{P(7, 7)}{2 \cdot 2}$ distinct words. We now have the answer to the original question: there are $\frac{P(7, 7)}{2 \cdot 2}$ distinct words formed from the letters in the word "college." This technique can be used in more complex situations, some of which are the topics of exercises (Exercises 30 to 33).

Exercises for Section 2.2

1. Evaluate the following numbers expressed in terms of the factorial symbol.
   (*a*) $6!$ (*b*) $\frac{6!}{5!}$ (*c*) $\frac{8!}{4!4!}$ (*d*) $\frac{52!}{48!4!}$
2. Evaluate the following numbers expressed in terms of the factorial symbol.
   (*a*) $(6-2)!$ (*b*) $6!-2!$ (*c*) $(5+3)!$ (*d*) $5!+3!$
3. Evaluate the following numbers expressed in terms of the permutation symbol.
   (*a*) $P(5, 2)$ (*b*) $P(7, 1)$ (*c*) $P(6, 6)$ (*d*) $P(20, 3)$
4. *Committees* A committee has 7 members. One member is to be selected as chairperson, and another member is to be selected as secretary. In how many ways can these selections be made?
5. *Task scheduling* A programmer has 6 programs to write, and he will write them in succession, in any order. In how many ways can he organize his work?
6. *College life* A group of 7 students is to make a presentation on 3 issues: parking fees, campus safety, and recreational sports facilities. One student is to be assigned primary responsibility for each issue. In how many ways can this assignment be made?
7. *Sports* A football team has 44 players, 22 on offense and 22 on defense. How many ways can 2 team captains be selected if 1 must be from the offense and 1 from the defense?
8. *Opinion surveys* Suppose that in Example 2.12 there are 4 districts in the city and 6 individuals available to carry out the surveys. How many different assignments of survey takers to districts can be made, assuming no one surveys more than one district?
9. How many 3-digit numbers can be formed with the digits 2, 4, 6, 8, and 9 if each digit is used at most once? How many of these numbers are smaller than 500?
10. *Sports* A horse race has 8 entries. In how many different ways can the horses finish in the payoff slots: win, place, and show? (Assume no ties occur.)
11. How many 2-letter words can be formed from the letters of the word "consider"?
12. In how many ways can 6 different books be removed from a shelf, one at a time?
13. *Government* A city has 8 districts, and each is to be assigned a district manager. If there are 10 managers available, how many ways can the assignments be made so that each district has a different manager?
14. *Education* Each student in a class of 25 is to be assigned a computer account. If there are 100 accounts available, how many ways can the assignments be made so that each student gets a different account?
15. *Committees* A committee has 9 members, and 3 different members are to be given special tasks: one is to serve as a delegate to a convention, one is to seek new members, and one is to seek donations. In how many different ways can the tasks be assigned?
16. *Committees* In the situation described in Exercise 15, suppose the committee consists of 5 men and 4 women. In how many ways can the tasks be assigned so that both men and women are given assignments?
17. *College life* Seven students meet to study for an economics examination, and they decide to take a break. They decide that one of them should go for pizza, one should go for cokes, one should copy some notes, one should send an e-mail message to the instructor, and the rest will relax. In how many ways can these tasks be assigned?

Theater casting 18. In Example 2.16, suppose that only 6 females and 5 males are trying out for the 4 female roles and 2 male roles, respectively. How many possible casts are there?

College life 19. A group of 6 students including Vania has been working on a research project for a course. The project has been completed, and a report is to be written. The report consists of an introduction, 2 chapters, and a conclusion, a total of 4 sections.
(*a*) In how many ways can the students divide up the task of writing the report if each section is to be written by a different student?
(*b*) In how many ways can the students divide up the report writing if each section is to be written by a different student and Vania writes the conclusion?

Travel 20. A student has job interviews in Atlanta, Boston, Chicago, and Detroit. In how many ways can the interviews be scheduled so that the Chicago interview precedes the Detroit interview?

Travel 21. A traveler can visit 3 of these cities: Amsterdam, Barcelona, Copenhagen, Rome, and Zurich. An itinerary for a trip is a list of the 3 cities in the order to be visited. How many different itineraries are there for the trip?

Travel 22. You are given the situation described in Exercise 21.
(*a*) How many of the itineraries include Copenhagen?
(*b*) In how many of these is Copenhagen the first city to be visited?
(*c*) How many of the itineraries include both Copenhagen and Rome?

Committees 23. A committee has 10 members including Alice. Three offices (chairperson, secretary, treasurer) must be filled.
(*a*) If each person can fill at most one office, in how many different ways can the offices be filled?
(*b*) In how many ways can the offices be filled so that Alice is one of the officers?
(*c*) In how many ways can the offices be filled so that Alice is chairperson?

Travel 24. A sales representative plans 1-week visits to Boston, Chicago, Detroit, and Los Angeles. If she visits Chicago and Detroit on consecutive weeks but in either order, how many possible schedules does she have?

Psychology 25. A psychologist plans an experiment that consists of 4 different activities which can be performed in any order. Instructions to the subject can be given in writing or orally, by either a male or a female. An experiment is defined by a method of instruction, an instructor, and the order of the 4 activities. How many different experiments are there?

Theater casting 26. There are 3 unfilled roles in a play at the community theater, 2 for females and 1 for a male. Auditioning for the female roles are 4 females including Susan; 3 males audition for the male roles. A cast consists of an assignment of specific people to specific roles.
(*a*) How many different casts are there?
(*b*) How many different casts include Susan?

27. Let $S = \{1, 2, 4, 5, 6, 7, 9\}$, and suppose a 4-digit number is formed by selecting digits, without replacement, from S.
(*a*) How many different such numbers can be formed?
(*b*) How many different such numbers can be formed if both even and odd digits must be used?

28. Let $S = \{1, 2, 3, 4, 6, 7, 8, 9\}$, and suppose a 3- or a 4-digit number is formed by selecting digits, without replacement, from S.

(*a*) How many different such numbers can be formed?
(*b*) How many different such numbers smaller than 5000 can be formed?

Entertainment

29. The time in a television variety show is utilized as indicated below, where C denotes commercial and S denotes skit.

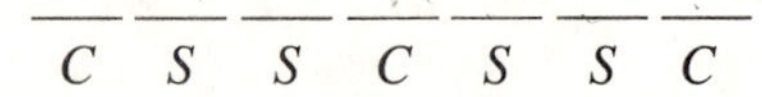

If the producer has 5 skits and 3 commercials, in how many ways can he create a television show? We assume here that a skit is never repeated, but (unfortunately) commercials may appear any number of times.

30. How many 6-letter words can be formed from the letters in the word "batter"?
31. How many 6-letter words can be formed from the letters in the word "bottom"?
32. How many 6-letter words can be formed from the letters in the word "feeder"?
33. How many 7-letter words can be formed from the letters in the word "breeder"?

*34. For which values of the integer n are both $P(n+1, 3)$ and $P(n-1, 4)$ defined and $P(n+1, 3) \geq P(n-1, 4)$?

*35. For which values of the integer n is it true that $P(n+2, 3)$ and $P(n, 5)$ are both defined and $2n \cdot P(n+2, 3) \geq P(n, 5)$?

2.3 COUNTING PARTITIONS: COMBINATIONS

In the preceding section we developed a method for determining the number of ways that r objects can be selected from a set of n objects and arranged in order. Suppose now that we are interested simply in selecting r objects from n but *not* in arranging these r objects in a particular order. Each such selection is called a *combination* of r objects selected from n. To study this new operation, it is convenient to view the process discussed in Section 2.2 as a two-step process consisting of *selecting* and *arranging*. That is, by the *multiplication principle*,

Number of ways r objects can be selected from n objects and arranged in order	=	number of ways r objects can be selected from n objects	$\times$	number of ways r objects can be arranged in order

The permutation principle can be used to evaluate the numbers represented by the quantities on the extreme right and left. We have

*Exercises denoted by an asterisk may require the use of a calculator or computer.

$$P(n, r) = \frac{n!}{(n-r)!} \qquad \text{and} \qquad P(r, r) = r!$$

and consequently

$$\frac{n!}{(n-r)!} = \boxed{\begin{array}{l}\text{number of ways}\\ r \text{ objects can be}\\ \text{selected from } n\\ \text{objects}\end{array}} \times r!$$

This expression leads immediately to a formula for the number of ways r objects can be selected from n objects. If we divide both sides of the equality by $r!$, the result is our next counting principle.

Combination Principle

A subset of r objects can be selected from a set of n distinct objects in $\dfrac{n!}{(n-r)!r!}$ different ways.

As an illustration of the ideas involved in the combination principle, consider the situation in which a retail merchant selects colored tags for sale merchandise. Suppose the merchant decides to discount some merchandise 10 percent, some 20 percent, and some 50 percent. Also suppose that he has tags colored blue (B), green (G), orange (O), red (R), and yellow (Y). He decides to color-code the sale so that all items discounted 10 percent are tagged one color, those discounted 20 percent are tagged another color, and those discounted 50 percent are tagged a third color. Thus, a color code for the sale may be represented by an ordered list of 3 letters, the colors of the tags for items discounted 10 percent, 20 percent, and 50 percent, in that order. The merchant must first select the 3 colors to be used from the 5 colors available and then decide how to use them. For instance, if the colors selected are blue, green, and red, then there are 6 different color codes which can be formed with these colors—the 6 permutations of B, G, and R: BGR, BRG, GBR, GRB, RBG, RGB. In fact, using the permutation principle, we know that there are $P(3, 3) = 3! = 6$ color codes for *each* choice of 3 colors. We also know that the total number of color codes is the number of permutations of 5 things taken 3 at a time, or $P(5, 3)$. Since there are $P(5, 3) = 5!/2! = 60$ color codes and 6 color codes for each choice of 3 colors, there must be $60/6 = 10$ ways to choose 3 colors from 5 colors. In the terminology of the combination principle:

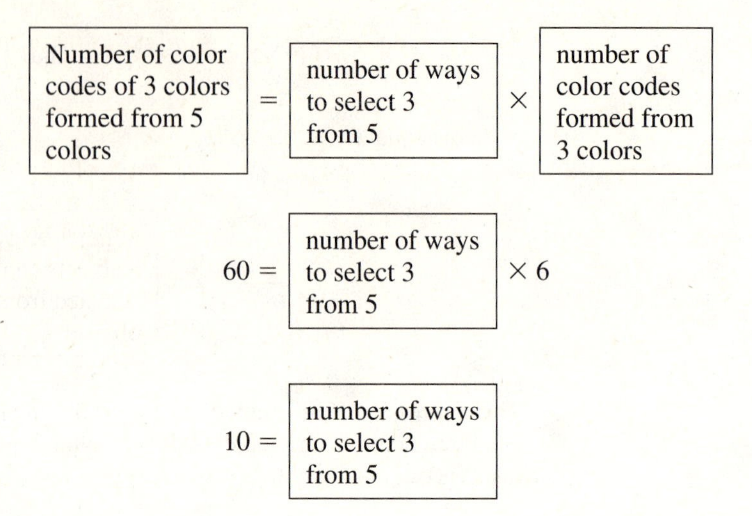

Example 2.19 An apartment complex has 12 smoke alarms, and 3 are to be selected for testing.

Product testing

Problem In how many ways can the selection be made? Notice that the question involves only *selection*, not selection and order.

Solution To emphasize the underlying ideas, we shall once again solve the problem in a way which illustrates the derivation of the combination principle, instead of simply applying the result. In later examples, after the underlying ideas are familiar, we apply the result directly.

The number of ways that 3 alarms can be selected from a set of 12 and arranged in order is $12 \times 11 \times 10$. Moreover, for each choice of 3 alarms, these 3 can be arranged (permuted) in $3 \times 2 \times 1$ different ways. Therefore, viewing the total process as selection and arrangement, we have

$$12 \times 11 \times 10 = \boxed{\begin{array}{l}\text{number of ways}\\ \text{3 alarms can be}\\ \text{selected from 12}\end{array}} \times (3 \times 2 \times 1)$$

Consequently, there are

$$\frac{12 \times 11 \times 10}{3 \times 2 \times 1} = 220$$

ways of selecting 3 alarms from the 12 available. ■

If we want to select 4 alarms for testing from the set of 12, then we can use the method of Example 2.19 except that we select 4 rather than 3. In particular, we can select 4 alarms and arrange them in order in $12 \times 11 \times 10 \times 9$ different ways. Next, for each choice of the 4 alarms, they can be arranged in $4 \times 3 \times 2 \times 1$ different ways. Consequently, there are

$$\frac{12 \times 11 \times 10 \times 9}{4 \times 3 \times 2 \times 1} = 495$$

ways of selecting 4 alarms for testing from the 12 alarms available.

Example 2.20 There are 8 engineers qualified to serve on a design team. If the team is to consist of 5 engineers, how many possible teams can be created?

Business teams

Solution This is a problem of selecting subsets of 5 elements each from a set of 8 elements. An application of the combination principle yields

$$\frac{8!}{(8-5)!5!} = \frac{8 \times 7 \times 6}{3 \times 2 \times 1} = 56$$

as the number of possible teams (note that $8! = 8 \times 7 \times 6 \times 5!$). ■

The expression $\frac{n!}{(n-r)!r!}$ will occur frequently in our study of probability, and it is useful to introduce a shorter notation for it.

The number of combinations of n things taken r at a time is

$$C(n, r) = \frac{n!}{(n-r)!r!} \tag{2.6}$$

The numbers $C(n, r)$ are also known as binomial coefficients, and they have many interesting and important properties. We list here several of these properties.

1. Each selection of r elements from a set of n elements identifies a subset of r elements (those selected). Thus, $C(n, r)$ denotes the number of different subsets of size r in a set of size n.
2. $C(n, r) = C(n, n - r)$. This fact, which is often useful in working problems, follows from the observation that to each r-element subset which is selected from a set of n elements, there is associated an $(n - r)$-element subset consisting of elements which are not selected. This can be directly verified by replacing r by $n - r$ in Equation (2.6).

3. $P(n, r) = r!C(n, r)$. This fact follows from the definitions of $P(n, r)$ and $C(n, r)$.
4. For computational purposes it is convenient to divide both the numerator and denominator of $C(n, r)$ by $(n - r)!$. The result is

$$C(n, r) = \frac{n(n-1)(n-2)\cdots(n-r+1)}{r(r-1)(r-2)\cdots 2\cdot 1}$$

Note that there are r factors in both the numerator and the denominator. This together with the fact that the first factor in the numerator is n and the first factor in the denominator is r makes the formula easy to remember. For example,

$$C(10, 4) = \frac{10\times 9\times 8\times 7}{4\times 3\times 2\times 1} = 210$$

An alternative notation for the binomial coefficient $C(n, r)$ is $\binom{n}{r}$, and many books use this symbol in place of $C(n, r)$.

In many of the problems which we solve by using counting techniques, it is necessary to combine counting principles in ways that vary from problem to problem. Problems which require combining counting techniques are often multistep problems, problems in which the eventual goal—solving the original problem—is achieved by first solving a series of (two or more) subproblems and then using the solutions of the subproblems to solve the original problem. Each subproblem may require its own counting technique, and the results may be combined by yet another technique. The following examples illustrate the use of the combination principle and the multiplication principle.

Example 2.21

Course selection

The core curriculum at Gigantic State University (GSU) requires that each student take 2 humanities courses from an approved set of 5 courses and 2 science courses from an approved set of 4 courses.

Problem Find the number of different course selections available to each student, assuming no schedule conflicts.

Solution We apply the combination principle to conclude that it is possible to select the 2 humanities courses in $C(5, 2)$ ways and the 2 science courses in $C(4, 2)$ ways. The process of pairing 2 humanities courses with 2 science courses to form a complete course selection can be treated as a two-stage experiment. The first stage consists of selecting 2 different humanities courses. There are $C(5, 2)$ results at the first stage. The second stage consists of selecting 2 different science courses. For each result at the first stage, there are $C(4, 2)$ results at the second stage. Thus, using the multiplication principle, we conclude that there are

$$C(5, 2) \times C(4, 2) = \frac{5 \times 4}{2 \times 1} \times \frac{4 \times 3}{2 \times 1} = 10 \times 6 = 60$$

ways of selecting 2 humanities courses and 2 science courses. That is, there are 60 different course selections available to each student. ■

Example 2.22 A student at GSU (see Example 2.21) plans to take 2 of her 4 required courses during the same term, and she prefers to take both either in humanities or in science.

Course selection

Problem Find the number of course selections available to her.

Solution A direct application of the combination principle shows that there are $C(5, 2)$ selections consisting of 2 humanities courses and $C(4, 2)$ selections of 2 science courses. Since no selection is to contain both science and humanities courses, we can think of the set of course selections as partitioned into two sets: one set consisting of 2 humanities courses and one set consisting of 2 science courses. Consequently, we can use the result from Chapter 1 which gives a method for counting the number of elements in a partition [Equation (1.2)]. We conclude that there are $C(5, 2) + C(4, 2) = 10 + 6 = 16$ different course selections available to her. ■

Example 2.23 A student has 7 coins; 5 are dimes and 2 are quarters.

Problem (*a*) In how many ways can 3 coins be selected from 7?
(*b*) In how many ways can 3 coins be selected if both dimes and quarters must be included?

Solution (*a*) This part of the problem can be solved with a direct application of the combination principle: The number of ways to select 3 objects from a set of 7 is $C(7, 3) = 35$.

(*b*) Since there are only 2 quarters, every selection of 3 coins must include at least one dime. However, an arbitrary selection of 3 coins could include only dimes; we seek to count the number of selections containing at least one quarter. If we compute the number of selections which include only dimes and subtract this number from the answer to part *a*, then we will have the answer to part *b*. The number of ways we can select 3 coins from 5 dimes is $C(5, 3) = 10$. Thus, the number of ways to select 3 coins including both dimes and quarters is $C(7, 3) - C(5, 3) = 35 - 10 = 25$. ■

Example 2.24

Management

A total quality management team consisting of 3 people from engineering, 2 people from marketing, and 1 person from finance is to be formed. There are 6 people from engineering, 5 from marketing, and 3 from finance qualified for the team.

Problem Find the number of different total quality management teams which can be formed.

Solution It is helpful to view the selection of a team as a three-stage experiment:

Stage 1: Select 3 from 6 people in engineering; $C(6, 3)$ results.
Stage 2: Select 2 from 5 people in marketing; $C(5, 2)$ results.
Stage 3: Select 1 from 3 people in finance; $C(3, 1)$ results.

We apply the multiplication principle to conclude that the number of different teams which can be formed is:

$$C(6, 3) \times C(5, 2) \times C(3, 1) = 20 \times 10 \times 3 = 600$$

■

Example 2.25

A psychologist plans an experiment in group dynamics which can be conducted with either 3, 4, or 5 subjects. There are 7 subjects available for the experiment.

Psychology

Problem With how many different groups can the experiment be conducted?

Solution We begin this problem by first noting that the entire set of groups can be partitioned into three subsets: subsets consisting of those groups with 3, 4, and 5 members, respectively. We proceed by counting the number of groups in each subset. Finally, we use Equation (1.2) and determine the total number by adding the numbers for each subset.

$$\begin{aligned}
\boxed{\begin{array}{l}\text{Number}\\ \text{of groups}\\ \text{with 3}\\ \text{members}\end{array}} + \boxed{\begin{array}{l}\text{number}\\ \text{of groups}\\ \text{with 4}\\ \text{members}\end{array}} + \boxed{\begin{array}{l}\text{number}\\ \text{of groups}\\ \text{with 5}\\ \text{members}\end{array}} &= C(7, 3) + C(7, 4) + C(7, 5) \\
&= 35 + 35 + 21 \\
&= 91
\end{aligned}$$

The experiment can be conducted with 91 different groups. ■

Examples 2.21 through 2.25 give an indication of the diversity of ways in which the basic counting principles can be applied to a problem divided into subproblems. This approach frequently leads to questions which are, in one form or another, about combining the answers to the subproblems and, more particularly, when the answers should be added and when they should be multiplied. This is the standard question: *When do I multiply and when do I add?*

It is difficult to give criteria which are easy to apply and which are always applicable. However, the following guidelines will help in essentially all cases. If the problem is subdivided so that each outcome of a subproblem is an outcome of the original problem, i.e., if the sample space of the original problem has been partitioned in some way, then the results of the subproblems should be added to give the answer to the original problem. This is illustrated in Examples 2.22 and 2.25. Alternatively, if the results of the subproblems are parts of the outcomes of the original problem, but not entire outcomes, then the multiplication principle should be applied. In this case the experiment is viewed as consisting of stages, the tree diagram is split into pieces, and the subtrees are considered separately. This is illustrated in Examples 2.21 and 2.24. The following summarizes these comments.

If the sample space is partitioned, then the numbers of outcomes of the subproblems are *added*; and if the experiment consists of stages, then the numbers of results at each stage are *multiplied.*

Finally, we repeat for emphasis the comment that an important part of many problems is distinguishing between situations in which there is selection only—and therefore the Combination Principle should be used—and situations in which there is both selection and arranging in order—and therefore the Permutation Principle should be used. The actual problem may be described in terms where the selection is not explicit, and a part of the solution is to determine which of these descriptions—selection only or selecting and arranging in order—fits the problem.

Pascal's Triangle

There are a well-known and very useful algorithm and diagram which provide a convenient method of obtaining the binomial coefficients $C(n, r)$ for different values of n and r. The algorithm is based on the fact (easily verified by substitution) that $C(n + 1, r) = C(n, r) + C(n, r - 1)$, for $r = 1, 2, \ldots, n$. The associated diagram is shown in Figure 2.2. In this figure, each row of the triangle represents one value of n in $C(n, r)$, and the entries in the row are the numbers $C(n, r)$, in order, for $r = 0, 1, 2, \ldots, n$. In each row the first and last entries are 1s [because $C(n, 0) = C(n, n) = 1$]; each remaining entry is the sum of the two nearest entries in the row just above. For example, the third entry in the row for

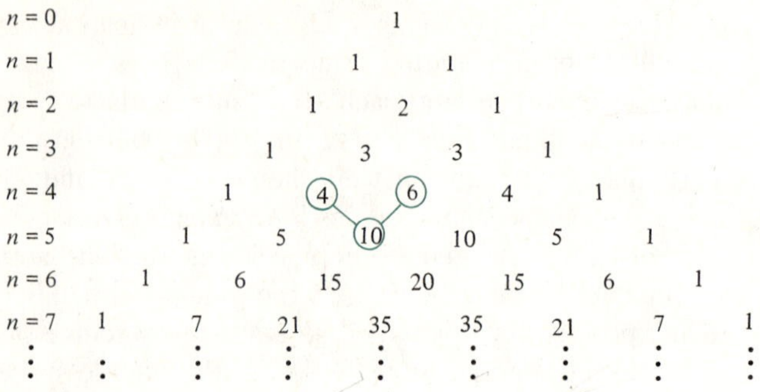

$n = 0$ 1
$n = 1$ 1 1
$n = 2$ 1 2 1
$n = 3$ 1 3 3 1
$n = 4$ 1 4 6 4 1
$n = 5$ 1 5 10 10 5 1
$n = 6$ 1 6 15 20 15 6 1
$n = 7$ 1 7 21 35 35 21 7 1
⋮ ⋮ ⋮ ⋮ ⋮ ⋮ ⋮ ⋮ ⋮

FIGURE 2.2

$n = 5$ is $C(5, 2) = 10$; and this is the sum of $C(4, 1) = 4$ and $C(4, 2) = 6$, the entries just above the 10. The same relationship holds for the entries in each row, and the triangle continues for all positive integers $n = 0, 1, 2, 3. \ldots$

Exercises for Section 2.3

1. Compute the values of $C(7, 3)$, $C(8, 3)$, and $C(9, 3)$.
2. Compute the values of $C(9, 2)$, $C(9, 3)$, and $C(9, 4)$.
3. Compute the following numbers.
   (*a*) $\frac{7!}{3!}$ (*b*) $C(7, 3)$ (*c*) $P(7, 3)$
   (*d*) $C(7, 0)$ (*e*) $C(3, 0) + C(3, 1) + C(3, 2) + C(3, 3)$
4. Which number is larger in each of the following pairs of numbers?
   (*a*) $C(9, 3)$, $C(10, 2)$ (*b*) $C(12, 4)$, $C(18, 3)$
   (*c*) $C(9, 3)$, $6!$ (*d*) 3^5, $5!$
5. A set S has 5 elements. How many subsets of S have *at least* 3 elements?
6. A set S has 9 elements. How many subsets of S have *at most* 3 elements?
7. A student has 4 pencils, 3 ballpoint pens, and 1 felt-tip pen in a cup.
   (*a*) In how many different ways can 2 of the 8 be selected?
   (*b*) In how many different ways can 2 items be selected so that not both are ballpoint pens?

Committees

8. In how many different ways can the chairperson of a 10-person committee select a subcommittee of 3 from the other 9 members of the committee?
9. Two coins are selected from 5 coins: 3 dimes and 2 quarters.
   (*a*) In how many ways can the selection be made?
   (*b*) In how many ways can the selection be made so that both are dimes?
   (*c*) In how many ways can the selection be made so that 1 dime and 1 quarter are selected?
   (*d*) In how many ways can the selection be made so that at least 1 coin is a dime?

Games of Chance 10. Four cards are drawn from a standard deck. In how many ways can they all be the same suit?

Games of Chance 11. Three coins are selected from 6 coins: 3 dimes, 2 nickels, and 1 quarter. In how many ways can the selection be made so that the value of the coins chosen is at least 25 cents?

Recreation 12. A camper has 4 jackets, 3 sleeping bags, and 10 different packages of freeze-dried food. An "outfit" for a camping trip consists of 1 jacket, 1 sleeping bag, and 3 packages of freeze-dried food. How many different outfits are there?

Music preferences 13. A student interested in jazz plans to buy 3 CDs. She is interested in 2 featuring piano, 3 featuring trumpet, and 4 featuring saxophone.
(*a*) In how many ways can the 3 CDs be selected?
(*b*) In how many ways can the selection be made if CDs featuring at least 2 different instruments are selected?

Food service 14. A delicatessen manager prepares snack trays, using 4 kinds of crackers and 3 kinds of cheese. Each tray contains 3 kinds of crackers and 2 kinds of cheese. How many different trays are there?

Food service 15. You have 3 types of vegetables and 5 types of fruits. A vegetable salad consists of a mixture of any 2 kinds of vegetables, and a fruit salad consists of a mixture of any 2 kinds of fruits.
(*a*) In how many different ways can you prepare a vegetable salad and a fruit salad?
(*b*) In how many ways can you prepare a vegetable salad or a fruit salad?
(*c*) You are to take 2 different kinds of salads on a picnic. In how many ways can you do this?

Course selection 16. A student is to take 1 mathematics course, 1 history course, 1 business course, and 2 electives. There are 3 courses in mathematics, 3 in history, and 5 in business for which she is prepared. Also, she is interested in 5 elective courses. How many course selections are available to her if there are no time conflicts?

Manufacturing 17. A candle manufacturer produces 5 types of round candles and 4 types of square candles. Gift packages contain 2 round candles and 2 square candles, all of different types. How many different gift packages are there?

Manufacturing 18. A candle manufacturer produces 5 types of round candles and 4 types of square candles. Gift packages contain 3 candles, all of different types. How many different gift packages are there? Note that a gift package may contain candles which are all of one shape or 2 candles of one shape and 1 of the other shape.

Committees 19. A senate committee consists of 6 Democrats and 4 Republicans. How many different subcommittees can be formed in the following situations?
(*a*) Each subcommittee must consist of exactly 3 Democrats and 2 Republicans.
(*b*) Each subcommittee must consist of either 3 Democrats and 2 Republicans or 4 Democrats and 1 Republican.

Business 20. A corporation car pool consists of 8 subcompact and 12 compact cars. The sales manager reserves 2 subcompact and 3 compact cars. In how many ways can the request be met?

Food selection 21. A pizza can be ordered with any of 3 different crusts (thin, regular, and deep-dish) and 4 toppings selected from the 6 available (sausage, pepperoni, olives, green peppers, mushrooms, and onions). How many different pizzas are there?

Food selection 22. In the setting of Exercise 21, suppose that either 3 or 4 toppings can be selected. How many different pizzas are there?

Sports 23. A neighborhood club has 8 girls and 6 boys. A basketball team of 5 members is to be selected.
(*a*) In how many ways can the selection be made?
(*b*) In how many ways can the selection be made so that the team contains both boys and girls?

Travel 24. A student has invitations for job interviews in Atlanta, Boston, Chicago, Detroit, and Seattle.
(*a*) In how many ways can he select 3 cities to visit?
(*b*) In how many ways can he select 3 cities to visit if both Atlanta and Seattle are included?
(*c*) In how many ways can he select 3 cities to visit if not both Atlanta and Seattle are included?

25. A bowl contains 6 red balls and 4 blue balls. In how many ways can 3 balls be selected so that both red and blue balls are obtained?

26. A bowl contains 3 red balls, 4 blue balls, and 5 green balls. In how many ways can 2 balls be selected so that at least 1 is blue?

Games of chance 27. A "hand" consists of a set of 5 cards selected from an ordinary deck of 52 cards.
(*a*) How many different hands contain exactly 3 spades?
(*b*) How many different hands contain exactly 3 cards of some suit (any of the 4 suits)?
(*c*) How many different hands contain cards of all 4 suits?

Games of chance 28. Refer to Exercise 27.
(*a*) How many different hands of cards contain exactly 4 aces?
(*b*) How many different hands contain cards of only 1 suit?

Committee officers 29. A political party has rules to the effect that the officers of every party committee (i.e., the chairperson, vice-chairperson, and secretary) must always include at least 1 woman and 1 black. A certain committee of 6 members consists of 2 black males, 1 black female, 2 white males, and 1 white female. In how many different ways can the officers of the committee be selected to obey party rules? Note that if individuals change offices, then a different selection exists.

Business 30. Five sales representatives are to be assigned to either telephone sales or travel.
(*a*) If 3 sales representatives are assigned to telephone sales and 2 to travel, in how many ways can the assignment be made?
(*b*) If each type of job must have at least 1 person assigned to it, then how many assignments are there?

Education 31. The faculty at GSU is divided into four schools: Humanities, Science, Music, and Athletics. The faculty council has 8 members, and it must contain at least 1 faculty member from each school. In how many different ways (in terms of school representatives) can the council be constituted? (For example, one possible way is to have 2 members from each of the 4 schools.)

32. For what values of the natural number n are both $C(2n, 4)$ and $C(n, 2)$ defined and equal?

33. Show that for $n = 3, 4, 5, \ldots$

$$C(n, 2) + C(n, 3) = C(n + 1, 3)$$

34. Show that for $n = 2, 3, 4, \ldots$ and $r = 1, 2, 3, \ldots, n$

$$C(n, r) + C(n, r-1) = C(n+1, r)$$

*35. For what values of n is it true that $C(100, 2n) > C(1000, n)$?

2.4 COMPUTING PROBABILITIES BY USING EQUALLY LIKELY OUTCOMES

We have used counting techniques to determine the number of outcomes of an experiment, where the term "experiment" has a fairly general meaning. We continue by using the same techniques to determine probabilities. If a sample space S contains n outcomes and if these outcomes are equally likely, then each should be assigned the same probability (weight), namely $1/n(S)$. In this special case of equally likely outcomes, to determine the probability of an event E (a subset of S), we need only count the number of elements in each of sets E and S and use the fact that the probability of any event is the sum of the probabilities of the outcomes in that event. Thus, if each outcome in event E has probability $1/n(S)$, then the probability of E, denoted $\Pr[E]$, is $\Pr[E] = n(E)/n(S)$. We repeat for emphasis that this technique should be used *only when the outcomes are equally likely*. In this section we illustrate the use of the four counting principles (partition, multiplication, permutation, and combination) to compute $n(E)$ and $n(S)$ and consequently $\Pr[E]$.

Example 2.26 A student has 5 coins—2 dimes and 3 quarters—and 2 coins are selected simultaneously and at random.

Problem Find the probability that both are quarters.

Solution Recall that the phrase "selected at random" means that equally likely outcomes are to be used. Since there are 5 coins, there are $C(5, 2) = 10$ ways of selecting 2 coins. Since there are 3 quarters, there are $C(3, 2) = 3$ ways of selecting 2 quarters. Therefore, if S is the sample space for the experiment of selecting 2 coins, then $n(S) = 10$; and if E is the event that 2 quarters are selected, then $n(E) = 3$. Consequently, $\Pr[E] = \frac{3}{10}$; the probability of selecting 2 quarters is $\frac{3}{10}$. ■

Example 2.27 An accounting firm regularly audits 4 firms each year. It has 4 chief auditors who direct these audits, and the assignment of chief auditors to the firms is done at random.

Task scheduling

Problem What is the probability that all 4 chief auditors will be assigned to the same firm this year as last year?

Solution Each assignment of auditors to firms corresponds to an arrangement of the names of the 4 chief auditors. Thus there are $P(4, 4) = 4!/0! = 24$ assignments. Only 1 of these 24 assignments will be the same as last year's. Hence, Pr[same assignment as last year] $= \frac{1}{24}$. ■

Example 2.28 A runner enters two races, both to be run on an 8-lane track. The runner is assigned a lane for each race.

Sports

Problem Assuming that the lane assignments are made at random, what is the probability that the runner has lane 1 (the inside lane) for at least one race?

Solution The runner could be assigned any one of 8 lanes for the first race and any one of 8 lanes for the second race. By the multiplication principle, there are $8 \times 8 = 64$ possible pairs of lane assignments for the two races. Let E be the event that the runner has lane 1 for at least one race, F the event that she has lane 1 for the first race, and G the event that she has lane 1 for the second race. Then $E = F \cup G$. Clearly $n(F) = 8$ since she could have lane 1 in the first race and any of the 8 lanes for the second race. Likewise, $n(G) = 8$ and $n(F \cap G) = 1$. Using Equation (1.6), we have

$$n(E) = n(F \cup G) = n(F) + n(G) - n(F \cap G)$$

and

$$n(E) = 8 + 8 - 1 = 15$$

Since $n(S) = 64$, it follows from Equation (2.4) that $\Pr[E] = \frac{15}{64}$. ■

Example 2.29 A 12-unit apartment building has 9 smoke alarms which pass inspection and 3 which do not. If 3 smoke alarms are selected at random and tested, what is the probability that all pass inspection?

Solution In Example 2.19 it was shown that there are $C(12, 3) = 220$ ways of selecting 3 smoke alarms from 12. Thus, $n(S) = 220$. Since 9 of the alarms pass inspection, there are $C(9, 3) = 84$ ways of selecting 3 alarms which pass. Therefore, $n(E) = 84$. Finally, the probability of selecting 3 alarms which pass inspection is $\Pr[E] = \frac{84}{220} = \frac{21}{55}$. ■

Example 2.30

Management

A total quality management team contains 8 people: 5 from customer service and 3 from data processing. Two team members are selected at random to present the results of the team's work.

Problem What is the probability of event E that 1 of the people selected is from customer service and 1 is from data processing?

Solution The two team members can be selected in $(8 \times 7)/2 = 28$ ways, so $n(S) = 28$. Event E is the complement of the event that both come from the same department. Since both can come from customer service in $(5 \times 4)/2 = 10$ ways, and since both can come from data processing in $(3 \times 2)/2 = 3$ ways, we see by the partition principle that $n(E') = 10 + 3 = 13$. Finally, $n(E) = n(S) - n(E') = 28 - 13 = 15$, and $\Pr[E] = \frac{15}{28}$. ■

Example 2.31

Course selection

A student must select 3 courses from a list of 7. The list includes 4 humanities courses and 3 science courses.

Problem If the selection is made at random, what is the probability that 3 humanities courses will be chosen?

Solution The number of ways to choose 3 courses from a list of 7 is

$$C(7, 3) = \frac{7 \times 6 \times 5}{3 \times 2 \times 1} = 35$$

The number of ways to choose 3 courses from a list of 4 humanities courses is

$$C(4, 3) = \frac{4 \times 3 \times 2}{3 \times 2 \times 1} = 4$$

Thus the probability of selecting all 3 courses from the humanities is $\frac{4}{35}$. ■

Example 2.32

Committees

Sam and Sally are members of a committee studying the effects of government regulation on business. The committee consists of 4 men and 3 women, and a subcommittee of 3 is to be chosen to study "paperwork." The subcommittee must include at least 1 woman, but it cannot consist entirely of women.

Problem If the subcommittee is selected at random from the set of all subcommittees which meet these conditions, what is the probability that the subcommittee selected will include Sam? What is the probability that it will include Sally?

Solution The subcommittee may consist of 1 man and 2 women or 2 men and 1 woman. The number of subcommittees with 1 man and 2 women is

$$C(4, 1) \times C(3, 2) = 4 \times 3 = 12$$

The number of subcommittees with 2 men and 1 woman is

$$C(4, 2) \times C(3, 1) = 6 \times 3 = 18$$

Thus there are $18 + 12 = 30$ possible subcommittees.

To answer the questions about Sally and Sam, we need to know the number of possible subcommittees which include them. If Sam is on a subcommittee, the remaining 2 members could be 2 women or 1 woman and 1 man. The number of ways 2 women could be chosen is $C(3, 2)$, and the number of ways 1 woman and 1 man could be chosen is $C(3, 1) \times C(3, 1)$. Therefore the number of such subcommittees is

$$C(3, 2) + C(3, 1) \times C(3, 1) = 3 + 3 \times 3 = 12$$

Thus the probability that the subcommittee selected will include Sam is

$$\tfrac{12}{30} = \tfrac{2}{5}$$

The analysis is similar for Sally. If she is on the subcommittee, then the other 2 members could be 2 men or 1 woman and 1 man. The number of such subcommittees is

$$C(4, 2) + C(4, 1) \times C(2, 1) = 6 + 4 \times 2 = 14$$

and the probability that the subcommittee selected will include Sally is

$$\tfrac{14}{30} = \tfrac{7}{15}$$ ■

Exercises for Section 2.4

1. Two coins are selected at random from 5 coins: 3 dimes and 2 quarters. What is the probability that both are dimes?
2. A group of 6 students—4 males and 2 females—selects 2 students at random to make a report. Find the probability that 2 females are selected.
3. Three cards are selected at random from a deck of 52 cards. What is the probability that all 3 are spades?
4. Four cards are selected at random from a deck of 52 cards. What is the probability that all 4 are aces?
5. A cooler contains 7 cans of cola: 4 regular colas and 3 diet colas. If 3 cans of cola are selected at random, what is the probability that all 3 are regular colas?

6. In the situation of Exercise 5, what is the probability that 2 are regular colas and 1 is a diet cola?
7. An experiment consists of rolling 1 red die, 1 white die, and 1 blue die and noting the result of each roll. If the dice are fair and all outcomes are equally likely, find the probability that should be assigned to each outcome.
8. Using the experiment of Exercise 7, what is the probability that the sum of the results on the 3 dice is 5?
9. *Employment* Five qualified individuals including a wife and husband apply for 3 vacant sales positions. If people are hired at random, what is the probability that both the husband and wife are hired? What is the probability that one is hired and one is not?
10. A fair coin is flipped 4 times. What is the probability that both heads and tails occur?
11. *Entertainment* Your friend has 8 tickets to a concert: 2 in the front row and 6 in the tenth row. She selects 2 tickets at random and gives them to you. Find the probability that both are in the front row.
12. Suppose that 3 digits are selected at random from the set $S = \{1, 2, 4, 5, 6\}$ and are arranged in random order. Find the probability that the resulting 3-digit number is less than 300.
13. *Task scheduling* A student has errands at the bookstore, the bank, and the music store. The student makes a random choice of the order in which to do the errands. Find the probability of each of the following.
    (*a*) The bank is visited first.
    (*b*) The bank is visited before the bookstore.
14. Two fair dice are rolled, and the sum is noted. Find the probability that the sum is less than 6.
15. *Biology* There are 5 black and 5 white mice available for an experiment which requires 4 mice. If a random selection of 4 mice is made from the set of 10, what is the probability that 2 black and 2 white mice are selected?
16. *Sports* Suppose there are 3 U.S. sprinters among the 8 finalists in the 100-meter dash at the Atlanta Olympics. If the 8 finalists are equally talented, find the probability that the U.S. sprinters finish first, second, and third.
17. *Travel* A student has job interviews in Albany, Buffalo, Cincinnati, and Denver. The interviews are scheduled at random. Find the probability that the Albany interview comes before both the Cincinnati and Denver interviews.
18. *Product testing* A computer technician knows that 2 chips are working properly and 2 chips are not. She tests chips one after another at random until both defective chips are found. Find the probability that 2 defective chips are identified after 2 tests.
19. *Food selection* A student selects 5 frozen dinners from 9 alternatives. The alternatives include 4 different pasta dinners, 3 different chicken dinners, and 2 different seafood dinners. If the selection is made at random, find the probability that *at least* 2 pasta dinners will be selected.
20. *Entertainment* A disk jockey for a jazz radio show is to prepare a program consisting of 4 tracks. Selections are to be made from 4 piano tracks and 5 saxophone tracks. The program, i.e., the tracks and the order in which they are to be played, is selected at random.
    (*a*) Find the probability that all tracks played are saxophone tracks.
    (*b*) Find the probability that the first 2 tracks are piano and the last 2 are saxophone.

College life 21. The Student Union Board has 10 members: 7 females and 3 males including Steve. Three tasks are to be assigned, including that of reserving a room for meetings. The tasks are assigned at random, at most one task per person.
(*a*) Find the probability that Steve is given a task.
(*b*) Find the probability that Steve is asked to reserve a room.

College life 22. You are given the situation in Exercise 21.
(*a*) Find the probability that both males and females are given tasks.
(*b*) Find the probability that Steve and at least 1 female are given tasks.

Committees 23. A congressional committee contains 5 Democrats and 4 Republicans. A subcommittee of 3 is to be randomly selected from all subcommittees of 3 which contain at least 1 Demo-crat. What is the probability that this new subcommittee will contain at least 2 Democrats?

24. Five students—Ann, Bill, Chuck, Debra and Ed—are randomly assigned seats from the seats numbered 1 through 10. What is the probability that Ann is assigned either seat 1 or seat 10?

Business 25. Employees at the Wonder Widget Corporation are assigned 3-digit identification numbers for payroll purposes. For data processing reasons, only the digits 1 through 9 are used, but each digit can be used any number of times. If all possible numbers have been assigned, what is the probability that a randomly selected employee has a payroll number which consists of 3 consecutive digits in their natural order (such as 345)?

Personal finance 26. Suppose your checkbook balance is at least $10.00 and no more than $99.99. Assuming that all balances are equally likely to occur, what is the probability that your balance is an exact number of dollars, for example, $21.00?

Sports 27. Suppose a basketball team is equally likely to win or lose each game. After 5 games the team has a "record," i.e., a sequence of wins and losses. What is the probability that there is a string of at least 3 consecutive wins in the team's record?

28. A student has 4 pencils, 3 ballpoint pens, and 1 felt-tip pen in a cup. If 2 of the 8 are selected at random, what is the probability that a ballpoint pen is *not* selected?

Entertainment 29. A magazine rack contains 3 different issues of *Newsweek* including the September 15 issue, 4 different issues of *Time*, and 2 different issues of *Sports Illustrated*. Three magazines are selected at random. Find the probability for each of the following.
(*a*) Exactly 1 is an issue of *Time*.
(*b*) One of the magazines is the September 15 issue of *Newsweek*.

Entertainment 30. In the situation described in Exercise 29, find the probability that
(*a*) One is an issue of *Newsweek* and 2 are issues of *Sports Illustrated*.
(*b*) At least 1 is an issue of *Sports Illustrated*.

Sports 31. The coach of the Gigantic State University football team must form a game plan by selecting 7 plays from his list of 14 plays. The list includes 8 pass plays and 6 running plays. If the selection is made at random, what is the probability that the game plan will include at least 2 running plays *and* at least 2 pass plays?

32. A group of 6 men enter a restaurant and check their coats. The checker puts all 6 coats on the same hook and gives one of the men a tag. When the men leave the restaurant, the checker hands each of the men a coat. If the checker hands out the coats at random, what is the probability that each man receives the correct coat?

*33. There are n cans of regular cola and 3 cans of diet cola in a cooler. Three cans are selected at random, and the type of cola, regular or diet, is noted. For what values of n is the probability of having all diet colas less than .01?

*34. In the situation described in Exercise 33, for what values of n is the probability of having 1 regular and 2 diet colas less than .01?

*35. There are n red balls and $2n$ blue balls in a box. Three balls are selected at random, and their colors are noted. For what values of n is the probability of having all blue balls less than .28?

IMPORTANT TERMS AND CONCEPTS

You should be able to describe, define, or give examples and use each of the following:

Relative frequency method of assigning probabilities
Deductive method of assigning probabilities
Experiment
Event
Assignment of probabilities
Probability of an event
Equally likely outcomes
Selected at random
Permutation principle
Combination principle
Binomial coefficient
Pascal's triangle

REVIEW EXERCISES

1. A bowl contains 3 red balls and 2 blue balls. An experiment consists of drawing 3 balls, in succession and without replacement, and noting the color of each ball drawn. Describe the sample space of this experiment. What is the event that *at least* 1 blue ball is drawn?

2. In the experiment described in Exercise 1, what is the event that *at most* 1 blue ball is drawn? What is the event that *exactly* 1 blue ball is drawn?

Biology

3. A researcher removes 3 mice from a cage one after the other and notes the sex of each mouse. What is the sample space for this experiment? What is the event that at least 2 of the mice selected are male?

4. A pack of cards is made up of all the hearts, diamonds, and clubs of a regular deck of cards. An experiment consists of drawing a card, noting its *color* and *rank*, and replacing it, then drawing a second card and noting its color and rank. How many outcomes are there for this experiment? How many outcomes are in the event "2 red cards are drawn"?

5. A die is weighted so that outcomes 3 and 4 are equally likely, and they are 3 times as likely as outcomes 1, 2, 5, and 6, which are all equally likely. What probabilities should be assigned to the outcomes to reflect these facts?

Management

6. A delicatessen manager plans to have a special once each week for 4 weeks. The special can be any one of these items: soup, salad, sandwich, beverage, fruit, or dessert. No item will be a special more than once during the 4 weeks. A plan consists of the items to be specials and their order. How many different plans are there?

Product codes 7. A manufacturer uses a code to indicate the place and month in which an item is built. The manufacturer has 4 plants, each of which operates 12 months. If the letters A through L are used to denote months and the letters W, X, Y, and Z are used to denote locations, how many different codes are there?

Product codes 8. In the setting of Exercise 7, suppose that 2 plants operate 12 months and 2 plants operate 6 months. How many different codes are there?

Food selection 9. Gwen & Harry's Ice Cream Shoppe has 3 types of cones and 14 flavors of ice cream.

(*a*) In how many different ways can you order 1 cone and 2 scoops of ice cream? Of course, vanilla on top of chocolate is different from chocolate on top of vanilla.

(*b*) In how many different ways can you order 1 cone and 2 scoops of ice cream which are not the same flavor?

Transportation 10. A motor pool contains 5 subcompact sedans, 4 compact sedans, and 3 midsize sedans. In how many ways can 3 cars be selected so that not all are the same size?

Education 11. There are 4 students who turn in their mathematics examinations without names. After grading, each of the examinations has a different score.

(*a*) In how many different ways can the examinations be returned to the 4 students?

(*b*) In how many different ways can the examinations be returned so that exactly 2 students receive the correct examinations?

Education 12. A student takes a quiz on which the possible grades are A, P, and F. In the student's view, a grade of A is twice as likely as an F and a P is 3 times as likely as an F. What weights should be assigned to the outcomes of the "experiment" of taking a quiz to reflect the student's assumptions?

Entertainment 13. There are 4 adjacent seats in a row in a theater. In how many different ways can 4 people be seated?

Education 14. First, second, and third prizes are to be awarded at a science fair in which 16 exhibits have been entered. In how many different ways can the prizes be awarded?

Committees 15. Manuel belongs to a committee which consists of 4 females and 3 males, including Manuel.

(*a*) In how many different ways can the names of the committee members be listed in a membership list?

(*b*) If the names of the committee members are listed in random order on the membership list, what is the probability that Manuel's name is last?

(*c*) A subcommittee of 3 members including both females and males is to be selected from the committee. In how many different ways can this be done?

16. Two fair dice are rolled. What is the probability of each of the following?

(*a*) A sum of 4, 6, or 8 is obtained.

(*b*) A sum of at least 10 is obtained.

17. Two fair dice are rolled, and the number on each is noted. Find the probability that one of the numbers is a 6 or the sum of the numbers is at least 9.

18. Find the number of different pairs of elements which can be selected from a set of 12 elements.

Psychology 19. A psychologist plans an experiment in group dynamics that requires subjects to play specific roles. There are 2 roles that must be played by females and 2 roles that can be played by either males or females. There are 5 females and 3 male subjects available. In how many ways can the experiment be set up?

Education

20. An economics exam has two parts. The first part consists of 10 true-false questions, and the second part consists of 5 multiple-choice questions where 1 answer in 4 possibilities is correct. How many different answer sheets can be submitted, assuming 1 answer is given for each question?

Marketing

21. A merchant has sale tags colored red, green, and yellow. She plans to mark down some merchandise 10 percent, some 20 percent, and some 50 percent. If the amount of discount and the colors of tags are matched randomly, find the probability that red tags are used to identify items with a 20 percent discount.

Health care

22. A large company requires yearly physical exams for its employees. Over a period of many years, it is found that each year about 12 employees out of every 500 examined will have emphysema and about 3 employees out of every 100 will have lung cancer. Moreover, about 1 employee per 1000 examined will have both emphysema and lung cancer. Consider an experiment which consists of examining an employee and which has these possible outcomes:
    (*a*) Emphysema is found but no lung cancer.
    (*b*) Lung cancer is found but no emphysema.
    (*c*) Both emphysema and lung cancer are found.
    (*d*) Neither emphysema nor lung cancer is found.
    Using the data given, what probabilities should be assigned to each of the 4 outcomes of this experiment?

Opinion survey

23. A staff member for a politician is conducting a survey, and there are questions on crime, defense, inflation, and unemployment. If the questions are arranged in random order on the questionnaire, find the probability of the following:
    (*a*) The question on inflation is first.
    (*b*) The question on inflation is first, and the question on crime is second.

Sports

24. A softball team has 6 aluminum bats and 9 magnesium bats. Three bats are selected at random from the bat rack.
    (*a*) What is the probability that all 3 are aluminum?
    (*b*) What is the probability that at least 1 bat is aluminum?

25. There are 24 cans of cola in a cooler; 6 are regular, and the remainder are diet. In how many ways can you select each of the following?
    (*a*) 3 cans from the cooler
    (*b*) 3 regular colas
    (*c*) 3 diet colas
    (*d*) 3 colas of which at least 2 are diet

26. In the setting of Exercise 25, find the probability of selecting 3 colas from the cooler such that
    (*a*) All are regular.
    (*b*) All are diet.
    (*c*) At least 2 are diet.
    (*d*) There are both regular and diet colas.

Games of chance

27. A poker hand consists of 5 cards selected at random from an ordinary deck. Find the probability that a poker hand contains
    (*a*) 4 aces
    (*b*) 4 of a kind (4 cards of the same rank)
    (*c*) 5 consecutive cards of the same suit (each ace can be used either as the first card, A2345, or as the last card, 10JQKA)

Games of chance 28. Find the probability that a poker hand contains
(*a*) Exactly 3 of a kind (*b*) Exactly 2 pairs (*c*) Exactly 1 pair

Production 29. A machine tool operator has a checklist of tasks that must be performed each day before production begins. The tasks are labeled *A*, *B*, *C*, *D*, and *E*. Task *A* must precede task *B*, and task *D* must precede *E*, but there are no other restrictions on the order in which the tasks are performed. In how many different orders can the tasks be performed subject to these restrictions?

Sports 30. Suppose that a bag containing 10 basketballs has 4 with defective valves. If 2 balls are selected at random from the bag, what is the probability that *at least* 1 will have a defective valve?

Sports 31. A coach has 6 stopwatches: 3 are accurate, 1 is fast, and 2 are equally slow. The coach makes a random selection of 2 watches to be used to time the first-place finisher in a race.
(*a*) What is the probability that the first-place finisher is timed accurately by both watches?
(*b*) What is the probability that the first-place finisher is timed accurately by at least 1 watch?
(*c*) What is the probability that the watches used to time the first-place finisher both show the same time?

Recreation 32. A student is planning a ski trip between semesters. She can go to Colorado or New England. She is free to go in either December or January, and there is space at each of 3 Colorado lodges either time. There is space at 4 New England lodges in December but at only 2 in January. A vacation plan is a location, a lodge, and a time. If she selects a vacation plan at random, find the probability of each of the following.
(*a*) She chooses Colorado.
(*b*) She chooses January.
(*c*) She chooses Colorado and January.

Management 33. A total quality management (TQM) team consists of 3 people from engineering, 2 people from marketing, and 1 person from finance. There are 6 people from engineering, 5 from marketing, and 3 from finance. One of the engineers cannot work productively with one of the marketing specialists and always quarrels with one of the people from finance. How many TQM teams can be formed so that this engineer is never on a team with either of these people?

34. A green die and a red die are rolled, and the numbers on each are noted. Assume both dice are fair.
(*a*) Find the probability that the numbers are the same on both dice.
(*b*) Find the probability that the sum of the numbers is even.
(*c*) Find the probability that the product of the numbers is less than 19.

Theater casting 35. A casting director is casting for a play with roles for 1 female, 1 child, and 2 males. There are 6 males, including Sam, 3 females, and 2 children, including Heidi, auditioning for the roles.
(*a*) In how many different ways can the roles be filled from those auditioning?
(*b*) In how many different ways can they be filled if exactly 1 of Sam and Heidi is given a role?
(*c*) If the roles are filled at random, what is the probability that both Heidi and Sam are given roles?

Quality control *36. A box of 50 widgets contains 5 defective ones. Three widgets are selected at random and tested. What is the most likely number of defective widgets in the sample?

*37. For what values of n does the number of ways of selecting n items from a set of 50 exceed the number of ways of selecting 10 items from a set of 100?

Quality control *38. There are n defective items in a box of 50. Three items are selected and tested, and 1 is found to be defective.

(*a*) Determine the probability of finding exactly 1 defective item in a random selection of 3 items. This probability will depend on n.

(*b*) For what value of n is the probability determined in (*a*) as large as possible?

CHAPTER

3 Probability

3.0 THE SETTING AND OVERVIEW

The outcomes of an experiment do not always occur with the same frequency. As a result, when using the concepts and methods of probability to study experiments, we cannot always assume that the outcomes are equally likely. In this chapter we shall see how the properties of probabilities which hold in general settings are related to those which hold when outcomes are equally likely. We shall use these properties to study the results of experiments and to compute the probabilities of various events. Also we shall study experiments for which we have partial advance information about the outcomes. In these cases we develop methods to determine probabilities in light of this information. We further develop the concept of a tree diagram as a way of representing certain types of experiments and as an aid in solving problems.

3.1 PROBABILITY MEASURES: AXIOMS AND PROPERTIES

Probability theory is a method of assigning numbers to events, i.e., to subsets of a sample space. In Chapter 2, we considered events which were subsets of a sample space of equally likely outcomes. However, many very interesting and important events do not consist of equally likely outcomes. For example, consider the experiment of checking computer memory chips until either a defective chip is found or three chips have been checked. If we let G represent a good chip and D represent a defective chip, then a natural sample space for this experiment is the set $S = \{D, GD, GGD, \text{and } GGG\}$. The outcomes in set S do not usually occur equally often, and hence they should not be assigned the same probability. This means that we need a more general method of assigning probabilities to events than the one used in Chapter 2, $\Pr[E] = n(E)/n(S)$, because in that method it was assumed that all outcomes had the same probability. Our more general method should include the situation of equally likely outcomes as a special case, and we begin with a brief review of this special case. First, however, it is useful to introduce some additional terminology.

Assigning probabilities to outcomes attaches a number to each outcome of an experiment. This probability is a measure of the likelihood of that outcome and is frequently referred to simply as the *weight* of that outcome. The result of all these assignments is a *probability measure*. In the special case in which all outcomes are equally likely, this is called the *equiprobable measure*.

If we use the equiprobable measure in a sample space S containing n outcomes, then we assign probability $1/n$ to each outcome, and we assign probability m/n to an event (subset of the sample space) which contains m outcomes. Several facts about this assignment are clear. First, the probability of each event is a nonnegative number. Second, since the sample space contains all outcomes, the probability assigned to the sample space is $n/n = 1$. Third, if we have two *disjoint* events E_1 and E_2 containing m_1 and m_2 outcomes, respectively, then there are $m_1 + m_2$ outcomes in the event $E_1 \cup E_2$, and

$$\Pr[E_1 \cup E_2] = \frac{m_1 + m_2}{n} = \frac{m_1}{n} + \frac{m_2}{n} = \Pr[E_1] + \Pr[E_2]$$

We require that any assignment of probabilities satisfy these three conditions. To use the terminology which is common in mathematics in such circumstances, we take these conditions as *axioms* for an assignment of probabilities.

Axioms for a Probability Measure

A *probability measure* assigns to each event E of a sample space S a number denoted by $\Pr[E]$ and called the *probability* of E: This assignment must satisfy

i. $0 \leq \Pr[E] \leq 1$ for each event E in S
ii. $\Pr[S] = 1$
iii. If E_1 and E_2 are disjoint events in S, then

$$\Pr[E_1 \cup E_2] = \Pr[E_1] + \Pr[E_2]$$

We note that a probability measure is an example of a function, a function defined on the subsets of a sample space S. To each subset E of S, this function assigns a number $\Pr[E]$. It is natural to refer to this function as Pr.

In Chapter 2 we defined the probability of an event by using the probabilities of outcomes. In particular:

For any event E, $E \neq \emptyset$,

$$\Pr[E] = \text{sum of probabilities of outcomes in } E$$

and $$\Pr[\emptyset] = 0$$

Given any assignment of probabilities to outcomes, this definition can be used to define the probability of any event. To confirm that it defines a probability measure, it is necessary to verify that axioms i to iii are fulfilled. This is a straightforward application of the properties (2.1) and (2.2). We omit the details.

Example 3.1

Education

The new class of first-year students at the main campus of Gigantic State University (GSU) consists of traditional students (those who just graduated from high school) and nontraditional students (those who did not go to college directly from high school). Both groups contain students who plan to major in the College of Arts and Sciences as well as students who plan to major in the School of Business. The selection of a student in this class can be thought of as an experiment with four possible outcomes. Using data about this class of students to determine the relative frequencies for these four outcomes yields the probabilities assigned in Table 3.1. Note that since the four outcomes listed are mutually exclusive and make up the entire sample space, the probabilities add to 1.

TABLE 3.1

Outcome		Probability
Traditional student in arts and sciences	(TA)	.40
Nontraditional student in arts and sciences	(NA)	.12
Traditional student in business	(TB)	.28
Nontraditional student in business	(NB)	.20

Problem Using the data from Table 3.1, determine each of the following probabilities: Pr[traditional student], Pr[nontraditional student], and Pr[traditional student or business major].

Solution We use the following notation for events:

T: a traditional student
N: a nontraditional student
A: a student who plans to major in arts and sciences
B: a student who plans to major in business

It follows that $T = \{TA, TB\}$ and $N = \{NA, NB\}$. Using the fact that the probability of an event is the sum of the probabilities of the outcomes in the event, we have

$$\Pr[T] = \Pr[TA] + \Pr[TB] = .40 + .28 = .68$$

Also events T and N form a partition of the sample space of all first-year students (every student is in either T or N, and no student is in $T \cap N$), so $\Pr[T] + \Pr[N] = 1$, and hence

$$\Pr[N] = 1 - \Pr[T] = 1 - .68 = .32$$

Also since a student in the event $T \cup B$ is either a traditional student or a prospective business major, we have

$$\begin{aligned}\Pr[\text{traditional student or business major}] &= \Pr[T \cup B] \\ &= \Pr[\{TA, TB, NB\}] \\ &= .40 + .28 + .20 \\ &= .88\end{aligned}$$

■

There are a number of useful properties of probability measures which can be deduced from the axioms.

Properties of a Probability Measure

1. For any event E, $\Pr[E'] = 1 - \Pr[E]$.
2. For any collection of pairwise disjoint events $E_1, E_2, \ldots, E_k$,

$$\Pr[E_1 \cup E_2 \cup \cdots \cup E_k] = \Pr[E_1] + \Pr[E_2] + \cdots + \Pr[E_k] \qquad (3.1)$$

3. For any events E and F,

$$\Pr[E \cup F] = \Pr[E] + \Pr[F] - \Pr[E \cap F] \qquad (3.2)$$

Equation (3.2), which holds for any probability measure, can be deduced from Equation (1.6) in the special case of equally likely outcomes. Indeed, Equation (1.6) for events E and F is

$$n(E \cup F) = n(E) + n(F) - n(E \cap F)$$

Dividing both sides of this equality by $n(S)$ and using the definition of probabilities for equally likely outcomes [Equation (2.4)], we have Equation (3.2).

Other properties can be deduced easily from these three. For instance, since the empty set $\varnothing$ is a subset of S, $\Pr[\varnothing]$ is defined. Also $S' = \varnothing$. It follows from axiom ii and property 1 that

$$\Pr[\varnothing] = \Pr[S'] = 1 - \Pr[S] = 1 - 1 = 0$$

The solution of many of the problems in applied probability considered in this book rests on properties 1 to 3 of a probability measure. In particular, we note that in Example 3.1 we used both properties 1 and 2.

Example 3.2 Let E and F be events in a sample space S with $\Pr[E] = .65$, $\Pr[F] = .4$, and $\Pr[E \cap F] = .3$.

Problem (*a*) Find $\Pr[E \cup F]$.

(*b*) Find the probability of event G, where G is the set of all outcomes which are in exactly one of events E or F.

Solution (*a*) We can apply property 3 directly. We have

$$\begin{aligned} \Pr[E \cup F] &= \Pr[E] + \Pr[F] - \Pr[E \cap F] \\ &= .65 + .4 - .3 = .75 \end{aligned}$$

(*b*) We recall from Chapter 1 that the set of all elements which are in exactly one of sets E or F can be written as the union of two sets: the set of elements which are in E but not in F and the set of elements which are in F but not in E. Therefore,

$$G = (E \cap F') \cup (E' \cap F)$$

Since $E \cap F'$ and $E' \cap F$ are disjoint, it follows from axiom 3 that

$$\begin{aligned} \Pr[G] &= \Pr[(E \cap F') \cup (E' \cap F)] \\ &= \Pr[E \cap F'] + \Pr[E' \cap F] \end{aligned}$$

Consequently, our goal is to determine $\Pr[E \cap F']$ and $\Pr[E' \cap F]$. To do so, it is useful to draw a Venn diagram and include the probabilities of appropriate subsets. Figure 3.1 shows sets E and F. We insert probabilities in Figure 3.1 in the same way that we inserted the numbers of elements in subsets of Venn diagrams in Section 1.3. Thus, in the set $E \cap F$, we write the probability .3. Since set E is

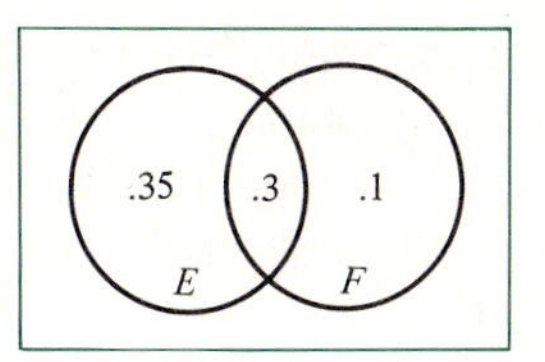

FIGURE 3.1

partitioned into $E \cap F$ and $E \cap F'$, $\Pr[E] = \Pr[E \cap F] + \Pr[E \cap F']$ and $.65 = .3 + \Pr[E \cap F']$, so that $\Pr[E \cap F'] = .35$. The number .35 has been inserted in Figure 3.1 in set $E \cap F'$. Likewise, we find $\Pr[E' \cap F]$ to be .1, and the number .1 has been inserted in Figure 3.1 in set $E' \cap F$. Using the information in Figure 3.1, we have

$$\Pr[G] = .35 + .1 = .45$$

■

Example 3.3 Let A, B, and C be events in a sample space S, with $A \cup B \cup C = S, A \cap (B \cup C) = \varnothing$, $\Pr[A] = .2$, $\Pr[B] = .5$, and $\Pr[C] = .7$.

Problem Find $\Pr[A']$, $\Pr[B \cup C]$, and $\Pr[B \cap C]$.

Solution Using property 1 and the information $\Pr[A] = .2$, we find

$$\Pr[A'] = 1 - \Pr[A] = .8$$

Next we are given that events A and $B \cup C$ are disjoint since $A \cap (B \cup C) = \varnothing$. Therefore, by property 2 and the fact that $\Pr[S] = 1$, we have $1 = \Pr[S] = \Pr[A \cup (B \cup C)] = \Pr[A] + \Pr[B \cup C] = .2 + \Pr[B \cup C]$. It follows that

$$\Pr[B \cup C] = .8$$

Finally, using property 3, we have

$$\begin{aligned} \Pr[B \cup C] &= \Pr[B] + \Pr[C] - \Pr[B \cap C] \\ .8 &= .5 + .7 - \Pr[B \cap C] \end{aligned}$$

from which we conclude that $\Pr[B \cap C] = .4$. ■

Note: The Venn diagram for Example 3.3 is shown in Figure 3.2. Although we did not need the diagram to answer the questions posed in the example, it is very useful to indicate the relationships between sets A, B, and C. Since $A \cap (B \cup C)$ is empty, sets A and $B \cup C$ do not overlap on the diagram. Also since $A \cup (B \cup C)$ is the entire sample space S, there is zero probability outside these sets. With this diagram it is clear that $\Pr[B \cup C] = .8$.

FIGURE 3.2

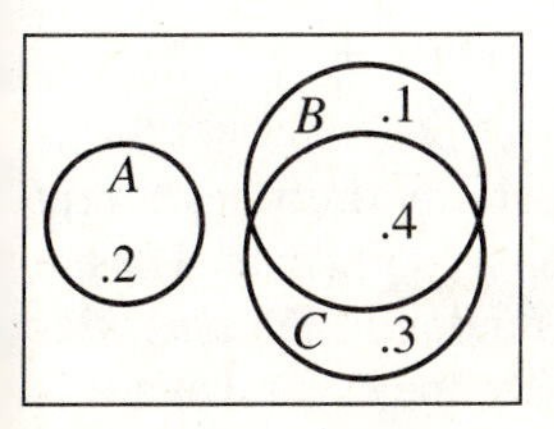

In Examples 3.2 and 3.3 the problems were solved by directly using the properties of a probability measure. Problems containing information given in descriptive form normally require an additional step. The events needed to answer the question must be identified, and the problem must be translated from words into symbols.

Example 3.4

Product testing

A quality control engineer inspects widgets on the production line of the Wonder Widget Company. A randomly selected widget can be acceptable or defective. Defective widgets are mismanufactured (too large or too small) or mislabeled (incorrect label or correct label in wrong place) or both. Probabilities based on past data are assigned (by the frequency method) as shown in Table 3.2. Note that since some widgets may be both mismanufactured and mislabeled, the sum of the probabilities in Table 3.2 is more than 1.

Problem Find the probability of each of the following events.
(*a*) A randomly selected widget is mismanufactured.
(*b*) A randomly selected widget is not mismanufactured.
(*c*) A randomly selected widget is mislabeled.

Solution We begin by labeling the events with the symbols F, G_1, G_2, H_1, and H_2 as shown in Table 3.2.

TABLE 3.2

Event		Probability
Acceptable	(F)	.72
Too large	(G_1)	.12
Too small	(G_2)	.08
Incorrect label	(H_1)	.15
Correct label in wrong place	(H_2)	.02

(*a*) The event that a randomly selected widget is mismanufactured is $G_1 \cup G_2$. Since $G_1 \cap G_2 = \varnothing$, we have

$$\Pr[G_1 \cup G_2] = \Pr[G_1] + \Pr[G_2] = .12 + .08 = .20$$

(*b*) The event that a widget is not mismanufactured is $(G_1 \cup G_2)'$, and

$$\Pr[(G_1 \cup G_2)'] = 1 - \Pr[G_1 \cup G_2] = 1 - .20 = .80$$

(*c*) The event that a widget is mislabeled is $H_1 \cup H_2$. Again, since $H_1 \cap H_2 = \varnothing$, we have

$$\Pr[H_1 \cup H_2] = \Pr[H_1] + \Pr[H_2] = .15 + .02 = .17$$

■

In assigning probabilities to events, we assign probability 1 to any event which always occurs and probability 0 to any event which never occurs. Thus in Example 3.4 we assign probability 1 to the event that a randomly selected widget is either acceptable or not acceptable, and we assign 0 to the event that a randomly selected widget is acceptable and not acceptable.

The computation of the desired probabilities in Example 3.4 is relatively direct. The solution of other problems may require several steps and the use of more than one of the properties. This is illustrated in our next example.

Example 3.5

Product testing

In the situation described in Example 3.4, find the probabilities of these events:

(*a*) A randomly selected widget is *either* mismanufactured *or* mislabeled (or both).

(*b*) A randomly selected widget is *both* mismanufactured *and* mislabeled.

Solution (*a*) Let $E_1 = G_1 \cup G_2$ denote the event that a randomly selected widget is mismanufactured, and let $E_2 = H_1 \cup H_2$ denote the event that a randomly selected widget is mislabeled. In Example 3.4 we determined $\Pr[E_1]$ and $\Pr[E_2]$. We have $E_1 \cup E_2 \cup F = S$, and clearly the event F (an acceptable widget is selected) is disjoint from $E_1 \cup E_2$. We conclude that $E_1 \cup E_2 = F'$, and therefore

$$\Pr[E_1 \cup E_2] = \Pr[F'] = 1 - \Pr[F] = 1 - .72 = .28$$

(*b*) The event that a randomly selected widget is both mismanufactured and mislabeled is $E_1 \cap E_2$, and we can determine $\Pr[E_1 \cap E_2]$ by using property 3, the results of Example 3.4, and part (*a*) of this example. Indeed, property 3 applied to events E_1 and E_2 gives

$$\Pr[E_1 \cup E_2] = \Pr[E_1] + \Pr[E_2] - \Pr[E_1 \cap E_2]$$

From Example 3.4, $\Pr[E_1] = .20$ and $\Pr[E_2] = .17$. Combining this with the result of part (*a*), we have

$$\Pr[E_1 \cap E_2] = \Pr[E_1] + \Pr[E_2] - \Pr[E_1 \cup E_2] = .37 - .28 = .09 \quad \blacksquare$$

Notice that the solution of part (*b*) uses the results of part (*a*) and the results of Example 3.4. If the problem of part (*b*) were posed as the only task, it would still be necessary to determine several other probabilities to solve it: It is a problem whose solution requires several steps.

Exercises for Section 3.1

1. Let A and B be events in the sample space S, and let $\Pr[A] = .45$, $\Pr[B] = .75$, and $\Pr[A \cap B] = .25$. Find the probability of each of the following events.
   (*a*) $B' \cap A$ (*b*) $A' \cap B$ (*c*) $A' \cap B'$
2. Let A and B be events in the sample space S, and let $\Pr[A] = .45$, $\Pr[B] = .75$, and $\Pr[A' \cap B] = .35$. Find the probability of each of the following events.
   (*a*) $B' \cap A$ (*b*) $A \cap B$ (*c*) $A' \cap B'$

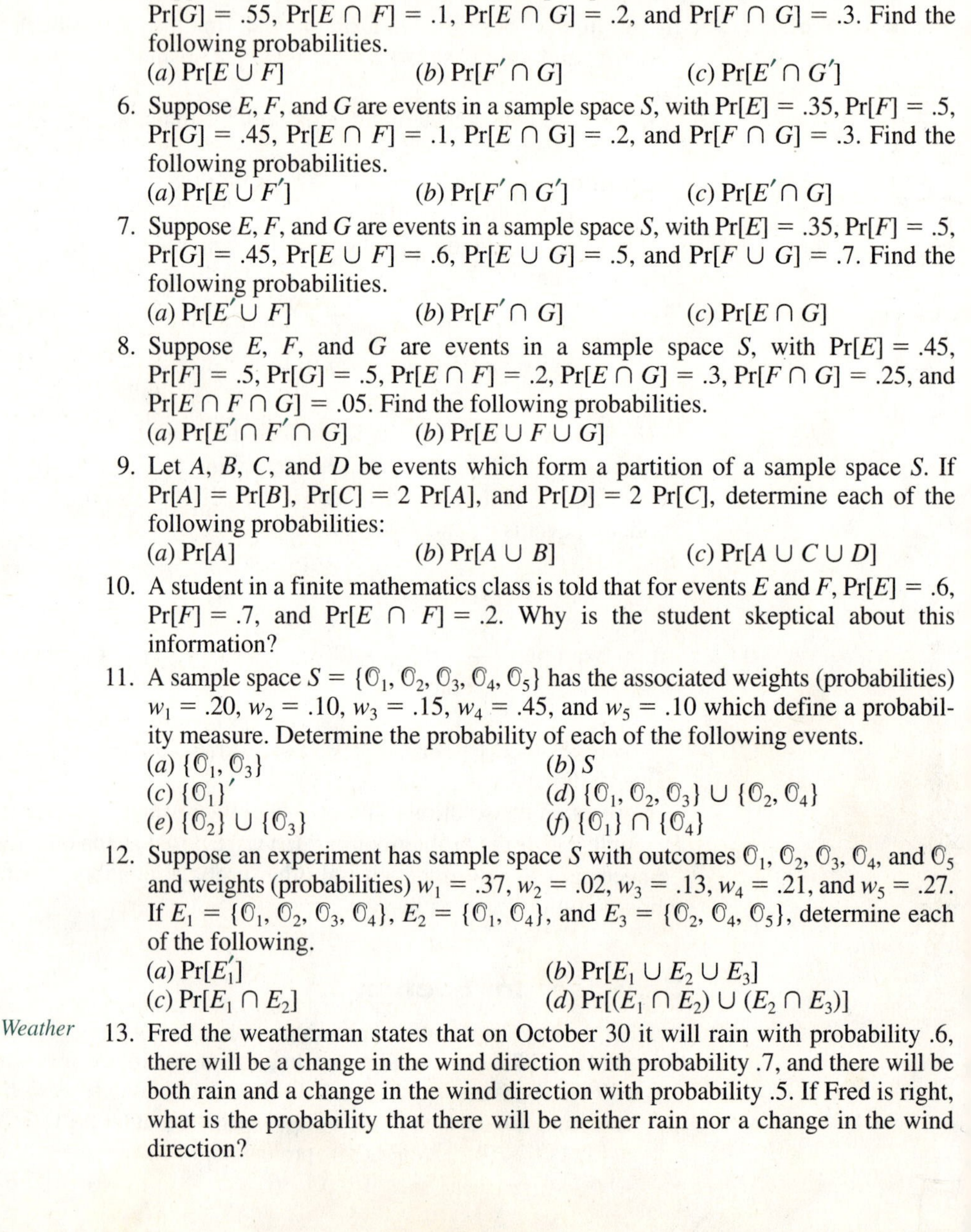

3. Let A and B be events in the sample space S, and let $\Pr[A'] = .45$, $\Pr[B'] = .35$, and $\Pr[A' \cap B'] = .15$. Find the probability of each of the following events.
   (*a*) $B' \cap A$ (*b*) $A \cap B$ (*c*) $A' \cap B$
4. Let A and B be events in a sample space S, with $\Pr[A] = .6$ and $\Pr[B] = .4$. Find $\Pr[A \cup B]$ if:
   (*a*) $B \subset A$ (*b*) $A' \cap B' = \varnothing$
5. Suppose E, F, and G are events in a sample space S, with $\Pr[E] = .25$, $\Pr[F] = .4$, $\Pr[G] = .55$, $\Pr[E \cap F] = .1$, $\Pr[E \cap G] = .2$, and $\Pr[F \cap G] = .3$. Find the following probabilities.
   (*a*) $\Pr[E \cup F]$ (*b*) $\Pr[F' \cap G]$ (*c*) $\Pr[E' \cap G']$
6. Suppose E, F, and G are events in a sample space S, with $\Pr[E] = .35$, $\Pr[F] = .5$, $\Pr[G] = .45$, $\Pr[E \cap F] = .1$, $\Pr[E \cap G] = .2$, and $\Pr[F \cap G] = .3$. Find the following probabilities.
   (*a*) $\Pr[E \cup F']$ (*b*) $\Pr[F' \cap G']$ (*c*) $\Pr[E' \cap G]$
7. Suppose E, F, and G are events in a sample space S, with $\Pr[E] = .35$, $\Pr[F] = .5$, $\Pr[G] = .45$, $\Pr[E \cup F] = .6$, $\Pr[E \cup G] = .5$, and $\Pr[F \cup G] = .7$. Find the following probabilities.
   (*a*) $\Pr[E' \cup F]$ (*b*) $\Pr[F' \cap G]$ (*c*) $\Pr[E \cap G]$
8. Suppose E, F, and G are events in a sample space S, with $\Pr[E] = .45$, $\Pr[F] = .5$, $\Pr[G] = .5$, $\Pr[E \cap F] = .2$, $\Pr[E \cap G] = .3$, $\Pr[F \cap G] = .25$, and $\Pr[E \cap F \cap G] = .05$. Find the following probabilities.
   (*a*) $\Pr[E' \cap F' \cap G]$ (*b*) $\Pr[E \cup F \cup G]$
9. Let A, B, C, and D be events which form a partition of a sample space S. If $\Pr[A] = \Pr[B]$, $\Pr[C] = 2\Pr[A]$, and $\Pr[D] = 2\Pr[C]$, determine each of the following probabilities:
   (*a*) $\Pr[A]$ (*b*) $\Pr[A \cup B]$ (*c*) $\Pr[A \cup C \cup D]$
10. A student in a finite mathematics class is told that for events E and F, $\Pr[E] = .6$, $\Pr[F] = .7$, and $\Pr[E \cap F] = .2$. Why is the student skeptical about this information?
11. A sample space $S = \{\mathcal{O}_1, \mathcal{O}_2, \mathcal{O}_3, \mathcal{O}_4, \mathcal{O}_5\}$ has the associated weights (probabilities) $w_1 = .20$, $w_2 = .10$, $w_3 = .15$, $w_4 = .45$, and $w_5 = .10$ which define a probability measure. Determine the probability of each of the following events.
   (*a*) $\{\mathcal{O}_1, \mathcal{O}_3\}$ (*b*) S
   (*c*) $\{\mathcal{O}_1\}'$ (*d*) $\{\mathcal{O}_1, \mathcal{O}_2, \mathcal{O}_3\} \cup \{\mathcal{O}_2, \mathcal{O}_4\}$
   (*e*) $\{\mathcal{O}_2\} \cup \{\mathcal{O}_3\}$ (*f*) $\{\mathcal{O}_1\} \cap \{\mathcal{O}_4\}$
12. Suppose an experiment has sample space S with outcomes $\mathcal{O}_1$, $\mathcal{O}_2$, $\mathcal{O}_3$, $\mathcal{O}_4$, and $\mathcal{O}_5$ and weights (probabilities) $w_1 = .37$, $w_2 = .02$, $w_3 = .13$, $w_4 = .21$, and $w_5 = .27$. If $E_1 = \{\mathcal{O}_1, \mathcal{O}_2, \mathcal{O}_3, \mathcal{O}_4\}$, $E_2 = \{\mathcal{O}_1, \mathcal{O}_4\}$, and $E_3 = \{\mathcal{O}_2, \mathcal{O}_4, \mathcal{O}_5\}$, determine each of the following.
   (*a*) $\Pr[E_1']$ (*b*) $\Pr[E_1 \cup E_2 \cup E_3]$
   (*c*) $\Pr[E_1 \cap E_2]$ (*d*) $\Pr[(E_1 \cap E_2) \cup (E_2 \cap E_3)]$

Weather

13. Fred the weatherman states that on October 30 it will rain with probability .6, there will be a change in the wind direction with probability .7, and there will be both rain and a change in the wind direction with probability .5. If Fred is right, what is the probability that there will be neither rain nor a change in the wind direction?

Weather 14. In the setting of Exercise 13, find the probability that it will rain but there will be no change in the wind direction.

Weather 15. In the setting of Exercise 13, find the probability that either it will rain or there will be a change in the wind direction, but not both.

16. Let A, B, and C be events in a sample space S, and suppose $\Pr[A] = .3$, $\Pr[B] = .6$, $\Pr[C] = .5$, $A \cap (B \cup C) = \varnothing$, and $A \cup B \cup C = S$. Find $\Pr[B \cap C]$.

17. Let $S = \{1, 2, 3, 4, 5, 6\}$, and suppose the weights assigned to these numbers are (in order) $\frac{1}{21}, \frac{2}{21}, \frac{3}{21}, \frac{4}{21}, \frac{5}{21}$, and $\frac{6}{21}$.
(*a*) Find Pr[the outcome is an even number].
(*b*) Find Pr[the outcome is greater than 2].

18. In Example 3.5 what is the probability of the event $E_1 \cup F$? Of $E_2 \cap F$?

Product performance 19. A sound system is such that when it is used, the microphone malfunctions with probability .1, the speakers malfunction with probability .05, and both malfunction with probability .01. What is the probability that neither the microphone nor the speakers malfunction?

20. In the setting of Exercise 19, what is the probability that either the microphone malfunctions or the speakers malfunction, but not both?

College life 21. Each Monday a student attends mathematics class with probability .6, skips accounting class with probability .3, and attends both with probability .5. Find the probability that she attends at least one class on Monday.

College life 22. In the setting of Exercise 21, find the probability that she attends exactly one class.

Investments 23. A securities analyst is reviewing the performance of a group of electric utilities. He finds that 65 percent have increased sales and 25 percent have increased earnings. Fifteen percent have increased neither sales nor earnings. A utility is selected at random from the group. What is the probability that it has increased both sales and earnings?

24. An experiment has a sample space $S = \{\mathbb{O}_1, \mathbb{O}_2, \mathbb{O}_3, \mathbb{O}_4\}$ with $\Pr[\{\mathbb{O}_1, \mathbb{O}_2\}] = .49$, $\Pr[\{\mathbb{O}_1, \mathbb{O}_3\}] = .51$, and $\Pr[\{\mathbb{O}_1, \mathbb{O}_2, \mathbb{O}_3\}] = .69$. Find the probabilities w_1, w_2, w_3, and w_4.

25. In Example 3.4 there are not enough data given to determine the probabilities of all events. For which of the following events are there insufficient data to determine the probability of the event?
(*a*) $G_1 \cup H_1$ (*b*) $F \cup G_1$ (*c*) $F \cap G_2$ (*d*) $G_2 \cap H_2$

Opinion survey 26. A consumer product-rating service sampled the owners of Arcticool air conditioners and asked questions about cost, ease of installation, and efficiency. The following data were obtained:

35 percent commented favorably about cost.
52 percent commented favorably about installation.
45 percent commented favorably about efficiency.
15 percent commented favorably about both cost and installation.
20 percent commented favorably about both cost and efficiency.
17 percent commented favorably about installation and efficiency.
5 percent commented favorably about all three.

A questionnaire is selected at random. Find the probability that:
(a) The owner commented favorably on exactly one item.
(b) The owner commented favorably on none of these three items.

Product performance

27. A television set which sometimes malfunctions has

A clear picture with probability .7
Good sound with probability .5
Good color with probability .6
Both a clear picture and good sound with probability .4
Both a clear picture and good color with probability .5
Both good sound and good color with probability .4
A clear picture, good sound, and good color with probability .3

Find the probability that the television has exactly one of these three characteristics (clear picture, good sound, good color).

Product evaluation

28. An inspector on an assembly line of a refrigerator plant classifies each refrigerator according to the properties of its enamel. Based on his data, the inspector assigns the probabilities listed in Table 3.3. Assume that no refrigerator has both too much and too little enamel and that the defects associated with E_1, E_2, E_3 are the only ones of concern to the inspector. What is the probability that a randomly selected refrigerator has each of the following?
(a) A paint defect
(b) A paint defect which includes an improper amount of paint
(c) A paint defect which results from an improper amount of paint and uneven application
(d) A paint defect which results from the proper amount of paint but uneven application

TABLE 3.3

Event		Observed Probability
E_1	Too much enamel	.04
E_2	Too little enamel	.12
E_3	Uneven application	.09
E_4	No defects noted	.82

29. Use the axioms of a probability measure to show that for any events E and F in a sample space S

$$\Pr[E \cup F] = \Pr[E] + \Pr[F] - \Pr[E \cap F]$$

30. Use the axioms of a probability measure to show that for any events E, F, and G in a sample space S

$$\Pr[E \cup F \cup G] = \Pr[E] + \Pr[F] + \Pr[G] - \Pr[E \cap F] - \Pr[E \cap G] - \Pr[F \cap G] + \Pr[E \cap F \cap G]$$

*31. Suppose the sample space S contains 100 disjoint events $E_1, E_2, \ldots, E_{100}$ and event E_n has probability K/n, where K is a constant, $n = 1, \ldots, 100$. Does there exist a value of K so that the events E_n form a partition of S? If such a K exists, find its value.

*32. In the setting of Exercise 31, suppose that event E_n has probability K/n^2, $n = 1, \ldots, 100$. Answer the questions posed in Exercise 31.

3.2 CONDITIONAL PROBABILITY AND INDEPENDENCE

The probability of an event is a number which indicates the likelihood that the event will occur. In certain situations we are naturally led to consider two events which are very similar, but which are distinguished in that for one we have more information on the outcomes than for the other. This additional information may mean that the two events have different probabilities.

Example 3.6 Suppose that we have an urn containing 3 red balls, 2 blue balls, and 4 white balls, and a ball is selected at random and its color is noted. Assuming that each ball is equally likely to be selected, i.e., each ball has probability $\frac{1}{9}$ of being selected, we conclude that

$$\Pr[\text{red ball}] = \tfrac{3}{9}$$
$$\Pr[\text{blue ball}] = \tfrac{2}{9}$$
$$\Pr[\text{white ball}] = \tfrac{4}{9}$$

Now suppose that we are given the additional information that the ball selected is *not white*. Since only 5 of the 9 balls are not white, we may think of the sample space as being limited to those 5 outcomes. With this information the probabilities become

$$\Pr[\text{red ball selected, given that the ball selected is not white}] = \tfrac{3}{5}$$
$$\Pr[\text{blue ball selected, given that the ball selected is not white}] = \tfrac{2}{5}$$
$$\Pr[\text{white ball selected, given that the ball selected is not white}] = \tfrac{0}{5} = 0$$

Indeed, in the latter case there are only 5 balls—3 red and 2 blue—which satisfy the condition of being not white, and the probabilities assigned to the outcomes are those determined by the equiprobable measure with 5 possible outcomes. ■

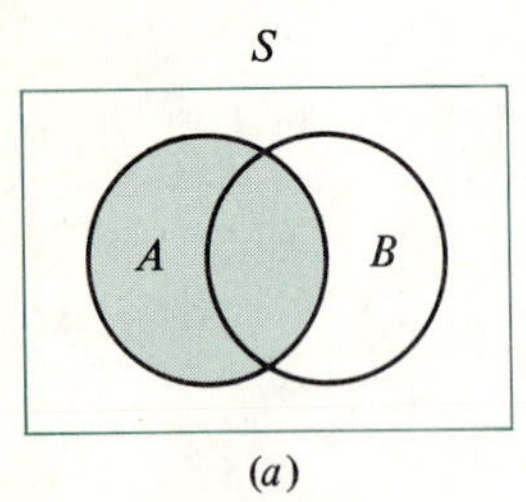

(a)

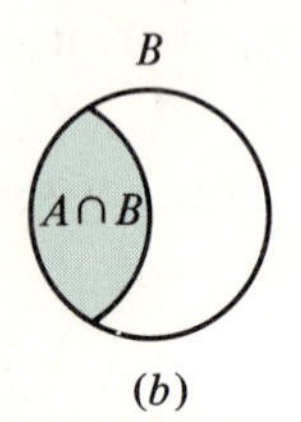

(b)

FIGURE 3.3

Example 3.6 illustrates the concept of *conditional probability*, i.e., the probability assigned to an event which is described in terms of two other events: the drawing of a red ball and the drawing of a ball which is not white. A precise definition is as follows.

> Let A and B be events in the sample space of an experiment with $\Pr[B] \neq 0$. The *conditional probability of event A given B*, written $\Pr[A \mid B]$, is
>
> $$\Pr[A \mid B] = \frac{\Pr[A \cap B]}{\Pr[B]}$$

In Example 3.6 we see that the conditional probability of an event is just the probability of that event in the smaller sample space specified by the condition. If A is the event "a red ball is selected" and B is the event "a white ball is not selected," then

$$\Pr[A \mid B] = \frac{\frac{3}{9}}{\frac{5}{9}} = \frac{3}{5}$$

This final ratio $\frac{3}{5}$ is simply the result of using the equiprobable measure in the sample space consisting of those outcomes in which the ball selected is not white.

In the general case we can illustrate the concept of conditional probability by using Venn diagrams. A sample space S and events A and B are shown in the Venn diagram of Figure 3.3*a*. The probability of A is the sum of the weights of all outcomes in A. Equivalently, since the weight of all outcomes in S is 1, this sum is the fraction of the total weight due to outcomes in A. The conditional probability of A given B corresponds to the situation in which B represents the sample space and $\Pr[A \mid B]$ is the weight of the outcomes in $A \cap B$ compared to the total weight of B. This is shown in Figure 3.3*b*.

Example 3.7 Let A and B be events in a sample space S, and suppose it is known that $\Pr[A] = \frac{2}{5}$, $\Pr[B] = \frac{1}{3}$, and $\Pr[A \cap B] = \frac{1}{5}$.

Problem Find $\Pr[A \mid B]$ and $\Pr[B \mid A]$.

Solution

$$\Pr[A \mid B] = \frac{\Pr[A \cap B]}{\Pr[B]} = \frac{\frac{1}{5}}{\frac{1}{3}} = \frac{3}{5}$$

$$\Pr[B|A] = \frac{\Pr[B \cap A]}{\Pr[A]} = \frac{\frac{1}{5}}{\frac{2}{5}} = \frac{1}{2}$$ ■

Remark Note that, in general, $\Pr[A|B] \neq \Pr[B|A]$. This fact is illustrated in Example 3.7. We also note that $\Pr[A|B]$ may be greater than, less than, or the same as $\Pr[A]$.

Example 3.8 *Product testing*

An inspector at a soft drink plant checks the quantity in each 12-ounce can. The data collected yield the probabilities shown in Table 3.4. Thus, the probability that a randomly selected can contains too much is .06.

Problem Suppose that the inspector knows that a certain can does not contain the correct amount. What is the probability that the can contains too much?

TABLE 3.4

Event	Probability
Too much (*M*)	.06
Too little (*L*)	.14
Correct amount (*C*)	.80

Solution Let events M, L, and C be as shown in Table 3.4. The given condition in this problem is that the can does not contain the correct amount. Since C is the event that the can does contain the correct amount, the given condition is the event C'. Our problem is to find the probability of too much (M), given C'. In other words, we seek $\Pr[M|C']$. Using the facts that $\Pr[C'] = .2$ and $M \cap C' = M$ (any can with too much is a can which does not have the correct amount), we obtain

$$\Pr[M|C'] = \frac{\Pr[M \cap C']}{\Pr[C']} = \frac{\Pr[M]}{\Pr[C']} = \frac{.06}{.20} = .3$$

Thus, if a can does not contain the correct amount, the probability that it has too much is .3. ■

Note: The situation which occurs in Example 3.8 is a common one: The event A is a subset of the event B so $A \cap B = A$. In Example 3.8, the event M is a subset of the event C', so $M \cap C' = M$.

Example 3.9 Suppose that an urn contains red balls marked 1, 2, 3; a blue ball marked 4; and white balls marked 5, 6, 7, 8. A ball is selected at random, and its color and number are noted.

Problem What is the probability that it is red? If the ball is known to have an even number, what is the probability that it is red?

Solution Let A be the event that the ball is red, and let B be the event that the ball has an even number. Then, since $n(A) = 3$ and $n(S) = 8$, we have $\Pr[A] = \frac{3}{8}$. This gives an answer to the first question. Next, $n(B) = 4$ and $n(A \cap B) = 1$. Therefore

$$\Pr[A\,|\,B] = \frac{\Pr[A \cap B]}{\Pr[B]} = \frac{n(A \cap B)/n(S)}{n(B)/n(S)} = \frac{\frac{1}{8}}{\frac{4}{8}} = \frac{1}{4}$$

and the answer to the second question is $\frac{1}{4}$. Clearly, *in this case*, $\Pr[A] \neq \Pr[A\,|\,B]$. Thus the knowledge that B has occurred can affect the probability assigned to A. ■

It can also happen that knowledge that event B occurs does not affect the probability that A occurs. Events for which this is the case are said to be *independent*. It is convenient to define the concept in slightly different terms and then to connect the definition with this interpretation.

Events A and B are said to be *independent* if

$$\Pr[A \cap B] = \Pr[A]\Pr[B] \tag{3.3}$$

The connection between this definition and the intuitive interpretation of the word "independent" is provided by the following observation: If A and B are independent events and $\Pr[B] \neq 0$, then

$$\Pr[A\,|\,B] = \frac{\Pr[A \cap B]}{\Pr[B]} = \frac{\Pr[A]\Pr[B]}{\Pr[B]} = \Pr[A]$$

Thus for some events A and B, in particular for independent events, we have $\Pr[A\,|\,B] = \Pr[A]$. In such cases we may say that knowing that B has occurred does not affect the probability that A occurs. Likewise, if A and B are independent and $\Pr[A] \neq 0$, then

$$\Pr[B\,|\,A] = \frac{\Pr[B \cap A]}{\Pr[A]} = \frac{\Pr[B]\Pr[A]}{\Pr[A]} = \Pr[B]$$

Thus, a knowledge of the occurrence of A does not affect the probability that B occurs.

It is important to recognize that independence (as we have defined the concept) has a precise *mathematical meaning*, and it may not be clear from the description of an experiment whether certain events are or are not independent. The only way to determine if events A and B are independent is to compute $\Pr[A]$, $\Pr[B]$, and $\Pr[A \cap B]$ and then check whether $\Pr[A \cap B] = \Pr[A]\Pr[B]$.

Example 3.10 Consider again the urn of Example 3.9 which contains red balls marked 1, 2, 3; a blue ball marked 4; and white balls marked 5, 6, 7, 8. Let A, B, and C be the following events:

A: A red ball is drawn.
B: A ball with an even number is drawn.
C: A white ball is drawn.

As before, the experiment consists of drawing a ball and noting its color and number.

Problem Decide whether A and B are independent and whether B and C are independent.

Solution We have (using equally likely outcomes)

$$\Pr[A] = \tfrac{3}{8} \qquad \Pr[B] = \tfrac{4}{8} \qquad \Pr[C] = \tfrac{4}{8}$$
$$\Pr[A \cap B] = \tfrac{1}{8} \qquad \Pr[B \cap C] = \tfrac{2}{8}$$

Thus

$$\Pr[A]\Pr[B] = \tfrac{3}{8} \cdot \tfrac{4}{8}$$

and

$$\Pr[A \cap B] \neq \Pr[A]\Pr[B]$$

Therefore the events A and B are *not* independent. However, since

$$\tfrac{1}{4} = \tfrac{2}{8} = \Pr[B \cap C] = \Pr[B]\Pr[C] = \tfrac{4}{8} \cdot \tfrac{4}{8} = \tfrac{1}{4}$$

events B and C are independent. ■

The question of whether two events are independent and the question of whether two events are disjoint are two quite different questions, even though both involve pairs of events. If A and B are disjoint, then $A \cap B = \varnothing$ and $\Pr[A \cap B] = 0$; if A and B are independent, then $\Pr[A \cap B] = \Pr[A]\Pr[B]$.

In general, two events which are independent are *not* disjoint. In fact, independent events which are also disjoint are very special: If two events A and B are both independent and disjoint, then $\Pr[A] = 0$ or $\Pr[B] = 0$ or both. You are asked to verify this fact in Exercise 15.

Example 3.11

Product performance

Adam's Audio purchases audiotapes from one manufacturer and tape decks from another. Approximately 2 percent of the audiotapes are defective, and the failure rate for tape decks is 8 percent.

Problem If the test results of tapes and tape decks are assumed to be independent, what is the probability that a random customer who purchases both an audiotape and a tape deck finds both defective?

Solution The probability that a randomly selected audiotape is defective is .02, and the probability of failure of a randomly selected tape deck is .08. Since these events are assumed to be independent, we have

$$\Pr[\text{defective tape and defective tape deck}] = .02 \times .08 = .0016.$$ ■

Exercises for Section 3.2

1. Let A and B be events such that $\Pr[A] = .7$, $\Pr[B] = .6$, and $\Pr[A \cap B] = .4$. Find $\Pr[A \mid B]$, $\Pr[B \mid A]$, and $\Pr[A \cup B]$.
2. Let A and B be events such that $\Pr[A] = .4$, $\Pr[B] = .5$, and $\Pr[A \cap B] = .3$. Find $\Pr[A \mid B]$, $\Pr[B \mid A]$, and $\Pr[A \cup B]$.
3. Let E and F be events in a sample space S. Suppose $\Pr[E] = \frac{5}{8}$, $\Pr[F] = \frac{3}{8}$, and $\Pr[(E \cup F)'] = 0$. Find $\Pr[E \mid F]$ and $\Pr[F \mid E]$.
4. Let E and F be events in a sample space S. Suppose $\Pr[E] = \frac{3}{8}$, $\Pr[F] = \frac{1}{2}$, and $\Pr[E \cap F'] = \frac{1}{8}$. Find $\Pr[E \mid F]$ and $\Pr[F \mid E]$.
5. Suppose A and B are events in a sample space S with $\Pr[A \mid B] = \frac{3}{4}$, $\Pr[A] = \frac{1}{2}$, and $\Pr[B'] = \frac{3}{4}$. Find $\Pr[B \mid A]$ and $\Pr[B \mid A']$.
6. Let A and B be events in a sample space S with $\Pr[A] = .6$, $\Pr[A \cap B'] = .2$, and $\Pr[B] = .5$. Find $\Pr[A \mid B]$ and $\Pr[B \mid A]$.
7. Let A and B be events such that $\Pr[A \mid B] = \frac{1}{2}$, $\Pr[B \mid A] = \frac{1}{3}$, and $\Pr[A \cap B] = \frac{1}{5}$. Are A and B independent?
8. Events E and F are independent in a sample space S with $\Pr[E] = .3$ and $\Pr[F] = .6$. Find $\Pr[E \mid F']$.
9. Let A and B be events such that $\Pr[A \cup B] = .8$ and $\Pr[A] = .6$. What is $\Pr[B]$ in the following cases:
   (*a*) A and B are independent.
   (*b*) A and B are disjoint events.
10. Let A and B be events such that $\Pr[A \cup B] = .9$ and $\Pr[A] = .5$. What is $\Pr[B]$ in the following cases:

(*a*) A and B are independent.
(*b*) A and B are disjoint events.

11. Suppose A and B are events with $\Pr[A \mid B] = \frac{3}{4}$, $\Pr[B \mid A] = \frac{3}{8}$, and $\Pr[A \cup B] = 1$. Find $\Pr[A]$, $\Pr[B]$, and $\Pr[A \cap B]$.
12. Suppose A and B are events with $\Pr[A \mid B] = \frac{2}{3}$, $\Pr[B \mid A] = \frac{1}{3}$, and $\Pr[A \cup B] = \frac{8}{9}$. Find $\Pr[A]$, $\Pr[B]$, and $\Pr[A \cap B]$.
13. In Example 3.9, are the following two events independent?

    F: A blue ball is drawn.
    B: An even-numbered ball is drawn.

14. Let $S = \{\mathcal{O}_1, \mathcal{O}_2, \mathcal{O}_3, \mathcal{O}_4, \mathcal{O}_5\}$ be a sample space with events $E = \{\mathcal{O}_1, \mathcal{O}_2, \mathcal{O}_3\}$, $F = \{\mathcal{O}_1, \mathcal{O}_5\}$, and $G = \{\mathcal{O}_2, \mathcal{O}_4\}$. Let probabilities be assigned to the outcomes as follows: $w_1 = \frac{1}{5}$, $w_2 = \frac{1}{10}$, $w_3 = \frac{1}{10}$, $w_4 = \frac{1}{5}$, and $w_5 = \frac{2}{5}$.
    (*a*) Are E and F independent?
    (*b*) Are F and G independent?
15. Let A and B be independent events. If A and B are also disjoint, show that either $\Pr[A] = 0$ or $\Pr[B] = 0$.
16. A box contains 1 blue, 2 red, and 3 white balls. A ball is selected at random, and its color is noted. Find the probability that it is not white given that it is not blue.
17. Two fair dice are rolled, and the numbers on the uppermost faces are noted.
    (*a*) What is the probability that exactly one die shows a 4 given that the sum of the numbers is 6?
    (*b*) What is the probability that the sum of the numbers is 6 given that exactly one die shows a 4?
    (*c*) What is the probability that the sum of the numbers is 6 given that at least one die shows a 4?
18. Two fair dice are rolled, and the numbers on the uppermost faces are noted.
    (*a*) What is the probability that exactly one die shows a 3 given that the sum of the numbers is 6?
    (*b*) What is the probability that the sum of the numbers is 6 given that exactly one die shows a 3?
    (*c*) What is the probability that the sum of the numbers is 6 given that at least one die shows a 3?

Committees

19. There are 5 Democrats and 3 Republicans on a committee. A subcommittee of 2 people is selected from the committee at random. Find the probability that both are Democrats given that at least one is a Democrat.

Entertainment

20. A student has 7 tickets to a play; 3 are in row A, 2 are in row B, and 2 are in row C. Two tickets are selected at random, and it is noticed that one is *not* in row A.
    (*a*) Find the probability that both are in row B.
    (*b*) Find the probability that both are in the same row.

Business

21. There are 3 Chevrolets and 4 Fords in the company motor pool. Two cars are selected at random and assigned to sales representatives. If they are both the same make, what is the probability that both are Fords?
22. There are 2 white balls, 2 red balls, and 3 blue balls in a box. Two balls are selected at random, and their colors are noted. Find the probability that neither is white given that neither is blue.

Committees 23. There are 4 men and 5 women on a committee. A subcommittee of 3 people is selected at random. Find the probability that all are male given that all are the same sex.

Committees 24. A subcommittee of 3 people is selected at random from a committee consisting of 6 women and 4 men. Find the probability that the committee consists of 2 men and 1 woman given that it contains both women and men.

College life 25. Three students are selected at random from a class consisting of 5 freshmen, 3 sophomores, and 2 juniors. Find the probability that 2 freshmen and 1 junior are selected given that at least 1 freshman is selected.

Course selection 26. Students at GSU register for courses, and then their schedules are prepared by computer. The computer is programmed to assign each student who registers for a course to a specific section of the course. There are 3 morning sections and 1 afternoon section of accounting and 1 morning section and 1 afternoon section of economics. All these sections are at different times, and the computer assigns students to sections at random. José registers for accounting and economics. If he is assigned 1 morning class and 1 afternoon class, what is the probability that he has accounting in the morning?

Biology 27. A biologist has 12 mice in a cage: 2 gray females, 3 gray males, 3 white females, and 4 white males. Two mice are selected simultaneously and at random, and their color and sex are noted.
   (*a*) Find the probability that both are females given that both are gray.
   (*b*) Find the probability that 1 mouse is female and 1 mouse is male given that both are gray.

Biology 28. The situation described in Exercise 27 is given.
   (*a*) Find the probability that at least 1 mouse is male given that 2 are white.
   (*b*) Find the probability that 1 mouse is female and 1 is male given that 1 is gray and 1 is white.

Biology 29. In the situation described in Exercise 27, what distribution of colors (2 gray or 2 white or 1 of each color) is most likely to result from the selection of 2 male mice?

30. An experiment consists of drawing a ball from an urn containing 2 red balls, 3 white balls, and 1 blue ball, noting the color, and then (without replacement) drawing a second ball and noting the color. Draw a tree diagram to represent this experiment, and assign conditional probabilities to the branches to reflect the likelihood of each selection.

31. Using the setting of Exercise 30, suppose that the first ball drawn is returned to the urn before the second ball is drawn. Again, draw the tree to represent this experiment, and assign conditional probabilities to the branches to reflect the likelihood of each selection.

*32. Suppose that 5 cards are randomly drawn from a standard deck. If it is known that 2 of them are aces, what is the probability that the other 3 cards are face cards?

*33. In the setting of Exercise 32 (i.e., it is known that 2 of the cards are aces), which of the following is most likely to be the other 3 cards: 3 cards of the same kind (such as kings) or 3 cards in a run (such as 5, 6, 7)?

3.3 STOCHASTIC PROCESSES AND TREES

Many experiments are naturally carried out in steps or stages, or they can be represented as being carried out in steps or stages. For these experiments, it is very useful to represent the steps and the outcomes by using a tree diagram with probabilities on the branches. We think of the main experiment as a sequence of subexperiments, one for each step or stage of the main experiment. Each of the subexperiments has a set of outcomes, and these outcomes have probabilities associated with them. Each outcome of the main experiment is determined by the results of the subexperiments, and the probabilities of these outcomes are determined in a simple way by the probabilities of the outcomes of the subexperiments. A natural method of representing the possible outcomes of both the subexperiments and the main experiment is to use a tree diagram.

An experiment which consists of a sequence of subexperiments is called a *stochastic process*.

Flipping a coin three times and noting the result of each flip is a simple example of a stochastic process. Most stochastic processes are more complicated than this, and it is useful to develop special techniques to use in studying them. The tree diagrams introduced in Chapter 1 are helpful, and our goal in this section is to refine and extend their use.

Example 3.12 Consider an experiment which consists of two steps. First a box is selected at random from a set of two boxes labeled a and b (see Figure 3.4), and then an urn is selected at random from the chosen box. Box a contains 3 urns, labeled A, B, and C, and box b contains 2 urns, labeled D and E.

Problem Draw a tree diagram to represent the outcomes of this experiment, and assign conditional probabilities to the tree to reflect the probabilities of the possible results at each stage.

FIGURE 3.4

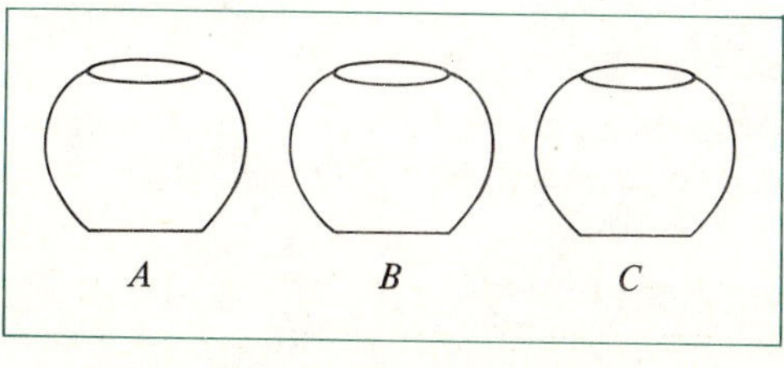

(a)

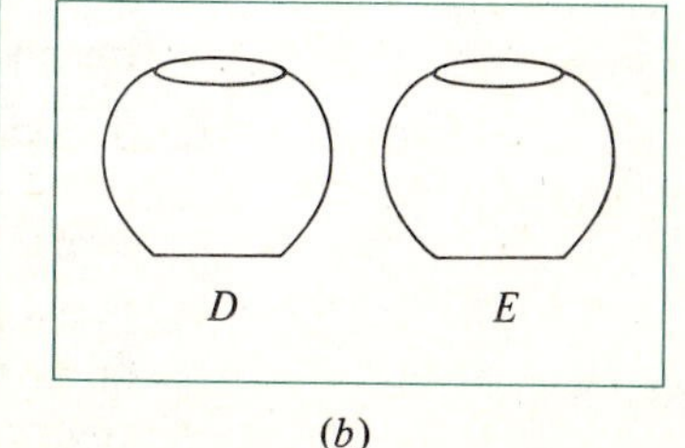

(b)

Solution At the first stage of the experiment there are 2 possible results (a or b). Hence the tree diagram should begin with a fork with 2 branches, one ending at a and the other at b. Since the boxes are selected at random (i.e., each choice is equally likely), we assign weights $\frac{1}{2}$ to each of these branches. After the first stage of the experiment is completed (and we have either box a or box b), the number of possible results for the second stage depends on the result at the first stage. This is represented in the tree diagram by having a fork with 3 branches at a and a fork with 2 branches at b. Again we use the fact that selections are made at random, and we assign conditional probabilities of $\frac{1}{3}$ to each of the 3 branches at a and $\frac{1}{2}$ to each of the 2 branches at b. The resulting tree with weights is shown in Figure 3.5. We have noted the outcomes of the experiment to the right of the tree, each outcome corresponding to a path which connects the left side of the tree ("Begin") to a result at each stage of the experiment. ■

The method used to obtain the tree diagram in Example 3.12 is typical of the method used to construct tree diagrams for any experiment which consists of a sequence of subexperiments. Each possible result of each subexperiment is represented by a branch of the tree. Each outcome of the complete experiment is represented pictorially by a path which connects the left of the tree (marked "Begin") to the successive results of the subexperiments in the order they occur. Each branch of the tree is assigned a number, the conditional probability associated with the result of the subexperiment represented by that branch. Thus the probability $\frac{1}{3}$ on the branch from a to A in Figure 3.5 is $\Pr[A \mid a]$; the $\frac{1}{2}$ on the branch from b to D is $\Pr[D \mid b]$.

FIGURE 3.5

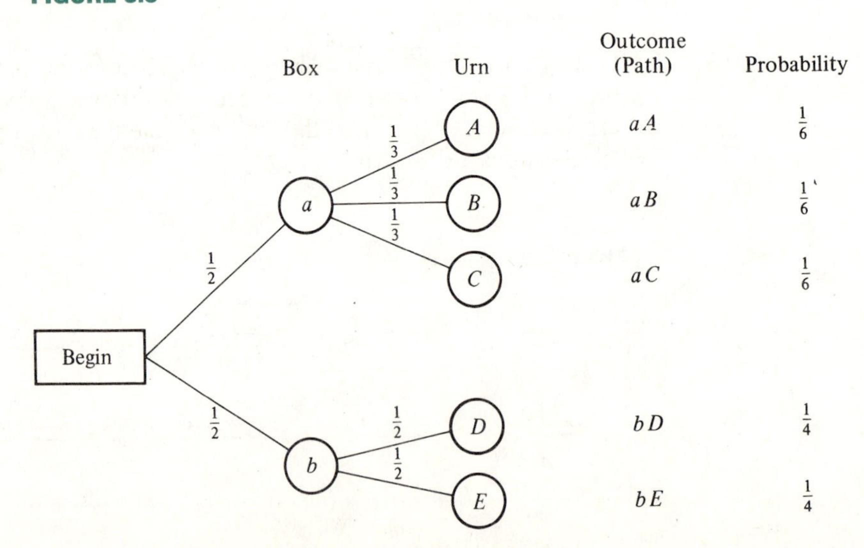

Before we continue our discussion of tree diagrams, it is necessary for us to look more carefully at our notation for the outcome of a sequential experiment. For example, in Figure 3.5 outcome aA means that a occurs at the first stage and A at the second stage. This outcome can be viewed as the intersection of two events. The first is the set of all outcomes in which a occurs at the first stage. Our shorthand notation for this event is simply a. The second event is the set of all outcomes in which A occurs at the second stage. Our shorthand notation for this is A. With this shorthand, the outcome aA can also be written $a \cap A$.

We use tree diagrams in studying stochastic processes to provide a simple method of finding an appropriate assignment of probabilities to the outcomes of the experiment. For example, in Figure 3.5 the outcome aA (which can also be written as $a \cap A$) is assigned probability $\frac{1}{6}$. This probability is the product $\frac{1}{2} \times \frac{1}{3}$ of the probability on the branch from "Begin" to a (that is, $\frac{1}{2}$) and the probability on the branch from a to A (that is, $\frac{1}{3}$). The reason this is the appropriate probability is that $\Pr[aA] = \Pr[a]\Pr[A \mid a]$. (This fact follows immediately from the definition of conditional probability: $\Pr[A \mid a] = \Pr[A \cap a] / \Pr[a]$.) This method of assigning a probability to an outcome, i.e., multiplying together the conditional probabilities on the branches of the path corresponding to the outcome, can be used for all tree diagrams. This method defines a probability measure.

Suppose we have a multistage experiment in which conditional probabilities can be assigned to each branch of the associated tree diagram. The *probability measure* of that experiment can be obtained by defining the probability of each outcome to be the product of the conditional probabilities assigned to the branches of the path in the tree diagram corresponding to that outcome.

Example 3.13 Consider the three-stage experiment which uses the boxes, urns, and colored balls of Figure 3.6 and which proceeds as follows: Select a box, then select an urn, then select a ball from that urn, and note the color of the ball. Suppose that all selections are random.

FIGURE 3.6

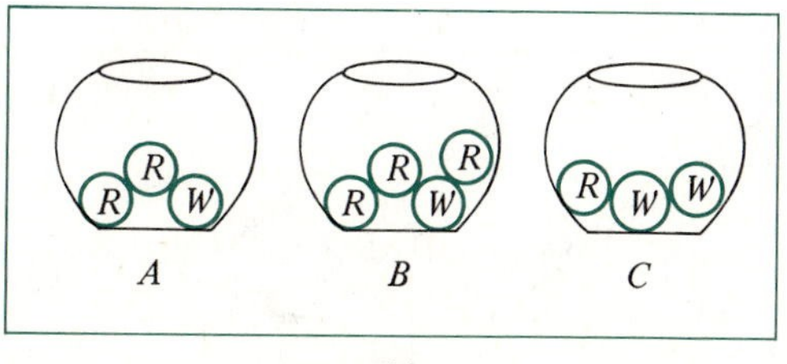

(a)

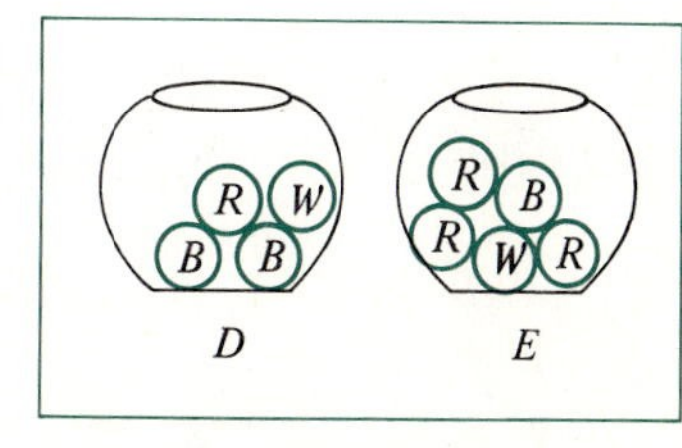

(b)

Problem Form a tree diagram for this experiment, and compute the probabilities for all outcomes.

Solution The first two stages of this experiment are identical to the two stages of the experiment in Example 3.12. Consequently, the first two stages of the tree diagram are identical to Figure 3.5. The third stage is formed by considering each of the urns in Figure 3.6 and placing one branch for each color which can be drawn from that urn at the fork of the tree representing that urn. For example, the fork for urn A has a branch for white and a branch for red since there are both red and white balls in urn A. There is only one branch for red even though there are 2 red balls. The branches at a fork correspond to distinct results. Since there are 2 red balls and only 1 white ball, the probability of drawing a red ball from urn A is $\frac{2}{3}$ and the probability of a white ball is $\frac{1}{3}$. These conditional probabilities are assigned to the branches from A to R and from A to W, respectively. In the same way we compute the other conditional probabilities, and the resulting tree is shown in Figure 3.7. The probability of any outcome is simply the product of the conditional probabilities for the branches in the path corresponding to that outcome. ■

Example 3.14 In the experiment described in Example 3.13, find the probabilities of these events:

$$F = \{\text{a red ball is drawn}\}$$
$$G = \{\text{a white ball is drawn}\}$$
$$H = \{\text{a blue ball is drawn}\}$$

Solution The complete tree for this experiment is shown in Figure 3.7. The probabilities shown at the right of this figure can be used to compute the probabilities of events F, G, and H.

Event F is the set of outcomes

$$F = \{aAR, aBR, aCR, bDR, bER\}$$

Using the definition of the probability assigned to an event, we have

$$\Pr[F] = \tfrac{2}{18} + \tfrac{3}{24} + \tfrac{1}{18} + \tfrac{1}{16} + \tfrac{3}{20} = \tfrac{121}{240}$$

Similarly,

$$\Pr[G] = \tfrac{1}{18} + \tfrac{1}{24} + \tfrac{2}{18} + \tfrac{1}{16} + \tfrac{1}{20} = \tfrac{77}{240}$$

and

$$\Pr[H] = \tfrac{2}{16} + \tfrac{1}{20} = \tfrac{42}{240}$$

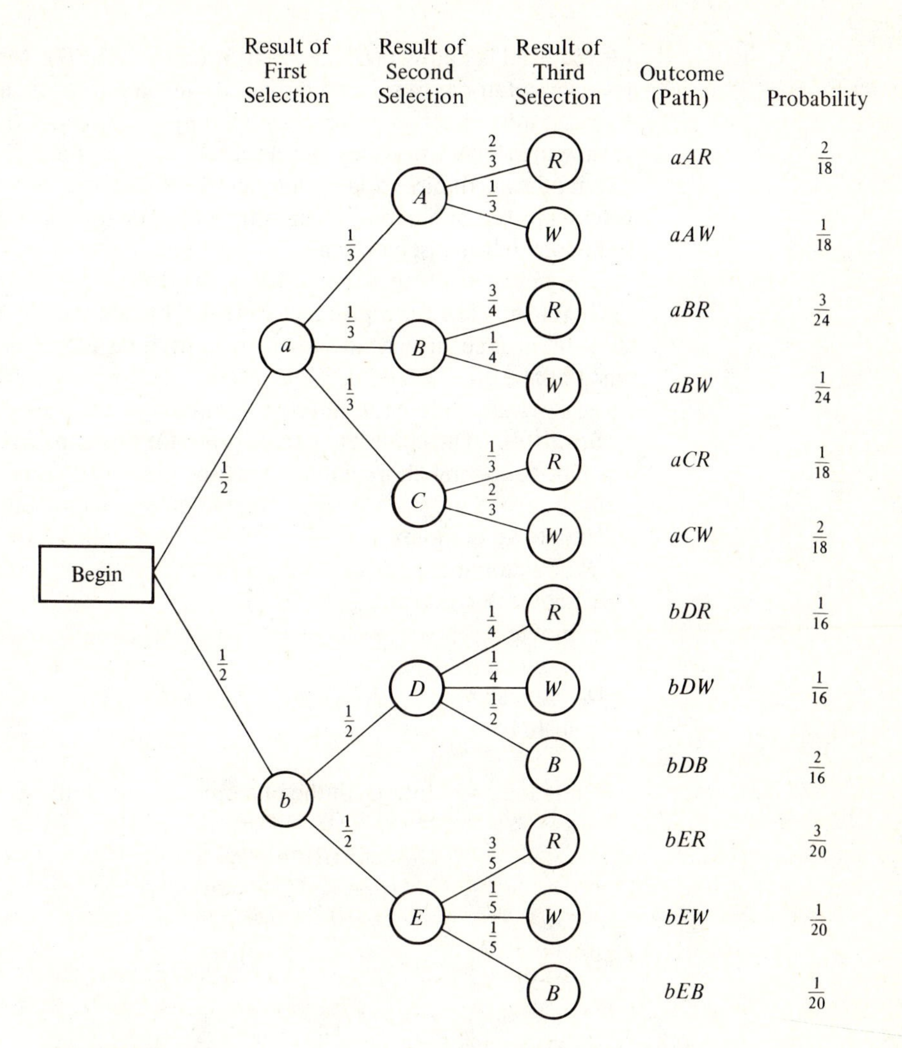

FIGURE 3.7

Note that since events F, G, and H form a partition of the sample space for this experiment, we have

$$\Pr[F] + \Pr[G] + \Pr[H] = \tfrac{121}{240} + \tfrac{77}{240} + \tfrac{42}{240} = \tfrac{240}{240} = 1 \qquad \blacksquare$$

Example 3.15

Recreation

When Harry met Sally, they decided to see a movie. They could go to either of two theater complexes, one uptown showing films A, B, and X, and another downtown, showing films A and B. They decide to flip a fair coin to pick a complex, a head is uptown (U) and a tail is downtown (D). After picking a complex, they roll a fair die to select a movie. If they are uptown and 1, 2, or 3 comes up, they see movie A; if 4 or 5 comes up, they see B; and if 6 comes up, they see X. If they are downtown, they see A with 1, 2, 3, or 4 and B with 5 or 6. After the movie, they will have a late-night snack. If they see movie A, they go to a delicatessen (De) for the snack; and if they see movie B or X, they are equally likely to go to a delicatessen or to a sushi bar (S).

Problem Find the probability that they see movie B, the probability that they eat at a sushi bar, and the probability that they see movie B and eat at a sushi bar.

Solution The complete tree diagram for this experiment is shown in Figure 3.8. The following abbreviations are used:

U: uptown complex
D: downtown complex
A: movie A
B: movie B
X: movie X
De: delicatessen
S: sushi bar

Using the probabilities of the outcomes shown at the right of the tree, we have

$$\Pr[B] = \tfrac{1}{12} + \tfrac{1}{12} + \tfrac{1}{12} + \tfrac{1}{12} = \tfrac{4}{12} = \tfrac{1}{3}$$
$$\Pr[S] = \tfrac{1}{12} + \tfrac{1}{24} + \tfrac{1}{12} = \tfrac{5}{24}$$

and

$$\Pr[B \cap S] = \tfrac{1}{12} + \tfrac{1}{12} = \tfrac{1}{6}$$ ■

Note that in Example 3.8, the events "see movie B" and "eat at a sushi bar" are not independent, because $\frac{1}{3} \cdot \frac{5}{24} \neq \frac{1}{6}$.

Exercises for Section 3.3

1. An experiment consists of flipping an unfair coin twice and noting the outcome (H or T) each time. Assume that $\Pr[H] = .2$, and draw a tree diagram for this experiment. Show all conditional probabilities on the branches of the tree, and compute the probabilities of the four outcomes of the experiment.
2. Repeat the experiment of Exercise 1. However, use a coin for which $\Pr[H] = .3$, and flip the coin 3 times.

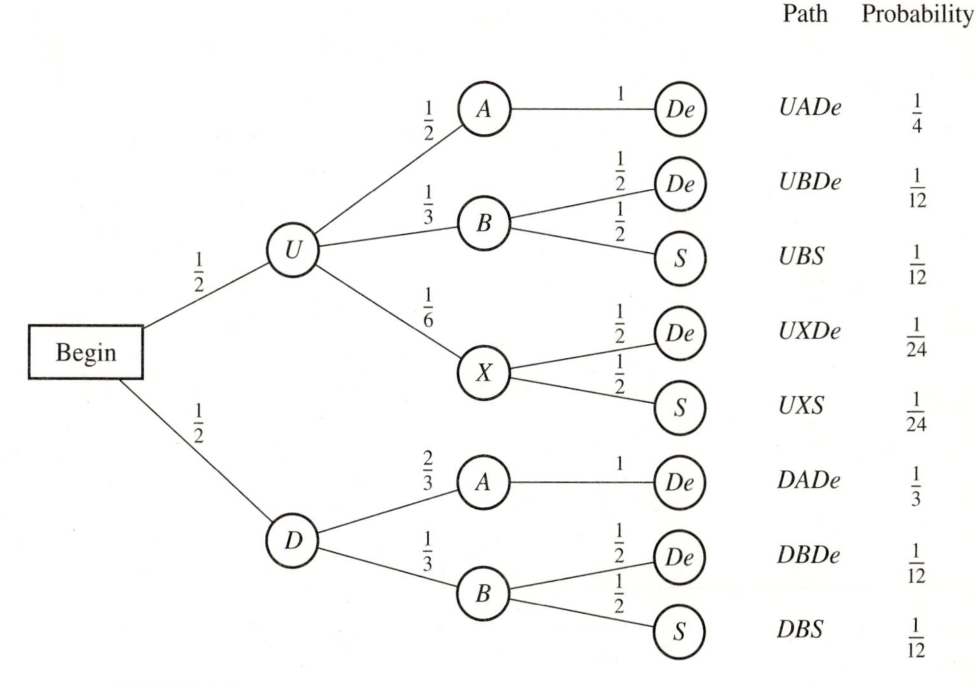

FIGURE 3.8

3. An experiment consists of first flipping a coin ($\Pr[H] = \frac{1}{3}$) and then randomly drawing a card from a reduced deck consisting of 4 aces, 3 kings, 2 queens, and 1 jack.
   (*a*) Draw a tree to represent this experiment, and compute the probabilities of the outcomes. For example, one outcome could be represented by HQ to indicate that the coin came up heads and the card drawn was a queen.
   (*b*) Find the probability of the event that either the coin came up heads or the card was a king (or both).
4. Suppose that in Exercise 3, whenever a tail comes up on the coin, then the aces are removed from the deck before the card is drawn. Answer the same two questions in this new setting.
5. An experiment consists of rolling a fair die 2 times. On the first roll, note whether the result is a small number (1, 2) or not (3, 4, 5, 6). If the first number is small, then on the second roll note whether the result is a 1 or not. If the first number is not small, then on the second roll note whether the number is small or not. Draw a tree diagram to represent this experiment, show the conditional probabilities, and compute the probabilities of the four outcomes.
6. Using the situation of Exercise 5, compute the probability that at least one of the two results is a small number.

7. There are 9 apples in a bag; 5 are red and 4 are yellow. Two apples are selected at random, one after another without replacement, and the color of each is noted.
   (*a*) Find the probability that the second is red.
   (*b*) Find the probability that at least 1 is red.

8. Repeat the experiment of Exercise 7, but in this case place the first apple back in the bag before selecting the second apple. Find the same two probabilities.

Education

9. At Gigantic State University, 20 percent of the students have taken calculus in high school. Of those who have taken calculus in high school, 60 percent plan to major in science. Of those who have not taken calculus in high school, 10 percent plan to major in science. What is the probability that a randomly selected student will plan to major in science?

10. For the setting of Exercise 9, what is the probability that a randomly selected student either will have taken calculus in high school or will not plan to major in science?

11. At the local Rent-A-Reck car rental lot, the fleet consists of Lexus, 20 percent; Infinity, 45 percent; and Mercedes, 35 percent. Half of the Mercedes, 70 percent of the Infinity, and 90 percent of the Lexus cars have a CD player.
   (*a*) If a car is selected at random from the lot, what is the probability that it has a CD player?
   (*b*) What is the probability that a car selected at random is either a Mercedes or has a CD player?

12. An unfair die, with $\Pr[1] = \Pr[2] = \Pr[3] = \frac{1}{12}$ and $\Pr[4] = \Pr[5] = \Pr[6] = \frac{1}{4}$, is rolled two times.
   (*a*) What is the probability that both numbers are even?
   (*b*) What is the probability that either one of the numbers is a 6 or both are odd?

13. There are 4 red balls and 3 blue balls in a box. Two balls are selected at random, one after the other without replacement, and the color of each ball is noted.
   (*a*) Find the probability that exactly 1 ball is blue.
   (*b*) Find the probability that the second ball is blue given that at least 1 of the balls is blue.

14. An unfair coin with $\Pr[H] = \frac{2}{3}$ is flipped until either 2 consecutive heads appear or a total of 4 flips has been made. Draw a tree diagram for this experiment, and determine the probability of the event "exactly 2 heads."

15. Suppose A and B are events in a sample space S, with $\Pr[A] = \frac{1}{3}$, $\Pr[B \mid A] = \frac{1}{6}$, and $\Pr[B \mid A'] = \frac{1}{3}$. Find $\Pr[B]$.

16. Suppose E and F are events in a sample space S, with $\Pr[E] = \frac{2}{5}$, $\Pr[F \mid E] = \frac{1}{2}$, and $\Pr[E' \cap F'] = \frac{3}{15}$. Find $\Pr[F]$ and $\Pr[E \cap F]$.

17. Find all missing probabilities on the tree diagram shown in Figure 3.9.

18. Find all missing probabilities on the tree diagram shown in Figure 3.10.

Biology

19. There are 8 mice in a cage: 4 white females, 3 white males, and 1 gray male. Two mice are selected one after the other without replacement, and their color and sex are noted. Draw a tree diagram for this experiment, and show all conditional probabilities on the tree. Assume random selections.
   (*a*) Compute the probabilities of all outcomes of the experiment.
   (*b*) Find the probability that 2 male mice are selected.

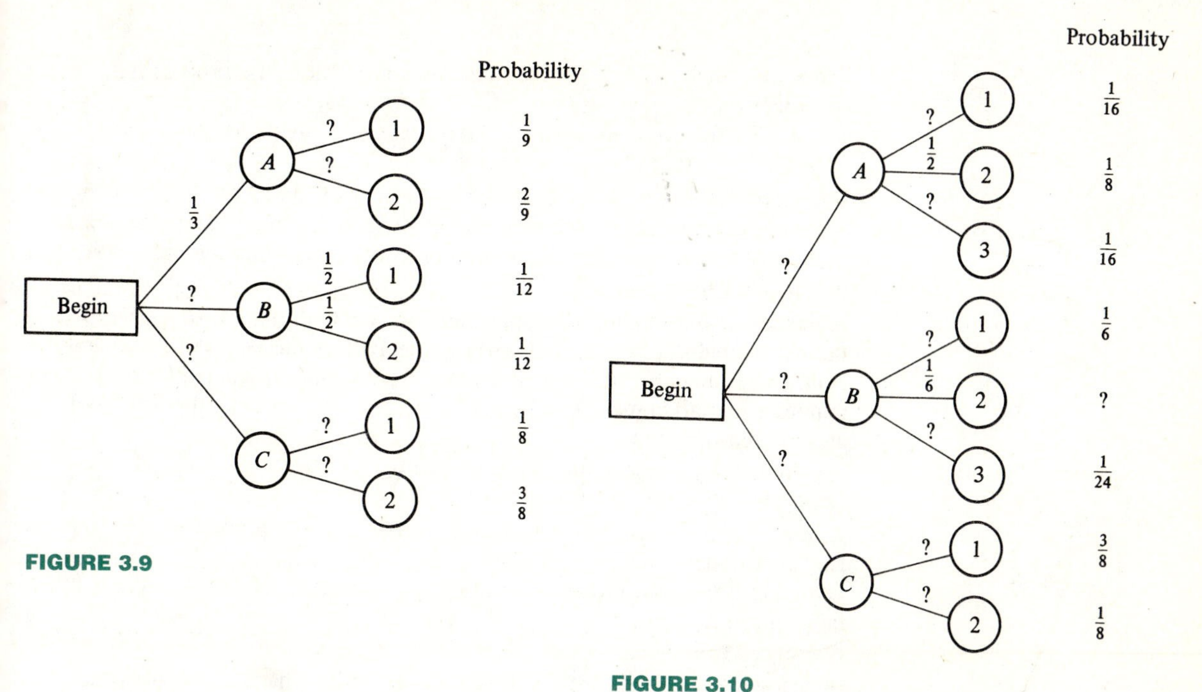

FIGURE 3.9

FIGURE 3.10

Biology 20. For the situation described in Exercise 19:
(*a*) Find the probability that the second mouse selected is gray.
(*b*) Find the probability that both gray and white mice are selected.

21. Consider 2 bags, each of which contains 3 colored balls. Bag *a* contains 2 red balls and 1 white ball, and bag *b* contains 1 red, 1 white, and 1 blue ball. An experiment consists of randomly drawing a ball from bag *a*, noting its color, placing the ball in bag *b*, and then randomly drawing a ball from bag *b* and noting its color. Draw a tree diagram to represent the outcomes of this experiment, i.e., the ordered pairs of colors obtained, and show all probabilities on the tree. What is the probability of obtaining 2 red balls? Also find Pr[2 red balls | first ball red].

Opinion survey 22. A group of people who had recently purchased new cars was surveyed about the purchases. Each person was asked to select the single most important factor in determining his or her choice and to indicate whether the purchase had proved satisfactory. Of this group, 35 percent cited price as the most important factor, 50 percent cited fuel economy, and the remainder cited styling. Fifty percent of those who cited price as the most important factor expressed satisfaction with their purchase, 80 percent of those who cited fuel economy were satisfied, and 30 percent of those who cited styling were satisfied. View this as an experiment whose outcomes are a reason for purchase (price, fuel economy, styling) and an evaluation of satisfaction (satisfied, not satisfied).

(*a*) Draw a tree diagram for this experiment, and find the probabilities for all outcomes.
(*b*) Let E be the event consisting of all outcomes in which the purchaser was satisfied. Find $\Pr[E]$.

23. There are 5 red and 3 white balls in box A and 2 red and 5 white balls in box B. A ball is selected at random from box A, its color is noted, and it is placed in box B. Then a ball is selected at random from box B. Find the probability that the second ball is red.

24. A box labeled H contains 1 red ball and 3 yellow balls, and a box labeled T contains 2 red balls, 1 yellow ball, and 1 green ball. A fair coin is flipped and if a head comes up, a ball is selected at random from the box labeled H and its color is noted. If a tail comes up, a ball is selected at random from the box labeled T and its color is noted.
(*a*) Draw a tree diagram for this experiment.
(*b*) Find the probability that a red ball is drawn.

Business 25. A new small business makes a profit the first year with probability .2. After the first year it makes a profit with probability .6 if it made a profit in the preceding year, and it makes a profit with probability .4 if it did not make a profit in the preceding year. Find the probability that it makes a profit for exactly 2 of the first 3 years.

Business 26. In the setting of Exercise 25, find the probability that the business makes a profit in at least 2 consecutive years of the first 3 years.

Sports 27. A basketball player is equally likely to make and to miss her first shot in each game. If she makes her first shot, then she is twice as likely to make future shots as to miss them. If she misses the first shot, then she is twice as likely to miss as to make future shots. Find the probability that the player makes exactly 2 of her first 3 shots.

28. A box contains 2 red, 3 blue, and 2 yellow balls. A ball is selected at random, and its color is noted. If it is yellow, it is replaced; otherwise, it is not. A second ball is selected, and its color is noted.
(*a*) Find the probability that the second ball is yellow.
(*b*) Find the probability that the second ball is red.

29. There are 5 blue and 2 yellow balls in a bucket. An experiment is described as follows. First, 2 balls are selected at random and the number of yellow balls is noted. The 2 balls just selected are set aside. Next, 2 additional balls are selected, one after the other without replacement, and the color of each is noted. Find the sample space and the probabilities of all outcomes for this experiment.

30. Let E and F be events in a sample space S with $\Pr[F] \neq 0$, and $\Pr[F] \neq 1$. Show that

$$\Pr[E] = \Pr[E|F]\Pr[F] + \Pr[E|F']\Pr[F']$$

*31. A special deck of cards consists of 10 groups of 10 cards, each labeled 1 through 10. These cards are mixed, and then random draws are made until the cards drawn include 5 consecutive numbers. What is the probability that only 5 cards will need to be drawn?

3.4 BAYES PROBABILITIES

Questions which involve conditional probabilities are very common in applications of probability theory, and one type of conditional probability problem is so common that it has a special name. In this type of problem, called a *Bayes probability* problem, it is easy to compute $\Pr[A|B]$, but the goal is to determine $\Pr[B|A]$. Our approach to these problems is to use both tree diagrams and the basic formula for computing conditional probabilities. We also give a general formula which can be used to compute Bayes probabilities.

Example 3.16

Quality control

Rap and Rock Records receives 70 percent of its inventory from Los Angeles and 30 percent from New York. Items from New York are defective 10 percent of the time, and items from Los Angeles are defective 5 percent of the time. Shipments arrive randomly, and the items are randomly distributed in the store.

Problem If a randomly selected item is found to be defective, what is the probability that it came from New York?

Solution We construct a tree diagram to represent the two-stage experiment of first noting the source of an item (*NY* or *LA*) and then noting whether it is defective (*D*) or acceptable (*A*). Using the data provided in the problem, we obtain the tree and the probabilities shown in Figure 3.11. Note, that we are using the notation $NY \cap D$ to indicate the path through *NY* and *D*, that is, to represent the outcome of a defective item from New York. Since the probability that an item is from New York is .3 and the conditional probability that an item from New York is defective is .1, the probability of outcome $NY \cap D$ is .03. Also, since the event "defective item" consists of the outcomes $NY \cap D$ and $LA \cap D$, we see that $\Pr[D] = \Pr[NY \cap D] + \Pr[LA \cap D] = .03 + .035 = .065$. We can now solve the problem posed. Using the definition of conditional probability, we have

$$\Pr[NY|D] = \frac{\Pr[NY \cap D]}{\Pr[D]} = \frac{.03}{.065} = \frac{6}{13} \quad \blacksquare$$

Example 3.16 has been solved by using a tree diagram and the definition of conditional probability. This approach is frequently the most direct for simple problems. However, there is another approach, based on the definition of conditional probability and axiom iii of a probability measure, which is sometimes useful. We illustrate the technique with an example.

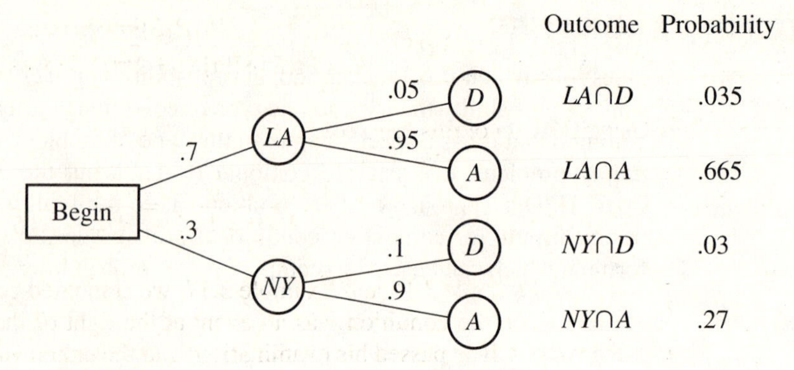

FIGURE 3.11

Example 3.17

College life

Clyde has a rather casual attitude toward his courses, and he studies for one-third of his examinations selected at random. In the past he has passed three-fourths of the examinations for which he has studied and one-fourth of those for which he has not studied.

Problem Clyde passed his mathematics examination. Find the probability that he studied for it.

Solution Let S and N denote study and not study, respectively, and let P and F denote pass and fail, respectively. The problem is to determine the conditional probability $\Pr[S|P]$. By definition this is equal to $\Pr[S \cap P]/\Pr[P]$ provided $\Pr[P] \neq 0$. To evaluate this ratio, we need $\Pr[S \cap P]$ and $\Pr[P]$. There is another expression for $\Pr[S \cap P]$ which involves only known quantities, namely, $\Pr[S \cap P] = \Pr[P|S]\Pr[S]$. It remains to determine $\Pr[P]$.

Since Clyde either studies or does not study for each examination, the set of examinations which he passes can be partitioned into $P \cap S$ and $P \cap N$. It follows from axiom iii of a probability measure that

$$\Pr[P] = \Pr[P \cap S] + \Pr[P \cap N]$$

Again using the definition of conditional probability, we have

$$\Pr[P \cap S] = \Pr[P|S]\Pr[S] \qquad \text{and} \qquad \Pr[P \cap N] = \Pr[P|N]\Pr[N]$$

Combining these relations, we have

$$\Pr[P] = \Pr[P|S]\Pr[S] + \Pr[P|N]\Pr[N]$$

We now have expressions for both $\Pr[S \cap P]$ and $\Pr[P]$ which involve only known quantities. Using these expressions, we have

$$\Pr[S \mid P] = \frac{\Pr[P \mid S]\Pr[S]}{\Pr[P \mid S]\Pr[S] + \Pr[P \mid N]\Pr[N]}$$

Using the data of this problem,

$$\Pr[S \mid P] = \frac{\frac{3}{4} \cdot \frac{1}{3}}{\frac{3}{4} \cdot \frac{1}{3} + \frac{1}{4} \cdot \frac{2}{3}} = \frac{\frac{1}{4}}{\frac{1}{4} + \frac{1}{6}} = \frac{3}{5}$$

■

In both Example 3.16 and Example 3.17, we computed conditional probabilities for which the condition was an event at the right of the tree (the item was defective or Clyde passed his examination) and the unknown probability was for an event at the left of the tree (the item came from New York or Clyde studied for the examination). Conditional probabilities such as these are known as *Bayes probabilities*. The formula derived in the solution of Example 3.17 is a general one, and it is an instance of Bayes' formula.

Bayes' Formula

Let S be the sample space of an experiment, and suppose that S is partitioned into subsets $S_1, S_2, \ldots, S_k$ such that $\Pr[S_i] > 0$ for $i = 1, 2, \ldots, k$. If A is any event such that $\Pr[A] > 0$, then

$$\Pr[S_i \mid A] = \frac{\Pr[A \mid S_i]\Pr[S_i]}{\Pr[A \mid S_1]\Pr[S_1] + \cdots + \Pr[A \mid S_k]\Pr[S_k]} \tag{3.4}$$

for $i = 1, 2, \ldots, k$.

In Example 3.17 the sample space (set of all examinations) was partitioned into two sets: those for which Clyde studied and those for which he did not study. Using the notation of Bayes' formula, formula (3.4), we have $k = 2$, and $S_1 = S$ (the set of examinations for which Clyde studied) and $S_2 = N$ (the set of examinations for which Clyde did not study).

There are situations in which the formula is simpler to use than a tree diagram, but most problems in this book can be solved more directly by using a tree diagram.

Example 3.18

Transportation

A faculty member at Gigantic State University regularly teaches an evening course in a neighboring city. Each time he makes the trip he is randomly assigned an automobile from the university motor pool. The motor pool consists of 50 percent Chevrolets, 30 percent Fords, and 20 percent Volkswagens. Some of the cars are air-conditioned, and others are not: 60 percent of the Chevrolets, 50 percent of the Fords, and 30 percent of the Volkswagens are air-conditioned.

Problem One afternoon the faculty member is assigned an air-conditioned car. What is the probability that it is a Chevrolet?

Solution We illustrate two methods of solving the problem: the use of a tree diagram and the use of Bayes' formula. Of course, the methods are not actually different; the formula simply expresses in a concise way the operations which are being carried out on a tree diagram. The tree diagram method is somewhat longer, but it provides a convenient means of organizing information.

(*a*) *Tree diagram*: A tree diagram is constructed from the information of the problem, and it is shown in Figure 3.12. The first branches represent the makes of automobiles, and the second branches distinguish between air-conditioned (A) and non-air-conditioned (N) vehicles. The event A that the faculty member receives an air-conditioned car can be partitioned into the disjoint events "air-conditioned Chevrolet," "air-conditioned Ford," and "air-conditioned Volkswagen." Using the data of the tree diagram, we have

$$\Pr[A] = \Pr[C \cap A] + \Pr[F \cap A] + \Pr[V \cap A] = .30 + .15 + .06 = .51$$

We can now solve the problem. We have

$$\Pr[C|A] = \frac{\Pr[C \cap A]}{\Pr[A]} = \frac{.30}{.51} = \frac{30}{51}$$

(*b*) *Bayes' formula*: The sample space can be partitioned into disjoint events C, F, and V (the faculty member is assigned a Chevrolet, Ford, or Volkswagen, respectively). Letting these events play the role of S_1, S_2 and S_3 in formula (3.4),

FIGURE 3.12

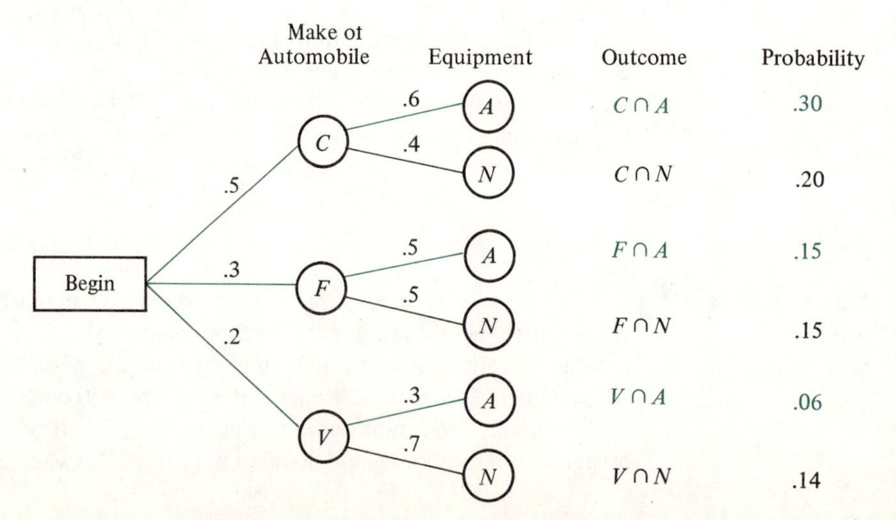

and letting A (the faculty member is assigned an air-conditioned car) be the A in formula (3.4), we have

$$\Pr[C|A] = \frac{\Pr[A|C]\Pr[C]}{\Pr[A|C]\Pr[C] + \Pr[A|F]\Pr[F] + \Pr[A|V]\Pr[V]}$$

$$= \frac{.6 \times .5}{.6 \times .5 + .5 \times .3 + .3 \times .2} = \frac{30}{51}$$ ■

Example 3.19 A bucket labeled H contains 3 yellow and 2 white balls, and a bucket labeled T contains 2 yellow balls and 1 white ball. An experiment consists of flipping a fair coin once, and according to the side which comes up, selecting two balls in succession and without replacement from the bucket labeled H (if a head) or T (if a tail). The color of each ball is noted as it is drawn: W is white and Y is yellow.

Problem (*a*) If exactly one white ball is drawn, what is the probability that the coin landed with a head up?

(*b*) If the second ball drawn is white, what is the probability that the coin landed with a head up?

Solution The tree diagram for this experiment is shown in Figure 3.13.

(*a*) We are interested in $\Pr[H|\text{exactly one } W]$. Here we are again using a shorthand notation to represent sets. Thus the expression "exactly one W"

FIGURE 3.13

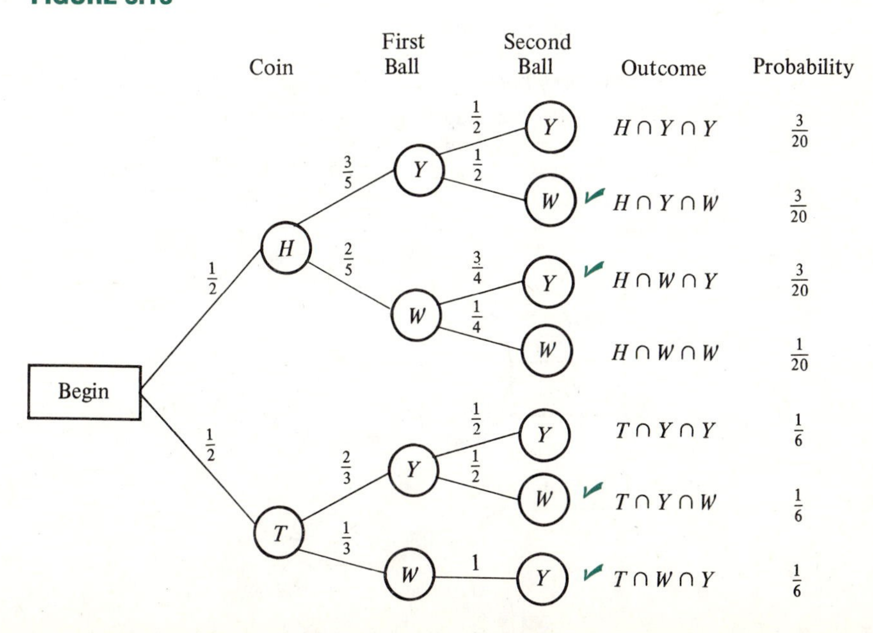

stands for the event consisting of all outcomes for which exactly one white ball is drawn. The outcomes in the event "exactly one W" are identified by a ✔ in the tree diagram. From the tree it is clear that

$$\Pr[H \cap \text{exactly one } W] = \tfrac{3}{20} + \tfrac{3}{20} = \tfrac{3}{10}$$
$$\Pr[\text{exactly one } W] = \tfrac{3}{20} + \tfrac{3}{20} + \tfrac{1}{6} + \tfrac{1}{6} = \tfrac{19}{30}$$

Using the definition of conditional probability, we find

$$\Pr[H|\text{exactly one } W] = \frac{\Pr[H \cap \text{exactly one } W]}{\Pr[\text{exactly one } W]} = \frac{\frac{3}{10}}{\frac{19}{30}} = \frac{9}{19}$$

(*b*) Our goal is to determine $\Pr[H|\text{second ball } W]$. Again using the tree diagram, we have

$$\Pr[H \cap \text{second ball } W] = \frac{3}{20} + \frac{1}{20} = \frac{1}{5}$$
$$\Pr[\text{second ball } W] = \frac{3}{20} + \frac{1}{20} + \frac{1}{6} = \frac{11}{30}$$

Combining these, we find

$$\Pr[H|\text{second ball } W] = \frac{\frac{1}{5}}{\frac{11}{30}} = \frac{6}{11}$$

■

Exercises for Section 3.4

1. The tree diagram for an experiment is shown in Figure 3.14. Compute the following probabilities.
   (*a*) $\Pr[B]$ (*b*) $\Pr[b]$ (*c*) $\Pr[b|B]$ (*d*) $\Pr[B|b]$
2. The tree diagram for an experiment is shown in Figure 3.15. Compute the following probabilities.
   (*a*) $\Pr[A]$ (*b*) $\Pr[X|A]$ (*c*) $\Pr[Y|B]$ (*d*) $\Pr[B|Y]$

FIGURE 3.14

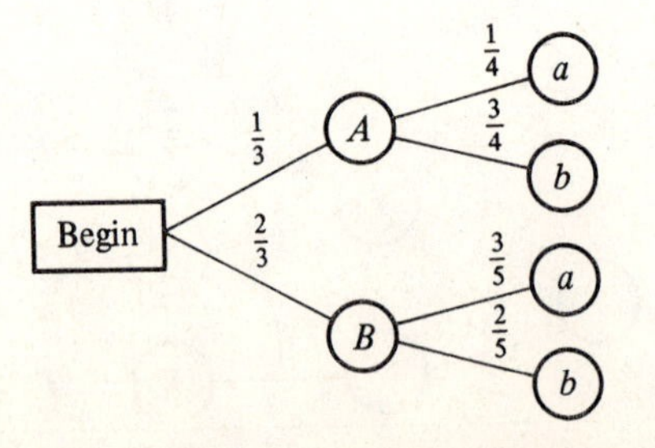

FIGURE 3.15

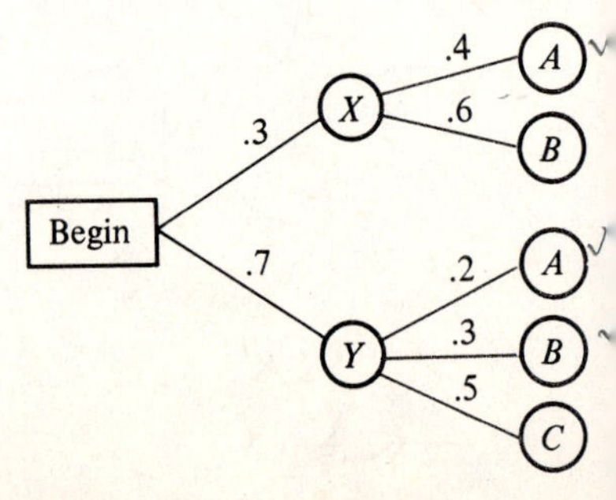

Quality control

3. In Example 3.16, concerning Rap and Rock Records, if it is known that an item is acceptable (A), what is the probability that it is from Los Angeles?
4. An experiment has the tree diagram shown in Figure 3.16. Find the following probabilities.
   (*a*) $\Pr[B \mid Y]$ (*b*) $\Pr[Y \mid B]$ (*c*) $\Pr[(X \text{ or } Y) \mid B]$
5. The tree diagram for an experiment is shown in Figure 3.17. Compute these probabilities:
   (*a*) $\Pr[(b \text{ and } B) \mid Y]$ (*b*) $\Pr[Y \mid (b \text{ and } B)]$

FIGURE 3.16 **FIGURE 3.17**

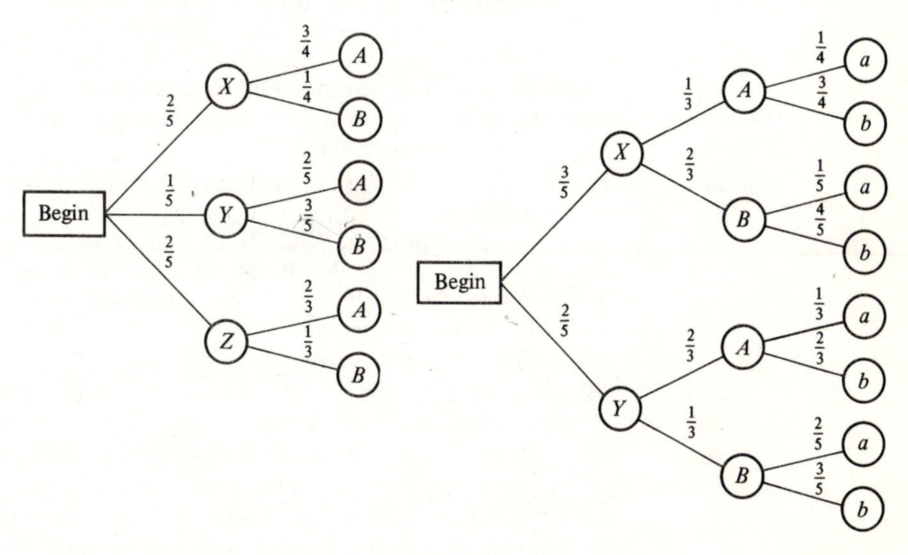

6. In Example 3.19, suppose both balls drawn are yellow (Y). What is the probability that the coin came up tails?
7. An experiment consists of making a random selection of one of the 3 urns shown in Figure 3.18 and then drawing a single ball and noting its color. If the ball is blue (B), what is the probability that urn X was selected? If the ball is either blue or white (W), what is the probability that urn Y was selected?

FIGURE 3.18

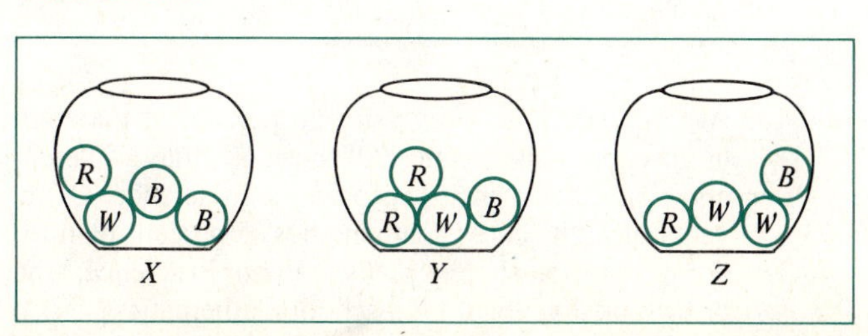

8. In the setting of Exercise 7, suppose two balls are drawn, in succession and without replacement. If both are the same color, what is the probability that urn X was selected?
9. Two balls are selected one after another and without replacement from a box which contains 2 red, 1 green, and 3 blue balls. Find the probability that the first ball was blue, given that the second was red. Assume random selections.
10. In the setting of Exercise 9, find the probability that the first was blue, given that the second was not blue.
11. Steve has two coins, one is fair and the other is weighted so that the probability of a head is .6. A coin is selected at random and flipped twice. The result of each flip is noted. Find the conditional probability that the fair coin was selected, given that there were two heads.
12. A random selection is made between two coins, and the selected coin is flipped twice. One of the coins is fair, and the other is unfair with $\Pr[H] = \frac{2}{3}$. If the result of the experiment is one head and one tail, what is the probability that the unfair coin was selected?

Education

13. A biology class consists of 50 percent freshmen, 25 percent sophomores, 15 percent juniors, and 10 percent seniors. Forty percent of the freshmen, 20 percent of the sophomores and juniors, and 10 percent of the seniors plan to go to medical school. A student is randomly selected from the class.
    (*a*) What is the probability that the student plans to go to medical school?
    (*b*) If the student selected plans to go to medical school, what is the probability that the student is a senior?

Education

14. In the setting described in Exercise 13, a student is randomly selected, and it is known that the student does not plan to go to medical school. What is the most likely class standing of the student: freshman, sophomore, junior, or senior?

Product distribution

15. A manufacturer of running shoes has plants in Korea, Australia, and Venezuela. Korea produces 60 percent of the shoes, Australia 20 percent, and Venezuela 20 percent. They make 2 types of shoes at each plant, a racing shoe and a training flat. The production at each plant is allocated as shown in Table 3.5. Suppose that these shoes are randomly distributed to stores in the United States and that you go into a store and buy a training flat. What is the probability that it came from Korea?

TABLE 3.5

Plant	Racing	Training
Korea	.50	.50
Australia	.25	.75
Venezuela	.40	.60

Education

16. At Gigantic State University 60 percent of the students have taken 4 years of high school mathematics. Of those who have taken 4 years of high school mathematics, 20 percent plan to major in science. Of those who have not taken 4 years of high school mathematics, 5 percent plan to major in science. If a randomly selected student plans to major in science, what is the probability that she or he took 4 years of high school mathematics?

Product distribution

17. The Fastback Cycle Company produces motorcycles at 3 plants, one each in Germany, Japan, and Brazil. Half the production comes from Germany with the other half split equally between Japan and Brazil. Each plant produces the same 3 different sizes of cycles, called the Mini, the Continental, and the Brute. Production of each size is allocated to the plants as shown in Table 3.6. Suppose that the cycles are distributed randomly to retail stores and someone buys a Continental cycle. What is the probability that the cycle was made in Germany?

TABLE 3.6

Plant	% Mini	% Continental	% Brute
Germany	30	40	30
Brazil	20	70	10
Japan	30	50	20

Product distribution

18. In Exercise 17, what size cycle should a person buy to have the highest probability of obtaining a cycle made in Japan?

Transportation

19. A rental car lot contains 35 percent BMWs, 45 percent Corvettes, and 20 percent Volvos. Fifty percent of the BMWs, 30 percent of the Corvettes, and 80 percent of the Volvos contain two airbags. If someone is randomly assigned a car and if the car has two airbags, what is the probability that the car is not a Corvette?

Transportation

20. In the rental car lot of Exercise 19, all of the Corvettes are white, half of the Volvos are white, and 20 percent of the BMWs are white. Also it is known that the property of being a white car is independent of the property of having two airbags. In this setting, if a randomly selected car has two airbags and is white, what is the probability that it is a Corvette?

Investments

21. The Discount Brokerage Company sells stocks and bonds. Half of the customers invest in stocks only, 20 percent in bonds only, and 30 percent in both. Of those who invest in stocks only, 20 percent make money; of those who invest in bonds only, 15 percent make money; and of those who invest in both, 30 percent make money.
   (*a*) A customer of the firm is selected at random. Find the probability that the customer makes money.
   (*b*) If a randomly selected customer makes money, find the probability that the customer invested in stocks only.

Product performance

22. A store stocks VCR tapes as follows: 50 percent are brand A, 20 percent are brand B, and the remainder are brand X. The store owner knows that 10 percent of brand A tapes are defective; also 5 percent of brand B and 20 percent of brand X are defective. A customer returns a defective tape. Find the probability that it is a brand X tape.

Education

23. At Gigantic State University 40 percent of the students are underclassmen (freshmen and sophomores), 35 percent are upperclassmen (juniors and seniors), and 25 percent are graduate students. Also 20 percent of the underclassmen are foreign students, 30 percent of the upperclassmen, and 50 percent of the graduate students are foreign students. A randomly selected student is not a foreign student. Find the probability that the student is an upperclassman.

Biology 24. There are 8 mice in a cage: 4 white females, 3 white males, and 1 gray male. Two mice are selected one after the other without replacement, and their color and sex are noted.
(*a*) Find the probability that the first was a female given that the second was a male.
(*b*) Find the probability that the first was gray given that the second was a male.

Forestry 25. Pine tree seeds of types A, B, and C are randomly scattered in a field. The seeds are 60 percent type A, 30 percent type B, and 10 percent type C. It is known that 30 percent of type A seeds will germinate, 40 percent of type B seeds will germinate, and 70 percent of type C seeds will germinate. If a randomly selected seed has germinated, what is the probability that it is type A?

Forestry 26. In the situation described in Exercise 25, suppose that a randomly selected seed has not germinated. Find the probability that it is either type A or type B.

27. There are 1 red, 2 green, and 3 yellow balls in an urn. A ball is selected at random, and its color is noted. If it is green, it is replaced; otherwise, it is not. A second ball is drawn and its color noted. If the second is green, what is the probability that the first was green also?

28. A box contains 2 red, 3 blue, and 2 yellow balls. A ball is selected at random, and its color is noted. If it is yellow, it is replaced; otherwise, it is not. A second ball is selected, and its color is noted. Find the probability that the first ball is yellow, given that the second ball is red.

Sports 29. A basketball player is equally likely to make and to miss his first shot in each game. If he makes his first shot, then he is twice as likely to make future shots as to miss them. If he misses the first shot, then he is twice as likely to miss as to make future shots.
(*a*) In a game in which the player makes his third shot, find the probability that he made his first shot as well.
(*b*) In a game in which the player makes his third shot, find the probability that he made all 3 shots.

Sports 30. In the setting of Exercise 29, suppose the player missed shot 2, but made shots 3 and 4. What is more likely to have happened on shot 1, he made it or missed it?

Travel 31. A time-conscious traveler is considering flying from New York to Los Angeles, and she reviews the on-time performance of Best Air, East-West Air, and Bi-Coastal Airlines. She finds that Best Air has 2 flights each day, East-West Air has 4 flights, and Bi-Coastal Airlines has 5 flights. Data show that Best Air is on time 50 percent of the time, East-West Air is on time 70 percent of the time, and Bi-Coastal Airlines is on time 40 percent of the time. She selects a flight at random, and she arrives on time. Find the probabilities that she flew on each of the three airlines.

Education 32. The Alumni Association of GSU has conducted a survey of 1990 graduates. That year the university awarded 5000 undergraduate degrees and 3000 graduate degrees, and no student received both a graduate and an undergraduate degree. Of the graduate degrees, 40 percent were in technical fields and 60 percent in nontechnical fields. Of the undergraduate degrees, 20 percent were in technical fields and 80 percent in nontechnical fields. Among those who earned graduate degrees, 98 percent of those with technical degrees and 95 percent of those with nontechnical degrees are employed, and the remainder are unemployed. Among those who earned undergraduate degrees, 95 percent of those with technical degrees and 90 percent of those with nontechnical degrees are employed, and the remainder are unemployed. Find the probability that a randomly selected employed graduate earned a graduate degree from GSU.

Education 33. In the setting of Exercise 32, which is more likely, that a randomly selected employed graduate earned a graduate degree or an undergraduate degree?

Education 34. The class standing and residency status of undergraduates at GSU are shown in Table 3.7. A randomly selected student is a state resident. Find the probability that the student is a freshman.

TABLE 3.7

Class Standing	Percentage in Class	% Resident
Freshman	26	75
Sophomore	24	73
Junior	22	73
Senior	28	70

3.5 BERNOULLI TRIALS

A stochastic process is an experiment defined as a sequence of subexperiments. In Section 3.3, we studied stochastic processes by using tree diagrams and by assigning probabilities to the branches. This works fine if there are only a small number of subexperiments and each has only a small number of possible outcomes. However, if the tree becomes large, then it may be difficult to draw and to analyze. In such cases, we need other tools to compute the probabilities of outcomes. Such tools are available for special stochastic processes, and this section considers one such process. We study the case where the process consists of independent repetitions of a single experiment with exactly two possible outcomes. The independence condition means that the probabilities of the two outcomes are the same on each repetition of the experiment.

An experiment with two possible outcomes is called a *Bernoulli*[1] *trial.* The two possible outcomes are usually called *success* (s) and *failure* (f). Examples of experiments with these properties are common: guessing on a true-false test, $s =$ correct and $f =$ incorrect; testing a light bulb, $s =$ it lights up and $f =$ it does not light. A *Bernoulli process* is a sequence of repetitions of the same Bernoulli trial. Since the repetitions are independent, the probability of success is the same at each trial.

[1]After Jakob Bernoulli (1654–1705), a Swiss mathematician and one of the founders of probability theory.

Example 3.20 Suppose an experiment consists of two repetitions of a Bernoulli trial with the probability of success equal to $\frac{1}{3}$ and the probability of failure equal to $\frac{2}{3}$. What is the probability that exactly one success is obtained in the two trials?

Solution We solve this problem with the aid of the tree diagram for the experiment (Figure 3.19). The assumption of independence is used in assigning the conditional probabilities on the tree. Thus we assume that the probabilities of the two possible outcomes (s and f) on the second trial are independent of the outcome (s or f) on the first trial. In other words, the second trial is identical (in terms of probabilities) to the first trial. Accordingly, each fork of the tree has two branches, one leading to s, which has probability $\frac{1}{3}$, and another leading to f, which has probability $\frac{2}{3}$.

The outcomes of the experiment which are of interest to us are sf and fs. Using the probabilities shown at the right of the tree diagram, we conclude that

$$\Pr[\text{exactly 1 success}] = \Pr[\{sf, fs\}] = \tfrac{2}{9} + \tfrac{2}{9} = \tfrac{4}{9}$$ ■

An example of an experiment which can be modeled by Example 3.20 is the following. A fair die is rolled twice, and the result is noted each time. On a single roll of the die, the result is said to be a success if a 1 or a 6 comes up, and the result is said to be a failure otherwise. Since the die is fair,

$$\Pr[s] = \Pr[\{1, 6\}] = \tfrac{1}{3} \qquad \text{and} \qquad \Pr[f] = \Pr[\{2, 3, 4, 5\}] = \tfrac{2}{3}$$

Two rolls of the die can be modeled as two repetitions of a Bernoulli trial, and this is precisely the experiment described in Example 3.20.

Throughout this section we will denote the probability of success in a single Bernoulli trial by p and the probability of failure by $q = 1 - p$. In general, we

FIGURE 3.19

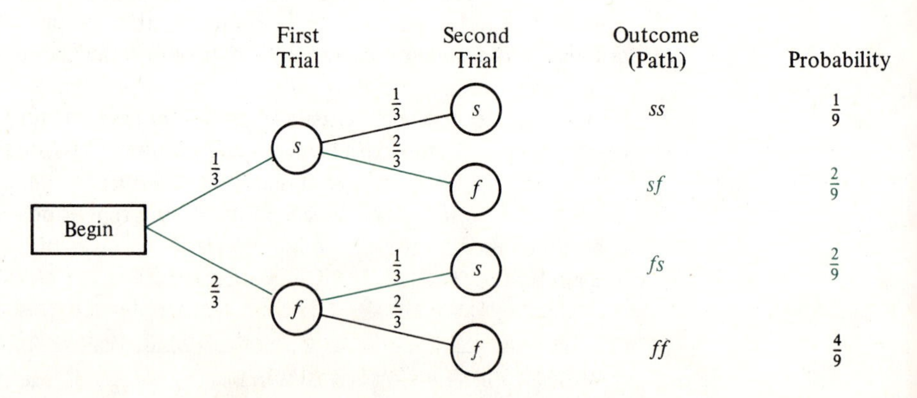

will be interested in computing the probability of obtaining a certain number of successes (say, r) in a given number of trials (say, n). We begin with a special case.

Example 3.21 An experiment consists of four Bernoulli trials. Compute the probabilities of obtaining exactly r successes for $r = 0, 1, 2, 3, 4$.

Solution In Figure 3.19 the Bernoulli process with two trials described in Example 3.20 is represented by a tree diagram. In this example the experiment consists of four Bernoulli trials instead of two, and hence the tree has four sets of forks instead of two. Otherwise, the tree for this example, shown in Figure 3.20, can be constructed in the same way as the tree in Figure 3.19. The conditional probabilities on the tree result from the fundamental assumption that the probability of success (s) is *always* p and the probability of failure (f) is *always* q. In particular, these probabilities do not depend on which trial in the sequence is being considered.

An examination of Figure 3.20 shows that for a Bernoulli process with 4 trials, the probability of any outcome which has r successes and $4 - r$ failures is $p^r q^{4-r}$. Indeed, each path which starts at "Begin" and includes a result for each of the 4 trials consists of r branches leading to an s, each with probability p, and $4 - r$ branches leading to an f, each with probability q. Consequently, to determine the probability that the experiment has an outcome consisting of r successes and $4 - r$ failures, it remains for us to count the number of such outcomes. We know the probability of each one.

For example, suppose we let $r = 1$. There are 4 paths with outcomes that consist of exactly 1 success and 3 failures. Each of these outcomes has probability pq^3. Hence,

$$\Pr[1 \text{ success and } 3 \text{ failures}] = 4pq^3$$

In a similar way (using Figure 3.20), we obtain the following:

$$\begin{aligned}
\Pr[0 \text{ successes and } 4 \text{ failures}] &= q^4 \\
\Pr[2 \text{ successes and } 2 \text{ failures}] &= 6p^2q^2 \\
\Pr[3 \text{ successes and } 1 \text{ failure}] &= 4p^3q \\
\Pr[4 \text{ successes and } 0 \text{ failures}] &= p^4
\end{aligned}$$

■

Although the method used in Example 3.21 is a relatively simple one (draw a tree and count outcomes), it is clear that for more than 4 trials, the tree and the method will become unwieldy. Fortunately, it is possible to use a rather simple formula to obtain the same results.

To justify our formula, let us compute the probability that we have exactly r successes in a Bernoulli process with n trials. First we note that we can represent

FIGURE 3.20

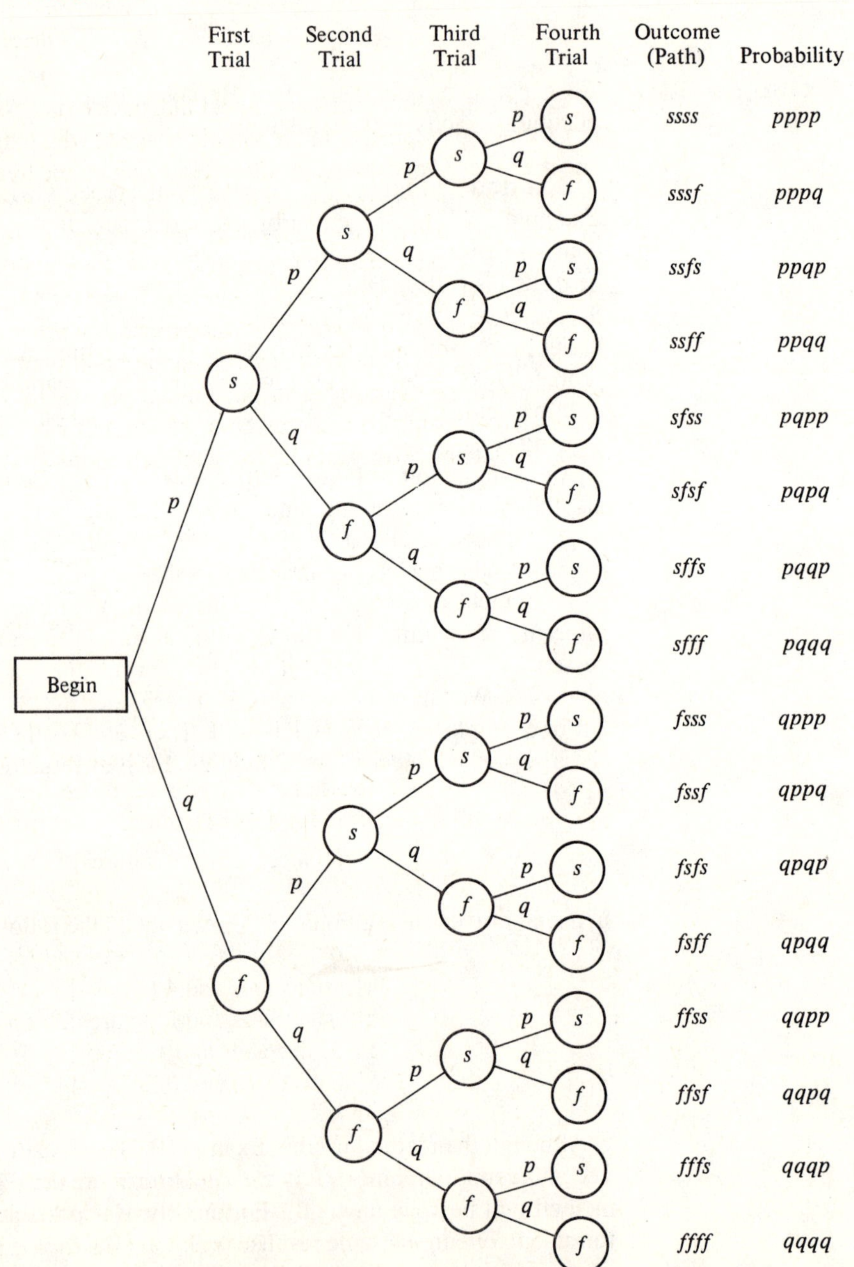

each element of the sample space by a list of n letters, each of which is an s or an f. For example, three elements of the sample space are

$$\underbrace{ff\ldots f}_{n \text{ times}}, \quad \underbrace{ss\ldots s}_{n \text{ times}}, \quad \text{and} \quad \underbrace{ss\ldots s}_{k \text{ times}}\underbrace{ff\ldots f}_{n-k \text{ times}}$$

Those outcomes of the experiment that have r successes and $n - r$ failures are represented by elements in the sample space in which the letter s appears r times and the letter f appears $n - r$ times. Note that on the tree diagram the path corresponding to such an outcome must have r branches with probability p (the successes) and $n - r$ branches with probability q (the failures). Since the probability of the outcome is the product of the conditional probabilities on the branches, the probability of this outcome is $p^r q^{n-r}$. Finally, since each outcome of r successes and $n - r$ failures has probability $p^r q^{n-r}$, we need only compute the number of such outcomes to find the probability of r successes in n trials.

To determine the number of outcomes with exactly r successes, we note that each such outcome corresponds to a selection of r positions from n in which to place the letter s. For example, two such selections are

$$\underbrace{ss\ldots s}_{\substack{r\\ \text{letters}\\ s}}\underbrace{ff\ldots f}_{\substack{n-r\\ \text{letters}\\ f}} \quad \text{and} \quad \underbrace{ff\ldots f}_{\substack{n-r\\ \text{letters}\\ f}}\underbrace{ss\ldots s}_{\substack{r\\ \text{letters}\\ s}}$$

and in general there are many more. Thus to count the number of ways to have r successes in n trials, we simply count the number of ways to select r objects from a set of n. We recognize this number to be $C(n, r)$ (recall Section 2.3). Combining these facts, we have this formula:

The probability of obtaining exactly r successes in a Bernoulli process consisting of n trials with success probability p and failure probability $q = 1 - p$ is

$$\Pr[r \text{ successes}] = C(n, r)p^r q^{n-r} \tag{3.5}$$

for $r = 0, 1, 2, \ldots, n$.

We illustrate formula (3.5) by applying it to the experiment described in Example 3.20. It consists of 2 Bernoulli trials with success probability $p = \frac{1}{3}$. Therefore $n = 2$, $p = \frac{1}{3}$, $q = \frac{2}{3}$. The problem asks for the probability of exactly 1 success. Therefore $r = 1$ in Equation (3.5). We have

$$\Pr[1 \text{ success}] = C(2,1)p^1q^1 = \frac{2!}{1!1!}\left(\frac{1}{3}\right)^1\left(\frac{2}{3}\right)^1$$
$$= 2 \cdot \tfrac{1}{3} \cdot \tfrac{2}{3} = \tfrac{4}{9}$$

as we obtained in Example 3.20.

Example 3.22 In a game of chance, you play by rolling a fair die 4 times, and you count the number of results which are 6s.

Problem What is the probability that in one play of the game you obtain exactly three 6s?

Solution We view a play of the game as a sequence of 4 Bernoulli trials. Since we are rolling a fair die, if we view a success as a 6, then $p = \frac{1}{6}$. We have 4 trials (rolls of the die), and we seek the probability of 3 successes. Using Equation (3.5), we have

$$\Pr[3 \text{ successes}] = C(4,3)\left(\frac{1}{6}\right)^3\left(\frac{5}{6}\right)^1 = \frac{4!}{3!1!}\left(\frac{1}{6}\right)^3\left(\frac{5}{6}\right)^1$$
$$= \frac{4 \cdot 5}{6^4} = \frac{20}{1296}$$

■

Example 3.23 Solve the problem posed in Example 3.21 by using Equation (3.5). Assume $p = \frac{1}{4}$, so $q = \frac{3}{4}$.

Solution In Example 3.21 we have $n = 4$. We are interested in the probabilities of 0, 1, 2, 3, and 4 successes. Using Equation (3.5) with $p = \frac{1}{4}$ and $q = \frac{3}{4}$, we find

$$\Pr[0 \text{ successes}] = C(4,0)\left(\frac{1}{4}\right)^0\left(\frac{3}{4}\right)^4 = \frac{4!}{4!0!} \cdot 1 \cdot \frac{3^4}{4^4} = 1 \cdot 1 \cdot \tfrac{81}{256} = \tfrac{81}{256}$$
$$\Pr[1 \text{ success}] = C(4,1)\left(\frac{1}{4}\right)^1\left(\frac{3}{4}\right)^3 = \frac{4!}{3!1!} \cdot \frac{1}{4} \cdot \frac{3^3}{4^3} = 4 \cdot \tfrac{1}{4} \cdot \tfrac{27}{64} = \tfrac{108}{256} = \tfrac{27}{64}$$
$$\Pr[2 \text{ successes}] = C(4,2)\left(\frac{1}{4}\right)^2\left(\frac{3}{4}\right)^2 = \frac{4!}{2!2!} \cdot \frac{1}{4^2} \cdot \frac{3^2}{4^2} = 6 \cdot \tfrac{1}{16} \cdot \tfrac{9}{16} = \tfrac{54}{256} = \tfrac{27}{128}$$
$$\Pr[3 \text{ successes}] = C(4,3)\left(\frac{1}{4}\right)^3\left(\frac{3}{4}\right)^1 = \frac{4!}{1!3!} \cdot \frac{1}{4^3} \cdot \frac{3}{4} = 4 \cdot \tfrac{1}{64} \cdot \tfrac{3}{4} = \tfrac{12}{256} = \tfrac{3}{64}$$
$$\Pr[4 \text{ successes}] = C(4,4)\left(\frac{1}{4}\right)^4\left(\frac{3}{4}\right)^0 = \frac{4!}{0!4!} \cdot \frac{1}{4^4} \cdot 1 = 1 \cdot \tfrac{1}{256} \cdot 1 = \tfrac{1}{256}$$

■

We note that in Example 3.23 the events 0 successes, 1 success, 2 successes, 3 successes, and 4 successes form a partition of the sample space S. Thus the sum of the probabilities for these events should be 1.

Indeed, we have

$$\begin{aligned}
\Pr[0 \text{ successes}] &+ \Pr[1 \text{ success}] + \Pr[2 \text{ successes}] \\
&+ \Pr[3 \text{ successes}] + \Pr[4 \text{ successes}] \\
&= \tfrac{81}{256} + \tfrac{108}{256} + \tfrac{54}{256} + \tfrac{12}{256} + \tfrac{1}{256} \\
&= \tfrac{256}{256} = 1
\end{aligned}$$

In fact, in any binomial experiment with n trials, the events 0 successes, 1 success, 2 successes, . . . , n successes form a partition of the sample space, and the sum of their probabilities must be 1.

> In any Bernoulli process with n trials,
>
> $$\Pr[0 \text{ successes}] + \Pr[1 \text{ success}] + \ldots + \Pr[n \text{ successes}] = 1 \qquad (3.6)$$

Formula (3.6) can frequently be used to reduce the amount of computation needed to solve a problem.

Example 3.24 *Sports*

A National Football League kicker knows that he makes one-fourth of his field goal attempts from more than 50 yards. He also knows that they are independent trials. In a certain game he attempts 5 field goals of more than 50 yards.

Problem What is the probability that he will make **at least** 2 of the 5 attempts?

Solution The probability of at least 2 successes (the kicker made at least 2 field goals) is the sum of the probabilities of the disjoint events: exactly 2 successes, exactly 3 successes, exactly 4 successes, and exactly 5 successes. That is, Pr[at least 2 made] = Pr[2 made] + Pr[3 made] + Pr[4 made] + Pr[5 made]. Thus, one method of obtaining the solution to this problem is to use formula (3.5) to compute Pr[2 successes], Pr[3 successes], Pr[4 successes], and Pr[5 successes] and then to add these 4 numbers. Using formula (3.5) with $n = 5$, $p = \frac{1}{4}$, $q = \frac{3}{4}$, and $r = 2, 3, 4, 5$, we obtain

$$\begin{aligned}
\Pr[\text{at least 2 made}] &= C(5, 2)\left(\frac{1}{4}\right)^2\left(\frac{3}{4}\right)^3 + C(5, 3)\left(\frac{1}{4}\right)^3\left(\frac{3}{4}\right)^2 \\
&\quad + C(5, 4)\left(\frac{1}{4}\right)^4\left(\frac{3}{4}\right)^1 + C(5, 5)\left(\frac{1}{4}\right)^5\left(\frac{3}{4}\right)^0 \\
&= \tfrac{270}{1024} + \tfrac{90}{1024} + \tfrac{15}{1024} + \tfrac{1}{1024} \\
&= \tfrac{376}{1024}
\end{aligned}$$

A second method of solving our problem is to use formula (3.6) for the case $n = 5$. Using a field goal made as a success, we have

$$\Pr[0 \text{ made}] + \Pr[1 \text{ made}] + \Pr[2 \text{ made}] + \Pr[3 \text{ made}] + \Pr[4 \text{ made}] + \Pr[5 \text{ made}] = 1$$

This can also be stated as

$$\Pr[0 \text{ made}] + \Pr[1 \text{ made}] + \Pr[\text{at least 2 made}] = 1$$

which gives

$$\Pr[\text{at least 2 made}] = 1 - \Pr[0 \text{ made}] - \Pr[1 \text{ made}]$$

Now if we use formula (3.5), we have

$$\begin{aligned}\Pr[\text{at least 2 made}] &= 1 - C(5, 0)\left(\frac{1}{4}\right)^0\left(\frac{3}{4}\right)^5 - C(5, 1)\left(\frac{1}{4}\right)^1\left(\frac{3}{4}\right)^4 \\ &= 1 - \tfrac{243}{1024} - \tfrac{405}{1024} = \tfrac{376}{1024}\end{aligned}$$

■

In the first method illustrated in Example 3.24, we partitioned the event "at least 2 made" into 4 events and used formula (3.5) 4 times. In the second method we partitioned the entire sample space into the disjoint events "0 made," "1 made," and "at least 2 made"; and [using formula (3.6)] we applied formula (3.5) only 2 times. The difference in the two methods would be even more striking if there were more trials. In 10 trials the probability of at least 2 successes can be obtained by using formula (3.6) and *two* applications of (3.5), or it can be obtained by directly applying (3.5) *nine times* and adding the results. Clearly, in some cases the method based on formula (3.6) is preferable.

In some interesting problems involving Bernoulli trials, the number of trials n is unknown. We solve such problems by a trial-and-error method.

Example 3.25

Investments

An investor makes money on approximately 70 percent of her investments. She would like to have at least 3 profitable investments in the coming year. How many investments should she make so that the probability of having *at least* 3 profitable investments is at least .8?

Solution We solve the problem by assuming that her investments are Bernoulli trials with $p = .7$. Here a success is a profitable investment. The precise mathematical question is the following: Find n such that

$$\Pr[3 \text{ or more successes in } n \text{ trials}] \geq .8$$

We proceed by using a "guess and test" strategy. Clearly there must be at least 3 investments.

If $n = 3$, then

$$\Pr[3 \text{ successes in } 3 \text{ trials}] = C(3, 3)(.7)^3(.3)^0 = .343$$

This is much too small, and consequently more than 3 investments are required.

If $n = 4$, then

$$\begin{aligned}\Pr[3 \text{ or more successes in } 4 \text{ trials}] &= \Pr[3 \text{ successes}] + \Pr[4 \text{ successes}] \\ &= C(4, 3)(.7)^3(.3)^1 + C(4, 4)(.7)^4(.3)^0 \\ &= .412 + .240 = .652\end{aligned}$$

If $n = 5$, then

$$\begin{aligned}&\Pr[3 \text{ or more successes in } 5 \text{ trials}] \\ &\quad = \Pr[3 \text{ successes}] + \Pr[4 \text{ successes}] + \Pr[5 \text{ successes}] \\ &\quad = C(5, 3)(.7)^3(.3)^2 + C(5, 4)(.7)^4(.3)^1 + C(5, 5)(.7)^5(.3)^0 \\ &\quad = .309 + .360 + .168 = .837\end{aligned}$$

We conclude that the investor should plan on making at least 5 investments to be sure that the probability of making money on 3 or more investments is at least .8. ■

Exercises for Section 3.5

1. For Bernoulli trials compute the following probabilities.
   (*a*) 1 success in 3 trials with $p = .2$
   (*b*) 2 successes in 3 trials with $p = .4$
   (*c*) 3 successes in 3 trials with $p = \frac{2}{3}$
2. For Bernoulli trials compute the following probabilities.
   (*a*) 2 successes in 4 trials with $p = .6$
   (*b*) 3 successes in 4 trials with $p = \frac{1}{3}$
   (*c*) 0 successes in 4 trials with $p = .3$
3. For Bernoulli trials compute the following probabilities.
   (*a*) 4 successes in 6 trials with $q = .4$
   (*b*) 2 failures in 5 trials with $q = .6$
4. For Bernoulli trials compute the following probabilities.
   (*a*) 3 successes in 6 trials with $q = .5$
   (*b*) 3 failures in 5 trials with $q = .7$
5. For Bernoulli trials compute the following probabilities.
   (*a*) At least 3 successes in 6 trials with $p = .2$
   (*b*) At least 3 failures in 5 trials with $p = .8$
6. For Bernoulli trials compute the following probabilities.
   (*a*) No more than 2 successes in 4 trials with $p = \frac{1}{3}$
   (*b*) No more than 3 failures in 5 trials with $p = \frac{2}{3}$

7. For a Bernoulli process with $n = 5$ and $p = \frac{2}{5}$, find the probability of at least 1 success and at least 1 failure.
8. For a Bernoulli process with $n = 7$ and $p = \frac{1}{2}$, find the probability of at least 1 success and at least 1 failure.
9. For a Bernoulli process with $n = 5$ and $p = \frac{3}{4}$, find the probability of exactly 2 successes or exactly 2 failures.
10. For a Bernoulli process with $n = 5$ and $p = .6$, find the probability of at least 4 successes or at least 3 failures.
11. A coin is weighted so that a head is twice as likely to occur as is a tail. The coin is flipped 4 times, and the result of each flip is noted. Find the probability that both heads and tails occur.
12. A coin is weighted so that a head is twice as likely to occur as is a tail. The coin is flipped 3 times, and the result of each flip is noted.
    (*a*) Find the probability that exactly 1 head occurs.
    (*b*) Find the probability that there is exactly 1 head given that both heads and tails occur.

Business

13. Of the telephone calls received by an airline reservation agent, 60 percent are requests for information and 40 percent are to make reservations. Assume the calls can be viewed as Bernoulli trials with success defined to be a call for a reservation. Six calls are received.
    (*a*) What is the probability that exactly 2 calls are for reservations?
    (*b*) What is the probability that at least 4 are for information?
14. A fair coin is flipped 6 times. Which is greater, the probability that heads and tails divide 3 and 3 or the probability that they divide 4 and 2 (either 4 heads and 2 tails or 4 tails and 2 heads)?

Educational testing

15. In a 20-question true-false test, what is the probability of answering exactly 18 questions correctly just by guessing?

Educational testing

16. In a 10-question true-false test, find the probability of answering at least 8 questions correctly just by guessing.

Educational testing

17. In an 8-question multiple-choice test, each question has 4 choices and you are told that one and only one is correct. If you guess at all 8 questions, what is the probability that you get *exactly* 2 right?

Educational testing

18. In the setting of Exercise 17, find the probability of answering at least 7 questions correctly just by guessing.

Product performance

19. In the inventory of Good Vehicles, Inc., 1 car in 10 is defective and will not start. At different times 3 individuals randomly select cars for test drives. What is the probability that at least 1 of them selects a car that will not start?

Biology

20. An orchard is infested with a large population of Mediterranean fruit flies. Suppose half are males and half are females. If 6 fruit flies are trapped, find the probability that more males are trapped than females. (*Hint*: Assume that the population is so large that the removal of a fly does not change the male-female ratio enough to alter the results.)

Sports

21. A deep-sea fisherman estimates that if he uses 4 lines, then the probability of making a catch on any line is .7; if he uses 5 lines, then the probability of making a catch on any line is .6; and if he uses 6 lines, then the probability of making a catch on any line is .5. If his objective is to catch at least 4 fish, how many lines should he use to maximize the probability of achieving his goal?

Product testing 22. A box contains 3 defective widgets and 7 acceptable widgets. Three inspectors sample widgets from the box in sequence as follows: Inspector 1 selects a widget from the box at random, tests it, and returns it to the box; inspector 2 does the same, followed by inspector 3, who also uses the same method. What is the probability that at least one inspector will select a defective widget?

Sports 23. A high school basketball player makes one-third of his three-point shots. If we assume that his shots are Bernoulli trials, how many must he shoot to have a probability .7 of making at least 1?

Sports 24. In the setting of Exercise 23, suppose the player only makes a three-point shot with probability $\frac{1}{4}$. Now how many must he shoot to have probability .75 of making at least 1 shot?

Sales 25. A used-car saleswoman estimates that each time she shows a customer a car, there is a probability .1 that the customer will buy the car. The saleswoman would like to sell at least 1 car per week. If showing a car is a Bernoulli trial, how many cars must the saleswoman show per week so that the probability is .95 of at least 1 sale?

Sales 26. In the setting of Exercise 23, suppose that the probability of a sale goes up from .1 to .2. Now how many cars must the saleswoman show to have probability .95 of at least 1 sale?

Surveys 27. In a large city 50 percent of the voters consider themselves Democrats, 25 percent consider themselves Independents, and 25 percent consider themselves Republicans. A poll taker asks 10 different people at random for their party affiliations. What is the probability that at least 7 are Democrats?

Sports 28. A basketball player has a .600 shooting average before a game. That is, the player has made 60 percent of the baskets attempted before the game. The player attempts 15 shots during the game. Assume the attempts are Bernoulli trials, and find the probability that exactly 9 of the shots are successful.

College life 29. A professor who intends to bring her briefcase to the office each morning forgets it one-quarter of the time. Assume that forgetting the briefcase is a Bernoulli trial, and find the probability that she forgets it at least twice a week (5 days).

Advertising 30. Advertisements for UP Airlines claim that its flights are on time 80 percent of the time. A business executive takes a trip which includes 5 flights on UP Airlines, and 2 of the flights are late. Assuming the airline's claim is correct, find the probability of this event.

Educational testing 31. A multiple-choice examination consists of 10 questions, each question with 5 choices for an answer, and each question has exactly 1 correct answer. Two students answer the questions by guessing. Assume the answer sheets are prepared independently, and find the probability that each student answers exactly 4 questions correctly.

Educational testing 32. In the setting of Exercise 31, find the probability that the 2 students answered *the same* 4 questions correctly just by guessing.

Recreation 33. North Fork River Expeditions, Inc., operates white-water rafting trips on the North Fork of the Trout River. Data from past years indicate that about 60 percent of the people who come on such trips prefer to paddle in rafts and the remainder prefer to paddle one-person kayaks. A specific trip has 20 people, and there are 6 kayaks. If the preferences of the people on the trip can be viewed as the result of a Bernoulli trials process, find the probability that everyone who prefers to paddle a kayak can be accommodated.

Recreation 34. In the setting of Exercise 33, how many kayaks would be needed so that everyone who prefers to paddle a kayak can do so with probability at least .9?

Recreation 35. In the setting of Exercise 33, suppose that data from past years indicate that 50 percent of the people prefer to paddle in rafts, 30 percent prefer to paddle a kayak, and 20 percent are willing to do either. Find the probability that everyone is satisfied with her or his paddling opportunity.

IMPORTANT TERMS AND CONCEPTS

You should be able to describe, define, or give examples of each of the following:

Axioms of a probability measure
Properties of a probability measure
Conditional probability
Independence
Probabilities on trees
Stochastic process
Bayes probabilities
Bernoulli trial
Bernoulli process

REVIEW EXERCISES

1. An experiment has the sample space $S = \{\mathcal{O}_1, \mathcal{O}_2, \mathcal{O}_3, \mathcal{O}_4, \mathcal{O}_5\}$ and probabilities $w_1 = .09$, $w_2 = .23$, $w_3 = .41$, $w_4 = .14$, and $w_5 = .13$. Define $E_1 = \{\mathcal{O}_1, \mathcal{O}_2, \mathcal{O}_4\}$, $E_2 = \{\mathcal{O}_3, \mathcal{O}_4, \mathcal{O}_5\}$, and $E_3 = \{\mathcal{O}_2, \mathcal{O}_3\}$. Find:
(*a*) $\Pr[E_1']$ (*b*) $\Pr[E_1 \cup E_2]$
(*c*) $\Pr[E_1 \cap E_3]$ (*d*) $\Pr[E_2 \cap E_3']$
2. A sample space contains events E and F with $\Pr[E] = .35$, $\Pr[F] = .30$, and $\Pr[E \cap F] = .20$. Find:
(*a*) $\Pr[E']$ (*b*) $\Pr[E \cup F]$ (*c*) $\Pr[E \cap F']$
3. A sample space contains events E, F, and G. Suppose $\Pr[E] = .5$, $\Pr[F] = .3$, $\Pr[G] = .35$, $\Pr[E \cup F] = .8$, and $\Pr[F \cap G] = .3$. Find
(*a*) $\Pr[E \cap F]$ (*b*) $\Pr[F \cup G]$ (*c*) $\Pr[E \cap F']$
4. Using the setting of Exercise 3, suppose we also know that $\Pr[E \cap G] = .05$. Find:
(*a*) $\Pr[E' \cap F']$ (*b*) $\Pr[E' \cap G]$ (*c*) $\Pr[E' \cap F' \cap G']$
5. Using the data of Exercises 3 and 4, find:
(*a*) $\Pr[E|G]$ (*b*) $\Pr[E|F]$ (*c*) $\Pr[G|E]$
6. Suppose that events E_1, E_2, E_3 form a partition of a sample space S and that $\Pr[E_1] = 2\Pr[E_2]$, $\Pr[E_3] = 3\Pr[E_1]$. Find the probabilities of events E_1, E_2, and E_3.
7. Suppose E and F are events in a sample space with $\Pr[E \cup F] = .8$, $\Pr[E \cap F] = .3$, and $\Pr[E'] = .4$.
(*a*) Find $\Pr[E]$. (*b*) Find $\Pr[F]$. (*c*) Find $\Pr[E|F]$.
8. A box contains 1 green, 2 blue, and 2 red balls. Two balls are selected simultaneously and at random, and their colors are noted. Find the probability that one is red and one is blue given that both are not the same color.

College life 9. Half of the students enrolled in a business course subscribe to *The Wall Street Journal*, and two-thirds of the students enrolled in this class subscribe to *The New York Times*. Subscriptions to these newspapers are independent. What is the probability that a randomly selected student enrolled in this class subscribes to exactly one of these newspapers?

10. A sample space contains the events E, F, and G, with $\Pr[E] = .3$, $\Pr[F] = .4$, and $\Pr[G] = .5$. If $\Pr[E \cap F] = .12$, $\Pr[E \cap G] = .1$, and $\Pr[F \cup G] = .7$, decide which (if any) of the following pairs are independent.
(*a*) E and F (*b*) E and G (*c*) F and G

11. Using the data of Exercise 10, which of the following pairs are independent?
(*a*) E' and F (*b*) E' and G' (*c*) F' and G'

12. Suppose E and F are independent events, with $\Pr[E] = .3$ and $\Pr[F] = .6$. What is $\Pr[E' \cap F']$?

13. Two fair dice are rolled, and the product of the two numbers is noted. What is the probability that the product is 12, given that at least one of the numbers is a 4?

14. In the setting of Exercise 13, what is the probability that one of the numbers is a 4, given that the product is 12?

Biology 15. There are 8 mice in a cage: 3 white males, 3 gray females, and 2 gray males. Two mice are selected simultaneously and at random, and their colors are noted. Find the probability that at least one mouse is a male, given that exactly one is gray.

Education 16. A student estimates that he will receive a B or better in English with probability .4 and a B or better in psychology with probability .6. If these are independent results, what is the probability that the student will receive at least one grade of B or better?

Product evaluation 17. The Human Factors Group at the Joltmobile Corporation is assigned the task of designing the interior of a new automobile. A preliminary design is tested by asking a sample of 100 people to evaluate various aspects. The data shown in Table 3.8 are collected. An evaluation sheet is selected at random. What is the probability that it evaluates favorably *exactly one* of the items—seats, instruments, pedals?

TABLE 3.8

Percentage Responding Favorably	Feature
35	Seats
42	Instruments
28	Pedals
22	Seats and instruments
18	Seats and pedals
8	All three
39	None of the three

18. A student has 3 pencils with erasers and 4 pencils without erasers in her desk

drawer. She selects 2 pencils simultaneously and at random. During a test she notices that the first one she takes out of her notebook does not have an eraser. Find the probability that the other does not have an eraser either.

19. A fair die is rolled 10 times. A roll is called a success if either a 1 or a 2 comes up. What is the probability that there are exactly 3 successes or exactly 3 failures in the 10 rolls?
20. Sam has an unfair coin with $\Pr[H] = \frac{1}{3}$, and Sally has an unfair coin with $\Pr[H] = \frac{1}{4}$. Sam flips his coin 3 times, and Sally flips hers 4 times. Who has the greater probability of having exactly 1 head?
21. A coin with $\Pr[H] = p > 0$ is flipped 10 times. If the probability of exactly 4 heads is equal to the probability of exactly 5 heads, what is p?
22. Sam has an unfair coin with $\Pr[H] = \frac{1}{6}$. How many times must Sam flip his coin to have the probability of at least 1 head exceed $\frac{1}{2}$?
23. A parking lot contains 2 Chevrolets, 3 Fords, and 1 Toyota. Two cars leave, one after the other; the make of each car is noted as it leaves. Draw a tree diagram to represent this, and find the probabilities of all outcomes.
24. In the situation described in Exercise 23, find:
    (*a*) The probability that the second car to leave is a Ford
    (*b*) The probability that the first car to leave is a Chevrolet given that the second car to leave is a Ford or Toyota
25. A fair coin is flipped once. If it comes up heads, it is flipped 1 more time. If the first flip is tails, it is flipped 2 more times. Find the probability that a total of 2 heads come up.
26. Two fair dice are rolled, and the result on each die is noted. Find the probability that:
    (*a*) The sum is 8, given that at least one of the numbers is even.
    (*b*) At least one of the numbers is even, given that the sum is 8.
    (*c*) At least one of the numbers is even, given that the sum is 7.
27. A 3-digit number is formed from the digits 1, 2, 4, 6, 9 by using 3 different digits. If the digits are selected and ordered at random, what is the probability that:
    (*a*) The number is even.
    (*b*) The number is less than 500.
    (*c*) The number is even, given that it is less than 500.
    (*d*) The number is less than 500, given that it is even.

Transportation

28. Two-thirds of the cars in a university motor pool are over 5 years old. Cars that are over 5 years old fail to start 10 percent of the time, and cars less than 5 years old fail to start 2 percent of the time. A faculty member is assigned a car at random. If the car starts the first 2 times, find the probability that it is more than 5 years old.

Transportation

29. There are 8 flights from Indianapolis to San Francisco, 3 direct and 5 with a stop. Of the direct flights, 2 are day flights and 1 is a night flight. Of the flights with a stop, 4 are day flights and 1 is a night flight. Daytime flights are on time 70 percent of the time, and night flights are on time 80 percent of the time. An executive makes a reservation at random on a flight from Indianapolis to San Francisco.
    (*a*) Find the probability that the flight is on time.
    (*b*) Find the probability that it is on time given that it is a direct flight.

(*c*) Find the probability that it is direct given that it is on time.
(*d*) Find the probability that it is direct given that it is on time and a daytime flight.

Product testing 30. Of the vehicles checked at an inspection station, 80 percent are passed, 18 percent are found to be defective but repairable, and 2 percent are found to be defective and unrepairable. Of those passed, three-fourths are less than 5 years old; of those found to be defective but repairable, one-half are less than 5 years old; and of those found to be defective and unrepairable, only 1 in 20 is less than 5 years old. What is the probability that your friend's 4-year-old car (about which you know only its age) will be passed by this inspection station?

College life 31. Each morning Tom decides whether to attend economics class. He attends randomly with probability .6, and each decision is independent of what he has done in the past. That is, his decision process can be viewed as a Bernoulli process. Find the probability that he attends at least 6 of 10 classes given that he attends at least 1 but not all of the 10 classes.

Education 32. At Gigantic State University 40 percent of the biology classes have laboratories, 30 percent have discussion sections, and 15 percent have both. A biology course is selected at random.
(*a*) Find the probability that it has neither a laboratory nor a discussion section.
(*b*) Find the probability that it has a laboratory, given that it has a discussion section.
(*c*) Find the probability that it has a laboratory, given that it has exactly one of these features.

Medical tests 33. Two percent of the laboratory tests for a certain disease give "false-positive" results. That is, the test indicates that the patient has the disease when, in fact, the patient does not. Five patients are tested, and we assume that the tests can be viewed as Bernoulli trials.
(*a*) Find the probability that exactly 2 tests give false-positive results.
(*b*) Find the probability that the first test and the last test give false-positive results.

Medical tests 34. In the setting of Exercise 33, find the probability that there are exactly 2 false–positives given that there are at least 2 false–positives.

Medical tests 35. In the setting of Exercise 33, find the probability that there are exactly 2 false–positives, given that the first test and the last test are false–positives.

CHAPTER 4

Random Variables, Averages, and Statistics

4.0 THE SETTING AND OVERVIEW

This chapter forms a bridge between the study of probability and the application of probability theory to the field of statistics. We will apply the methods of Chapters 2 and 3 to study several topics which are very useful in business, the social and life sciences, and everyday life. One such topic is the values (or costs) which are sometimes associated with the outcomes of experiments. A common problem in such situations, a problem which occurs frequently in decision making, is to find an average value to associate with an experiment. To solve such problems, we use the concepts of random variable, mean, and standard deviation. The normal random variable, which occurs in many statistical applications, will be introduced and studied, as will the binomial random variable. We conclude the chapter by studying the close relationship between the normal random variable and the binomial random variable.

4.1 RANDOM VARIABLES AND PROBABILITY DENSITY FUNCTIONS

In many situations that involve uncertainty, we are interested in a numerical quantity whose value depends on the outcome of an experiment. For instance, gamblers are interested in the amount of money won or lost on each play of a game; business managers are interested in the day-to-day demand for products; and students are interested in semester-to-semester grade-point averages. Each of these quantities (winnings, demands, grade-point averages) is a number, and in general it is not known in advance what the number will be. We will use probability to help us understand such situations.

Example 4.1

Lotteries

Consider the following simple lottery. You pay \$1 to play, and then you select an integer in the set $\{0, 1, 2, 3, 4, 5, 6, 7, 8, 9\}$. Next the person running the lottery randomly picks an integer from the same set. If the number picked is the same as yours, you win \$5; if not, you win nothing. If you play this game many times, how much should you expect to win (or lose)?

Solution In playing this game, you will sometimes be lucky and win \$5, but most of the time you will not win anything. Since it costs you \$1 to play, in most cases you lose that \$1. In a few cases you gain \$4 (the \$5 you win, minus the \$1 you paid to play). In addition to how much you win or lose, you need to know how often you will win or lose. Once you select a number in the given set of 10 integers, your chance of winning is $\frac{1}{10}$, because your number is 1 of 10 and in a random selection it has 1 chance in 10 of being selected by the person running the lottery. Thus, in playing 10 times, you would expect to win 1 time and lose 9 times. Of course, this is just an average. You might not win at all, or you might win more than once, but on average you win 1 time in 10 tries. When you win, you end up \$4 ahead; and when you lose, you end up short \$1. Thus, on average, when you play 10 times, you lose \$9 and gain \$4, for a net loss of \$5. In other words, your average loss is \$.50 per play. ■

In Example 4.1, we are not able to predict the result of the lottery on any given play of the game. However, we know that if we play many times, then we lose an average of 50 cents per play. In many situations in life, we cannot predict a specific outcome, but we can compute an "average" value. Of course, we need to be precise about our meaning of the term "average." We work one more example and then develop the general setting.

Example 4.2

Games of chance

Suppose that the following game is proposed by a friend: A fair die is to be rolled, and if an outcome of 2, 3, 4, or 5 occurs, then your friend will pay you \$1.50. If either a 1 or a 6 occurs, then you pay your friend an amount in dollars equal to the outcome. That is, you pay \$1.00 if a 1 occurs and \$6.00 if a 6 occurs.

Problem If you play the game many times, how much gain (or loss) should you expect?

Solution We begin an analysis of this game by viewing each play as an experiment. We construct a table that shows the outcomes of the experiment, the gain or loss to you associated with each outcome, and the probability that each outcome occurs. Table 4.1 contains this information.

It is impossible to predict with certainty the outcome of any specific play of this game. Sometimes you will lose \$1.00, sometimes you will lose \$6.00, and sometimes you will win \$1.50. Although it is impossible to predict a single outcome of the game, it is possible to make predictions about the results of many plays of the game. If there are many plays of the game (rolls of the die), then we expect each of about one-sixth of these plays (rolls) to result in a loss of \$1.00, each of another one-sixth to result in a loss of \$6.00, and each of the remaining four-sixths of the outcomes to result in a win of \$1.50. These values reflect the probabilities of the various outcomes given in Table 4.1.

TABLE 4.1

Outcome	Gain (+) or Loss (−)	Probability
1	−1.0	$\frac{1}{6}$
2	+1.5	$\frac{1}{6}$
3	+1.5	$\frac{1}{6}$
4	+1.5	$\frac{1}{6}$
5	+1.5	$\frac{1}{6}$
6	−6.0	$\frac{1}{6}$

Suppose that you play this game 600 times and each outcome of the die occurs exactly 100 times. Then 100 times you lose \$1.00, 100 times you lose \$6.00, and on each of the other 400 times you win \$1.50. Your losses would add up to \$700, and your gains would add up to \$600; so your net return would be −\$100 (a loss of \$100). Since you played 600 times, your average return per play would be − \$100/600 = −(\$1)/6. Summarizing these computations, we see that the average return is

$$V = \frac{100(-\$1.00) + 400(+\$1.50) + 100(-\$6.00)}{600}$$

$$= \frac{-\$100 + \$600 - \$600}{600} = -\frac{\$1}{6}$$

In 600 actual plays of this game, it is unlikely that each outcome on the die will occur exactly 100 times. However, it is likely that each outcome will occur about 100 times, and thus a net return of −\$1/6 is about what you would expect for a fair die. Without reference to the number of plays, we expect that each outcome on the die will occur one-sixth of the time. Then the net return per play, called the *expected return*, is (in dollars)

$$\begin{aligned} V &= (-1.00)(\tfrac{1}{6}) + (1.50)(\tfrac{1}{6}) + (1.50)(\tfrac{1}{6}) + (1.50)(\tfrac{1}{6}) \\ &\quad + (1.50)(\tfrac{1}{6}) + (-6.00)(\tfrac{1}{6}) \\ &= \frac{-7+6}{6} \\ &= -\tfrac{1}{6} \end{aligned}$$

Therefore, in any number of plays of this game, you expect to lose one-sixth of a dollar per play of the game. Your "friend" has a definite advantage! ■

The method of Example 4.2 can be used in any experiment where numbers are attached to outcomes. An assignment of numbers to outcomes is a *random variable*. In Example 4.2 the elements in the sample space are the outcomes of the die (1, 2, 3, 4, 5, 6), and the random variable attaches -1 to outcome 1, -6 to outcome 6, and 1.5 to each of the outcomes 2, 3, 4, and 5. A general definition follows:

A *random variable* on a sample space S is an assignment of real numbers to the elements of S, exactly one number assigned to each outcome.

In mathematical terms a random variable on S is a real-valued function with domain S.

To illustrate further the idea of a random variable, consider the experiment of flipping a coin four times and noting each time whether it lands heads or tails. If to each outcome of this experiment we assign the *number of heads* obtained in the four flips, then we have defined a random variable. The *values* of the random variable are the numbers assigned to the outcomes. In this case the value of the random variable can be 0, 1, 2, 3, or 4, since these are the possible numbers of heads obtained in four flips. If we think of obtaining a head as a success in the Bernoulli trial of flipping a coin, then our random variable counts the total number of successes in four trials. A random variable of this type occurs often, and it has its own name.

A random variable which assigns to each outcome of a Bernoulli process the number of successes is called a *binomial random variable*.

Random variables are usually represented by uppercase letters such as X and Y. The values of the random variable X will be denoted by the lowercase letters $x_1, x_2, \ldots, x_k$. Following convention, we will list these values in increasing order: $x_1 < x_2 < \cdots < x_k$. Also, in most problems we will specify a random variable by using a table in which each row corresponds to a single value of the random variable.

We illustrate this with two examples: first, the binomial random variable which assigns the number of heads to each outcome of four flips of a fair coin. The values, labeled by our convention, are $x_1 = 0$, $x_2 = 1$, $x_3 = 2$, $x_4 = 3$, and $x_5 = 4$. The appropriate table is Table 4.2. The first and third columns of this table will be discussed after we give another example.

TABLE 4.2

Event E_j	Value of Random Variable X, x_j	Probability $\Pr[E_j]$
No heads	0	$\frac{1}{16}$
One head	1	$\frac{4}{16}$
Two heads	2	$\frac{6}{16}$
Three heads	3	$\frac{4}{16}$
Four heads	4	$\frac{1}{16}$

As a second example, we use the random variable of Example 4.2. The random variable X has three values (-6, -1, and 1.5), and using our convention, we set $x_1 = -6$, $x_2 = -1$, and $x_3 = 1.5$. Forming a table in which each value of X is listed once, we represent the data of Table 4.1 in Table 4.3.

It is important to note that in Table 4.3 the column at the left consists of events and *not* of single outcomes, as in Table 4.1. This is necessary because we are using only one row of the table for each value of the random variable, and it is often the case (as in Example 4.2) that more than one outcome is assigned the same value of the random variable.

The technique of grouping together all outcomes for which a random variable assumes a specific value is an important and useful one. We next introduce terminology and notation which describe the process of grouping more precisely. We concentrate on sample spaces which contain a finite number of elements, and consequently on random variables which assume a finite number of values. A different setting will be needed in Section 4.3 where we study the normal distribution.

TABLE 4.3

Event E_j	Value of Random Variable X, x_j	Probability $\Pr[E_j]$
{6}	-6.0	$\frac{1}{6}$
{1}	-1.0	$\frac{1}{6}$
{2, 3, 4, 5}	$+1.5$	$\frac{4}{6}$

Suppose that the values assumed by a random variable X are $x_1, x_2, \ldots, x_k$. For $j = 1, 2, \ldots, k$ let E_j be the event consisting of all outcomes to which the random variable X assigns the number x_j. That is, E_j is the event such that X takes the value x_j for each outcome in E_j and it takes the value x_j for no other outcomes. In the game of Example 4.2, the random variable takes the three values -6, -1, and 1.5, so $k = 3$. If we let

$$x_1 = -6 \qquad x_2 = -1 \qquad x_3 = 1.5$$

then

$$E_1 = \{6\} \qquad E_2 = \{1\} \qquad E_3 = \{2, 3, 4, 5\}$$

Notice that x_j is the value of the random variable assigned to each outcome in event E_j, which is a set of outcomes. The first column in tables such as Tables 4.2 and 4.3 lists the events E_j corresponding to the distinct values x_j of the random variable.

The third column in Table 4.2 lists the probabilities assigned to events E_j. Since E_j consists of those outcomes for which X takes the value x_j, we have

$$\Pr[X = x_j] = \Pr[E_j]$$

Thus we have a correspondence between values x_j and probabilities p_j defined by $p_j = \Pr[X = x_j]$. This correspondence is of sufficient interest and importance to have a name of its own.

The *probability density function,* or simply *density function,* of a random variable X assigns probabilities p_j to values of x_j of X, where p_j is the probability of event E_j on which the random variable X takes the value x_j, $j = 1, 2, \ldots, k$.

We note that since the events E_j are always a partition of the sample space S,

$$p_1 + p_2 + \cdots + p_k = 1$$

In the setting described in Example 4.2 and shown in Table 4.3, we have $p_1 = \frac{1}{6}$, $p_2 = \frac{1}{6}$, and $p_3 = \frac{4}{6}$. Thus, even though the values x_j are always different for different values of j, it is possible that some of the probabilities p_j are the same.

Remark In Example 4.1, we considered a simple lottery, and we computed the average loss per play without using the concepts of random variable or density function. However, it is natural to define a random variable X to be the net return on each play of the lottery. Then X takes values of -1 (you lose your dollar) and 4 (you win \$5, but paid \$1 to play). Also, the probability of the value

-1 is $\frac{9}{10}$, and the probability of the value 4 is $\frac{1}{10}$. Hence the density function for X is as shown in Table 4.4.

TABLE 4.4

Value of X	Probability
-1	$\frac{9}{10}$
4	$\frac{1}{10}$

Example 4.3

Product testing

A production process that uses obsolete equipment produces defective audiotapes 10 percent of the time. An experiment consists of selecting a sample of 3 tapes, testing the tapes, and noting how many are defective. A random variable X is defined by assigning to each outcome the number of defective tapes.

Problem Assume that this is a Bernoulli process, and find the density function of X.

Solution We define a success to be the selection of a defective tape. Since 10 percent of the tapes are defective, we conclude that $p = .1$. The probability of the event for which X takes the value 0, that is, the set of outcomes with no defective tapes, can be obtained by using (3.5) with $n = 3$, $r = 0$, and $p = .1$. We have

$$\Pr[X = 0] = C(3, 0)(.1)^0(.9)^3 = .729$$

Likewise,

$$\Pr[X = 1] = C(3, 1)(.1)^1(.9)^2 = .243$$
$$\Pr[X = 2] = C(3, 2)(.1)^2(.9)^1 = .027$$
$$\Pr[X = 3] = C(3, 3)(.1)^3(.9)^0 = .001$$

The values of X and the associated values of the density function for X are shown in Table 4.5. ■

TABLE 4.5

Value of X	Probability
0	.729
1	.243
2	.027
3	.001

Example 4.4

Committees

A congressional committee consists of 4 Democrats and 3 Republicans. Suppose that a subcommittee of 3 is selected at random, and a random variable X is defined to be the number of Republicans on the subcommittee.

Problem Find the density function of X.

Solution The random variable X takes the values 0, 1, 2, and 3, so we set $x_1 = 0$, $x_2 = 1$, $x_3 = 2$, and $x_4 = 3$. The probabilities of the associated events are obtained by the methods of Chapter 2. Thus, we have

TABLE 4.6

Value of X	Probability
0	$\frac{4}{35}$
1	$\frac{18}{35}$
2	$\frac{12}{35}$
3	$\frac{1}{35}$

$$p_1 = \Pr[E_1] = \Pr[X = x_1] = \frac{C(4, 3)}{C(7, 3)} = \frac{4}{35}$$

$$p_2 = \Pr[E_2] = \Pr[X = x_2] = \frac{C(4, 2) \times C(3, 1)}{C(7, 3)} = \frac{18}{35}$$

$$p_3 = \Pr[E_3] = \Pr[X = x_3] = \frac{C(4, 1) \times C(3, 2)}{C(7, 3)} = \frac{12}{35}$$

$$p_4 = \Pr[E_4] = \Pr[X = x_4] = \frac{C(3, 3)}{C(7, 3)} = \frac{1}{35}$$

The density function of X is shown in Table 4.6. ■

In addition to using tables to represent random variables and density functions, it is often useful to represent them graphically. This is done by using a coordinate system for which the horizontal coordinates represent the values of the random variable and the vertical coordinates are the probabilities associated with these values. In Figure 4.1, we display the graphs associated with the random variables of Table 4.4 (part (*a*) of the graph) and Table 4.6 (part (*b*) of the graph).

Exercises for Section 4.1

1. A fair coin is flipped 2 times, and a random variable X is defined to be the number of heads minus the number of tails. Find the density function of X.
2. A fair coin is flipped 3 times, and a random variable X is defined to be 3 times the number of heads minus 2 times the number of tails. Find the density function of X.
3. A fair die is rolled, and a random variable X is defined to be 0 if the number on the die is even and the number on the die if the number is odd. Find the density function of X.
4. A fair die is rolled, and a random variable X is defined to be the number on the die if the number is even and the negative of the number on the die if the number is odd. Find the density function of X.

FIGURE 4.1

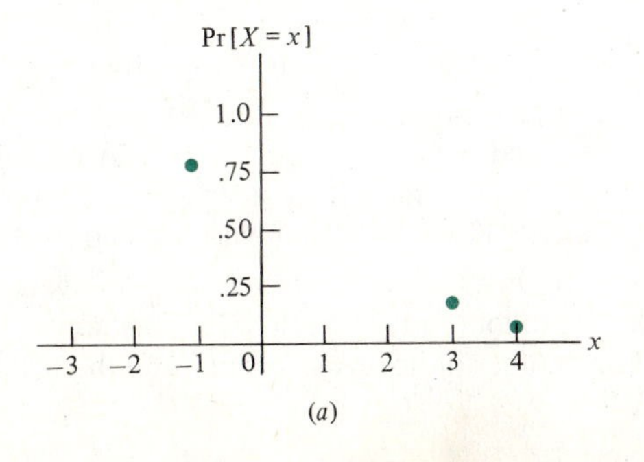

(*a*)

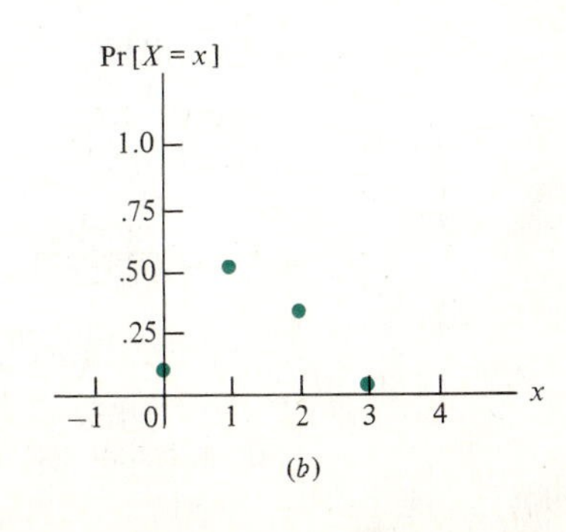

(*b*)

Committees

5. A student committee has 6 members, 4 females and 2 males. Two students are selected at random, and a random variable X is defined to be the number of females selected. Find the density function of X.

Committees

6. The student committee of Exercise 5 is enlarged to include 5 females and 3 males. Three students are selected at random, and a random variable Y is defined to be the number of females minus the number of males. Find the density function of Y.
7. An experiment consists of rolling two dice and noting the product of the two numbers. Let the random variable X be defined as -1 if the product is an even number and 2 if the product is an odd number. Find the density function of X.
8. Repeat Exercise 7, but now the random variable X is defined as 1 if the product is less than 11, as 2 if the product is between 11 and 19, and as 3 if the product exceeds 19. Find the density function of X.

Product performance

9. A vending machine yields the item selected 80 percent of the time and no item at all 20 percent of the time. Three individuals attempt to use the machine. Let the random variable X be defined as the number of individuals who obtain the item selected. Find the values taken by X and the density function of X, under the assumption that the attempts to use the machine form a Bernoulli process.
10. A coin is weighted so that $\Pr[H] = \frac{1}{3}$. The coin is flipped 3 times. A random variable X is defined by assigning to each outcome the number of heads obtained in the 3 flips. Find the values assumed by the random variable and the density function of X.
11. An experiment consists of randomly drawing 2 cards from a standard deck (without replacement). The random variable X is defined to be the number of aces drawn. Find the density function of X.
12. Using the same experiment as in Exercise 11, the random variable Y is defined to be 5 times the number of aces drawn. Find the density function of Y.
13. An experiment consists of randomly drawing 3 cards from a standard deck (without replacement). The random variable Z is defined to be 5 times the number of aces plus 2 times the number of face cards which are not aces (jacks, queens, or kings). Find the density function of Z.
14. Using the same experiment as in Exercise 13, suppose that the random variable W is the random variable Z minus 3 times the number of cards which are not face cards. Find the density function of W.
15. An experiment consists of selecting 2 coins simultaneously and at random from a collection consisting of 2 nickels and 3 dimes. A random variable is defined which assigns to each outcome the value of the coins in cents. Find the values assumed by this random variable and its density function.

Product testing

16. In a box containing 12 hand calculators, 9 work and 3 do not work. A sample of 5 calculators is selected at random from the box. A random variable is defined which assigns to each outcome the number of calculators which do not work. Find the values assumed by this random variable and its density function.

Entertainment

17. A student has 3 volumes of short stories and 2 novels on a bookshelf. She selects 3 books at random to take home over vacation. A random variable X is defined to be the number of novels selected. Find the values assumed by X and its density function.

18. There are 20 balls labeled 1 through 20 in a box. A ball with an even number is worth \$1, and a ball with an odd number is worth \$2. Three balls are selected simultaneously and at random. A random variable X is defined as the total value (in dollars) of the 3 balls selected. Find the probability density function of X.

Games of chance

19. Five cards are selected at random from an ordinary deck, and a random variable X is defined as the number of aces. Find the probability density function of X.

Sports

20. A track coach has 8 stopwatches: 5 are accurate, 1 is slow, and 2 are fast. Three stopwatches are selected at random and tested for accuracy. A random variable X is defined to be the number of accurate stopwatches in the sample. Find the values assumed by X and its density function.

Sports

21. In the situation described in Exercise 20, a random variable X is defined to be the number of stopwatches in the sample which are fast. Find the values assumed by X and its density function.

22. An unfair coin is weighted so that $\Pr[H] = \frac{2}{3}$. The coin is flipped until a head appears or 4 consecutive tails appear. The random variable X is defined to be the total number of flips of the coin. Find the values assumed by X and the density function of X.

Product performance

23. A motor pool contains 10 automobiles, 2 of which have defective fuel gauges. A saleswoman is assigned an automobile at random on 4 occasions. A random variable X is defined to be the number of times she receives an automobile with a defective gauge. Find the values of X and the density function of X.

Education

24. A kindergarten teacher has a class with 9 boys and 8 girls. He picks 4 students' names at random (without replacement). A random variable X is defined to be the number of girls selected. Find the density function of X.

25. An experiment consists of rolling a fair die until the sum of all numbers which have appeared exceeds 3. A random variable X is defined to be the total number of rolls. Find the density function of X.

26. A fair coin is flipped until either 3 heads have appeared or 3 tails have appeared. A random variable X is defined to be the total number of flips. Find the density function of X.

27. Repeat Exercise 26 with an unfair coin such that $\Pr[H] = \frac{1}{3}$.

Games of chance

28. A game is played by drawing 5 cards at random from a completely shuffled deck. You receive \$10 if you have exactly 1 pair (exactly 2 cards of the same value), \$50 if you have exactly 2 pairs, \$100 if you have exactly 3 of a kind, and \$5000 if you have 4 of a kind. It costs you \$1 to play the game. Define a random variable X to be your net gain per play, and find the density function of X.

29. You have two coins: One is fair and the other is unfair with $\Pr[H] = \frac{2}{3}$. An experiment consists of selecting a coin at random and flipping it twice, noting the result of each flip. Define a random variable X to be the number of heads. Find the density function of X.

30. You have 3 coins: 2 are fair and 1 is unfair with $\Pr[H] = \frac{1}{3}$. An experiment consists of selecting a coin at random and flipping it twice, noting the result of each flip. Define a random variable X to be the number of heads. Find the density function of X.

31. An experiment consists of randomly choosing two 4-digit numbers from the set of consecutive integers from 0000 through 9999. The selection is with replace-

ment, and we assume the selections are independent. Thus, possible selections are 0135 and 2553. If the two numbers selected are the same, you win \$500; and if they are not the same, you lose \$1. Find the density function of the random variable which gives your gain or loss.

32. An experiment consists of randomly selecting two sets of six 2-digit numbers from the set of consecutive integers from 01 through 45. Thus, one possible selection is {11, 23, 24, 25, 44, 45}. We assume the selections are with replacement and independent, and we define a random variable to have value \$1,000,000 if the sets selected are the same and to have value −\$1 if the selections are not the same. Find the density function of this random variable.

4.2 EXPECTED VALUES AND STANDARD DEVIATIONS OF RANDOM VARIABLES

In the discussion of Example 4.1, we concluded that it was reasonable to expect to lose \$.50 per play of the game. We obtained this average loss by noting that in each 10 plays of the game, we expect to win 1 time (\$4) and lose 9 times ($9 \times \$1 = \$9$). Thus, in 10 plays we lose \$5 or, on average, \$.50 per play. Similarly, in Example 4.2, we decided the average return per play is −\$1/6. This followed from the fact that a loss of \$6.00 occurred with probability $\frac{1}{6}$, a loss of \$1.00 with probability $\frac{1}{6}$, and a gain of \$1.50 with probability $\frac{4}{6}$:

$$\text{Expected gain (in dollars)} = (-6)(\tfrac{1}{6}) + (-1)(\tfrac{1}{6}) + (1.5)(\tfrac{4}{6}) = -\tfrac{1}{6}$$

This concept of an average or expected gain can be extended to any random variable.

Let X be a random variable with values $x_1, x_2, \ldots, x_k$. Also, let $p_i = \Pr[X = x_i]$ for $i = 1, 2, \ldots, k$. The expected value of X, $E[X]$, is defined to be

$$E[X] = x_1p_1 + x_2p_2 + \cdots + x_kp_k \tag{4.1}$$

The expected value of X is also called the *mean* of X.

The probabilities p_i are given by the probability density function for the random variable X. Thus a necessary step in determining the expected value of X is to find the density function of X.

Example 4.5

Sales

The sale table for compact disks at Tone Down Sounds has 3 premium labels and 5 discount specials. On sale, each premium label disk sells for \$10 and each discount special sells for \$6. Two disks are selected at random. A random variable X is defined to be the total cost of the CDs selected.

Problem Find the expected value of the random variable X.

Solution In general we will apply formula (4.1) by using a table. In this first example, however, we apply the formula directly to emphasize the meaning of each term in the formula. We first determine the values assumed by X. If 2 discount specials are selected, then the cost in dollars is \$12; if 1 discount special is selected along with 1 premium label, then the cost is \$16; and if 2 premium labels are selected, then the cost is \$20. Thus we set $x_1 = 12$, $x_2 = 16$, and $x_3 = 20$. We determine the density function of X (using the techniques of Chapter 2) to be

$$p_1 = \Pr[X = x_1] = \frac{C(5,2)}{C(8,2)} = \frac{(5 \cdot 4)/(1 \cdot 2)}{(8 \cdot 7)/(1 \cdot 2)} = \frac{20}{56}$$
$$p_2 = \Pr[X = x_2] = \frac{C(5,1)C(3,1)}{C(8,2)} = \frac{(5/1)(3/1)}{(8 \cdot 7)/(1 \cdot 2)} = \frac{30}{56}$$
$$p_3 = \Pr[X = x_3] = \frac{C(3,2)}{C(8,2)} = \frac{(3 \cdot 2)/(1 \cdot 2)}{(8 \cdot 7)/(1 \cdot 2)} = \frac{6}{56}$$

Applying the definition of expected value, we have

$$E[X] = x_1p_1 + x_2p_2 + x_3p_3 = 12(\tfrac{20}{56}) + 16(\tfrac{30}{56}) + 20(\tfrac{6}{56}) = \tfrac{840}{56} = 15$$

In terms of the original setting, the expected value of the cost of the two CDs selected is \$15.

Remarks

1. In Example 4.5, we computed the expected value of a cost. In such a case we often shorten the phrase "expected value of the cost" to "expected cost." Similarly, we say "expected grade-point average" for "expected value of the grade-point average," etc.
2. Notice that in Example 4.5 the expected value of X (which is 15) is *not* one of the values taken by X (which are 12, 16, 20). This is often, but not always, the case.

TABLE 4.7

Value of X x_i	Probability p_i	Product x_ip_i
12	$\frac{20}{56}$	$\frac{240}{56}$
16	$\frac{30}{56}$	$\frac{480}{56}$
20	$\frac{6}{56}$	$\frac{120}{56}$
		$E[X] = \frac{840}{56} = 15$

In Section 4.1 we saw how to form a table (Table 4.5) giving the density function of a random variable. It is easy and frequently useful to modify such a table for use in finding the expected value of a random variable. Indeed, such a table already contains the values of X and their associated probabilities. We add another column labeled "product" in which we multiply each value of X by its associated probability. The expected value of X is, by Equation (4.1), the sum of these products. Using this idea in Example 4.5, we have Table 4.7. The table is completed by adding the entries in the column labeled "product." This sum is the expected value of X.

Example 4.6

Games of chance

A friend invites you to play the following game: 2 coins are to be selected at random from 6 coins: 2 nickels, 3 dimes, and 1 quarter. If the sum of the values of the coins is 10, 20, or 30 cents, your friend pays you 25 cents. Otherwise you must pay your friend the value of the coins. What is your expected gain per play of the game?

Solution Let N, D, and Q denote nickel, dime, and quarter, respectively. Since the order in which the coins are drawn is unimportant, the sample space of the experiment can be represented as $\{DQ, DN, DD, NQ, NN\}$. The gain to you is -35 cents for DQ, -15 cents for DN, and $+25$ cents for the other three outcomes. Let X denote the random variable which assigns to each outcome its value to you in cents. The density function of X is given in Table 4.8, and we have

$$E[X] = -\tfrac{105}{15} - \tfrac{90}{15} + \tfrac{150}{15} = -\tfrac{45}{15} = -3$$

You should expect to lose 3 cents to your friend on each play of the game. ■

TABLE 4.8

Value	Probability	Product
-35	$\dfrac{C(3,1)C(1,1)}{C(6,2)} = \dfrac{3}{15}$	$-\dfrac{105}{15}$
-15	$\dfrac{C(2,1)C(3,1)}{C(6,2)} = \dfrac{6}{15}$	$-\dfrac{90}{15}$
$+25$	$\dfrac{C(2,2)}{C(6,2)} + \dfrac{C(3,2)}{C(6,2)} + \dfrac{C(2,1)C(1,1)}{C(6,2)} = \dfrac{6}{15}$	$+\dfrac{150}{15}$
		$E[X] = -\frac{45}{15} = -3$

Example 4.7 *Sales*

A sales representative makes sales to 20 percent of the customers on which he calls. His commissions are \$100 each for the first 3 sales each week, \$200 each for every sale above 3. In a certain week the sales representative calls on 8 customers.

Problem Find the expected value of the total income of the sales representative that week.

Solution Define a random variable X to be the total income of the sales representative. We view the calls made by the sales representative as Bernoulli trials. There are 8 trials, and each has success probability $p = .2$. There is no income for no sales. The total income for 1, 2, or 3 sales is \$100, \$200, or \$300, respectively. Each sale above 3 adds \$200 to the total income. Thus for 4, 5, 6, 7, and 8 sales the total income is \$500, \$700, \$900, \$1100, and \$1300, respectively. The values and probability density function for X (taken from Appendix A) are shown in Table 4.9. Using the definition of expected value, we add the entries in the product column, and we conclude that the expected value of X is (in dollars)

$$E[X] = 166.80$$

Thus the expected total income of the sales representative for that week is \$166.80. ■

TABLE 4.9

Value, \$	Probability	Product
0	.1678	0
100	.3355	33.55
200	.2936	58.72
300	.1468	44.04
500	.0459	22.95
700	.0092	6.44
900	.0011	.99
1100	.0001	.11
1300	.0000	.00
		$E[X] = 166.80$

In computing the mean or expected value of a random variable, it is usually necessary to use the process just described. However, for the important special case of a binomial random variable (introduced in Section 4.1) there is a formula which greatly simplifies the task.

Consider a Bernoulli process consisting of n trials, and let X be the random variable which assigns to each outcome the number of successes. Suppose p is the probability of success on a single trial. The random variable X takes the values 0, 1, 2, . . . , n. The probability associated with each of these values was determined in formula (3.5); it is

$$\Pr[X = r] = C(n, r)p^r(1 - p)^{n-r}$$

for $r = 0, 1, 2, \ldots, n$. According to the definition of expected value, we have

$$\begin{aligned} E[X] &= 0 \cdot \Pr[X = 0] + 1 \cdot \Pr[X = 1] + 2 \cdot \Pr[X = 2] \\ &\quad + \cdots + n \cdot \Pr[X = n] \\ &= 0 \cdot C(n, 0)p^0(1 - p)^n + 1 \cdot C(n, 1)p(1 - p)^{n-1} \\ &\quad + 2 \cdot C(n, 2)p^2(1 - p)^{n-2} + \cdots + n \cdot C(n, n)p^n \end{aligned}$$

Although the formula for $E[X]$ appears to be a complicated sum, it can be written in a very simple form. We illustrate the simplification for $n = 1$ and $n = 2$. First, suppose $n = 1$. Then

$$E[X] = 0 \cdot p^0(1 - p)^1 + 1 \cdot p^1(1 - p)^0 = p$$

Next, if $n = 2$, then

$$E[X] = 0 \cdot p^0(1 - p)^2 + 1 \cdot 2 \cdot p^1(1 - p)^1 + 2 \cdot 1 \cdot p^2(1 - p)^0 = 2p$$

The argument can be continued and an analogous result can be established for any value of n by using mathematical induction.

Expected Value of a Binomial Random Variable

If X is a binomial random variable for a Bernoulli process consisting of n trials with success probability p, then

$$E[X] = np \tag{4.2}$$

Example 4.8 A political poll is to be taken by mailing 1000 questionnaires. It is known that the probability that any given questionnaire will be returned is .14.

Opinion surveys

Problem Find the expected number of questionnaires that will be returned (assume that each return or nonreturn of a questionnaire is a Bernoulli trial).

Solution We use formula (4.2). We define a success to be a returned questionnaire, so $p = .14$. Since $n = 1000$, the expected number of returned questionnaires is $1000 \times .14 = 140$. ■

Example 4.9 *Sports*

An NBA basketball player makes 30 percent of his 3-point shots and 45 percent of his 2-point shots. If each year he attempts 300 three-point shots and 800 two-point shots, what is the average number of points he makes per year on field goals? You may assume that his 3-point shots and his 2-point shots are both Bernoulli trial experiments.

Solution The total number of points on field goals is the sum of the points from 3-point shots and the points from 2-point shots. If we consider the 3-point shots as a Bernoulli process with 300 trials and $p = .30$, the expected number of points from 3-point shots is $3 \times 300 \times .30 = 270$. Similarly, the expected number of points from 2-point shots is $2 \times 800 \times .45 = 720$. Thus, the expected total number of points from field goals is $270 + 720 = 990$. ■

Remark In the solution of Example 4.9, we used a powerful fact about the expected value of a random variable. In that problem we began with two random variables—the number of points from 3-point shots made in a year and the number of points from 2-point shots made in a year. We were interested in a third random variable, namely, the total number of points from field goals. Using the formula for the expected value of a random variable (4.2), we obtained the expected number of successful 3-point shots ($300 \times .3 = 90$) and the expected number of successful 2-point shots ($800 \times .45 = 360$). This gave us $270 = 3 \times 90$ expected points from 3-point shots and $720 = 2 \times 360$ expected points from 2-point shots. Finally, we added these to obtain the expected number of total points, $990 = 270 + 720$. The powerful result that we are using is that the steps combining these numbers are legal. In particular, if a random variable Z is equal to $aX + bY$, where X and Y are random variables and a and b are numbers, then

$$E[Z] = aE[X] + bE[Y]$$

In our case, X is the number of 3-point shots made, Y is the number of 2-point shots made, $a = 3$, and $b = 2$.

Example 4.10 Suppose we have experiments labeled A, B, C and corresponding random variables X_A, X_B, and X_C. Moreover, suppose that the probability functions of these random variables are as shown in Table 4.10.

Problem Find the mean of each of these random variables and draw graphs of their density functions. To draw a graph, you plot the points (value, probability of that value) on a coordinate system.

TABLE 4.10

X_A		X_B		X_C	
Value	**Probability**	**Value**	**Probability**	**Value**	**Probability**
+5	.5	+1	.25	+5	.1
−5	.5	0	.5	0	.8
		−1	.25	−5	.1

Solution The graphs of the density functions are shown in Figure 4.2. Using Equation (4.1), we find

$$E[X_A] = (5)(.5) + (-5)(.5) = 0$$
$$E[X_B] = (1)(.25) + (0)(.5) + (-1)(.25) = 0$$
$$E[X_C] = (5)(.1) + (0)(.8) + (-5)(.1) = 0$$

■

In Example 4.10 each of the three random variables X_A, X_B, and X_C has mean 0. However, it is clear from Table 4.10 and Figure 4.2 that these random variables are quite different. For instance, X_A never takes the value 0 and in fact takes each of the values +5 and −5 with probability .5. On the other hand, X_B takes only values relatively close to 0 and in fact takes the value 0 with probability .5. Finally, X_C also takes the values +5 and −5, but it is far more likely to take the value 0. Since these quite different random variables have the same mean, it is clear that another measure is needed to help distinguish one from another. One approach is to measure the dispersion of the values of X about the

FIGURE 4.2

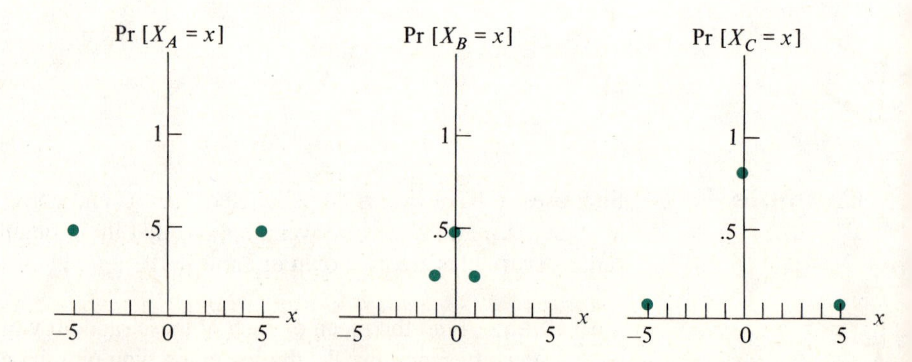

mean, i.e., how the values of X are spread out. To do so, we shall use the *variance* and *standard deviation*. It is common to adopt notation using the Greek letters μ (mu) and σ (sigma) for the mean and standard deviation, respectively.

Let X be a random variable which takes the values $x_1, x_2, \ldots, x_n$ with probabilities $p_1, p_2, \ldots, p_n$, respectively. Let $\mu = E[X]$ denote the *mean* of X. The *variance* of X is the number

$$\text{Var}[X] = (x_1 - \mu)^2 p_1 + (x_2 - \mu)^2 p_2 + \cdots + (x_n - \mu)^2 p_n \tag{4.3}$$

The *standard deviation* of X is the number

$$\sigma = \sqrt{\text{Var}[X]} \tag{4.4}$$

Equation (4.4) simply says that once we have the variance of X, we obtain the standard deviation by taking the square root of the variance.

Example 4.11 Determine the variance and standard deviation of the random variables X_A, X_B, and X_C of Example 4.10.

Solution Using definitions (4.3) and (4.4), we have

$$\begin{aligned}
\text{Var}[X_A] &= (5 - 0)^2(.5) + (-5 - 0)^2(.5) = 25 \\
\text{Var}[X_B] &= (1 - 0)^2(.25) + (0 - 0)^2(.5) + (-1 - 0)^2(.25) = .5 \\
\text{Var}[X_C] &= (5 - 0)^2(.1) + (0 - 0)^2(.8) + (-5 - 0)^2(.1) = 5
\end{aligned}$$

and therefore for X_A we have $\sigma = 5$, for X_B we have $\sigma = \sqrt{.5} = .71$, and for X_C we have $\sigma = \sqrt{5} = 2.2$, where the square roots are correct to two digits. ■

Notice that in this example the variances of X_A, X_B, and X_C correspond in a qualitative way to the distribution of the values about the mean. That is, the random variable with the largest variance, X_A, takes values relatively far from the mean (0) with high probability, and the random variable with the smallest variance, X_B, takes only values relatively close to the mean.

Just as there is a simple formula for the mean of a binomial random variable, there is a corresponding formula for the variance.

Variance of a Binomial Random Variable

If X is a binomial random variable for a Bernoulli process consisting of n trials with success probability p, then the variance of X, Var[X], is

$$\text{Var}[X] = np(1 - p) \tag{4.5}$$

and the standard deviation of X, $\sigma[X]$, is

$$\sigma[X] = \sqrt{np(1 - p)}$$

Example 4.12

Product performance

Sales information provided by O. A. Rowe Seed Company asserts that the germination rate for its best-selling Robust Red Tomato seeds is 98 percent. Suppose 2000 seeds are planted. Let X denote a random variable which assigns to each outcome, for the "experiment" of planting 2000 seeds, the number of seeds which germinate. Assume that this is a Bernoulli process and that the assertion is correct.

Problem Find the expected value and variance of X.

Solution The random variable X is a binomial random variable with $n = 2000$ and $p = .98$. (A success is a germinated seed.) Using (4.2) and (4.5), we have

$$E[X] = 2000(.98) = 1960 \qquad \text{and} \qquad \text{Var}[X] = 2000(.98)(.02) = 39.2 \quad \blacksquare$$

Exercises for Section 4.2

1. A random variable X has the density function shown below. What is the expected value of X?

Value of X	Probability
1	$\frac{3}{6}$
2	$\frac{2}{6}$
3	$\frac{1}{6}$

2. A random variable X has the density function shown below. What is the expected value of X?

Value of X	Probability
−2	$\frac{2}{10}$
−1	$\frac{2}{10}$
0	$\frac{1}{10}$
1	$\frac{2}{10}$
2	$\frac{3}{10}$

3. A random variable X has the density function shown below. What is the expected value of X?

Value of X	Probability
1	$\frac{1}{21}$
2	$\frac{2}{21}$
3	$\frac{3}{21}$
4	$\frac{4}{21}$
5	$\frac{5}{21}$
6	$\frac{6}{21}$

4. A random variable X has the density function shown below. What is the expected value of X?

Value of X	Probability
−1	$\frac{1}{21}$
2	$\frac{2}{21}$
−3	$\frac{3}{21}$
4	$\frac{4}{21}$
−5	$\frac{5}{21}$
6	$\frac{6}{21}$

5. A random variable X has the density function shown on the following page. What is the expected value of X?

Value of X	Probability
−2	.05
−1	.10
0	.15
1	.30
2	.40

6. A random variable X has the density function shown below. What is the expected value of X?

Value of X	Probability
−20	.05
−10	.10
0	.15
1	.30
2	.40

7. A random variable X has the density function table shown below. Find the missing entries in the table and the expected value of X.

Value of X	Probability	Product
−1	?	−.3
−2	?	−.4
?	.2	.3
2	.3	.6

8. A random variable X has the density function table shown below. Find the missing entries in the table.

Value of X	Probability	Product
?	?	?
1	.2	.2
?	.3	.6
3	?	.3
4	?	1.2
		$E[X]=2.1$

9. A drawer contains 3 nickels, 2 dimes, and 4 quarters. One coin is randomly selected from the drawer. What is the expected value of the coin selected?

10. Suppose 2 coins are simultaneously selected from the drawer in Exercise 9. What is the expected value of the coins selected?

Education

11. A student estimates that her chances of getting various grades in a finite mathematics class are as shown in Table 4.11 (points per grade are also shown). What is the expected value of the number of grade points she will earn in the course?

TABLE 4.11

Grade	Grade Points	Probability
A	4.0	.05
A−	3.7	.1
B+	3.3	.2
B	3.0	.3
B−	2.7	.2
C+	2.3	.05
C	2.0	.05
C−	1.7	.05

Sports

12. For a certain baseball player, hitting against left-handed pitchers is a Bernoulli process with the probability of a hit equal to .3, while hitting against a right-handed pitcher is a Bernoulli process with the probability of a hit equal to .25. One month this player bats 20 times against left-handed pitchers and 70 times against right-handed pitchers. What is the expected number of hits the batter obtains that month?

13. A coin is weighted so that $\Pr[H] = \frac{2}{5}$. The coin is flipped 10 times. A random variable X is defined by assigning to each outcome the number of heads obtained in the 10 flips. Find $E[X]$.

Product testing

14. Defective widgets are produced at random with probability .2. An experiment consists of examining widgets one after another as they come off the production line until either a defective widget is found or 4 widgets are examined. Find the expected number of widgets examined in the experiment.

15. You have two coins; one is fair and the other is unfair with $\Pr[H] = \frac{2}{3}$. An experiment consists of selecting a coin at random and flipping it twice, noting the result of each flip. Find the expected number of heads.

16. You have three coins; two are fair and one is unfair with $\Pr[H] = \frac{1}{3}$. An experiment consists of selecting a coin at random and flipping it twice, noting the result of each flip. Find the expected number of heads.

17. A fair coin is flipped repeatedly until there are either 2 consecutive tails or a total of 4 flips. Find the expected number of times the coin is flipped.

Sports

18. A golfer makes 60 percent of her putts when they are less than 10 feet and 30 percent when they are 10 feet or more in length. Assume that her putts are Bernoulli trials and in one tournament she takes 32 putts from less than 10 feet and 78 from 10 feet or more. What is the expected number of putts made?

Lotteries

19. The probability of winning \$10,000 in the lottery is .00002, the probability of winning \$100,000 is .000004, and the probability of winning \$1,000,000 is

.0000001. A ticket costs \$10. Find your expected gain or loss on each ticket you buy.

Psychology

20. A psychologist has designed a word-matching experiment as follows. A subject is presented with a word as a stimulus and a set of 4 words from which a response is to be selected. Exactly one of the responses is correct. The subject selects responses one after another, and after each selection he is told whether the selection is correct. Assume that responses are selected at random and that once the subject knows a response is incorrect, that response will not be selected again. A trial of the experiment ends with the selection of the correct response. Find the expected number of responses in each trial of the experiment.
21. An experiment consists of rolling a pair of fair dice. The random variable X assigns to each roll the *product* of the results. What is the expected value of the random variable?

Games of chance

22. At a local carnival a game can be played with a fishpond containing 100 fish: 90 are white, 9 are red, and 1 is blue. A contestant randomly catches a fish and receives payment as follows:

White:	\$.30
Red:	\$1.00
Blue:	\$10.00

If it costs \$.60 to play this game, how much (on the average) does the carnival gain on each play?

23. There are two urns a and b. Urn a contains 2 red balls and 1 blue ball; urn b contains 1 red ball and 1 blue ball. An experiment consists of drawing a ball at random from urn a, noting its color, placing it in urn b, and then drawing a ball at random from urn b and noting its color. A random variable X is defined by assigning to each outcome the total number of red balls drawn. Find the expected value of X.
24. An experiment consists of rolling a fair die and noting the number which comes up. A random variable X is defined by assigning the value 1 to any outcome that is an odd number and by assigning the outcome value to any outcome that is even. Find $E[X]$.

Educational testing

25. An examination consists of 2 true-false questions and 2 multiple choice questions (5 options per question), each with exactly 1 correct answer. If a student selects answers at random, one answer for each question, what is the expected number of correct answers?
26. In a Bernoulli process with $n = 1000$ and $p = .1$, find the expected number of successes and the standard deviation of the number of successes.

Production

27. A foundry worker is assigned bonus points for producing castings without defects. In a typical day the worker produces 10 castings, and he expects that 90 percent of the time he can produce a casting without defects. The reward system assigns 100 points if none of the 10 castings has defects, 80 points of 1 or 2 castings have defects, 50 points if 3 or 4 castings have defects, and no bonus points if 5 or more castings have defects. Find the expected number of bonus points per day for the worker. (*Hint*: View this as a Bernoulli process and use Table A.1 in Appendix A.)

Production

28. In Exercise 27 what happens to the expected number of bonus points if the percentage of defect-free castings drops to 80 percent?
29. A Bernoulli process consists of 10 trials with $p = .2$. A random variable X is defined by assigning to each outcome 3 times the number of successes minus the number of failures. Find the expected value of X. (*Hint*: Use Table A.1.)
30. Find the variance and standard deviation of the random variable of Exercise 1.
31. Find the variance and standard deviation of the random variable of Exercise 3.
32. Find the variance and standard deviation of the random variable of Exercise 6.

4.3 NORMAL RANDOM VARIABLES AND THE NORMAL APPROXIMATION TO THE BINOMIAL

In this section we consider a topic which is not strictly a part of finite mathematics. In particular, we study a class of random variables, normal random variables, which assume infinitely many values. We do so for two reasons. First, many important applications of probability and statistics involve normal random variables. Second, normal random variables are often useful as an approximation to binomial random variables, and we have seen earlier the usefulness of binomial random variables. For instance, if X is a binomial random variable for a Bernoulli process with $n = 100$ and $p = .2$, then the direct computation of $\Pr[X \geq 22]$ is very cumbersome. Alternative methods are needed.

We are concerned here with experiments which can have infinitely many different outcomes. In some situations of this sort, it is possible and useful to present the probabilities of various events in a different way than we have seen before. For instance, suppose that we are interested in the length of telephone calls made by the sales staff in an office. The length of a call (in minutes) can be any number between, say, 2 and 40. Using data which have been collected for a large number of calls, we can determine a curve, called a *density curve* (Figure 4.3), which plays a role similar to that played by the density function discussed in preceding sections. (The details of how one uses data to determine a density curve involve concepts from statistics which are outside the scope of this discussion and are omitted.) The density curve can be used to solve problems such as determining the probability that a random call lasts at least 5 minutes but not more than 18 minutes. The density curve enables us to solve this problem in the following way. The basic property of a density curve is that it relates probabilities to areas. In fact, the probability that a random call lasts between 5 and 18 minutes is the area under the density curve between 5 and 18. This area is shown in Figure 4.3.

Using a density curve for a random variable in this way, we can determine the probability that the random variable takes a value in any specified interval. This use is similar to the use of the density functions we defined in Section 4.1 to find the probability of events. Since in this case the random variable can take all values in some interval, it is customary to refer here to *continuous* random variables, and in the previous case (Section 4.1) to *discrete* random variables.

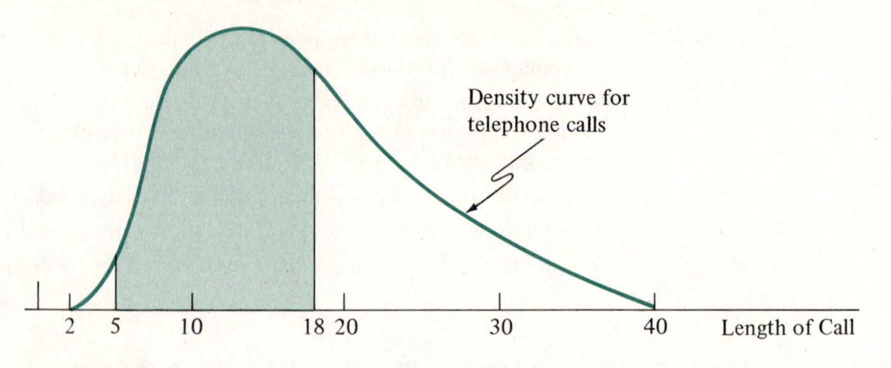

FIGURE 4.3

In general, each random variable has its own density curve. However, a particular density curve which occurs frequently in applications is called the *standard normal curve*,[1] and it is pictured in Figure 4.4.

Example 4.13

Product performance

A direct-drive record turntable (used to digitize classic albums) is designed to operate at 33 rpm (revolutions per minute). A quality control engineer finds that the manufacturing process actually produces turntables whose speeds deviate from design speed. The deviations are distributed according to a standard normal curve. That is, the probability that a randomly selected turntable has a speed between 32 and 34 rpm is the area under the standard normal curve between -1 and $+1$. ■

In Figure 4.4*a* the area under the curve between $z = a$ and $z = b$ (the area shaded in the figure) represents the probability that a random variable takes a value between a and b. In general, such areas are difficult to compute, but for the standard normal curve the areas have been computed and are available in tables.

The random variable associated with the standard normal curve is called the *standard normal random variable* and is denoted by Z. The properties of this random variable which are most helpful in solving problems are summarized next. The notation $\Pr[Z \leq z]$ denotes the probability that Z takes a value less than or equal to some specific value z, and $\Pr[a \leq Z \leq b]$ denotes the probability that Z takes a value between a and b, inclusive.

[1]The standard normal curve is the graph described by the equation $y = \dfrac{1}{\sqrt{2\pi}}e^{-z^2/2}$, for z any number. Here e is a number whose value is (to 5 decimal places) 2.71828. This number occurs in many different ways, some quite unexpected, throughout mathematics. Figure 4.4*a* shows the graph of this function for $-3 \leq z \leq 3$.

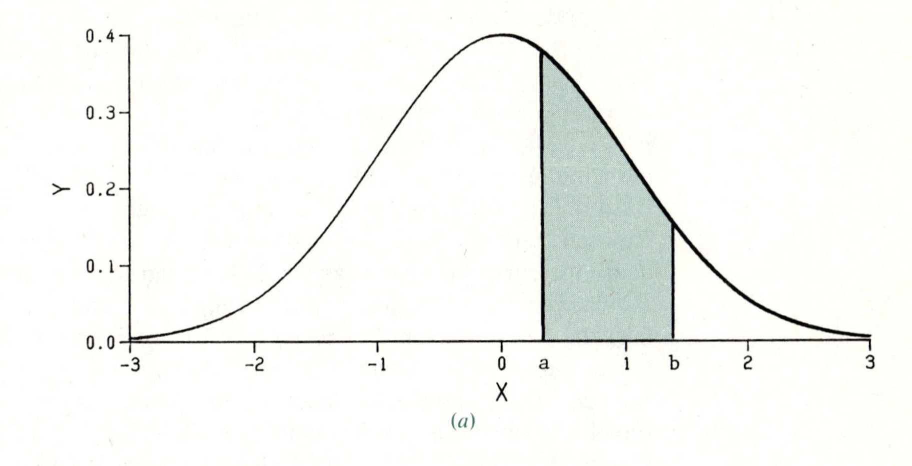

(*a*)

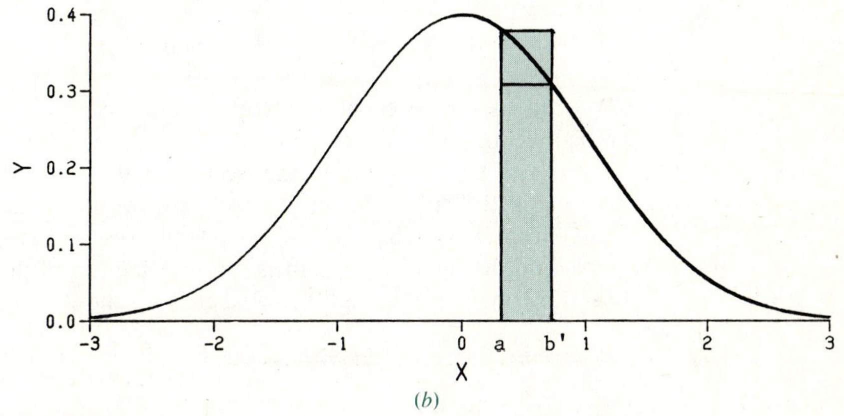

(*b*)

FIGURE 4.4

Properties of the Standard Normal Random Variable

Let Z denote the standard normal random variable.

1. The probability that Z takes a value between a and b, $a < b$, is the area under the standard normal curve between $z = a$ and $z = b$.
2. $\Pr[Z \geq 0] = .5$ and $\Pr[Z \leq 0] = .5$ (4.6)
3. For each real number a, $a > 0$,

$$\Pr[-a \leq Z \leq 0] = \Pr[0 \leq Z \leq a] \tag{4.7}$$

4. For each real number a,

$$\Pr[Z = a] = 0 \tag{4.8}$$

Property 1 is just a restatement of the relation between areas and probabilities. Properties 2 and 3 reflect the symmetry of the standard normal curve about $z = 0$. Property 4, which states that the probability that Z takes any *specific* value is zero, is quite different from the properties of any of the random variables studied in our earlier sections. However, it is natural in view of the relation between probabilities and areas. Indeed, suppose that in Figure 4.4a, the point a is held fixed and b is moved toward a. As b approaches a, the area under the curve between a and b gets smaller and smaller. For instance, in Figure 4.4b the area under the curve between a and b' is less than the area of the larger rectangle shown with the line segment having endpoints a and b' as base, and it is more than the area of the smaller rectangle shown with the same base. As point b' moves closer and closer to a, the areas of both rectangles become smaller and smaller. In fact, the areas of both rectangles approach zero. Since areas correspond to probabilities in this situation, this is the motivation for property 4. It follows from property 4 that for each number z we have

$$\Pr[Z \leq z] = \Pr[Z < z] \qquad \text{and} \qquad \Pr[Z \geq z] = \Pr[Z > z]$$

We will use these equalities frequently and without further comment in the remainder of this chapter.

Most of the problems in this section will require the use of Table B.1 in Appendix B. The entries in that table are the areas under the standard normal curve from 0 to t, where the units and tenths digits of t are given to the left of the table and the hundredths digit is given at the top of the table. For example, if $t = 1.57$, then we look in the row labeled 1.5 and the column labeled .07:

	.06	↓ **.07**	**.08**
1.4	.4279	.4292	.4306
→1.5	.4406	.4418	.4429
1.6	.4608	.4525	.4535

We find that the area under the standard normal curve from 0 to 1.57 is .4418.

Example 4.14 Let Z be the standard normal random variable. Find the following probabilities:

(a) $\Pr[Z$ is less than or equal to $.43] = \Pr[Z \leq .43]$
(b) $\Pr[Z$ is at least $1.7] = \Pr[Z \geq 1.7]$

Solution We use Appendix B and Equations (4.6) and (4.7) to find these probabilities.

(*a*) We must determine the area under the standard normal curve to the left of .43. We think of this area as consisting of two pieces (Figure 4.5*a*): the area to the left of 0 and the area from 0 to .43. We use Equation (4.6) to determine the former (dark shading in Figure 4.5*a*), and we use the table to determine the latter (light shading in Figure 4.5*a*). By Equation (4.6) we have $\Pr[Z \le 0] = .5$. To determine $\Pr[0 \le Z \le .43]$, we use the row labeled .4 and the column labeled .03 in Table B.1. The area under the standard normal curve from 0 to .43 is .1664. That is,

$$\Pr[0 \le Z \le .43] = .1664$$

Finally,

$$\begin{aligned}\Pr[Z \le .43] &= \Pr[Z \le 0] + \Pr[0 \le Z \le .43] \\ &= .5000 + .1664 = .6664\end{aligned}$$

FIGURE 4.5

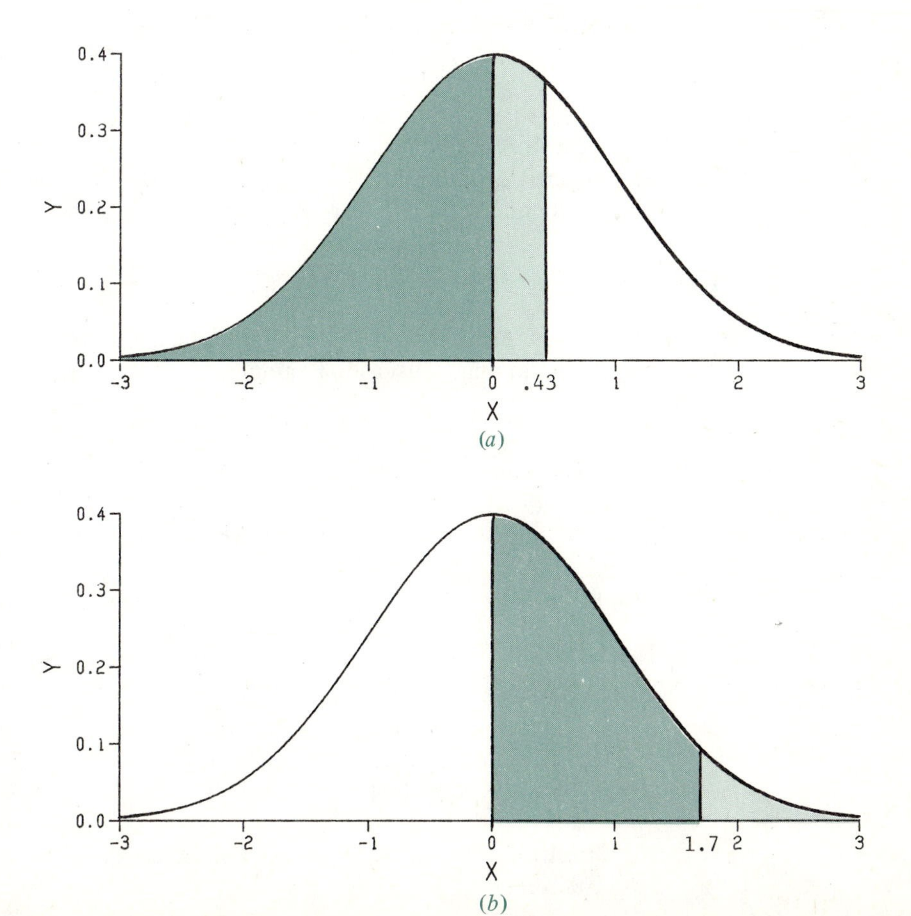

(*b*) Our goal is to determine $\Pr[Z \geq 1.7]$ or, equivalently, $\Pr[1.7 \leq Z]$. We begin with $\Pr[Z \geq 0] = .5$. We then divide the area to the right of $z = 0$ into two pieces at $z = 1.7$, shown with dark and light shading in Figure 4.5*b*. Since the two shaded areas together comprise the entire area to the right of $z = 0$, their total area is .5. Therefore we have

$$\Pr[0 \leq Z \leq 1.7] + \Pr[1.7 \leq Z] = .5$$

By Table B.1 the term $\Pr[0 \leq Z \leq 1.7]$ is equal to .4554. That is, the portion of Figure 4.5*b* with the dark shading has area .4554. Therefore,

$$.4554 + \Pr[1.7 \leq Z] = .5000$$

or

$$\Pr[1.7 \leq Z] = .0446$$

The portion of Figure 4.5*b* with the light shading has area .0446. ■

Example 4.15 Let Z be the standard normal random variable. Find

$$\Pr[Z \text{ is less than } -1.2] = \Pr[Z < -1.2] = \Pr[Z \leq -1.2]$$

Solution Here we use property 3, the fact that the standard normal curve is symmetric about the vertical line through 0. Consequently the area under the curve to the left of -1.2 is equal to the area under the curve to the right of $+1.2$:

$$\Pr[Z \leq -1.2] = \Pr[Z \geq 1.2]$$

These areas are shaded in Figure 4.6, and the area to the right of 1.2, $\Pr[Z \geq 1.2]$, can be determined just as in Example 4.14*b*. We have

FIGURE 4.6

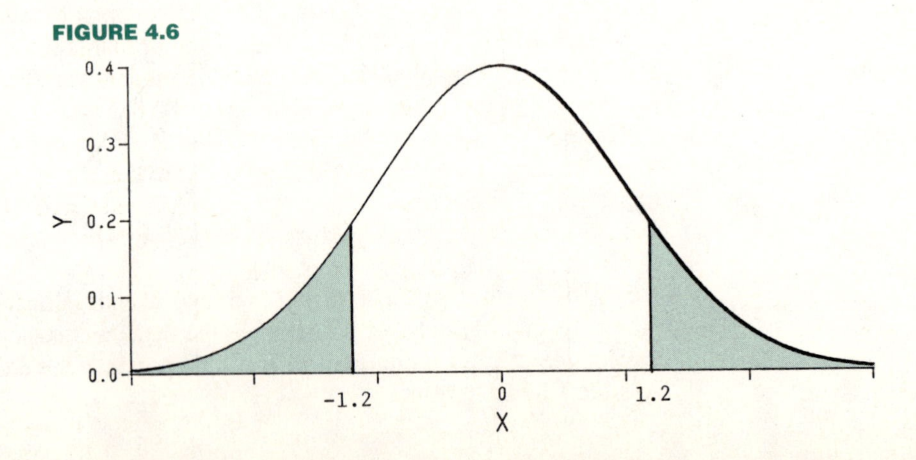

$$\Pr[0 \leq Z \leq 1.2] + \Pr[1.2 \leq Z] = .5$$
$$.3849 + \Pr[1.2 \leq Z] = .5000$$

and

$$\Pr[1.2 \leq Z] = .1151$$

We conclude that $\Pr[Z \leq -1.2] = .1151$. ■

Example 4.16 Let Z be the standard normal random variable. Find
(*a*) Pr[Z is between .2 and 1.2] $= \Pr[.2 \leq Z \leq 1.2]$
(*b*) Pr[Z is between -1.2 and .2] $= \Pr[-1.2 \leq Z \leq .2]$

Solution (*a*) To find the area under the standard normal curve from .2 to 1.2, we use Table B.1 to find the area from 0 to 1.2 and from 0 to .2, and then we subtract the latter from the former. That is,

$$\begin{aligned}\Pr[.2 \leq Z \leq 1.2] &= \Pr[0 \leq Z \leq 1.2] - \Pr[0 \leq Z \leq .2] \\ &= .3849 - .0793 = .3056\end{aligned}$$

(*b*) To find the area under the curve between -1.2 and .2, we compute the area between -1.2 and 0 and add to this the area from 0 to .2. Using Equation (4.7), we have

$$\begin{aligned}\Pr[-1.2 \leq Z \leq .2] &= \Pr[-1.2 \leq Z \leq 0] + \Pr[0 \leq Z \leq .2] \\ &= \Pr[0 \leq Z \leq 1.2] + \Pr[0 \leq Z \leq .2] \\ &= .3849 + .0793 = .4642\end{aligned}$$ ■

According to properties 2 and 3, the value 0 plays a special role for the standard normal random variable Z. According to property 2, in fact, Z is as likely to take a value greater than 0 as it is to take a value less than 0. Also, according to property 3, it is as likely to take a value in the interval between $-a$ and 0 as it is to take a value in the interval between 0 and a. In this sense the value 0 is special. There are many applications involving random variables in which the special value for the random variable is not 0, but is some other number. For instance, if we consider a random variable X which associates with a record turntable its actual speed (see Example 4.13), then the speed 33 rpm plays the same role for X as 0 does for Z. In this case X is related to Z in a simple way. To explore this relation, next we discuss one way in which a new random variable can be defined in terms of a given one.

Suppose we are given a random variable X and numbers μ and σ, $\sigma > 0$. (Later we will assume that μ and σ are the mean and standard deviation of X, respectively, but for the moment this is unimportant.) We define a new random variable Y by the formula

$$Y = \frac{X - \mu}{\sigma}$$

That is, associated with each value x of X there is a value $(x - \mu)/\sigma$ of Y. If X takes values in an interval with endpoints a and b, $a < b$, then we can determine the interval in which Y takes values. Indeed, $a \leq X \leq b$ is equivalent to $a \leq X$ and $X \leq b$. Subtracting μ from both inequalities and dividing by σ, we have (since $\sigma > 0$)

$$\frac{a - \mu}{\sigma} \leq \frac{X - \mu}{\sigma} \quad \text{and} \quad \frac{X - \mu}{\sigma} \leq \frac{b - \mu}{\sigma}$$

These two inequalities can be combined into

$$\frac{a - \mu}{\sigma} \leq \frac{X - \mu}{\sigma} \leq \frac{b - \mu}{\sigma} \quad \text{or} \quad \frac{a - \mu}{\sigma} \leq Y \leq \frac{b - \mu}{\sigma} \tag{4.9}$$

We conclude that if X takes values in an interval with endpoints a and b, then Y takes values in an interval with endpoints $(a - \mu)/\sigma$ and $(b - \mu)/\sigma$.

Example 4.17

Psychology

A study of reaction times is conducted by repeating an experiment several times with each subject. Let a random variable X be defined by assigning a time (in seconds) to each repetition of the experiment.

Problem Suppose X takes values between 1.28 and 2.54, inclusive. Find the corresponding interval in which the random variable $Y = (X - 1.7)/.42$ takes values.

Solution The inequality $1.28 \leq X \leq 2.54$ is the same as $X \geq 1.28$ and $X \leq 2.54$. Therefore, for the quantity $X - 1.7$, we have

$$X - 1.7 \geq -.42 \quad \text{and} \quad X - 1.7 \leq .84$$

Finally, for the quantity $(X - 1.7)/.42$, we have

$$\frac{X - 1.7}{.42} \geq -1 \quad \text{and} \quad \frac{X - 1.7}{.42} \leq 2$$

Therefore, the random variable Y takes values between -1 and 2. ■

A random variable X is said to be *normal* if there are numbers μ and σ, $\sigma > 0$, such that the random variable Z defined by $Z = (X - \mu)/\sigma$ is the standard normal random variable. If X is a normal random variable, then there is only one pair of numbers μ, σ such that $(X - \mu)/\sigma$ is the standard normal random variable. The number μ is the *mean* of X, and σ is the *standard deviation* of X. The standard normal random variable has mean 0 and standard deviation 1.

Just as for the random variables studied in Sections 4.1 and 4.2, the constants μ and σ are measures of an "average value" of X and the dispersion of the values of X about this average, respectively.

The set of values of X described by $a \leq X \leq b$ is *exactly the same as the set of values of* $(X - \mu)/\sigma$ described by

$$\frac{a - \mu}{\sigma} \leq \frac{X - \mu}{\sigma} \leq \frac{b - \mu}{\sigma}$$

In fact, the corresponding probabilities of the random variables X and $Z = (X - \mu)/\sigma$ are equal.

Let X be a normal random variable with mean μ and standard deviation σ, and let Z be the standard normal random variable. For every a and b, with $a < b$, we have

$$\Pr[a \leq X \leq b] = \Pr\left[\frac{a - \mu}{\sigma} \leq Z \leq \frac{b - \mu}{\sigma}\right] \tag{4.10}$$

The relationship (4.10) and the probabilities in Table B.1 can be used to compute the probability that a normal random variable takes values in any interval. The technique is illustrated in the next example.

Example 4.18 Let X be a normal random variable with mean 5 and standard deviation 10.

Problem Find the following probabilities for X:
(*a*) $\Pr[5 \leq X \leq 15]$
(*b*) $\Pr[-10 \leq X \leq 10]$

Solution Since the random variable Z defined by $Z = (X - 5)/10$ is the standard normal random variable, it has the probabilities given in Table B.1.

(*a*) We are interested in values of X in the interval $5 \leq X \leq 15$. If X has values in this interval, $X - 5$ has values in the interval $0 \leq X - 5 \leq 10$, and $(X - 5)/10$ has values in $0 \leq (X - 5)/10 \leq 1$. Therefore, by Equation (4.10),

$$\Pr[5 \leq X \leq 15] = \Pr[0 \leq Z \leq 1]$$

where Z is the standard normal random variable. From Table B.1, $\Pr[0 \leq Z \leq 1] = .3413$, and consequently

$$\Pr[5 \leq X \leq 15] = .3413$$

(*b*) Proceeding as in part *a*, we have

$$\begin{aligned} \Pr[-10 \leq X \leq 10] &= \Pr[-1.5 \leq Z \leq .5] \\ &= \Pr[-1.5 \leq Z \leq 0] + \Pr[0 \leq Z \leq .5] \\ &= \Pr[0 \leq Z \leq 1.5] + \Pr[0 \leq Z \leq .5] \\ &= .4332 + .1915 = .6247 \end{aligned}$$

■

Example 4.19 Suppose that the reaction times described in Example 4.17 are given by a normal random variable X with mean 1.7 and standard deviation .42.

Psychology

Problem Find the probability that X takes values between 1.28 and 2.54.

Solution Using the results of Example 4.17, we have

$$\Pr[1.28 \leq X \leq 2.54] = \Pr[-1 \leq Z \leq 2]$$

The probability on the right-hand side can be written as

$$\Pr[-1 \leq Z \leq 0] + \Pr[0 \leq Z \leq 2]$$

or, by Equation (4.7), as

$$\Pr[0 \leq Z \leq 1] + \Pr[0 \leq Z \leq 2]$$

Using Table B.1 to evaluate the latter probabilities, we have

$$\begin{aligned} \Pr[0 \leq Z \leq 1] &= .3413 \\ \Pr[0 \leq Z \leq 2] &= .4772 \end{aligned}$$

and consequently,

$$\Pr[1.28 \leq X \leq 2.54] = .3413 + .4772 = .8185$$

■

The techniques introduced in this section can also be used to solve problems which are in a sense the reverse of those considered previously. That is, if you know a certain probability for a normal random variable, it may be possible to deduce information about the mean or standard deviation. In turn, this additional information can be used to solve other problems.

Example 4.20

Product testing

The thermometers produced at Accutemp Corporation are tested for accuracy at 40°C. The testing is done on an automatic testing device which rejects any thermometer whose registered temperature differs from 40°C by more than 1°C. After efforts at quality control are intensified, the rejection rate is 17.7 percent. That is, with probability .823 a randomly selected thermometer will show a temperature between 39 and 41°C when tested.

Problem Assume that the manufacturing process produces thermometers whose registered temperature at 40°C is a normal random variable X with mean 40 (in degrees Celsius) and standard deviation σ.

(*a*) Find the standard deviation σ.

(*b*) Find the probability that when a randomly selected thermometer is tested it will show a temperature which differs from 40°C by more than 2°C.

Solution (*a*) Let σ denote the standard deviation of X. Then $Z = (X - 40)/\sigma$ is the standard normal random variable. Since we know $\Pr[39 \leq X \leq 41]$, we need to find the interval of values of Z which correspond to the values of X in the interval from 39 to 41.

If X takes the value 39, then Z takes the value $(39 - 40)/\sigma = -1/\sigma$.

If X takes the value 41, then Z takes the value $(41 - 40)/\sigma = 1/\sigma$.

Now, using Equation (4.10), we have

$$\Pr\left[\frac{-1}{\sigma} \leq Z \leq \frac{1}{\sigma}\right] = \Pr[39 \leq X \leq 41] = .8230$$

Taking advantage of the symmetry of Z (property 3), we have

$$\Pr\left[\frac{-1}{\sigma} \leq Z \leq \frac{1}{\sigma}\right] = 2\Pr\left[0 \leq Z \leq \frac{1}{\sigma}\right]$$

so

$$2\Pr\left[0 \leq Z \leq \frac{1}{\sigma}\right] = .8230$$

and

$$\Pr\left[0 \leq Z \leq \frac{1}{\sigma}\right] = .4115$$

To complete the problem, we use Table B.1. This time, however, we know the probability .4115, and we seek the interval. Since

$$.4115 = \Pr[0 \le Z \le 1.35]$$

we have $1/\sigma = 1.35$. Therefore, $\sigma = 1/1.35 = .7407$. The standard deviation of X is .7407.

(*b*) Now that we know σ, we can solve this part. Note that part *b* is likely to be the question of real interest, but it cannot be answered until we know σ. Restated in a slightly different form, (*b*) is the following: X is a normal random variable with mean 40 and standard deviation .7407; find the probability that X takes a value less than 38 or larger than 42. Since we have

$$\Pr[X < 38] + \Pr[38 \le X \le 42] + \Pr[X > 42] = 1$$

the probability we seek

$$\Pr[X < 38] + \Pr[X > 42]$$

is equal to

$$1 - \Pr[38 \le X \le 42]$$

We turn now to the computation of $\Pr[38 \le X \le 42]$. The random variable $Z = (X - 40)/.7407$ is the standard normal random variable, and the inequality $38 \le X \le 42$ for X becomes the inequality $-2.700 \le Z \le 2.700$ for Z. Thus

$$\Pr[38 \le X \le 42] = \Pr[-2.70 \le Z \le 2.70]$$

and by property 3 of the standard normal random variable,

$$\Pr[38 \le X \le 42] = 2\Pr[0 \le Z \le 2.70]$$

Finally, using Table B.1, we have

$$\Pr[38 \le X \le 42] = 2(.4965) = .993$$

Thus the desired probability is

$$\Pr[X < 38] + \Pr[X > 42] = 1 - .993 = .007$$

Less than 1 percent of the thermometers will show testing errors of more than 2°C. ■

In Section 4.2 we saw that two random variables with the same mean can have quite different values and density functions. We now consider a related question. Namely, given the mean and the standard deviation of a random variable, what can be said about the values and the density function of the random variable? In particular, what is the probability that the random variable will assume a value within 1 standard deviation of the mean? Two standard deviations? Three? We answer the question for normal random variables, and then we indicate how to estimate the answer for many binomial random variables.

The following result for a normal random variable can be obtained by using the fact that $(X - \mu)/\sigma$ is a standard normal random variable, together with Table B.1.

If X is a normal random variable with mean μ and standard deviation σ, then

$$\begin{aligned} \Pr[\mu - \sigma \leq X \leq \mu + \sigma] &= .6826 \\ \Pr[\mu - 2\sigma \leq X \leq \mu + 2\sigma] &= .9554 \\ \Pr[\mu - 3\sigma \leq X \leq \mu + 3\sigma] &= .9974 \end{aligned} \tag{4.11}$$

The relation between these probabilities and areas under a normal curve is shown in Figure 4.7.

Both Equations (4.11) and Figure 4.7 illustrate the fact that about 68 percent of the time (actually 68.26 percent) a normal random variable takes a value within 1 standard deviation of the mean, 95 percent of the time within 2 standard deviations, and over 99 percent of the time within 3 standard deviations.

Example 4.21

Anthropology

An experiment consists of randomly selecting a male student at Gigantic State University and measuring his height. A random variable X is defined by associating with each student his height. Suppose that X is a normal random variable with mean 5 feet 10 inches and standard deviation 2 inches.

Problem Find the percentage of male students with heights between 5 feet (ft) 6 inches (in) and 6 feet 2 inches.

Solution We know that for the random variable X, $\sigma = 2$ inches. Since

$$5 \text{ ft } 10 \text{ in} - 2(2 \text{ in}) = 5 \text{ ft } 6 \text{ in}$$

and

$$5 \text{ ft } 10 \text{ in} + 2(2 \text{ in}) = 6 \text{ ft } 2 \text{ in}$$

we have

$$\begin{aligned} \Pr[5 \text{ ft } 6 \text{ in} \leq X \leq 6 \text{ ft } 2 \text{ in}] &= \Pr[\mu - 2\sigma \leq X \leq \mu + 2\sigma] \\ &= .95 \end{aligned}$$

We conclude that 95 percent of males are between 5 feet 6 inches and 6 feet 2 inches. ■

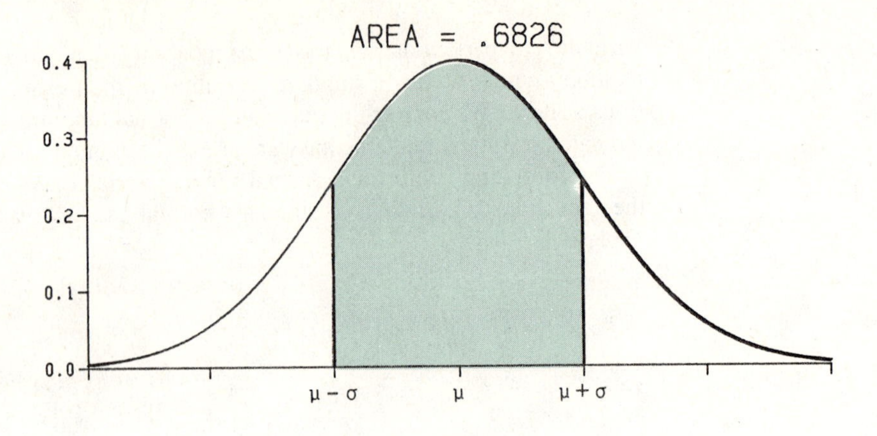

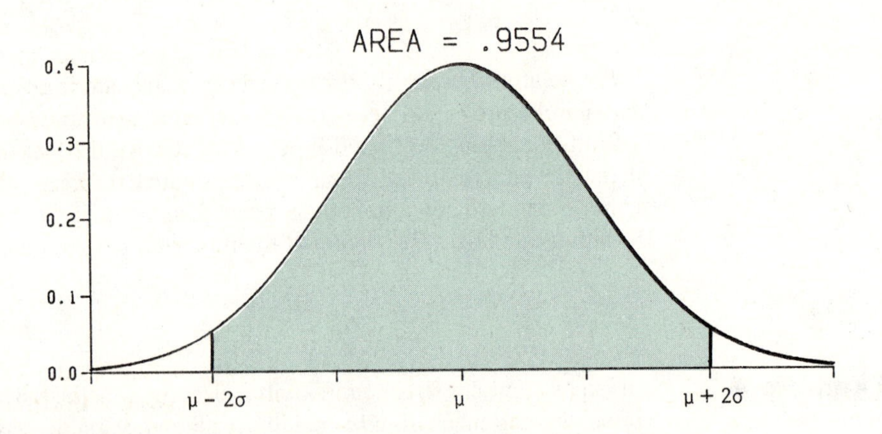

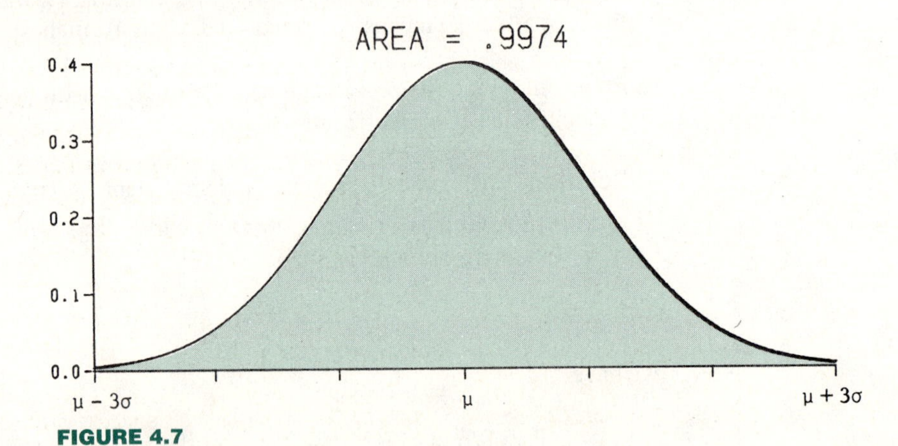

FIGURE 4.7

The information provided by Equations (4.11) is information about normal random variables, and in general it cannot be expected to hold for other random variables. However, there are special classes of random variables (in particular many binomial random variables) which have density curves similar to those of normal random variables. In such cases, Equations (4.11) and the probabilities of Appendix B can be used. This is a very significant advantage since the task of determining the probability that a random variable takes values within 1, 2, or 3 standard deviations of the mean is, in general, very difficult. For example, consider the computations involved in finding $\Pr[\mu - \sigma \leq X \leq \mu + \sigma]$, where X is a binomial random variable with $n = 1000$ and $p = .4$. For this random variable $\mu = np = 400$ and $\sigma = \sqrt{np(1-p)} = \sqrt{(1000)(.4)(.6)} = 15.49$ (with accuracy to two decimal places). Thus

$$\begin{aligned}\Pr[\mu - \sigma \leq X \leq \mu + \sigma] &= \Pr[384.51 \leq X \leq 415.49] \\ &= \Pr[X = 385] + \Pr[X = 386] \\ &\quad + \cdots + \Pr[X = 414] + \Pr[X = 415]\end{aligned}$$

Each entry in this sum involves a very large number of multiplications. Even with the use of a large computer, complex algebraic computations of this sort can result in inaccurate results, a consequence of round-off errors. Thus a simpler method of computing $\Pr[\mu - \sigma \leq X \leq \mu + \sigma]$ and similar probabilities would be very helpful. Such a method is our next topic.

The Normal Approximation to a Binomial Random Variable

In Section 3.5 we encountered problems similar to the following: A Bernoulli process consists of 100 trials with $p = .2$. What is the probability that there are at least 20 successes and at least 20 failures? The only difference between this problem and the ones we solved in Section 3.5 is the number of outcomes in the event we are to consider. However, this is an important difference. Solving the problem posed above with the techniques of Section 3.5 is not realistic, even if a calculator is used to aid in doing the arithmetic. We now develop a technique using normal random variables and Table B.1 in Appendix B to solve many probability problems involving binomial random variables.

In our discussion of normal random variables, we introduced a connection between probabilities and areas under a curve. The same idea can be carried over to binomial random variables.

Example 4.22 A Bernoulli process consists of 10 trials with success probability .4. Let X be the binomial random variable for the experiment. That is, X counts the number of successes which occur.

Problem Compute the values of the probability function for X, and represent these probabilities as areas under a curve.

TABLE 4.12

x	$\Pr[X = x]$
0	.0060
1	.0403
2	.1209
3	.2150
4	.2508
5	.2007
6	.1115
7	.0425
8	.0106
9	.0016
10	.0001

Solution Using $\Pr[X = r] = C(10, r)(.4)^r(.6)^{10-r}$, $r = 0, 1, 2, \ldots, 10$, we obtain the probabilities shown in Table 4.12. Next, on a horizontal axis mark off equally spaced points labeled 0, 1, 2, . . . , 10. Also mark the points $-.5$, .5, 1.5, . . . , 9.5, and 10.5. Using the segment with endpoints $-.5$ and .5 as a base, construct a rectangle with height $\Pr[X = 0]$. Note that the point 0 is the midpoint of the base of this rectangle. Similarly, using the segment with endpoints .5 and 1.5 as a base, construct a rectangle with height $\Pr[X = 1]$. Continue the process until 11 rectangles have been constructed. The result is shown in Figure 4.8.

In Figure 4.8 note that each value x_j of X is the center of the base of one rectangle. Since the length of the base of each rectangle is 1, the area of the rectangle is $1 \cdot \Pr[X = x_j] = p_j$. The rectangle corresponding to $x_j = 4$ is shaded in Figure 4.8. Each rectangle has area equal to the probability that X takes the value of the midpoint of the base. That is, $\Pr[X = k]$ is the area of the rectangle with base from $k - .5$ to $k + .5$. If we view the polygonal line which bounds the collection of 11 rectangles from above as a "curve," then again we have probabilities as areas under a curve. ■

It is worthwhile to compare Figure 4.8 with the curve corresponding to a normal random variable with the same mean and standard deviation. The mean of the binomial random variable X with $p = .4$ and $n = 10$ is 4, and its standard deviation is $\sqrt{10(.4)(.6)} = \sqrt{2.4}$. The curve for the normal random variable with $\mu = 4$ and $\sigma = \sqrt{2.4}$ is shown in Figure 4.9, where it is superimposed on Figure 4.8. Sometimes one curve is higher, sometimes the other. However, it is not the heights of the curves at any single point which are of interest to us; instead it is the area under the curves between points such as 1.5 and 2.5, which are endpoints for the base of one of the rectangles. These areas represent the val-

FIGURE 4.8

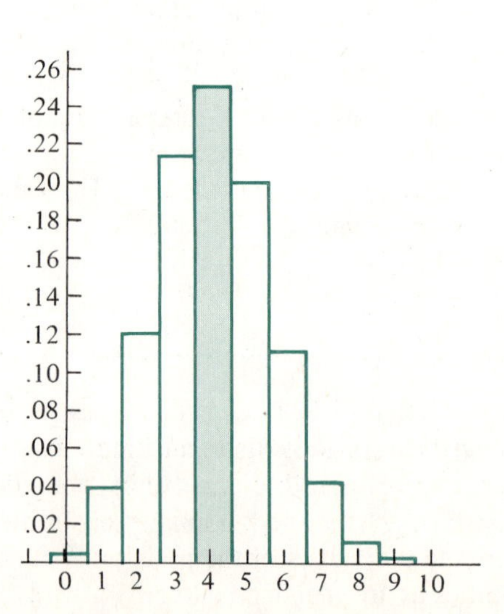

FIGURE 4.9

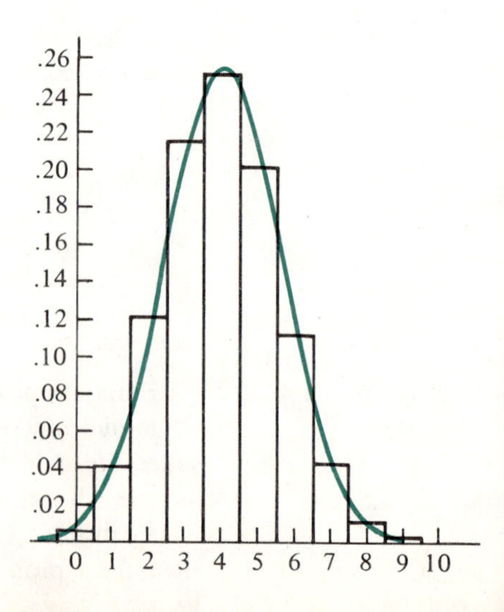

ues of the binomial probability function, and these areas are often quite close in value to the areas under the normal curve between the same points. For example, using Appendix B, we can obtain the areas under the normal curve with mean 4 and standard deviation $\sqrt{2.4}$. The areas under the curve and between values of the form $k - .5$ to $k + .5$ are shown in Table 4.13, where they are compared with the areas under the curve corresponding to the binomial random variable. The values in Table 4.13 indicate that for some intervals the areas are very close, while others (interval 2.5 to 3.5, for example) are not quite so close.

Whether probabilities computed by using a normal curve are acceptable approximations for binomial probabilities depends, of course, on your definition of acceptability. Approximations which are acceptable for one purpose may well be unacceptable for another purpose. However, there are commonly agreed upon notions of acceptability and corresponding criteria involving n and p for deciding whether an approximation can be used. We use the following simple condition.

The normal approximation to the binomial may be used to compute probabilities for a Bernoulli process consisting of n trials with success probability p whenever

$$np > 5 \quad \text{and} \quad n(1 - p) > 5 \tag{4.12}$$

TABLE 4.13

Interval	Area under Binomial Curve	Area under Normal Curve
(−.5, +.5)	.0060	.0101
(.5, 1.5)	.0403	.0390
(1.5, 2.5)	.1209	.1156
(2.5, 3.5)	.2150	.2032
(3.5, 4.5)	.2508	.2606
(4.5, 5.5)	.2007	.2032
(5.5, 6.5)	.1115	.1156
(6.5, 7.5)	.0425	.0390
(7.5, 8.5)	.0106	.0101
(8.5, 9.5)	.0016	.0016
(9.5, 10.5)	.0001	.0002

This criterion provides a means of distinguishing those values of n and p for which the approximate probabilities computed by using the normal approximation are sufficiently close to the true binomial probabilities to be used in many applications.

If X is a binomial random variable with n and p for which condition (4.12) holds, then we say that the normal approximation to X is *valid*.

Our next task is to describe the approximation method. Suppose we have a binomial random variable X for which the normal approximation is valid. Let C be the curve (consisting of horizontal and vertical line segments) constructed for X as described in Example 4.22. Using the technique introduced in that example, we have for any integers k and l, $0 \le k \le l \le n$,

$$\Pr[k \le X \le l] = \text{area under } C \text{ from } k - .5 \text{ to } l + .5$$

This follows from the fact that

$$\Pr[k \le X \le l] = \Pr[X = k] + \Pr[X = k + 1] + \cdots + \Pr[X = l]$$

and

$$\begin{aligned} \Pr[X = k] &= \text{area under } C \text{ from } k - .5 \text{ to } k + .5 \\ \Pr[X = k + 1] &= \text{area under } C \text{ from } k + .5 \text{ to } k + 1.5 \\ &\vdots \\ \Pr[X = l] &= \text{area under } C \text{ from } l - .5 \text{ to } l + .5 \end{aligned}$$

We now approximate the area under C by the area under a normal curve. The normal random variable we use here must have the same mean $(=np)$ and standard deviation $\left[= \sqrt{np(1 - p)}\,\right]$ as X. Finally, we use the method developed earlier in this section to compute the area under the normal curve from $k - .5$ to $l + .5$. These steps can be summarized as follows.

Approximation Method

Let X be a binomial random variable for which the normal approximation is valid. To find an approximate value for $\Pr[k \le X \le l]$ for any k and l, $0 \le k \le l \le n$, determine

$$\Pr[k - .5 \le Y \le l + .5] \tag{4.13}$$

where Y is a normal random variable with the same mean and standard deviation as X.

We illustrate the method with examples.

Example 4.23 A Bernoulli process consists of 100 trials with success probability $p = .2$.

Problem Find the probability that the binomial random variable takes a value between 25 and 29, inclusive.

Solution Let X denote the binomial random variable with $n = 100$ and $p = .2$. The mean and standard deviation of X are $np = 20$ and $\sqrt{np(1-p)} = \sqrt{16} = 4$, respectively. Let Y be the normal random variable with mean 20 and standard deviation 4. Since $np = 100(.2) = 20 > 5$ and $n(1-p) = 100(.8) = 80 > 5$, condition (4.12) is fulfilled and the normal approximation is valid.

The problem is to find $\Pr[25 \le X \le 29]$. Using Equation (4.13), we see that our task reduces to finding

$$\Pr[25 - .5 \le Y \le 29 + .5]$$

or

$$\Pr[24.5 \le Y \le 29.5]$$

This latter probability can be evaluated by using the method described earlier. Since Y has mean 20 and standard deviation 4, the random variable defined by

$$Z = \frac{Y - 20}{4}$$

is the standard normal random variable. The inequality

$$24.5 \le Y \le 29.5$$

for Y can be transformed into an inequality on Z by using the techniques introduced above. Indeed, subtracting $\mu = 20$ from all parts of the inequality, we have

$$24.5 - 20 \le Y - 20 \le 29.5 - 20$$

or

$$4.5 \le Y - 20 \le 9.5$$

Next, dividing all parts of the inequality by $\sigma = 4$, we obtain

$$\frac{4.5}{4} \le \frac{Y - 20}{4} \le \frac{9.5}{4}$$

In terms of Z this inequality is

$$1.12 \le Z \le 2.38$$

to two-decimal-place accuracy. We have

$$\Pr[24.5 \leq Y \leq 29.5] = \Pr[1.12 \leq Z \leq 2.38]$$

The probability on the right-hand side of this equation can be evaluated by using Appendix B. Since

$$\Pr[1.12 \leq Z \leq 2.38] = \Pr[0 \leq Z \leq 2.38] - \Pr[0 \leq Z \leq 1.12]$$
$$\Pr[0 \leq Z \leq 1.12] = .3686 \qquad \text{and} \qquad \Pr[0 \leq Z \leq 2.38] = .4913$$

it follows that

$$\Pr[1.12 \leq Z \leq 2.38] = .4913 - .3686 = .1227$$

The answer to our original problem, namely, the probability that X takes a value between 25 and 29, inclusive, is approximately .1227. Increased accuracy in the determination of $\Pr[1.12 \leq Z \leq 2.38]$ can be obtained by retaining three decimal places in the values of the normal random variables and using interpolation in Table B.1. Remember, however, that the approximation method imposes a limitation to the accuracy which can be obtained in determining $\Pr[25 \leq X \leq 29]$. ■

Example 4.24

Opinion surveys

A survey indicates that 40 percent of the voters in Metroburg support a bond issue. At random, 150 different voters are sampled and asked questions about the bond issue. Assume that the population of Metroburg is large enough that the sampling can be viewed as a Bernoulli process.

Problem Find the probability that *at least* 45 of those sampled support the bond issue.

Solution This is a Bernoulli process consisting of 150 trials with probability of support $p = .4$. Since $np = 150(.4) = 60 > 5$ and $n(1 - p) = 150(.6) = 90 > 5$, the normal approximation is valid.

Let X denote the random variable which associates with each sample of 150 voters the number who support the bond issue. The mean of X is $np = 150(.4) = 60$, and the standard deviation is $\sqrt{np(1-p)} = \sqrt{36} = 6$. We are to determine $\Pr[X \geq 45]$. Using our approximation method with Y a normal random variable with mean 60 and standard deviation 6, we seek $\Pr[Y \geq 44.5]$. Subtracting 60 from both sides of this inequality, we have

$$Y - 60 \geq -15.5$$

and dividing by 6 gives

$$\frac{Y - 60}{6} \geq \frac{-15.5}{6} = -2.5833$$

Now, since $Z = (Y - 60)/6$ is a standard normal random variable, we determine $\Pr[Y \geq 44.5]$ by using the techniques developed earlier.

$$\Pr[Y \geq 44.5] = \Pr[Z \geq -2.5833]$$

and

$$\begin{aligned}\Pr[Z \geq -2.5833] &= \Pr[0 \leq Z \leq 2.5833] + \Pr[Z \geq 0] \\ &= .4951 + .5000 \\ &= .9951\end{aligned}$$

Therefore the probability that at least 45 of the 150 voters surveyed support the bond issues is approximately .9951. In qualitative terms, it is nearly certain that at least 45 of those surveyed support the bond issue. ■

In the last two examples we have provided many of the details in illustrating the use of the normal approximation method of determining binomial probabilities. We did so to provide some insight into the method. In practice, however, it is common to use an abbreviated version of the method. This abbreviated version can be summarized as follows:

Let X be a binomial random variable with parameters n (number of trials) and p (success probability), and suppose $np > 5$ and $n(1 - p) > 5$ so that the normal approximation is valid. Then

$$\Pr[k \leq X \leq l] \simeq \Pr\left[\frac{k - .5 - np}{\sqrt{np(1-p)}} \leq Z \leq \frac{l + .5 - np}{\sqrt{np(1-p)}}\right] \tag{4.14}$$

where $\simeq$ means "is approximately equal to" and Z is the standard normal random variable.

Two comments are in order. First, probabilities involving the standard normal random variable can be computed by using Appendix B. Second, it may happen that instead of the probability of a value being between k and l, inclusive, we are interested in the probability of either being less than or equal to l or being greater than or equal to k. In such a case Equation (4.14) becomes

$$\Pr[X \leq l] \simeq \Pr\left[Z \leq \frac{l + .5 - np}{\sqrt{np(1-p)}}\right] \tag{4.15}$$

or

$$\Pr[k \leq X] \simeq \Pr\left[\frac{k - .5 - np}{\sqrt{np(1-p)}} \leq Z\right] \tag{4.16}$$

respectively. We employ the abbreviated method, and in fact formula (4.15) in our last example.

Example 4.25

Educational testing

Data collected by the Admissions Testing Program in 1994 show that about 90 percent of college-bound high school seniors have taken at least 4 years of high school English. Dean Jones of Gigantic State University suspects that the percentage at GSU is higher than this. At her request, the admissions office at GSU reviews 200 randomly selected student records and finds that 15 entering freshmen have taken fewer than 4 years of high school English.

Problem Assume that the data collected by the Admissions Testing Program accurately reflect the amount of English taken by GSU students. Find the probability that a random sample of 200 students will contain 15 or fewer students with less than 4 years of high school English.

Solution Consider the review to be a Bernoulli process with $n = 200$ and $p = .1$. That is, the selection of a student who has *fewer* than 4 years of English is viewed as a success. Let X denote the associated binomial random variable. The mean number of successes is $\mu = np = 20$, and the standard deviation is $\sigma = \sqrt{np(1-p)} = \sqrt{18} = 4.24$ to two decimal places. Since $np = 20 > 5$ and $n(1-p) = 180 > 5$, the normal approximation is valid.

We wish to compute $\Pr[X \leq 15]$. Using Equation (4.15), we find that the probability we seek is approximately equal to

$$\Pr\left[Z \leq \frac{15 + .5 - 20}{4.24}\right] = \Pr[Z \leq -1.06]$$

Using Appendix B, we find

$$\begin{aligned}\Pr[Z \leq -1.06] &= \Pr[Z \geq 1.06] \\ &= .5 - \Pr[0 \leq Z \leq 1.06] \\ &= .5 - .3554 \\ &= .1446\end{aligned}$$

We conclude that if the data accurately describe GSU students, then with probability approximately equal to .1446 there will be 15 or fewer students with fewer than 4 years of high school English in a random sample of 200 entering freshmen. That is, assuming that the statement of the Admissions Testing Program is true, then the results of the survey at GSU are rather unlikely. Perhaps Dean Jones has a point. ■

Exercises for Section 4.3

In Exercises 1 through 4, Z denotes the standard normal random variable.

1. Find (*a*) $\Pr[0 \leq Z \leq 1]$ (*b*) $\Pr[-1 \leq Z \leq 0]$.
2. Find (*a*) $\Pr[-2 \leq Z \leq 1]$ (*b*) $\Pr[-1 \leq Z \leq 2]$.
3. Find (*a*) $\Pr[0 \leq Z \leq .7]$ (*b*) $\Pr[-.3 \leq Z \leq 0]$.
4. Find (*a*) $\Pr[-.3 \leq Z \leq 1.5]$ (*b*) $\Pr[-.8 \leq Z \leq .8]$.

5. Let X be a random variable which takes values between 2 and 8. A random variable Y is defined by the formula $Y = (X - 2)/2$. Find numbers a and b such that Y takes values between a and b.
6. Let X be a random variable which takes values larger than 10. A random variable Y is defined by the formula $Y = X + 20$. What can you say about the values of Y?
7. Let X be a normal random variable with $\mu = 5$ and $\sigma = 2$. Find
(*a*) $\Pr[5 \leq X \leq 8]$ (*b*) $\Pr[6 \leq X \leq 8]$.
8. Let X be as in Exercise 7. Find
(*a*) $\Pr[X \geq 0]$ (*b*) $\Pr[X \leq 10]$.
9. Let X be a normal random variable with mean -2 and standard deviation 2. Find the probability that X takes values between -1 and 3.
10. Consider a Bernoulli process with $n = 6$ and $p = .5$.
(*a*) Find the density function of the random variable X which associates with each outcome the number of successes. Give your answer in a table like Table 4.12.
(*b*) Represent these probabilities as areas under a curve, as in Figure 4.8.
11. Repeat Exercise 10 with $n = 8$ and $p = .5$.
12. Construct a table similar to Table 4.13 for a Bernoulli process with $n = 10$ and $p = .2$. Is the normal approximation in this case better or worse than in Table 4.13?
13. For each of the following pairs n, p, let X be the binomial random variable for the Bernoulli process with that n and p. In which cases is the normal approximation to X valid?
(*a*) $n = 30, p = .1$ (*b*) $n = 50, p = .2$
(*c*) $n = 30, p = .3$ (*d*) $n = 50, p = .7$
14. Repeat Exercise 13 with the following:
(*a*) $n = 12, p = .4$ (*b*) $n = 14, p = .4$
(*c*) $n = 16, p = .3$ (*d*) $n = 18, p = .3$
15. A Bernoulli process consists of 160 trials with $p = .25$. Use the normal approximation to the binomial to approximate the probability that the number of successes X satisfies $35 \leq X \leq 45$.

Product performance

16. Memory chips by Ultrachip Inc. have a mean time to failure of 10,000 hours and a standard deviation of 1000 hours. If time to failure is a normal random variable, what is the probability that a randomly selected chip will last at least 8000 hours?

Biology

17. Suppose that the weight (in ounces) of perch fry grown in a fish hatchery is a normal random variable X with mean 2 (in ounces). Which is larger, $\Pr[X \geq 3.5]$ or $\Pr[X \leq 1]$?

Production

18. Suppose that a certain widget-making machine produces widgets in such a way that the length of the widget is a normal random variable with $\mu = 6$ inches and $\sigma = .3$ inch. What is the probability that a randomly selected widget will have a length between 5.4 and 6.6 inches?

Psychology

19. An experiment in a psychology laboratory is designed to measure reaction times. Suppose that for a specific group of subjects the reaction times are normally distributed with mean .2 (in seconds). If half the subjects had reaction times in the interval from .16 to .24 second, find the standard deviation of the reaction times.

Product performance

20. The Ultra-Fi reel-to-reel tape recorder is designed to operate at a tape speed of 1.75 inches per second. Suppose that the tape speeds of recorders coming off an assembly line are described by a normal random variable with mean 1.75 and standard deviation .25. What is the probability that a randomly selected recorder has a tape speed which differs from the design speed by less than .1 inch per second?

Meteorology

21. Weather records for Metroburg lead the local weather forecaster to suggest that the high temperature for May 15 is a normal random variable with mean 22°C and standard deviation 5°C. Using this information, find the probability that next May 15 the high temperature will be between 20 and 25°C.

22. A bag with 1600 fair pennies is dumped onto a table. Use the normal approximation to the binomial to estimate the following probabilities.
    (*a*) The number of heads is between 760 and 840.
    (*b*) The number of heads is at least 780.

Anthropology

23. Suppose that the average height of 1000 senior male students is 5 feet 10 inches and the standard deviation is 3 inches. Suppose also that the students' heights closely follow a normal distribution. How many of these students would you expect to be 6 feet 2 inches or taller? How many 5 feet 4 inches or shorter?

Educational testing

24. On an examination in an economics class, the mean score was 70 and the standard deviation was 10. Suppose that the scores were approximately normally distributed and grades were assigned as follows: over 88 = A, from 75 to 88 = B, from 60 to 74 = C, from 48 to 59 = D, below 48 = F. What percentage of the students received each grade? (*Hint*: Represent the probability of a specific score as an area under a curve.)

Surveys

25. A recent survey showed that about one-quarter of all workers were employed in jobs that they had found in the last 12 months. What is the probability that in a random sample of 300 workers over 100 of them are working in jobs they have held for over a year?

Psychology

26. Eighty percent of the students at GSU have good depth perception. If 100 GSU students are selected at random, what is the probability that between 70 and 90, inclusive, have good depth perception?

Product testing

27. Shortshocks Electronics, Inc., manufactures hand calculators. Company records indicate that even when the machinery is operating normally, defective circuit boards are produced at random with a frequency of about 1 in 5. The company selects 100 circuit boards per day and tests them. If 30 or more defective boards are found, the production process is halted while the machinery is adjusted. Find the probability that the company will shut down production even though the machinery is operating normally, i.e., is averaging only 20 percent defective circuit boards overall.

Product testing

28. In Exercise 27, suppose only 80 units are tested, and if the machinery is operating normally, then 1 in 4 is defective. Production is stopped when 25 defective boards are found. What is the probability that production is stopped even though the machinery is operating normally?

Careers

29. Ten percent of the students at a certain college are science majors. A reporter for the student paper would like to interview at least 5 science majors. If the reporter randomly selects 100 names from the student directory and begins contacting them, what is the probability that she will contact at least 5 science majors before she runs out of names?

Careers *30. In Exercise 29, what is the probability that the reporter will contact at least 5 science majors in the first 60 names? In the first 25 names?

IMPORTANT TERMS AND CONCEPTS

You should be able to describe, define, or give examples and use each of the following:

Random variable
Binomial random variable
Density function
Mean or expected value
Variance
Standard deviation
Standard normal curve
Standard normal random variable
Normal random variable
Normal approximation to a binomial random variable

REVIEW EXERCISES

1. A fair die is rolled, and a random variable X is defined to be -1 if the result is a 1, 2 if the result is a 2 or a 3, and 5 if the result is a 4, 5, or 6. Find the density function of X.
2. A fair coin is flipped 3 times, and a random variable is defined to be the number of heads plus twice the number of tails. Find the values assumed by X and the density function of X.
3. A box contains 3 red balls and 2 blue balls. Three balls are selected simultaneously and at random, and a random variable is defined to be the number of blue balls. Find the density function of X.
4. A box contains 3 red balls and 2 blue balls. Three balls are selected one after another without replacement, and the color of each ball is noted. A random variable is defined to be 5 if the second ball selected is blue and 10 if the second ball selected is red. Find the density function of X.
5. An experiment consists of flipping a fair coin repeatedly until there are either 2 consecutive tails or a total of 5 flips. A random variable X is defined as the number of times the coin is flipped.
(*a*) Find the values taken by X.
(*b*) Find the probability density function of X.

College life

6. Clarence attends class at random 50 percent of the time. The instructor takes attendance 3 times and awards 10 points to students who are there at least twice and no points otherwise. A random variable X is defined by associating with each attendance record the number of points awarded. Find the density function of the random variable X associated with Clarence.

Games of chance

7. A carnival game consists of selecting 3 balls simultaneously and at random from a box containing 5 red balls and 3 green balls. Each red ball selected pays \$.20, and each green ball pays \$.50. It costs \$1.00 to play the game. A random variable X is defined by associating with each play of the game the *net* payoff to the player. Find the density function of X.
8. A random variable X has values 5, 10, and 20 and associated probabilities .5, .3, and .2. Find the expected value of X.
9. A random selection is made from a collection of coins containing 4 nickels, 3 dimes, and 2 quarters. What is the expected value of the coin selected?

Sports

10. A basketball player makes 80 percent of her free throws, 50 percent of her 2-point field goal attempts, and 30 percent of her 3-point field goal attempts. In a typical game the player attempts 10 free throws, 12 two-point field goals, and 5 three-point field goals. If we assume that each type of shot is a Bernoulli process, find the expected number of points the player will score.

Biology

11. An aquarium contains 7 fish: 3 of the fish weigh 200 grams each, 2 weigh 150 grams each, and the remaining 2 weigh 100 grams each. A sample of 3 fish is selected at random. A random variable Y is defined by associating with each outcome the total weight of the 3 fish.
   (*a*) Find the density function of Y. (*b*) Find the mean of Y.
   (*c*) Find the variance and standard deviation of Y.

x	$\Pr[X = x]$
−1	.20
0	.35
1	.10
2	.15
4	.20

12. Let X be a random variable whose density function is given at the left.
   (*a*) Find the mean of X.
   (*b*) Find the variance of X.
   (*c*) Find the standard deviation of X.
   (*d*) Find the probability that X takes a value in the interval from $\mu - \sigma$ to $\mu + \sigma$.

Customer service

13. My morning paper is delivered late 40 percent of the time. Suppose deliveries can be described as a Bernoulli process, and let a random variable X be defined by counting the number of late deliveries in a year (365 days).
   (*a*) Find the mean of X.
   (*b*) Find the standard deviation of X.
14. Let X be a binomial random variable with $n = 100$ and $p = .25$. Find the mean and standard deviation of X.
15. Repeat Exercise 14 with $p = .05$.

Production

16. An automatic counter on a machine is supposed to reset itself each time the machine is turned off. However, it fails to do so at random 35 percent of the time. If it fails to reset, then it must be reset by hand. The machinery is turned off at the end of each day and only then. Assume the reset failures can be viewed as Bernoulli trials, and find how many days must pass before the expected number of hand resets is at least 10.
17. A binomial random variable X has $n = 100$ and $p = .6$. Another binomial random variable Y has $n = 150$.
   (*a*) What must be the value of p for Y so that the two random variables have the same mean?
   (*b*) Using this value of p for Y, decide which random variable has the larger variance.
18. For each of the following pairs n, p, check if the normal approximation to the binomial is valid.
   (*a*) $n = 20, p = .1$ (*b*) $n = 15, p = .7$
   (*c*) $n = 30, p = .2$ (*d*) $n = 25, p = .9$
19. Let X be a random variable, and let Y be defined by the formula $Y = (X - 100)/15$. If X takes values between 55 and 145, inclusive, find the corresponding interval in which Y takes values.
20. Let X be a normal random variable with a mean of 100 and a standard deviation of 15. Find:
   (*a*) $\Pr[70 \le X \le 110]$ (*b*) $\Pr[80 \le X \le 125]$

21. Let Z be the standard normal random variable. Find:
 (*a*) $\Pr[-.92 \le Z \le 1.35]$ (*b*) $\Pr[Z \le -.40]$
 (*c*) $\Pr[Z \le 1.05]$

22. Let X be a normal random variable with mean 10 and standard deviation 4. Find:
 (*a*) $\Pr[6 \le X \le 12]$ (*b*) $\Pr[X \ge 16]$
 (*c*) $\Pr[E]$, where E is the event on which X takes values less than 7 or greater than 15

23. Let Y be a binomial random variable with $n = 162$ and $p = \frac{1}{3}$.
 (*a*) Find the mean μ and standard deviation σ of Y.
 (*b*) Show that the normal approximation to Y is valid.
 (*c*) Use the normal approximation to evaluate $\Pr[45 \le Y \le 66]$.

Product testing 24. An inspector checks the seals on the doors of microwave ovens on an assembly line. Experience shows that doors with defective seals occur at random and with probability .2. How many doors must the inspector check so that the expected number of ovens with nondefective door seals is at least 1000? Assume that each oven is checked exactly once.

Sports 25. A major league baseball player has a batting average (number of hits divided by number of times at bat) of .200. If the player's turns at bat are viewed as Bernoulli trials with $p = .2$, what is the probability that in 200 times at bat he has at least 50 hits?

Production 26. One-fifth of the holes punched on a high-speed punching machine in an automobile plant are more than 1 millimeter out of place. If the positions of 400 randomly selected holes are checked, what is the probability that the number of holes more than 1 millimeter out of place is no more than 60?

Educational testing 27. A student takes a 100-question true-false test. A passing grade is 70 or more correct answers.
 (*a*) What is the probability that the student passes the test by guessing? When the student is guessing, the probability of a correct answer on any question is .5.
 (*b*) If the student studies and has probability .8 of answering each question correctly, what is the probability that she does *not* pass the test?

Product performance 28. A report on Wik cigarette lighters states that the average number of lights from a Wik lighter is 500 and the variance is 25. If the report is accurate, find a number M such that the following claim is legitimate: "Ninety-nine percent of the Wik lighters sold will give more than M lights."

Sales 29. A sales representative estimates that on each call there is a probability $\frac{1}{4}$ of making a sale. A bonus of \$100 is paid for the first and second sale each day, and a bonus of \$200 is paid for each additional sale. If it is possible to make 3 calls each day, what is the expected income from bonuses? What is the standard deviation of income from bonuses?

Games of chance 30. A carnival game has a tank of water containing 6 plastic fish. Each fish has a payoff marked on the bottom. Three of the fish are worth \$0.25, 2 are worth \$0.50, and 1 is worth \$2.00. It costs \$1.00 to play the game, and it is played as follows: A fish is selected at random. If it is worth \$0.50 or \$2.00, the game ends. If it is worth \$0.25, then another fish is selected (the first is not returned to the tank) and the game ends. You receive the total amount on the fish you select.

Find your expected *net* gain (the amount you receive minus the amount you pay) per play of the game.

31. A box contains 3 red balls and 2 blue balls. Three balls are selected one after another without replacement, and the color of each ball is noted. A random variable X is defined to be 5 if the first ball is blue and 0 otherwise; a random variable Y is defined to be the number of red balls drawn on the second and third selections. Find the expected value of the random variable $X + Y$.

Psychology 32. A cage of rats in a psychology laboratory contains 8 rats of which 3 are fully trained, 1 is partially trained, and the rest are untrained. A sample of 4 rats is selected at random. A random variable X is defined by associating with each outcome the number of fully trained rats.
(*a*) Find the density function of X. (*b*) Find the mean of X.
(*c*) Find the variance of X.

33. A biased coin with Pr[heads] = .6 is flipped 3 times, and the result of each flip is noted. A random variable X is defined to be 5 if there are exactly 2 heads, 20 if there are 3 heads, and 0 otherwise. Find the expected value and standard deviation of X.

Demographics 34. The current population of Metroburg is 120,000. Projections are that in 10 years the population will increase 10 percent with probability .6, it will increase 5 percent with probability .3, and it will decline 2 percent with probability .1. Find the expected population of Metroburg in 10 years.

Lotteries *35. A person plays a state lottery by selecting 5 numbers in the set of integers from 1 through 50, with no repetitions allowed. The selections are not ordered, so {10, 20, 30, 40, 50} and {50, 20, 30, 40, 10} are the same selection. The winning numbers are selected at random by a computer. In a given week, 500,000 people play this lottery, and we assume that each person selects 5 numbers at random. It costs $2 to play this lottery, and there is a prize of $1,000,000 if you match all 5 numbers. Find the expected gain to the state.

PART
2

Linear Models

CHAPTER

5

Systems of Linear Equations

5.0 THE SETTING AND OVERVIEW: LINEAR MODELS

The first step in most applications of mathematics involves determining an appropriate mathematical model of the system being studied. A mathematical model consists of both symbolic notation (i.e., symbols and an agreement as to what the symbols mean) and relations among the symbols. The relations we propose depend on what we are willing to assume about the situation being studied.

Many of the situations in which mathematical concepts and methods are applied in the management, social, and life sciences involve quantities which are assumed to be related to each other through one or more linear equations. To illustrate the ideas, suppose we have two quantities, and let them be denoted by x and y. A linear expression or linear function of the variables x and y is an expression of the form $Ax + By$, where A and B are numbers. We say that x and y satisfy a *linear equation* if there are numbers A, B, and C such that $Ax + By = C$. For instance, $4x - 3y$ is a linear expression, and $3x + 2y = 6$ is a linear equation. On the other hand, $4x^2 - 2y$ and $x + (3/y)$ are not linear expressions in variables x and y. Mathematical models in which the assumptions lead to linear expressions and linear equations are included in a class known as *linear models*. The term "linear" refers to the geometric concept of a line.

As an example, suppose we have a manufacturing plant which uses electricity in the manufacturing process as well as in the administrative office. Suppose that the office uses 6000 kilowatthours each month and production requires 400 kilowatthours for each unit of product produced. Then if x denotes the number of units of product produced in a month and y denotes the amount of electricity used in that month, then x and y are related by $y = 400x + 6000$. Since this is equivalent to $-400x + y = 6000$, which is of the form $Ax + By = C$ with $A = -400$, $B = 1$, and $C = 6000$, we see that x and y satisfy a linear equation.

Notice that we have made several implicit assumptions in this discussion. First, we are assuming that the amount of electricity used in the office does not depend on the amount of manufacturing done. This is reasonable if the level of activity in the office is approximately the same no matter how much product is produced. Such an assumption is likely to be legitimate over a range of manufacturing levels, but is unlikely to be valid for all levels. For instance, if the manufacturing level is very low for an extended period, then the office staff may be reduced and electricity use would decline. Also, if the manufacturing level increases beyond that which can be supported by the present office staff, then new staff would be added, and this would involve a new level of office electricity use. Also, we are assuming that the same amount of electricity is needed to produce one unit of product independent of the level of production. This may not always be a legitimate assumption. For example, the manufacturing process may use less electricity to produce a unit of product when the process is operating at one level than at other levels. Indeed, it is common for manufacturing processes to have so-called optimal operating levels which are more efficient than levels either above or below. Another implicit assumption is that we know exactly how much electricity will be used by the office and in the manufacturing process. In practice, it would be rare indeed to have no variation in these quantities from one month to another. The fact that we take them to be constants is an assumption that needs to be carefully considered in light of the data actually available. In this example the relation $y = 400x + 6000$ is a mathematical model because it represents an idealized situation, a situation using precise notation and described by specific assumptions.

Another situation in which linear equations are common involves linear, or straight-line, depreciation. Suppose that you have a building whose original value is \$250,000 and whose useful life is 30 years and whose salvage value, i.e., the value after its useful life is over, is \$10,000. The depreciable value of the building is \$250,000 − \$10,000 = \$240,000. By using linear depreciation, one of several options available in most cases, the depreciation which can be claimed each year is the depreciable value of the building divided by its useful life, \$240,000 divided by 30, or \$8000. It follows that if we let y denote the depreciated value of the building measured in dollars after x years, then y and x are related by $y = 250{,}000 - 8000x$.

As we shall see, it is frequently useful to find values of variables which satisfy several linear equations simultaneously. Each such equation describes a specific relation among the variables, and satisfying several equations means that several relations hold simultaneously.

5.1 REVIEW OF EQUATIONS AND GRAPHS OF LINES

The basic expressions used in this chapter are linear equations and their graphs. In this section we review briefly the concepts for expressions with two variables, and we introduce similar ideas for expressions with three variables in Section 5.3.

For expressions with two variables, we use the familiar cartesian coordinate system shown in Figure 5.1*a* to associate pairs of numbers with points in a plane. If P is the label of the point (x, y), then the numbers x and y are the *coordinates* of point P. For example, for the point $P = (2, 1)$, 2 is the x coordinate and 1 is the y coordinate. Several other examples are shown in Figure 5.1*b*. We are especially interested in pairs (x, y) where x and y satisfy a linear equation, as illustrated in the following example.

Example 5.1 *Sociology*

A sociologist studying the relationship between years of formal education and income level is looking for a simple formula to relate these two numbers. After looking at several data sets, she proposes that the average income level (in thousands of dollars) is related to the average number of years of formal education by the formula

$$\text{Average income level} = 8 + (\tfrac{3}{2})(\text{average number of years of formal education})$$

Problem Using the formula given above, graph the average income level for 6, 9, 12, 16, 18, and 21 years of formal education. Note that graduation from high school corresponds to 12 years, an undergraduate degree from college corresponds to 16 years, and a 5-year Ph.D. program corresponds to 21 years.

FIGURE 5.1

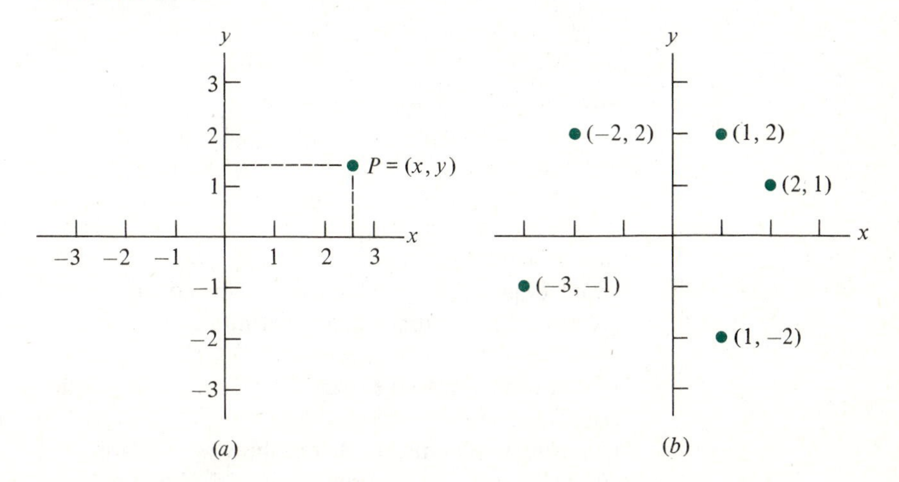

Solution We use a standard cartesian coordinate system, and we let

$$x = \text{number of years of formal education}$$
$$y = \text{income level in thousands of dollars}$$

Thus, for $x = 12$ years, the formula predicts an income level of $8 + (\frac{3}{2})(12) = 8 + 18 = 24$. The points representing the pairs (x, y), for $x = 6, 9, 12, 16, 18, 21$, are shown in Figure 5.2. ■

Note that the formula used in Example 5.1 ($y = 8 + 1.5x$) is an example of a linear model with one variable, in this case x, the number of years of formal education. This model may or may not make good predictions; it depends on the data used to create the model and on the current state of the economy. Also, the predictions involve averages (means), and thus the formula may not apply to individuals. There may be individuals with very little formal education and very high income and other individuals with lots of formal education and very low income, and yet the model (the equation) may be correct for averages.

The formula given in Example 5.1 is called a linear model because the graph of this equation is a line. To make our discussion precise, we define the term "line" in an algebraic way as a set of points which satisfy a certain type of equation. The geometric representation is the graph of this set on a cartesian coordinate system.

For any numbers A, B, and C with A and B not both zero, the set points

$$\{(x, y): \quad Ax + By = C\}$$

is a *line*. If (x_1, y_1) and (x_2, y_2) are two points on this line, i.e., if $Ax_1 + By_1 = C$ and $Ax_2 + By_2 = C$, then the line is referred to as the *line through* (x_1, y_1) *and* (x_2, y_2).

We emphasize that the point (x_1, y_1) lies on the line $Ax + By = C$ if $Ax_1 + By_1 = C$ and that the point does not lie on the line if $Ax_1 + By_1 \neq C$.

Example 5.2 The set of points (x, y) which satisfy the equation $x - 3y = 6$ is a line (take $A = 1, B = -3, C = 6$ in the above definition).

Problem Graph the line on a cartesian coordinate system.

Solution We recall that a line is determined by two points, and our approach to solving the problem is to find two points which lie on the line. That

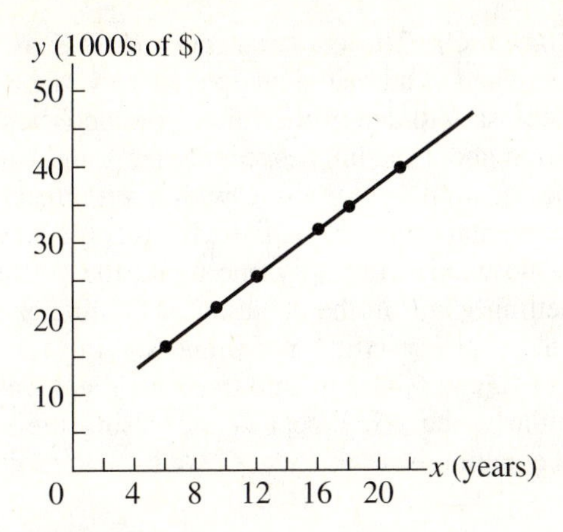

FIGURE 5.2

is, we find two points each of which satisfies the equation $x - 3y = 6$. To find points which satisfy the equation, we select a value of x and then find the value of y which satisfies $x - 3y = 6$. We have

$$y = \frac{x - 6}{3}$$

To illustrate the method, we take $x = 0$ and $x = 3$. For $x = 0$ we have $y = -2$, and for $x = 3$ we have $y = -1$. The points $(0, -2)$ and $(3, -1)$ and the line through them are shown on Figure 5.3. ■

It is customary to refer to the equation $Ax + By = C$ as a *line*, and we shall retain this custom where it is convenient. We mean, of course, the set of points

FIGURE 5.3

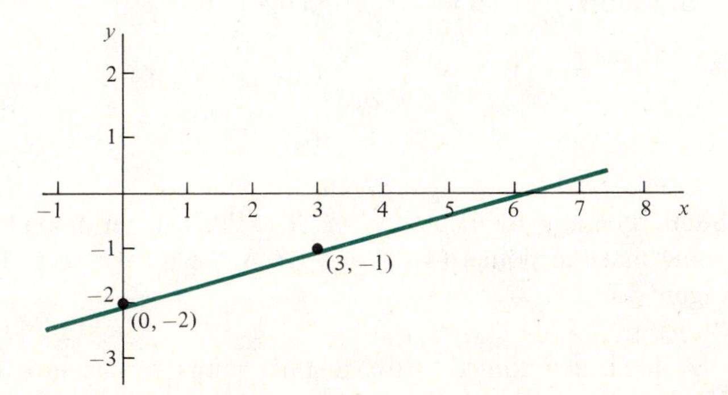

$\{(x, y): Ax + By = C\}$. The equation $Ax + By = C$ is known as the *general equation of a line* and as a linear equation in two variables.

It is often necessary to determine values for coefficients A, B, and C by using given information about the line, especially the x and y *intercepts* and the *slope*. If a line has the equation $Ax + By = C$ with $A \neq 0$, then it intersects the x axis in exactly one point, and that point is the x intercept. Likewise, if $B \neq 0$, then the line intersects the y axis in exactly one point, the y intercept. We obtain the x intercept by setting $y = 0$ in the equation of the line and solving for x, and we obtain the y intercept by setting $x = 0$ and solving for y. In Example 5.2, we have the line $x - 3y = 6$, and to find the x intercept we set $y = 0$. The x intercept is 6. Similarly, the y intercept is -2. Using the same method on a line whose general equation is $Ax + By = C$, we have the following result.

> The line $Ax + By = C$ has an x intercept if $A \neq 0$ and the x intercept is C/A. The line has a y intercept if $B \neq 0$, and the y intercept is C/B.

If $C = 0$ and both intercepts exist ($A \neq 0$ and $B \neq 0$), then both intercepts are 0. If $C \neq 0$ and $A = 0$, then the line does not intersect the x axis and therefore has no x intercept. If $C \neq 0$ and $B = 0$, then the line does not intersect the y axis and therefore has no y intercept.

Note The intercept of a line on an axis can be denoted by a single number, such as the x intercept is 2, or by the pair (2, 0) giving the point of intersection in the plane.

Example 5.3 Find the x and y intercepts of the line $3x + 2y = 12$, and graph the line.

Solution We set $y = 0$ to find the x intercept:

$$\begin{aligned} 3x + 2(0) &= 12 \\ 3x &= 12 \\ x &= 4 \end{aligned}$$

Similarly, the y intercept is $y = 6$. The line is determined by two points, and we know that the points (4, 0) and (0, 6) lie on the line. The line is shown in Figure 5.4. ■

A line is determined by two distinct points, for instance, (x_1, y_1) and (x_2, y_2) in Figure 5.5. It is also determined by one point and its inclination or slope.

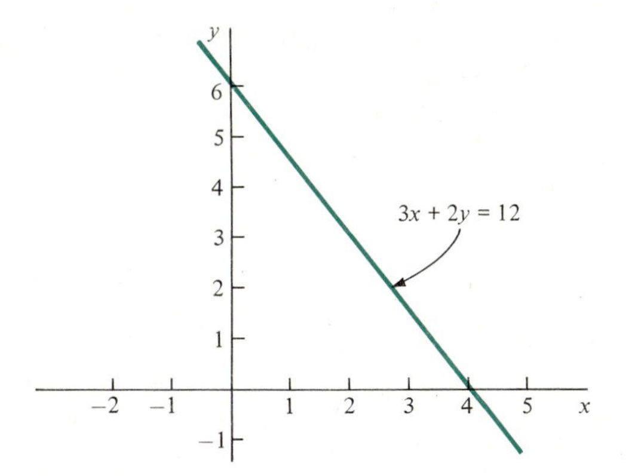

FIGURE 5.4

The *slope* of a line is the ratio of the difference of the y coordinates to the difference of the x coordinates of any two distinct points on the line, whenever the latter difference is not zero. That is, if (x_1, y_1) and (x_2, y_2) are two distinct points on a line, with $x_1 \neq x_2$, then the slope m of the line is

$$m = \frac{y_2 - y_1}{x_2 - x_1} \tag{5.1}$$

If $x_2 = x_1$ and $y_2 \neq y_1$, then the line is vertical and its slope is not defined. The slope m of a line does not depend on the choice of the points (x_1, y_1) and (x_2, y_2) used in formula (5.1).

FIGURE 5.5

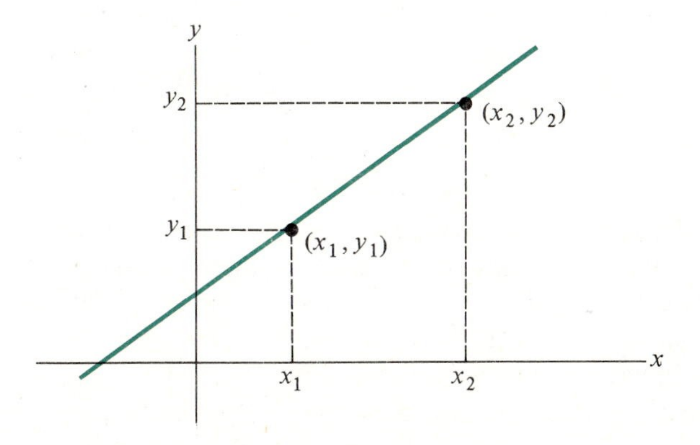

Example 5.4 Find the slope of the line with the equation $x + 2y = 4$.

Solution The intercepts of $x + 2y = 4$ are (4, 0) and (0, 2). Thus, using these two points in Equation (5.1), we compute the slope to be

$$m = \frac{0-2}{4-0} = -\frac{1}{2}$$ ■

In Example 5.4 we showed that the line whose equation is $x + 2y = 4$ has slope $-\frac{1}{2}$. This number can be obtained directly from the equation by taking $-A/B$, where $A = 1$ and $B = 2$. Such a technique always works, and we have

> If $B \neq 0$, the slope of the line $Ax + By = C$ is $m = -A/B$. If $B = 0$, the slope is not defined.

Both methods of determining the slope of a line (using two points and using the formula $-A/B$) always give the same value.

Example 5.5 Find the slope of the line $x - 3y = 6$
(*a*) By using the formula $m = -A/B$
(*b*) By using Equation (5.1) and the two points $(1, -\frac{5}{3})$ and $(2, -\frac{4}{3})$

Solution (*a*) In the equation $x - 3y = 6$, $A = 1$ and $B = -3$. Hence the slope is $m = (-1)/(-3) = \frac{1}{3}$.
(*b*) Using the two points specified and Equation (5.1), we get

$$\begin{aligned} m &= \frac{y_2 - y_1}{x_2 - x_1} \\ &= \frac{-\frac{5}{3} - (-\frac{4}{3})}{1-2} = \frac{-\frac{1}{3}}{-1} = \frac{1}{3} \end{aligned}$$ ■

Another useful form of the equation of a line is the *slope-intercept* form $y = mx + b$. To derive this form of the equation, we begin with the general equation $Ax + By = C$ and assume $B \neq 0$, so the slope is defined. Dividing the equation by B, we obtain $(A/B)x + y = C/B$. Next, subtracting $(A/B)x$ from both sides, we have

$$y = -\frac{A}{B}x + \frac{C}{B}$$

Notice here that $-A/B$ is the slope of the line and C/B is the y intercept. Thus we have

$$y = -\frac{A}{B}x + \frac{C}{B} = mx + b$$

In the equation $y = mx + b$, the number m is always the slope of the line and b is always the y intercept.

We emphasize that there is exactly one line through two points and there is exactly one line through a point with a specified slope. Another way of saying this is that a line is uniquely determined by two points or by one point and a slope.

We use the concept of slope to define the term "parallel."

> Two lines are *parallel* if they have the same slope or if the slope is undefined for both.

In Section 5.2 we will relate this definition of parallel to the geometric notion that parallel lines do not cross. If two lines are not parallel, then they intersect. The problem of determining the point (or points) of intersection is considered in the next section.

To complete our study of the slope of a line, we must consider the case of a line $Ax + By = C$ when either $A = 0$ or $B = 0$. According to the definition of a line, at least one of A and B must be different from 0 (Exercise 24 of this section). The equation of a line with $A = 0$ is $By = C$ with $B \neq 0$. Every point of

FIGURE 5.6

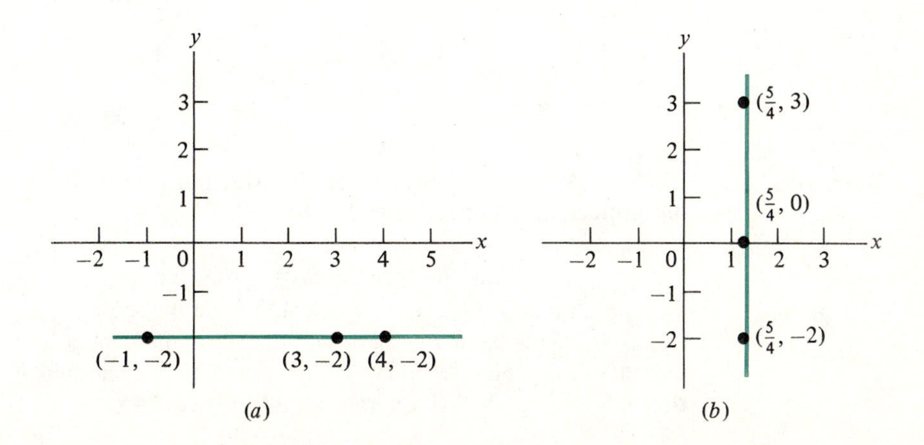

the form $(x, C/B)$ satisfies this equation. Thus if the equation is $-3y = 6$, then $(-1, -2)$, $(3, -2)$, and $(4, -2)$ are all solutions. The graph determined by the equation $-3y = 6$ is a line parallel to the x axis which intersects the y axis at the point -2 (Figure 5.6*a*). The slope is 0.

It remains to consider the equation $Ax = C$, $A \neq 0$. Every point of the form $(C/A, y)$ satisfies the equation. Thus if the equation is $4x = 5$, the points $(\frac{5}{4}, -2)$, $(\frac{5}{4}, 0)$, and $(\frac{5}{4}, 3)$ are solutions of the equation. The graph determined by $4x = 5$ is the line parallel to the y axis which intersects the x axis at the point $\frac{5}{4}$ (Figure 5.6*b*). Since $B = 0$, the slope is not defined.

We conclude our discussion of lines with three typical problems.

Example 5.6 Find the equation of the line through the points $(3, 2)$ and $(-1, 0)$.

Solution Using the given points, we find that the line has the slope

$$m = \frac{2 - 0}{3 - (-1)} = \frac{2}{4} = \frac{1}{2}$$

Therefore, in slope-intercept form, the equation of the line is

$$y = \tfrac{1}{2}x + b$$

To find b, the y intercept, we use the fact that the point $(3, 2)$ is on the line. That is, $x = 3$ and $y = 2$ must satisfy $y = \frac{1}{2}x + b$. It follows that

$$2 = \tfrac{1}{2}(3) + b$$

and therefore $b = \frac{1}{2}$. The desired equation is

$$y = \tfrac{1}{2}x + \tfrac{1}{2}$$

Other forms of the same equation are $2y = x + 1$ and $x = 2y - 1$. ■

Example 5.7 Find the general equation of the line which is parallel to the line $4x - 2y = 3$ and which has x intercept equal to $\frac{1}{2}$.

Solution Using the formula $m = -A/B$, we find that the slope of the line $4x - 2y = 3$ is $(-4)/(-2) = 2$. Thus the line we seek also has slope 2, and the equation has the form $y = 2x + b$. Since the x intercept is $\frac{1}{2}$, the point $(\frac{1}{2}, 0)$ is on the line. Using this, we find $0 = 2(\frac{1}{2}) + b = 1 + b$, and thus $b = -1$. The equation is $y = 2x - 1$. Written in general form, it is $-2x + y = -1$. ■

Example 5.8 Suppose that the cost of an airline ticket is related to the distance traveled by a linear equation. The cost of a 200-mile flight is \$76, and the cost of a 350-mile flight is \$100.

Travel

Problem Find the equation relating cost and distance, and find the cost of a 275-mile flight.

Solution Let

$$x = \text{distance of the flight in miles}$$
$$y = \text{cost of the flight in dollars}$$

The slope of the line through the points (200, 76) and (350, 100) is $\frac{4}{25}$, and therefore the slope-intercept form of the equation of the line through these points is $y = \frac{4}{25}x + b$. Using the fact that (200, 76) satisfies this equation, we can evaluate b. We find $b = 44$. Since y = cost (in dollars) and x = distance (in miles),

$$\text{Cost} = \tfrac{4}{25}(\text{distance}) + 44$$

The cost of a 275-mile flight (in dollars) is

$$\tfrac{4}{25}(275) + 44 = 88$$

■

Exercises for Section 5.1

1. Graph the line whose equation is
   (*a*) $3x - 4y = 12$ (*b*) $2x + 6y = 6$
2. Graph the line whose equation is
   (*a*) $-5x - 2y = 20$ (*b*) $x - 3y = 0$
3. *Foreign currency* A liter of gasoline purchased in Canada is 1.057 quarts. Write an equation to convert the price of gasoline in Canada (in Canadian dollars per liter) to the price of gasoline in the United States (in U.S. dollars per gallon). You may assume 80 cents U.S. = \$1.00 Canadian. What is the U.S. equivalent price of a Canadian price of 65 cents per liter?
4. On a cartesian coordinate system, draw the line through the points $(-4, 9)$ and $(4, -3)$.
5. On a cartesian coordinate system, draw the line through the points $(-3, 3)$ and $(1, 1)$. Also draw the line through the points $(-3, -3)$ and $(1, 1)$.
6. Find the x and y intercepts of each of the following lines and graph the lines.
   (*a*) $2x - y = 1$ (*b*) $2x - y = 0$
   (*c*) $y = 3x + 7$ (*d*) $x = 5$
7. Find the x and y intercepts of each of the following lines and graph the lines.
   (*a*) $-y + x = 2$ (*b*) $y = 3$
   (*c*) $x = 2y - 9$ (*d*) $x = 0$
8. *Metric conversion* Write an equation to convert a speed expressed in miles per hour to the same speed expressed in kilometers per minute.

9. Find the slope of the line through each of the following pairs of points.
   (*a*) $(1, -1), (3, 2)$ (*b*) $(-3, 7), (-1, 9)$
10. Find the slope of the line through each of the following pairs of points.
   (*a*) $(7, 3), (-1, 2)$ (*b*) $(3, -5), (1, 2)$
11. Find the equation of the line through each of the following pairs of points and graph each line.
   (*a*) $(1, 3), (4, 1)$ (*b*) $(-3, -1), (1, 3)$
12. Find the slope, x intercept, and y intercept of the line whose equation is $5x - 3y = -15$.
13. Find the slope, x intercept, and y intercept of the line whose equation is $x = 3y - 6$.
14. State the equation of the line with
   (*a*) Slope -3 and containing the point $(1, 0)$
   (*b*) Slope 0 and containing the point $(1, -2)$
   (*c*) Slope not defined and containing the point $(3, 2)$
15. Find the equation of the line through the point $(-2, -1)$ and parallel to the line through the points $(2, 3)$ and $(4, 1)$. Graph both lines.
16. State the equation of the line with
   (*a*) Slope -3, y intercept $= 4$
   (*b*) Slope 5, y intercept $= -2$

Production

17. Production records indicate that the number of defective basketballs produced on a certain production line is related by a linear equation to the total number produced. Suppose this is true. If 10 defective balls are produced in a total of 300 in one day and 15 defective balls are produced in a total of 425 on another day, then how many defective basketballs will be expected on a day when the total production is 500?

Production

18. Using the data of Exercise 17, answer the following question. The production supervisor wants to produce as many basketballs as possible without having more than 25 defective ones produced. What total production should be scheduled?

Temperature conversion

19. Find the equation of the line through the points $(32, 0)$ and $(212, 100)$, and graph this line. What is the slope of this line? This graph can be related to temperature scales. How should the axes be labeled to illustrate this relationship?

Business

20. The following data give the year-by-year resale value of a car which cost \$10,000 when it was new. Decide whether these data are linear, i.e., decide whether these pairs of points (age, resale value) lie on a line:

Age, years	1	2	3	4	5
Resale value, \$	8000	6400	5200	4400	4000

Travel

21. A certain car can be rented at either of two rates:
   (*a*) \$40 per day and \$0.30 per mile driven
   (*b*) \$30 per day and \$0.50 per mile driven

   Write equations which describe the costs (in dollars) of driving x miles in 1 day under each of these rates. Which rate is less expensive for someone who plans to drive 30 miles on a single day?

22. Using the data of Exercise 21, which rate is less expensive for someone who plans to drive 225 miles in 3 days?
23. For the situation in Example 5.8, how many miles could you fly for \$524?
24. Using the general equation of a line, suppose $A = B = 0$.
    (*a*) If $C = 0$, show that the solution set of $Ax + By = C$ is the entire xy plane.
    (*b*) If $C \neq 0$, show that the solution set of $Ax + By = C$ is the empty set.
    (Thus in neither case is the solution set a line.)
25. Suppose that a line is defined by the equation $Ax + By = C$ with $B \neq 0$. Show that if (x_1, y_1) and (x_2, y_2) are two distinct points on this line, then $x_1 \neq x_2$. (*Hint*: Each of the points must satisfy the equation of the line. Write down the two equations and subtract one from the other.)

Sports

26. A jogger is able to run long distances at a pace of 2 minutes 15 seconds for each quarter mile. Write an equation which gives the distance covered in terms of the time the jogger has been running. How long will it take this jogger to complete a marathon?

5.2 FORMULATION AND SOLUTION OF SYSTEMS OF LINEAR EQUATIONS IN TWO VARIABLES

An important part of the skill of using mathematics to solve problems is the ability to translate verbal expressions into mathematical expressions, usually equations. In this section, we consider several problems which can be translated into linear equations in two variables. We formulate these problems, and then we solve them. We begin with the following example.

Example 5.9

Manufacturing

Willie's Waterbeds decides to make giant rectangular beds for which the length is 20 percent longer than the width and for which the perimeter is 330 inches. What are the length and width of these beds?

Solution The length and width of the beds are unknowns, so it is natural to use these as our variables. Thus, let

$$x = \text{the length of the waterbed in inches}$$

and

$$y = \text{the width of the waterbed in inches}$$

Next, since the length x is to be 20 percent longer than the width y, we have the equation

$$x = y + .2y \tag{5.2}$$

Also, the perimeter of the bed $2x + 2y$ is known to be 330 inches, so we have the second equation

$$2x + 2y = 330 \tag{5.3}$$

Together, Equations (5.2) and (5.3) form a system of two equations in the two variables x and y. If we rewrite Equation (5.2) by adding the two terms in y, then an equivalent form of the two equations is

$$x = 1.2y \tag{5.4}$$

$$2x + 2y = 330 \tag{5.5}$$

We are interested in the solution of this system of equations, i.e., a pair of numbers (x, y) which satisfies both equations. Since each equation is the equation of a line, we are interested in a point on both lines, i.e., the point of intersection of the lines. A direct way of finding this point is to substitute Equation (5.4) into Equation (5.5) to obtain a single equation in the variable y and then solve for y. We obtain the following:

$$\begin{aligned} 2(1.2y) + 2y &= 330 \\ 4.4y &= 330 \\ y &= 75 \end{aligned}$$

Since $y = 75$, using Equation (5.4), we obtain $x = (1.2)(75) = 90$. Thus the length of the giant waterbed is 90 inches, and the width is 75 inches. ■

A graph of the lines in Example 5.9 is shown in Figure 5.7.

FIGURE 5.7

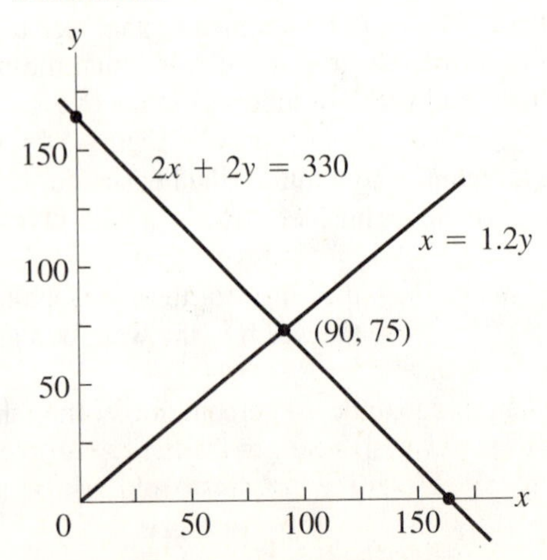

Example 5.10

Resource allocation

Murphy's Muffin Shoppe makes two sizes of raisin muffins using prepackaged dough and raisins. Each large muffin uses 5 ounces of dough and 2 ounces of raisins, and each small muffin uses 2 ounces of dough and 1 ounce of raisins. Each day the shop receives 450 ounces of dough and 200 ounces of raisins. How many large muffins and small muffins should be baked each day to use up all the dough and all the raisins?

Solution The unknowns in this problem are the number of large muffins to be baked and the number of small muffins to be baked. Thus, we specify variables x and y as follows:

$$x = \text{the number of large muffins to be baked each day}$$
$$y = \text{the number of small muffins to be baked each day}$$

Next, we describe the conditions that x and y must satisfy due to the limited amounts of dough and raisins. Each large muffin uses 5 ounces of dough, so x large muffins require $5x$ ounces of dough. Also, each small muffin uses 2 ounces of dough, so y small muffins require $2y$ ounces of dough. Thus, the total amount of dough used by x large muffins and y small muffins is $5x + 2y$. Since 450 ounces of dough are available each day, x and y must satisfy the equation

$$5x + 2y = 450 \tag{5.6}$$

Reasoning in the same way for raisins, we see that x large muffins and y small muffins use a total of $2x + y$ ounces of raisins. There are 200 ounces of raisins available each day, so x and y must satisfy

$$2x + y = 200 \tag{5.7}$$

Equations (5.6) and (5.7) must be satisfied at the same time, and since they are both equations of lines, we are looking for a point, i.e., a pair (x, y), which lies on both lines. To find the values of x and y at the intersection of these two lines, we once again express one of the variables in terms of the other and then substitute to obtain a single equation in one variable. Using (5.7), we have $y = 200 - 2x$, and substituting this into Equation (5.6) gives $5x + 2(200 - 2x) = 450$ or, equivalently, $5x + 400 - 4x = 450$, which in turn gives $x = 50$. Finally, using Equation (5.7) with $x = 50$, we have $y = 200 - 2(50) = 100$. Thus, Murphy should bake 50 large muffins and 100 small muffins in order to use all the dough and all the raisins. ■

Note In Example 5.10, the values obtained for variables x and y are integers, 50 and 100. However, it is easy to modify the problem so that this is not the case, and in such situations the answer would not be a practical one for the baker.

Example 5.11 *Sports*

Two joggers begin running together at the same time on an oval 400-meter track. One jogger runs at the rate of 400 meters every 2 minutes, and the other runs at the rate of 400 meters every 2 minutes 15 seconds.

Problem How far has the slower runner run at the time when the faster runner first catches up with the slower runner?

Solution The faster runner covers 400 meters every 2 minutes, or 400/2 meters/minute, so a formula which relates distance covered in meters to time of running in minutes, for the faster runner, is

$$d = \frac{400}{2}t \tag{5.8}$$

where t is measured in minutes and d is measured in meters. A similar formula for the slower runner gives

$$d = \frac{400}{2.25}t \tag{5.9}$$

In Equation (5.9), the term 400/2.25 occurs because the slower runner completes 400 meters in 2 minutes 15 seconds, which is 2.25 minutes, and thus the runner completes 400/2.25 meters in one minute.

To solve the problem posed, we need to find the distance d covered by the faster runner when that distance is exactly 400 meters beyond the distance for the slower runner. To find this distance, we first find the time that the runners have been running when the faster runner catches the slower runner. To do this, we add 400 meters to formula (5.9), the distance covered by the slower runner plus one lap, and then we set this equal to the distance run by the faster runner. This gives

$$\frac{400}{2.25}t + 400 = \frac{400}{2}t$$

Solving this equation for t gives $t = 18$ minutes. To complete the solution, we use $t = 18$ minutes in Equation (5.8), and this gives $d = (400/2)18 = 3600$ meters. Thus, the faster runner must run about $2\frac{1}{4}$ miles before first catching the slower runner. The slower runner has run $3600 - 400 = 3200$ meters. ■

In Examples 5.9, 5.10, and 5.11, we used basic algebraic techniques to solve two equations in two variables. In each case, we obtained a single value for x and a single value for y, and we noted that since each equation represented a line, we had found the point of intersection of the two lines. However, not all pairs of lines intersect in a single point. If the lines are parallel but different, then they do not intersect–they have no points in common; and if the two algebraic equations represent the same geometric line (set of points), then the two equations have infinitely many points in common. Our goal is to develop a method to solve all systems of linear equations, including those with more than two variables and/or more than two equations. Here by the word "solve" we mean that we find the set of points which satisfy all the equations of the system. Of course, one possibility is that this set is empty, in which case we say that the system has no solution. In this section we consider the case of two variables, and in Section 5.3 we consider the general case.

There are several methods for solving systems of equations in two variables, and we used one of them above in solving the systems of Examples 5.9, 5.10, and 5.11. This method is called the *substitution method* since it is based on using one equation to obtain an expression for one variable in terms of the other variable and then substituting that expression into the other equation to obtain a new equation with only one variable. The substitution method is simple and easy to use, and most students have used it in high school. However, the method is not a general one, and it is of limited value for the more general systems studied later in this chapter and in Chapters 8 and 10. Therefore, we also present a second method, called the *reduction method*, which can be used for all linear systems. First, however, we illustrate the substitution method with two more examples.

Example 5.12 Solve the system of equations

$$\begin{aligned} x - y &= 6 \\ -2x + 2y &= 5 \end{aligned} \tag{5.10}$$

and graph the lines described by these equations.

Solution We solve the first equation for x in terms of y. This gives

$$x = 6 + y$$

That is, if (x, y) is a solution of the system, then x and y must be related as $x = 6 + y$. Now use this expression for x in the left-hand side of the second equation:

$$-2x + 2y = -2(6 + y) + 2y = -12 - 2y + 2y = -12$$

We have shown that if (x, y) satisfies the first equation, then for that (x, y) the left-hand side of the second equation is -12. But the right-hand side of the second equation is 5. Therefore, we have shown that if (x, y) satisfies the first equation, it *cannot* satisfy the second $(-12 \neq 5)$. The system has no solution (no point of intersection), and the lines described by Equations (5.10) are parallel. They are shown in Figure 5.8. ■

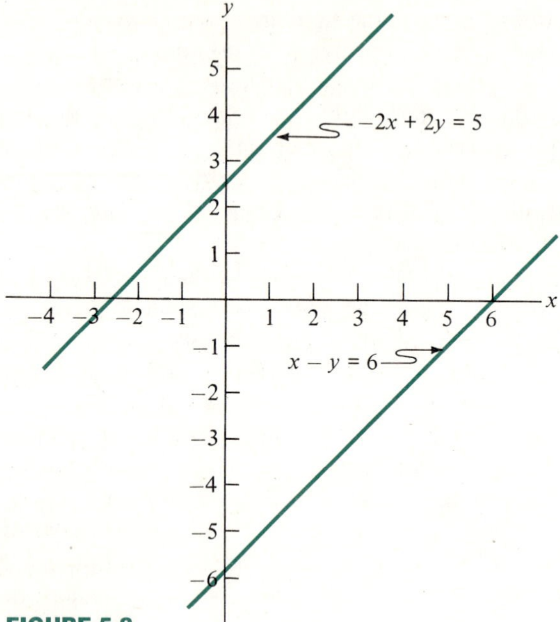

FIGURE 5.8

We note that if Equations (5.10) are written in slope-intercept form, we have

$$y = x - 6$$
$$y = x + \tfrac{5}{2}$$

Thus both lines have slope $m = 1$, and we see immediately that they are parallel.
There is one more case that can arise.

Example 5.13 Solve the system of equations

$$x - y = 3$$
$$-3x + 3y = -9$$

Solution We solve the first equation for x:

$$x = 3 + y$$

Substituting this expression for x into the left-hand side of the second equation, we have

$$-3x + 3y = -3(3 + y) + 3y = -9 - 3y + 3y = -9$$

That is, if (x, y) satisfies the first equation, then the left-hand side of the second equation is -9 and the second equation is automatically satisfied. Therefore, all points which satisfy the first equation also satisfy the second equation. This fact also follows from the observation that the second equation is simply -3 times the first equation. We conclude that for *every value of* y the pair $(x, y) = (3 + y, y)$ is a solution of both equations. Thus there are infinitely many solutions, and the two equations describe the same line. It is shown in Figure 5.9. ■

To summarize our last three examples, we have now shown that there are at least three types of results that arise when we solve a system of two equations in two variables. In fact, these are the only possible results. First, the system may describe two lines which intersect in a single point (as in Example 5.11). This point is the solution of the system. Second, the system may describe two parallel lines, line which do not meet (as in Example 5.12). In this case the system is said to be *inconsistent*, and the solution set is empty. Third, the system may consist of two equations which describe the same line (as in Example 5.13). The solution set in this case consists of infinitely many points, namely, all points on this line. In the first and third cases (one point common and infinitely many points common to the two lines), the system is said to be *consistent*.

The substitution method is useful primarily for systems of two equations in two variables. For more general systems, another method is needed. The *reduction method*, which is developed in Section 5.3 for systems with more than two variables, can also be used to solve systems of two equations in two variables. The basic idea of the method is to transform the given system into one new system after another until we obtain a system whose solution is obvious. In particular, our goal will be a system in which each equation contains a single variable. Since the number of variables in each equation has been reduced, our method is called the reduction method.

FIGURE 5.9

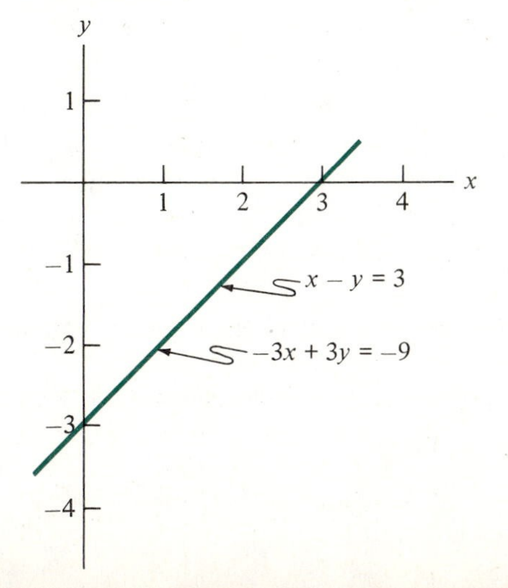

To carry out the reduction method, we use two basic operations:

1. Multiply any equation (both sides of the equals sign) by a number different from zero.
2. Replace any equation by the sum of that equation and any multiple of another equation.

That is, operation 1 means that if E is an equation in a system, then E can be replaced by kE, where k is any number different from 0. Operation 2 means that if E and F are equations in a system, then F can be replaced by $F + kE$, where k is any number. These operations form the basis of the methods discussed in Section 5.3 and in Chapter 10.

The reduction method works because the use of these two operations does not change the set of solutions of a system of equations.

Example 5.14 Solve the system of equations

$$\begin{aligned} 2x + 4y &= 6 \\ 3x - 6y &= 1 \end{aligned} \tag{5.11}$$

Solution We solve this system by converting it into one new system after another, using operations 1 and 2. Our goal is to reach a system whose solution is obvious, i.e., one in which each equation contains a single variable.

First we obtain a system in which the second equation contains a single variable. We begin our conversion process by multiplying the first equation of (5.11) by $\frac{1}{2}$, an operation of type 1. Our goal is to replace the original system by one in which the coefficient of x (the number multiplied by x) in the first equation is 1. We have

$$\begin{aligned} x + 2y &= 3 \\ 3x - 6y &= 1 \end{aligned}$$

Next, we multiply the first equation in this new system by -3 and add the result to the second equation. We obtain an equation which contains only the variable y:

$$-12y = -8$$

We obtain a new system by replacing the second equation by $-12y = -8$.

$$\begin{aligned} x + 2y &= 3 \\ -12y &= -8 \end{aligned}$$

This second replacement has been accomplished by an operation of type 2. We have succeeded in converting the original system to a new system in which the second equation contains a single variable. Now we divide both sides of the second equation in the last system by -12 to obtain the system

$$\begin{aligned} x + 2y &= 3 \\ y &= \tfrac{2}{3} \end{aligned}$$

Now we convert the last system to one in which the first equation also contains a single variable. To do so, we multiply the second equation by -2 and add the result to the first equation. We obtain an equation which contains only the variable x:

$$x = \tfrac{5}{3}$$

Finally, we replace the first equation by $x = \frac{5}{3}$, and hence we have converted the original system to one in which each equation contains a single variable:

$$\begin{aligned} x &= \tfrac{5}{3} \\ y &= \tfrac{2}{3} \end{aligned}$$

The solution of this system is obvious. The point $(x, y) = (\frac{5}{3}, \frac{2}{3})$ is the only solution of the original system. In other words, there is only one point common to the two lines, and it is $P = (\frac{5}{3}, \frac{2}{3})$. The lines (5.11) and the point P are shown in Figure 5.10. ■

FIGURE 5.10

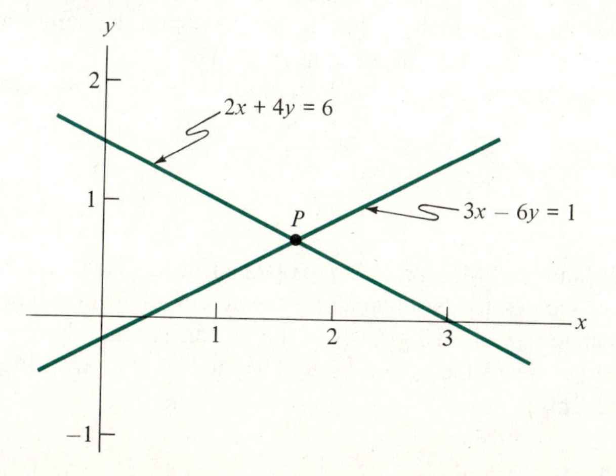

The reduction method also enables us to identify those systems which have no solutions.

Example 5.15 Solve the system of equations

$$\begin{aligned} 8x - 4y &= 16 \\ 2x - y &= 0 \end{aligned} \tag{5.12}$$

Solution We begin our conversion of the system (5.12) by dividing the first equation by 8 (the coefficient is x) and using the result as a new first equation. The resulting system is

$$\begin{aligned} x - \tfrac{1}{2}y &= 2 \\ 2x - y &= 0 \end{aligned} \tag{5.13}$$

The next step is to eliminate the variable x from the second equation in the system (5.13). We do this by multiplying the first equation by -2, adding it to the second equation, and using this sum as a new second equation. We have

$$\begin{aligned} x - \tfrac{1}{2}y &= 2 \\ 0 &= -4 \end{aligned} \tag{5.14}$$

At this stage we stop the process because $0 \neq -4$, and we conclude that the system (5.14) *has no solutions*. Since the solution set of the system remains unchanged during the conversions from one system to another, we conclude that the original system has no solutions. The lines described by the system (5.12) are parallel. ■

Finally, the reduction method also enables us to identify those systems which have infinitely many solutions and to obtain the solution sets for such systems. (See Exercise 6 of this section.)

We conclude with a final example of the use of the reduction method.

Example 5.16 *Small business*

William and Margaret plan to open a fruit-drink stand. They have 15 lemons and 30 oranges to use in making two types of drinks. They use 2 lemons and 1 orange to make 10 glasses of the tart drink and 1 lemon and 3 oranges to make 10 glasses of the sweet drink. They use mountain spring water as an all-natural mixer.

Problem Find the number of glasses of each type of drink they should make to exactly use their supply of fruit.

Solution There are two unknown quantities to determine: the number of glasses of each of the two drinks. Let

$$x = \text{the number of glasses of tart drink}$$
$$y = \text{the number of glasses of sweet drink}$$

Since 10 glasses of tart drink require 2 lemons, each single glass requires $\frac{2}{10}$ lemon. Likewise, each single glass of sweet drink requires $\frac{1}{20}$ lemon. Thus the total number of lemons used to make x glasses of tart drink and y glasses of sweet drink is $\frac{2}{10}x + \frac{1}{10}y$.

There are 15 lemons available, and if all are to be used,

$$\tfrac{2}{10}x + \tfrac{1}{10}y = 15$$

Using the same reasoning for oranges, we have

$$\tfrac{1}{10}x + \tfrac{3}{10}y = 30$$

Thus the problem is to find values of x and y which satisfy the system of equations

$$\begin{aligned} \tfrac{2}{10}x + \tfrac{1}{10}y &= 15 \\ \tfrac{1}{10}x + \tfrac{3}{10}y &= 30 \end{aligned}$$

From the setting, we are interested only in solutions for which $x \geq 0$ and $y \geq 0$.

Using the reduction method, we convert the system successively as follows: Multiply the first equation by $\frac{10}{2}$ and use the result as a new first equation.

$$\begin{aligned} x + \tfrac{1}{2}y &= 75 \\ \tfrac{1}{10}x + \tfrac{3}{10}y &= 30 \end{aligned}$$

Multiply the first equation by $-\frac{1}{10}$, add it to the second equation, and use the sum as a new second equation:

$$\begin{aligned} x + \tfrac{1}{2}y &= 75 \\ \tfrac{5}{20}y &= \tfrac{225}{10} \end{aligned}$$

Multiply the second equation by $\frac{20}{5}$, and use the result as a new second equation:

$$\begin{aligned} x + \tfrac{1}{2}y &= 75 \\ y &= 90 \end{aligned}$$

Finally, multiply the second equation by $-\frac{1}{2}$, add it to the first, and use the sum as a new first equation:

$$\begin{aligned} x &= 30 \\ y &= 90 \end{aligned}$$

We conclude that Margaret and William should make 30 glasses of tart drink and 90 glasses of sweet drink to exactly use the fruit available. ■

Exercises for Section 5.2

For Exercises 1 through 4, formulate the problem as a system of linear equations in two variables. Be sure to define your variables.

Manufacturing

1. Suppose that Willie's Waterbeds makes a rectangular children's bed for which the length is 30 percent longer than the width and the sum of the length plus the width is 115 inches. Formulate a system of equations which can be used to find the length and width of the bed.

Resource allocation

2. Suppose that Murphy's Muffin Shoppe decides to make both large and small bran muffins. Each large muffin uses 4 ounces of dough and 2 ounces of bran, while a small muffin uses 1 ounce of dough and 1 ounce of bran. Suppose also that there are 300 ounces of dough available each day and 160 ounces of bran. Formulate a system of equations to determine how many muffins of each size should be baked each day to use up all the dough and all the bran.

Resource allocation

3. Using the setting of Exercise 2, suppose that the bran muffins also contain raisins and that a large muffin has 1 ounce of raisins while a small muffin has $\frac{1}{3}$ ounce of raisins. Assume there are 75 ounces of raisins available each day, and formulate a system of equations to determine the number of muffins to bake each day to use up all the dough, bran, and raisins.

Sports

4. Sally runs 10 percent faster than Mary. Assume that they start off running together on a 400-meter oval track and that Mary runs each lap of 400 meters in 2 minutes. Formulate a system of equations to use to find the distance that Sally has run when she catches up to Mary.

For Exercises 5 through 10, solve each of the two systems of equations and graph the lines.

5. (*a*) $\begin{aligned} 3x &= 12 \\ 5x + 2y &= 14 \end{aligned}$ (*b*) $\begin{aligned} 2x - y &= 7 \\ 2y &= 6 \end{aligned}$

6. (*a*) $\begin{aligned} x - 3y &= -1 \\ 3x - 9y &= -3 \end{aligned}$ (*b*) $\begin{aligned} x + 2y &= 10 \\ x - 2y &= 10 \end{aligned}$

7. (*a*) $\begin{aligned} 4x + 2y &= 3 \\ -4x - 2y &= 6 \end{aligned}$ (*b*) $\begin{aligned} -x + y &= 2 \\ x - 2y &= 8 \end{aligned}$

8. (*a*) $\begin{aligned} 13x + 5y &= 60 \\ 3x - y &= 9 \end{aligned}$ (*b*) $\begin{aligned} -x + y &= 7 \\ x - y &= -8 \end{aligned}$

9. (*a*) $\begin{aligned} 10x + 5y &= 20 \\ 6x + 3y &= 12 \end{aligned}$ (*b*) $\begin{aligned} -x + y &= 7 \\ 2x - y &= -8 \end{aligned}$

10. (*a*) $3x - y = -1$
$-x + 3y = 1$

(*b*) $3x - y = -1$
$6x - 2y = -2$

11. Solve the problem formulated in Exercise 1.

12. Solve the problem formulated in Exercise 2.

13. Solve the problem formulated in Exercise 3.

14. Solve the problem formulated in Exercise 4.

Demographics 15. A rapidly growing suburb has a population of 50,000 and is growing at the rate of 7500 people per year. The adjacent declining city has a population of 1,000,000 and is decreasing at a rate of 125,000 people per year. If these rates continue, in how many years will the population of the suburb equal the population of the city?

Demographics 16. In the setting of Exercise 15, suppose that the suburb is growing at the rate of r people per year. What value must r have in order that the suburb grow to half the size of the city in exactly 6 years?

Investments 17. A lottery winner plans to invest part of her $1,000,000 in utility bonds paying 12 percent per year and the rest in a savings account paying 8 percent per year. How much should be allocated to each investment if the yearly incomes from the two investments are to be the same?

Investments 18. In the situation described in Exercise 17, how much should be allocated to each investment if the income from utility bonds is to be twice the income from the savings account?

Science 19. The formula for converting Fahrenheit temperatures to Celsius is $C = \frac{5}{9}(F - 32)$. At what temperatures (if any) do the two temperature scales have the same value?

Sales 20. The sales of ABC Corporation are now $5,000,000 per year and are increasing by $350,000 per year. The sales of XYZ Incorporated are now $2,000,000 per year and are increasing by $950,000 per year. In how many years will the sales of the two companies be equal?

Investments 21. Fred wins $100,000 in the lottery, and he invests all the money in two stocks. At the end of the year, he sells both stocks and his profit is 10 percent in the first and 20 percent in the second. If total profit is $17,000, how much did he invest in each stock?

Investments 22. Suppose that Fred (see Exercise 21) invests $117,000 in year 2 and he again invests it all in two stocks. At the end of year 2, he sells both stocks, and he notices that in year 2 he lost 10 percent in stock 1 and 20 percent in stock 2. If his total loss was $17,000, how much did he invest in each stock?

Investments 23. Janet and Michael decide to pool their money for investments. They will invest a total of $10,000,000, and it is all to be in stocks, bonds, and real estate. Suppose that the amount in stocks must be twice the amount in bonds, and the amount in stocks and bonds together must be 3 times the amount in real estate. How should their money be distributed among the three areas?

24. For what numbers a and b is this system of equations satisfied for $x = b$ and $y = a$?

$$x - 3y = a$$
$$2x + 7y = 5$$

25. For what numbers a and b is this system of equations satisfied by $x = 1$ and $y = b$?

$$\begin{aligned} ax + y &= 3 \\ -x + 2y &= 4 \end{aligned}$$

26. The system of equations

$$\begin{aligned} 3x - 2y &= -4 \\ -6x + ay &= 8 \end{aligned}$$

has a unique solution for all but one value of the number a. What is this exceptional value of a? How many solutions does the system have for this value of a?

Criminology

27. Tom, Dick, and Harry are being questioned by the police after a robbery at a local fruit market. All three fit the descriptions of the fruit thief, and all three were seen at the market. The police ask each what he bought and how much he paid. Tom says he bought 20 apples and 20 oranges and paid \$8.00. Dick says he bought 15 apples and 5 oranges and paid \$6.75. Harry says he bought 10 apples and 25 oranges and paid \$11.00. After a short pause for computational purposes, the police released Dick and Harry but kept Tom for additional questioning. Why?

Resource allocation

28. Murphy's Muffin Shoppe makes large and small apple-raisin muffins. A large apple-raisin muffin requires 5 ounces of dough, 2 ounces of apples, and 0.5 ounce of raisins. A small apple-raisin muffin requires 3 ounces of dough, 1 ounce of apples, and 0.3 ounce of raisins. Murphy has 270 ounces of dough, 100 ounces of apples, and 30 ounces of raisins. How many muffins of each type can Murphy make to use all the dough and apples? If he does so, how many ounces of raisins remain unused?

Resource allocation

29. In the setting of Exercise 28, suppose Murphy has only 25 ounces of raisins. Will he have enough raisins to use all the dough and apples? If not, find the number of large and small muffins he should make to use all the raisins.

5.3 FORMULATION AND SOLUTION OF SYSTEMS OF LINEAR EQUATIONS IN THREE OR MORE VARIABLES

Most problems involving linear equations have more than two unknown quantities. Consequently, it is important to develop methods to solve systems of equations with more than two variables. Our methods will be primarily algebraic, though we do discuss the geometry of the three-variable situation. In this section we formulate some problems with more than two variables, and we develop an algorithm (i.e., a step-by-step method) which will produce all solutions of a system of linear equations when there are solutions.

Example 5.17

Landscaping

Nan's Nursery sells bushes, trees, and perennial flowers. Each bush costs \$20, each tree costs \$40, and each flower costs \$2. To plant these bushes, trees, and flowers, Nan's charges \$10 per bush, \$10 per tree, and \$1 per flower. Also, Nan will insure the items she plants for 1 year at a charge of \$4 per bush, \$6 per tree, and \$0.20 per flower. A new homeowner has budgeted \$1000 for purchases, \$400 for planting the purchases, and \$150 for insuring purchases from Nan's. How many bushes, trees, and flowers should the homeowner buy to exactly use up the money budgeted?

Problem Formulate the homeowner's problem as a system of linear equations.

Solution Since the homeowner is interested in the number of bushes, trees, and flowers to buy, we let

x = the number of bushes to buy
y = the number of trees to buy
z = the number of flowers to buy

Next, since \$1000 is budgeted for purchases, and since bushes, trees, and flowers cost \$20, \$40, and \$2, respectively, we know that variables x, y, and z must satisfy the equation

$$20x + 40y + 2z = \$1000 \tag{5.15}$$

Similarly, since the planting budget is \$400, and since planting each bush, tree, and flower costs \$10, \$10, and \$4, respectively, we also know that variables x, y, and z must satisfy

$$10x + 10y + z = \$400 \tag{5.16}$$

Finally, since \$150 is budgeted for insurance for the first year, we have

$$4x + 6y + .2z = \$150 \tag{5.17}$$

The variables x, y, and z must satisfy Equations (5.15), (5.16), and (5.17) simultaneously, and these three equations form the desired system. Also, because of the meanings of x, y, and z, we are only interested in integer solutions which satisfy $x \geq 0$, $y \geq 0$, and $z \geq 0$. ■

Example 5.18

Recreation

The Hiker and Biker Outfitting Shop makes packages of snacks to order. One customer likes peanuts, raisins, and chocolate chips but wants only two ingredients per snack. In peanut-raisin mixes she likes twice as many peanuts as raisins, in peanut–chocolate chip mixes she likes twice as many chocolate chips as peanuts, and in raisin–chocolate chip mixes she likes equal amounts of the two foods. She buys 4 ounces of peanuts, 6 ounces of raisins, and 12 ounces of chocolate chips.

Problem Formulate a mathematical problem whose solution will give the amounts of the various mixes which will meet her conditions and completely use up the foods she has purchased.

Solution Let

$$
\begin{aligned}
x &= \text{amount (in ounces) of peanut-raisin mixture} \\
y &= \text{amount (in ounces) of peanut–chocolate chip mixture} \\
z &= \text{amount (in ounces) of raisin–chocolate chip mixture}
\end{aligned}
$$

Now consider the information on peanuts. Since she likes a 2-to-1 mixture in the peanut-raisin combination, each ounce of that combination requires $\frac{2}{3}$ ounce of peanuts. Similarly, each ounce of peanut–chocolate chip mixture requires $\frac{1}{3}$ ounce of peanuts. Since there are 4 ounces of peanuts available, x and y are to be selected so that

$$\tfrac{2}{3}x + \tfrac{1}{3}y = 4$$

Next we use the information on raisins. The 2-to-1 ratio in the peanut-raisin mixture means that each ounce of mix requires $\frac{1}{3}$ ounce of raisins, and the 1-to-1 ratio in the raisin–chocolate chip mixture means that each ounce of that mixture requires $\frac{1}{2}$ ounce of raisins. There are 6 ounces of raisins available, and consequently we have the equation

$$\tfrac{1}{3}x + \tfrac{1}{2}z = 6$$

By using the same type of argument, the information on chocolate chips leads to the equation

$$\tfrac{2}{3}y + \tfrac{1}{2}z = 12$$

The mathematical problem whose solution gives the amounts of the various mixtures which meet her conditions and use all the food is the following: Find x, y, and z which satisfy

$$
\begin{aligned}
\tfrac{2}{3}x + \tfrac{1}{3}y &= 4 \\
\tfrac{1}{3}x + \tfrac{1}{2}z &= 6 \\
\tfrac{2}{3}y + \tfrac{1}{2}z &= 12
\end{aligned}
\tag{5.18}
$$

Again, because of the meanings of x, y, and z, we are only interested in solutions which satisfy $x \geq 0$, $y \geq 0$, and $z \geq 0$. ■

In Examples 5.17 and 5.18, the formulation of the problem resulted in a system of three linear equations with three variables. In Section 5.2, when we discussed solving linear systems of equations with two variables, we noted that the equations represented lines in the plane, and, in seeking a solution, we were really seeking a point common to these lines. A similar situation holds with equations in three variables, but now the equations represent planes instead of lines and we are seeking a point common to these planes. Since the idea of a plane is also important in Chapter 7, we now introduce the idea of graphing planes in three dimensions.

Graphing Planes in Three-Dimensional Space

Many of the algebraic and graphical techniques we reviewed for linear equations with two variables can be used equally well for equations with three variables. Although we shall not exploit graphical methods for problems with three variables in the same systematic way as for problems with two variables, it is useful to have available the rudiments of graphing in three dimensions. Three-dimensional space with a cartesian coordinate system replaces two-dimensional space (the plane) for problems with three variables. In this case, we identify points in three-dimensional space with ordered triples of numbers (x, y, z). Figure 5.11 shows several points plotted on a cartesian coordinate system with the axes labeled x, y, and z. The x axis should be visualized as pointing out of the page, i.e., as perpendicular to the plane determined by the y and z axes.

A linear equation with three variables is an equation of the form $Ax + By + Cz = D$, where A, B, C, and D are numbers. The graph of a linear

FIGURE 5.11

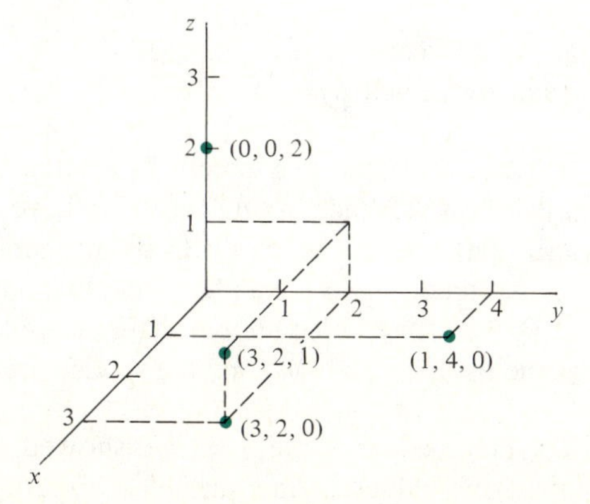

equation with three variables is a plane in three-dimensional space. Conversely, any plane in three-dimensional space has an equation of the form $Ax + By + Cz = D$ for some numbers A, B, C, and D. We define a plane in a manner analogous to our definition of a line.

For any numbers A, B, C, and D with A, B, and C not all zero, the set of points

$$\{(x, y, z): \quad Ax + By + Cz = D\}$$

is a *plane*. If (x_1, y_1, z_1), (x_2, y_2, z_2), and (x_3, y_3, z_3) are three points on this plane, i.e., each point satisfies the equation $Ax + By + Cz = D$, then the plane is referred to as a *plane through these three points*. If the three points do not lie on a straight line, then there is only one plane through them.

The fact that we know the form of the graph of a linear equation in three variables makes the actual construction of the graph relatively simple. The x, y, and z intercepts of a plane can be defined as follows:

The plane $Ax + By + Cz = D$ has an x intercept if $A \neq 0$, and the x intercept is D/A.
The plane has a y intercept if $B \neq 0$, and the y intercept is D/B.
The plane has a z intercept if $C \neq 0$, and the z intercept is D/C.

The x intercept can be determined by applying the formula of the definition, or equivalently, by setting $y = 0$ and $z = 0$ in the equation of the plane and solving for x. The y and z intercepts can be determined in analogous ways.

Three points not in a straight line determine a plane, and if the intercepts exist (i.e., if A, B, and C are all nonzero), then these intercepts are normally used to graph the plane.

Example 5.19 Find the x, y, and z intercepts of the plane $3x + 2y + 4z = 12$, and graph the plane on a cartesian coordinate system.

Solution We determine the x intercept by setting $y = 0$ and $z = 0$ in the equation $3x + 2y + 4z = 12$ and solving for x. We have $3x = 12$ and $x = 4$; the x intercept is 4. Similarly, the y intercept is 6, and the z intercept is 3. These three points—the three intercepts—define a plane, and that portion of the plane in the set $\{(x, y, z): x \geq 0, y \geq 0, z \geq 0\}$ is shown as the shaded triangle in Figure 5.12. Of course, the plane itself extends beyond the shaded area. ■

In graphing the plane of Example 5.19, we showed only that portion of the plane with nonnegative values of the variables (Figure 5.12). To simplify our

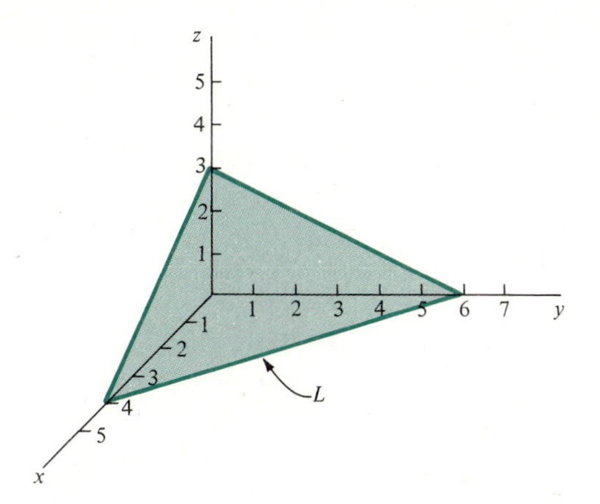

FIGURE 5.12

three-dimensional figures, we restrict our attention to planes which intersect the set $\{(x, y, z): x \geq 0, y \geq 0, z \geq 0\}$ and we show only the intersection of the plane with that set.

The equation $z = 0$ defines a plane, the xy coordinate plane. Likewise, the equation $x = 0$ defines a plane, the yz coordinate plane; and $y = 0$ defines the xz coordinate plane. The intersection of the plane $3x + 2y + 4z = 12$ of Example 5.13 and the plane $z = 0$ is the line $3x + 2y = 12$, the line labeled L in Figure 5.12.

The equation $2x + 5z = 10$ also defines a plane, a plane which does not intersect the y axis. This plane is shown in Figure 5.13. In fact any equation of the form $Ax + By + Cz = D$ with one of the numbers A, B, or C equal to 0 and $D \neq 0$ defines a plane which does not intersect the coordinate axis associated with the zero coefficient.

FIGURE 5.13

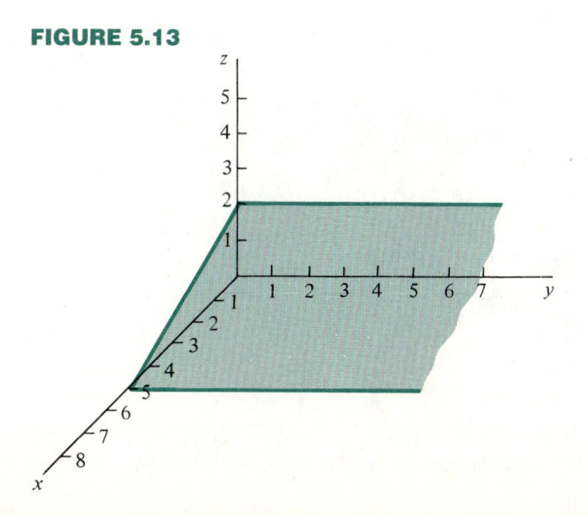

Example 5.20 Graph the two planes

$$\begin{aligned} 3x + 2y + 4z &= 12 \\ 2x + 5z &= 10 \end{aligned}$$

on the same coordinate system, and identify the intersection of the two graphs.

Solution We begin with the observation that if we think about the possible intersections of two planes in three-dimensional space, we anticipate that the intersection will be either a line, a plane, or empty. The graphs of the two planes are shown in Figure 5.14. The intersection of each of the two planes with the plane $y = 0$ is a line, and these two lines intersect at point P. Likewise, the intersection of each of the two planes with the plane $x = 0$ is a line, and these two lines intersect at point Q. The intersection of the two original planes is a line through points P and Q. (The segment connecting P and Q is shown as the dark line in Figure 5.14.) We developed methods of finding the coordinates of points P and Q in Section 5.2, and we will develop methods for finding the equation of the dark line in the figure in this section. ■

There is a notion of orientation for planes which corresponds to the concept of slope for a line. Also, the concepts developed in this section can be given meaning for linear functions of more than three variables. For instance, the set of points $(x_1, x_2, \ldots, x_n)$ which satisfy an equation of the form $A_1x_1 + A_2x_2 + \cdots + A_nx_n = B$ is called a *hyperplane* in n-dimensional space. However, to introduce these ideas would take us astray from the main goals of this chapter, and we omit them.

The primary goal of this section is to develop methods of solving problems like those posed in Examples 5.17 and 5.18. Since we also want to solve similar

FIGURE 5.14

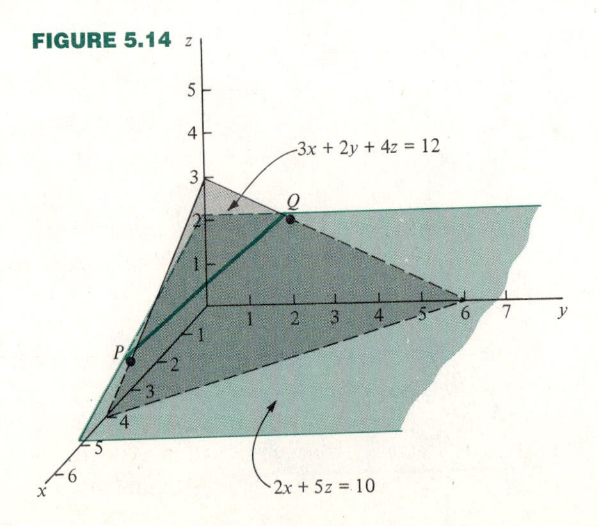

systems with more than three variables, it is useful to begin by introducing a general notation.

In Examples 5.17 and 5.18, the problems were formulated using the three variables x, y, and z. In general, if the problem involves n variables, it is customary to label them $x_1, x_2, \ldots, x_n$, particularly when $n > 3$. With this notation, a system of m linear equations in n variables can be written in the form

$$\begin{aligned} a_{11}x_1 + a_{12}x_2 + \cdots + a_{1n}x_n &= b_1 \\ a_{21}x_1 + a_{22}x_2 + \cdots + a_{2n}x_n &= b_2 \\ \vdots \qquad \vdots \qquad\quad \vdots \quad &\quad \vdots \\ a_{m1}x_1 + a_{m2}x_2 + \cdots + a_{mn}x_n &= b_m \end{aligned} \tag{5.19}$$

It is assumed that the a_{ij}'s and b_j's are known numbers. That is, the values of these numbers are given in the problem. The values of the variables are to be determined.

A solution of system (5.19) is an ordered set of n numbers which when substituted into Equations (5.19) for $x_1, x_2, \ldots, x_n$, respectively, results in actual equalities.

Example 5.21 Show that $x = \frac{6}{5}$, $y = \frac{48}{5}$, $z = \frac{56}{5}$ is a solution of the system (5.18).

Solution On the left-hand side of system (5.18) we replace x by $\frac{6}{5}$, y by $\frac{48}{5}$, and z by $\frac{56}{5}$. We have

$$\begin{aligned} \tfrac{2}{3}x + \tfrac{1}{3}y &= \tfrac{2}{3}(\tfrac{6}{5}) + \tfrac{1}{3}(\tfrac{48}{5}) = \tfrac{4}{5} + \tfrac{16}{5} = \tfrac{20}{5} = 4 \\ \tfrac{1}{3}x + \tfrac{1}{2}z &= \tfrac{1}{3}(\tfrac{6}{5}) + \tfrac{1}{2}(\tfrac{56}{5}) = \tfrac{2}{5} + \tfrac{28}{5} = \tfrac{30}{5} = 6 \\ \tfrac{2}{3}y + \tfrac{1}{2}z &= \tfrac{2}{3}(\tfrac{48}{5}) + \tfrac{1}{2}(\tfrac{56}{5}) = \tfrac{32}{5} + \tfrac{28}{5} = \tfrac{60}{5} = 12 \end{aligned}$$

After making the substitution, we see that the left-hand side of each expression is equal to the right-hand side. In other words, the substitution yields three actual equalities. ■

In Examples 5.17 and 5.18, the labels we used to denote the unknown quantities, i.e., the variables, are unimportant. We used x, y, z. We could equally well have used x_1, x_2, x_3 or u, v, w. The important information of system (5.18) is contained in the coefficients of the variables in the equations. This information can be extracted and represented in an array of numbers.

In particular, we write the coefficients of the variables x, y, z in the first equation in order in the first row of the array. We write the coefficients of x, y, z in the second equation as the second row of the array; the coefficients of the third equation form the third row, and we enclose the array in brackets.

$$\begin{bmatrix} \frac{2}{3} & \frac{1}{3} & 0 \\ \frac{1}{3} & 0 & \frac{1}{2} \\ 0 & \frac{2}{3} & \frac{1}{2} \end{bmatrix} \tag{5.20}$$

Notice that the first entry in each row is a coefficient of variable x, the second entry in each row is a coefficient of variable y, and the third entry in each row is a coefficient of variable z. Each equation must be viewed as containing *all* variables; a missing variable is considered to have a zero coefficient. For example, the first equation in (5.18) has only x and y terms, but it can be written equivalently as

$$\tfrac{2}{3}x + \tfrac{1}{3}y + 0z = 4$$

The array of numbers in (5.20) with the coefficients of each variable aligned in a column is an example of a *coefficient matrix*. In general:

The array

$$\begin{bmatrix} a_{11} & a_{12} & \cdots & a_{1n} \\ a_{21} & a_{22} & \cdots & a_{2n} \\ \cdot & \cdot & & \cdot \\ \cdot & \cdot & & \cdot \\ \cdot & \cdot & & \cdot \\ a_{m1} & a_{m2} & \cdots & a_{mn} \end{bmatrix} \tag{5.21}$$

in which the entries are the coefficients that appear in system (5.19) is called the *coefficient matrix* of the system. The entries in a horizontal line in this matrix form a *row*, and the entries in a vertical line form a *column*. For instance, $[a_{21} \quad a_{22} \quad \cdots \quad a_{2n}]$ is the second row, and the third column is

$$\begin{bmatrix} a_{13} \\ a_{23} \\ \cdot \\ \cdot \\ \cdot \\ a_{m3} \end{bmatrix}$$

In determining the coefficient matrix of a system of equations, it is important to remember that the variables must appear in the same order in every equation and that every variable must have a coefficient, possibly zero, in each equation. The array (5.20) is the coefficient matrix of the system (5.18).

It is convenient to have a shorthand notation for the coefficient matrix of a system of equations. We use the boldface letter **A** to denote the matrix (5.21).

Example 5.22 Find the coefficient matrix **A** of the following system of equations:

$$\begin{aligned} 3x_2 - 4x_1 - x_4 &= 6 \\ x_3 + x_2 - x_1 &= 2 \end{aligned}$$

Solution First we rewrite the system of equations, placing the variables in the same order in both equations and inserting missing variables with zero coefficients:

$$\begin{aligned} -4x_1 + 3x_2 + 0x_3 - x_4 &= 6 \\ -x_1 + x_2 + x_3 + 0x_4 &= 2 \end{aligned}$$

From this it is easy to find the coefficient matrix:

$$\mathbf{A} = \begin{bmatrix} -4 & 3 & 0 & -1 \\ -1 & 1 & 1 & 0 \end{bmatrix}$$ ■

We now turn to the task of describing an algorithm which will enable us to solve all systems of the form (5.19) which have solutions. For notational convenience we write the system (5.19) symbolically as $\mathbf{AX} = \mathbf{B}$. Here **A**, **B**, and **X** are just symbols. In Chapter 6 we will attach a meaning to the symbols and see that the representation $\mathbf{AX} = \mathbf{B}$ is more than just notation.

As we saw in Section 5.2, there are systems of equations which have a unique solution, there are systems which have infinitely many solutions, and there are systems which have no solutions.

A system of equations that has at least one solution is said to be *consistent*, and a system that has no solution is said to be *inconsistent*.

The solution algorithm introduced here is based on the following idea: We successively transform the original system into simpler and simpler systems *without changing the set of solutions* until we obtain a system whose solutions are clear. The technique is the same as that used in Section 5.2 to solve systems of two equations in two variables. The method rests on the following important fact about systems of linear equations.

Theorem on Transforming Systems of Equations

If a system of equations is transformed into a new system by any one of the following operations

1. Interchange two equations.
2. Multiply any equation by a nonzero number.
3. Replace any equation by the sum of that equation and a multiple of any other equation.

then the set of solutions of the transformed system is the same as the set of solutions of the original system.

Our goal in transforming a system is to obtain a new system in which each equation contains only one variable. For such a system the solution can be obtained by reading the value of each variable from the equation containing only that variable. We saw in Section 5.2, however, that even in the special case of two equations and two variables, we could not always achieve this goal. The best that we can do is to always obtain a new system with a special form, called the *reduced form*, that enables us to find the solutions of that system easily. The coefficient matrix we seek has the following properties:

A. The entry in the first row and first column of the coefficient matrix is a 1, and the first nonzero entry in each row is 1. These entries are called *leading ones* (1s).
B. If a column of the coefficient matrix contains a leading 1, then all other entries in that column are 0.
C. As you move from left to right through the columns of the coefficient matrix, the leading 1s occur in successive rows. In particular, all rows with only 0s are at the bottom of the matrix.

Condition C is sometimes phrased in descriptive terms as "The leading 1s march downward and to the right." Matrix **A** shown below is an example of a matrix which satisfies conditions A, B, and C.

$$\mathbf{A} = \begin{bmatrix} 1 & 0 & 0 & 0 \\ 0 & 1 & 1 & 0 \\ 0 & 0 & 0 & 1 \\ 0 & 0 & 0 & 0 \end{bmatrix}$$

Notice that the leading 1s occur successively in row 1, row 2, and row 3.

A coefficient matrix which has the form described by conditions A, B, and C is said to be in *reduced form.*

Example 5.23 Which of the following matrices are in reduced form? If a matrix is not in reduced form, determine which of the conditions A, B, or C are violated.

$$(a)\begin{bmatrix} 1 & 2 & 2 & 0 \\ 0 & 1 & 3 & 0 \\ 0 & 0 & 0 & 1 \end{bmatrix} \qquad (b)\begin{bmatrix} 1 & 0 & 0 & 0 \\ 0 & 0 & 1 & 3 \\ 0 & 1 & 0 & 0 \end{bmatrix}$$

$$(c)\begin{bmatrix} 1 & 0 & 0 & 2 \\ 0 & 1 & 0 & -1 \\ 0 & 0 & 1 & 3 \end{bmatrix} \qquad (d)\begin{bmatrix} 1 & 2 & 0 & 0 \\ 0 & 0 & 1 & 0 \\ 0 & 0 & 0 & 1 \end{bmatrix}$$

Solution (*a*) This matrix is not in reduced form. The first nonzero entry in each row is a 1; however, in the second column there is a leading 1 and a nonzero entry, the 2 immediately above the 1.

(*b*) This matrix is not in reduced form. Conditions A and B are met, but condition C is violated. Indeed, as you move from left to right through the columns of the matrix, the leading 1s appear in row 1, then row 3, then row 2. They do *not* appear in successive rows.

(*c*) and (*d*) Both are in reduced form. ■

The first step in our method of solving a system of equations is to transform it into a new system with a 1 as the upper left entry in the coefficient matrix, as required by condition A. We continue to transform the system until all of the conditions A, B, and C are satisfied.

Example 5.24 Solve the system of equations

$$\begin{aligned} 2x_1 + 4x_2 + 2x_3 &= -4 \\ 3x_1 + 6x_2 + 4x_3 &= -5 \\ x_2 + 2x_3 &= 1 \end{aligned} \tag{5.22}$$

Solution We begin by transforming the system into one in which the coefficient of x_1 in the first equation is 1, so that condition A is satisfied for the first row. To accomplish this, we multiply the first equation by $\frac{1}{2}$ (an operation of type 2). We obtain

$$\begin{aligned} x_1 + 2x_2 + x_3 &= -2 \\ 3x_1 + 6x_2 + 4x_3 &= -5 \\ x_2 + 2x_3 &= 1 \end{aligned}$$

To satisfy condition B, we must transform the system into one in which the coefficients of x_1 in the second and third equations are zero. Since the coefficient of x_1 in the third equation is already 0, we need to consider only the second equation. We multiply the first equation by -3, add it to the second, and use the result to replace the second equation (an operation of type 3). We now have

$$\begin{aligned} x_1 + 2x_2 + x_3 &= -2 \\ x_3 &= 1 \\ x_2 + 2x_3 &= 1 \end{aligned}$$

The leading coefficient in each row is now a 1; however, these leading coefficients do not march down and to the right. To obtain such a form, we need to interchange the second and third equations. After the interchange we have

$$\begin{aligned} x_1 + 2x_2 + x_3 &= -2 \\ x_2 + 2x_3 &= 1 \\ x_3 &= 1 \end{aligned}$$

Our system of equations has now been transformed into one which satisfies two of the three conditions for reduced form (A and C). To satisfy condition B, we wish to keep the coefficient of x_3 as a 1 in the third equation and eliminate x_3 (obtain a zero coefficient) in the other two equations. To accomplish this, we multiply the third equation by -2, add it to the second equation, and use the result as a new second equation. We have

$$\begin{aligned} x_1 + 2x_2 + x_3 &= -2 \\ x_2 \qquad &= -1 \\ x_3 &= 1 \end{aligned}$$

Next, we multiply the third equation by -1, add it to the first, and use the result as a new first equation. We have

$$\begin{aligned} x_1 + 2x_2 \qquad &= -3 \\ x_2 \qquad &= -1 \\ x_3 &= 1 \end{aligned}$$

Finally, we multiply the second equation by -2, add it to the first, and use the result as a new first equation. We have

$$\begin{aligned} x_1 \qquad\qquad &= -1 \\ x_2 \qquad &= -1 \\ x_3 &= \ \ 1 \end{aligned} \tag{5.23}$$

The coefficient matrix of this last system is

$$\begin{bmatrix} 1 & 0 & 0 \\ 0 & 1 & 0 \\ 0 & 0 & 1 \end{bmatrix}$$

which clearly satisfies conditions A, B, and C. That is, it is in reduced form.

A solution (in fact, the only solution) of system (5.23) is obvious: $x_1 = -1$, $x_2 = -1$, $x_3 = 1$. According to the theorem, this is also a solution (and again the only solution) of the original system (5.22). ■

It is important to realize that the system of Example 5.24 could have been solved by using other sequences of transformations. However, each sequence of transformation (if correctly carried out) will lead to the same set of solutions.

It is clear that in Example 5.24 the variables x_1, x_2, and x_3 were just "carried along for the ride." That is, the entire process can be carried out by working with only the entries in the coefficient matrix and the numbers on the right-hand side, the numbers b_j in Equation (5.19). To take advantage of this simplification we introduce the idea of an augmented matrix.

The *augmented matrix* of a system of equations $\mathbf{AX} = \mathbf{B}$ of the form (5.19) is the array of numbers

$$\left[\begin{array}{cccc|c} a_{11} & a_{12} & \cdots & a_{1n} & b_1 \\ a_{21} & a_{22} & \cdots & a_{2n} & b_2 \\ \cdot & \cdot & \cdot & \cdot & \cdot \\ \cdot & \cdot & \cdot & \cdot & \cdot \\ \cdot & \cdot & \cdot & \cdot & \cdot \\ a_{m1} & a_{m2} & \cdots & a_{mn} & b_m \end{array}\right]$$

The vertical line between the two columns on the right indicates that the column on the far right is not a part of the coefficient matrix, but instead represents the right-hand side of Equation (5.19).

Example 5.25 Find the augmented matrix for system (5.22).

Solution

$$\left[\begin{array}{ccc|c} 2 & 4 & 2 & -4 \\ 3 & 6 & 4 & -5 \\ 0 & 1 & 2 & 1 \end{array}\right]$$

■

In working the next example, we show each step in two forms: the form displaying the variables and the form utilizing the augmented matrix.

Example 5.26 Solve the system posed in Example 5.17.

$$\begin{aligned} 20x + 40y + 2z &= 1000 \\ 10x + 10y + z &= 400 \\ 4x + 6y + .2z &= 150 \end{aligned} \tag{5.24}$$

Solution The augmented matrix for system (5.24) is

$$\left[\begin{array}{ccc|c} 20 & 40 & 2 & 1000 \\ 10 & 10 & 1 & 400 \\ 4 & 6 & .2 & 150 \end{array}\right]$$

We begin by multiplying the first equation in system (5.24) by $\frac{1}{20}$, in order to make the first coefficient in the first equation a 1. In matrix terms, we multiply the first row by $\frac{1}{20}$. The new system and the corresponding augmented matrix are

$$\begin{aligned} 1x + 2y + .1z &= 50 \\ 10x + 10y + z &= 400 \\ 4x + 6y + .2z &= 150 \end{aligned} \qquad \left[\begin{array}{ccc|c} 1 & 2 & .1 & 50 \\ 10 & 10 & 1 & 400 \\ 4 & 6 & .2 & 150 \end{array}\right]$$

Next, we replace the second equation—row 2 in the augmented matrix—by the sum of the second equation and -10 times the first equation. This yields a second equation with a first coefficient of 0. The new system and augmented matrix are

$$\begin{aligned} 1x + 2y + .1z &= 50 \\ 0x + (-10)y + 0z &= -100 \\ 4x + 6y + .2z &= 150 \end{aligned} \qquad \left[\begin{array}{ccc|c} 1 & 2 & .1 & 50 \\ 0 & -10 & 0 & -100 \\ 4 & 6 & .2 & 150 \end{array}\right]$$

We now replace the third equation (row 3 of the matrix) by the sum of the third equation and -4 times the first equation. This gives the following system and augmented matrix:

$$\begin{aligned} 1x + \quad 2y + \quad .1z &= 50 \\ 0x + (-10)y + \quad 0z &= -100 \\ 0x + \ (-2)y + (-.2)z &= -50 \end{aligned} \qquad \left[\begin{array}{ccc|c} 1 & 2 & .1 & 50 \\ 0 & -10 & 0 & -100 \\ 0 & -2 & -.2 & -50 \end{array}\right]$$

At this stage of the reduction process we have completed work on the first column of the augmented matrix. In terms of the equations, there is only one equation, the first one, that contains variable x. In the second and third equations the coefficient of x is 0. Next, we need to form a leading 1 in the second row of the augmented matrix. Since the first nonzero entry in the second row is -10, we multiply the second row (equation) by $-\frac{1}{10}$. This gives the following system and augmented matrix:

$$\begin{aligned} 1x + \quad 2y + \quad .1z &= 50 \\ 0x + \quad 1y + \quad 0z &= 10 \\ 0x + (-2)y + (-.2)z &= -50 \end{aligned} \qquad \left[\begin{array}{ccc|c} 1 & 2 & .1 & 50 \\ 0 & 1 & 0 & 10 \\ 0 & -2 & -.2 & -50 \end{array}\right]$$

Since we now have a leading 1 in the second row, we need to create 0s above and below this entry. Accordingly, we replace the first row by the sum of the first row and -2 times the second row, and we replace the third row by the sum of the third row and 2 times the second row. This gives the following:

$$\begin{aligned} 1x + 0y + \quad .1z &= 30 \\ 0x + 1y + \quad 0z &= 10 \\ 0x + 0y + (-.2)z &= -30 \end{aligned} \qquad \left[\begin{array}{ccc|c} 1 & 0 & .1 & 30 \\ 0 & 1 & 0 & 10 \\ 0 & 0 & -.2 & -30 \end{array}\right]$$

The first and second rows now have leading 1s, and we need to form a leading 1 in the third row. Since the first nonzero entry in the third row is $-.2$, we multiply row 3 by $-1/.2 = -5$ to obtain

$$\begin{aligned} 1x + 0y + .1z &= 30 \\ 0x + 1y + 0z &= 10 \\ 0x + 0y + 1z &= 150 \end{aligned} \qquad \left[\begin{array}{ccc|c} 1 & 0 & .1 & 30 \\ 0 & 1 & 0 & 10 \\ 0 & 0 & 1 & 150 \end{array}\right]$$

The final step in obtaining reduced form for this augmented matrix is to transform all the entries above the leading 1 in the third column into 0s. The only such entry is in the first row, so we replace row 1 by the sum of row 1 and $-.1$ times row 3. This gives the reduced form

$$\begin{aligned} 1x + 0y + 0z &= 15 \\ 0x + 1y + 0z &= 10 \\ 0x + 0y + 1z &= 150 \end{aligned} \qquad \left[\begin{array}{ccc|c} 1 & 0 & 0 & 15 \\ 0 & 1 & 0 & 10 \\ 0 & 0 & 1 & 150 \end{array}\right]$$

The solution of the problem posed in Example 5.17 is given by this last system of equations, namely,

$$\begin{aligned} x \qquad\qquad &= 15 \\ y \qquad &= 10 \\ z &= 150 \end{aligned}$$

To check our work, we verify that this solution satisfies the original system (5.24). We have

$$\begin{aligned} 20(15) + 40(10) + 2(150) &= 1000 \\ 10(15) + 10(10) + 1(150) &= 400 \\ 4(15) + 6(10) + .2(150) &= 150 \end{aligned}$$

This verifies that our solution is correct, and thus the homeowner should buy 15 bushes, 10 trees, and 150 flowers. ■

If a system of equations has any solutions, then the technique illustrated in Examples 5.24 and 5.26 will produce all solutions. Also, if a system has no solutions, then the technique will show that fact as well.

Example 5.27 Find the solutions (if any) of the system of equations

$$\begin{aligned} 3x_1 - x_2 &= 4 \\ 6x_1 - 2x_2 &= 3 \end{aligned}$$

Solution To illustrate the method, we again proceed by using both the equations and the augmented matrix. The augmented matrix for the given system is

$$\left[\begin{array}{cc|c} 3 & -1 & 4 \\ 6 & -2 & 3 \end{array}\right]$$

We multiply the first equation by -2, add it to the second equation, and use the result as a new second equation. We have

$$\begin{aligned} 3x_1 - x_2 &= 4 \\ 0 &= -5 \end{aligned} \qquad \left[\begin{array}{cc|c} 3 & -1 & 4 \\ 0 & 0 & -5 \end{array}\right]$$

Thus there is an equation of the form $0x_1 + 0x_2 = -5$ in the transformed system. Since this is true for *no choice* of x_1 and x_2, we conclude that the original system has no solution. ■

Inconsistent Systems

If in using this technique it ever happens that a transformed system has an equation with zero on the left-hand side and a nonzero number on the right, then the original system has no solutions, i.e., it is inconsistent.

In terms of the augmented matrix (or its transform) this condition is as follows: If there is a row with only 0s to the left of the vertical line and a nonzero entry in that row to the right of the vertical line, then the system has no solutions.

The remainder of this section is concerned with systems which have more than one solution. A system of linear equations has *no solutions*, *exactly one solution*, or *infinitely many solutions*. There are no other possibilities.

The solution set of a single linear equation in three variables is a plane in three-dimensional space. The set of points which are simultaneously the solution of two linear equations in three variables will be empty if the planes associated with the two equations are parallel; it will be a line if the planes intersect but are not coincident; and it will be a plane if the planes are coincident. The set of points that are simultaneously the solution of three linear equations in three variables can be empty, a single point, a line, or a plane. In the first case, the system has no solution; in the second case, a unique solution; and in the third and fourth cases, infinitely many solutions—either all points on a line or all points in a plane. Thus, there are infinitely many solutions associated with intersections which are either lines or planes. We conclude that there can be different kinds of solution sets with infinitely many elements. We look further into this situation after our next example, and we make the ideas precise in a theorem.

A graph of one of the cases, three equations which have a unique solution, is shown in Figure 5.15. Suppose we have three equations, denoted Equation 1, Equation 2, and Equation 3, in three-dimensional space where the coordinate axes are denoted by x_1, x_2, and x_3. The graph of the solution set of Equation 1 might appear as in Figure 5.15*a*. The solution set of the single equation is a plane. The graphs of the solutions of Equations 1 and 2 might appear as in Figure 5.15*b*. Here the solution set of the two equations is a line. Finally, the solution set of all three equations, a point, might appear as in Figure 5.15*c*.

Example 5.28 Find all solutions of the system

$$\begin{aligned} x_1 + 2x_2 - x_3 &= 5 \\ 2x_1 + 3x_2 - 3x_3 &= 8 \\ x_2 + x_3 &= 2 \end{aligned} \tag{5.25}$$

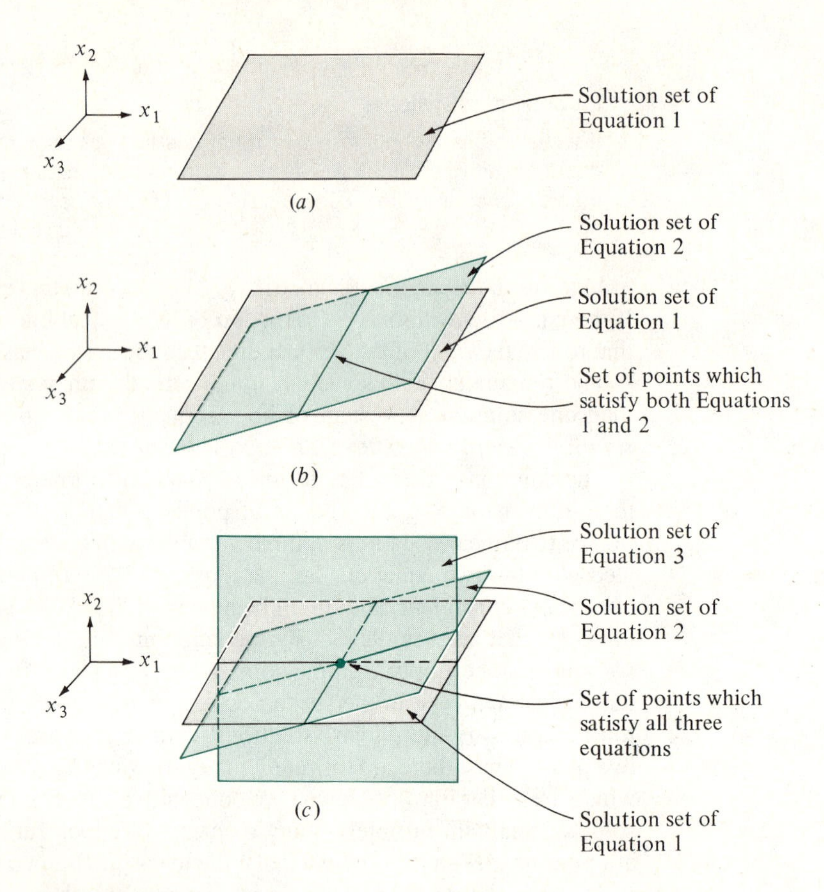

FIGURE 5.15

Solution In this example, and in general, we will work only with the augmented matrix. We begin by interchanging the second and third rows to obtain a leading 1 as the entry in the second row and second column of the matrix. We have

$$\left[\begin{array}{rrr|r} 1 & 2 & -1 & 5 \\ 0 & 1 & 1 & 2 \\ 2 & 3 & -3 & 8 \end{array}\right]$$

Next, we multiply the first row by -2, add it to the third row, and use the result as a new third row. We have

$$\left[\begin{array}{rrr|r} 1 & 2 & -1 & 5 \\ 0 & 1 & 1 & 2 \\ 0 & -1 & -1 & -2 \end{array}\right]$$

Next, we add the second row to the third row and use the result as a new third row. We have

$$\left[\begin{array}{ccc|c} 1 & 2 & -1 & 5 \\ 0 & 1 & 1 & 2 \\ 0 & 0 & 0 & 0 \end{array}\right]$$

Finally, we multiply the second row by -2, add it to the first row, and use the result as a new first row. We have

$$\left[\begin{array}{ccc|c} 1 & 0 & -3 & 1 \\ 0 & 1 & 1 & 2 \\ 0 & 0 & 0 & 0 \end{array}\right] \tag{5.26}$$

The augmented matrix is now in reduced form.

The system of equations with the augmented matrix (5.26) is

$$\begin{aligned} x_1 \qquad - 3x_3 &= 1 \\ x_2 + \ x_3 &= 2 \end{aligned}$$

Suppose $x_3 = 0$. Then the system becomes

$$\begin{aligned} x_1 &= 1 \\ x_2 &= 2 \end{aligned}$$

Thus one solution of the system (5.25) is $x_1 = 1$, $x_2 = 2$, $x_3 = 0$. Likewise, if $x_3 = -1$, then the system becomes

$$\begin{aligned} x_1 - 3(-1) &= 1 \\ x_2 + \ (-1) &= 2 \end{aligned}$$

with the solution $x_1 = -2$, $x_2 = 3$, $x_3 = -1$. The same method works when x_3 is set equal to any number. That is, for any choice of x_3 we can determine x_1 and x_2. We have

$$\begin{aligned} x_1 - 3x_3 &= 1 \\ x_2 + \ x_3 &= 2 \end{aligned} \qquad \text{or} \qquad \begin{aligned} x_1 &= 1 + 3x_3 \\ x_2 &= 2 - \ x_3 \end{aligned}$$

For each number x_3, a solution of (5.25) is given by $x_1 = 1 + 3x_3$, $x_2 = 2 - x_3$, x_3. Since there are infinitely many choices for x_3—any number will do—and since each choice of x_3 gives a different solution of the system, there are infinitely many solutions of the system (5.25). Moreover, every solution of the system is given in this form for some value of x_3, and consequently the solution set is specified by

$$x_1 = 1 + 3x_3, \qquad x_2 = 2 - x_3, \qquad x_3 \text{ arbitrary}$$

■

In Example 5.28 one of the variables (x_3) could be assigned an arbitrary value. The geometry in this case is that the solution set of system (5.25) is a line in three-dimensional space. It may happen that two or more of the variables can be assigned arbitrary values. For example, if the solution set of a system of three equations in three variables is a plane in three-dimensional space, then two of the variables can be specified arbitrarily. There is a means of determining how many variables can be specified arbitrarily.

Theorem on the Solution Set of a System of Linear Equations

Suppose that a consistent system of linear equations in n variables has been transformed into reduced form with k nonzero rows. Then there are solutions in which $n - k$ of the variables can be specified arbitrarily. In particular, each of the $n - k$ variables associated with a column in the reduced coefficient matrix which does *not* contain a leading 1 can be specified arbitrarily. If $n > k$, then there are infinitely many solutions.

Example 5.29 Find all the solutions of the system

$$\begin{aligned} x_1 - x_2 - 2x_4 &= 3 \\ 2x_1 - 2x_2 + 2x_3 - 2x_4 &= 4 \\ x_3 + x_4 &= -1 \\ x_1 - x_2 + 2x_3 &= 1 \end{aligned}$$

Solution There are four variables in the system, so $n = 4$. To find k, we must transform the system to reduced form. Again we work only with the augmented matrix. The original augmented matrix and the transformed matrix obtained by multiplying row 1 by -2, adding it to row 2, and using the result as a new row 2 are

$$\left[\begin{array}{cccc|c} 1 & -1 & 0 & -2 & 3 \\ 2 & -2 & 2 & -2 & 4 \\ 0 & 0 & 1 & 1 & -1 \\ 1 & -1 & 2 & 0 & 1 \end{array}\right] \quad \text{and} \quad \left[\begin{array}{cccc|c} 1 & -1 & 0 & -2 & 3 \\ 0 & 0 & 2 & 2 & -2 \\ 0 & 0 & 1 & 1 & -1 \\ 1 & -1 & 2 & 0 & 1 \end{array}\right]$$

Next, multiply row 1 by -1, add it to row 4, and use the result as a new row 4. We have

$$\left[\begin{array}{cccc|c} 1 & -1 & 0 & -2 & 3 \\ 0 & 0 & 2 & 2 & -2 \\ 0 & 0 & 1 & 1 & -1 \\ 0 & 0 & 2 & 2 & -2 \end{array}\right]$$

Next, we multiply row 2 by -1 and add it to row 4 to obtain a new row 4; also we multiply row 2 by $-\frac{1}{2}$, add it to row 3, and use the result as a new row 3. We have

$$\left[\begin{array}{cccc|c} 1 & -1 & 0 & -2 & 3 \\ 0 & 0 & 2 & 2 & -2 \\ 0 & 0 & 0 & 0 & 0 \\ 0 & 0 & 0 & 0 & 0 \end{array}\right]$$

Finally, multiply row 2 by $\frac{1}{2}$ to obtain the reduced-form augmented matrix

$$\left[\begin{array}{cccc|c} 1 & -1 & 0 & -2 & 3 \\ 0 & 0 & 1 & 1 & -1 \\ 0 & 0 & 0 & 0 & 0 \\ 0 & 0 & 0 & 0 & 0 \end{array}\right]$$

We see from this reduced-form matrix that there are two nonzero rows, and hence $k = 2$. Moreover, according to the theorem, the variables x_2 and x_4 can be specified arbitrarily. The equations corresponding to the last augmented matrix are

$$\begin{aligned} x_1 - x_2 - 2x_4 &= 3 \\ x_3 + x_4 &= -1 \end{aligned} \qquad \text{or} \qquad \begin{aligned} x_1 &= x_2 + 2x_4 + 3 \\ x_3 &= -x_4 - 1 \end{aligned}$$

This form of the system shows that x_1 and x_3 depend on x_2 and x_4 and that the solution set of the original system can be expressed as

$$x_1 = x_2 + 2x_4 + 3, \; x_3 = -x_4 - 1, \; x_2 \text{ arbitrary and } x_4 \text{ arbitrary} \tag{5.27}$$

■

Example 5.30 Find the specific solution of the system of Example 5.29 with $x_2 = 2$ and $x_4 = -1$.

Solution We use the result (5.27) of Example 5.29. Setting $x_2 = 2$ and $x_4 = -1$, we have the specific solution

$$x_1 = 3, \qquad x_2 = 2, \qquad x_3 = 0, \qquad x_4 = -1$$

■

As our final example we return to the situation introduced in Example 5.20 and pictured in Figure 5.14.

Example 5.31 Find the intersection of the planes

$$\begin{aligned} 3x + 2y + 4z &= 12 \\ 2x + 5z &= 10 \end{aligned} \tag{5.28}$$

and find the intersection of this line with the coordinate planes $x = 0$ and $y = 0$.

Solution The augmented matrix for the system is

$$\left[\begin{array}{ccc|c} 3 & 2 & 4 & 12 \\ 2 & 0 & 5 & 10 \end{array}\right]$$

Using the techniques developed in this section, we transform the augmented matrix successively to

$$\left[\begin{array}{ccc|c} 1 & \frac{2}{3} & \frac{4}{3} & 4 \\ 2 & 0 & 5 & 10 \end{array}\right]$$

then

$$\left[\begin{array}{ccc|c} 1 & \frac{2}{3} & \frac{4}{3} & 4 \\ 0 & -\frac{4}{3} & \frac{7}{3} & 2 \end{array}\right]$$

then

$$\left[\begin{array}{ccc|c} 1 & \frac{2}{3} & \frac{4}{3} & 4 \\ 0 & 1 & -\frac{7}{4} & -\frac{3}{2} \end{array}\right]$$

and finally

$$\left[\begin{array}{ccc|c} 1 & 0 & \frac{5}{2} & 5 \\ 0 & 1 & -\frac{7}{4} & -\frac{3}{2} \end{array}\right]$$

Consequently, the solution set of the system (5.28) is

$$x = 5 - \tfrac{5}{2}z \qquad y = -\tfrac{3}{2} + \tfrac{7}{4}z \qquad z \text{ arbitrary} \tag{5.29}$$

The equations (5.29) define a line in three-dimensional space. We complete the solution by finding the intersection of this line with the coordinate planes $x = 0$ and $y = 0$. To find the intersection with the coordinate plane $x = 0$, we set $x = 0$ in (5.29). We find that z must be equal to 2. This in turn gives $y = 2$. Consequently, the point of intersection of the line (5.29) with the plane $x = 0$ is $(0, 2, 2)$. This is the point labeled Q on Figure 5.14. Similarly, the point of intersection of the line (5.29) with the plane $y = 0$ is $(\frac{20}{7}, 0, \frac{6}{7})$. This is the point labeled P in Figure 5.14. ■

We close this section with an explicit statement of the method we have developed to solve systems of linear equations.

Algorithm for Solving a System of Linear Equations

1. Form the augmented matrix.
2. Use row operations to transform the augmented matrix into reduced form.
3. Interpret the reduced form to obtain information about the solution set.
   a. If there is a row with zeros to the left of the vertical line and a nonzero entry to the right, then the system has no solutions.
   b. If the system has n equations in n variables and the reduced form has a leading 1 in each row, then there is a unique solution.
   c. If a consistent system has n variables and the reduced form has k, $k < n$, nonzero rows, then there are infinitely many solutions in which $n - k$ of the variables can be specified arbitrarily.

Exercises for Section 5.3

For Exercises 1 through 4, formulate the problem as a system of linear equations. Be sure to define your variables carefully.

Manufacturing

1. Ted's Toys makes toy airplanes, boats, and cars. The materials used are plastic, wood strips, and steel. Each airplane uses 100 grams of plastic, 10 inches of wood strips, and 200 grams of steel. Each boat uses 50 grams of plastic, 100 inches of wood strips, and 50 grams of steel; and each car uses 50 grams of plastic and 150 grams of steel. If Ted's has on hand 10,500 grams of plastic, 1500 inches of wood strips, and 25,500 grams of steel, how many planes, boats, and cars should be made to use up all these supplies?

Investments

2. Rachael, Stephanie, and Tina are investing for retirement, using stocks, bonds, and money market funds. They use the following guidelines: Rachael wants half of her money in stocks and the rest split equally in bonds and money market funds. Stephanie wants her money split equally among all three areas, and Tina wants her money split equally between stocks and bonds. If the annual return on stocks is 9 percent, on bonds 6 percent, and on money market funds 3 percent, what is the total that each woman should invest in order that each of them gain $10,000 from her investment?

Investments

3. Suppose that in Exercise 2 each woman decides to invest $100,000 and that Rachael expects a gain of $9000, Stephanie expects a gain of $6000 and Tina expects a gain of $8000. For what set of returns on stocks, bonds, and money market funds will each of these investors be satisfied?

Resource allocations

4. Raskins and Bobbins Ice Cream Shop makes three kinds of ice cream using skim milk, cream, vanilla, and cacao. Each gallon of Deluxe Vanilla uses 3 quarts of milk, 1 quart of cream, and 2 ounces of vanilla. Each gallon of Regular Vanilla uses 3.5 quarts of milk, 0.5 quart of cream, and 1 ounce of vanilla. Each gallon of Deluxe Chocolate uses 3.25 quarts of milk, 0.75 quart of cream, and 2 ounces of cacao. How many gallons of each type of ice cream should be made in order to use up 100 gallons of milk, 25 gallons of cream, 5 pounds of vanilla, and 10 pounds of cacao?

5. Decide which of the following matrices are in reduced form and which are not. If a matrix is not in reduced form, tell why it is not.
(*a*) $\begin{bmatrix} 0 & 1 \\ 1 & 1 \end{bmatrix}$ (*b*) $\begin{bmatrix} 1 & 1 \\ 0 & 0 \end{bmatrix}$ (*c*) $\begin{bmatrix} 1 & 1 \\ 1 & 1 \end{bmatrix}$

6. Decide which of the following matrices are in reduced form and which are not. If a matrix is not in reduced form, tell why it is not.
(*a*) $\begin{bmatrix} 1 & -2 & 0 \\ 0 & 0 & 1 \\ 0 & 0 & 0 \end{bmatrix}$ (*b*) $\begin{bmatrix} 1 & -2 & 0 \\ 0 & 0 & 1 \\ 0 & 0 & 1 \end{bmatrix}$ (*c*) $\begin{bmatrix} 1 & 0 & 2 & 3 \\ 0 & 0 & 2 & 7 \\ 0 & 1 & 4 & 9 \\ 0 & 0 & 0 & 1 \end{bmatrix}$

7. Find the augmented matrix for each of the following systems of equations.
(*a*) $\begin{aligned} 2x_1 - x_2 - x_3 &= 2 \\ x_1 + x_2 + 2x_3 &= 4 \end{aligned}$ (*b*) $\begin{aligned} 3x &= 4 + 2y \\ x + 3y + 1 &= 0 \end{aligned}$

8. Find the augmented matrix for each of the following systems of equations.
(*a*) $\begin{aligned} 3x_1 + 2x_2 - x_3 &= 8 \\ x_1 + x_2 + x_3 &= 2 \\ 2x_1 + x_2 - x_3 &= 5 \end{aligned}$ (*b*) $\begin{aligned} 2x_2 - x_1 &= 5 \\ 3x_3 - x_1 &= 2 \\ 2x_1 - x_3 &= 6 \end{aligned}$

9. Verify that $x_1 = 1, x_2 = 2, x_3 = -1$ is a solution of Exercise 8*a*.

10. Suppose that

$$\left[\begin{array}{rrr|r} 2 & 3 & 1 & 6 \\ 6 & -2 & -8 & 7 \\ 8 & 5 & -3 & 17 \end{array}\right]$$

is the augmented matrix of a system of equations.
(*a*) Write the system in the form (5.19).
(*b*) Verify that $x_1 = \frac{5}{2}, x_2 = 0, x_3 = 1$ is a solution of the system.
(*c*) Verify that $x_1 = \frac{3}{2}, x_2 = 1, x_3 = 0$ is a solution of the system.

In Exercises 11 through 20, find all solutions of the given system of equations.

11. $\begin{aligned} 2x + y + 3z &= 1 \\ 3x - 2y + 4z &= -1 \\ 2x - 4y + 2z &= -2 \end{aligned}$

12. $\begin{aligned} -x + 3y + 2z &= 2 \\ 4x + 2y + z &= 4 \\ -2x + 4y - z &= 6 \end{aligned}$

13. $\begin{aligned} 2x - 4y + z &= 7 \\ 4y + 2z &= -3 \end{aligned}$

14. $\begin{aligned} 2x - 8y + 3z &= 8 \\ 3x - 9y + 6z &= 12 \\ x - 3y + 2z &= 4 \end{aligned}$

15. $\begin{aligned} x - 2y &= 3 \\ x + 2y + 2z &= 3 \\ x + 6y + 4z &= 3 \end{aligned}$

16. $\begin{aligned} 3x_1 + 5x_2 - x_3 + 2x_4 &= 2 \\ 2x_1 - x_3 &= 0 \\ 2x_2 - x_4 &= -3 \\ x_1 - 3x_2 - x_3 &= 1 \end{aligned}$

17. $\begin{aligned} x - y + z &= 3 \\ 2x - 2y + z &= 7 \\ x - 2y + 3z &= 3 \end{aligned}$

18. $\begin{aligned} x_1 + x_3 &= 2 \\ 2x_1 - x_2 + x_3 &= 3 \\ 2x_2 + 3x_3 &= 2 \end{aligned}$

19. $\begin{aligned} 2x + 2y - z &= 8 \\ -2x + y + z &= 4 \\ 4x - y + 2z &= -4 \end{aligned}$

20. $\begin{aligned} x - y + 2z &= 2 \\ x - 5y + 5z &= -2 \\ x + 3y &= 2 \end{aligned}$

Folklore 21. Robin makes bows and arrows using wood, string, and feathers. Each bow uses 5 feet of wood and 4 feet of string, while each arrow uses 3 feet of wood and 4 feathers. If Robin has 100 feet of wood and 32 feet of string, how many feathers does he need so that he can use up all the wood, string, and feathers making bows and arrows?

Folklore 22. Using the setting of Exercise 21, suppose that Robin has 120 feathers and 100 feet of wood. How much string does he need so that he can use up all the wood, string, and feathers making bows and arrows? Answer the same question if he has 100 feet of wood and 140 feathers?

23. Solve the problem formulated in Exercise 1.

24. Solve the problem formulated in Exercise 2.

25. Solve the problem formulated in Exercise 3.

26. Solve the problem formulated in Exercise 4.

Management 27. A small businessman allocates his time among sales (both new clients and old clients), office management, and long-range planning. He decides that he should devote half his time to sales and twice as much time to old clients as to new clients. Also, he decides to devote twice as much time to new clients as to long-range planning. Assume he works 40 hours each week.
 (*a*) Find a system of four equations in four variables that represents this information in mathematical form.
 (*b*) Solve the system and determine how the businessman should allocate his time to meet his goals.

Education 28. A student is trying to decide how to allocate her study time among mathematics, English, biology, and economics. She decides to spend a total of 45 hours per week studying and to spend twice as much time on mathematics and biology combined as on English and economics combined. Also, she will spend twice as much time on economics as on English and the same amount of time on mathematics as on biology. How much time will she spend on each subject per week?

29. (*a*) Find all solutions of the system

$$\begin{aligned} 3x_1 - 2x_2 + x_3 + 2x_4 &= 6 \\ 2x_1 + 4x_2 - 2x_3 + 3x_4 &= 2 \end{aligned}$$

 (*b*) Find all solutions with $x_1 = 0$.

30. (*a*) Find all solutions of the system

$$\begin{aligned} 3x - 2y + z - w &= 8 \\ 2x + y - z &= 6 \\ x - 3y + w &= 2 \end{aligned}$$

 (*b*) Find all solutions with $w = 0$.
 (*c*) Find all solutions with $y = 0$.

31. Find all solutions of the system

$$\begin{aligned} 3x - 2y + z - w &= 8 \\ 4x - 5y + w &= 8 \\ 2x + y - z &= 6 \\ x - 3y + w &= 2 \end{aligned}$$

32. (*a*) Find all solutions of the system

$$\begin{aligned} x_1 - x_2 + 2x_3 - x_4 &= 7 \\ x_1 + x_2 - 4x_3 + x_4 &= 3 \\ x_1 - x_3 &= 5 \\ 2x_2 + 6x_3 - 2x_4 &= 4 \end{aligned}$$

(*b*) Find a specific solution with $x_1 = 0$ and $x_2 = 0$.
(*c*) Are there any solutions with $x_1 = 0$ and $x_3 = 0$?

33. (*a*) Solve the following system of equations to find the intersection of the corresponding planes:

$$\begin{aligned} 5x + 5y + 2z &= 20 \\ 3x + y &= 6 \end{aligned}$$

(*b*) Find the intersection of this line with the coordinate planes $y = 0$ and $z = 0$.
(*c*) Draw a figure similar to Figure 5.14 for this situation.

34. (*a*) Solve the following system of equations to find the intersection of the corresponding planes:

$$\begin{aligned} 6x + 2y + z &= 10 \\ 3y + 4z &= 20 \end{aligned}$$

(*b*) Find the intersection of this line with the coordinate planes $x = 0$ and $y = 0$.
(*c*) Draw a figure similar to Figure 5.14 for this situation.

IMPORTANT TERMS AND CONCEPTS

You should be able to describe, define, or give examples of and use each of the following:

Coordinates in the plane and in three-dimensional space
Line
General equation of a line
Intercept
Slope
Slope-intercept equation of a line
Equation of a plane
System of linear equations
Consistent system
Inconsistent system
Solution of a system of linear equations
Algorithm for solving a system of linear equations
Coefficient matrix
Augmented matrix
Reduced form

REVIEW EXERCISES

Economics

1. An economist reviews recent economic growth and inflation, and she concludes that both economic growth and inflation appear to be linear functions of time. Economic growth is now 3.0 percent per year, and it is increasing 0.6 percent per year. Inflation is now 2.0 percent per year, and it is increasing 1 percent per year. If the economist's assumptions continue to hold, when will economic growth and inflation be equal?

Economics

2. In the setting of Exercise 1, suppose that economic growth is increasing at the rate of 2.5 percent per year. If the assumptions continue to hold, when will economic growth be twice the rate of inflation?

3. Find the slope, the x intercept, and the y intercept of each of the lines.
   (*a*) $3x - 6y = 18$ (*b*) $-3x + 5y = -7$

4. Find the equation of the line through the point $(-3, 5)$ and parallel to the line through the points $(-2, -1)$ and $(3, -4)$.

5. Find all solutions of the system of equations

$$\begin{aligned} -3x + 5y &= 15 \\ -6x - 4y &= 2 \end{aligned}$$

6. Find all solutions of the system of equations

$$\begin{aligned} 3x + 2y &= -1 \\ 5x + 4y &= 0 \end{aligned}$$

7. Find all solutions of the system of equations

$$\begin{aligned} -2x - 4y &= 6 \\ 6x + 6y &= -3 \end{aligned}$$

Resource allocation

8. Every left-handed widget uses 5 ounces of plastic and 10 ounces of steel, while every right-handed widget uses 4 ounces of plastic and 11 ounces of steel. What is the total number of widgets that can be made from 300 ounces of plastic and 750 ounces of steel?

Sports

9. Rachael, Stephanie, and Tina begin running together on a 400-meter oval track. Rachael runs at a pace of 2 minutes per lap, Stephanie at a pace of 2 minutes 15 seconds per lap, and Tina at a pace of 2 minutes 30 seconds per lap. How far has Rachael run when Stephanie first catches up to Tina?

Depreciation

10. A car which cost \$18,900 when new is depreciated linearly for a period of 10 years. At the end of 10 years, the salvage value is \$350.
    (*a*) What is the depreciated value (called the *book value*) of the car after 4 years?
    (*b*) If v is the book value of the car and t is the number of years since the car was purchased, find a formula which expresses the relationship between v and t.

Depreciation

11. A data processing company has a computer which cost \$1,000,000 when new and which depreciates linearly to a salvage value of \$50,000 over 10 years. The company also has a computer which cost \$2,225,000 when new and which depreciates linearly to a salvage value of \$75,000 over a period of 10 years. If the computers were purchased at the same time, when will the book value of the more expensive computer be exactly twice the book value of the less expensive computer?

12. Find the coefficient matrix of each of the following systems of equations.

(*a*)
$$\begin{aligned} x_1 + x_2 + 3x_4 &= 7 \\ x_2 &= 10 \\ x_1 - x_3 &= 20 \end{aligned}$$

(*b*)
$$\begin{aligned} x_1 + x_2 &= 5 - x_2 \\ 3x_1 - 3x_2 &= 2x_1 \end{aligned}$$

13. Verify that $x_1 = 3, x_2 = 1$ is a solution of the system of equations in Exercise 12*b*.

14. Verify that $x_1 = 0$, $x_2 = 10$, $x_3 = -20$, $x_4 = -1$ is a solution of the system of equations in Exercise 12*a*.

15. Find all solutions of the system of equations

$$\begin{aligned} 2x_1 + 4x_2 &= 8 \\ 4x_1 - 2x_2 &= 4 \\ 6x_1 + 2x_2 &= 12 \end{aligned}$$

16. Find all solutions of the system of equations

$$\begin{aligned} 2x_1 + 2x_2 - 4x_3 &= 12 \\ x_1 + x_2 + x_3 &= 6 \\ 3x_1 + 3x_2 - 3x_3 &= 3 \end{aligned}$$

17. Find all solutions of the system of equations

$$\begin{aligned} 2x - 5y &= 0 \\ x - 3y - z &= -1 \\ -x + 2y - z &= -1 \end{aligned}$$

18. Find all solutions of the system of equations

$$\begin{aligned} x_1 + 5x_2 + 2x_3 &= 0 \\ 2x_1 + 7x_2 + x_3 &= -3 \\ 2x_2 + 3x_3 &= 3 \end{aligned}$$

19. Find all solutions of the system of equations

$$\begin{aligned} x_1 - x_3 &= 2 \\ 2x_1 - x_2 - x_3 &= 3 \\ 2x_2 - 3x_3 &= 4 \end{aligned}$$

20. Find all solutions of the system of equations

$$\begin{aligned} x - 3y &= 2 \\ x + y + 2z &= 2 \\ x + 5y + 5z &= 4 \end{aligned}$$

21. (*a*) Find all solutions of the system

$$\begin{aligned} 2x_3 + 3x_4 &= 5 \\ x_1 + x_2 - x_3 &= 2 \\ 2x_2 - 2x_3 - x_4 &= 1 \end{aligned}$$

(*b*) Find the solutions with $x_2 = 0$.

(*c*) Find the solutions with $x_2 = 2$.

22. (*a*) Find all solutions of the system

$$\begin{aligned} 2x_1 - x_2 - 3x_3 &= 6 \\ x_1 + 2x_3 &= 1 \\ x_1 + x_2 + 3x_3 &= -3 \end{aligned}$$

(*b*) Are there any solutions with $x_1 = 1$?

Business 23. The owner of a fleet of trucks is looking for a simple formula to help him predict the maintenance costs for each year of use of a truck. He believes that the maintenance cost for a year is determined by the age of the truck at the start of the year and the miles the truck is driven that year. He proposes that the cost for a year, in dollars, is given by the formula

$$\text{Cost} = 50 + 20 \times (\text{miles driven in 1000s}) + 70 \times (\text{age in years})$$

Plot this function in three dimensions and indicate the cost line that corresponds to an almost-new truck (1 year old) and the cost line for a truck driven 10,000 miles. Assume that all trucks are at most 10 years old and that they are driven at most 20,000 miles in a year. At what point do these two lines intersect?

24. Using the setting of Exercise 23, suppose that the owner of the fleet wishes to keep maintenance costs under $500 for each of his trucks. Show, in a figure, how this condition constrains the miles that certain trucks can be driven and how it constrains the possible ages for trucks that are driven a certain number of miles.

In Exercises 25 through 29, the augmented matrix shown has been obtained by a sequence of row operations. In each case determine which of the following statements is true about the associated system of equations.

(*a*) The system has a unique solution.

(*b*) The system has no solution.

(*c*) The system has an infinite number of solutions in which one variable can be selected arbitrarily.

(*d*) The system has an infinite number of solutions in which two variables can be selected arbitrarily.

25. $\left[\begin{array}{rrr|r} 1 & 0 & 5 & 5 \\ 0 & 1 & -3 & 4 \\ 0 & 2 & -6 & -8 \end{array}\right]$

26. $\left[\begin{array}{rrr|r} 1 & 0 & 4 & 4 \\ 0 & 1 & -3 & 4 \\ 0 & 2 & -6 & 8 \end{array}\right]$

27. $\left[\begin{array}{rrr|r} 1 & 0 & 3 & 3 \\ 0 & 1 & -3 & 4 \\ 0 & 2 & -5 & 8 \end{array}\right]$

28. $\left[\begin{array}{rrrr|r} 1 & 0 & 0 & 0 & 2 \\ 0 & 1 & -3 & 4 & 0 \\ 0 & 2 & -6 & -8 & 1 \\ 0 & 0 & 1 & 1 & 2 \end{array}\right]$

29. $\left[\begin{array}{rrrr|r} 1 & 0 & 0 & 0 & 2 \\ 0 & 1 & -3 & 4 & 1 \\ 0 & 2 & -6 & 8 & 2 \\ 0 & 0 & 1 & 1 & 2 \end{array}\right]$

30. A system of equations is given by

$$\begin{aligned} 3x + y - z &= 6 \\ 2x \phantom{{}+y} - z &= 2 \\ y + kz &= 3 \end{aligned}$$

where k is a number. For what values of k (if any) does the system have infinitely many solutions?

31. Graph the planes $2x + 5y = 10$ and $3y + 4z = 8$ on a three-dimensional coordinate system. Restrict your attention to $x \geq 0, y \geq 0, z \geq 0$.

32. The planes defined in Exercise 31 intersect in a line.
    (*a*) Solve the system of equations of Exercise 31.
    (*b*) Use the solution you obtained in (*a*) to find the coordinates of the points of intersection of the line with the planes $x = 0$ and $y = 0$.

CHAPTER

6

Matrix Algebra and Applications

6.0 THE SETTING AND OVERVIEW

Matrices are mathematical objects which are useful in the study of systems of linear equations, the topic of Chapter 5, as well as in many other areas of mathematics. Matrices are arrays of numbers, and they can be manipulated and combined in much the same way as we handle numbers. In this chapter we develop an "algebra" of matrices, i.e., a method of adding, subtracting, multiplying, and dividing matrices when these operations make sense. Also, we illustrate the use of the ideas in this chapter combined with those in Chapter 5 in an important model from economic analysis. This model is useful in theoretical studies as well as in very practical situations.

6.1 MATRIX NOTATION AND ALGEBRA

A *matrix* is a rectangular array. If the array has m (horizontal) rows and n (vertical) columns, the matrix is said to be of dimension $m \times n$, read "m by n." The entry in the ith row and the jth column of the matrix is said to be the (i, j) entry in the matrix. Matrix **A** with entries a_{ij} is

$$\mathbf{A} = \begin{bmatrix} a_{11} & a_{12} & \cdots & a_{1n} \\ a_{21} & a_{22} & \cdots & a_{2n} \\ \cdot & \cdot & & \cdot \\ \cdot & \cdot & & \cdot \\ \cdot & \cdot & & \cdot \\ a_{m1} & a_{m2} & \cdots & a_{mn} \end{bmatrix}$$

We also use the shorthand notation $\mathbf{A} = [a_{ij}]$.

Example 6.1

$$\mathbf{A} = \begin{bmatrix} 1 & 2 & 3 \\ 4 & 5 & 6 \end{bmatrix} \quad \mathbf{B} = \begin{bmatrix} -1 & 0 \\ .5 & 2 \end{bmatrix} \quad \mathbf{C} = \begin{bmatrix} 1 \\ -2 \\ 5 \end{bmatrix}$$

$$\mathbf{D} = [-.2 \quad .4 \quad .6 \quad 1.8]$$

Problem Find the dimensions of **A**, **B**, **C**, and **D**. Also find the (2, 1) entry of **A**, the (1, 2) entry of **B**, the (3, 1) entry of **C**, and the (1, 1) entry of **D**.

Solution

A is 2×3; the (2, 1) entry of **A** is 4.
B is 2×2; the (1, 2) entry of **B** is 0.
C is 3×1; the (3, 1) entry of **C** is 5.
D is 1×4; the (1, 1) entry of **D** is $-.2$. ■

A matrix with one column is a *column vector*, and a matrix with one row is a *row vector*. The entries in a vector are called *coordinates*. A column (or row) vector with n coordinates is referred to as a *column* (or *row*) *n-vector*, and n is the *dimension* of the vector. We use boldface letters such as **X**, **Y**, and **U** to denote vectors.

A row vector **U** with m coordinates can be represented as

$$\mathbf{U} = [u_1 \quad u_2 \quad \cdots \quad u_m]$$

and a column vector **X** with n coordinates can be represented as

$$\mathbf{X} = \begin{bmatrix} x_1 \\ x_2 \\ \cdot \\ \cdot \\ \cdot \\ x_n \end{bmatrix}$$

The number of coordinates or the dimension of a vector will always be clear from the way it arises.

Example 6.2 $\mathbf{W} = [1 \quad 3 \quad 8 \quad -5 \quad -2]$

$$\mathbf{X} = \begin{bmatrix} -1 \\ 1 \\ 5 \end{bmatrix} \qquad \mathbf{Y} = \begin{bmatrix} 1 & 2 \\ 3 & -1 \\ 0 & 2 \end{bmatrix} \qquad \mathbf{Z} = [1]$$

Problem Decide whether **W**, **X**, **Y**, and **Z** are vectors; if so, give their dimensions. Also find the second coordinate of each vector.

Solution

W is a row vector of dimension 5, and the second coordinate of **W** is 3.

X is a column vector of dimension 3, and its second coordinate is 1.

Y is neither a row vector nor a column vector.

Z is both a row vector and a column vector of dimension 1, and **Z** has no second coordinate. ■

Example 6.3

Sales

The Inshur & Bshur Insurance Agency records the sales made by each agent on a large bulletin board. The agency has a sales staff of five, and it offers life insurance, automobile insurance, and home insurance. The bulletin board can be organized as shown in Table 6.1. The data in that table are for October.

TABLE 6.1

	Salesperson				
Type of Policy	Alice	Barbara	Charles	David	Ellen
Life	8	10	6	3	5
Automobile	3	4	12	12	10
Home	7	6	3	2	8

If we agree to remember the meaning of the entry in each location, then we can represent the sales figures for October in a 3×5 matrix; call it $\mathbf{A}$:

$$\mathbf{A} = \begin{bmatrix} 8 & 10 & 6 & 3 & 5 \\ 3 & 4 & 12 & 12 & 10 \\ 7 & 6 & 3 & 2 & 8 \end{bmatrix}$$

The corresponding matrix for another month would be viewed as the same as that for October only if all its entries were the same as the entries for October. If the corresponding matrix for November is

$$\mathbf{B} = \begin{bmatrix} 4 & 6 & 5 & 6 & 4 \\ 5 & 2 & 2 & 6 & 8 \\ 6 & 8 & 5 & 2 & 6 \end{bmatrix}$$

then the matrix which represents the combined sales for October and November is

$$\begin{bmatrix} 12 & 16 & 11 & 9 & 9 \\ 8 & 6 & 14 & 18 & 18 \\ 13 & 14 & 8 & 4 & 14 \end{bmatrix}$$

This new matrix represents the sum of $\mathbf{A}$ and $\mathbf{B}$. The entry 9 in the upper right-hand corner, the (1, 5) entry, is the sum of the sales of life insurance policies by Ellen in October and November. We denote this matrix, the matrix formed by adding the respective entries of matrices $\mathbf{A}$ and $\mathbf{B}$, by $\mathbf{A} + \mathbf{B}$. If the sales manager sets a goal for December for each salesperson which is twice the November sales for each type of policy, then the goal can be represented by the matrix

$$\begin{bmatrix} 8 & 12 & 10 & 12 & 8 \\ 10 & 4 & 4 & 12 & 16 \\ 12 & 16 & 10 & 4 & 12 \end{bmatrix}$$

Since the November sales matrix is denoted by $\mathbf{B}$, it is reasonable to denote the matrix which gives the December goals as $2\mathbf{B}$. ■

Example 6.3 illustrates the following definitions of *equality of matrices*, *matrix addition*, and *scalar multiplication*.

Equality of Matrices

Let $\mathbf{A} = [a_{ij}]$ and $\mathbf{B} = [b_{ij}]$ be two $m \times n$ matrices. Matrix **A** is said to be *equal* to matrix **B**, written $\mathbf{A} = \mathbf{B}$, if $a_{ij} = b_{ij}$ for $1 \le i \le m$, $1 \le j \le n$.

That is, two matrices **A** and **B** are equal if they are the same size and if their corresponding entries are all equal.

Addition of Matrices

With **A** and **B** as above, the sum $\mathbf{A} + \mathbf{B}$ of matrices **A** and **B** is defined to be the matrix $\mathbf{C} = [c_{ij}]$, where $c_{ij} = a_{ij} + b_{ij}$, $1 \le i \le m$, $1 \le j \le n$.

That is, the sum of two $m \times n$ matrices is the $m \times n$ matrix whose entries are each the sum of the corresponding entries of **A** and **B**. Addition of matrices is defined only for matrices of the same size, i.e., with the same number of rows and the same number of columns.

The sum of three or more $m \times n$ matrices, **A**, **B**, **C**, . . . can be computed by first computing the sum $\mathbf{A} + \mathbf{B}$, then $(\mathbf{A} + \mathbf{B}) + \mathbf{C}$, and so on. In fact, the parentheses are unnecessary: The sum $(\mathbf{A} + \mathbf{B}) + \mathbf{C}$ is the same matrix as the sum $\mathbf{A} + (\mathbf{B} + \mathbf{C})$. Therefore there is no ambiguity in writing simply $\mathbf{A} + \mathbf{B} + \mathbf{C}$ for the sum, and we shall do so in the future.

Scalar Multiplication

If c is a number and **A** is an $m \times n$ matrix, then the *scalar multiple* $c\mathbf{A}$ is defined to be the matrix $\mathbf{D} = [d_{ij}]$, where $d_{ij} = ca_{ij}$, $1 \le i \le m$, $1 \le j \le n$.

That is, the scalar multiple of a matrix **A** is the matrix whose entries are each the corresponding entry of **A** multiplied by the scalar.

Example 6.4 Let

$$\mathbf{A} = \begin{bmatrix} 3 & 2 \\ 1 & -1 \\ 0 & 2 \end{bmatrix} \qquad \mathbf{B} = \begin{bmatrix} 2 & 1 \\ 0 & 2 \\ -1 & -3 \end{bmatrix} \qquad \mathbf{C} = \begin{bmatrix} a & 2 \\ b & -1 \\ 0 & c \end{bmatrix}$$

Problem Find $\mathbf{A} + \mathbf{B}$, $2\mathbf{A}$, $(-1)\mathbf{B}$, and the conditions on a, b, and c such that $\mathbf{A} = \mathbf{C}$.

Solution

$$\mathbf{A} + \mathbf{B} = \begin{bmatrix} 3 & 2 \\ 1 & -1 \\ 0 & 2 \end{bmatrix} + \begin{bmatrix} 2 & 1 \\ 0 & 2 \\ -1 & -3 \end{bmatrix} = \begin{bmatrix} 5 & 3 \\ 1 & 1 \\ -1 & -1 \end{bmatrix}$$

$$2\mathbf{A} = 2\begin{bmatrix} 3 & 2 \\ 1 & -1 \\ 0 & 2 \end{bmatrix} = \begin{bmatrix} 6 & 4 \\ 2 & -2 \\ 0 & 4 \end{bmatrix}$$

$$(-1)\mathbf{B} = (-1)\begin{bmatrix} 2 & 1 \\ 0 & 2 \\ -1 & -3 \end{bmatrix} = \begin{bmatrix} -2 & -1 \\ 0 & -2 \\ 1 & 3 \end{bmatrix}$$

$$\mathbf{A} = \mathbf{C} \qquad \text{or} \qquad \begin{bmatrix} 3 & 2 \\ 1 & -1 \\ 0 & 2 \end{bmatrix} = \begin{bmatrix} a & 2 \\ b & -1 \\ 0 & c \end{bmatrix}$$

if and only if $a = 3$, $b = 1$, and $c = 2$. ■

There are several features of matrix addition and scalar multiplication which will be useful to us.

Properties of Matrix Algebra

If **A** and **B** are $m \times n$ matrices and c and d are numbers, then

$$\mathbf{A} + \mathbf{B} = \mathbf{B} + \mathbf{A}$$
$$c\mathbf{A} + d\mathbf{A} = (c + d)\mathbf{A}$$
$$c\mathbf{A} + c\mathbf{B} = c(\mathbf{A} + \mathbf{B})$$
$$(cd)\mathbf{A} = c(d\mathbf{A})$$

In the work which follows in this and later chapters, we use these properties freely without specifically mentioning them. At times more than one may be combined into a single step in a computation. We write $-\mathbf{A}$ for the matrix $(-1)\mathbf{A}$ and $\mathbf{A} - \mathbf{B}$ for $\mathbf{A} + (-\mathbf{B})$.

Matrix Multiplication

We now turn to the topic of multiplying matrices. We begin with a special case which is important in the study of systems of equations, and then we continue by using this special case to define general matrix multiplication. First we define the product of a row n-vector **U** by a column n-vector **X**, to obtain **UX**. Then we use this idea to define the product of an $m \times n$ matrix **A** and an $n \times k$ matrix **B**, to obtain **AB**.

Row-by-Column Multiplication

If

$$\mathbf{U} = [u_1 \quad u_2 \quad \cdots \quad u_n]$$

is a row n-vector and

$$\mathbf{X} = \begin{bmatrix} x_1 \\ x_2 \\ \cdot \\ \cdot \\ \cdot \\ x_n \end{bmatrix}$$

is a column n-vector, then the product **UX** is defined and

$$\mathbf{UX} = u_1x_1 + u_2x_2 + \cdots + u_nx_n \tag{6.1}$$

The row vector in (6.1) is always written to the left of the column vector. The product of a row k-vector and a column n-vector with $n \neq k$ is not defined.

Example 6.5 Let

$$\mathbf{U} = [1 \quad -1 \quad 3] \quad \text{and} \quad \mathbf{X} = \begin{bmatrix} 4 \\ 2 \\ -1 \end{bmatrix}$$

Find **UX**.

Solution

$$\mathbf{UX} = 1(4) + (-1)(2) + 3(-1) = -1$$

■

Example 6.6 Let

$$\mathbf{U} = [2 \quad 3 \quad -4 \quad 5] \quad \text{and} \quad \mathbf{X} = \begin{bmatrix} x_1 \\ x_2 \\ x_3 \\ x_4 \end{bmatrix}$$

Find **UX**.

Solution In the column vector **X**, the entries x_1, x_2, x_3, and x_4 are variables which represent unknown numbers. By formula (6.1) the product **UX** has the form

$$\mathbf{UX} = 2x_1 + 3x_2 - 4x_3 + 5x_4$$

Thus, **UX** is a linear expression in the variables x_1, x_2, x_3, and x_4. ■

Matrix Multiplication

If $\mathbf{A} = [a_{ij}]$ is an $m \times n$ matrix and $\mathbf{B} = [b_{ij}]$ is an $n \times k$ matrix, then the *matrix product* **AB** is defined to be the $m \times k$ matrix $\mathbf{C} = [c_{ij}]$, where

$$c_{ij} = a_{i1}b_{1j} + a_{i2}b_{2j} + \cdots + a_{in}b_{nj} \qquad 1 \le i \le m,\ 1 \le j \le k \tag{6.2}$$

Notice that the product **AB** is defined only between matrices **A** and **B** where the number of columns of **A** is equal to the number of rows of **B**. Since the (i, j) entry of the product **AB** is determined by the ith row of matrix **A** and the jth column of matrix **B**, the (i, j) entry is said to be obtained through *row-by-column* multiplication. For example,

$$c_{32} = [\text{third row of } \mathbf{A}] \begin{bmatrix} \text{second} \\ \text{column} \\ \text{of } \mathbf{B} \end{bmatrix}$$

$$= a_{31}b_{12} + a_{32}b_{22} + \cdots + a_{3n}b_{n2}$$

Example 6.7 Find the matrix products **AB** and **BC**, where

$$\mathbf{A} = \begin{bmatrix} 2 & 3 & -1 \\ -1 & 0 & 2 \end{bmatrix} \quad \mathbf{B} = \begin{bmatrix} 1 & -1 \\ -1 & 0 \\ 1 & 2 \end{bmatrix} \quad \mathbf{C} = \begin{bmatrix} 1 & -1 \\ -3 & 2 \end{bmatrix}$$

Solution Since $\mathbf{A}$ is 2×3 and $\mathbf{B}$ is 3×2, the product $\mathbf{AB}$ is defined and is a 2×2 matrix. Likewise, $\mathbf{BC}$ is defined and is a 3×2 matrix.

$$\mathbf{AB} = \begin{bmatrix} 2(1) + 3(-1) + (-1)(1) & 2(-1) + 3(0) + (-1)(2) \\ (-1)(1) + 0(-1) + 2(1) & (-1)(-1) + 0(0) + 2(2) \end{bmatrix} = \begin{bmatrix} -2 & -4 \\ 1 & 5 \end{bmatrix}$$

$$\mathbf{BC} = \begin{bmatrix} 1(1) + (-1)(-3) & 1(-1) + (-1)(2) \\ (-1)(1) + 0(-3) & (-1)(-1) + 0(2) \\ 1(1) + 2(-3) & 1(-1) + 2(2) \end{bmatrix} = \begin{bmatrix} 4 & -3 \\ -1 & 1 \\ -5 & 3 \end{bmatrix}$$

■

Multiplication of three or more matrices can be defined by grouping the matrices and computing the product in steps by multiplying two matrices at a time. For example, the product of an $m \times n$ matrix $\mathbf{A}$, an $n \times r$ matrix $\mathbf{B}$, and an $r \times s$ matrix $\mathbf{C}$ (in that order) can be defined as $(\mathbf{AB})\mathbf{C}$. This is to be interpreted as the result of multiplying the $m \times r$ matrix $\mathbf{AB}$ (which is defined by the definition of multiplication of two matrices) by the $r \times s$ matrix $\mathbf{C}$. It can be shown that this is the same as the product of the $m \times n$ matrix $\mathbf{A}$ and the $n \times s$ matrix $\mathbf{BC}$. That is, $(\mathbf{AB})\mathbf{C} = \mathbf{A}(\mathbf{BC})$, and in the future we write simply $\mathbf{ABC}$ since there is no possible ambiguity.

In contrast to the freedom we have in grouping matrices in a product in different ways, we cannot in general change the order of the factors in a matrix product. If the product $\mathbf{AB}$ of matrices $\mathbf{A}$ and $\mathbf{B}$ in that order is defined, it may be that the product of these matrices in the reverse order is not defined. Even if $\mathbf{AB}$ and $\mathbf{BA}$ are both defined, they may not be equal.

Example 6.8 If

$$\mathbf{A} = \begin{bmatrix} 2 & 0 \\ 1 & 3 \\ 0 & -1 \end{bmatrix} \quad \text{and} \quad \mathbf{B} = \begin{bmatrix} 1 & 1 \\ -1 & -1 \end{bmatrix}$$

then $\mathbf{AB}$ is defined since $\mathbf{A}$ is 3×2 and $\mathbf{B}$ is 2×2. The product $\mathbf{AB}$ is 3×2. However, the product $\mathbf{BA}$ is not defined since $\mathbf{B}$ has 2 columns and $\mathbf{A}$ has 3 rows.

With $\mathbf{B}$ as above and $\mathbf{C} = \begin{bmatrix} 2 & 2 \\ 1 & 1 \end{bmatrix}$, we have both $\mathbf{CB}$ and $\mathbf{BC}$ defined, but

$$\mathbf{CB} = \begin{bmatrix} 0 & 0 \\ 0 & 0 \end{bmatrix} \quad \text{and} \quad \mathbf{BC} = \begin{bmatrix} 3 & 3 \\ -3 & -3 \end{bmatrix}$$

This shows that in general $\mathbf{AB} \neq \mathbf{BA}$. It also illustrates that the product of a matrix $\mathbf{A}$ that is not all zeros and a matrix $\mathbf{B}$ that is not all zeros can be a matrix with all zeros. Thus the familiar result "if $ab = 0$, then either $a = 0$ or $b = 0$," which holds for numbers, does *not* hold for matrices.

There is a matrix $\mathbf{I}$ which plays the same role in matrix multiplication that the number 1 plays with respect to multiplication of numbers.

The $n \times n$ matrix $\mathbf{I}$ defined by

$$\mathbf{I} = \underbrace{\begin{bmatrix} 1 & 0 & 0 & \cdots & 0 \\ 0 & 1 & 0 & \cdots & 0 \\ 0 & 0 & 1 & \cdots & 0 \\ \cdot & \cdot & \cdot & & \cdot \\ \cdot & \cdot & \cdot & & \cdot \\ \cdot & \cdot & \cdot & & \cdot \\ 0 & 0 & 0 & \cdots & 1 \end{bmatrix}}_{n \text{ columns}} \left.\vphantom{\begin{bmatrix} 1\\0\\0\\ \cdot\\ \cdot\\ \cdot\\0 \end{bmatrix}}\right\} n \text{ rows}$$

is said to be the $n \times n$ *identity* matrix.

We shall not distinguish notationally between identity matrices of different sizes. When we write $\mathbf{IA}$, we assume that $\mathbf{I}$ has the correct size for the product to be defined.

For every matrix $\mathbf{A}$ and the identity matrices $\mathbf{I}$ for which the products $\mathbf{AI}$ and $\mathbf{IA}$ are defined, we have

$$\mathbf{AI} = \mathbf{A} \quad \text{and} \quad \mathbf{IA} = \mathbf{A}$$

We use $\mathbf{A} = \begin{bmatrix} 1 & 3 \\ 2 & -1 \end{bmatrix}$ and $\mathbf{I} = \begin{bmatrix} 1 & 0 \\ 0 & 1 \end{bmatrix}$ to illustrate the equality $\mathbf{AI} = \mathbf{A}$. We have

$$\begin{aligned} \mathbf{AI} = \begin{bmatrix} 1 & 3 \\ 2 & -1 \end{bmatrix}\begin{bmatrix} 1 & 0 \\ 0 & 1 \end{bmatrix} &= \begin{bmatrix} 1(1) + 3(0) & 1(0) + 3(1) \\ 2(1) + (-1)(0) & 2(0) + (-1)(1) \end{bmatrix} \\ &= \begin{bmatrix} 1 & 3 \\ 2 & -1 \end{bmatrix} \\ &= \mathbf{A} \end{aligned}$$

Example 6.9 Let

$$\mathbf{A} = \begin{bmatrix} 2 & 3 & 4 & 5 \\ -1 & 2 & 0 & 4 \\ 3 & -2 & 1 & -3 \end{bmatrix} \quad \text{and} \quad \mathbf{X} = \begin{bmatrix} x_1 \\ x_2 \\ x_3 \\ x_4 \end{bmatrix}$$

Problem Find **AX**.

Solution Since **A** is 3×4 and **X** is 4×1, the product **AX** is defined and is a 3×1 matrix. As in Example 6.6, the entries in the column vector **X** represent variables, and using definition (6.2), we obtain

$$\mathbf{AX} = \begin{bmatrix} 2x_1 + 3x_2 + 4x_3 + 5x_4 \\ -x_1 + 2x_2 + 4x_4 \\ 3x_1 - 2x_2 + x_3 - 3x_4 \end{bmatrix}$$

We see that each entry in the 3×1 matrix **AX** is a linear expression in the variables x_1, x_2, x_3, and x_4. ■

Using the ideas of Example 6.9 , we now show that systems of linear equations, such as those studied in Chapter 5, can be conveniently represented in matrix form. If a system of m equations and n variables has the form

$$\begin{array}{cccc} a_{11}x_1 + a_{12}x_2 + \cdots + a_{1n}x_n = b_1 \\ a_{21}x_1 + a_{22}x_2 + \cdots + a_{2n}x_n = b_2 \\ \vdots \\ a_{m1}x_1 + a_{m2}x_2 + \cdots + a_{mn}x_n = b_m \end{array} \tag{6.3}$$

then, as in Chapter 5, we let **A** represent the coefficient matrix. Thus,

$$\mathbf{A} = \begin{bmatrix} a_{11} & a_{12} & \cdots & a_{1n} \\ a_{21} & a_{22} & \cdots & a_{2n} \\ \vdots & \vdots & & \vdots \\ a_{m1} & a_{m2} & \cdots & a_{mn} \end{bmatrix}$$

We also define the column vectors **X** and **B** as follows:

$$\mathbf{X} = \begin{bmatrix} x_1 \\ x_2 \\ \cdot \\ \cdot \\ \cdot \\ x_n \end{bmatrix} \qquad \mathbf{B} = \begin{bmatrix} b_1 \\ b_2 \\ \cdot \\ \cdot \\ \cdot \\ b_m \end{bmatrix}$$

Then, using the definitions of matrix multiplication and matrix equality, we see that the system (6.3) is identical to the matrix equation $\mathbf{AX} = \mathbf{B}$. This justifies the shorthand notation used in Chapter 5.

Exercises for Section 6.1

1. Find the dimension of each of the following matrices, and decide which are vectors.

$$(a) \begin{bmatrix} 1 & 2 & 3 \\ 0 & 1 & 5 \\ 2 & -1 & 6 \end{bmatrix} \qquad (b) \begin{bmatrix} -1 & 0 \\ .2 & .8 \\ .5 & 2 \end{bmatrix} \qquad (c) \begin{bmatrix} -1 \\ 2 \\ .5 \end{bmatrix}$$

2. Find the dimension of each of the following matrices, and decide which are vectors.

$$(a) \begin{bmatrix} 5 & 0 & 1 & -3 \\ 2 & 10 & 0 & -2 \\ 1 & 16 & 8 & 0 \end{bmatrix} \qquad (b) \begin{bmatrix} 15 & 10 & 40 & 50 \\ 20 & 35 & 30 & 65 \end{bmatrix} \qquad (c) \; [.2 \;\; 0 \;\; .2 \;\; .6]$$

3. Let

$$\mathbf{A} = \begin{bmatrix} 5 & 2 \\ 1 & 0 \end{bmatrix} \quad \text{and} \quad \mathbf{B} = \begin{bmatrix} 1 & 4 \\ 2 & 3 \end{bmatrix}$$

Find $\mathbf{A} + \mathbf{B}$ and $\mathbf{A} - \mathbf{B}$.

4. Let

$$\mathbf{A} = \begin{bmatrix} 2 & 3 \\ 1 & -2 \end{bmatrix} \quad \text{and} \quad \mathbf{B} = \begin{bmatrix} 2 & -1 \\ 4 & -3 \end{bmatrix}$$

Find $2\mathbf{A} + \mathbf{B}$ and $\mathbf{A} - 2\mathbf{B}$.

5. Let

$$\mathbf{A} = \begin{bmatrix} 3 & -1 & 3 \\ 0 & 2 & 1 \\ 4 & -2 & -2 \end{bmatrix} \quad \text{and} \quad \mathbf{B} = \begin{bmatrix} -1 & 2 & -1 \\ 2 & -1 & 0 \\ 0 & -4 & 3 \end{bmatrix}$$

Find $3\mathbf{A} + 2\mathbf{B}$ and $2\mathbf{A} - \mathbf{B}$.

6. Find the product **AB** in each of the following cases.
   (*a*) **A** and **B** defined as in Exercise 3
   (*b*) **A** and **B** defined as in Exercise 4
   (*c*) **A** and **B** defined as in Exercise 5
7. Let

$$\mathbf{A}=\begin{bmatrix}2 & -1 & 0\\ 4 & -2 & 3\end{bmatrix}\qquad \mathbf{B}=\begin{bmatrix}1 & -2 & -1\\ 2 & 1 & 0\\ 3 & 2 & -3\end{bmatrix}\qquad \mathbf{C}=\begin{bmatrix}-1 & 0 & -1\\ 2 & -2 & 0\end{bmatrix}$$

   Decide which of the following operations are defined, and carry out those which are defined.
   (*a*) **A** + **B** (*b*) **AB** (*c*) **BC** (*d*) (**A** + **C**)**B**
8. For matrices **A**, **B**, and **C** given in Exercise 7, decide which of the following operations are defined, and carry out those which are defined.
   (*a*) 2**A** (*b*) **AC** (*c*) **AB** + **C** (*d*) **CB**
9. For matrices **A** and **C** given in Exercise 7, find a matrix **D** such that 2**A** − **D** = **C**.
10. For matrices **A** and **C** given in Exercise 7, find a matrix **D** such that **A** + 2**D** = **C**.
11. In the situation described in Example 6.3, a goal is set for sales in the quarter January through March to be equal to the sum of the sales in October and twice those in November. Find the goals for the quarter January through March for each salesperson.
12. (Continuation of Exercise 11) In the situation described in Example 6.3, suppose that the sales in January are given by the matrix

$$\begin{bmatrix}2 & 4 & 3 & 5 & 6\\ 6 & 4 & 2 & 2 & 5\\ 4 & 9 & 8 & 4 & 3\end{bmatrix}$$

   Find the sales in the 2-month period of February and March necessary to achieve the goal determined in Exercise 11.
13. Let

$$\mathbf{A}=\begin{bmatrix}2 & 2\\ 3 & 1\end{bmatrix}\qquad \mathbf{B}=\begin{bmatrix}3 & 1\\ -1 & 0\\ 4 & 2\end{bmatrix}\qquad \mathbf{C}=\begin{bmatrix}4 & 0 & 1\\ 1 & -1 & 2\\ 3 & 1 & 1\end{bmatrix}\qquad \mathbf{D}=\begin{bmatrix}-1 & 0\\ 2 & 1\end{bmatrix}$$

   Find:
   (*a*) 2**B** (*b*) **BA** (*c*) **CBA**
14. With matrices **A**, **B**, **C**, and **D** defined as in Exercise 13, find:
   (*a*) 3**A** − 2**D** (*b*) **CB** (*c*) **CBA** − **BD**
15. Let matrices **A**, **B**, and **C** be defined by

$$\mathbf{A}=\begin{bmatrix}2 & 0 & -3\\ 1 & 4 & 6\end{bmatrix}\qquad \mathbf{B}=\begin{bmatrix}-1 & 2 & 0\\ 1 & 4 & 3\end{bmatrix}\qquad \mathbf{C}=\begin{bmatrix}3 & 1\\ 2 & 0\end{bmatrix}$$

Decide which of the following operations are defined, and carry out those which are defined.

(*a*) $3\mathbf{A} - \mathbf{B}$ (*b*) $\mathbf{CA}$

(*c*) $(\mathbf{A} - 2\mathbf{B})\mathbf{C}$ (*d*) $\mathbf{ABC}$

16. Let matrices **A**, **B**, and **C** be as in Exercise 15, and let

$$\mathbf{X} = \begin{bmatrix} 1 \\ 0 \\ 4 \end{bmatrix} \qquad \mathbf{Y} = \begin{bmatrix} 2 \\ -1 \end{bmatrix}$$

Decide which of the following are defined, and evaluate those which are defined.

(*a*) $2\mathbf{X} - \mathbf{Y}$ (*b*) $\mathbf{AX}$

(*c*) $\mathbf{AX} + \mathbf{CY}$ (*d*) $\mathbf{CAX}$

17. Let

$$\mathbf{A} = \begin{bmatrix} 3 & 1 \\ 2 & 1 \end{bmatrix}$$

For what value of c does the matrix

$$\mathbf{B} = \begin{bmatrix} 1 & -1 \\ -2 & c \end{bmatrix}$$

satisfy $\mathbf{AB} = \mathbf{I}$?

18. For what values of c and d does the following matrix equation hold?

$$c\begin{bmatrix} 3 \\ 1 \end{bmatrix} + d\begin{bmatrix} -1 \\ 2 \end{bmatrix} = \begin{bmatrix} 7 \\ 0 \end{bmatrix}$$

19. Let

$$\mathbf{A} = \begin{bmatrix} 2 & -1 \\ 4 & 0 \end{bmatrix} \quad \text{and} \quad \mathbf{B} = \begin{bmatrix} -2 & 3 \\ -1 & 2 \end{bmatrix}$$

(*a*) Find a matrix **C** which satisfies $2\mathbf{A} + \mathbf{C} = \mathbf{B}$.

(*b*) Find a matrix **D** which satisfies $\mathbf{A} + \mathbf{B} + \mathbf{D} = \mathbf{I}$.

20. Let **A** and **B** be as in Exercise 19. Find $\mathbf{AB} - \mathbf{BA}$.

21. Find three 2×2 matrices, **A**, **B**, and **C** (**C** not all zeros) such that $\mathbf{AC} = \mathbf{BC}$ but $\mathbf{A} \neq \mathbf{B}$.

22. Find a 2×2 matrix **A** whose entries are all different from zero but for which **AA** has all zero entries.

23. For a square matrix **A** (equal numbers of rows and columns), the product **AA** is always defined. This product is denoted $\mathbf{A}^2$. Likewise, $\mathbf{A}^3 = \mathbf{AAA}$, and for any positive integer n, the product of $\mathbf{AAA} \cdots \mathbf{A}$ with n factors is denoted $\mathbf{A}^n$. For

$$\mathbf{A} = \begin{bmatrix} 2 & 1 \\ 1 & 0 \end{bmatrix}$$

find $\mathbf{A}^2$ and $\mathbf{A}^3$.

24. Let $\mathbf{P} = \begin{bmatrix} .5 & .5 \\ 1 & 0 \end{bmatrix}$. Using the definitions of Exercise 23, find $\mathbf{P}^2$, $\mathbf{P}^4$, and $\mathbf{P}^8$.

25. Let $\mathbf{P}$ be as in Exercise 24, and let $\mathbf{X} = [2 \quad 1]$. Show that $\mathbf{XP} = \mathbf{X}$ and $\mathbf{XP}^2 = \mathbf{X}$.

26. Let $\mathbf{P}$ be as in Exercise 24, and let $\mathbf{X} = \begin{bmatrix} .5 \\ .5 \end{bmatrix}$. Show that $\mathbf{PX} = \mathbf{X}$ and $\mathbf{P}^2\mathbf{X} = \mathbf{X}$.

27. Let

$$\mathbf{P} = \begin{bmatrix} 0 & 1 & 0 \\ 0 & 0 & 1 \\ \frac{2}{3} & 0 & \frac{1}{3} \end{bmatrix}$$

Using the definitions of Exercise 23, find $\mathbf{P}^2$ and $\mathbf{P}^4$.

28. Let $\mathbf{P}$ be as in Exercise 27, and let $\mathbf{X} = [1 \quad 1 \quad \frac{3}{2}]$. Show that $\mathbf{XP} = \mathbf{X}$. In fact, by using mathematical induction, one can show $\mathbf{XP}^n = \mathbf{X}$ for any positive integer n. Show that $\mathbf{XP}^4 = \mathbf{X}$.

29. Let $\mathbf{P}$ be as in Exercise 27, and let $\mathbf{X}$ satisfy $\mathbf{XP} = \mathbf{X}$.
    (*a*) Show that if $\mathbf{Y} = k\mathbf{X}$ for any number k, then $\mathbf{Y}$ satisfies $\mathbf{YP} = \mathbf{Y}$.
    (*b*) Find a vector $\mathbf{Y}$ the sum of whose coordinates is 1 which satisfies $\mathbf{YP} = \mathbf{Y}$. (*Hint*: Use Exercise 28.)

30. Let $\mathbf{P} = \begin{bmatrix} 0 & 1 \\ .5 & .5 \end{bmatrix}$. Find a vector $\mathbf{X}$ which satisfies $\mathbf{XP} = \mathbf{X}$.

31. Let $\mathbf{P} = \begin{bmatrix} 0 & 1 & 0 \\ 0 & 0 & 1 \\ \frac{1}{4} & \frac{1}{4} & \frac{1}{2} \end{bmatrix}$. Find a vector $\mathbf{X}$ which satisfies $\mathbf{XP} = \mathbf{X}$. (*Hint*: Use the techniques of Chapter 5.)

32. Write the following systems of equations in the matrix form $\mathbf{AX} = \mathbf{B}$. Thus, find $\mathbf{A}$, $\mathbf{X}$, and $\mathbf{B}$.
    (*a*) $\begin{aligned} 3x_1 + 4x_2 &= 5 \\ -x_1 + 2x_2 &= 3 \end{aligned}$
    (*b*) $\begin{aligned} -2x_1 + x_3 &= 7 \\ 4x_2 + 9x_3 &= 19 \\ 4x_1 - x_2 &= 29 \end{aligned}$

33. Let

$$\mathbf{A} = \begin{bmatrix} 1 & 2 & 1 \\ 2 & 3 & 1 \\ -2 & 4 & 6 \end{bmatrix} \qquad \mathbf{X} = \begin{bmatrix} 2+t \\ 1-t \\ t \end{bmatrix} \qquad \mathbf{B} = \begin{bmatrix} 4 \\ 7 \\ 0 \end{bmatrix}$$

Use the definition of matrix multiplication to show that for each value of the variable t, $\mathbf{X}$ is a solution of the matrix equation $\mathbf{AX} = \mathbf{B}$.

34. Find a 2×2 matrix $\mathbf{B}$ such that $\mathbf{AB} = \mathbf{I}$, where $\mathbf{A} = \begin{bmatrix} 1 & 2 \\ 3 & 7 \end{bmatrix}$ and $\mathbf{I} = \begin{bmatrix} 1 & 0 \\ 0 & 1 \end{bmatrix}$.
(*Hint*: Solve two systems, each with two equations in two variables.)
35. Find a 2×2 matrix $\mathbf{B}$ such that $\mathbf{AB} = \mathbf{C}$, where

$$\mathbf{A} = \begin{bmatrix} 1 & -2 \\ -3 & 4 \end{bmatrix} \quad \text{and} \quad \mathbf{C} = \begin{bmatrix} 10 & -7 \\ -22 & 15 \end{bmatrix}$$

(*Hint*: Solve two systems of two equations in two variables.)

6.2 MATRIX INVERSES

Linear equations in a single scalar variable are especially easy to solve. For instance, to find x for which $5x = 3$, we multiply both sides of the equation by $\frac{1}{5}$ (the reciprocal of 5) to obtain

$$\tfrac{1}{5}(5x) = 1 \cdot x = \tfrac{1}{5} \cdot 3 = \tfrac{3}{5}$$

or $x = \frac{3}{5}$. Superficially, a system of equations written in matrix form as $\mathbf{AX} = \mathbf{B}$ appears much like the above scalar equation $5x = 3$. It would be useful to have a reciprocal of matrix $\mathbf{A}$ which could be used to solve the equation $\mathbf{AX} = \mathbf{B}$ in the same way as $\frac{1}{5}$ was used to solve the scalar equation $5x = 3$. Unfortunately, in general, a matrix $\mathbf{A}$ will not have a reciprocal which is analogous to the reciprocal of a nonzero number. There is, however, a special case in which the method described above can be applied to matrix equations.

Inverse of a Matrix

An $n \times n$ matrix $\mathbf{A}$ is said to be *invertible* if there is an $n \times n$ matrix $\mathbf{B}$ such that $\mathbf{BA} = \mathbf{AB} = \mathbf{I}$. The matrix $\mathbf{B}$ with this property is said to be the *inverse* of $\mathbf{A}$ and is written $\mathbf{A}^{-1}$.

Not all square matrices are invertible; some are and others are not. If a matrix is invertible, then *there is only one* matrix $\mathbf{A}^{-1}$ such that $\mathbf{A}^{-1}\mathbf{A} = \mathbf{I} = \mathbf{AA}^{-1}$; that is, the inverse of a square matrix is unique.

If $\mathbf{A}$ is invertible, then the solution of the system of equations $\mathbf{AX} = \mathbf{B}$ is straightforward. Indeed, if $\mathbf{A}^{-1}$ is the inverse of $\mathbf{A}$, then

$$\mathbf{A}^{-1}\mathbf{AX} = \mathbf{IX} = \mathbf{X}$$

Therefore, if we multiply the equation $\mathbf{AX} = \mathbf{B}$ on the left by $\mathbf{A}^{-1}$, then we obtain

$$\mathbf{A}^{-1}\mathbf{AX} = \mathbf{A}^{-1}\mathbf{B} \qquad \text{or} \qquad \mathbf{X} = \mathbf{A}^{-1}\mathbf{B}$$

Since $\mathbf{A}^{-1}$ and $\mathbf{B}$ are known matrices, this provides a vector $\mathbf{X}$ which satisfies the equation $\mathbf{AX} = \mathbf{B}$.

In general, one would not solve a single system $\mathbf{AX} = \mathbf{B}$ by computing the inverse of $\mathbf{A}$ since this computation frequently involves more labor than the direct solution of $\mathbf{AX} = \mathbf{B}$ by the methods described in Section 5.3. However, if one is asked to solve several systems with the same coefficient matrix, e.g.,

$$\mathbf{AX} = \mathbf{B} \qquad \mathbf{AX} = \mathbf{B}' \qquad \mathbf{AX} = \mathbf{B}''$$

then it may be desirable to compute $\mathbf{A}^{-1}$. Indeed, once we have $\mathbf{A}^{-1}$, we can obtain the solutions of these equations immediately as $\mathbf{A}^{-1}\mathbf{B}$, $\mathbf{A}^{-1}\mathbf{B}'$, and $\mathbf{A}^{-1}\mathbf{B}''$, respectively. Also there are important uses of the inverse matrix other than its use in solving systems of equations. One such use will be discussed in our study of Markov chains (Chapter 8).

Our method for computing the inverse of a matrix is essentially the same as the method described in Section 5.3 for solving a system of linear equations. This method consists of systematically transforming a certain augmented matrix into reduced form. Since it may seem somewhat mysterious that this method should be useful in computing inverse matrices, first we describe briefly *why* the method works, then we show *how* it works.

Given a matrix $\mathbf{A}$, our problem is to find a matrix $\mathbf{B}$ such that $\mathbf{AB} = \mathbf{I}$. If we write $\mathbf{B}^{(1)}, \mathbf{B}^{(2)}, \ldots, \mathbf{B}^{(n)}$ for the columns of $\mathbf{B}$ in order, then the equation $\mathbf{AB} = \mathbf{I}$ is equivalent to n systems of equations for unknown vectors $\mathbf{B}^{(1)}, \mathbf{B}^{(2)}, \ldots, \mathbf{B}^{(n)}$:

$$\mathbf{AB}^{(1)} = \begin{bmatrix} 1 \\ 0 \\ \cdot \\ \cdot \\ \cdot \\ 0 \end{bmatrix}, \quad \mathbf{AB}^{(2)} = \begin{bmatrix} 0 \\ 1 \\ \cdot \\ \cdot \\ \cdot \\ 0 \end{bmatrix}, \quad \ldots, \quad \mathbf{AB}^{(n)} = \begin{bmatrix} 0 \\ 0 \\ \cdot \\ \cdot \\ \cdot \\ 1 \end{bmatrix} \tag{6.4}$$

For example, given the matrix

$$\mathbf{A} = \begin{bmatrix} 3 & 2 \\ 4 & 3 \end{bmatrix}$$

the problem of finding the inverse $\mathbf{B} = [\mathbf{B}^{(1)}, \mathbf{B}^{(2)}]$ of $\mathbf{A}$ is equivalent to solving

$$\mathbf{AB} = \mathbf{I} = \begin{bmatrix} 1 & 0 \\ 0 & 1 \end{bmatrix} \qquad \text{or} \qquad \mathbf{AB}^{(1)} = \begin{bmatrix} 1 \\ 0 \end{bmatrix} \qquad \mathbf{AB}^{(2)} = \begin{bmatrix} 0 \\ 1 \end{bmatrix}$$

In the general case, if we solve the n systems (6.4) for the vectors $\mathbf{B}^{(1)}$, $\mathbf{B}^{(2)}, \ldots, \mathbf{B}^{(n)}$, then we have matrix **B**, that is, the inverse of **A**. If **A** is invertible, then each of the systems (6.4) can be solved by the method of Section 5.3. In particular, we can use the operations of types 1, 2, and 3 on the augmented matrices

$$\left[\begin{array}{c|c} & 1 \\ & 0 \\ \mathbf{A} & \vdots \\ & 0 \end{array}\right], \quad \left[\begin{array}{c|c} & 0 \\ & 1 \\ \mathbf{A} & \vdots \\ & 0 \end{array}\right], \quad \ldots, \quad \left[\begin{array}{c|c} & 0 \\ & 0 \\ \mathbf{A} & \vdots \\ & 1 \end{array}\right]$$

However, since the coefficient matrices are the same in each case, we can solve all n systems simultaneously! We simply set up the "augmented" matrix

$$\left[\begin{array}{c|cccc} & 1 & 0 & \cdots & 0 \\ \mathbf{A} & 0 & 1 & \cdots & 0 \\ & \cdots & \cdots & \cdots & \cdots \\ & 0 & 0 & \cdots & 1 \end{array}\right] \quad \text{or} \quad [\mathbf{A} \mid \mathbf{I}] \tag{6.5}$$

and then we use the operations of types 1, 2, and 3 to successively replace this by simpler systems until the $n \times n$ matrix on the left of (6.5) is the identity matrix. Now the matrix on the right of the new augmented matrix is $\mathbf{A}^{-1}$.

Method of Determining Invertibility and Computing the Inverse of a Square Matrix A

1. Form the matrix $[\mathbf{A} \mid \mathbf{I}]$.
2. Use operations of types 1, 2, and 3 to successively transform this matrix.
   a. If $[\mathbf{A} \mid \mathbf{I}]$ can be so transformed into $[\mathbf{I} \mid \mathbf{B}]$, then **A** is invertible and $\mathbf{A}^{-1} = \mathbf{B}$.
   b. If $[\mathbf{A} \mid \mathbf{I}]$ cannot be transformed into the form $[\mathbf{I} \mid \mathbf{B}]$, then **A** is not invertible. Matrix **A** is not invertible if and only if one of the matrices obtained from $[\mathbf{A} \mid \mathbf{I}]$ by these transformations has a row consisting entirely of zeros to the left of the vertical line.

Example 6.10 Find the inverse (if it exists) of the matrix

$$\mathbf{A} = \begin{bmatrix} 2 & 1 \\ 5 & 3 \end{bmatrix}$$

Solution We form the augmented matrix $[\mathbf{A}\,|\,\mathbf{I}]$:

$$\left[\begin{array}{cc|cc} 2 & 1 & 1 & 0 \\ 5 & 3 & 0 & 1 \end{array}\right]$$

Our goal is to transform the left-hand side of this augmented matrix into an identity matrix. First, we multiply the first row by $\frac{1}{2}$ to obtain

$$\left[\begin{array}{cc|cc} 1 & \frac{1}{2} & \frac{1}{2} & 0 \\ 5 & 3 & 0 & 1 \end{array}\right]$$

Next, we multiply the first row by -5, add it to the second row, and use the result as a new second row:

$$\left[\begin{array}{cc|cc} 1 & \frac{1}{2} & \frac{1}{2} & 0 \\ 0 & \frac{1}{2} & -\frac{5}{2} & 1 \end{array}\right]$$

Continuing, we multiply the second row by 2:

$$\left[\begin{array}{cc|cc} 1 & \frac{1}{2} & \frac{1}{2} & 0 \\ 0 & 1 & -5 & 2 \end{array}\right]$$

Finally, we multiply the second row by $-\frac{1}{2}$, add it to the first row, and use the result as a new first row:

$$\left[\begin{array}{cc|cc} 1 & 0 & 3 & -1 \\ 0 & 1 & -5 & 2 \end{array}\right]$$

The left-hand side is now an identity matrix. We conclude that the matrix $\mathbf{A}$ is invertible and that

$$\mathbf{A}^{-1} = \begin{bmatrix} 3 & -1 \\ -5 & 2 \end{bmatrix}$$ ■

Example 6.11 Find the inverse (if it exists) of the matrix

$$\mathbf{A} = \begin{bmatrix} 2 & 0 & 1 \\ 2 & 1 & -1 \\ 3 & 1 & -1 \end{bmatrix}$$

Solution We form the augmented matrix

$$\left[\begin{array}{ccc|ccc} 2 & 0 & 1 & 1 & 0 & 0 \\ 2 & 1 & -1 & 0 & 1 & 0 \\ 3 & 1 & -1 & 0 & 0 & 1 \end{array}\right]$$

First, we multiply the first row by $\frac{1}{2}$ to obtain

$$\left[\begin{array}{ccc|ccc} 1 & 0 & \frac{1}{2} & \frac{1}{2} & 0 & 0 \\ 2 & 1 & -1 & 0 & 1 & 0 \\ 3 & 1 & -1 & 0 & 0 & 1 \end{array}\right]$$

Next we multiply the first row by -2 and add it to the second row and multiply the first row by -3 and add it to the third row. The next matrix is

$$\left[\begin{array}{ccc|ccc} 1 & 0 & \frac{1}{2} & \frac{1}{2} & 0 & 0 \\ 0 & 1 & -2 & -1 & 1 & 0 \\ 0 & 1 & -\frac{5}{2} & -\frac{3}{2} & 0 & 1 \end{array}\right]$$

The next step is to multiply the second row by -1 and add it to the third row. The resulting matrix is

$$\left[\begin{array}{ccc|ccc} 1 & 0 & \frac{1}{2} & \frac{1}{2} & 0 & 0 \\ 0 & 1 & -2 & -1 & 1 & 0 \\ 0 & 0 & -\frac{1}{2} & -\frac{1}{2} & -1 & 1 \end{array}\right]$$

Multiplying the third row by -2, we obtain

$$\left[\begin{array}{ccc|ccc} 1 & 0 & \frac{1}{2} & \frac{1}{2} & 0 & 0 \\ 0 & 1 & -2 & -1 & 1 & 0 \\ 0 & 0 & 1 & 1 & 2 & -2 \end{array}\right]$$

Finally, multiplying the third row by 2 and adding it to the second row and multiplying the third row by $-\frac{1}{2}$ and adding it to the first row, we obtain

$$\left[\begin{array}{ccc|ccc} 1 & 0 & 0 & 0 & -1 & 1 \\ 0 & 1 & 0 & 1 & 5 & -4 \\ 0 & 0 & 1 & 1 & 2 & -2 \end{array}\right]$$

We conclude that the matrix **A** is invertible and that

$$\mathbf{A}^{-1} = \begin{bmatrix} 0 & -1 & 1 \\ 1 & 5 & -4 \\ 1 & 2 & -2 \end{bmatrix}$$

Although it is not a necessary part of the method, we verify our work in this example by showing that $\mathbf{A}^{-1}\mathbf{A} = \mathbf{I}$. We have

$$\begin{bmatrix} 0 & -1 & 1 \\ 1 & 5 & -4 \\ 1 & 2 & -2 \end{bmatrix} \begin{bmatrix} 2 & 0 & 1 \\ 2 & 1 & -1 \\ 3 & 1 & -1 \end{bmatrix}$$

$$= \begin{bmatrix} 0(2)+(-1)(2)+1(3) & 0(0)+(-1)(1)+1(1) & 0(1)+(-1)(-1)+1(-1) \\ 1(2)+5(2)+(-4)(3) & 1(0)+5(1)+(-4)(1) & 1(1)+5(-1)+(-4)(-1) \\ 1(2)+2(2)+(-2)(3) & 1(0)+2(1)+(-2)(1) & 1(1)+2(-1)+(-2)(-1) \end{bmatrix}$$

$$= \begin{bmatrix} 1 & 0 & 0 \\ 0 & 1 & 0 \\ 0 & 0 & 1 \end{bmatrix}$$ ■

It is worthwhile to consider an example which illustrates how we can infer from the failure of the method proposed here that a matrix is not invertible.

Example 6.12 Find the inverse (if it exists) of the matrix

$$\mathbf{A} = \begin{bmatrix} 2 & 0 & 1 \\ 3 & 1 & -1 \\ -2 & -2 & 4 \end{bmatrix}$$

Solution We form the augmented matrix

$$\left[\begin{array}{rrr|rrr} 2 & 0 & 1 & 1 & 0 & 0 \\ 3 & 1 & -1 & 0 & 1 & 0 \\ -2 & -2 & 4 & 0 & 0 & 1 \end{array}\right]$$

We multiply the first row by $\frac{1}{2}$ to obtain

$$\left[\begin{array}{rrr|rrr} 1 & 0 & \frac{1}{2} & \frac{1}{2} & 0 & 0 \\ 3 & 1 & -1 & 0 & 1 & 0 \\ -2 & -2 & 4 & 0 & 0 & 1 \end{array}\right]$$

Next we multiply the first row by -3 and add it to the second row, and we multiply the first row by 2 and add it to the third row. The new matrix is

$$\left[\begin{array}{ccc|ccc} 1 & 0 & \frac{1}{2} & \frac{1}{2} & 0 & 0 \\ 0 & 1 & -\frac{5}{2} & -\frac{3}{2} & 1 & 0 \\ 0 & -2 & 5 & 1 & 0 & 1 \end{array}\right]$$

We continue by multiplying the second row by 2 and adding it to the third row. We have

$$\left[\begin{array}{ccc|ccc} 1 & 0 & \frac{1}{2} & \frac{1}{2} & 0 & 0 \\ 0 & 1 & -\frac{5}{2} & -\frac{3}{2} & 1 & 0 \\ 0 & 0 & 0 & -2 & 2 & 1 \end{array}\right]$$

Notice the third row of this matrix. It has only zeros to the left of the vertical line. Therefore, $\mathbf{A}$ is not invertible. In particular, since there is a nonzero entry in the (3, 4) spot of the augmented matrix, we conclude that the system of equations

$$\mathbf{AX} = \begin{bmatrix} 1 \\ 0 \\ 0 \end{bmatrix} \tag{6.6}$$

has no solutions. In fact, in this case neither do the systems

$$\mathbf{AX} = \begin{bmatrix} 0 \\ 1 \\ 0 \end{bmatrix} \quad \text{and} \quad \mathbf{AX} = \begin{bmatrix} 0 \\ 0 \\ 1 \end{bmatrix}$$

■

It is important to remember the connection between the invertibility of a matrix $\mathbf{A}$ and the solvability of a system of equations $\mathbf{AX} = \mathbf{B}$. We have seen that if $\mathbf{A}$ has an inverse, then $\mathbf{AX} = \mathbf{B}$ always has a unique solution, given by $\mathbf{X} = \mathbf{A}^{-1}\mathbf{B}$. On the other hand, if $\mathbf{A}$ does not have an inverse, then $\mathbf{AX} = \mathbf{B}$ may or may not have a solution; it depends on the particular $\mathbf{A}$ and $\mathbf{B}$. For instance, with matrix $\mathbf{A}$ of Example 6.12, which we have just shown to be noninvertible, the equation

$$\mathbf{AX} = \begin{bmatrix} 1 \\ 2 \\ -2 \end{bmatrix}$$

is solvable, and the vector

$$\mathbf{X} = \begin{bmatrix} 1 \\ -2 \\ -1 \end{bmatrix}$$

is a solution. We saw that Equation (6.6) with the same matrix $\mathbf{A}$ had no solutions.

Example 6.13 Let

$$\mathbf{A} = \begin{bmatrix} 2 & -1 \\ 4 & 3 \end{bmatrix}$$

Problem Use the inverse of matrix $\mathbf{A}$ to solve the equations

$$\mathbf{AX} = \begin{bmatrix} 1 \\ 3 \end{bmatrix} \qquad \mathbf{AX} = \begin{bmatrix} 3 \\ 2 \end{bmatrix} \qquad \mathbf{AX} = \begin{bmatrix} 1 \\ -1 \end{bmatrix}$$

Solution We begin by computing $\mathbf{A}^{-1}$. First we form the augmented matrix $[\mathbf{A}\,|\,\mathbf{I}]$:

$$\left[\begin{array}{cc|cc} 2 & -1 & 1 & 0 \\ 4 & 3 & 0 & 1 \end{array}\right]$$

We proceed to use row reduction operations to determine $\mathbf{A}^{-1}$.

$$\left[\begin{array}{cc|cc} 1 & -\frac{1}{2} & \frac{1}{2} & 0 \\ 4 & 3 & 0 & 1 \end{array}\right]$$

$$\left[\begin{array}{cc|cc} 1 & -\frac{1}{2} & \frac{1}{2} & 0 \\ 0 & 5 & -2 & 1 \end{array}\right]$$

$$\left[\begin{array}{cc|cc} 1 & -\frac{1}{2} & \frac{1}{2} & 0 \\ 0 & 1 & -\frac{2}{5} & \frac{1}{5} \end{array}\right]$$

$$\left[\begin{array}{cc|cc} 1 & 0 & \frac{3}{10} & \frac{1}{10} \\ 0 & 1 & -\frac{2}{5} & \frac{1}{5} \end{array}\right]$$

Therefore,

$$\mathbf{A}^{-1} = \begin{bmatrix} \frac{3}{10} & \frac{1}{10} \\ -\frac{2}{5} & \frac{1}{5} \end{bmatrix}$$

The solution of $\mathbf{AX} = \begin{bmatrix} 1 \\ 3 \end{bmatrix}$ is

$$\mathbf{X} = \mathbf{A}^{-1}\begin{bmatrix} 1 \\ 3 \end{bmatrix} = \begin{bmatrix} \frac{3}{10} & \frac{1}{10} \\ -\frac{2}{5} & \frac{1}{5} \end{bmatrix}\begin{bmatrix} 1 \\ 3 \end{bmatrix} = \begin{bmatrix} \frac{3}{5} \\ \frac{1}{5} \end{bmatrix}$$

The solution of $\mathbf{AX} = \begin{bmatrix} 3 \\ 2 \end{bmatrix}$ is

$$\mathbf{X} = \mathbf{A}^{-1}\begin{bmatrix} 3 \\ 2 \end{bmatrix} = \begin{bmatrix} \frac{11}{10} \\ -\frac{4}{5} \end{bmatrix}$$

and the solution of $\mathbf{AX} = \begin{bmatrix} 1 \\ -1 \end{bmatrix}$ is

$$\mathbf{X} = \mathbf{A}^{-1}\begin{bmatrix} 1 \\ -1 \end{bmatrix} = \begin{bmatrix} \frac{1}{5} \\ -\frac{3}{5} \end{bmatrix}$$

■

Example 6.14 Let

$$\mathbf{A} = \begin{bmatrix} 2 & 1 \\ 5 & 3 \end{bmatrix} \quad \text{and} \quad \mathbf{C} = \begin{bmatrix} 2 & 1 \\ 1 & -1 \end{bmatrix}$$

Problem Find a matrix $\mathbf{B}$ which satisfies $\mathbf{AB} = \mathbf{C}$.

Solution If $\mathbf{A}$ is invertible, then the equation $\mathbf{AB} = \mathbf{C}$ is the same as $\mathbf{B} = \mathbf{A}^{-1}\mathbf{C}$. (Why?) Therefore, if we can invert $\mathbf{A}$, then we can solve the problem. But we have already found (in Example 6.10) that $\mathbf{A}$ is invertible and that

$$\mathbf{A}^{-1} = \begin{bmatrix} 3 & -1 \\ -5 & 2 \end{bmatrix}$$

Consequently we have

$$\mathbf{B} = \mathbf{A}^{-1}\mathbf{C} = \begin{bmatrix} 3 & -1 \\ -5 & 2 \end{bmatrix}\begin{bmatrix} 2 & 1 \\ 1 & -1 \end{bmatrix} = \begin{bmatrix} 5 & 4 \\ -8 & -7 \end{bmatrix}$$

as the desired matrix. ■

Exercises for Section 6.2

1. Let $\mathbf{A} = \begin{bmatrix} 1 & 0 \\ 2 & 1 \end{bmatrix}$ and $\mathbf{B} = \begin{bmatrix} 1 & 0 \\ -2 & 1 \end{bmatrix}$. Decide whether $\mathbf{B} = \mathbf{A}^{-1}$ by computing $\mathbf{AB}$ and $\mathbf{BA}$.
2. Let $\mathbf{A} = \begin{bmatrix} 1 & 2 \\ 3 & 4 \end{bmatrix}$ and $\mathbf{B} = \begin{bmatrix} -2 & 1 \\ \frac{3}{2} & -\frac{1}{2} \end{bmatrix}$. Decide whether $\mathbf{B} = \mathbf{A}^{-1}$ by computing $\mathbf{AB}$ and $\mathbf{BA}$.

3. Let $\mathbf{A} = \begin{bmatrix} 1 & 0 & 1 \\ 0 & 1 & 0 \\ 1 & 1 & 1 \end{bmatrix}$ and $\mathbf{B} = \begin{bmatrix} 1 & 1 & 0 \\ -1 & 0 & 1 \\ 0 & -1 & 0 \end{bmatrix}$. Decide whether $\mathbf{B} = \mathbf{A}^{-1}$ by computing $\mathbf{AB}$ and $\mathbf{BA}$.

4. Let $\mathbf{A} = \begin{bmatrix} 1 & 1 & 0 \\ 0 & 1 & 0 \\ 1 & 1 & 1 \end{bmatrix}$ and $\mathbf{B} = \begin{bmatrix} 1 & -1 & 0 \\ 0 & 1 & 0 \\ -1 & 0 & 1 \end{bmatrix}$. Decide whether $\mathbf{B} = \mathbf{A}^{-1}$ by computing $\mathbf{AB}$ and $\mathbf{BA}$.

5. Let $\mathbf{A} = \begin{bmatrix} 3 & 2 \\ 4 & 3 \end{bmatrix}$. Verify that $\mathbf{A}^{-1} = \begin{bmatrix} 3 & -2 \\ -4 & 3 \end{bmatrix}$.

6. Let $\mathbf{A} = \begin{bmatrix} 2 & 0 & 1 \\ 2 & 1 & -1 \\ 3 & 1 & -1 \end{bmatrix}$. Verify that $\mathbf{A}^{-1} = \begin{bmatrix} 0 & -1 & 1 \\ 1 & 5 & -4 \\ 1 & 2 & -2 \end{bmatrix}$.

7. Find $\mathbf{C}^{-1}$ and $\mathbf{D}^{-1}$ for

$$\mathbf{C} = \begin{bmatrix} 1 & -1 \\ 2 & 3 \end{bmatrix} \quad \text{and} \quad \mathbf{D} = \begin{bmatrix} .3 & .4 \\ .5 & .8 \end{bmatrix}$$

8. Find $\mathbf{P}^{-1}$ and $\mathbf{Q}^{-1}$ for

$$\mathbf{P} = \begin{bmatrix} 3 & 2 \\ 2 & 1 \end{bmatrix} \quad \text{and} \quad \mathbf{Q} = \begin{bmatrix} .4 & .6 \\ .1 & .9 \end{bmatrix}$$

9. Let $\mathbf{A} = \begin{bmatrix} 2 & 0 \\ 3 & 1 \end{bmatrix}$. Find the inverse of $\mathbf{A}$ and use it to solve the systems of equations

$$\mathbf{AX} = \begin{bmatrix} 1 \\ 0 \end{bmatrix} \quad \text{and} \quad \mathbf{AX} = \begin{bmatrix} 1 \\ -1 \end{bmatrix}$$

10. Let $\mathbf{A} = \begin{bmatrix} 2 & 3 \\ 5 & 9 \end{bmatrix}$. Find the inverse of $\mathbf{A}$ and use it to solve the systems of equations

$$\mathbf{AX} = \begin{bmatrix} 3 \\ 2 \end{bmatrix} \quad \text{and} \quad \mathbf{AX} = \begin{bmatrix} 4 \\ -1 \end{bmatrix}$$

11. Find the inverse (if it exists) of the matrix

$$\mathbf{A} = \begin{bmatrix} 3 & -2 \\ -6 & 4 \end{bmatrix}$$

12. Find the inverse (if it exists) of the matrix

$$\mathbf{A} = \begin{bmatrix} 0 & 1 & 0 \\ 0 & 0 & 1 \\ .2 & .8 & 0 \end{bmatrix}$$

13. Find the inverse (if it exists) of the matrix

$$\mathbf{A} = \begin{bmatrix} 0 & 1 & 1 \\ 2 & 0 & 1 \\ 0 & 0 & 1 \end{bmatrix}$$

14. Find the inverse (if it exists) of the matrix

$$\mathbf{A} = \begin{bmatrix} 0 & 2 & 4 \\ 1 & 0 & 1 \\ 3 & 0 & 1 \end{bmatrix}$$

15. (*a*) Express the following system of equations in matrix form:

$$\begin{aligned} 2x_1 + 4x_2 &= -2 \\ x_1 + 3x_2 &= 4 \end{aligned}$$

(*b*) Use the inverse of the coefficient matrix in part *a* to solve the system of equations.

16. (*a*) Express the following system of equations in matrix form:

$$\begin{aligned} 2x_1 - 2x_2 &= 2 \\ 3x_1 - 2x_3 &= 4 \\ 2x_2 - x_3 &= -2 \end{aligned}$$

(*b*) Find the inverse of the coefficient matrix of (*a*).
(*c*) Use the inverse of the coefficient matrix determined in (*b*) to solve the system of equations.

17. Let $\mathbf{A} = \begin{bmatrix} 5 & 2 \\ 2 & 1 \end{bmatrix}$. Find $\mathbf{A}^{-1}$ and then the inverse of $\mathbf{A}^{-1}$, denoted by $(\mathbf{A}^{-1})^{-1}$, using the method of this section.

18. Let $\mathbf{A} = \begin{bmatrix} 5 & 2 \\ 2 & 1 \end{bmatrix}$ and $\mathbf{B} = \begin{bmatrix} 1 & 3 \\ 2 & 5 \end{bmatrix}$. Find $\mathbf{A}^{-1}$, $\mathbf{B}^{-1}$, $(\mathbf{AB})^{-1}$, and $(\mathbf{BA})^{-1}$.

19. Using the results of Exercise 18, find $\mathbf{A}^{-1}\mathbf{B}^{-1}$ and $\mathbf{B}^{-1}\mathbf{A}^{-1}$. What is the relationship among $(\mathbf{AB})^{-1}$, $(\mathbf{BA})^{-1}$, $\mathbf{A}^{-1}\mathbf{B}^{-1}$, and $\mathbf{B}^{-1}\mathbf{A}^{-1}$?

20. Let **A**, **B**, and **C** be 2×2 matrices, and suppose

$$\mathbf{A} = \begin{bmatrix} 2 & 1 \\ -3 & 1 \end{bmatrix} \quad \text{and} \quad \mathbf{C} = \begin{bmatrix} 2 & 6 \\ 1 & -3 \end{bmatrix}$$

(*a*) Find **B** such that $\mathbf{AB} = \mathbf{C}$.
(*b*) Find **B** such that $\mathbf{BA} = \mathbf{C}$.

21. Let $\mathbf{A} = \begin{bmatrix} .2 & 0 & -.6 \\ 0 & .5 & -.8 \\ -.2 & 0 & .4 \end{bmatrix}$.
   (*a*) Find $\mathbf{A}^{-1}$.
   (*b*) Evaluate $\mathbf{A}\mathbf{A}^{-1}\mathbf{A}$.

22. Using matrices **A** and **B** of Exercise 18 and the results of that exercise, decide whether $(\mathbf{A} + \mathbf{B})^{-1}$ is equal to $\mathbf{A}^{-1} + \mathbf{B}^{-1}$.

23. Find the inverse (if it exists) of the matrix

$$\mathbf{A} = \begin{bmatrix} 1 & -\frac{1}{4} & 0 \\ -\frac{1}{4} & 1 & -\frac{1}{4} \\ 0 & -\frac{1}{4} & 1 \end{bmatrix}$$

24. Find the inverse (if it exists) of the matrix

$$\mathbf{A} = \begin{bmatrix} 0 & 1 & 2 & 1 \\ 1 & 0 & -1 & -1 \\ 1 & 2 & 0 & 0 \\ 1 & 0 & -2 & 0 \end{bmatrix}$$

25. Find the inverse of matrix **A** and use it to solve the equations

$$\mathbf{AX} = \begin{bmatrix} 1 \\ 3 \\ 1 \end{bmatrix} \quad \text{and} \quad \mathbf{AX} = \begin{bmatrix} 2 \\ 1 \\ 0 \end{bmatrix}$$

where

$$\mathbf{A} = \begin{bmatrix} 2 & 1 & 0 \\ -1 & 1 & 1 \\ 0 & -1 & 1 \end{bmatrix}$$

26. Find the inverse of matrix **A** and use it to solve the equations

$$\mathbf{AX} = \begin{bmatrix} 1 \\ 1 \\ 1 \end{bmatrix} \quad \text{and} \quad \mathbf{AX} = \begin{bmatrix} 1 \\ -1 \\ 1 \end{bmatrix}$$

where

$$\mathbf{A} = \begin{bmatrix} 2 & 0 & 1 \\ 2 & 1 & -1 \\ 3 & 1 & -1 \end{bmatrix}$$

27. Use the concept of a matrix inverse to solve the following systems of equations:

(a)
$$\begin{aligned} x + y + z &= 15 \\ y - z &= 1 \\ x \quad + z &= 10 \end{aligned}$$

(b)
$$\begin{aligned} x + y + z &= 9 \\ y - z &= 0 \\ x \quad + z &= 6 \end{aligned}$$

(c)
$$\begin{aligned} x + y + z &= 0 \\ y - z &= 1 \\ x \quad + z &= 0 \end{aligned}$$

28. Let $\mathbf{A} = \begin{bmatrix} 1 & 1 & 1 & 1 \\ 2 & 0 & 0 & 1 \\ 0 & 1 & 1 & 0 \\ 0 & 0 & 1 & 1 \end{bmatrix}$

(a) Find $\mathbf{A}^{-1}$.

(b) Use $\mathbf{A}^{-1}$ to solve the system of equations $\mathbf{AX} = \mathbf{B}$ where

$$\mathbf{B} = \begin{bmatrix} 1 \\ -1 \\ 1 \\ -2 \end{bmatrix}$$

29. Let $\mathbf{A} = \begin{bmatrix} 1 & 3 \\ 1 & 5 \end{bmatrix}$. Compute $\mathbf{AA} = \mathbf{A}^2$, $\mathbf{A}^{-1}$, $(\mathbf{A}^{-1})(\mathbf{A}^{-1})$, and $(\mathbf{AA})^{-1}$. Is $(\mathbf{A}^2)^{-1}$ the same as $(\mathbf{A}^{-1})^2$?

30. (Continuation of Exercise 29) Use the definition of a matrix inverse to show that for any square matrix $\mathbf{A}$, if $\mathbf{A}$ has an inverse, then

$$(\mathbf{A}^n)^{-1} = (\mathbf{A}^{-1})^n \qquad \text{for} \qquad n = 1, 2, \ldots$$

31. Let $\mathbf{A} = \begin{bmatrix} 1 & b \\ 0 & 1 \end{bmatrix}$, where $b \neq 0$. Find $\mathbf{A}^{-1}$.

32. Let $\mathbf{A} = \begin{bmatrix} 1 & b \\ c & 1 \end{bmatrix}$, where $b \neq 0$, $c \neq 0$, and $bc \neq 1$. Find $\mathbf{A}^{-1}$.

33. Let $\mathbf{A} = \begin{bmatrix} a & b \\ c & d \end{bmatrix}$, where $ad \neq bc$. Find $\mathbf{A}^{-1}$.

34. Let $\mathbf{A} = \begin{bmatrix} a & 0 & 0 \\ 0 & b & 0 \\ 0 & 0 & c \end{bmatrix}$, where $a \neq 0$, $b \neq 0$, $c \neq 0$. Find $\mathbf{A}^{-1}$.

35. Let $\mathbf{A} = \begin{bmatrix} 1 & a & b \\ 0 & 1 & c \\ 0 & 0 & 1 \end{bmatrix}$. Find $\mathbf{A}^{-1}$.

6.3 A LINEAR ECONOMIC MODEL

In 1973 Professor Wassily Leontief was awarded the Nobel Prize for Economics in recognition of his development of mathematical methods to study economic systems in terms of inputs and outputs. In practice, these models are quite complicated because of the very large number of variables involved and because of modifications made to increase the accuracy of the predictions. In their simplest form, however, Leontief input-output models are very straightforward.

As an introduction, suppose we have an economy with a single good.

Example 6.15

Production

Consider an economy with a single good which we take to be lumber. We suppose here, and in the general case as well, that some of the good being produced is consumed in the production process. For this example, suppose .2 unit of lumber is consumed in the production of 1 unit of lumber. Finally, suppose we have an order for 10 units of lumber.

Problem How many units of lumber must be produced to fill this order?

Solution To answer the question, let x denote the number of units of lumber produced by the economy. Then, for each unit produced, .2 unit is consumed. Net production is $x - .2x$. The problem is to find x such that

$$x - .2x = 10 \qquad \text{or} \qquad .8x = 10$$

We conclude that $x = 12.5$. The production process must produce 12.5 units of lumber to fill an order for 10 units. ■

The ideas introduced in this simple one-good example can be applied in more general situations. We consider first the case of two goods and then the case of n goods.

Economies with Two Goods

Suppose we have an economy with two industries, each of which produces a single good, and suppose the economy operates over a specific time period (e.g., a year). Consider a situation where each of the industries may use some of both goods in the production processes. For example, if the two goods are lumber and steel, then some lumber and some steel are used in the production of lumber, and likewise both goods are used in the production of steel. Let x_1 be the amount of good 1 produced in this economy, and let x_2 be the amount of good 2 produced. In studying this simple economy, we also need to know—and represent—the amounts of each good used in the production of the two goods. To represent this information, we let

a_{11} = the amount of good 1 used to produce one unit of good 1
a_{12} = the amount of good 1 used to produce one unit of good 2
a_{21} = the amount of good 2 used to produce one unit of good 1
a_{22} = the amount of good 2 used to produce one unit of good 2

Also suppose there is an external demand d_1 for good 1 and an external demand d_2 for good 2. By an *external demand* we mean an order for goods from outside the production processes.

Next, since it takes a_{11} units of good 1 to produce one unit of good 1, we assume that it takes $a_{11}x_1$ units of good 1 to produce x_1 units of good 1. Also we assume it will take $a_{12}x_2$ units of good 1 to produce x_2 units of good 2. Thus, the total amount of good 1 used in production is $a_{11}x_1 + a_{12}x_2$. Since there are x_1 units of good 1 produced and d_1 units are needed for external demand, we have the following equation

$$x_1 = a_{11}x_1 + a_{12}x_2 + d_1 \tag{6.7}$$

Similarly, for good 2, we have the equation

$$x_2 = a_{21}x_1 + a_{22}x_2 + d_2 \tag{6.8}$$

Equations (6.7) and (6.8) form a system of two equations in two variables (x_1 and x_2); and if we let

$$\mathbf{A} = \begin{bmatrix} a_{11} & a_{12} \\ a_{21} & a_{22} \end{bmatrix} \qquad \mathbf{X} = \begin{bmatrix} x_1 \\ x_2 \end{bmatrix} \qquad \mathbf{D} = \begin{bmatrix} d_1 \\ d_2 \end{bmatrix}$$

then this system of equations can be written in matrix form as

$$\mathbf{X} = \mathbf{AX} + \mathbf{D} \tag{6.9}$$

The matrix **A** in (6.9) is called the *technology matrix* for the economy, **X** is called the *production schedule*, and **D** is called the *demand vector*.

Example 6.16 *Production*

Consider a Leontief input-output model for an economy with two goods, steel and coal. Suppose .1 unit of steel and 2 units of coal are required to produce 1 unit of steel, and .3 unit of steel is required to produce 1 unit of coal.

Problem Find the technology matrix **A**.

Solution Let x_1 and x_2 denote the number of units of steel and coal, respectively, produced in the economy. The amount of steel necessary to produce 1 unit of steel is .1, and the amount of steel required to produce 1 unit of coal is .3. Therefore, we have

$$x_1 = .1x_1 + .3x_2$$

Likewise, the amount of coal necessary to produce 1 unit of steel is 2, and no coal is required in the production of coal. Therefore, we have

$$x_2 = 2x_1 + 0x_2 = 2x_1$$

The technology matrix **A** is the matrix which arises when these equations are written in matrix form. We have

$$\begin{bmatrix} x_1 \\ x_2 \end{bmatrix} = \begin{bmatrix} .1 & .3 \\ 2 & 0 \end{bmatrix} \begin{bmatrix} x_1 \\ x_2 \end{bmatrix}$$

and consequently the technology matrix **A** is

$$\mathbf{A} = \begin{bmatrix} .1 & .3 \\ 2 & 0 \end{bmatrix}$$

■

Warning

A common error which occurs in forming technology matrices is to mix up the rows and columns of the matrix. To avoid this error you should be careful that, when you form the system of equations, you are representing only one good per equation. For example, in the equation $x_1 = .1x_1 + .3x_2$ (of Example 6.16), the left side is the amount of steel produced and the right side is the sum of the amount of steel needed for production of steel ($.1x_1$) and the amount of steel needed for production of coal ($.3x_2$). Both sides of the equation represent amounts of steel.

In most applications of Leontief models, one is interested in the relationship between the demand vector **D** and the production **X**. Using the matrix algebra tools of Sections 6.1 and 6.2, we can convert (6.9) to the form

$$(\mathbf{I} - \mathbf{A})\mathbf{X} = \mathbf{D} \qquad \text{where } \mathbf{I} = \begin{bmatrix} 1 & 0 \\ 0 & 1 \end{bmatrix}$$

Therefore, if the square matrix $\mathbf{I} - \mathbf{A}$ has an inverse, then we can express **X** in terms of **D** as follows:

$$\mathbf{X} = (\mathbf{I} - \mathbf{A})^{-1}\mathbf{D} \tag{6.10}$$

Equation (6.10) is especially useful when one is interested in the effect of various demands on production. For example, if a government has a goal of stimulating the economy, then it can use (6.10) to evaluate the effect of specific stimulus actions, different demand vectors **D**, on the level of economic activity. For this reason it is common to determine the production schedule **X** by using (6.10). However, it is also possible to solve (6.9) for **X** by using the techniques of Chapter 5. In Example 6.17 we illustrate both methods; from then on we use (6.10).

Example 6.17 *Production*

Consider a Leontief input-output model for an economy with two goods, steel and coal, and the technology matrix (as derived in Example 6.16)

$$\mathbf{A} = \begin{bmatrix} .1 & .3 \\ 2 & 0 \end{bmatrix}$$

Recall that the first row (and column) is identified with steel and the second row (and column) with coal. Suppose that the external demand vector is $\begin{bmatrix} 12 \\ 3 \end{bmatrix}$.

Problem Find the production schedule which meets this demand.

Solution We illustrate both methods of finding the production schedule: First we solve the system (6.9), and then we find $(\mathbf{I} - \mathbf{A})^{-1}$ and use (6.10).

To use the techniques of Chapter 5 on Equation (6.9), we write it in the form $(\mathbf{I} - \mathbf{A})\mathbf{X} = \mathbf{D}$, and we form the augmented matrix for this system. The left-hand side of the augmented matrix is determined by computing $\mathbf{I} - \mathbf{A}$:

$$\mathbf{I} - \mathbf{A} = \begin{bmatrix} 1 & 0 \\ 0 & 1 \end{bmatrix} - \begin{bmatrix} .1 & .3 \\ 2 & 0 \end{bmatrix} = \begin{bmatrix} .9 & -.3 \\ -2 & 1 \end{bmatrix}$$

Since $\mathbf{D} = \begin{bmatrix} 12 \\ 3 \end{bmatrix}$, the augmented matrix is

$$\left[\begin{array}{cc|c} .9 & -.3 & 12 \\ -2 & 1 & 3 \end{array}\right]$$

Using elementary row operations, we have successively

$$\left[\begin{array}{cc|c} 1 & -\frac{1}{3} & \frac{40}{3} \\ -2 & 1 & 3 \end{array}\right]$$

$$\left[\begin{array}{cc|c} 1 & -\frac{1}{3} & \frac{40}{3} \\ 0 & \frac{1}{3} & \frac{89}{3} \end{array}\right]$$

$$\left[\begin{array}{cc|c} 1 & 0 & 43 \\ 0 & 1 & 89 \end{array}\right]$$

From this we conclude that $\begin{bmatrix} 41 \\ 89 \end{bmatrix}$ is the production schedule required to produce the demand vector $\begin{bmatrix} 12 \\ 3 \end{bmatrix}$.

Next, we use Equation (6.10) to determine the demand vector, and to do so, we must determine $(\mathbf{I} - \mathbf{A})^{-1}$. Using the methods of Section 6.2, we find

$$(\mathbf{I} - \mathbf{A})^{-1} = \begin{bmatrix} \frac{10}{3} & 1 \\ \frac{20}{3} & 3 \end{bmatrix}$$

Using (6.10), we have

$$\mathbf{X} = (\mathbf{I} - \mathbf{A})^{-1}\mathbf{D} = \begin{bmatrix} \frac{10}{3} & 1 \\ \frac{20}{3} & 3 \end{bmatrix} \begin{bmatrix} 12 \\ 3 \end{bmatrix} = \begin{bmatrix} 43 \\ 89 \end{bmatrix}$$

Interpreting these results in the original setting, we see that it is necessary to produce 43 units of steel and 89 units of coal to meet the external demand. We know from the demand vector that 12 units of steel and 3 units of coal are used outside the production system, and therefore we see that $43 - 12 = 31$ units of steel and $89 - 3 = 86$ units of coal are used within the economy. ■

Economies with *n* Goods

The discussion preceding Examples 6.16 and 6.17 described a Leontief model for an economy in which two goods were produced. Such a model consists of assumptions and equations. We turn now to a similar model for an economy in which many goods are produced.

Suppose we have an economy in which there are n industries, each of which produces a single good. We assume that each industry utilizes the goods produced by some or all of the industries in the production process.

Let x_i denote the amount of the ith good (measured in appropriate units) produced in unit time. We suppose that this output is used in the production of the ith good, in the production of other goods, and in the satisfaction of the demand external to the economic system being considered. Thus

$$\begin{aligned} x_i = {} & \text{amount of } i\text{th good used to produce goods} \\ & + \text{amount used to satisfy external demand} \end{aligned} \tag{6.11}$$

That is, a portion of the amount of good i produced goes outside the system (it satisfies the external demand), and a portion remains in the system (it is used in the production processes). The amount of good i used in the production processes can be further subdivided into the amounts used to produce good 1, good 2, . . . , good n, respectively. We repeat for emphasis that the situation in which good i is used in the production of good i is included in the model.

We set a_{ij} equal to the amount of good i used to produce one unit of good j, and we assume that the amount of good i needed to produce x_j units of good j is $a_{ij}x_j$. We also assume that the various production processes are unrelated in the sense that the amount of good i used to produce x_1 units of good 1 and x_2 units of good 2 and . . . and x_n units of good n is the sum of the amounts of good i needed to produce those quantities of the goods individually:

$$a_{i1}x_1 + a_{i2}x_2 + \cdots + a_{in}x_n$$

These assumptions enable us to use the methods of this chapter to study the model. Now, if we let d_i denote the external demand for good i, then the representation of x_i given in Equation (6.11) has the form

$$x_i = a_{i1}x_1 + a_{i2}x_2 + \cdots + a_{in}x_n + d_i$$

We assume that all the goods produced in the economy can be described by similar equations. It follows that we have a system of linear equations in the variables $x_1, x_2, \ldots, x_n$:

$$\begin{aligned}
x_1 &= a_{11}x_1 + a_{12}x_2 + \cdots + a_{1n}x_n + d_1 \\
x_2 &= a_{21}x_1 + a_{22}x_2 + \cdots + a_{2n}x_n + d_2 \\
&\ \ \vdots \\
x_n &= a_{n1}x_1 + a_{n2}x_2 + \cdots + a_{nn}x_n + d_n
\end{aligned}$$

In matrix-vector notation, this system of equations has the form

$$\mathbf{X} = \mathbf{AX} + \mathbf{D}$$

where $\mathbf{A}$ is the $n \times n$ matrix whose (i, j) entry is a_{ij}, and $\mathbf{X}$ and $\mathbf{D}$ are column n-vectors whose coordinates are $x_1, \ldots, x_n$ and $d_1, \ldots, d_n$, respectively.

In a Leontief input-output model represented by the equation $\mathbf{X} = \mathbf{AX} + \mathbf{D}$, the matrix $\mathbf{A}$ is the *technology* or *input-output* matrix, the vector $\mathbf{D}$ is the *external demand* vector, and the vector $\mathbf{X}$ is a *production schedule* (a list of amounts of goods to be produced).

Since the coordinates of a production schedule are the amounts of various goods produced, each of these coordinates must be nonnegative.

As a result of our assumption that each industry produces a single good (the jth industry produces the jth good only), it follows that $a_{jk} \geq 0$ for $j \neq k$. Indeed, if there were industries j and k with $a_{jk} < 0$, then the effect of producing x_k units of good k would be to reduce the amount of good j required by an amount $|a_{jk}x_k|$. Thus, if this were the case, the production of good k would be equivalent to the production of good j. We assume that this is not the case.

Finally, we assume that all the coordinates of the demand vector **D** are nonnegative. This assumption reflects the fact that we are assuming **D** to be a genuine demand vector, and demands ought not to decrease the amount of a good required.

The problem we consider is that of finding a production schedule **X** which satisfies the demand **D** with a given technology matrix **A**. Thus we seek **X** with nonnegative coordinates such that

$$\mathbf{X} = \mathbf{AX} + \mathbf{D} \qquad \text{or} \qquad (\mathbf{I} - \mathbf{A})\mathbf{X} = \mathbf{D}$$

We assume that $\mathbf{I} - \mathbf{A}$ is invertible. This assumption can be given economic justification. Indeed, the failure of $\mathbf{I} - \mathbf{A}$ to be invertible would imply a particular and exact relationship among the coefficients a_{ij} of **A**. Since the coefficients of **A** are usually estimated and consequently not known precisely, it is not unduly restrictive to assume that $\mathbf{I} - \mathbf{A}$ is invertible.

For the remainder of this discussion we shall assume that **A** has nonnegative entries, that $\mathbf{I} - \mathbf{A}$ is invertible, and that $(\mathbf{I} - \mathbf{A})^{-1}$ has nonnegative entries. We can make additional assumptions about the nature of the economic system that will guarantee that $(\mathbf{I} - \mathbf{A})^{-1}$ has nonnegative entries, but we shall not discuss this assumption in economic terms. With these assumptions, the equation $(\mathbf{I} - \mathbf{A})\mathbf{X} = \mathbf{D}$ can be solved for the production schedule **X** by multiplying both sides by $(\mathbf{I} - \mathbf{A})^{-1}$. We conclude that

$$\mathbf{X} = (\mathbf{I} - \mathbf{A})^{-1}\mathbf{D} \tag{6.12}$$

Since **D** and $(\mathbf{I} - \mathbf{A})^{-1}$ contain only nonnegative entries, it follows that **X** has only nonnegative entries, and consequently **X** is a legitimate production schedule.

Example 6.18 *Production*

Suppose an economy has three goods—cement, electricity, and steel—and a technology matrix

$$\mathbf{A} = \begin{bmatrix} .2 & .6 & .2 \\ 0 & .1 & .4 \\ .6 & 0 & .3 \end{bmatrix}$$

In **A**, the first row (and column) corresponds to cement, the second to electricity, and the third to steel.

Problem Find the external demand **D** which will be met by the production schedule

$$\mathbf{X} = \begin{bmatrix} 30 \\ 20 \\ 40 \end{bmatrix}$$

Solution The technology matrix **A**, demand **D**, and production schedule **X** are related by

$$\mathbf{X} = \mathbf{AX} + \mathbf{D}$$

Since we are to determine **D** given **A** and **X**, it is helpful to rewrite this equation in the form

$$\mathbf{D} = \mathbf{X} - \mathbf{AX}$$

Using the values given for **A** and **X**, we have

$$\mathbf{D} = \begin{bmatrix} 30 \\ 20 \\ 40 \end{bmatrix} - \begin{bmatrix} .2 & .6 & .2 \\ 0 & .1 & .4 \\ .6 & 0 & .3 \end{bmatrix} \begin{bmatrix} 30 \\ 20 \\ 40 \end{bmatrix} = \begin{bmatrix} 4 \\ 2 \\ 10 \end{bmatrix}$$ ■

Example 6.19

Production

Consider an economy which produces three goods: lumber, steel, and coal. Suppose the production of 1 unit of lumber requires .5 unit of lumber, .2 unit of steel, and 1 unit of coal. Also the production of 1 unit of steel requires .8 unit of steel and .4 unit of coal, and the production of 1 unit of coal requires .2 unit of lumber and .12 unit of steel.

Problem Find the technology matrix for this economy, and find a production schedule which will satisfy an external demand for 5 units of lumber, 3 units of steel, and 4 units of coal.

Solution Since we are interested in the amounts of lumber, steel, and coal to be produced, we let

x_1 = the number of units of lumber produced
x_2 = the number of units of steel produced
x_3 = the number of units of coal produced

Then, using the information given about the amount of lumber used in producing lumber, steel, and coal, and the external demand for lumber, we have

$$x_1 = .5x_1 + 0x_2 + .2x_3 + 5$$

Similarly, we obtain the equations

$$\begin{aligned} x_2 &= .2x_1 + .8x_2 + .12x_3 + 3 \\ x_3 &= 1x_1 + .4x_2 + 0x_3 + 4 \end{aligned}$$

for steel and coal, respectively. To use matrix-vector notation, we let

$$\mathbf{X} = \begin{bmatrix} x_1 \\ x_2 \\ x_3 \end{bmatrix} \qquad \mathbf{A} = \begin{bmatrix} .5 & 0 & .2 \\ .2 & .8 & .12 \\ 1 & .4 & 0 \end{bmatrix} \qquad \mathbf{D} = \begin{bmatrix} 5 \\ 3 \\ 4 \end{bmatrix}$$

and we have the equation

$$\mathbf{X} = \mathbf{AX} + \mathbf{D}$$

Again, we proceed by determining $(\mathbf{I} - \mathbf{A})^{-1}$ and using Equation (6.12). In this case

$$\mathbf{I} - \mathbf{A} = \begin{bmatrix} .5 & 0 & -.2 \\ -.2 & .2 & -.12 \\ -1 & -.4 & 1 \end{bmatrix}$$

and

$$(\mathbf{I} - \mathbf{A})^{-1} = \begin{bmatrix} 7.6 & 4 & 2 \\ 16 & 15 & 5 \\ 14 & 10 & 5 \end{bmatrix}$$

The production vector $\mathbf{X}$ can be obtained from (6.12):

$$\mathbf{X} = (\mathbf{I} - \mathbf{A})^{-1}\mathbf{D} = \begin{bmatrix} 7.6 & 4 & 2 \\ 16 & 15 & 5 \\ 14 & 10 & 5 \end{bmatrix} \begin{bmatrix} 5 \\ 3 \\ 4 \end{bmatrix} = \begin{bmatrix} 58 \\ 145 \\ 120 \end{bmatrix}$$

To satisfy the external demand, it is necessary to produce 58 units of lumber, 145 units of steel, and 120 units of coal. ■

Example 6.20 *Production*

Consider an economy which has two goods, lumber and steel, and suppose that Table 6.2 gives the goods required in the production processes. For example, the entry .5 indicates that the production of 1 unit of lumber requires .5 unit of steel.

TABLE 6.2

	Number of Units Needed to Produce 1 Unit of	
Good	**Lumber**	**Steel**
Lumber	.6	.6
Steel	.5	.2

Problem Find the associated technology matrix **A**, and determine whether the inverse of **I** − **A** has all positive entries.

Solution Let

$$\begin{aligned} x_1 &= \text{the number of units of lumber to be produced} \\ x_2 &= \text{the number of units of steel to be produced} \\ d_1 &= \text{the external demand for lumber} \\ d_2 &= \text{the external demand for steel} \end{aligned}$$

Then, using Table 6.2, we have

$$\begin{aligned} x_1 &= .6x_1 + .6x_2 + d_1 \\ x_2 &= .5x_1 + .2x_2 + d_2 \end{aligned}$$

Hence the technology matrix **A** is given by

$$\mathbf{A} = \begin{bmatrix} .6 & .6 \\ .5 & .2 \end{bmatrix} \quad \text{and} \quad \mathbf{I} - \mathbf{A} = \begin{bmatrix} .4 & -.6 \\ -.5 & .8 \end{bmatrix}$$

Using the techniques of Section 6.2, we determine the inverse of **I** − **A** by converting the following matrix to reduced form:

$$\left[\begin{array}{cc|cc} .4 & -.6 & 1 & 0 \\ -.5 & .8 & 0 & 1 \end{array}\right]$$

We obtain

$$\left[\begin{array}{cc|cc} 1 & 0 & 40 & 30 \\ 0 & 1 & 25 & 20 \end{array}\right]$$

Therefore, the inverse of **I** − **A** is the matrix $\begin{bmatrix} 40 & 30 \\ 25 & 20 \end{bmatrix}$, and we see that it has all positive entries. ■

Exercises for Section 6.3

1. Using the technology matrix of Example 6.16, compute the production schedule for each of the following demand vectors.

   (*a*) $\mathbf{D} = \begin{bmatrix} 15 \\ 5 \end{bmatrix}$ (*b*) $\mathbf{D} = \begin{bmatrix} 1000 \\ 125 \end{bmatrix}$

2. Using the technology matrix of Example 6.16, find the external demand which is satisfied by each of the following production schedules.

   (*a*) $\mathbf{X} = \begin{bmatrix} 20 \\ 50 \end{bmatrix}$ (*b*) $\mathbf{X} = \begin{bmatrix} 100 \\ 240 \end{bmatrix}$

3. Using the technology matrix of Example 6.19, compute the production schedule for each of the following demand vectors.

   (*a*) $\mathbf{D} = \begin{bmatrix} 20 \\ 20 \\ 8 \end{bmatrix}$ (*b*) $\mathbf{D} = \begin{bmatrix} 100 \\ 120 \\ 50 \end{bmatrix}$

4. Using the technology matrix of Example 6.19, find the external demand which is satisfied by each of the following production schedules.

   (*a*) $\mathbf{X} = \begin{bmatrix} 9 \\ 22 \\ 20 \end{bmatrix}$ (*b*) $\mathbf{X} = \begin{bmatrix} 60 \\ 140 \\ 120 \end{bmatrix}$

5. In the setting of Example 6.19, find the production schedule which meets an external demand for 100 units of lumber and 60 units of steel. There is no external demand for coal.

6. In the situation described in Exercise 5, how much coal must be produced to meet the needs of the production processes?

7. Let $\mathbf{A} = \begin{bmatrix} .2 & .4 \\ .5 & .3 \end{bmatrix}$. If $\mathbf{A}$ is the technology matrix for a Leontief model, can the methods of this section be used to determine production schedules to meet external demands? That is, (*a*) is $(\mathbf{I} - \mathbf{A})$ invertible, and (*b*) if so, does $(\mathbf{I} - \mathbf{A})^{-1}$ have all positive entries?

8. Repeat Exercise 7 for the matrix $\mathbf{A} = \begin{bmatrix} .2 & .4 \\ .8 & .6 \end{bmatrix}$.

In Exercises 9 through 12, check whether the assumptions of this section hold. That is, does $(\mathbf{I} - \mathbf{A})^{-1}$ exist and have nonnegative entries?

9. $\mathbf{A} = \begin{bmatrix} .4 & .2 \\ .9 & .2 \end{bmatrix}$

10. $\mathbf{A} = \begin{bmatrix} .2 & .8 \\ .6 & .4 \end{bmatrix}$

11. $\mathbf{A} = \begin{bmatrix} .2 & .4 & 0 \\ .1 & 0 & .1 \\ 0 & .1 & .2 \end{bmatrix}$

12. $\mathbf{A} = \begin{bmatrix} .4 & .1 & .3 \\ .3 & .7 & .1 \\ .9 & .5 & .1 \end{bmatrix}$

13. Consider an economy with two goods and a technology matrix **A** as in Exercise 9.
    (*a*) Find the production schedule which meets an external demand for 24 units of good 1 and 15 units of good 2.
    (*b*) Find the production schedule which meets an external demand for 48 units of good 1 and 30 units of good 2.
    (*c*) What is the relation between the production schedules determined in (*a*) and (*b*)?
14. Consider an economy with two goods and a technology matrix **A** as in Exercise 9. If **X** is the production schedule which meets external demand **D**, what production schedule will meet external demand 2**D**? (*Hint*: Use the result of Exercise 13.) What about external demand 10**D**?

In Exercises 15 through 22, find the production schedule for the given technology matrix and demand vector.

15. $\mathbf{A} = \begin{bmatrix} .5 & .4 \\ .75 & .2 \end{bmatrix}$ $\qquad \mathbf{D} = \begin{bmatrix} 2 \\ 1 \end{bmatrix}$

16. $\mathbf{A} = \begin{bmatrix} .2 & .6 \\ .6 & .05 \end{bmatrix}$ $\qquad \mathbf{D} = \begin{bmatrix} 8 \\ 10 \end{bmatrix}$

17. $\mathbf{A} = \begin{bmatrix} .7 & .3 \\ .5 & .4 \end{bmatrix}$ $\qquad \mathbf{D} = \begin{bmatrix} 6 \\ 5 \end{bmatrix}$

18. $\mathbf{A} = \begin{bmatrix} \frac{3}{5} & \frac{3}{10} \\ \frac{1}{3} & \frac{1}{2} \end{bmatrix}$ $\qquad \mathbf{D} = \begin{bmatrix} 30 \\ 20 \end{bmatrix}$

19. $\mathbf{A} = \begin{bmatrix} .5 & 0 & .3 \\ 0 & .8 & .4 \\ 0 & .2 & .4 \end{bmatrix}$ $\qquad \mathbf{D} = \begin{bmatrix} 3 \\ 1 \\ 2 \end{bmatrix}$

20. $\mathbf{A} = \begin{bmatrix} .2 & .4 & 0 \\ 0 & .4 & .5 \\ 0 & .1 & .5 \end{bmatrix}$ $\qquad \mathbf{D} = \begin{bmatrix} 8 \\ 2 \\ 3 \end{bmatrix}$

21. $\mathbf{A} = \begin{bmatrix} .6 & .2 & 0 \\ 0 & .4 & .1 \\ .1 & .1 & .6 \end{bmatrix}$ $\qquad \mathbf{D} = \begin{bmatrix} 54 \\ 18 \\ 90 \end{bmatrix}$

22. $\mathbf{A} = \begin{bmatrix} .2 & 0 & 0 & .2 \\ 0 & .4 & 0 & .2 \\ 0 & 0 & .2 & .4 \\ 0 & 0 & .4 & .5 \end{bmatrix}$ $\qquad \mathbf{D} = \begin{bmatrix} 50 \\ 40 \\ 20 \\ 100 \end{bmatrix}$

23. An economy has two goods: lumber and steel. The production of 1 unit of steel requires .2 unit of steel and .5 unit of lumber, and the production of 1 unit of lumber requires .6 unit of steel and .6 unit of lumber. Find the production schedule which meets an external demand for 20 units of steel and 10 units of lumber.
24. Consider an economy in which the goods are grain, steel, and coal. Also suppose that the units of measurement are such that Table 6.3 describes the relationships between the units. Find the associated technology matrix, and decide whether this matrix satisfies the assumptions of this section.

TABLE 6.3

	Number of Units Needed to Produce 1 Unit of		
Good	**Grain**	**Steel**	**Coal**
Grain	.2	.4	0
Steel	0	.4	.5
Coal	.8	.1	.5

25. In the setting of Exercise 24, can an external demand for grain, steel, and coal of 20, 30, and 50 units, respectively, be met? If so, what is the production schedule which meets this demand?
26. In the setting of Exercise 24, find all production schedules which provide no grain for an external demand and which provide twice as much coal as steel for the external market.
27. An economy has three goods: coal, lumber, and steel. The production of 1 unit of coal requires .4 unit of coal, .5 unit of lumber, and .3 unit of steel. The production of 1 unit of lumber requires no coal, .2 unit of lumber, and .6 unit of steel. Finally, the production of 1 unit of steel requires .2 unit of coal, .5 unit of lumber, and .2 unit of steel. Find the production schedule which meets an external demand for 48 units of coal, 24 units of lumber, and 18 units of steel.
28. In the setting of Exercise 27, what production schedules provide only lumber for the external market? That is, the amounts of coal and steel produced are exactly the amounts used in the production process.
29. For what values of c does the technology matrix $\mathbf{A} = \begin{bmatrix} .2 & c \\ .6 & .4 \end{bmatrix}$ satisfy the assumptions of this chapter?
30. Let $\mathbf{A} = \begin{bmatrix} .5 & .5 \\ .5 & a \end{bmatrix}$. For which positive values of variable a does $\mathbf{I} - \mathbf{A}$ have an inverse with all positive entries?
31. Repeat Exercise 30 for

$$\mathbf{A} = \begin{bmatrix} .2 & a \\ .6 & .5 \end{bmatrix}$$

32. Repeat Exercise 30 for

$$\mathbf{A} = \begin{bmatrix} .2 & .4 \\ .6 & a \end{bmatrix}$$

33. Repeat Exercise 30 for

$$\mathbf{A} = \begin{bmatrix} .8 & a \\ a & .2 \end{bmatrix}$$

34. Let $\mathbf{A} = \begin{bmatrix} .5 & a \\ b & .5 \end{bmatrix}$. For which values of variables a and b does $\mathbf{I} - \mathbf{A}$ have an inverse with all positive entries?

35. An economy with two goods has the technology matrix

$$\mathbf{A} = \begin{bmatrix} .2 & k \\ .4 & .5 \end{bmatrix}$$

where k is a number, $0 < k < 1$. There is an external demand for 20 units of good 1 and 10 units of good 2.

(*a*) Find the amount of good 2 which must be produced to meet the external demand when $k = .1$.

(*b*) Repeat part *a* for $k = .2, .3, \ldots, .9$.

(*c*) Graph the amount of good 2 as a function of k. That is, on a coordinate system with k on the horizontal axis and the amount of good 2 on the vertical axis, plot the amount of good 2 for each value of k, $k = .1, .2, \ldots, .9$.

IMPORTANT TERMS AND CONCEPTS

You should be able to describe, define, or give examples of and use each of the following:

Matrix
Dimension of a matrix
The (i, j) entry in a matrix
Column vector
Row vector
Coordinates of a vector
Equality of matrices
Matrix addition
Scalar multiplication
Row-by-column multiplication
Matrix multiplication
Inverse of a matrix
Leontief model
Technology matrix
External demand vector
Production schedule

REVIEW EXERCISES

1. Find the dimensions of each of the following matrices.

(*a*) $\begin{bmatrix} 50 & -30 \\ -25 & 120 \\ 0 & 85 \\ 15 & 60 \end{bmatrix}$ (*b*) $\begin{bmatrix} 65 & 15 & 80 \\ 90 & 15 & 105 \end{bmatrix}$

(*c*) $[.2 \quad 0 \quad .2 \quad .6]$ (*d*) $\begin{bmatrix} 85 \\ 240 \\ 490 \end{bmatrix}$

2. Give an example of a 1×5 vector and a 4×1 vector.

3. Let

$$\mathbf{A} = \begin{bmatrix} -1 & 7 \\ 0 & 15 \\ 12 & -8 \end{bmatrix} \qquad \mathbf{B} = \begin{bmatrix} 5 & -1 \\ 12 & 8 \\ 5 & 2 \end{bmatrix}$$

Find $5\mathbf{A} + \mathbf{B}$ and $\mathbf{A} - 2\mathbf{B}$.

4. Let

$$\mathbf{A} = \begin{bmatrix} 10 & 3 \\ 0 & 5 \\ -2 & 4 \end{bmatrix} \qquad \mathbf{X} = \begin{bmatrix} -1 \\ 5 \end{bmatrix} \qquad \mathbf{B} = \begin{bmatrix} 5 \\ 25 \\ 22 \end{bmatrix}$$

Show that $\mathbf{AX} = \mathbf{B}$.

5. Using vectors $\mathbf{X} = \begin{bmatrix} -1 \\ 5 \\ -10 \end{bmatrix}$ and $\mathbf{B}$ as in Exercise 4, find vectors $\mathbf{Y}$ and $\mathbf{Z}$ such that $\mathbf{X} + 2\mathbf{Y} = \mathbf{B}$ and $2\mathbf{X} + \mathbf{Z} = \mathbf{B}$.

6. Let

$$\mathbf{A} = \begin{bmatrix} 1 & 0 & 3 \\ 1 & 1 & 1 \\ -3 & 2 & 1 \end{bmatrix} \qquad \mathbf{B} = \begin{bmatrix} 2 & -1 & 1 \\ 0 & 3 & -1 \\ -2 & -1 & -3 \end{bmatrix}$$

Find $2\mathbf{A} + 3\mathbf{B}$.

7. Using matrices $\mathbf{A}$ and $\mathbf{B}$ of Exercise 6, find p and q such that

$$p\mathbf{A} + q\mathbf{B} = \begin{bmatrix} -3 & 1 & -4 \\ -1 & -4 & 0 \\ 5 & -1 & 2 \end{bmatrix}$$

8. Let $\mathbf{A}$ be 3×5, $\mathbf{B}$ be 5×3, $\mathbf{C}$ be 3×3, and $\mathbf{D}$ be 5×5. Decide which of the following products are defined; if the product is defined, determine the size of the product.
   (*a*) **ABC** (*b*) **BAD**
   (*c*) **ABD** (*d*) **DBAC**

9. Matrices $\mathbf{A}$, $\mathbf{B}$, $\mathbf{C}$, and $\mathbf{X}$ are defined as follows:

$$\mathbf{A} = \begin{bmatrix} 10 & 3 \\ 0 & 5 \\ -2 & 4 \end{bmatrix} \qquad \mathbf{B} = \begin{bmatrix} 3 & 6 \\ 1 & 5 \end{bmatrix} \qquad \mathbf{C} = \begin{bmatrix} -2 & 4 \\ 3 & 0 \\ 2 & 2 \end{bmatrix} \qquad \mathbf{X} = \begin{bmatrix} 3 \\ 2 \end{bmatrix}$$

For each of the following, perform the indicated operations where possible, and express your answer as a single matrix. If the indicated operations are not defined, answer "not defined."
   (*a*) **2B** (*b*) **A + B** (*c*) **A + C**
   (*d*) **AB** (*e*) **AC** (*f*) **AX**

10. Matrices **A**, **B**, **C**, and **X** are defined as follows:

$$\mathbf{A} = \begin{bmatrix} 2 & -1 \\ 3 & 1 \end{bmatrix} \quad \mathbf{B} = \begin{bmatrix} 4 & 2 \\ 1 & -1 \\ 3 & 0 \end{bmatrix} \quad \mathbf{C} = \begin{bmatrix} -1 & 0 \\ 2 & 5 \\ 3 & -2 \end{bmatrix} \quad \mathbf{X} = \begin{bmatrix} -2 \\ 3 \end{bmatrix}$$

In each of the following, perform the indicated operations where possible, and express your answer as a single matrix. If the indicated operations are not defined, answer "not defined."

(*a*) 3**A** (*b*) 4**B** − **C** (*c*) **AB**
(*d*) **BA** (*e*) **BX** (*f*) **CX**

11. Matrices **A**, **B**, **C**, and **D** are defined as follows:

$$\mathbf{A} = \begin{bmatrix} 1 & 2 \\ -2 & 3 \end{bmatrix} \quad \mathbf{B} = \begin{bmatrix} 1 & 2 & -1 \\ 0 & -1 & 2 \end{bmatrix}$$

$$\mathbf{C} = \begin{bmatrix} 4 & 0 & 1 \\ 2 & 1 & 3 \end{bmatrix} \quad \mathbf{D} = \begin{bmatrix} 2 & 0 \\ 3 & -1 \\ -2 & 0 \end{bmatrix}$$

In each of the following, perform the indicated operations where possible, and express your answer as a single matrix. If the indicated operations are not defined, answer "not defined."

(*a*) 2**D** (*b*) 2**B** − **C** (*c*) **CA**
(*d*) **AB** (*e*) **BC** (*f*) **BD**

12. Matrices **A**, **B**, **C**, and **Y** are defined as follows:

$$\mathbf{A} = \begin{bmatrix} -5 & -3 \\ 1 & 2 \\ 2 & -4 \end{bmatrix} \quad \mathbf{B} = \begin{bmatrix} 1 & 5 & -2 \\ 5 & -2 & 3 \end{bmatrix}$$

$$\mathbf{C} = \begin{bmatrix} 2 & -3 \\ -1 & 2 \\ 0 & 4 \end{bmatrix} \quad \mathbf{Y} = \begin{bmatrix} 1 \\ 4 \\ 5 \end{bmatrix}$$

In each of the following, perform the indicated operations where possible, and express your answer as a single matrix. If the indicated operations are not defined, answer "not defined."

(*a*) −**A** (*b*) **A** + **B** (*c*) **AY**
(*d*) **BY** (*e*) **AC** (*f*) **BC**

13. Let

$$\mathbf{A} = \begin{bmatrix} 1 & -2 \\ 5 & 3 \end{bmatrix} \quad \mathbf{B} = \begin{bmatrix} -1 & -1 & 3 \\ 4 & 2 & 0 \end{bmatrix}$$

Find $\mathbf{A}^2$, **AB**, **A**(**AB**), and $\mathbf{A}^2\mathbf{B}$.

14. Let

$$\mathbf{A} = \begin{bmatrix} 1 & -2 \\ 4 & 0 \\ 5 & 3 \end{bmatrix} \qquad \mathbf{B} = \begin{bmatrix} -1 & -1 & 3 \\ 4 & 2 & 0 \end{bmatrix}$$

Find **AB**, **BA**, and **BAB**.

15. Let

$$\mathbf{A} = \begin{bmatrix} 3 & -1 \\ -2 & 8 \end{bmatrix} \qquad \mathbf{B} = \begin{bmatrix} 2 & 1 \\ 0 & 4 \end{bmatrix} \qquad \mathbf{C} = \begin{bmatrix} 2 & 4 \\ 1 & 2 \end{bmatrix}$$

Find **AC** and **BC**. What do you conclude from this exercise?

16. Let **A** and **B** be as in Exercise 15. Find a matrix **D** such that $\mathbf{B} + 2\mathbf{D} = \mathbf{A}$.

17. Let $\mathbf{A} = \begin{bmatrix} 7 & 4 \\ 5 & 3 \end{bmatrix}$. Find $\mathbf{A}^{-1}$.

18. Let $\mathbf{A} = \begin{bmatrix} 7 & 4 \\ 4 & 2 \end{bmatrix}$. Find $\mathbf{A}^{-1}$.

19. Let $\mathbf{A} = \begin{bmatrix} 1 & 1 & 1 \\ 2 & 3 & 2 \\ 3 & 3 & 4 \end{bmatrix}$. Find $\mathbf{A}^{-1}$.

20. Let $\mathbf{A} = \begin{bmatrix} 1 & -.5 & 0 \\ 0 & 1 & -.5 \\ -.2 & -.2 & 1 \end{bmatrix}$. Find $\mathbf{A}^{-1}$.

21. Let

$$\mathbf{A} = \begin{bmatrix} 3 & 2 \\ 5 & 4 \end{bmatrix} \qquad \mathbf{B} = \begin{bmatrix} -2 & 8 \\ 4 & 5 \end{bmatrix}$$

(*a*) Find a matrix **C** such that $\mathbf{AC} = \mathbf{B}$.
(*b*) Find a matrix **D** such that $\mathbf{DA} = \mathbf{B}$.

22. Decide whether the following matrix has an inverse. If it does, find it.

$$\begin{bmatrix} 1 & -1 & 2 \\ 2 & 3 & -2 \\ 3 & 2 & 0 \end{bmatrix}$$

23. For which value(s) of the constant k does the matrix shown below not have an inverse?

$$\mathbf{A} = \begin{bmatrix} 3 & k \\ -6 & 12 \end{bmatrix}$$

24. Let

$$\mathbf{A} = \begin{bmatrix} 1 & 2 & 3 \\ 2 & 4 & 7 \\ 3 & 8 & 9 \end{bmatrix} \qquad \mathbf{X} = \begin{bmatrix} x_1 \\ x_2 \\ x_3 \end{bmatrix} \qquad \mathbf{B} = \begin{bmatrix} 10 \\ 0 \\ -10 \end{bmatrix}$$

Find $\mathbf{A}^{-1}$ and use it to solve the system $\mathbf{AX} = \mathbf{B}$.

25. Let **A** be the matrix of Exercise 19. Solve the systems of equations

$$\mathbf{AX} = \begin{bmatrix} -1 \\ 2 \\ 3 \end{bmatrix} \qquad \mathbf{AX} = \begin{bmatrix} -1 \\ 0 \\ 1 \end{bmatrix} \qquad \mathbf{AX} = \begin{bmatrix} 1 \\ 2 \\ 3 \end{bmatrix}$$

26. Let **A** be the matrix of Exercise 19, and let

$$\mathbf{C} = \begin{bmatrix} 5 & -1 \\ 4 & 1 \\ -3 & 3 \end{bmatrix} \qquad \mathbf{D} = \begin{bmatrix} 10 & 10 \\ 25 & 40 \\ 60 & 35 \end{bmatrix}$$

(*a*) Find a matrix **B** such that $\mathbf{AB} = \mathbf{C}$.
(*b*) Find a matrix **H** such that $\mathbf{HA} = \mathbf{D}$.

27. Let $\mathbf{A} = \begin{bmatrix} .2 & a \\ a & .2 \end{bmatrix}$. For which values of a does $\mathbf{I} - \mathbf{A}$ have an inverse with all positive entries?

28. Decide whether the following technology matrices **A** and **B** satisfy the assumptions of the Leontief model in Section 6.3.

$$\mathbf{A} = \begin{bmatrix} .4 & .6 \\ .6 & .4 \end{bmatrix} \qquad \mathbf{B} = \begin{bmatrix} .8 & .4 \\ .2 & .8 \end{bmatrix}$$

29. Find the production vector for the following technology matrix **A** and demand vector **D**.

$$\mathbf{A} = \begin{bmatrix} .5 & .1 \\ .2 & .6 \end{bmatrix} \qquad \mathbf{D} = \begin{bmatrix} 4 \\ 11 \end{bmatrix}$$

30. Given the technology matrix **A** of Exercise 29, what demand vector **D** will be met by the production schedule $\mathbf{X} = \begin{bmatrix} 30 \\ 60 \end{bmatrix}$?

31. Find the production schedule for the following technology matrix and demand vector.

$$\mathbf{A} = \begin{bmatrix} .5 & .8 & .1 \\ 0 & .2 & .4 \\ .5 & 0 & .4 \end{bmatrix} \qquad \mathbf{D} = \begin{bmatrix} 3 \\ 8 \\ 2 \end{bmatrix}$$

32. Using the technology matrix **A** of Exercise 31, find the production schedule **X** which gives the demand vector **D** shown below.

$$\mathbf{D} = \begin{bmatrix} 0 \\ d \\ 0 \end{bmatrix} \qquad \text{where } d \text{ is a positive integer}$$

33. Decide whether the following matrix **A** is such that **I** − **A** has an inverse with all positive entries.

$$\mathbf{A} = \begin{bmatrix} .4 & .6 & .6 \\ .3 & .7 & .6 \\ .5 & .5 & .5 \end{bmatrix}$$

34. An economy has three goods—electricity, gravel, and timber—and the goods required in the production processes are specified in Table 6.4.

TABLE 6.4

	Amount Required to Produce 1 Unit of		
	Electricity	**Gravel**	**Timber**
Electricity	.20	.15	.25
Gravel	.20	.25	.30
Timber	.15	.35	.30

(*a*) Find the technology matrix for this economy.
(*b*) Find the production schedule which fills an external demand consisting of 45 units of electricity, 60 units of gravel, and 110 units of timber.

35. An economy with four goods has the technology matrix

$$\mathbf{A} = \begin{bmatrix} .1 & .25 & 0 & .2 \\ 0 & .2 & .2 & .3 \\ 0 & .3 & .3 & .1 \\ .15 & .25 & 0 & .3 \end{bmatrix}$$

Find the production schedule which meets the external demand

$$\mathbf{D} = \begin{bmatrix} 1200 \\ 840 \\ 660 \\ 1620 \end{bmatrix}$$

CHAPTER 7

Linear Programming: Modeling and Graphical Solution

7.0 THE SETTING AND OVERVIEW: LINEAR OPTIMIZATION MODELS

Mathematics and physics have been tightly tied together for hundreds of years, and developments in one of these two fields often led to developments in the other. The links between mathematics and fields such as the management sciences and the social and life sciences came much later, but these links are now well established and are very important in current developments in these fields. One major use of mathematics in the management sciences and, to a lesser extent, in the life and social sciences is to formulate and solve optimization problems. In general, an optimization problem is one in which the goal is to select values of variables in such a way that some quantity that depends on the variables is as large or as small as possible. Normally, the permissible choices of the variables are restricted in some way. For example, in linear optimization problems, the quantity to be optimized and the restrictions are represented by using linear expressions, expressions such as $3x + 5y - 8z$. Linear optimization problems of the types considered in this book are referred to as *linear programming* problems.

A good case can be made that linear programming is one of the most widely applied and effective mathematical tools used in the social, life, and management sciences. The importance of linear programming became clear in the complex scheduling and resource allocation problems arising during World War II. The techniques are now routinely applied to a variety of problems arising throughout business—problems from marketing, manufacturing, finance, distribution, transportation, etc. In addition, the ideas and methods are used in both theoretical and applied economics and in unexpected ways in situations that seem far from standard business problems. For instance, the question as to whether a specific baseball team has been eliminated from contention for a playoff spot, a much more complex question than it first appears, can be answered by using linear programming.

Linear programming problems typically arise outside mathematics, and the first step in solving these problems is to correctly formulate them using linear expressions, such as those studied in Chapter 5. To illustrate this process of formulation, we use an example which has great historical importance as a setting for our modeling discussion; it is a version of one of the first problems solved by using linear programming. The problem is one of selecting foods which satisfy certain dietary requirements in a way that minimizes the cost of the food.

We make the situation specific and yet keep the discussion fairly simple by considering only two foods and two dietary requirements; the techniques developed in this chapter will enable us to handle more complicated situations. Suppose that we are to use broccoli and milk—hardly an appetizing diet, but a nutritious one—to meet dietary needs for calcium and iron and that we are to select amounts of these two foods that meet our needs at the least possible cost. The following data are available to us: The diet must contain at least .8 gram of calcium and 10 milligrams of iron. Each serving of broccoli contains .13 gram of calcium and 1.3 milligrams of iron, and each serving of milk contains .28 gram of calcium and .2 milligram of iron. Broccoli costs \$0.80 per serving, and milk costs \$0.50 per serving.

To describe a diet, we need to specify the amount of broccoli and the amount of milk. The first step in building a model consists of introducing appropriate notation. We let x denote the amount of broccoli and y denote the amount of milk to be consumed, with x and y each measured in units of servings. If there is .13 gram of calcium in one serving of broccoli, then there are $.13x$ grams in x servings; and if there are 1.3 milligrams of iron in one serving of broccoli, then there are $1.3x$ milligrams in x servings. Likewise, if there is .28 gram of calcium in one serving of milk, then there are $.28y$ grams in y servings; and if there is .2 milligram of iron in one serving of milk, then there are $.2y$ milligrams in y servings. Next, if we consume both x servings of broccoli and y servings of milk, then we have $.13x + .28y$ grams of calcium. The requirement that the diet provide at least .8 gram of calcium is, in our notation, $.13x + .28y \geq .8$.[1] A completely analogous argument leads to the inequality $1.3x + .2y \geq 10$, which represents the requirement that the diet contain at least 10 milligrams of iron.

[1]The inequality sign $\geq$ means that the quantity $.8x + .28y$ is to be *greater than or equal to* .8. If we want to restrict $.8x + .28y$ to be greater than .8, we write $.8x + .28y > .8$. The latter expression is referred to as a *strict inequality*. Similarly, the symbols $\leq$ and $<$ mean "less than or equal to" and "less than," respectively.

Looking at the cost aspect of the problem, we see that since each serving of broccoli costs \$0.80, x servings should cost \$0.80x; and since each serving of milk costs \$0.50, y servings of milk cost \$0.50y. Therefore, the diet of x servings of broccoli and y servings of milk costs \$0.80x + \$0.50y. To find a least-cost diet, we should make $.80x + .50y$ as small as possible.

Although this all seems straightforward, it is important to realize that there are some unstated assumptions:

The first implicit assumption is that we actually know the calcium content of broccoli and milk. There is likely to be variation in the vitamin and mineral content of foods depending on how they are grown, processed, and stored. In some cases this variation can be quite large.

Second, we are assuming that if we get .13 gram of calcium from one serving of broccoli, then we get $.13x$ gram from x servings. This is a legitimate assumption for some values of x, but not for other values of x. As x becomes large, it is impossible for a human to extract all nutrients available in the food.

Next, we are assuming that if there is a certain amount of calcium available in the broccoli and another amount in the milk, then the sum of those two amounts is available in the combination of the two foods. This would be the case if the two foods were consumed independently; however, in some cases the presence of one food may affect the ability of the body to extract nutrients from other foods.

With regard to the cost assumptions, the statement that x servings of broccoli cost \$0.80x has the implicit assumption that there are no bulk discounts. Although the statement may be true for the amounts of broccoli that an individual normally buys, there should be some savings if the purchaser is buying broccoli by the boxcar load!

The statement that the cost of the diet is the sum of the costs of broccoli and of milk means that there are no discounts for purchasing both items. Frequently it is possible to negotiate a better price if you buy several items from one supplier.

It is clear that in formulating mathematical descriptions of situations, even fairly straightforward ones, we must be sensitive to the implications of our assumptions. The mathematical relations we develop incorporate the assumptions; and consequently, when we deduce conclusions from these relations, our conclusions will depend on the assumptions. If in some circumstances the assumptions are not completely fulfilled, then we should expect that the conclusions are not entirely valid.

This simple example illustrates the formulation of a linear programming problem and the many assumptions that underlie such a formulation. Although we will not emphasize the modeling aspect in the chapter which follows, we will formulate many mathematical problems from situations described in words. Indeed, the first section of this chapter is devoted entirely to the formulation of linear programming problems. In that discussion we will freely use the implicit assumptions that we have identified here. Remember that these are assumptions, and the reliability of the answers you obtain depends on the legitimacy of the assumptions.

7.1 FORMULATION OF LINEAR PROGRAMMING PROBLEMS

The first task in solving an optimization problem is to model the situation and to formulate a precise mathematical problem. We will model all situations discussed in this chapter as linear programming problems. The formulation phase involves taking a situation described in words and translating the information presented in words into mathematical form, i.e., into symbols and relations. The technique of translation can best be described by examples.

Example 7.1

Manufacturing

Ted's Toys makes toy cars and toy trucks using plastic and steel. Each car requires 4 ounces of plastic and 3 ounces of steel, while each truck requires 3 ounces of plastic and 6 ounces of steel. Each day Ted has 30 pounds of plastic and 45 pounds of steel to use in making toy cars and trucks, and he can sell all the cars and trucks he makes with these materials. His profit is $5 per car and $4 per truck. Ted is motivated by financial goals, and he would like to know how many cars and trucks he should make in order to maximize the total profit from the sale of these toys.

Problem Formulate a mathematical problem whose solution gives the numbers of cars and trucks Ted should make to maximize his profit.

Solution Ted's goal is to decide how many cars and trucks to make, so it is natural to introduce variables, called *decision variables*, which represent these quantities. Thus, we set

x = number of cars to be made each day

and y = number of trucks to be made each day

Next, since each car requires 4 ounces of plastic, x cars require $4x$ ounces of plastic. Also, each truck requires 3 ounces of plastic, and y trucks require $3y$ ounces of plastic. Therefore, x cars and y trucks together require $4x + 3y$ ounces of plastic. Each day Ted has 30 pounds of plastic available, that is, $30 \cdot 16 = 480$ ounces, and the variables x and y are constrained by the requirement that at most 480 ounces of plastic can be used. Since $4x + 3y$ ounces of plastic are required to produce x cars and y trucks, the mathematical constraint is

$$4x + 3y \leq 480$$

In the same way, x cars require $3x$ ounces of steel and y trucks require $6y$ ounces of steel, and there are 45 pounds, or $45 \cdot 16 = 720$ ounces, of steel available. Therefore, another constraint on the number of cars and trucks which can be produced is

$$3x + 6y \leq 720$$

Finally, the variables x and y cannot be negative; a negative number of cars and trucks cannot be produced. Collecting all these constraints, we find that the variables x and y must satisfy the conditions

$$\begin{gathered} x \geq 0 \qquad y \geq 0 \\ 4x + 3y \leq 480 \\ 3x + 6y \leq 720 \end{gathered} \tag{7.1}$$

The constraints (7.1) must be satisfied by any choice of x and y that Ted makes. However, Ted's goal is to choose x and y so that his profit is a maximum. Since his profit is \$5 for each car, it is (\$5)x when he produces x cars. Likewise, his profit for each truck is \$4, and therefore it is (\$4)y when he produces y trucks. It follows that his total profit is (in dollars) $5x + 4y$ when he produces x cars and y trucks. We have now formulated a mathematical problem which describes the decision Ted's Toys must make:

Find x and y which satisfy (7.1) and which make the profit $5x + 4y$ as large as possible [i.e., in comparison with the profit for any choice of x and y which satisfies (7.1)]. ■

Example 7.2

Nutrition

A hiker is planning her trail food, which is to include a snack mix of peanuts and raisins. Each day she wants 600 calories and 90 grams of carbohydrates from this mix. Each gram of raisins contains .8 gram of carbohydrates and 3 calories and costs 4 cents. Each gram of peanuts contains .2 gram of carbohydrates and 6 calories and costs 5 cents.

Problem Formulate a mathematical problem whose solution gives the number of grams of each food which will meet the hiker's needs at the smallest cost per day.

Solution First we identify the decision variables in the situation, i.e., the quantities which we wish to determine. Let

$$\begin{aligned} x &= \text{number of grams of raisins} \\ y &= \text{number of grams of peanuts} \end{aligned}$$

From the meaning of x and y we have the inequalities

$$x \geq 0 \qquad y \geq 0$$

Since each gram of raisins contains .8 gram of carbohydrate and each gram of peanuts contains .2 gram of carbohydrate, the total amount of carbohydrate is $.8x + .2y$; and consequently to meet the goal of the hiker regarding carbohydrate intake, we impose the constraint

$$.8x + .2y \geq 90$$

Likewise, the requirement on total calories leads to the constraint

$$3x + 6y \geq 600$$

Finally, the cost of x grams of raisins is $4x$ (in cents), and the cost of y grams of peanuts is $5y$ (in cents). Consequently, in cents,

$$\text{Total cost} = 4x + 5y$$

The mathematical problem is the following: Find the pair (or pairs) of values x and y satisfying the set of inequalities

$$\begin{aligned} x \geq 0 \qquad y &\geq 0 \\ .8x + .2y &\geq 90 \\ 3x + 6y &\geq 600 \end{aligned} \tag{7.2}$$

for which $4x + 5y$ is as small as possible. ■

Some special terminology is useful in discussing linear programming problems.

Constraints, Feasible Sets, and Objective Functions

The inequalities that specify permissible values of the variables are the *constraints* of the problem. The constraints considered in this book will always be inclusive inequalities; they will be expressed with $\leq$ or $\geq$.

The set of points satisfying the constraints of the problem is known as the *feasible set* for the problem.

The function to be maximized or minimized is known as the *objective function* for the problem.

The constraints in Example 7.2 are the inequalities given in (7.2), and the feasible set is the set described by these inequalities. The objective function in Example 7.2 is the total cost, $4x + 5y$.

The formulation of a linear programming problem involves three steps:

1. Specify the variables.
2. Specify the constraints using the variables.
3. Specify the objective function using the variables.

Example 7.3

Task scheduling

An account executive divides his time between sales and support activities, primarily paperwork and reading up on new products. Keeping up to date on new products requires that he spend at least 5 hours each week reading trade newspapers and magazines. In addition, each hour he devotes to sales generates .1 hour of paperwork. He prefers sales, and he wants to devote at least half his time to that activity, but there is enough to do in support activities that any time not devoted to sales can be used for that purpose. He plans to devote at most 50 hours per week to his job. Finally, he estimates that the time he devotes to sales is worth $15 per hour and that the time he devotes to support activities is worth $10 per hour. (It would cost $10 per hour to hire a staff person to do that work.)

Problem Formulate a mathematical problem whose solution gives the account executive an allocation of time that maximizes the value of his activities to his employer.

Solution Let x and y denote the number of hours devoted to sales and support activities, respectively. We have

$$x \geq 0 \qquad y \geq 0$$

The total number of hours worked is $x + y$, and the constraint that he work at most 50 hours per week gives

$$x + y \leq 50$$

Also, the constraint that at least half his effort be devoted to sales is

$$x \geq .5(x + y)$$

The latter inequality is equivalent to[2]

$$x \geq .5x + .5y$$

or

$$.5x - .5y \geq 0$$

Next, each hour devoted to sales generates 0.1 hour of paperwork, and an additional 5 hours of support are required each week. Consequently, y must be at least as large as the sum of the paperwork generated by sales ($.1x$) and the normal support activities (5). We have the inequality

$$y \geq 5 + .1x$$

or

$$-.1x + y \geq 5$$

Finally, the businessman's contribution can be measured (in dollars) as

$$15x + 10y$$

In summary, the mathematical problem is the following:

Maximize $$15x + 10y$$

subject to
$$\begin{aligned} x \geq 0 \qquad y &\geq 0 \\ x + y &\leq 50 \\ .5x - .5y &\geq 0 \\ -.1x + y &\geq 5 \end{aligned}$$

■

Each of these first three examples has two unknown quantities or variables. Of course, most problems actually arising in applications have more than two variables, in some cases several hundreds or thousands of variables. We continue the formulation discussion with two examples involving more than two variables.

Example 7.4

Resource allocation

The Plant Power Fertilizer Company makes three types of fertilizer: 20-8-8 for lawns, 4-8-4 for gardens, and 4-4-2 for general purposes. The numbers in each case refer to the percentage by weight of nitrate, phosphate, and potash, respectively, in a sack of fertilizer. The company has 6000 pounds of nitrate, 10,000 pounds of phosphate, and 4000 pounds of potash on hand. The profit is \$3 per 100 pounds of lawn fertilizer, \$8 per 100 pounds of garden fertilizer, and \$6 per 100 pounds of general-purpose fertilizer.

---

[2]Here and in the remaining sections of this chapter we freely use the following properties of inequalities:

If $a \leq b$ and $c \leq d$, then $a + c \leq b + d$.
If $a \leq b$ and $\lambda > 0$, then $\lambda a \leq \lambda b$.
If $a \leq b$ and $\lambda < 0$, then $\lambda a \geq \lambda b$.

Problem Formulate a linear programming problem whose solution will give the number of pounds of each type of fertilizer that should be produced to yield maximum profit.

Solution Introduce the variables

x = number of hundreds of pounds of 20-8-8 fertilizer
y = number of hundreds of pounds of 4-8-4 fertilizer
z = number of hundreds of pounds of 4-4-2 fertilizer

Here the units (hundreds of pounds) are selected to simplify the equations that express the resource constraints. The equations that express the constraints on nitrate, phosphate, and potash are, respectively,

$$\begin{aligned} 20x + 4y + 4z &\leq 6{,}000 \\ 8x + 8y + 4z &\leq 10{,}000 \\ 8x + 4y + 2z &\leq 4{,}000 \end{aligned} \tag{7.3}$$

The objective function is

$$p = \text{profit} = 3x + 8y + 6z$$

It follows from the definition that each of the variables must be nonnegative.
Accordingly, the mathematical problem is:

Maximize $3x + 8y + 6z$

subject to $x \geq 0 \quad y \geq 0 \quad z \geq 0$
and the inequalities (7.3) ■

Example 7.5

Task scheduling

The office manager of an accounting firm must allocate the time of the office staff each week among three activities: auditing, business accounting, and tax accounting. Each hour billed as auditing requires 15 minutes of an accountant's time and 30 minutes of clerical time. Each hour billed as business accounting requires 20 minutes of accountant time, 60 minutes of clerical time, and 6 minutes of computer time. Each hour billed as tax accounting requires 30 minutes of accountant time, 45 minutes of clerical time, and 3 minutes of computer time. The net profit to the firm from 1 hour of auditing is \$4, and from business accounting and tax accounting the net profits are \$10 and \$6, respectively. This week the staff available can provide 80 hours of accountant time, 180 hours of clerical time, and 30 hours of computer time.

Problem Formulate a linear programming problem whose solution gives the allocation of time which provides the maximum net profit to the firm.

Solution We begin by identifying the variables. In this case, we let x, y, and z denote the number of hours billed as auditing, business accounting, and tax accounting, respectively.

In discussing the allocation of accountant, clerical, and computer time among these activities, it is convenient to express the times in minutes. The constraint resulting from the condition that accountant time not exceed 80 hours can be stated in terms of the variables x, y, and z by the equation

$$15x + 20y + 30z \leq 80(60) = 4800$$

Similarly, the constraint resulting from the limitation on clerical time is

$$30x + 60y + 45z \leq 180(60) = 10{,}800$$

and the constraint resulting from the limitation on computer time is

$$6y + 3z \leq 30(60) = 1800$$

The net profit that results from x hours billed as auditing, y hours billed as business accounting, and z hours billed as tax accounting is (in dollars)

$$4x + 10y + 6z$$

Finally, there is the nonnegativity restriction $x \geq 0$, $y \geq 0$, $z \geq 0$ which follows from the definitions of x, y, and z. Thus for the original allocation problem we have the following mathematical formulation:

Find numbers x, y, and z such that

$$p = \text{profit} = 4x + 10y + 6z$$

is a maximum for all x, y, z satisfying

$$\begin{aligned} x \geq 0 \qquad y \geq 0 \qquad z &\geq 0 \\ 15x + 20y + 30z &\leq 4{,}800 \\ 30x + 60y + 45z &\leq 10{,}800 \\ 6y + 3z &\leq 1{,}800 \end{aligned}$$

■

Exercises for Section 7.1

In each of the following exercises, formulate a linear programming problem for the situation described. Be sure to identify the variables, the constraints, and the objective function.

Resource allocation

1. Sam's Deli makes sandwiches using bread and meat. Each regular sandwich uses 6 inches of bread and 2 ounces of meat, while each large sandwich uses 10 inches of bread and 4 ounces of meat. The profit on a small sandwich is $0.80 and on a large sandwich is $1.20. Each day the deli has 110 feet of bread and 30 pounds of meat. How many sandwiches of each size should be made to maximize profit?

Resource allocation

2. At a plant of Bigoil Corporation, the manager has a supply of 2400 barrels of crude oil each day to use in the production of two distillates A and B. It takes 2 barrels of crude oil and 250 cubic feet of natural gas (for energy) to produce 1 barrel of distillate A and 3 barrels of crude oil and 150 cubic feet of natural gas to produce 1 barrel of distillate B. Each day the manager has 150,000 cubic feet of natural gas available. Also, due to restrictions on the emission of by-products into the atmosphere, at most 400 barrels of distillate A can be produced. The profit on 1 barrel of distillate A is $3, and on 1 barrel of distillate B is $2. How many barrels of each distillate should the manager produce to maximize profit?

Resource allocation

3. The Natural Fertilizer Company produces 100-pound sacks of two types of fertilizer: 25-10-5 for lawns and 8-10-10 for gardens. The numbers in each case refer to the percentage by weight of nitrate, phosphate, and potash, respectively, in a sack of fertilizer. The company has 6 tons of nitrate, 5 tons of phosphate, and 3.5 tons of potash on hand. The profit per 100 pounds of lawn fertilizer is $7, and the profit per 100 pounds of garden fertilizer is $5. How much of each fertilizer should the company produce to maximize profit?

Manufacturing

4. A manufacturer makes two in-line skate models, the California and the Florida. Both models require two operations in production: finishing the frame and installing and balancing the wheels. These operations require the following amounts of time (in minutes):

	Frame	Wheels
California	15	5
Florida	10	20

There are 120 hours of labor per day available for each job. The profit per California model is $15, and the profit per Florida model is $18. How should daily production be scheduled to maximize profit?

Purchasing

5. Tony CD, Inc., purchases a portion of the material used in its compact disks (CDs) from an independent supplier and the remainder from a wholly owned subsidiary. The company needs 45,000 pounds of material in the next month, and the subsidiary can produce at most 35,000 pounds. The cost of this material is $0.80 per pound from the subsidiary and $1.00 per pound from the independent supplier. How much material should be ordered from the subsidiary and how much from the independent supplier to meet the needs at minimum cost?

Purchasing

6. In Exercise 5, suppose that in addition to the other conditions Tony CD agrees to an antitrust ruling which requires that it purchase at least one-third of its material from an independent supplier. Formulate the cost minimization problem in this case.

Purchasing

7. In Exercise 5, suppose that Tony CD is required to purchase at least one-third of the total material used from an independent supplier and that the independent supplier allocates its production so that no customer receives more than 20,000 pounds per month. If all other conditions are the same, formulate the cost minimization problem.

Resource allocation

8. An office furniture manufacturer has available 18 tons of sheet steel to be used to make desks and filing cabinets. It requires 50 pounds of steel and 3 hours of labor to make a filing cabinet and 75 pounds of steel and 2 hours of labor to make a desk. There are 1500 hours of labor available, but only enough tops for 400 desks. The net profit is $20 on each desk and $15 on each filing cabinet. How many desks and how many filing cabinets should be produced to provide maximum profit?

Resource allocation

9. Glen and Barrys blends skim milk, sugar, and cream to make both regular and low-calorie ice cream. Each gallon of regular ice cream requires .6 gallon of skim milk, 1 pound of sugar, and .4 gallon of cream. Each gallon of low-calorie ice cream requires .7 gallon of skim milk, .3 pound of sugar, and .4 gallon of cream. Each day there are 800 gallons of skim milk, 400 pounds of sugar, and 400 gallons of cream available. The profit per gallon is $1.00 for regular ice cream and $1.20 for low-calorie ice cream. How many gallons of each type of ice cream should be produced to maximize profit?

Resource allocation

10. The Natural Fertilizer Company makes three types of fertilizer: 20-5-5 for lawns, 10-15-10 for gardens, and 5-5-5 for trees. The numbers in each case refer to the percentage by weight of nitrates, phosphates, and potash, respectively. The fertilizer is packed in 100-pound sacks. The company has the following amounts of raw materials on hand: 7 tons of nitrates, 4 tons of phosphates, and 3 tons of potash. The profit per 100 pounds of lawn fertilizer is $6, the profit per sack of garden fertilizer is $4, and the profit per sack of tree fertilizer is $3. The company has a contract to supply at least 1 ton of garden fertilizer. How much of each fertilizer should the company produce to maximize profit?

Resource allocation

11. A psychologist plans to conduct an experiment which involves subjects who perform activities. After data have been collected, the data are to be analyzed by a team of (highly paid) expert consultants. The psychologist has 15 subject hours available, and she will need to use at least 6 of them. She has funds for a maximum of 200 minutes of consultant time, and each hour of subject time requires at least 30 minutes of consultant time to analyze the data. Depending upon the depth of the analysis, up to 50 minutes of consultant time per subject hour can be profitably used. The information which the psychologist obtains from the experiment depends upon the number of subject hours and the amount of analysis. She estimates that, in appropriate units, 1 unit of information is obtained from each subject hour and 1 unit is obtained from each 25 minutes of consultant analysis. How should the experiment be organized (i.e., how many subject hours and how much consultant analysis) to give the maximum information?

Purchasing

12. A purchasing agent for a college finds it necessary to decide how much hard and soft chalk should be purchased each month. He knows that a typical instructor will use two pieces of soft chalk or one piece of hard chalk for each class. In addition, he has observed that hard chalk always amounts to at least one-quarter of the total used. Finally, his supplier has limited him to a purchase of at most 60 gross (8640 pieces) of soft chalk. There are 3600 classes to be

taught each month. If soft chalk is \$1.50 per gross and hard chalk is \$3.50 per gross, how much of each should be purchased to meet the needs and to keep costs to a minimum?

Health care

13. The government has mobilized to inoculate the student population against sleeping sickness. There are 200 doctors and 450 nurses available. An inoculation team can consist of either 1 doctor and 3 nurses (called a *full team*) or 1 doctor and 2 nurses (called a *half team*). On average, a full team can inoculate 180 people per hour, while a half team can inoculate 100 people per hour. How many teams of each type should be formed to maximize the number of inoculations per hour?

Manufacturing

14. The Mount Cycle Company makes two mountain bike models, Starstreak and Superstreak. Both models require three operations in production: (1) frame assembly, (2) installing the wheels, and (3) decorating. These operations require the following amounts of time (in minutes):

	Frame	Wheels	Decoration
Starstreak	20	10	14
Superstreak	10	15	18

There are 120 hours of labor available per day for frame assembly, 90 hours for attaching the wheels, and 75 for decorating. Also, each Starstreak model brings a profit of \$15 and each Superstreak model a profit of \$21. How should daily production be scheduled to maximize profit?

Resource allocation

15. Brown Brothers Box Company produces both standard and heavy-duty shipping containers. One standard container requires 1 square foot of 100-pound test cardboard and 3 square feet of liner board. One heavy-duty container requires 5 square feet of 100-pound test cardboard and 1 square foot of liner board. Each week the company has 5000 square feet of 100-pound test cardboard and 4500 square feet of liner board available to produce shipping containers. There is a commitment to produce at least 500 standard containers each week. If the profit is \$0.30 for each standard container and \$0.40 for each heavy-duty container, how many of each type should be produced to meet the commitment and maximize profit?

Resource allocation

16. In the situation described in Exercise 15, suppose that in addition it is required that both of the following conditions be met:
    (*a*) The number of heavy-duty containers produced does not exceed the number of standard containers.
    (*b*) There is a commitment to produce at least 200 heavy-duty containers each week.

    How many containers of each type should be produced to meet the commitments and maximize profit?

Resource allocation

17. A toy company makes three monster dolls: Scary Harry, Horrible Harriet, and The Glob. The manufacturing of these dolls is a three-step process: (1) the body is molded from plastic, (2) clothes are put on, and (3) special monster features are added. The amounts of time and material for each step vary from doll to doll, and consequently each doll has its own production cost and associated profit. Data of the manufacturing process are shown in Table 7.1. How many dolls of each type should be manufactured each hour to maximize profit?

Resource allocation

18. Repeat Exercise 17 for the data given in Table 7.2.

Resource allocation

19. Is the vector which corresponds to the production of 5 Scary Harry, 5 Horrible Harriet, and 10 The Glob dolls feasible for the problem of Exercise 18? How about the vector which corresponds to 10 Scary Harry, 3 Horrible Harriet, and 2 The Glob dolls?

TABLE 7.1

Doll	Plastic, Ounces	Time for Clothes, Minutes	Time for Special Features, Minutes	Profit per Doll, $
Scary Harry	4	3	2	1.00
Horrible Harriet	3	4	4	1.25
The Glob	9	1	3	1.50
Available time or material per hour of operation	160	50	50	

TABLE 7.2

Doll	Plastic, Ounces	Time for Clothes, Minutes	Time for Special Features, Minutes	Profit per Doll, $
Scary Harry	5	2	3	1.10
Horrible Harriet	3	4	4	1.30
The Glob	10	1	6	2.00
Available time or material per hour of operation	192	55	45	

Resource allocation

20. Brown Brothers Box Company recently lost a contract to produce price signs for gasoline stations, and it finds itself with excess capacity. It has 300 pounds of heavy-duty liner board, 120 pounds of finish cardboard, and 10 hours of labor available each day. It can use these resources to produce shipping boxes, mailing tubes, and boxes for retail use. Each 100 shipping boxes use 150 pounds of heavy-duty liner board, 30 pounds of finish cardboard, and 2 hours of labor. Each 600 mailing tubes use 50 pounds of heavy-duty liner board and 30 pounds of finish cardboard and require 2 hours of labor. Finally, each 100 retail boxes use 60 pounds of heavy-duty liner board, 40 pounds of finish cardboard, and 5 hours of labor. The net profit is \$0.10 per retail box, \$0.01 per mailing tube, and \$0.04 per shipping box. How should the resources be allocated to produce maximum profit?

Sports

21. The Bait Shop sells three kinds of bait packages. The packages contain worms, minnows, and grasshoppers in the amounts shown in Table 7.3. The profit per package is \$1.00 for *A*, \$0.75 for *B*, and \$1.25 for *C*. There are 1000 worms available, 250 minnows, and 300 grasshoppers. How many packages of each type should be made to maximize profit?

Sports

22. Refer to the situation described in Exercise 21.
    (*a*) There are labels for 30 packages of type *A* bait and for 20 packages of type *B* bait. Formulate the situation as a linear programming problem with these additional constraints. Assume ample labels for type *C* bait.
    (*b*) The owner knows that she will sell at least twice as many packages of type *A* bait as type *C*. Formulate the situation as a linear programming problem with this additional constraint.

Resource allocation

23. Formulate the following situation as a linear programming problem. The Wonder Widget Company produces small, medium, and large widgets with locks and small widgets without locks. There is an assembly line with three stages: assembly, painting, and installation of locks; and there are 8 hours of assembly time, 9 hours of painting time, and 2 hours of lock installation time to be assigned. The time used (in hours) for each 100 widgets at each stage is shown in Table 7.4. The net profit is \$0.02 per widget without lock and \$0.10, \$0.11, and \$0.20 for small, medium, and large widgets with locks, respectively. How should production be scheduled (i.e., how many of each type of widget should be produced) to yield maximum net profit?

TABLE 7.3

Type	Number of Worms	Number of Minnows	Number of Grasshoppers
A	25	10	10
B	10	15	25
C	50	5	5

TABLE 7.4

Type of Widget	Time for Assembly, Hours	Time for Painting, Hours	Time for Lock Installation, Hours
Small without lock	1	1	0
Small with lock	2	5	3
Medium with lock	3	4	1
Large with lock	6	8	4

Travel

24. Columbus Cruiselines offers one-week cruises using three ships: the Nina, the Pinta, and the Santa Maria. Each ship has both regular cabins and deluxe cabins. The Nina has 500 regular rooms and 200 deluxe, the Pinta has 400 regular and 400 deluxe, and the Santa Maria has 800 regular and 500 deluxe rooms. The cost to run each ship for a week is $100,000 for the Nina, $120,000 for the Pinta, and $180,000 for the Santa Maria. There is demand for 12,000 one-week cruises in regular rooms and 8000 one-week cruises in deluxe rooms. How many weeks should each ship be scheduled to meet the demand at minimum cost?

Resource allocation

25. The California Dried Fruit Company prepares three types of dried-fruit packages for sale during the holiday season. The Deluxe Pack contains 16 ounces of dates, 24 ounces of apricots, and 12 ounces of candied fruit. The Special Pack contains 20 ounces of dates, 12 ounces of apricots, and 3 ounces of candied fruit. The Standard Pack contains 16 ounces of dates and 8 ounces of apricots. The company has 1200 ounces of dates, 900 ounces of apricots, and 360 ounces of candied fruit. If the net profit is $3.00 for each Deluxe Pack, $2.00 for each Special Pack, and $1.50 for each Standard Pack, how many packages of each type should be produced to yield maximum profit?

Resource allocation

26. Explain how the problem posed in Exercise 25 changes if both the following conditions are added:
    (*a*) The company has an order for 20 Special Packs which must be filled.
    (*b*) There are shipping cartons for only 30 Deluxe Packs.

Landscape design

27. A greenhouse operator plans to bid for the job of providing flowers for the city parks. He will use tulips, daffodils, and flowering shrubs in three types of layouts. A type 1 layout uses 30 tulips, 20 daffodils, and 4 flowering shrubs. A type 2 layout uses 10 tulips, 40 daffodils, and 3 flowering shrubs. A type 3 layout uses 20 tulips, 50 daffodils, and 2 flowering shrubs. The net profit is $50 for each type 1 layout, $30 for each type 2 layout, and $60 for each type 3 layout. He has 1000 tulips, 800 daffodils, and 100 flowering shrubs. How many layouts of each type should be used to yield the maximum net profit?

Landscape design

28. Explain how the problem formulated in Exercise 27 changes if both the following constraints are imposed.
    (*a*) The number of type 1 layouts cannot exceed the number of type 2 layouts.
    (*b*) There must be at least 5 layouts of each type.

Forestry 29. A timber company has forest land with trees suitable for furniture, plywood, and pulpwood. It requires 100 hours of labor and 20 hours of machine time to harvest 1 unit of timber for furniture; it requires 80 hours of labor and 30 hours of machine time to harvest 1 unit of timber for plywood; and it requires 50 hours of labor and 30 hours of machine time to harvest 1 unit of timber for pulpwood. The net profit is $500 per unit for furniture timber, $400 per unit for plywood timber, and $200 per unit for pulpwood timber. The company has 1000 hours of labor, 500 hours of machine time, and ample supplies of all three types of timber. How much of each type of timber should be harvested to yield the maximum net profit?

Farming 30. A farmer mixes three types of food—foods 1, 2, and 3—for her livestock. The foods must provide at least 300 units of nutrient A, 20 units of nutrient B, and 100 units of nutrient C. The costs of the foods and the amounts of each type of nutrient that each food provides are shown in Table 7.5. How much of each type of food should be used to provide the minimum nutrient needs at the lowest cost?

TABLE 7.5

		Amount of Nutrient per Unit of Food		
Food	**Cost per Unit, $**	***A***	***B***	***C***
1	40	100	5	20
2	20	60	8	10
3	50	100	12	30

7.2 SYSTEMS OF LINEAR INEQUALITIES IN TWO VARIABLES

The examples of Section 7.1 illustrate the way linear inequalities arise in linear programming problems. Indeed, the feasible sets for linear programming problems are defined by systems of linear inequalities, and developing a way to obtain points in feasible sets is important in the solution of linear programming problems. In Chapter 5 we developed ways to solve systems of linear equations, and we discussed the relation between algebraic expressions and the geometry of systems of linear equations. In this section we develop geometric methods to find points in sets described by systems of linear inequalities in two variables, and in Chapter 10 we develop algebraic methods for the same problem with systems of linear inequalities in any number of variables.

Example 7.6 Represent on a graph the set of points (x, y) for which $3x + 2y \leq 6$.

Solution The line specified by the equation $3x + 2y = 6$ defines two subsets, called *half planes*, of the cartesian plane. One half plane consists of the points (x, y) such that $3x + 2y \leq 6$. It is the shaded set shown in Figure 7.1 together with the set of points on the line itself. The line, i.e., the set of points which satisfy $3x + 2y = 6$, is called the *boundary* of the half plane. The other half plane consists of the points (x, y) such that $3x + 2y \geq 6$. This half plane consists of the unshaded area in Figure 7.1 together with the boundary line. We determine which of the two half planes is the one described by the given inequality $3x + 2y \leq 6$ by testing a single point not on the boundary line. For example, we can test the point $(1, 1)$. We have $3(1) + 2(1) = 5 < 6$. Therefore, the point $(1, 1)$ is included in the set described by the inequality $3x + 2y \leq 6$. ■

There are two points worth noting in this example.

1. The same technique can be applied to graph any set described by an inequality of the form $Ax + By \leq C$ (or $Ax + By \geq C$).
2. We used the point $(1, 1)$ to determine which half plane was described by $3x + 2y \leq 6$ and which by $3x + 2y \geq 6$. We could have used *any* point not on the line $3x + 2y = 6$. The point $(0, 0)$ is usually the easiest point to test, and it can be used whenever the line does not pass through the origin. Using the point $(0, 0)$ in this example, we have $3(0) + 2(0) = 0 < 6$, so $(0, 0)$ belongs to the half plane described by $3x + 2y \leq 6$.

FIGURE 7.1

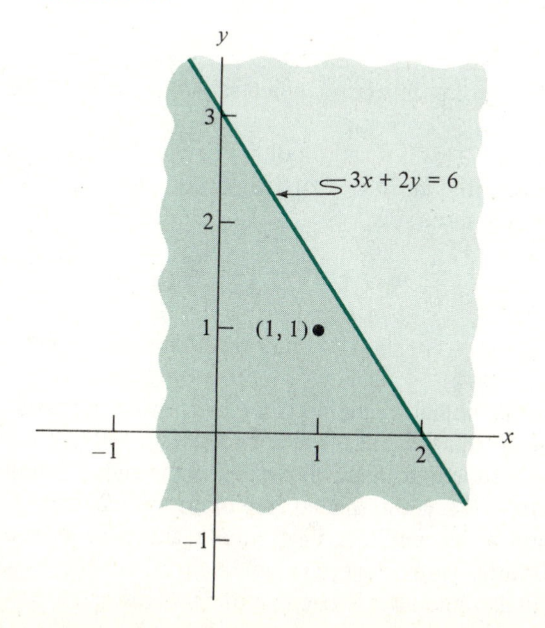

Example 7.7 *Nutrition*

A hiker believes that on a long hike she will need snacks with at least 600 calories, and she plans to take chocolate and raisins. The chocolate she plans to take has 150 calories per ounce, and the raisins have 80 calories per ounce.

Problem Find a system of inequalities that describes those combinations of food which satisfy her requirements.

Solution Let x denote the number of ounces of chocolate she plans to take, and let y denote the number of ounces of raisins. This definition implies that $x \geq 0$ and $y \geq 0$. Since 1 ounce of chocolate contains 150 calories, x ounces of chocolate contain $150x$ calories. Likewise y ounces of raisins contain $80y$ calories. Also, the total calories obtained from the two foods is the sum of the number of calories from chocolate and the number of calories from raisins. That is,

$$\text{Total calories} = 150x + 80y$$

She believes that she needs at least 600 calories (the total calories must equal 600 or more). This leads to the condition

$$150x + 80y \geq 600$$

In summary, if x = number of ounces of chocolate and y = number of ounces of raisins, then her requirements are met by any (x, y) satisfying

$$\begin{array}{c} x \geq 0 \qquad y \geq 0 \\ 150x + 80y \geq 600 \end{array} \tag{7.4}$$

■

Example 7.8 Graph the set of points (x, y) which satisfy the inequalities (7.4).

Solution This is a set of points (x, y) which satisfy three inequalities. Since these inequalities must be satisfied simultaneously, the desired set is the intersection of the three sets:

$$\begin{aligned} A &= \{(x, y): x \geq 0\} \\ B &= \{(x, y): y \geq 0\} \\ C &= \{(x, y): 150x + 80y \geq 600\} \end{aligned}$$

The set A is the half plane bounded by $x = 0$ (the y axis) and to the right of that line. The set B is the half plane bounded by $y = 0$ (the x axis) and above that line. The set C is the half plane above and to the right of the line $150x + 80y = 600$. The set $A \cap B \cap C$ is the shaded area in Figure 7.2. ■

Examples 7.7 and 7.8 have illustrated a general technique for graphing inequalities.

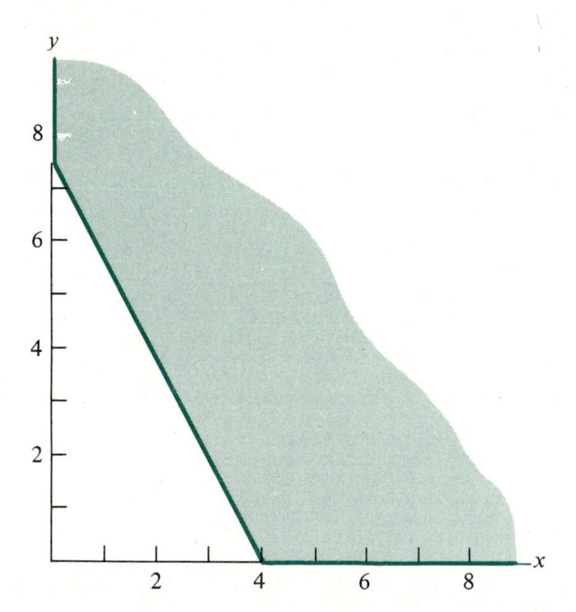

FIGURE 7.2

Method for Finding the Set Described by Linear Inequalities

1. For each inequality, graph the boundary line and test a point not on that line to determine which half plane satisfies the inequality.
2. The desired set is the intersection of all the half planes determined in step 1.

Example 7.9 Graph the set of points which satisfy the inequalities

$$\begin{aligned} x &\geq 1 \\ x &\leq 2y \\ 3x + 4y &\leq 12 \end{aligned}$$

Solution The graph of the line whose equation is $x = 1$ is the vertical line through the point (1, 0). The half plane A described by $x \geq 1$ is the half plane to the right of this line and is shown in Figure 7.3*a*.

The graph of the line $x = 2y$ or $x - 2y = 0$ is the line through the points (0, 0) and (2, 1). The half plane B described by $x \leq 2y$ is above and to the left of this line. The half plane is shown in Figure 7.3*b*. To verify that the half plane is as shown, check that the point (1, 1) satisfies the condition $x \leq 2y$.

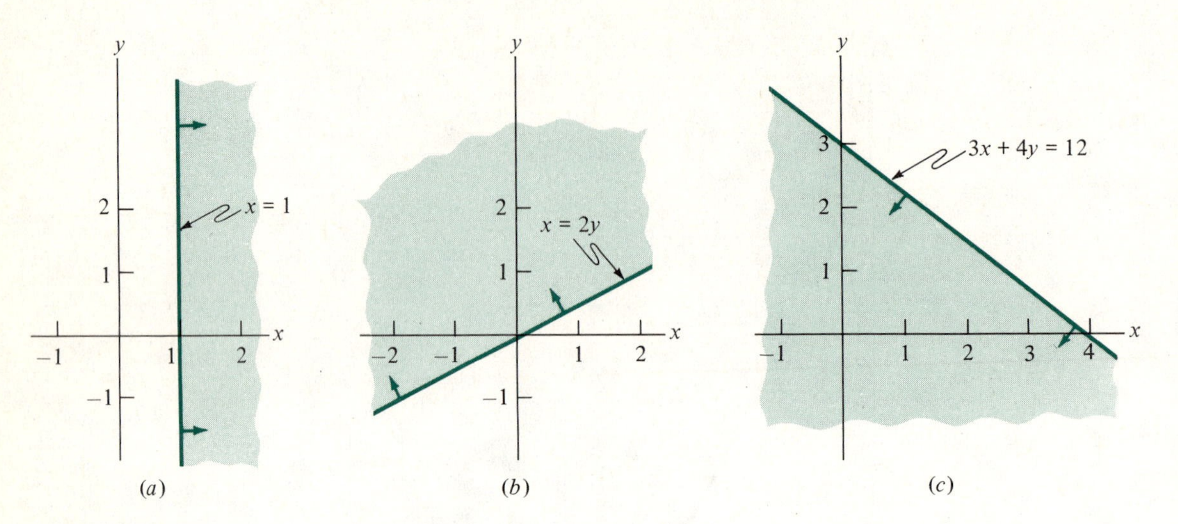

FIGURE 7.3

The graph of the line $3x + 4y = 12$ is the line through the points (4, 0) and (0, 3). The half plane C described by $3x + 4y \leq 12$ is below and to the left of this line. The half plane is shown in Figure 7.3*c*.

The set which we want to graph is $A \cap B \cap C$. This set is shown as the shaded area in Figure 7.4. ■

Notice that in Figure 7.3 we have shown small arrows from each boundary line directed into the half plane described by the relevant inequality. Such arrows are an alternative to shading and are frequently useful in identifying feasible sets

FIGURE 7.4

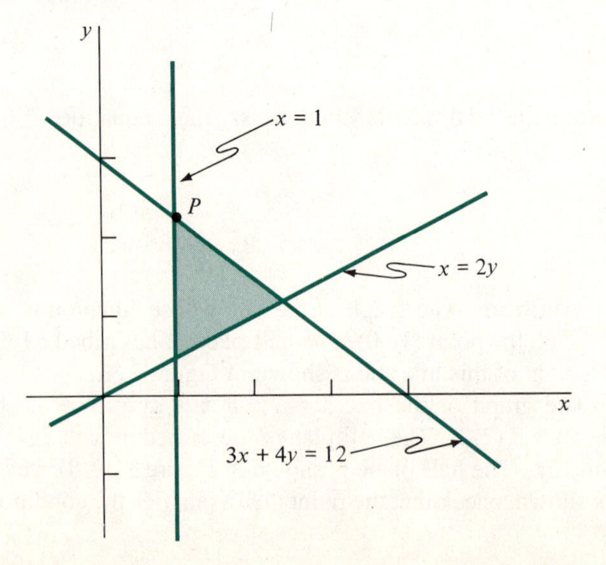

in situations where there are several inequalities. To keep the figures in this book as simple as possible, we will not show such arrows in subsequent figures.

Example 7.10 Find the coordinates of the corner point P shown in Figure 7.4.

Solution The coordinates of the point P can be obtained by solving the system of equations

$$\begin{aligned} x &= 1 \\ 3x + 4y &= 12 \end{aligned}$$

This system is easily solved, and we find $P = (1, \frac{9}{4})$. ■

Example 7.11 Graph the set of points which satisfy the inequalities

$$\begin{aligned} x + y - 10 &\leq 0 \\ x + y + 10 &\geq 0 \\ 2x - y - 10 &\leq 0 \\ x - 3y + 15 &\geq 0 \end{aligned} \tag{7.5}$$

Solution We begin by drawing the boundary line for the first of the four half planes described by the system (7.5). The equation for the first boundary line is $x + y - 10 = 0$. The x intercept is 10, and the y intercept is 10. The boundary line is shown as the line labeled $L1$ in Figure 7.5. Since the point (0, 0) does not lie on the boundary line, it can be used to test which side of that line is described by the inequality. The point (0, 0) satisfies the inequality $x + y - 10 \leq 0$, so the half plane described by the inequality contains the origin. The boundary lines for the sets described by the remaining inequalities in the system (7.5) are shown as lines labeled $L2$, $L3$, and $L4$ in Figure 7.5. In each case the origin belongs to the set described by the inequality. The set described by the system (7.5) is shown as the shaded area in Figure 7.5. ■

Exercises for Section 7.2

In Exercises 1 through 10, graph and shade the set of points which satisfy the given system of inequalities.

1. $5x + 3y \leq 15, \quad x - y \leq 3, \quad x \geq 1$
2. $x + 3y \geq 3, \quad y \leq 2, \quad x \geq 0$
3. $x \leq 4, \quad -x + 4y \geq 8, \quad y - x \leq 4$
4. $2x + 5y \leq 10, \quad x \leq 3, \quad y \geq -1$
5. $x - y \geq 0, \quad x + y \geq 2, \quad 4x + y \leq 4$

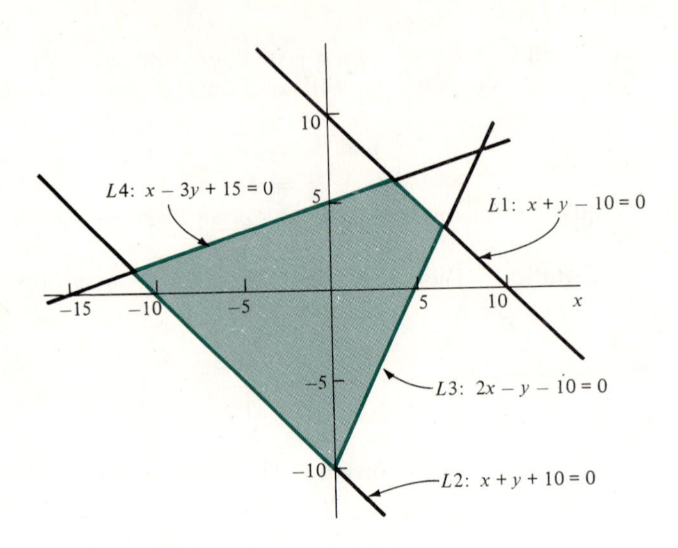

FIGURE 7.5

6. $x + y \geq 0, \quad x - y + 2 \geq 0, \quad 5x - 2y \leq 10$
7. $2x + 3y \geq 6, \quad y - x \leq 1, \quad x + y \leq 6$
8. $x + y \leq 2, \quad x - y \leq 2, \quad -x + y \leq 2, \quad -x - y \leq 2$
9. $5y - 2x - 20 \leq 0, \quad 4y + 3x + 24 \geq 0, \quad 6y - 7x + 21 \geq 0$
10. $3y + 2x + 12 \geq 0, \quad x + 4 \geq 0, \quad y + 2 \geq 0, \quad y - 2 \leq 0$
11. Find the corner points of the set described by the inequalities in Exercise 3.
12. Find the corner points of the set described by the inequalities in Exercise 5.
13. Find the corner points of the set described by the inequalities in Exercise 7.
14. A set of points is described by the inequalities

$$\begin{aligned} x &\geq 0 \\ y &\geq 1 \\ x + 2y &\geq 4 \\ x + y &\leq 6 \end{aligned}$$

  (*a*) Graph this set of points.
  (*b*) Find the vertices of the polygon which is described by the inequalities.

15. Find the corner points of the set described by the system (7.5) in the first quadrant.
16. Suppose that in the system (7.5) the inequality $x + y - 10 \leq 0$ is replaced by the inequality $x + y - 10 \leq 10$, with all other inequalities remaining the same. Graph the set described by the inequalities, and find the corner point(s) in the first quadrant.

17. A set in the plane is described by the inequalities

$$\begin{aligned} 2x - 5y &\leq 14 \\ 9x - 2y &\geq -6 \\ y &\leq 2 \end{aligned}$$

Graph the set and find the coordinates of the corner point in the first quadrant.

18. A set of points is described by the inequalities

$$\begin{aligned} x - y &\geq -4 \\ 2x - 4y &\leq 8 \\ 5x + 2y &\leq 10 \end{aligned}$$

Graph this set of points, and find the vertices of the polygon which is described by these inequalities.

19. A set of points is described by the inequalities

$$\begin{aligned} x + y &\geq -2 \\ 2x + 4y &\leq 8 \\ 5x - 2y &\leq -10 \end{aligned}$$

Graph this set of points, and find the vertices of the polygon which is described by these inequalities.

20. A set of points is described by the inequalities

$$\begin{aligned} 2y - x - 4 &\leq 0 \\ y - 1 &\geq 0 \\ 4x - y - 5 &\leq 0 \\ x &\geq 0 \end{aligned}$$

(*a*) Graph this set of points.
(*b*) Find the vertices of the polygon which is described by the inequalities.

21. A set of points is described by the inequalities

$$\begin{aligned} x &\geq 0 \\ x &\geq y \\ x &\leq y + 2 \\ 3y + x &\leq 6 \end{aligned}$$

(*a*) Graph this set of points.
(*b*) Find the vertices of the polygon which is described by the inequalities.

Investments

22. An investor has \$100,000 to invest in common and preferred stock. She determines that no more than three-fourths of the total should be invested in either type and that the amount invested in common stock should be at least as large as the amount invested in preferred stock. Graphically represent the set of choices available to the investor.

Production 23. The manager of a paper-box plant is scheduling the work for one production line for a week. He can produce standard and heavy-duty boxes. Each standard box requires 2 pounds of kraft paper, and each heavy-duty box requires 4 pounds. Also it requires 8 hours of labor to produce 100 standard boxes and 3 hours of labor to produce 100 heavy-duty boxes. (The machine which produces heavy-duty boxes is more efficient.) There are 50 tons of kraft paper and 2400 hours of labor available during the week. Finally, the manager has a contract which requires him to deliver 10,000 heavy-duty boxes at the end of the week. Graphically represent the set of choices available to the manager.

Sports 24. A National Basketball Association player is about to take forced retirement (bad knee), and he has \$1 million to invest for retirement income. He has decided to divide the money between a high-yield junk bond fund paying 10 percent per year and a government securities fund paying 6 percent per year. If he wants to be sure that his income is at least \$75,000 per year, what allocations of funds are possible for him? Graph the set of possible investments.

Biology 25. A biologist is planning to conduct two experiments on 3 acres of a salt marsh. She plans to conduct one experiment on one portion of the 3 acres and the other experiment on all or part of the remainder. She can collect samples for the first experiment at a rate such that it would take 2 days to collect samples from 1 acre. For the second experiment, samples are much harder to collect, and it would require 8 days to collect samples from 1 acre. Laboratory work for the samples from the first experiment would take 12 hours to complete for the samples from one acre, and it would take 4 hours to complete the laboratory work for the samples collected from 1 acre for the second experiment. The biologist has 16 days to collect data from the entire 3 acres, and her budget allows for 24 hours of laboratory work. Graphically represent the set of alternatives available to the biologist.

Investments 26. The manager of a pension fund has \$1 million to distribute among common stocks, bonds, and treasury notes. Guidelines for managing the fund call for no more than 60 percent of the assets to be invested in any single type of security.

(*a*) Graphically represent the set of options available to the fund manager. (*Note*: If x = number of dollars allocated to stocks and y = number of dollars allocated to bonds, then $1{,}000{,}000 - x - y$ = number of dollars allocated to treasury notes.)

(*b*) Suppose that stocks yield 6 percent, bonds yield 11 percent, and treasury notes yield 9 percent. If the goal of the manager is to maximize yield, find the objective function.

27. On a three-dimensional cartesian coordinate system, graph the feasible set for the following linear programming problem:

Maximize $\qquad p = x + 2y + 3z$

subject to $\qquad x \geq 0 \quad y \geq 0 \quad z \geq 0$

$$5x + 3y + 4z \leq 60$$

28. On a three-dimensional cartesian coordinate system, graph the feasible set for the following linear programming problem:

Maximize $$p = x + 2y + 3z$$

subject to
$$\begin{gathered} x \geq 0 \qquad y \geq 0 \qquad z \geq 0 \\ 5x + 3y + 4z \leq 60 \\ x - y \leq 0 \\ x - y \geq 0 \end{gathered}$$

29. Graph the feasible set and find the corner points for the following linear programming problem:

Maximize $$p = 2x + y + 3z$$

subject to
$$\begin{gathered} x \geq 0 \qquad y \geq 0 \qquad z \geq 0 \\ 3x + 4y + 5z \leq 60 \\ 2x + 3y \leq 0 \\ x - y \geq 0 \end{gathered}$$

30. Solve the linear programming problem posed in Exercise 29. In this special case the problem can be solved directly: Examine the feasible set, and evaluate the objective function on this feasible set.
31. Graph the feasible set and find the corner points for the following linear programming problem:

Minimize $$p = 4x + 7y$$

subject to
$$\begin{gathered} x \geq 0 \qquad y \geq 0 \\ 4x + 4y \geq 16 \\ 2x + 6y \geq 12 \end{gathered}$$

Evaluate the objective function at the corner points.

7.3 GRAPHICAL SOLUTION OF LINEAR PROGRAMMING PROBLEMS WITH TWO VARIABLES

In this section we develop a method to solve linear programming problems with two variables similar to those formulated in Section 7.1. The method is based on the techniques of Section 7.2, namely, graphing the feasible set and finding the corner points. We begin with an example and then present the general method.

Example 7.12

Manufacturing

(Continuation of Example 7.1) The problem formulated in Example 7.1 is to find the maximum of $5x + 4y$ for (x, y) satisfying

$$\begin{aligned} x \geq 0 \qquad y \geq 0 \\ 4x + 3y \leq 480 \\ 3x + 6y \leq 720 \end{aligned} \tag{7.6}$$

Recall that we are seeking x and y, where x is the number of toy cars and y is the number of toy trucks that Ted's Toys should produce in order to maximize profit.

Problem Solve the problem posed above; i.e., find the values of x and y which maximize profit $= 5x + 4y$.

Solution The feasible set for the problem, the set of (x, y) satisfying the inequalities (7.6), is shown as the shaded region in Figure 7.6. Since our goal is to make $5x + 4y$ as large as possible, we begin by considering possible values for this quantity. Suppose we denote the value of $5x + 4y$ by C. Then the set of all (x, y) for which $5x + 4y = C$ is a line. For instance, the lines on which $C = 120$ and $C = 240$ are shown on Figure 7.7. For some values of C the line intersects the feasible set, for example, $C = 240$, and for some values of C the line does not intersect the feasible set, for example, $C = 700$. If for a specific value of C, say $C = C'$, the line does intersect the feasible set, then there are values of x and y that satisfy the inequalities (7.5) and for which $5x + 4y = C'$. Since we are interested in finding the point (x, y) in the feasible set for which $5x + 4y$ is as large as possible, we are seeking the largest value of C for which the line $5x + 4y = C$ intersects the feasible set. We denote that value of C by C^*.

FIGURE 7.6

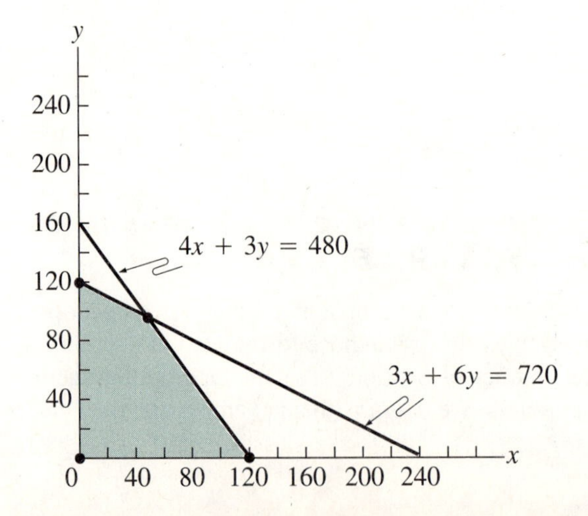

(It is common practice to use an asterisk to denote a maximum or a minimum value, and we use that convention in this section.)

The boundary of the feasible set is shown by the solid polygonal line in Figure 7.7 connecting points marked O, P, Q, R. Imagine a line with the equation $5x + 4y = C$ moving in the direction of the large arrow in Figure 7.7. The lines corresponding to any two values of C are parallel (they have the same slope), and the one with the larger value of C is "farther out" in the direction of the arrow. It is clear from Figure 7.7 that the value C^* corresponds to the line through the point labeled Q. The coordinates of the corner point Q can be obtained by solving the system of equations

$$4x + 3y = 480$$
$$3x + 6y = 720$$

The solution of the system is $x = 48$, $y = 96$. Therefore,

$$C^* = 5(48) + 4(96) = 624$$

The problem posed in Example 7.1 has now been solved. The maximum value of $5x + 4y$, for (x, y) satisfying (7.6), is 624. This value is attained for $x = 48$, $y = 96$. Stated differently, for any point (x, y) satisfying (7.1), we have $5x + 4y \leq 624$, and for $(x, y) = (48, 96)$, we have $5x + 4y = 624$.

In terms of the original problem, Ted's Toys should make 48 toy cars and 96 toy trucks to maximize profit. Notice that all the steel and all the plastic are used in the production of these toys and trucks. This follows because

$$4(48) + 3(96) = 480$$
$$3(48) + 6(96) = 720$$

■

FIGURE 7.7

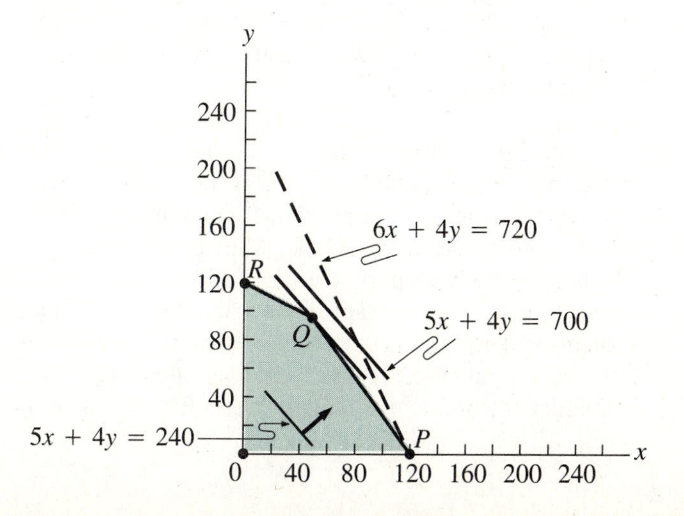

The last comment in the solution of Example 7.12 touches on an important issue. Suppose we have a linear programming problem in which the goal is to allocate resources to yield maximum profit. If the solution (x, y) of the problem satisfies a constraint $Ax + By \leq D$ as an equality, that is, $Ax + By = D$, then *all* the resources which are represented by that constraint are used when production is at the level (x, y). Geometrically this means that the boundary line of the half plane defined by the resource constraint contains the point (x, y). On the other hand, if the constraint is satisfied as a strict inequality $Ax + By < D$, then *not all* the resources represented by that constraint are used when production is at the level (x, y). Geometrically, the boundary line of the half plane does not contain the point (x, y). The amount of unused resources is $D - (Ax + By)$.

Example 7.13

Manufacturing

Suppose in Example 7.1 (solved in Example 7.12) that the profit per car is changed from \$5 to \$6. If all other conditions remain the same, how many toy cars and trucks should be produced to maximize profit? Also, are all the plastic and all the steel used in this new situation?

Solution Since only the objective function (the profit) has changed, the feasible set for this example is the same as the one in Example 7.12, namely, that shown in Figures 7.6 and 7.7. However, in this case, the profit line is $6x + 4y = C$, and we need to find the largest value of C such that this new line intersects the feasible set. Arguing as in Example 7.12, i.e., thinking of moving the line $6x + 4y = C$ by increasing the value of C, we conclude that the largest value of C for which the line meets the feasible set is $C = 720$. This line is shown as the dashed line in Figure 7.7. Since $C = 720$, the largest possible profit is \$720, and we see that this occurs at point P, where $x = 120$, $y = 0$. Thus, Ted's Toys should make 120 cars and no trucks. Also, note that for this level of production all the steel is used, because $4(120) + 3(0) = 480$, but not all the plastic is used, because $3(120) + 6(0) = 360 < 720$. Indeed, there are $720 - 360 = 360$ ounces of plastic left over. The amount of extra plastic is called the *slack* in the constraint for plastic. ■

Notice that in both Examples 7.12 and 7.13 the maximum value of the objective function was attained at one of the corner points of the feasible set. This is true for a large class of problems which includes all linear programming problems considered in this book. Although Figure 7.7 provides some geometric intuition which supports the conclusion in the special cases of Examples 7.12 and 7.13, it is by no means obvious that the conclusion must always hold. It is in situations such as this that mathematical theorems are useful. We observed a fact in two special cases. If it were true in all cases of interest to us, it would greatly simplify the problem-solving process. Its truth under certain carefully specified hypotheses is asserted in a *theorem* and is demonstrated in a *proof*. We shall state carefully conditions under which the assertion is valid, but we omit the proof. We consider only those sets that arise as feasible sets for linear programming problems.

Bounded and Unbounded Sets

A set of points in the plane is said to be *bounded* if it is contained in some circle centered at (0, 0). Otherwise it is said to be *unbounded.*

The set shown in Figure 7.8*a* is bounded; the set in Figure 7.8*b* is unbounded.

Corner Point

A point T is a *corner point* (or an *extreme point*) of a feasible set for a linear programming problem if every line segment that is contained in the set and that contains T has T as one of its endpoints.

In Figure 7.8*a*, points T_1 and T_2 are corner points, but points P and T_3 are not. This definition makes precise the intuitive notion of a corner point used in Examples 7.12 and 7.13. As in those examples, a corner point can always be determined by solving a system of linear equations.

Using these definitions, we can state a mathematical theorem which provides the basis for the method we use to solve linear programming problems.

FIGURE 7.8

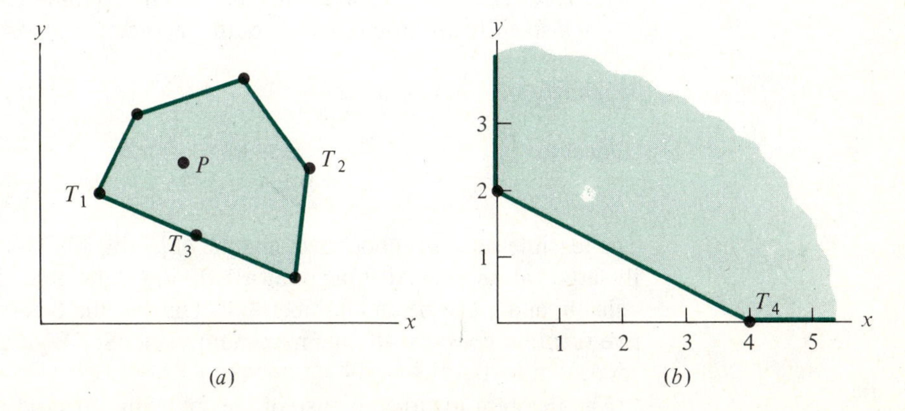

Theorem on Solutions of Linear Programming Problems

Let F be the feasible set for a linear programming problem, $F \neq \varnothing$, and let f be the objective function.

1. If F is bounded, then f attains its maximum value at a corner point of F and its minimum value at a corner point of F.
2. If F is unbounded and has at least one corner point, then exactly one of the following holds:
   (*a*) f attains its maximum at a corner point of F.
   (*b*) f takes arbitrarily large positive values on F.

The statement for unbounded sets and minimum values is as follows:

2′. If F is unbounded and has at least one corner point, then exactly one of the following holds:
   (*a*) f attains its minimum at a corner point of F.
   (*b*) f takes arbitrarily large negative values on F.

Stated in somewhat less formal terms, the first part of the theorem says that if the feasible set for a linear programming problem is bounded (such a set is shown shaded in Figure 7.8*a*), and if we seek to maximize the objective function, then the maximum value is attained at a corner point. Consequently, if we are to maximize the objective function in a linear programming problem and if the feasible set is bounded, then we need only examine the values of the objective function at the corner points and take the largest. This is the maximum value of the objective function on the entire feasible set. A similar statement holds for problems seeking the minimum of the objective function.

The assertions of the theorem regarding unbounded sets are necessarily phrased somewhat differently from the assertion regarding bounded sets. Indeed, for an unbounded feasible set F (e.g., the set shown in Figure 7.8*b*), the objective function need not attain a maximum or a minimum value.

For a specific instance of case 2 of the theorem, consider the problem:

Maximize $\quad x + y$

subject to $\quad x \geq 0 \quad y \geq 0$

$$x + 2y \geq 4$$

The feasible set F is unbounded, and the objective function $x + y$ takes arbitrarily large values: Indeed at the point $x = n$, $y = 0$ the objective function takes the value n, and n can be any number ≥ 4. Thus, as the theorem asserts, the objective function does not attain a maximum value. See Figure 7.8*b* for the feasible set F.

The theorem justifies the use of the following method of solving linear programming problems in two variables.

Solution Method for Linear Programming Problems

1. Graph the feasible set for the problem.
2. Determine the coordinates of each of the corner points of the feasible set.
3. Evaluate the objective function at each corner point.
4. Select the largest of these values if the problem is to maximize the objective function and the smallest of these values if the problem is to minimize the objective function.
   (*a*) If the feasible set is bounded, the value selected is the solution of the problem.
   (*b*) If the feasible set is unbounded and if the problem has a solution, the value selected is the solution.

Comments on the Method

Step 1 is important because without the aid of a graph it may be difficult to determine which pairs of the constraints determine corner points.

Since each corner point is the intersection of two lines, it can be determined by solving a pair of linear equations. Thus step 2 is carried out by solving pairs of linear equations, one pair for each corner point.

The information required in step 3 is easily presented in a table with the coordinates of the corner points forming one column and the values of the objective function forming the other. (See Table 7.6 for an example.)

If the feasible set is bounded, then the selection described in step 4 solves the problem. If the feasible set is unbounded, then a separate argument is needed to determine whether the problem has a solution. Concentrate for a moment on maximum problems. Either the problem has a solution, in which case the maximum value of the objective function is attained at a corner point, or the objective function takes arbitrarily large values and the problem has no solution. Similar comments hold for minimum problems.

We illustrate the case of a bounded feasible set in Example 7.14 and the case of an unbounded feasible set in Example 7.15.

Example 7.14 Find the largest value of $x + y$ for x and y satisfying the constraints

$$\begin{aligned} x \geq 0 \qquad y &\geq 0 \\ x + 2y &\geq 6 \\ x - y &\geq -4 \\ 2x + y &\leq 8 \end{aligned}$$

Solution Beginning with step 1 of our solution method, we graph the feasible set. This set is shown as the colored area in Figure 7.9. Note that the constraint $y \geq 0$ is redundant in this problem. That is, the feasible set is the same whether or not the constraint $y \geq 0$ is used to determine it.

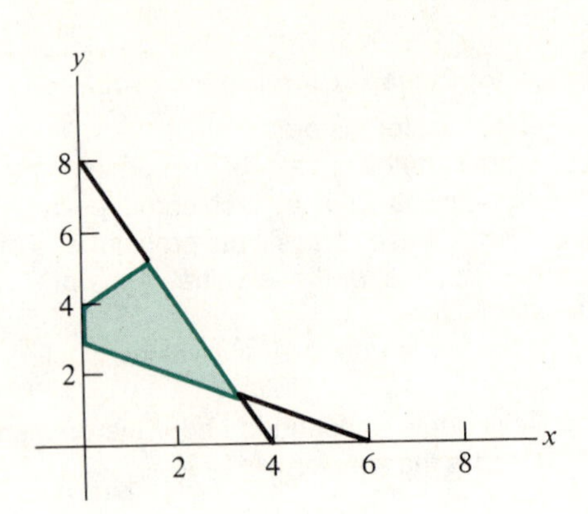

FIGURE 7.9

TABLE 7.6

Corner Point	Value of $x + y$
(0, 4)	4
$(\frac{4}{3}, \frac{16}{3})$	$\frac{20}{3}$
$(\frac{10}{3}, \frac{4}{3})$	$\frac{14}{3}$
(0, 3)	3

Continuing with step 2, we determine the corner points. They are shown in Table 7.6 along with the values of the objective function $x + y$ at the corner points. We select the value $\frac{20}{3}$ as the largest value taken by the objective function at a corner point. Since the feasible set is bounded, we conclude that the maximum value of the objective function $x + y$ on the feasible set is $\frac{20}{3}$, and this value is taken at the corner point $(\frac{4}{3}, \frac{16}{3})$. ■

Example 7.15 Solve the linear programming problem posed in Example 7.2.

Nutrition

Solution The equations defining the feasible set are

$$x \geq 0 \qquad y \geq 0$$
$$.8x + .2y \geq 90$$
$$3x + 6y \geq 600$$

and we wish to minimize the objective function $3x + 5y$. The feasible set, graphed in Figure 7.10, has three corner points denoted T_1, T_2, and T_3 in that figure. The coordinates of each corner point can be determined by solving the pair of equations corresponding to the lines which intersect at the corner point. Thus to determine the coordinates of T_1, T_2, and T_3, we solve the respective equations

$$T_1: \quad x = 0, \quad .8x + .2y = 90$$
$$T_2: \quad .8x + .2y = 90, \quad 3x + 6y = 600$$
$$T_3: \quad y = 0, \quad 3x + 6y = 600$$

We conclude that $T_1 = (0, 450)$, $T_2 = (100, 50)$, and $T_3 = (200, 0)$. Evaluating the objective function $3x + 5y$ at each corner point, we obtain the data in Table

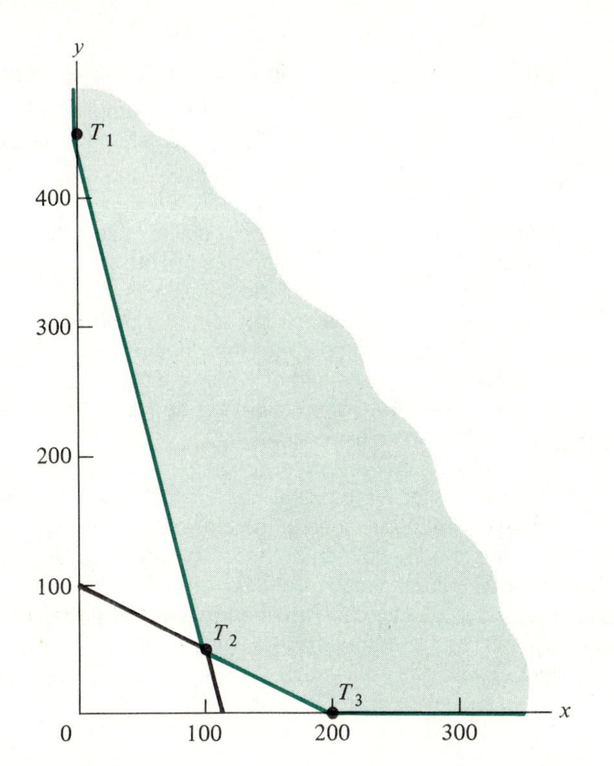

FIGURE 7.10

TABLE 7.7

Corner Point	Value of $3x + 5y$
(0, 450)	2250
(100, 50)	550
(200, 0)	600

7.7. Since the problem is to minimize the objective function (recall that the original problem was to minimize cost), we select the smallest value of the objective function. This is 550, which is attained at $T_2 = (100, 50)$. Therefore, since the feasible set is unbounded, we know that if the objective function does not attain arbitrarily large negative values, then the smallest value of the objective function is 550. In fact, this is the smallest value of the objective function on the feasible set since $3x + 5y$ takes only nonnegative values. (See 2′ of the theorem on p. 328.)

In terms of the original problem, the optimal mixture of peanuts and raisins is 100 grams of raisins and 50 grams of peanuts. The cost of this mixture is 550 cents, or $5.50, and this is the minimum cost of a mixture meeting the conditions set in the problem. ■

As we have seen, in some cases where the feasible set is unbounded special arguments can be used to show that a problem has a solution or that it has no solution. However, it is important to have a method that can be used in all cases of unbounded feasible sets. Such a method is the following.

Solutions of Problems with Unbounded Feasible Sets

Suppose the feasible set for a linear programming problem is unbounded. Determine the corner points of the feasible set and evaluate the objective function at each corner point.

(a) If the problem is a maximization problem, select the corner point (or points) at which the objective function has the largest value. At each such corner point, two boundary lines of the feasible set meet. On each such boundary line, select a feasible point, not a corner point, and evaluate the objective function. If these values are all less than or equal to the largest value of the objective function attained at a corner point, then the problem has a solution, and it is the largest value attained at a corner point. Otherwise, the problem has no solution.

(b) If the problem is a minimum problem, select the corner point (or points) at which the objective function has the smallest value. At each such corner point, two boundary lines of the feasible set meet. On each such boundary line, select a feasible point, not a corner point, and evaluate the objective function. If these values are all greater than or equal to the smallest value of the objective function attained at a corner point, then the problem has a solution, and it is the smallest value attained at a corner point. Otherwise, the problem has no solution.

Example 7.16 Although this technique is somewhat complicated to explain, it is quite easy to apply, as the following example shows.

Minimize $5x - 3y$

subject to
$$x \geq 0 \qquad y \geq 0$$
$$.8x + .2y \geq 90$$
$$3x + 6y \geq 600$$

Solution The constraints of this problem are the same as those of Example 7.15, and consequently the feasible sets are the same. Therefore, the feasible set for this problem is shown in Figure 7.10. The corner points and the values of the objective function at the corner points are shown in Table 7.8. The smallest value of the objective function is -1350, and it is attained at the corner point (0, 450). Following step (b) of the method, we select one point on each of the two boundary lines intersecting at (0, 450). We select (0, 500) and (25, 350). At the point (0, 500) the objective function takes the value -1500, and at (25, 350) it takes the value -925. Since -1500 is smaller than -1350, the problem has no solution. ■

TABLE 7.8

Corner Point	Value of $5x - 3y$
(0, 450)	-1350
(100, 50)	350
(200, 0)	1000

The geometric method requires an additional step when the feasible set is unbounded. It would be nice to have a method in which the distinction between bounded and unbounded feasible sets is unimportant. Such a method is the topic of Chapter 10.

We conclude this section, and the chapter, with an example which illustrates in a slightly different way the geometry of linear programming problems. We utilize the graphical techniques introduced in Chapter 5 to represent the feasible set in the xy plane and the value of the objective function in the direction perpendicular to that plane.

Example 7.17 A linear programming problem is defined as follows:

Maximize $$3x + y$$

subject to
$$\begin{aligned} x + y &\geq 3 \\ 3x - y &\geq -1 \\ x &\leq 2 \end{aligned}$$

Problem Graph the feasible set, find the value of $p = 3x + y$ for each corner point (x, y), and graph the value of p corresponding to each corner point on a three-dimensional coordinate system.

Solution The graph of the feasible set in the xy plane is shown in Figure 7.11. The values of the objective function at the corner points are shown in Table 7.9.

In Figure 7.12 the feasible set is shown as a dark-shaded area in the xy plane in a three-dimensional coordinate system. We denote the direction perpendicular to the xy plane as the p direction. For each corner point (x, y), we plot the point (x, y, p), where $p = 3x + y$. That is, for each corner point we plot the value of the objective function in the vertical direction. The relation between x, y, and p is $p = 3x + y$, or $p - 3x - y = 0$, and the graph of this linear equation is a plane in xyp space. If you consider only those points (x, y, p) on this plane for which (x, y) is in the feasible set, then you have the light-shaded area in Figure 7.12. In this example the maximum value of the objective function is 13, and this value is attained at the corner point $(2, 7)$. It is clear from the figure that the plane $p - 3x - y = 0$ slopes in such a way that its maximum height from the xy plane is attained at the point $(2, 7)$.

TABLE 7.9

Corner Point	Value of $p = 3x + y$
$(\frac{1}{2}, \frac{5}{2})$	4
(2, 1)	7
(2, 7)	13

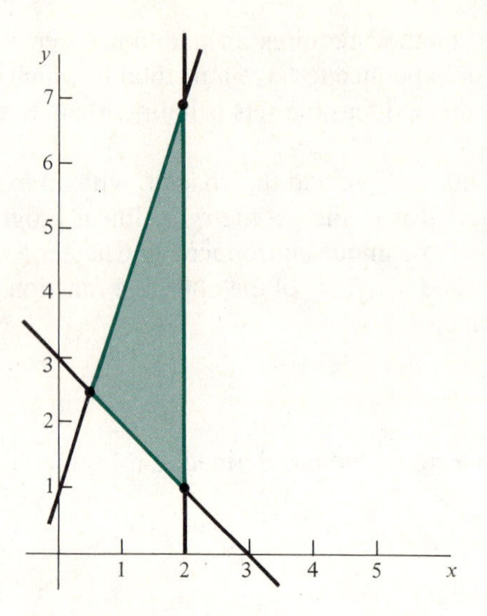

FIGURE 7.11

This situation prevails in general: If the feasible set is bounded, then the plane defined by

$$p - \text{objective function} = 0$$

attains its maximum height from the xy plane at a corner point. This corner point gives the solution of the linear programming problem. ■

FIGURE 7.12

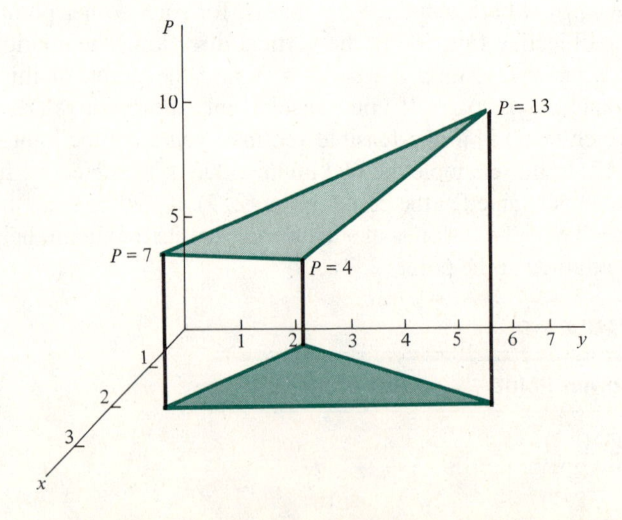

Exercises for Section 7.3

Each of the systems of inequalities in Exercises 1 through 5 describes a set in the xy plane. Graph that set and find the corner points. In each case decide if the set is bounded or unbounded.

1. $x - y \leq 3, \quad x + 3y \leq 3, \quad 3x + y \geq -3$
2. $x + y \leq 4, \quad x - y \leq 0, \quad x - y \geq -2$
3. $x \geq 0, y \geq 0, \quad x + y \leq 3, \quad x - y \leq 1$
4. $3x + 2y \leq 10, \quad y \leq 3, \quad x - y \leq 1, \quad x \geq 0$
5. $x - y \leq -3, \quad x - y \geq 1, \quad y \leq 3$
6. A set S in the xy plane is described by the inequalities

$$\begin{aligned} y &\geq -1 \\ x &\leq 3 \\ y &\leq 3x + 6 \\ 5y + 2x &\leq 10 \end{aligned}$$

   (*a*) Graph this set.
   (*b*) Find the corner points of the set.
   (*c*) Find the maximum value of $3x + 2y$ for $(x, y) \in S$.

7. Find the maximum value of $2x + y$ for (x, y) satisfying $x \geq 1$, $y \geq 1$, $x \leq 5$, $y \leq 5$, and $x + y \leq 6$. Also find the minimum value of $x - 2y$ on this set.
8. Find the minimum value of $x/2 + y$ for (x, y) satisfying $y \geq 2, x \geq 0, x \geq y$, and $x + y \leq 6$. Also find the maximum value of $-x + 2y$ on this set.
9. Find the maximum value and the minimum value of $3x + 5y$ for (x, y) satisfying $x + 5y \leq 50$ and $x - 5y \geq -25$.
10. Find the maximum and minimum values of $2x - 5y$ for (x, y) satisfying $x + 3y \leq 30, x \geq 0, 3x - 4y \geq -27$, and $y \geq 0$.
11. Find the minimum and maximum values of $5x - 3y$ for (x, y) satisfying the constraints in Exercise 9.
12. Find the minimum and maximum values of $-x + 4y$ for (x, y) satisfying the constraints in Exercise 10.
13. Maximize $\quad 3x - 2y$

   subject to

$$\begin{aligned} x + y &\geq 0 \\ x - y &\leq 0 \\ -2x + 4y &\leq 5 \end{aligned}$$

14. Find the minimum of the function $3x + 2y$ on the set described by the constraints of Exercise 13.
15. Maximize $\quad x - 3y$

   subject to

$$\begin{aligned} x \geq 0 \quad & y \geq 0 \\ 2x - 3y &\leq 6 \\ -x + 4y &\leq 4 \end{aligned}$$

16. Let S be the set described by the inequalities $y - x \leq 1$, $2y + x \geq 5$, $y + x \leq 5$, and $x \leq 3$.
    (*a*) Find the maximum value of $y + 2x$ for $(x, y) \in S$.
    (*b*) Find the minimum value of $2y - 3x$ for $(x, y) \in S$.

Solve the linear programming problems formulated in the following exercises from Section 7.1.

Production

17. Exercise 1
18. Exercise 2
19. Exercise 3

Purchasing

20. Exercise 4
21. Exercise 8
22. Exercise 9

Psychology

23. Exercise 11
24. Exercise 5
25. Exercise 13
26. Exercise 14
27. Exercise 15
28. Exercise 16

Food service

29. A bakery makes two types of cakes each day: poppy seed and German chocolate. The profit to the bakery is \$2 on each poppy seed cake and \$4 on each German chocolate cake. A poppy seed cake requires 400 grams of flour, 200 grams of butter, and 100 grams of poppy seeds. A German chocolate cake requires 600 grams of flour, 100 grams of butter, and 150 grams of chocolate. There are 9600 grams of flour, 2400 grams of butter, 1500 grams of poppy seeds, and 2100 grams of chocolate.
    (*a*) How many cakes of each type should be made to yield maximum profit? What is the maximum profit?
    (*b*) When the baker produces cakes to yield maximum profit, does any flour remain unused at the end of the day? How much?
    (*c*) Repeat (*b*) for butter.

Food service

30. In the situation described in Exercise 29, suppose the profit for each poppy seed cake is \$3 and the profit for each German chocolate cake is \$4. Answer questions (*a*), (*b*), and (*c*) of Exercise 29 for these profit levels.
31. Let F be the feasible set defined by the constraints $2x + y \geq 12$, $x + 2y \geq 12$, $-5x + 5y \leq 25$, and $5x - 5y \leq 25$.
    (*a*) Find the minimum and the maximum of $x - 2y$ over the set F.
    (*b*) Find the minimum and the maximum of $x + 2y$ over the set F.
    (*c*) Find the minimum and the maximum of $x + 6y$ over the set F.
    (*d*) Find the minimum and the maximum of $-2x + y$ over the set F.
32. Let F be the feasible set defined by the constraints $x \geq 1$, $y \geq 1$, $2x + 3y \geq 6$, and $3x - 2y \geq -6$.
    (*a*) Find the minimum and the maximum of $2x - y$ over the set F.
    (*b*) Find the minimum and the maximum of $-2x + 3y$ over the set F.
    (*c*) Find the minimum and the maximum of $2y$ over the set F.
    (*d*) Find the minimum and the maximum of $-x$ over the set F.

Nutrition

33. Sam's Snacks prepares packages of trail mix which contain raisins, peanuts, and dried apple slices. A box of regular mix contains 4 ounces of raisins, 8 ounces of peanuts, and 12 ounces of apple slices. A box of deluxe mix contains 6 ounces of raisins, 6 ounces of peanuts, and 8 ounces of apple slices. There are 24 pounds of raisins, 36 pounds of peanuts, and 60 pounds of apple slices available. The profit per box is \$2.00 for regular mix and \$2.50 for deluxe mix. How many boxes of each type should be prepared to maximize profit?
34. Solve the following linear programming problem:

Minimize $\quad 3x - 2y$

subject to
$$\begin{aligned} y &\geq 0 \\ 2x + y &\geq 2 \\ 2x - y + 2 &\geq 0 \\ x - y &\leq 4 \end{aligned}$$

35. Let F be the feasible set defined by the constraints $x + y \geq 4$, $4x - y \geq 1$, and $x - 4y \leq 0$.
(*a*) Find the maximum of $2x - 10y$ for (x, y) in F.
(*b*) Find the maximum of $2x - 6y$ for (x, y) in F.
(*c*) Find the maximum of $2x - 8y$ for (x, y) in F.
36. Let F be the feasible set defined by the constraints $3x + 2y \geq 15$, $3x - y \geq 6$, and $x + 4y \leq 10$.
(*a*) Find the minimum of $2x + 8y$ for (x, y) in F.
(*b*) Find the minimum of $2x + 3y$ for (x, y) in F.
(*c*) Find the minimum of $2x + 10y$ for (x, y) in F.
37. Find the minimum of the function $2x - 5y$ for (x, y) satisfying $x \geq 0$, $y \geq 0$, $3x + 2y \geq 30$, $5x - 2y \geq 18$, and $x - 2y \leq 20$.
38. Find the maximum of the function $3x - 5y$ for (x, y) satisfying $x \geq 0$, $y \geq 0$, $2x + 3y \geq 30$, $5x - 2y + 30 \geq 0$, and $x - 2y \leq 20$.

IMPORTANT TERMS AND CONCEPTS

You should be able to describe, define, or give examples of and use each of the following:

Linear programming problem
Constraint
Feasible set
Objective function
Bounded (unbounded) set
Corner point
Graphical solution method for linear programming problems

REVIEW EXERCISES

In Exercises 1 through 8, formulate (but do not solve) each situation as a linear programming problem.

Resource allocation

1. Ted's Toys decides to manufacture and sell two types of toy airplanes. The planes are made out of steel, plastic, and wood strips. Each passenger plane uses 4 ounces of steel, 1 ounce of plastic, and 8 inches of wood strips; and each cargo plane uses 5 ounces of steel, 3 ounces of plastic, and 12 inches of wood strips. Ted has available 20 pounds of steel, 15 pounds of plastic, and 20 feet of wood strips. If the profit per passenger plane is $3 and the profit per cargo plane is $4, how many of each type of plane should be made to maximize profit?

Resource allocation

2. A manufacturer of basketball shoes makes shoes using leather, nylon, rubber, and labor. The company makes two models, the Dunker and the Flyer. Each pair of Dunker shoes requires 6 ounces of leather, .6 square feet of nylon, 10 ounces of rubber, and 7 minutes of labor. Each pair of Flyer shoes requires 8 ounces of leather, .4 square foot of nylon, 12 ounces of rubber, and 9 minutes of labor. The company has 20 pounds of leather, 150 square feet of nylon, 50 pounds of rubber, and 20 hours of labor. The profit per pair is $16 for the Dunker and $24 for the Flyer. How many pairs of each type of shoe should the company produce to achieve maximum profit?

Farming

3. A farmer owns a 2000-acre farm and can plant any combination of two crops, A and B. Crop A requires 1 person-day of labor and $90 of capital for each acre planted, while crop B requires 2 person-days and $60 of capital for each acre planted. Crop A produces $170 in revenue per acre, and crop B produces $190 in revenue per acre. The farmer has $150,000 of capital and 3000 person-days of labor available for the year. How many acres of each crop should the farmer plant in order to maximize the total revenue?

Recreation

4. White Water Rapids, Inc., offers two types of white-water adventure day trips, one using rubber rafts and one using kayaks, on the Trout and Salamander rivers. On the Trout River they can handle 50 rafts and 20 kayaks per day at a cost of $1000. On the Salamander River they can handle 60 rafts and 40 kayaks per day at a cost of $1200. During the season they have a demand for 1800 raft trips and 1000 kayak trips. How many days should they operate on each river to meet all the demand at minimum cost?

Resource allocation

5. Murphy's Muffin Shoppe makes both large and small bran muffins, using dough and bran. Each large muffin uses 4 ounces of dough and 2 ounces of bran, and each small muffin uses 1 ounce of dough and 1 ounce of bran. There are 300 ounces of dough and 160 ounces of bran available each day, and the profit per muffin is $.25 for a large muffin and $.10 for a small muffin. How many muffins of each size should be made each day to maximize profits?

Production

6. The Trimetal Mining Company operates two mines from which gold, silver, and copper are mined. The Alpha Mine costs $8000 per day to operate, and it yields 30 ounces of gold, 200 ounces of silver, and 400 pounds of copper each day. The Omega Mine costs $12,000 per day to operate, and it yields 40 ounces of gold, 400 ounces of silver, and 300 pounds of copper each day. The company has a contract to supply at least 300 ounces of gold, 3000 ounces of silver, and 2000 pounds of copper. How many days should each mine be operated so that the contract can be filled at minimum cost?

Resource allocation

7. The Plant Power Fertilizer Company produces three fertilizer brands: Standard, Special, and Super. One sack (100 pounds) of each brand contains the nutrients nitrogen, phosphorus, and potash in the amounts shown below. The profit per sack is shown at the right, and the available material (in pounds) is shown at the bottom. How many sacks of each brand should be produced to maximize profit?

	Amount per Sack, Pounds			
Brand	**Nitrogen**	**Phosphorus**	**Potash**	**Profit, $**
Standard	20	5	5	10
Special	10	10	10	5
Super	5	15	10	15
Available	2000	1500	1500	

Resource allocation

8. Ms. Li produces widgets. For each 100 left-handed widgets she uses 1 pound of metal and 5 pounds of fiberglass. For each 100 right-handed widgets she uses 2 pounds of metal and 3 pounds of fiberglass. Each week Ms. Li has 65 pounds of metal and 150 pounds of fiberglass available. She makes a profit of $2.50 on each right-handed widget and a profit of $2.00 on each left-handed widget. How many widgets of each type should Ms. Li produce each week to maximize profit?

Resource allocation

9. Solve the problem formulated in Exercise 1. Are all the resources consumed?

Resource allocation

10. Solve the problem formulated in Exercise 2. Are all the resources consumed?

Farming

11. Solve the problem formulated in Exercise 3. Are all the resources consumed?

Recreation

12. Solve the problem formulated in Exercise 4.

Resource allocation

13. Solve the problem formulated in Exercise 5. Are all the resources consumed?

Production

14. Solve the problem formulated in Exercise 6.

Resource allocation

15. Solve the problem formulated in Exercise 7. Are all the resources consumed?

Resource allocation

16. Solve the problem formulated in Exercise 8. Are all the resources consumed?

Resource allocation

17. In the situation described in Exercise 8, suppose that Ms. Li decides to produce only one type of widget. What type and how many should she produce to maximize profit?

Resource allocation

18. In the situation described in Exercise 8, suppose that the production of 100 widgets (either type) requires 10 kilowatthours of electricity. Also suppose 350 kilowatthours are available for this process. How many widgets of each type should be produced each week to maximize profit?

19. A set in the plane is described by the inequalities

$$\begin{aligned} x - y &\le 3 \\ x + 3y &\le 0 \\ 3x + y &\ge -3 \end{aligned}$$

Graph the set and find its corner points.

20. A set in the plane is described by the inequalities

$$\begin{aligned} 4x - 6y &\geq 12 \\ 3x + 2y &\leq 14 \\ y + 2 &\geq 0 \end{aligned}$$

(*a*) Graph the set.
(*b*) Find the corner points in the first quadrant ($x \geq 0, y \geq 0$).

21. A set S is defined by the inequalities

$$\begin{aligned} x - y - 4 &\leq 0 \\ x - 6 &\leq 0 \\ y - 6 &\leq 0 \\ 2x + 3y &\geq 0 \end{aligned}$$

(*a*) Graph S.
(*b*) Find the corner points of S.
(*c*) Find the maximum value of the function $2x - 5y$ on S.

22. Find the maximum and minimum values of $2x + 3y$ for (x, y) satisfying the inequalities

$$\begin{aligned} y + 1 &\geq 0 \\ x &\leq 2 \\ 2x + y &\leq 5 \\ y - 3x &\leq 10 \end{aligned}$$

23. A set S is described by

$$\begin{aligned} x - y &\geq 0 \\ x + y + 1 &\geq 0 \\ y &\leq 2 \end{aligned}$$

Find the maximum value and the minimum value of the function $3x + 5y$ on S.

24. A set S is described by

$$\begin{aligned} x - y &\leq 0 \\ x + y + 1 &\geq 0 \\ y - 2x &\leq 8 \end{aligned}$$

Find the maximum and the minimum values of the function $2x - y$ on S.

25. Find the maximum and the minimum values of the function $5x + 3y$ on the feasible set defined by

$$\begin{aligned} x \geq 0 \qquad y &\geq 0 \\ 3x + 2y &\geq 6 \\ 2x + 3y &\geq 6 \end{aligned}$$

26. Find the maximum and minimum values of the function $2x + y$ on the feasible set defined by

$$\begin{aligned} y &\geq -1 \\ x &\leq 2 \\ 2x + 3y &\leq 5 \\ y - 3x &\leq 10 \end{aligned}$$

27. Find the maximum and minimum values of the function $2x + 3y$ on the feasible set defined by

$$\begin{aligned} x - y + 2 &\geq 0 \\ x - 4y &\leq 4 \\ x + y &\leq 5 \end{aligned}$$

28. (*a*) Graph the following system of inequalities:

$$\begin{aligned} y - 3x &\leq 2 \\ 2y + 3x &\geq 12 \\ y &\leq 8 \\ y + x &\leq 14 \\ 2x - y &\leq 16 \\ 2x - 5y &\leq 8 \end{aligned}$$

(*b*) Find the corner points of the set described in (*a*).
(*c*) Find the maximum of $2x + 3y$ over this set.
(*d*) Find the minimum of $4y - 3x$ over this set.

29. A set S is defined by the inequalities:

$$\begin{aligned} x + y + 2 &\geq 0 \\ y + 1 &\geq 0 \\ x - y &\leq 2 \\ 3x - y + 5 &\geq 0 \end{aligned}$$

(*a*) Find the maximum value of $x - 2y$ for $(x, y) \in S$.
(*b*) Find the minimum value of $x - 2y$ for $(x, y) \in S$.

30. Let S be the set defined in Exercise 29.
(*a*) Find the maximum value of $2x - y$ for $(x, y) \in S$.
(*b*) Find the minimum value of $2x - y$ for $(x, y) \in S$.

PART

3

Applications and Related Topics

CHAPTER 8

Markov Chains

8.0 THE SETTING AND OVERVIEW

As we have illustrated in earlier chapters, many situations in the social, life, and management sciences which involve uncertainty can be modeled by using multistage experiments. We introduced the idea of a multistage experiment in Chapter 1, including the helpful representation with tree diagrams, and we continued the study in Chapter 3. In Section 3.5 we considered the special case of experiments which can be modeled by using Bernoulli trials. In this chapter we study another special case of multistage stochastic processes, *Markov chains*.

Markov chains have several advantages as mathematical models: They are general enough to provide useful models for many situations; they have been intensively studied, and many results are known; and they are special enough to be easy to use. The diversity of application can be illustrated with a few examples: psychology—learning and memory; sociology—social mobility and demography; biology—ecological systems and nutrient flows; business—decision making under uncertainty. The ease of use is a consequence, in part, of using matrix algebra to simplify the computations.

Ideas introduced in earlier chapters play an important role in our study of Markov chains. For instance, the concepts and methods from our earlier discussions of tree diagrams, conditional probability, multistage experiments, matrices and matrix inverses, systems of equations, and the expected value of a random variable all play a role in this chapter.

8.1 STATES, TRANSITIONS, TRANSITION DIAGRAMS, AND TRANSITION MATRICES

As we have seen in earlier chapters, it is frequently useful to view a stochastic process as an experiment consisting of a number of successive steps or stages, and we introduced tree diagrams to help us study such processes. In general, the probabilities of the results at one stage depend on the results of preceding stages, and this dependence can take many forms. For example, tree diagrams for three experiments, each consisting of repetitions of a subexperiment with results labeled X and Y, are shown in Figure 8.1. In each case the result of the initial observation is X. In Figure 8.1*a* the probabilities of results X and Y are .2 and .8, respectively, at every stage of the experiment and independent of the result of the preceding stage. Thus, the experiment whose tree diagram is Figure 8.1*a* is a Bernoulli trial process. Figure 8.1*b* illustrates a process in which the probabilities of results X and Y are not always the same. However, it is the case that

FIGURE 8.1

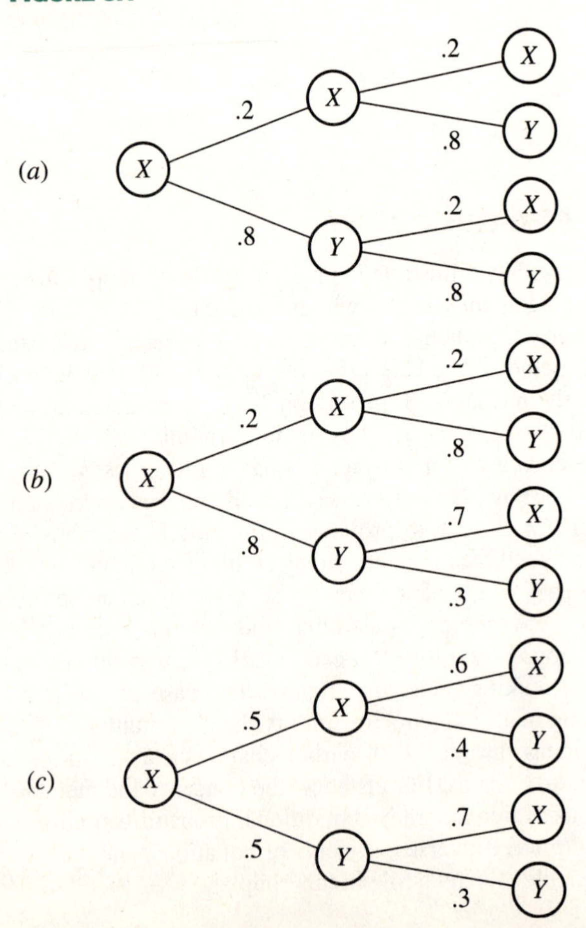

whenever the result at one stage is X, then the probabilities of X and Y at the next stage are *always* .2 and .8, respectively; and whenever the result at one stage is Y, then the probabilities of X and Y at the next stage are *always* .7 and .3, respectively. That is, probabilities of the possible results at the next stage depend *only* on the current result. Finally, Figure 8.1*c* illustrates a process in which the probabilities of results X and Y vary with both the stage of the experiment and the result of the current stage. For instance, the tree diagram shows the result of the first stage is X and the probability of result X at the second stage is .5. But if the result is X at the second stage, then the probability of result X at the third stage is .6.

Although our basic approach to the study of multistage experiments is through the use of tree diagrams, when the experiment has many stages, such analyses become very complicated and we seek alternative methods. In Chapter 3 we developed techniques, other than tree diagrams, to study one special type of multistage experiment, Bernoulli processes. We now consider another special type, namely, processes with tree diagrams similar to Figure 8.1*b*. We distinguish among multistage experiments according to the way in which the probabilities of results at one stage depend on the results at earlier stages. For example, in a Bernoulli process the results at one stage do *not* depend at all on the results at other stages. We now distinguish another special type of multistage experiment.

Markov Chain

A *Markov chain* is a stochastic process which satisfies the conditions:

1. At each stage the result is one of a fixed number of states.
2. The conditional probability of a transition from any given state to any other state depends on only the two states.

Example 8.1 *Employment*

A freelance computer network consultant is employed only when she has a contract for work, and each of her contracts is for 1 week of work. Each week she is either employed (E) or unemployed (U), and her records support the following assumptions about the conditional probabilities:

(*a*) If she is employed this week, then next week she will be employed with probability .8 and unemployed with probability .2.

(*b*) If she is unemployed this week, then next week she will be employed with probability .6 and unemployed with probability .4.

Problem She is employed this week. Find the probability that she will be employed 2 weeks from now.

Solution A tree diagram for this situation is shown in Figure 8.2. Since we know that she is employed this week, the "Begin" box of the tree diagram is replaced by E. From the tree diagram we conclude that the probability that she is employed 2 weeks from now is

$$.64 + .12 = .76$$ ■

The solution of this problem involved the use of techniques from Chapter 3. However, if we had asked for the probability that she will be employed in 5 weeks or in 10 weeks or "in the long run," then the techniques of Chapter 3 would be cumbersome or impossible to use. To develop techniques which will be more effective in such problems, we need to have new terminology and notation.

We will be concerned with systems which can be in any one of N possible *states*. The systems are observed successively, and *transitions* between states are noted. In Example 8.1 the system is our consultant, and the states describe her employment status: employed or unemployed. Our consultant is observed each week, and her state is noted.

The information given in Example 8.1(a) and (b) can be represented concisely in a *transition diagram* such as Figure 8.3. In Figure 8.3 we have indicated states E and U and, on arrows connecting the states, the probabilities of being in successive states on successive observations. For instance, the .2 on the arrow directed from E to U means that if the system is in state E on one observation, then it is in state U on the next observation with probability .2.

As we noted in the definition of Markov chains, the fundamental property which distinguishes a Markov chain from other sequential probabilistic processes can be described as follows:

FIGURE 8.2

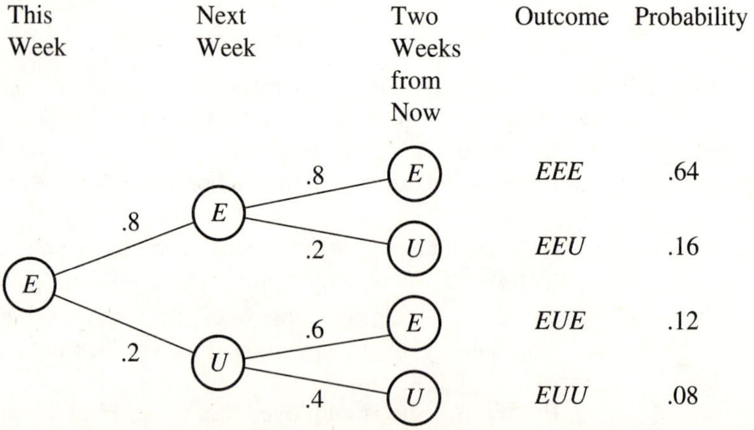

This Week	Next Week	Two Weeks from Now	Outcome	Probability
E	.8 → E	.8 → E	EEE	.64
		.2 → U	EEU	.16
	.2 → U	.6 → E	EUE	.12
		.4 → U	EUU	.08

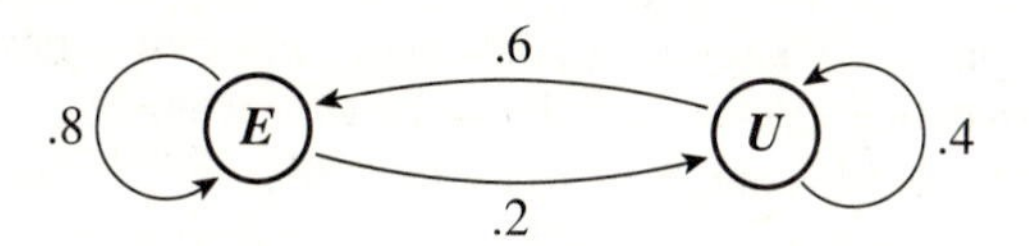

FIGURE 8.3

Markov Property

If a system is in state i on one observation, then the conditional probability that it is in state j on the next observation depends on only i and j (and not on what happened before the system reached state i or on the stage of the experiment). This probability will be denoted by p_{ij}, the probability of making a direct transition from i to j.

One way that this property can be viewed intuitively is to think of Markov chains as mathematical descriptions for systems without memories. That is, the probability that the system makes a transition from one state (say, state i) to another state (or even back into state i) depends on the two states and not on the number of transitions or the states occupied before the system reached state i.

The transition probabilities p_{ij} are the numbers on the arrows of the transition diagram. Thus, in Example 8.1, if we label state E by 1 and state U by 2, then $p_{11} = .8$ and $p_{12} = .2$.

Transition Matrix

Consider a Markov chain with N states. Let p_{ij} be the probability of making a direct transition from state i to state j, $1 \leq i \leq N$, $1 \leq j \leq N$. The matrix

$$\mathbf{P} = [p_{ij}]$$

is the *one-step transition matrix*, or simply the *transition matrix*, for the Markov chain.

Example 8.2 Find the transition matrix for the process described in Example 8.1.

Employment

Solution Let E and U be labeled as states 1 and 2, respectively. The transition probabilities are given on the transition diagram in Figure 8.2. We have

$$\mathbf{P} = \begin{bmatrix} .8 & .2 \\ .6 & .4 \end{bmatrix}$$

■

Note that if we had labeled the states differently, then we would have obtained a different transition matrix **P**. However, providing the identification of states with rows and columns is consistent, then all probabilities computed for the process will be the same, regardless of the transition matrix used.

Example 8.3 A tree diagram for a stochastic process with two states is shown in Figure 8.4.

Problem Determine whether this is the tree diagram of a Markov chain.

Solution The states are labeled A and B in Figure 8.4. Note that the probability of making a transition from A to A at stage 1 is $\frac{1}{3}$, the probability of making a transition from A to A at the second stage is $\frac{1}{4}$, and the probability of making a transition from A to A at the third stage is $\frac{1}{5}$. Thus, the probability of making a transition from A to A *does* depend on the stage, and consequently this is *not* the tree diagram of a Markov chain. ■

Example 8.4 *Scheduling*

The service specialists of Metroburg Heating and Air Conditioning Company make their calls with a fleet of radio-dispatched service trucks. Established procedure is that when a call requesting service is received by the office, the dispatcher sends the next available truck to respond to the call. One of the service specialists keeps records of his service calls, and his data are summarized in Table 8.1.

FIGURE 8.4

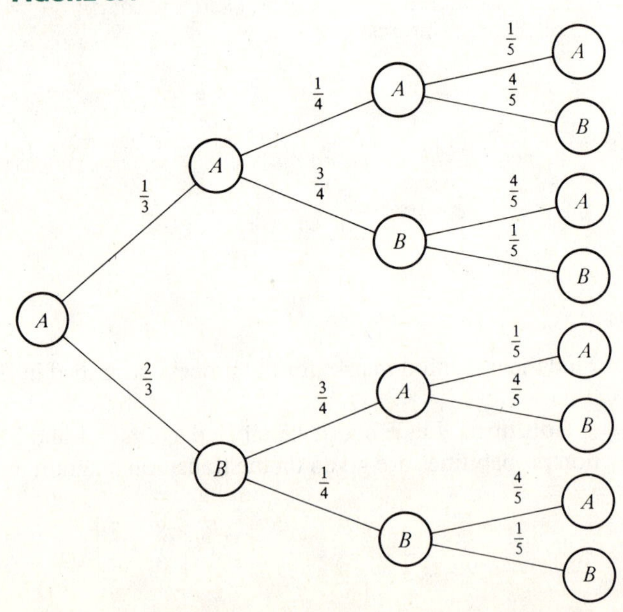

TABLE 8.1

District of Current Call	District of Next Call	Percentage of Calls
East	East	50
	Central	40
	West	10
Central	East	10
	Central	60
	West	30
West	East	30
	Central	60
	West	10

Problem Formulate this situation as a Markov chain. Find the transition diagram and the one-step transition matrix.

Solution To formulate this as a Markov chain, we must identify the states of the system. We define state 1 to be a service call in the east district, state 2 a call in the central district, and state 3 a call in the west district. Using the data of Table 8.1, we have the transition diagram shown in Figure 8.5 and the transition matrix **P**.

$$\mathbf{P} = \begin{bmatrix} .5 & .4 & .1 \\ .1 & .6 & .3 \\ .3 & .6 & .1 \end{bmatrix}$$ ■

FIGURE 8.5

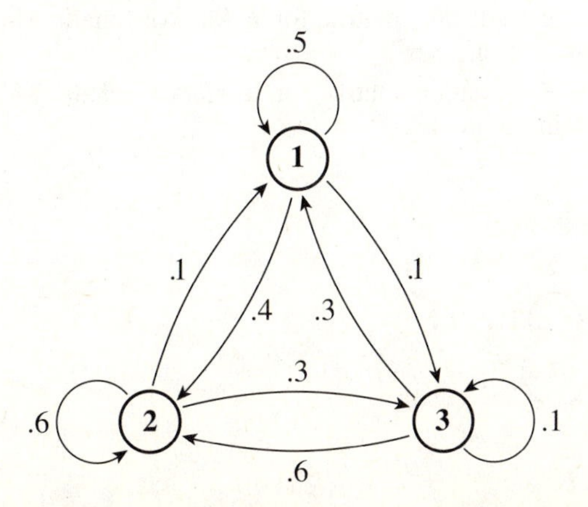

Remark The method of dispatching service trucks used in this example is not the only one which could be used. Other methods may be more efficient in terms of travel time and expense, but less efficient in terms of customer waiting time. Many businesses are faced with this problem, and they solve it in different ways.

Exercises for Section 8.1

1. A Markov chain has the transition matrix **P** shown below. Find the transition diagram for this Markov chain.

$$\mathbf{P} = \begin{bmatrix} .3 & .7 \\ .9 & .1 \end{bmatrix}$$

2. A Markov chain has the transition matrix given in Exercise 1. On the first observation the system is in state 1.
   (*a*) Find the probability that it is in state 1 on the second observation.
   (*b*) Find the probability that it is in state 1 on the third observation.
3. A Markov chain has the transition matrix given in Exercise 1. On the first observation it is in state 2. What state is it most likely to occupy on the third observation?
4. A Markov chain has the transition matrix given in Exercise 1. If the system begins in state 2, find the probability that it alternates between states 1 and 2 for the first four observations. That is, it occupies state 2, then state 1, then state 2, and finally state 1 again.
5. A Markov chain has the transition matrix **P** shown below. Find the transition diagram for this Markov chain.

$$\mathbf{P} = \begin{bmatrix} .2 & .8 \\ 1 & 0 \end{bmatrix}$$

6. A Markov chain has the transition matrix given in Exercise 5. On the first observation it is in state 2.
   (*a*) Find the probability that it is in state 2 on the third observation.
   (*b*) In what state is it most likely to be on the fourth observation?
7. Find the transition matrix for a Markov chain whose transition diagram is shown in Figure 8.6.
8. Find the transition matrix for a Markov chain whose transition diagram is shown in Figure 8.7.

FIGURE 8.6

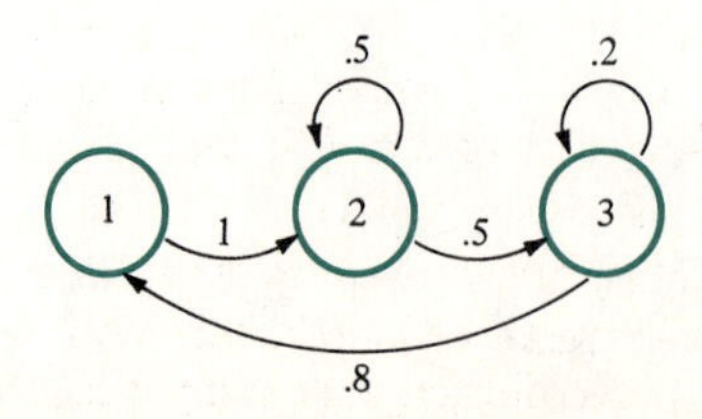

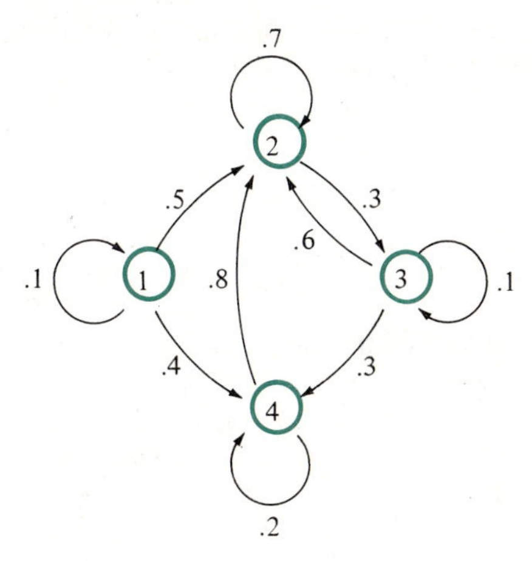

FIGURE 8.7

9. A Markov chain has the transition matrix **P** shown below. Find the transition diagram for this Markov chain.

$$\mathbf{P} = \begin{bmatrix} .4 & 0 & .6 \\ 1 & 0 & 0 \\ 0 & 1 & 0 \end{bmatrix}$$

10. A Markov chain has the transition matrix **P** shown below. Find the transition diagram for this Markov chain.

$$\mathbf{P} = \begin{bmatrix} .3 & .2 & .5 \\ .4 & 0 & .6 \\ 1 & 0 & 0 \end{bmatrix}$$

11. A Markov chain has the transition matrix given in Exercise 10. On the first observation it is in state 1.
    (*a*) Find the probability that it is in state 1 on the next observation.
    (*b*) What state is it most likely to occupy on the next observation?
12. A Markov chain has the transition matrix given in Exercise 10. On the first observation it is in state 1. Find the probability that it is in state 1 on the third observation. (*Hint*: Use a tree diagram.)
13. A Markov chain has the transition matrix given in Exercise 9. On the first observation it is in state 1. In what state is it most likely to be on the third observation?
14. A Markov chain has the transition matrix given in Exercise 10. If it begins in state 2, find the probability that on the next four observations it successively occupies states 3, 1, 2, and 1 (in that order).

Business

15. Profits at a brokerage firm are determined by the volume of securities sold, and this volume fluctuates from week to week. Each week volume is classified as high or low, and information collected over many weeks leads to the following assumptions:

i) If volume is high this week, then next week it will be high with probability .7 and low with probability .3.

ii) If volume is low this week, then next week it will be high or low with equal probability.

Formulate a Markov chain model for this situation; i.e., find the states and the transition matrix.

Business 16. In the situation described in Exercise 15, suppose that the volume is high this week.

(*a*) Find the probability that the volume will be high two weeks from now.

(*b*) Find the probability that volume will be high for three consecutive weeks.

Ecology 17. A small animal lives in a territory that can be divided into areas described as meadow and woods, and it moves randomly from one area to another. If the animal is in the woods on one observation, then it is twice as likely to be in the woods as the meadow on the next observation. Likewise, if the animal is in the meadow on one observation, then it is twice as likely to be in the meadow as the woods on the next observation.

(*a*) Identify appropriate states.

(*b*) Find the transition diagram.

(*c*) Find the transition matrix.

Ecology 18. In the situation described in Exercise 17, suppose the animal is initially in the woods, and find the probability that it is in the woods on the next three observations. If it is initially in the woods, what is the probability that it is in the meadow on the next three observations?

Demographics 19. People living in a small town are classified according to employment into three groups: employed in industry, employed in small business, and self-employed. Data on employment are collected for several years, and the results are summarized in Table 8.2. Formulate a Markov chain model for this situation.

(*a*) Identify appropriate states.

(*b*) Find the transition diagram.

(*c*) Find the transition matrix.

TABLE 8.2

Employment Last Year	Employment This Year	Percentage
Industry	Industry	70
	Small business	20
	Self-employed	10
Small business	Industry	30
	Small business	50
	Self-employed	20
Self-employed	Industry	30
	Small business	30
	Self-employed	40

Sports

20. A football player is practicing field goal kicks. He finds that if he makes a kick, then he makes the next kick 85 percent of the time, but if he misses a kick, then he makes the next kick only 60 percent of the time.
    (*a*) Identify appropriate states, and formulate this situation as a Markov chain.
    (*b*) Find the transition diagram for the Markov chain formulated in (*a*).
    (*c*) Find the transition matrix for the Markov chain formulated in (*a*).

Games of chance

21. A game is played on the board shown in Figure 8.8 where the arrows indicate "forward" motion. A marker is moved around the board according to the following rules. A fair die is rolled at each play of the game, and:
    (*a*) If the marker is in location red, then it moves forward one step if the result on the die is even and it moves backward one step if the result is odd.
    (*b*) If the marker is in any other location, then it moves forward two steps if the result on the die is a 6 and one step if the result on the die is a 4 or 5. It does not move at all if the result on the die is anything else.

    Formulate this situation as a Markov chain, and find the transition matrix.

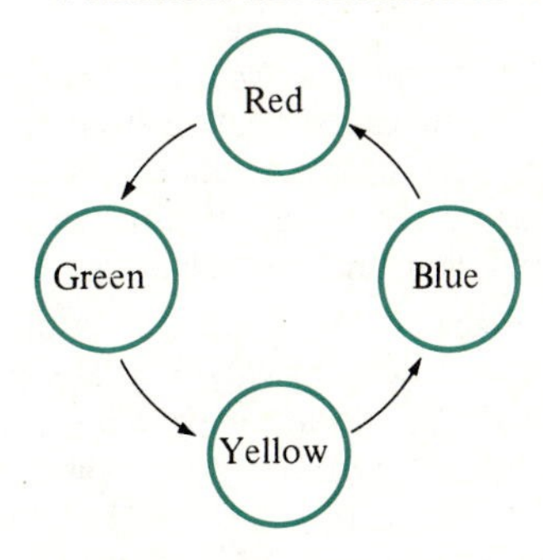

FIGURE 8.8

College life

22. A not-so-enthusiastic student often misses class on Friday afternoon. If he attends class on a certain Friday, then the next Friday he is twice as likely to be absent as to attend. On the other hand, if he misses on a certain Friday, then the next week he is 3 times as likely to attend as to miss class again. Formulate this situation as a Markov chain, and find the transition matrix.

Consumer choices

23. A student eats in a restaurant every Sunday evening. Each week she chooses among Chinese, Greek, and Italian food, and she never eats the same kind of food for 2 consecutive weeks. If she eats at a Chinese restaurant one week, then she is twice as likely to have Greek as Italian food the next week. If she has Greek food one week, then she is equally likely to have Chinese and Italian food the next week. Finally, if she has Italian food one week, then she is 5 times as likely to have Chinese as Greek food the next week. Formulate this situation as a Markov chain, and find the transition matrix.

Consumer choices

24. A consumer purchases an automobile every 2 years. She is twice as likely to purchase one from the maker of her present car as from the most likely alternative. If she makes a change from a GM car, then she is equally likely to buy a Ford, Chrysler, Honda, or VW product. If she makes a change from a non-GM car, then she is twice as likely to buy a GM car as one from another builder. Formulate this situation as a Markov chain, and find the transition matrix.

Sports 25. A nervous basketball player finds that her success at shooting a free throw depends on what happened when she last shot a free throw. If she made the last one, her probability of making the next one is .8, while if she missed the last one, her probability of making the next one is .4. Formulate this situation as a Markov chain, and find the transition matrix.

College life 26. As the semester progresses, the not-so-enthusiastic student (see Exercise 22) acquires another habit: attending class and reading a newspaper. Suppose that he attends class and pays attention, attends class and reads a newspaper, or misses class. If he attends and pays attention one Friday, then the next Friday he is equally likely to attend and pay attention, to attend and read a newspaper, and to miss class. If he attends and reads a newspaper one Friday, then the next Friday he is twice as likely to attend and pay attention as to miss, and he never attends and reads a newspaper two consecutive weeks. If he misses one week, then he attends the next week and he is 3 times as likely to pay attention as to read a newspaper. Formulate this situation as a Markov chain, and find the transition matrix.

Investments 27. A stock broker believes that the stock market can be described as a Markov chain. Each day she evaluates the relative strength of stocks and bonds, and she decides whether stocks are stronger than bonds, bonds are stronger than stocks, or they are equally strong. The data are summarized in Table 8.3. Formulate this situation as a Markov chain, and find the transition matrix.

Product performance 28. A soft drink vending machine dispenses cola drinks, lemon-lime drinks, and diet drinks in a relatively random way. If it last produced a cola drink, then it next produces a cola drink half the time and a lemon-lime drink half the time. If it last produced a lemon-lime drink, it next produces a lemon-lime drink two-thirds of the time and a diet drink one-third of the time. And finally, if it last produced a diet drink, it next produces a diet drink three-fourths of the time and a cola drink one-fourth of the time. Formulate this situation as a Markov chain, and find the transition matrix.

TABLE 8.3

Relative Strength Last Week	Relative Strength This Week	Percentage of Weeks
Stocks stronger	Stocks stronger	60
	Bonds stronger	30
	Equally strong	10
Bonds stronger	Stocks stronger	30
	Bonds stronger	50
	Equally strong	20
Equally strong	Stocks stronger	40
	Bonds stronger	40
	Equally strong	20

College life 29. Using the setting of Exercise 22, suppose that the student's attendance on Friday afternoon is described by a Markov chain with transition matrix

$$\begin{array}{c c} & \text{Attends} \quad \text{Misses} \\ \begin{array}{r}\text{Attends}\\ \text{Misses}\end{array} & \begin{bmatrix} 0 & 1 \\ .5 & .5 \end{bmatrix}\end{array}$$

Also suppose that the student attends class the first Friday of the semester. Find the expected number of times the student attends class on Friday afternoon during the first 4 weeks of the semester.

Sports 30. Suppose that the basketball player of Exercise 25 has four free throws every game and she always makes the first. What is the expected number of free throws made each game?

8.2 BASIC PROPERTIES OF MARKOV CHAINS

We have seen that information about transitions in a Markov chain—which transitions are possible and their probabilities—can be given in a transition diagram or a transition matrix. For computational purposes, the transition matrix is often the most useful. The transition matrix for a Markov chain with N states, introduced in Section 8.1, is an $N \times N$ matrix whose (i, j) entry is the probability of a transition from state i to state j in one step. There are corresponding probabilities for transitions from one state to another in k steps; these are usually called *k-step transition probabilities*.

The conditional probability of making a transition from state i to state j in exactly k steps is denoted by $p_{ij}(k)$. The matrix whose (i, j) entry is $p_{ij}(k)$ is denoted by $\mathbf{P}(k)$ and will be called the *k-step transition matrix for a Markov chain.*

Example 8.5 The transition matrix for a two-state Markov chain is

$$\mathbf{P} = \begin{bmatrix} .8 & .2 \\ .6 & .4 \end{bmatrix}$$

Problem Use tree diagrams to find the two-step transition matrix $\mathbf{P}(2)$.

Solution To determine the first row of $\mathbf{P}(2)$, we use a tree diagram which represents a two-stage experiment in which the system is initially in state 1. Such a tree diagram is shown in Figure 8.9*a*. From this tree diagram we see that a transition from state 1 to state 1 in two steps can occur in either of two ways, and the probability $p_{11}(2)$ is the sum

$$p_{11}(2) = .64 + .12 = .76$$

Similarly,

$$p_{12}(2) = .16 + .08 = .24$$

To determine the second row of $\mathbf{P}(2)$, we use a tree diagram in which the system is initially in state 2, Figure 8.9*b* in this case. We have

$$p_{21}(2) = .48 + .24 = .72 \qquad p_{22}(2) = .12 + .16 = .28$$

FIGURE 8.9

First Observation Second Observation Third Observation Probability

1 → .8 → 1 → .8 → 1: .64
1 → .8 → 1 → .2 → 2: .16
1 → .2 → 2 → .6 → 1: .12
1 → .2 → 2 → .4 → 2: .08

(*a*)

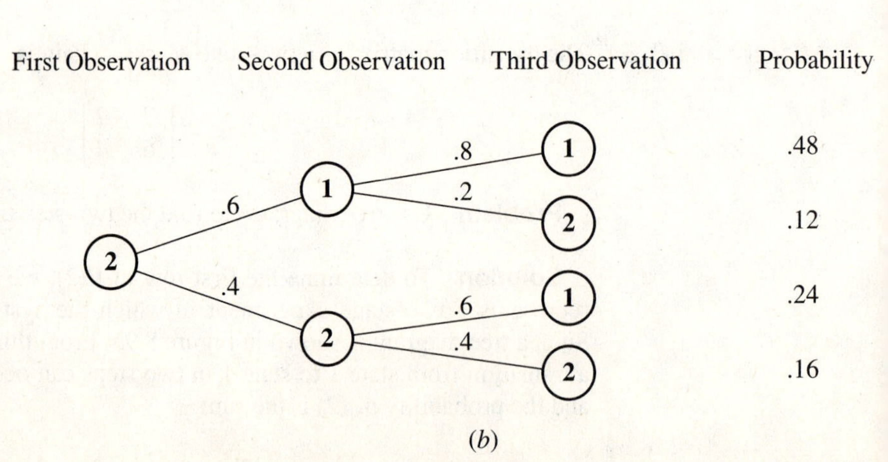

(*b*)

Therefore,

$$\mathbf{P}(2) = \begin{bmatrix} p_{11}(2) & p_{12}(2) \\ p_{21}(2) & p_{22}(2) \end{bmatrix} = \begin{bmatrix} .76 & .24 \\ .72 & .28 \end{bmatrix}$$ ■

The technique illustrated in Example 8.5 can be used to construct **P**(2) for any Markov chain for which the transition matrix **P** can be determined. However, for large matrices the process is cumbersome, and one of the very useful properties of Markov chains is that there is a simple method of finding **P**(2) from **P** without using tree diagrams. The idea behind the method can be seen by looking more carefully at the calculation of $p_{12}(2)$ in Example 8.5. Using the tree diagram, we found

$$p_{12}(2) = .8(.2) + .2(.4) = p_{11}p_{12} + p_{12}p_{22}$$

This last expression is exactly the (1,2) entry in the matrix product **PP**.

Now let us look at a similar argument in the case of a Markov chain with N states and transition matrix **P**. We find an expression for $p_{ij}(2)$: the probability that if the system is in state i on one observation, then it is in state j on the second subsequent observation. The system must be in some state on the intervening observation, and consequently we have the situation depicted in Figure 8.10. The transition probabilities between the various states are shown on the lines connecting those states. The system can move from state i to state j in two steps by moving from i to 1 to j. This happens with probability $p_{i1}p_{1j}$. Recall that the probability that the system makes a transition from state 1 to state j in one step is independent of the states it occupied before state 1. Likewise, the system can

FIGURE 8.10

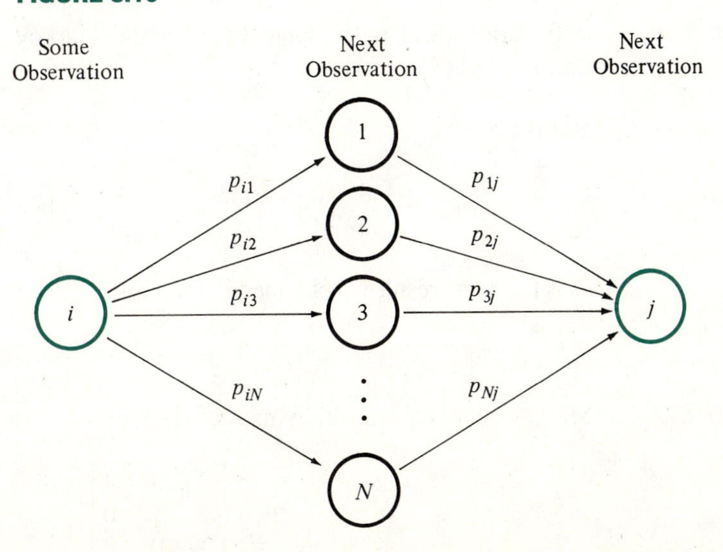

move from state i to state j through any of states 2, 3, . . . , N. These events happen with probabilities $p_{i2}p_{2j}, p_{i3}p_{3j}, \ldots, p_{iN}p_{Nj}$, respectively. Since the system must move from state i to state j through exactly one intermediate state, we have

$$p_{ij}(2) = p_{i1}p_{1j} + p_{i2}p_{2j} + p_{i3}p_{3j} + \cdots + p_{iN}p_{Nj}$$

This expression for $p_{ij}(2)$ is exactly the (i, j) entry in the matrix product **PP**. We have the result

$$\mathbf{P}(2) = \mathbf{PP} = \mathbf{P}^2$$

This is a special case of the following more general result.

Let **P** be the (one-step) transition matrix for a Markov chain. Then the matrix $\mathbf{P}(k)$ of k-step transition probabilities is

$$\mathbf{P}(k) = \underbrace{\mathbf{PP}\cdots\mathbf{P}}_{k \text{ factors}} = \mathbf{P}^k \qquad (8.1)$$

This result can be justified by using the special case $k = 2$, $\mathbf{P}(2) = \mathbf{P}^2$ (which was verified above), and mathematical induction.

Example 8.6 Use Equation (8.1) to compute $\mathbf{P}(2)$ for the Markov chain whose transition matrix is matrix **P** of Example 8.5.

Solution

$$\mathbf{P}(2) = \begin{bmatrix} .8 & .2 \\ .6 & .4 \end{bmatrix}^2 = \begin{bmatrix} .8 & .2 \\ .6 & .4 \end{bmatrix}\begin{bmatrix} .8 & .2 \\ .6 & .4 \end{bmatrix} = \begin{bmatrix} .76 & .24 \\ .72 & .28 \end{bmatrix}$$

This is the same result we obtained in Example 8.5 (as it must be). ■

Example 8.7 A Markov chain has transition matrix

$$\mathbf{P} = \begin{bmatrix} .3 & .3 & .4 \\ .5 & .5 & 0 \\ 1 & 0 & 0 \end{bmatrix}$$

Problem (*a*) Find the two-step transition matrix **P**(2).
(*b*) If the system is initially observed in state 1, what is the probability that it is in state 1 two observations later?

Solution (*a*) Applying Equation (8.1), we have

$$\mathbf{P}(2) = \mathbf{P}^2 = \begin{bmatrix} .3 & .3 & .4 \\ .5 & .5 & 0 \\ 1 & 0 & 0 \end{bmatrix}^2 = \begin{bmatrix} .64 & .24 & .12 \\ .4 & .4 & .2 \\ .3 & .3 & .4 \end{bmatrix}$$

(*b*) We are asked to find $p_{11}(2)$. This is the (1, 1) entry of **P**(2), and consequently the answer is .64. ■

Example 8.8 The transition diagram for a Markov chain is shown in Figure 8.11.

Problem Find the matrix of three-step transition probabilities for this Markov chain.

Solution Using Figure 8.11, we find that the transition matrix for this Markov chain is

$$\mathbf{P} = \begin{bmatrix} \frac{1}{4} & \frac{3}{4} & 0 \\ 0 & 0 & 1 \\ \frac{1}{2} & 0 & \frac{1}{2} \end{bmatrix}$$

It follows from Equation (8.1) that

FIGURE 8.11

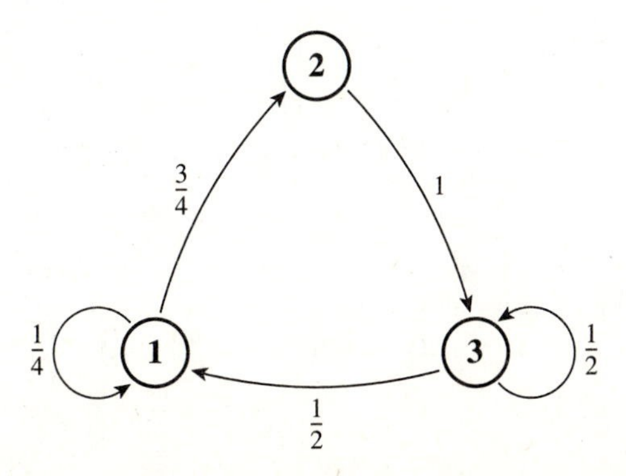

$$\mathbf{P}(3) = \mathbf{P}^3 = \mathbf{P}\mathbf{P}^2$$
$$= \begin{bmatrix} \frac{1}{4} & \frac{3}{4} & 0 \\ 0 & 0 & 1 \\ \frac{1}{2} & 0 & \frac{1}{2} \end{bmatrix} \begin{bmatrix} \frac{1}{16} & \frac{3}{16} & \frac{3}{4} \\ \frac{1}{2} & 0 & \frac{1}{2} \\ \frac{3}{8} & \frac{3}{8} & \frac{1}{4} \end{bmatrix}$$
$$= \begin{bmatrix} \frac{25}{64} & \frac{3}{64} & \frac{36}{64} \\ \frac{3}{8} & \frac{3}{8} & \frac{1}{4} \\ \frac{14}{64} & \frac{18}{64} & \frac{1}{2} \end{bmatrix}$$

■

In Example 8.8, notice that the 0 in the (2, 2) entry of **P** means that it is impossible to make a direct transition from state 2 to state 2. Likewise, the 0 in the (2, 2) entry of $\mathbf{P}^2$ means that it is impossible to make a transition from state 2 to state 2 in two steps. Since the (2, 2) entry of $\mathbf{P}^3$ is not zero (it is equal to $\frac{3}{8}$), it is possible to make a transition from state 2 to state 2 in three steps. In fact, beginning in state 2, there is a positive probability of making each of the transitions $2 \to 3$, $3 \to 1$, and $1 \to 2$.

Example 8.9 (Continuation of Examples 8.1 and 8.2) Suppose that the consultant is unemployed this week.

Employment

Problem Find the probability that she will be unemployed 4 weeks from now.

Solution Adopting the notation introduced in Example 8.2, that is, labeling states E and U as 1 and 2, respectively, we have the transition matrix **P**:

$$\mathbf{P} = \begin{bmatrix} .8 & .2 \\ .6 & .4 \end{bmatrix}$$

We seek $p_{22}(4)$. Using Equation (8.1), we have $\mathbf{P}(4) = \mathbf{P}^4$. Also, $\mathbf{P}^4 = \mathbf{P}^2\mathbf{P}^2$, so first we find $\mathbf{P}^2$.

$$\mathbf{P}^2 = \begin{bmatrix} .8 & .2 \\ .6 & .4 \end{bmatrix} \begin{bmatrix} .8 & .2 \\ .6 & .4 \end{bmatrix} = \begin{bmatrix} .76 & .24 \\ .72 & .28 \end{bmatrix}$$

Since we are interested in the (2, 2) entry of **P**(4), we need to compute only that entry and not the entire matrix **P**(4). That is, we need only to evaluate $[.72 \quad .28] \begin{bmatrix} .24 \\ .28 \end{bmatrix}$. We have

$$p_{22}(4) = .72(.24) + .28(.28) = .2512$$

Therefore, if the consultant is unemployed this week, then the conditional probability that she will be unemployed 4 weeks from now is .2512. ■

The entries $p_{i1}, p_{i2}, \ldots, p_{iN}$ in the ith row of the transition matrix **P** are the probabilities that the system moves from state i to states $1, 2, \ldots N$, respectively, in one step. Now, since at each step the system must first move from state i to some state (perhaps to i itself), it follows that the sum of the entries in the ith row of matrix **P** is 1. Vectors such as the ith row of **P** have a special name.

A *probability vector* is a vector with nonnegative coordinates for which the sum of the coordinates is 1.

Using this terminology, we have the following fact.

Each row of a transition matrix **P** of a Markov chain is a probability vector.

Likewise, for each k the rows of matrix $\mathbf{P}(k)$ are probability vectors. Indeed, the entries of the ith row of $\mathbf{P}(k)$ are the probabilities $p_{i1}(k), p_{i2}(k), \ldots, p_{iN}(k)$, and the system must move from state i to some state (perhaps i itself) in k steps. It is easily seen that each row of matrix **P** of Example 8.7 is a probability vector, that each row of matrices $\mathbf{P}(2)$ of Examples 8.6 and 8.7 is a probability vector, and that each row of matrix $\mathbf{P}(3)$ of Example 8.8 is a probability vector.

Example 8.10

Psychology

Many experiments concerned with the behavior of animal subjects have been modeled by using Markov chains. We consider a very simple example of such an experiment and the behavior of a subject, say a mouse, in the setting shown in Figure 8.12. The physical apparatus consists of four compartments: a nest, a compartment containing food, a compartment containing water, and a compartment containing an exercise device known as a squirrel cage. An experiment is designed to study the mouse's moves among the compartments. The mouse is initially released in one compartment, and its movements after that time are observed.

Suppose that the behavior of the mouse can be described by a Markov chain. That is, suppose that the behavior of the mouse depends only on the compartment it currently occupies and not on where the mouse was earlier. Transitions can occur in two ways: the mouse can move to a neighboring compartment, and the mouse can stay in the same compartment for 2 minutes. In the latter case, we say that a transition into the same compartment has taken place.

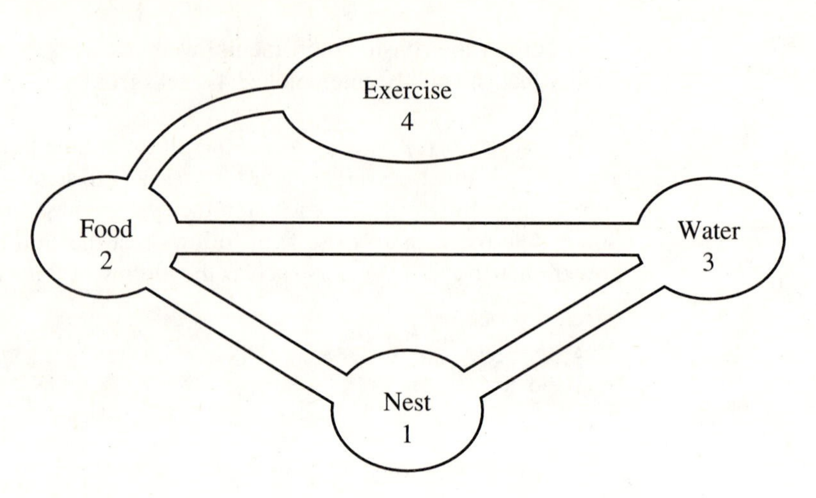

FIGURE 8.12

Problem (*a*) Find the transition matrix under the assumptions that the mouse is twice as likely to remain in the same compartment as to move and that if it moves, it is equally likely to make any of the possible moves.

(*b*) Under the assumptions of part *a*, find the probability that the mouse is in the nest after two transitions given that it began in compartment 2, the compartment with food.

Solution (*a*) Label the states as shown in Figure 8.12. Since the mouse is observed each time it moves from one compartment to another, only transitions from a compartment to itself or to an immediately adjacent compartment have nonzero probability. Consider the system when the mouse is in state 3 (the compartment with food). The mouse is twice as likely to remain in compartment 3 as to move, and if it does move, then the mouse is equally likely to move to compartments 1 and 2. Thus, using the techniques of Chapter 2, we see that the probabilities of transitions from state 3 to states 1, 2, 3, and 4 are $\frac{1}{6}$, $\frac{1}{6}$, $\frac{2}{3}$, and 0, respectively. The remaining transition probabilities are determined similarly, and the transition matrix is:

$$\mathbf{P} = \begin{bmatrix} \frac{2}{3} & \frac{1}{6} & \frac{1}{6} & 0 \\ \frac{1}{9} & \frac{2}{3} & \frac{1}{9} & \frac{1}{9} \\ \frac{1}{6} & \frac{1}{6} & \frac{2}{3} & 0 \\ 0 & \frac{1}{3} & 0 & \frac{2}{3} \end{bmatrix}$$

(*b*) If the mouse is released in compartment 2, then the probability that it is in compartment 1 after two transitions is the (2, 1) entry of the two-step transition matrix $\mathbf{P}(2) = \mathbf{P}^2$. The desired probability is $p_{21}(2) = \frac{1}{6}$. ■

Exercises for Section 8.2

1. A Markov chain has the transition matrix

$$\mathbf{P} = \begin{bmatrix} .5 & .5 \\ 1 & 0 \end{bmatrix}$$

(*a*) Find the two-step transition matrix $\mathbf{P}(2)$.
(*b*) Find $p_{12}(2)$ and $p_{21}(2)$.

2. A Markov chain has the transition matrix

$$\mathbf{P} = \begin{bmatrix} 0 & 1 \\ .2 & .8 \end{bmatrix}$$

(*a*) Find the two-step transition matrix $\mathbf{P}(2)$.
(*b*) Find $p_{11}(2)$ and $p_{22}(2)$.

3. For the transition matrix $\mathbf{P}$ of Exercise 1, find $\mathbf{P}(3)$ and $\mathbf{P}(4)$.
4. A Markov chain has the transition matrix

$$\mathbf{P} = \begin{bmatrix} .6 & .4 \\ .4 & .6 \end{bmatrix}$$

(*a*) Find the two-step transition matrix $\mathbf{P}(2)$.
(*b*) Find the three-step transition matrix $\mathbf{P}(3)$.

5. Find the three-step transition matrix for the Markov chain of Example 8.5.
6. Find the three-step transition probability $p_{13}(3)$ for the Markov chain of Example 8.8.
7. A Markov chain has the transition matrix

$$\mathbf{P} = \begin{bmatrix} 0 & 1 & 0 \\ 0 & 0 & 1 \\ 1 & 0 & 0 \end{bmatrix}$$

(*a*) Find the two-step transition matrix $\mathbf{P}(2)$.
(*b*) Find the three-step transition matrix $\mathbf{P}(3)$.

8. Find the three-step transition probability $p_{32}(3)$ for the Markov chain of Example 8.7.
9. A Markov chain has the transition matrix

$$\mathbf{P} = \begin{bmatrix} 0 & .8 & .2 \\ 0 & 1 & 0 \\ .5 & 0 & .5 \end{bmatrix}$$

(*a*) Find $\mathbf{P}(2)$, $\mathbf{P}(3)$, and $\mathbf{P}(4)$.
(*b*) Find $p_{22}(2)$, $p_{22}(3)$, and $p_{22}(4)$.
(*c*) What can you say about $p_{22}(k)$ for any positive integer k?

10. Let $\mathbf{P}$ be the transition matrix of Exercise 9.
    (*a*) Show that the rows of $\mathbf{P}(2)$, $\mathbf{P}(3)$, and $\mathbf{P}(4)$ are probability vectors.
    (*b*) What can you say about the relative sizes of $p_{12}(2)$, $p_{12}(3)$, and $p_{12}(4)$?
    (*c*) What can you say about the relative sizes of $p_{32}(2)$, $p_{32}(3)$, and $p_{32}(4)$?

11. Let $\mathbf{P}$ be the transition matrix for a Markov chain where

$$\mathbf{P} = \begin{bmatrix} 0 & 1 & 0 \\ 0 & 0 & 1 \\ .5 & 0 & .5 \end{bmatrix}$$

    (*a*) Suppose the system is in state 2 on an initial observation. Draw a tree diagram for a sequential experiment consisting of three observations of the system, including the initial observation.
    (*b*) Determine the second row of matrix $\mathbf{P}(2)$, using the tree diagram.
    (*c*) Determine matrix $\mathbf{P}(2)$, using formula (8.1).

12. Let $\mathbf{P}$ (given below) be the transition matrix for a Markov chain, and suppose that the system is initially in state 2.
    (*a*) Draw the tree diagram for a sequential experiment consisting of three observations of the system, including the initial observation.
    (*b*) Determine the second row of matrix $\mathbf{P}(2)$, using the tree diagram.
    (*c*) Determine matrix $\mathbf{P}(2)$, using formula (8.1).

$$\mathbf{P} = \begin{bmatrix} \frac{1}{2} & \frac{1}{2} & 0 \\ \frac{1}{3} & 0 & \frac{2}{3} \\ \frac{1}{4} & \frac{1}{2} & \frac{1}{4} \end{bmatrix}$$

13. Determine $\mathbf{P}(2)$ and $\mathbf{P}(3)$ for a Markov chain with transition matrix

$$\mathbf{P} = \begin{bmatrix} .2 & .2 & .6 \\ .4 & .2 & .4 \\ .5 & 0 & .5 \end{bmatrix}$$

    Verify that the rows of $\mathbf{P}(2)$ and $\mathbf{P}(3)$ are probability vectors.

Business

14. Profits at a brokerage firm are determined by the volume of securities sold, and this volume fluctuates from week to week (see Exercise 15, Section 8.1): If volume is high this week, then next week it will be high with probability .7 and low with probability .3. If volume is low this week, then next week it will be high or low with equal probability.
    (*a*) If the volume is high this week, find the probability that it will be high 4 weeks from now.
    (*b*) If the volume is high this week, use the answer to part *a* to determine the probability that the volume will be low 4 weeks from now.

Sports

15. Suppose that a basketball player's success in free-throw shooting can be described with a Markov chain with transition matrix

$$\mathbf{P} = \begin{array}{c} \\ \text{Make} \\ \text{Miss} \end{array} \begin{array}{c} \begin{array}{cc} \text{Make} & \text{Miss} \end{array} \\ \begin{bmatrix} .8 & .2 \\ .4 & .6 \end{bmatrix} \end{array}$$

That is, if the player makes a free throw, then he will make the next one with probability .8 and he will miss it with probability .2. Also, if the player misses a free throw, then he will make the next one with probability .4 and miss it with probability .6.

(*a*) If the player makes his first free throw, what is the probability that he also makes the third one?

(*b*) If the player misses his first free throw, what is the probability that he makes the third?

College life

16. A not-so-enthusiastic student tends to miss class on Friday afternoons. She is always in one of the two states, absent or present, and transitions occur according to matrix **P**, where

$$\mathbf{P} = \begin{array}{c} \\ \text{Absent} \\ \text{Present} \end{array} \begin{array}{c} \begin{array}{cc} \text{Absent} & \text{Present} \end{array} \\ \begin{bmatrix} .3 & .7 \\ .6 & .4 \end{bmatrix} \end{array}$$

If the student is absent on a given Friday afternoon, find the probability she will be present 3 weeks later.

Nutrition

17. An overweight student is always in one of the three states: fast, balanced diet, sweet binge. Transitions between these stages occur weekly according to the probabilities shown in matrix **P**:

$$\mathbf{P} = \begin{array}{r} \\ \text{Fast} \\ \text{Balanced diet} \\ \text{Sweet binge} \end{array} \begin{array}{c} \begin{array}{ccc} \text{Fast} & \text{Balanced diet} & \text{Sweet binge} \end{array} \\ \begin{bmatrix} .2 & .2 & .6 \\ .3 & .4 & .3 \\ .5 & .4 & .1 \end{bmatrix} \end{array}$$

Find the two-step transition matrix **P**(2) and the transition probability of going from sweet binge to sweet binge in three steps.

Ecology

18. A small animal lives in a territory that can be divided into two areas described as meadow and woods, and it moves randomly from one area to another (see Exercise 17, Section 8.1). If it is in the woods on one observation, then it is twice as likely to be in the woods as the meadow on the next observation. Likewise, if it is in the meadow on one observation, then it is twice as likely to be in the meadow as the woods on the next observation. If the animal is in the woods on the first observation, find the probability that it is in the woods on the fourth observation.

Scheduling

19. Consider the situation described in Example 8.4. If the service specialist is currently in the east district, what is the probability that after two more calls he will be in the west district? In what district is he most likely to be?

Scheduling

20. Consider the situation described in Example 8.4. Beginning in what district is it most likely that the service specialist will be in the east district on the second subsequent call?

Demographics

21. Employment shifts in a certain city follow the pattern given in Table 8.4. Formulate a Markov chain model for these shifts.
    (*a*) Find the two-step transition matrix for this Markov chain.
    (*b*) Find the probability that an individual employed in industry at one time is employed in small business 2 years later.

Demographics

22. (Continuation of Exercise 21) If an individual is self-employed this year, what is her most likely employment status 2 years from now?

Demographics

23. (Continuation of Exercise 22) A person would like to be self-employed 2 years from now. What current employment status gives her the greatest likelihood of achieving her goal?

Consumer choices

24. A wealthy computer scientist buys a new car every year. It is always a Cadillac, Lexus, or Mercedes. She is 3 times as likely to buy a different model from the one she last purchased as she is to buy the same model, and she is equally likely to buy each of the other two models. Formulate this situation as a Markov chain, and find the transition matrix. If she buys a Cadillac in 1994, what is the probability that she buys another Cadillac in 1998?

Consumer choices

25. (Continuation of Exercise 24) If she buys a Cadillac in 1994, what model is she most likely to buy in 1998?

26. A Markov chain has the transition matrix **P** shown below. If the system begins in state 4, find the probability that it is in state 4 after four steps.

$$\mathbf{P} = \begin{bmatrix} 0 & 1 & 0 & 0 \\ 0 & 0 & 1 & 0 \\ 0 & 0 & 0 & 1 \\ .5 & 0 & 0 & .5 \end{bmatrix}$$

TABLE 8.4

Employment Last Year	Employment This Year	Percentage
Industry	Industry	90
	Small business	5
	Self-employed	5
Small business	Industry	10
	Small business	80
	Self-employed	10
Self-employed	Industry	10
	Small business	30
	Self-employed	60

27. Using the Markov chain of Exercise 26, suppose the system begins in state 4. In what state is it most likely to be after four steps?

28. A Markov chain has the transition matrix

$$\mathbf{P} = \begin{bmatrix} 0 & 1 & 0 & 0 & 0 \\ 0 & 0 & 1 & 0 & 0 \\ 0 & 0 & 0 & 1 & 0 \\ 0 & 0 & 0 & 0 & 1 \\ .5 & 0 & 0 & 0 & .5 \end{bmatrix}$$

What is the smallest positive integer k for which it is possible to go between any two states in k steps?

29. A Markov chain has the transition matrix

$$\mathbf{P} = \begin{bmatrix} .4 & .6 & 0 & 0 \\ .4 & .4 & .2 & 0 \\ 0 & 0 & 0 & 1 \\ 0 & 0 & .5 & .5 \end{bmatrix}$$

(*a*) Compute $\mathbf{P}(2)$, $\mathbf{P}(4)$, and $\mathbf{P}(8)$.
(*b*) Is it possible to go from state 3 to state 1 in 2, 4, or 8 steps?
(*c*) Is it possible to go from state 3 to state 1 in any number of steps? (*Hint*: Draw a transition diagram for this Markov chain.)

30. A Markov chain has the transition matrix $\mathbf{P}$ shown below. Without computing the powers of the matrix, give an argument to show that no power of matrix $\mathbf{P}$ has all positive entries.

$$\mathbf{P} = \begin{bmatrix} 0 & 1 & 0 & 0 \\ 0 & 0 & 1 & 0 \\ 0 & 0 & 0 & 1 \\ 1 & 0 & 0 & 0 \end{bmatrix}$$

8.3 REGULAR MARKOV CHAINS

In the preceding sections we introduced the basic property of Markov chains, and we developed the formula $\mathbf{P}(k) = \mathbf{P}^k$, which shows that the k-step transition matrix equals the kth power of the 1-step transition matrix. In this section and the next, we continue by considering briefly two special types of Markov chains which are especially useful in applications, and we give such an application in the final section of the chapter.

Since Markov chains are stochastic processes, we do not ordinarily know what will happen at each stage, and we must describe the system in terms of probabilities.

State Vector

Consider a Markov chain with N states. A state vector for the Markov chain is a probability N vector $\mathbf{X} = [x_1 \quad x_2 \quad \cdots \quad x_N]$. The ith coordinate x_i of the state vector $\mathbf{X}$ is to be interpreted as the probability that the system is in state i. We write a state vector as a row vector, i.e., as a $1 \times N$ matrix.

Example 8.11

Scheduling

The Markov chain model for Metroburg Heating and Air Conditioning Company formulated in Example 8.4 has three states. In this situation the state vector $[.7 \quad 0 \quad .3]$ is to be interpreted as follows. With probability .7 the current call is to a location in the east district; the current call is not in the central district; and with probability .3 the current call is to a location in the west district. ■

If the system is known to be in a specific state, the state vector has a particularly simple form. If the system is in the ith state, the ith coordinate of the state vector is 1 and all other coordinates are 0. For example, if in the Markov chain defined in Example 8.4 the system is known to be in state 1, the current call is to a location in the east district, and the associated state vector is $[1 \quad 0 \quad 0]$.

The behavior of a Markov chain (i.e., the states it can enter and how likely it is to enter them) can be described with a sequence of state vectors. The initial state of the system can be described with a state vector which we denote by $\mathbf{X}_0$. After one transition, the system can again be described with a state vector which we call $\mathbf{X}_1$. After two transitions it can be described by another state vector which we call $\mathbf{X}_2$. In general, after k transitions the system can be described by a state vector which we call $\mathbf{X}_k$. The relation between these state vectors can be represented with vector-matrix multiplication. We summarize the facts in a theorem.

Theorem on State Vectors for a Markov Chain

If $\mathbf{X}_k$ and $\mathbf{X}_{k+1}$ denote the state vectors which describe a Markov chain after k and $k + 1$ transitions, respectively, then $\mathbf{X}_{k+1} = \mathbf{X}_k\mathbf{P}$, where $\mathbf{P}$ is the transition matrix of the chain. In particular, $\mathbf{X}_1 = \mathbf{X}_0\mathbf{P}$, $\mathbf{X}_2 = \mathbf{X}_1\mathbf{P} = \mathbf{X}_0\mathbf{P}^2, \ldots,$ and in general $\mathbf{X}_k = \mathbf{X}_0\mathbf{P}^k$. That is, the state vector $\mathbf{X}_k$ which describes the system after k transitions is the product of the initial state vector and the kth power of the transition matrix.

The theorem can be verified by examining each coordinate of the product $\mathbf{X}_k\mathbf{P}$ and using the meaning of $\mathbf{X}_k$ and $\mathbf{P}$. The conclusion is that the jth coordinate of $\mathbf{X}_k\mathbf{P}$ is exactly the probability that the system is in state j after $k + 1$ transitions, i.e., the jth coordinate of $\mathbf{X}_{k+1}$. Note that if the chain has N states, so that $\mathbf{X}_k$ is an N-vector and $\mathbf{P}$ is an $N \times N$ matrix, then $\mathbf{X}_k\mathbf{P}$ is also an N-vector.

Example 8.12 The Markov chain model for Metroburg Heating has the transition matrix (Example 8.4)

Scheduling

$$\mathbf{P} = \begin{bmatrix} .5 & .4 & .1 \\ .1 & .6 & .3 \\ .3 & .6 & .1 \end{bmatrix}$$

Problem If the initial state vector is $\mathbf{X}_0 = [.7 \quad .0 \quad .3]$, find the state vector after one transition.

Solution Using the theorem, we find that the state vector after one transition is

$$\begin{aligned} \mathbf{X}_1 = \mathbf{X}_0\mathbf{P} &= [.7 \quad 0 \quad .3] \begin{bmatrix} .5 & .4 & .1 \\ .1 & .6 & .3 \\ .3 & .6 & .1 \end{bmatrix} \\ &= [.44 \quad .46 \quad .10] \end{aligned}$$

■

Example 8.13 A Markov chain has the transition matrix

$$\mathbf{P} = \begin{bmatrix} .5 & .5 \\ .8 & .2 \end{bmatrix}$$

Problem If the system begins in state 2, find the state vector after two transitions.

Solution The initial state vector is $[0 \quad 1]$. According to the theorem, the state vector after two transitions is

$$\begin{aligned} \mathbf{X}_2 = \mathbf{X}_0\mathbf{P}^2 &= [0 \quad 1] \begin{bmatrix} .5 & .5 \\ .8 & .2 \end{bmatrix} \begin{bmatrix} .5 & .5 \\ .8 & .2 \end{bmatrix} \\ &= [0 \quad 1] \begin{bmatrix} .65 & .35 \\ .56 & .44 \end{bmatrix} \\ &= [.56 \quad .44] \end{aligned}$$

■

We turn now to the primary concern of this section. There are several different ways of classifying Markov chains, and we choose one which distinguishes between chains on the basis of their long-run behavior, i.e., on the nature of the state vector after many transitions. As we will see in our examples, the long-run behavior of a stochastic process may be very important in applications. Although the kth-state vector can be determined for any k, it may be a lot of work to do so even for moderately large k. Also, after you have an $\mathbf{X}_k$ for some k, in general you do not know much about $\mathbf{X}_{k+1}$ or $\mathbf{X}_{k+2}$ unless you actually compute them. Thus, if you are interested in studying a stochastic process over many transitions, then it is appropriate to study its long-run behavior. For instance, suppose that in Example 8.4 the first call is to a location in the east district. Then we can easily determine the probabilities that calls will be to locations in the east district after 1, 2, or 3 calls, that is p_{11}, $p_{11}(2)$, and $p_{11}(3)$. The task is more difficult if we are interested in the probabilities of calls to locations in the east district after 10, 20, or 30 calls; that is $p_{11}(10)$, $p_{11}(20)$, and $p_{11}(30)$. The long-run behavior of the probabilities $p_{11}(k)$ gives the likelihood that if the day begins with a call to the east district, then there is a call to the east district after many transitions. Alternatively, these long-run probabilities can also be viewed as the fraction of calls (over the long run) to locations in the east district, given a start in the east district.

In the general case of a Markov chain with N states, the probability that the system is in the jth state after k trials depends upon the state in which it started. Thus $p_{1j}(k)$ is the probability that the system is in state j after k trials if it is initially in state 1. There are similar meanings attached to $p_{2j}(k), p_{3j}(k), \ldots, p_{Nj}(k)$. There is no reason to have (or expect) equality among all these probabilities. However, for some Markov chains there is a positive probability q_j associated with the jth state such that the k-step transition probabilities $p_{ij}(k)$ all become close to q_j for large k. That is, the likelihood that the system is in state j after k transitions is, for large k, nearly the same for all starting states. We study such Markov chains in some detail because they form an important special class. We begin by defining the class. The definition does not directly refer to the long-run behavior of the chain; however, the connection with that behavior will be explored below. We use our definition because it is easier to check that a chain satisfies the definition than that it has certain long-run behaviors.

Regular Markov Chain

A Markov chain with transition matrix **P** is *regular* if there is a positive integer k such that $\mathbf{P}^k$ has all positive entries.

Example 8.14 Each of the following matrices is the transition matrix for a Markov chain.

$$\mathbf{P}_1 = \begin{bmatrix} \frac{1}{2} & \frac{1}{2} \\ 1 & 0 \end{bmatrix} \qquad \mathbf{P}_2 = \begin{bmatrix} 0 & 1 \\ 1 & 0 \end{bmatrix}$$

Problem Which of the Markov chains associated with these matrices are regular?

Solution For matrix $\mathbf{P}_1$ we have

$$\mathbf{P}_1^2 = \begin{bmatrix} \frac{1}{2} & \frac{1}{2} \\ 1 & 0 \end{bmatrix} \begin{bmatrix} \frac{1}{2} & \frac{1}{2} \\ 1 & 0 \end{bmatrix} = \begin{bmatrix} \frac{3}{4} & \frac{1}{4} \\ \frac{1}{2} & \frac{1}{2} \end{bmatrix}$$

Thus, the second power of $\mathbf{P}_1$ has all positive entries, and by the definition given above, the associated Markov chain is regular.

For matrix $\mathbf{P}_2$ we have

$$\mathbf{P}_2^2 = \begin{bmatrix} 0 & 1 \\ 1 & 0 \end{bmatrix} \begin{bmatrix} 0 & 1 \\ 1 & 0 \end{bmatrix} = \begin{bmatrix} 1 & 0 \\ 0 & 1 \end{bmatrix} = \mathbf{I}$$

$$\mathbf{P}_2^3 = \mathbf{P}_2\mathbf{P}_2^2 = \begin{bmatrix} 0 & 1 \\ 1 & 0 \end{bmatrix} \begin{bmatrix} 1 & 0 \\ 0 & 1 \end{bmatrix} = \begin{bmatrix} 0 & 1 \\ 1 & 0 \end{bmatrix} = \mathbf{P}_2$$

and

$$\mathbf{P}_2^4 = \mathbf{P}_2\mathbf{P}_2^3 = \begin{bmatrix} 0 & 1 \\ 1 & 0 \end{bmatrix} \begin{bmatrix} 0 & 1 \\ 1 & 0 \end{bmatrix} = \begin{bmatrix} 1 & 0 \\ 0 & 1 \end{bmatrix} = \mathbf{I}$$

We see at this stage that $\mathbf{P}_2 = \mathbf{P}_2^3$ and $\mathbf{P}_2^2 = \mathbf{P}_2^4 = \mathbf{I}$. This means that every even power of $\mathbf{P}_2$ will be the identity matrix $\mathbf{I}$ since every even power is a power of $\mathbf{P}_2^2 = \mathbf{I}$. Also, every odd power of $\mathbf{P}_2$ will be $\mathbf{P}_2$ since it is just some power of $\mathbf{P}_2^2 = \mathbf{I}$ times $\mathbf{P}_2$. Since both $\mathbf{I}$ and $\mathbf{P}_2$ contain zeros, we see that no power of $\mathbf{P}_2$ will have all positive entries. In other words, the Markov chain for $\mathbf{P}_2$ is *not regular*. ■

It is important to connect our definition of regular with the long-run behavior of a Markov chain. The next example is a step in that direction.

Example 8.15 A Markov chain has the transition matrix

$$\mathbf{P} = \begin{bmatrix} .5 & .5 & 0 \\ 0 & .5 & .5 \\ .75 & .25 & 0 \end{bmatrix}$$

Problem (*a*) For which state i is p_{i3} largest? For which state i is it smallest?
(*b*) For which state i is $p_{i3}(2)$ largest, and for which is it smallest?
(*c*) For which state i is $p_{i3}(3)$ largest, and for which is it smallest?

Solution (*a*) The probabilities p_{i3}, $i = 1, 2, 3$, form the third column of matrix $\mathbf{P}$. Therefore, $p_{23} = .5$ is the largest, and $p_{13} = p_{33} = 0$ are the smallest.

We have

$$\mathbf{P}(2) = \begin{bmatrix} .25 & .50 & .25 \\ .375 & .375 & .25 \\ .375 & .50 & .125 \end{bmatrix} \quad \text{and} \quad \mathbf{P}(3) = \begin{bmatrix} .3125 & .4375 & .2500 \\ .3750 & .4375 & .1875 \\ .28125 & .46875 & .2500 \end{bmatrix}$$

(*b*) The probabilities $p_{13}(2)$, $p_{23}(2)$, and $p_{33}(2)$ form the third column of matrix $\mathbf{P}(2)$. Therefore $p_{13}(2) = p_{23}(2) = .25$ are the largest, and $p_{33}(2) = .125$ is the smallest.

(*c*) The probabilities $p_{13}(3)$, $p_{23}(3)$, and $p_{33}(3)$ form the third column of matrix $\mathbf{P}(3)$. Therefore, $p_{13}(3) = p_{33}(3) = .25$ are the largest, and $p_{23}(3) = .1875$ is the smallest.

Notice that the two-step transition probabilities $p_{13}(2) = .25$, $p_{23}(2) = .25$, and $p_{33}(2) = .125$ are closer together than the one-step transition probabilities. The three-step transition probabilities $p_{13}(3) = .25$, $p_{23}(3) = .1875$, and $p_{33}(3) = .25$ are still closer. Indeed, the probabilities that the system is in state 3 after three transitions given that it began in states 1, 2, and 3 differ from each other by at most .0625, whereas the differences after one transition can be as large as .5. As we shall see in a moment, the probabilities $p_{13}(k)$, $p_{23}(k)$, and $p_{33}(k)$ all approach $\frac{2}{9} = .2222$ as k becomes large. ■

The definition of a regular chain (although stated in terms of the powers of **P**) has the following important consequence. For each j and for k sufficiently large, each of the transition probabilities $p_{1j}(k), p_{2j}(k), \ldots, p_{Nj}(k)$ is close to the same number, call it q_j. That is, each of the entries in the jth column of the k-step transition matrix $\mathbf{P}(k)$ is close to q_j. Another way of saying this is that for large values of k, the k-step transition matrix

$$\mathbf{P}(k) = \begin{bmatrix} p_{11}(k) & p_{12}(k) & \cdots & p_{1N}(k) \\ p_{21}(k) & p_{22}(k) & \cdots & p_{2N}(k) \\ \cdot & \cdot & & \cdot \\ \cdot & \cdot & & \cdot \\ \cdot & \cdot & & \cdot \\ p_{N1}(k) & p_{N2}(k) & \cdots & p_{NN}(k) \end{bmatrix}$$

is very close to a matrix that has all rows identical,

$$\begin{bmatrix} \mathbf{W} \\ \mathbf{W} \\ \cdot \\ \cdot \\ \cdot \\ \mathbf{W} \end{bmatrix} = \begin{bmatrix} w_1 & w_2 & \cdots & w_N \\ w_1 & w_2 & \cdots & w_N \\ \cdot & \cdot & & \cdot \\ \cdot & \cdot & & \cdot \\ \cdot & \cdot & & \cdot \\ w_1 & w_2 & \cdots & w_N \end{bmatrix}$$

where $W = [w_1 \quad w_2 \quad \cdots \quad w_N]$.

Example 8.16 Consider the transition matrix $\mathbf{P}$ of Example 8.12.

$$\mathbf{P} = \begin{bmatrix} .5 & .4 & .1 \\ .1 & .6 & .3 \\ .3 & .6 & .1 \end{bmatrix}$$

A straightforward computation of the k-step transition matrices (best carried out on a computer) gives

$$\mathbf{P}(2) = \begin{bmatrix} .32 & .50 & .18 \\ .20 & .58 & .22 \\ .24 & .54 & .22 \end{bmatrix}$$

$$\mathbf{P}(4) = \begin{bmatrix} .2456 & .5472 & .2072 \\ .2328 & .5552 & .2120 \\ .2376 & .5520 & .2104 \end{bmatrix}$$

$$\mathbf{P}(8) = \begin{bmatrix} .2369 & .5526 & .2105 \\ .2368 & .5527 & .2105 \\ .2369 & .5526 & .2105 \end{bmatrix}$$

where the entries have been rounded off to the four decimal places shown. It is now clear that the rows of $\mathbf{P}(8)$ are essentially equal. This illustrates the assertion that as k increases, the k-step transition matrix $\mathbf{P}(k)$ becomes closer and closer to a matrix all of whose rows are equal to the same vector $\mathbf{W}$. ■

The rows in a transition matrix $\mathbf{P}$, and those of its powers $\mathbf{P}^k$, are all probability vectors. For regular chains the rows in $\mathbf{P}^k$ all become closer and closer to the same probability vector as k increases. This special probability vector is determined by $\mathbf{P}$ and is called a *stable vector* for $\mathbf{P}$. To be precise:

Theorem on Stable Probabilities

Let $\mathbf{P}$ be the transition matrix for a regular Markov chain. There is a unique probability vector $\mathbf{W} = [w_1 \;\; w_2 \;\; \cdots \;\; w_N]$ such that for each state j the difference $|p_{ij}(k) - w_j|$ can be made as small as we choose by selecting k sufficiently large. The vector $\mathbf{W}$ is known as a *stable* vector, and its coordinates are known as *stable probabilities* for the Markov chain.

In a regular Markov chain, the probabilities $p_{ij}(k)$ are for all large values of k nearly equal to the stable probabilities w_j. This assertion holds for each initial state i, $i = 1, 2, \ldots, N$. The stable probabilities w_j can be obtained from the vector $\mathbf{W}$, which is closely approximated by any row of $\mathbf{P}(k)$ for large values of k.

However, obtaining **W** from **P**(k) usually requires computing $\mathbf{P}^k$ for several values of k, a method that is frequently impractical. Fortunately there is an alternative method of obtaining the stable probabilities.

Theorem on a Method of Obtaining Stable Probabilities

Let **P** be the transition matrix of a regular Markov chain. Then there is a unique probability vector **W** which satisfies **WP** = **W**. The coordinates of this vector are the stable probabilities for the Markov chain.

This theorem (whose proof we omit) provides a direct method of obtaining the stable probabilities. Indeed, we need only solve a system of linear equations.

Example 8.17 The matrix $\mathbf{P} = \begin{bmatrix} .25 & .75 \\ .60 & .40 \end{bmatrix}$ is the transition matrix of a regular Markov chain.

Problem Determine the vector **W** of stable probabilities for this Markov chain.

Solution We make use of the theorem quoted above. That is, we find the probability vector **W** which satisfies the system of equations

$$\mathbf{WP} = \mathbf{W} \qquad \text{or equivalently} \qquad \mathbf{W}(\mathbf{P} - \mathbf{I}) = \mathbf{0}$$

If $\mathbf{W} = [w_1 \quad w_2]$, then the condition that **W** be a probability vector requires that $w_1 + w_2 = 1$, and the system $\mathbf{W}(\mathbf{P} - \mathbf{I}) = \mathbf{0}$ is

$$[w_1 \quad w_2]\begin{bmatrix} -.75 & .75 \\ .60 & -.60 \end{bmatrix} = [0 \quad 0]$$

$$-.75w_1 + .60w_2 = 0$$

$$.75w_1 - .60w_2 = 0$$

Therefore, the complete system to be solved is

$$w_1 + w_2 = 1$$

$$-.75w_1 + .60w_2 = 0$$

$$.75w_1 - .60w_2 = 0$$

Using the techniques of Chapter 5, gaussian elimination, we find

$$w_1 = \tfrac{4}{9} \qquad \text{and} \qquad w_2 = \tfrac{5}{9}$$

The vector of stable probabilities for the Markov chain whose transition matrix is **P** is $\mathbf{W} = [\tfrac{4}{9} \quad \tfrac{5}{9}]$. ■

Example 8.18 (Continuation of Example 8.15) Find the stable probabilities for the Markov chain whose transition matrix is

$$\mathbf{P} = \begin{bmatrix} .5 & .5 & 0 \\ 0 & .5 & .5 \\ .75 & .25 & 0 \end{bmatrix}$$

Solution Since $N = 3$, there are three states; $\mathbf{W}$ is a 3-vector, $\mathbf{W} = [w_1 \quad w_2 \quad w_3]$. The system of equations $\mathbf{WP} = \mathbf{W}$ or $\mathbf{W}(\mathbf{P} - \mathbf{I}) = \mathbf{0}$ is

$$[w_1 \quad w_2 \quad w_3] \begin{bmatrix} -.5 & .5 & 0 \\ 0 & -.5 & .5 \\ .75 & .25 & -1 \end{bmatrix} = [0 \quad 0 \quad 0] \tag{8.2}$$

The condition that $\mathbf{W}$ be a probability vector gives

$$w_1 + w_2 + w_3 = 1$$

Therefore, the system consisting of the probability vector condition together with Equation (8.2) is

$$\begin{aligned} w_1 + \quad w_2 + \quad w_3 &= 1 \\ -.5w_1 \qquad\qquad + .75w_3 &= 0 \\ .5w_1 - .5w_2 + .25w_3 &= 0 \\ .5w_2 - \quad w_3 &= 0 \end{aligned}$$

Using the techniques of Chapter 5, we conclude that the vector of stable probabilities is $[\frac{3}{9} \quad \frac{4}{9} \quad \frac{2}{9}]$. This confirms the assertion made in Example 8.15 that as k becomes large, $p_{i3}(k)$ approaches $\frac{2}{9}$ for each initial state $i = 1, 2, 3$. ■

Note that in each of the last two examples we wrote the condition that the sum of the coordinates of $\mathbf{W}$ is 1 as the *first* equation of our system. The remaining equations came from the system $\mathbf{WP} = \mathbf{W}$. Writing the equations in this order, i.e., with the probability condition first, makes the use of the techniques of Chapter 5 somewhat easier.

Example 8.19 (Continuation of Example 8.4) The Markov chain used to describe the calls made by the service specialist for Metroburg Heating is a regular Markov chain.

Scheduling

Problem Find the long-run probabilities that the service specialist will be in each of the three districts.

Solution We know that one of the interpretations of the stable probabilities is that they give the likelihoods that, in the long run, the system will be in each of the states. Thus we can solve the problem by determining the vector **W** of the stable probabilities. Using the fact that **WP** = **W**, we see that our task is to find a probability vector **W** which satisfies

$$\mathbf{WP} = \mathbf{W} \qquad \text{with} \quad \mathbf{P} = \begin{bmatrix} .5 & .4 & .1 \\ .1 & .6 & .3 \\ .3 & .6 & .1 \end{bmatrix}$$

We set $\mathbf{W} = [w_1 \quad w_2 \quad w_3]$, and we write $\mathbf{WP} = \mathbf{W}$ as $\mathbf{W}(\mathbf{P} - \mathbf{I}) = \mathbf{0}$. The resulting system, including the probability condition, is

$$\begin{aligned} w_1 + \quad w_2 + \quad w_3 &= 1 \\ -.5w_1 + .1w_2 + .3w_3 &= 0 \\ .4w_1 - .4w_2 + .6w_3 &= 0 \\ .1w_1 + .3w_2 - .9w_3 &= 0 \end{aligned}$$

Using the techniques of Chapter 5, we find that

$$\mathbf{W} = [\tfrac{9}{38} \quad \tfrac{21}{38} \quad \tfrac{8}{38}]$$

Thus, in the long run the service specialist will be in the east district with probability $\frac{9}{38}$, in the central district with probability $\frac{21}{38}$, and in the west district with probability $\frac{8}{38}$. ■

We now have a means of computing the vector of stable probabilities for any regular Markov chain. To show that a Markov chain is regular, we must be able to show that some power of the transition matrix has all positive entries. It is important to note that we do not need to know the actual entries of the power of the matrix. We only need to know that the entries are all positive.

Example 8.20 Show that the matrix **P** given below is the transition matrix of a regular Markov chain.

$$\mathbf{P} = \begin{bmatrix} \frac{1}{3} & \frac{2}{3} & 0 & 0 \\ \frac{1}{4} & \frac{1}{4} & \frac{1}{2} & 0 \\ 0 & \frac{1}{4} & \frac{1}{4} & \frac{1}{2} \\ 0 & 0 & \frac{1}{2} & \frac{1}{2} \end{bmatrix}$$

Solution Since we are only interested in knowing which entries are positive and which ones are zero, we adopt the convention that + and 0 denote positive and zero entries, respectively, in $\mathbf{P}^k$. For $k = 1$, we have

$$\mathbf{P} = \begin{bmatrix} + & + & 0 & 0 \\ + & + & + & 0 \\ 0 & + & + & + \\ 0 & 0 & + & + \end{bmatrix}$$

For $k = 2$ and $k = 3$, we have

$$\mathbf{P}(2) = \mathbf{P}^2 = \begin{bmatrix} + & + & + & 0 \\ + & + & + & + \\ + & + & + & + \\ 0 & + & + & + \end{bmatrix} \text{ and } \mathbf{P}(3) = \mathbf{P}^3 = \begin{bmatrix} + & + & + & + \\ + & + & + & + \\ + & + & + & + \\ + & + & + & + \end{bmatrix}$$

We conclude that $\mathbf{P}(3)$ contains only positive entries, and therefore $\mathbf{P}$ is the transition matrix of a regular Markov chain. ■

We comment that the techniques of using pluses and zeros as described in Example 8.20 is useful only when we are dealing with nonnegative matrices.

How many $\mathbf{P}^k$ might you be required to test to determine whether a Markov chain is regular? The answer depends upon the size of matrix $\mathbf{P}$. If $\mathbf{P}$ is the transition matrix for a regular Markov chain with N states, then one of the first $(N - 1)^2 + 1$ powers of $\mathbf{P}$ contains only positive entries. [In most regular chains one has to test fewer than $(N - 1)^2 + 1$ powers to find one with only positive entries.] Thus, if each of the first $(N - 1)^2 + 1$ powers of the transition matrix contains at least one zero entry, the Markov chain is not regular.

In general, it is not efficient to compute $\mathbf{P}^2$, then $\mathbf{P}^3$, then $\mathbf{P}^4$, etc. Instead it is preferable to compute $\mathbf{P}^2$, then multiply $\mathbf{P}^2$ by $\mathbf{P}^2$ to obtain $\mathbf{P}^4$, then multiply $\mathbf{P}^4$ by $\mathbf{P}^4$ to obtain $\mathbf{P}^8$, and so on. This method enables you to determine whether $\mathbf{P}^8$ has only positive entries with only three matrix multiplications. We do not have to worry about "missing" a power which had all positive entries, say, $\mathbf{P}^7$, because any time one power has only positive entries, all higher powers also have only positive entries. A zero can never return after all zeros are gone.

Exercises for Section 8.3

1. Which of the following are probability vectors?
   (*a*) [.3 0 0 .2 .5] (*b*) [.1 .2 0 0 .6]
   (*c*) [.1 .4 .5 0 .1] (*d*) [0 .1 0 .1 .8]
2. In each case find a value of the number x for which the vector is a probability vector.
   (*a*) $[.2 \quad x \quad 0 \quad 0 \quad .5]$ (*b*) $[x \quad 0 \quad x \quad 0 \quad .2]$
   (*c*) $[x \quad x \quad x \quad 0 \quad x]$ (*d*) $[2x \quad 3x \quad 3x \quad 2x \quad x]$
3. A state vector $\mathbf{X}$ for a three-state Markov chain is described as follows: The system is equally likely to be in states 1 and 2, and it is twice as likely to be in state 3 as in state 1. Find the state vector $\mathbf{X}$.

4. A state vector **X** for a four-state Markov chain is described as follows: The system is twice as likely to be in state 1 as in state 3, it is never in state 4, and it is in state 2 with probability .4. Find the state vector **X**.

Business

5. (Continuation of Exercise 15, Section 8.1) Profits at a brokerage firm are determined by the volume of securities sold, and this volume fluctuates from week to week. Each week volume is classified as high or low, and information collected over many weeks leads to the following assumptions:

    (*i*) If volume is high this week, then next week it will be high with probability .7 and low with probability .3.
    (*ii*) If volume is low this week, then next week it will be high or low with equal probability.

    The manager estimates that volume is twice as likely to be high as low this week.
    (*a*) Find the state vector which represents the manager's estimate.
    (*b*) Using the estimate as an initial state vector, find the probability of high volume 2 weeks from now.
    (*c*) Repeat (*b*) for 3 weeks from now.

Business

6. In the situation described in Exercise 5, the volume this week is low. How many weeks must pass before a week comes along in which the probability of high volume is at least .6?

Ecology

7. (Continuation of Exercise 17, Section 8.1) A small animal lives in a territory that can be divided into areas described as meadow and woods, and it moves randomly from one area to another. If it is in the woods on one observation, then the animal is twice as likely to be in the woods as the meadow on the next observation. Likewise, if it is in the meadow on one observation, then it is twice as likely to be in the meadow as the woods on the next.
    (*a*) If the animal is 3 times as likely to be in the meadow as the woods, find the state vector which represents this fact.
    (*b*) Using the state vector determined in (*a*) as an initial state vector, find the probability that the animal is in the meadow on the third subsequent observation.

Ecology

8. Consider the situation described in Exercise 7. If the probability that the animal is in the meadow is .1 at a specific time, how many subsequent observations must be made before the probability that it is in the meadow exceeds .3?

9. Find the vector of stable probabilities for the situation described in Exercise 5.

10. The matrix $\mathbf{P} = \begin{bmatrix} 0 & 1 \\ .5 & .5 \end{bmatrix}$ is the transition matrix of a regular Markov chain.

    Determine $\mathbf{P}^2$, $\mathbf{P}^4$, and $\mathbf{P}^8$. Use this information to estimate the vector of stable probabilities.

11. Find the vector of stable probabilities for the Markov chain whose transition matrix is

$$\begin{bmatrix} \frac{1}{2} & \frac{1}{2} \\ \frac{2}{3} & \frac{1}{3} \end{bmatrix}$$

12. Find the vector of stable probabilities for the Markov chain whose transition matrix is

$$\mathbf{P} = \begin{bmatrix} 0 & 1 & 0 \\ .1 & 0 & .9 \\ 1 & 0 & 0 \end{bmatrix}$$

For Exercises 13 through 17, decide whether the Markov chain with the transition matrix shown is regular. If it is, determine the vector of stable probabilities.

13. $\begin{bmatrix} 0 & 1 \\ \frac{1}{3} & \frac{2}{3} \end{bmatrix}$

14. $\begin{bmatrix} 0 & 1 & 0 \\ 0 & 0 & 1 \\ \frac{1}{3} & 0 & \frac{2}{3} \end{bmatrix}$

15. $\begin{bmatrix} 1 & 0 & 0 \\ \frac{1}{2} & \frac{1}{2} & 0 \\ 0 & \frac{2}{3} & \frac{1}{3} \end{bmatrix}$

16. $\begin{bmatrix} 0 & \frac{1}{3} & 0 & \frac{2}{3} \\ \frac{1}{3} & 0 & \frac{2}{3} & 0 \\ 0 & \frac{1}{2} & 0 & \frac{1}{2} \\ \frac{1}{2} & 0 & \frac{1}{2} & 0 \end{bmatrix}$

17. $\begin{bmatrix} .6 & .4 & 0 & 0 \\ 0 & .5 & .5 & 0 \\ 0 & 0 & .6 & .4 \\ .5 & 0 & .5 & 0 \end{bmatrix}$

18. Consider a Markov chain whose transition matrix is

$$\begin{bmatrix} \frac{1}{3} & 0 & \frac{2}{3} & 0 \\ \frac{1}{4} & 0 & \frac{3}{4} & 0 \\ \frac{1}{3} & \frac{1}{3} & 0 & \frac{1}{3} \\ 0 & \frac{1}{2} & 0 & \frac{1}{2} \end{bmatrix}$$

In which state is the system most likely to be in the long run?

Sports 19. Find the long-run free-throw shooting probabilities for a basketball player who has the properties that if she makes a shot, then she makes the next one with probability .85, but if she misses a shot, then she makes the next one with probability .6.

Sports 20. A fisherman has either good luck or poor luck each time he goes fishing. He notices that if he has good luck one time, then he has good luck the next time with probability .6 and if he has poor luck one time, then he has good luck the next time with probability .8. What fraction of the time in the long run does the fisherman have good luck?

College life 21. How often (in the long run) does our not-so-enthusiastic student attend class on Friday afternoon? Use the transition matrix

$$\mathbf{P} = \begin{array}{c} \\ \text{Absent} \\ \text{Present} \end{array} \begin{array}{c} \begin{array}{cc} \text{Absent} & \text{Present} \end{array} \\ \begin{bmatrix} .3 & .7 \\ .6 & .4 \end{bmatrix} \end{array}$$

Demographics 22. People in a small town shift jobs according to Table 8.2 (reproduced below). How will people be distributed in the various employment groups over the long run?

Employment Last Year	Employment This Year	Percentage
Industry	Industry	70
	Small business	20
	Self-employed	10
Small business	Industry	30
	Small business	50
	Self-employed	20
Self-employed	Industry	30
	Small business	30
	Self-employed	40

Psychology 23. Consider the situation described in Example 8.10.

(*a*) Determine $\mathbf{P}^4$ and $\mathbf{P}^8$.

(*b*) What is the largest difference between elements of the first column of $\mathbf{P}^4$? Of $\mathbf{P}^8$?

(*c*) Find the vector of stable probabilities for this situation.

(*d*) In which state is the mouse most likely to be in the long run?

Psychology 24. Consider a mouse moving in a set of compartments connected as shown in Figure 8.13. Suppose that the movements of the mouse can be modeled as a Markov chain. Assume that the mouse is twice as likely to move as not and that if it moves, then it is equally likely to move to any adjacent compartment. (See also Example 8.10.) In what compartment is the mouse most likely to be in the long run, and what is the probability that the mouse is in that compartment?

FIGURE 8.13

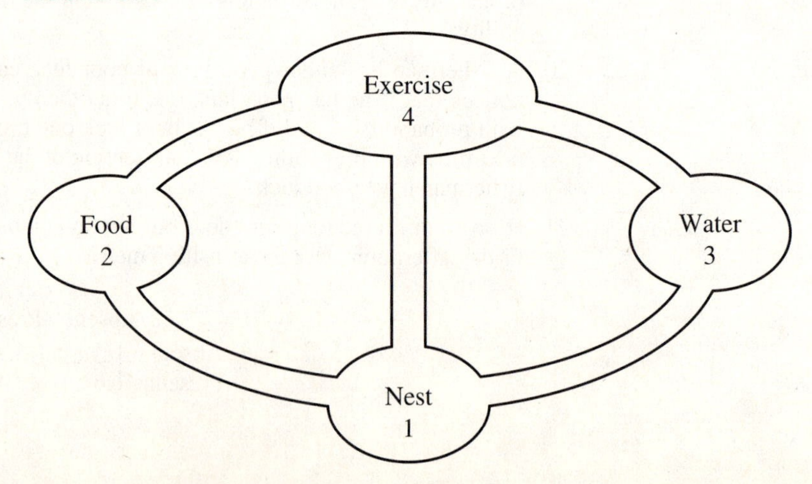

25. Two four-state Markov chains have the transition matrices shown below.

$$\mathbf{P}_1 = \begin{bmatrix} 0 & 1 & 0 & 0 \\ 0 & 0 & 1 & 0 \\ 0 & 0 & 0 & 1 \\ .5 & .5 & 0 & 0 \end{bmatrix} \qquad \mathbf{P}_2 = \begin{bmatrix} 0 & 1 & 0 & 0 \\ 0 & 0 & 1 & 0 \\ 0 & 0 & 0 & 1 \\ .5 & 0 & 0 & .5 \end{bmatrix}$$

(*a*) Find the transition diagram for each Markov chain.
(*b*) In each case find the smallest integer k for which the kth power of the transition matrix has only positive entries.

26. A four-state Markov chain has the transition matrix

$$\mathbf{P} = \begin{bmatrix} 0 & 1 & 0 & 0 \\ 0 & 0 & 1 & 0 \\ 0 & 0 & 0 & 1 \\ .5 & 0 & .5 & 0 \end{bmatrix}$$

(*a*) Find the transition diagram for this Markov chain.
(*b*) Show that there is a positive probability of making a transition from state 1 to state 3 in two steps, in four steps, in six steps, in eight steps, etc.
(*c*) Show that it is impossible to go from state 1 to state 3 in any odd number of steps.

27. A five-state Markov chain has the transition matrix $\mathbf{P}$.

$$\mathbf{P} = \begin{bmatrix} 0 & 1 & 0 & 0 & 0 \\ 0 & 0 & 1 & 0 & 0 \\ 0 & 0 & 0 & 1 & 0 \\ 0 & 0 & 0 & 0 & 1 \\ x & y & 0 & z & 0 \end{bmatrix}$$

Show that for any nonnegative values of x, y, and z the eight-step transition matrix $\mathbf{P}(8) = \mathbf{P}^8$ has at least one zero entry.

28. With the transition matrix $\mathbf{P}$ of Exercise 27, find values of x, y, and z for which the 16th power of the matrix $\mathbf{P}$ has at least one zero and the 17th power contains only positive entries. (*Hint*: Work Exercises 25 and 26 first.)

29. Let the transition matrix for a regular Markov chain be

$$\mathbf{P} = \begin{bmatrix} a & 1-a \\ b & 1-b \end{bmatrix}$$

where $0 < a < 1$ and $0 < b < 1$. Show that the vector of stable probabilities is

$$\begin{bmatrix} \dfrac{b}{1+b-a} & \dfrac{1-a}{1+b-a} \end{bmatrix}$$

30. Use the result of Exercise 29 to compute the stable probabilities for the matrix

$$\mathbf{P} = \begin{bmatrix} \frac{3}{4} & \frac{1}{4} \\ \frac{1}{3} & \frac{2}{3} \end{bmatrix}$$

31. Find the vector of stable probabilities for the Markov chain with the transition matrix

$$\mathbf{P} = \begin{bmatrix} 0 & 1 \\ b & 1-b \end{bmatrix} \qquad 0 < b < 1$$

32. Find the vector of stable probabilities for the Markov chain with transition matrix

$$\mathbf{P} = \begin{bmatrix} 0 & 1 & 0 \\ 0 & 0 & 1 \\ a & b & 1-a-b \end{bmatrix}$$

where $0 < a < 1$, $0 < b < 1$, and $a + b < 1$.

8.4 ABSORBING MARKOV CHAINS

The behavior of a Markov chain which is not regular can differ from that of a regular chain in several ways. For a regular chain, some power of the transition matrix contains only positive entries, and consequently it is possible to make transitions between *any* two states in some number of steps. One way in which this can fail is for there to be a state, or a subset of states, from which transitions are impossible. This is illustrated in the next example.

Example 8.21 *Psychology*

Consider the arrangement of connected compartments introduced in Example 8.10 (and shown in Figure 8.12), and suppose that the mouse can enter compartment 3 but cannot leave it. This can be accomplished, e.g., by making one-way doors into compartment 3.

Problem If the transitions are governed by the assumptions of Example 8.10, find the transition matrix of the resulting Markov chain.

Solution Since the transitions from compartments 1, 2, and 4 are unaffected by making it impossible to leave compartment 3, rows 1, 2, and 4 of the transition matrix **P** are as in Example 8.10. The third row must be changed to represent the fact that the only possible transition from state 3 is back to state 3. We have

$$\mathbf{P} = \begin{bmatrix} \frac{2}{3} & \frac{1}{6} & \frac{1}{6} & 0 \\ \frac{1}{9} & \frac{2}{3} & \frac{1}{9} & \frac{1}{9} \\ 0 & 0 & 1 & 0 \\ 0 & \frac{1}{3} & 0 & \frac{2}{3} \end{bmatrix}$$

■

The situation illustrated in Example 8.21 is typical of a type of Markov chain called an *absorbing Markov chain*. To define absorbing Markov chains, we introduce the notion of an *absorbing state*. Heuristically, an absorbing state is a state which it is impossible to leave. We give a precise definition using the concept of a unit vector.

The ith unit vector is a probability vector with a 1 as the ith coordinate.

Absorbing States

The ith state of a Markov chain is said to be an *absorbing state* if $p_{ii} = 1$ and $p_{ij} = 0$ for $j \neq i$. That is, state i is absorbing if the ith row of the transition matrix is the ith unit vector.

In the Markov chain whose transition matrix is shown below, the second state is absorbing. Note, however, that the fifth state is not absorbing. Even though the fifth row contains a single 1, it is not in the fifth column, so the fifth row is not the fifth unit vector.

$$\begin{bmatrix} \frac{1}{3} & 0 & \frac{1}{3} & 0 & \frac{1}{3} \\ 0 & 1 & 0 & 0 & 0 \\ \frac{1}{2} & 0 & \frac{1}{2} & 0 & 0 \\ \frac{1}{10} & \frac{2}{10} & \frac{3}{10} & \frac{4}{10} & 0 \\ 0 & 0 & 0 & 1 & 0 \end{bmatrix} \tag{8.3}$$

Absorbing chains must have absorbing states, but that is not enough. It must also be possible to go from nonabsorbing states to absorbing states.

Absorbing Markov Chain

A Markov chain is said to be *absorbing* if

(*a*) There is at least one absorbing state, and

(*b*) For every nonabsorbing state i there is some absorbing state j and a positive integer k such that the probability of a transition from state i to state j in k steps is positive.

Notice that there are two parts to the definition of an absorbing chain. For a chain to be absorbing, *both* conditions must hold. For example, the transition matrix (8.3) is the transition matrix of an absorbing chain. Indeed, condition (*a*) is satisfied since the second state is absorbing, and condition (*b*) is satisfied since transitions can occur from each of the other states to state 2; that is,

$$p_{12}(3) > 0 \qquad p_{32}(4) > 0 \qquad p_{42} > 0 \qquad p_{52}(2) > 0$$

To verify this, it is convenient to use the technique introduced at the end of Section 8.3 or the transition diagram for the chain.

A chain is not absorbing if *either* of the conditions of the definition fails to hold. For instance, the transition matrix shown below is not the transition matrix for an absorbing Markov chain since condition (*b*) fails to hold for states 3 and 4. State 2 is the only absorbing state, and the probability of a transition in any number of steps from either state 3 or state 4 to state 2 is zero.

$$\begin{bmatrix} \frac{1}{2} & \frac{1}{2} & 0 & 0 \\ 0 & 1 & 0 & 0 \\ 0 & 0 & \frac{1}{3} & \frac{2}{3} \\ 0 & 0 & \frac{3}{5} & \frac{2}{5} \end{bmatrix}$$

The analysis of situations modeled by absorbing Markov chains will be easier if the transition matrices are written in a particular form. Since the absorbing states play a special role, it is conventional to collect them together and to label them as states 1 through k—we suppose that there are k absorbing states. The remaining states are labeled $k + 1$ through N. In relabeling states, it is important to note that both the rows and columns of the transition matrix are altered. A transition matrix with the states labeled in this way is said to be in *canonical form*. A technique for writing a transition matrix in canonical form is illustrated in the next example.

Example 8.22 Matrix **P** is the transition matrix of a Markov chain.

$$\mathbf{P} = \begin{bmatrix} \frac{1}{3} & 0 & \frac{2}{3} \\ 0 & 1 & 0 \\ 0 & \frac{1}{2} & \frac{1}{2} \end{bmatrix}$$

Problem Show that this is the transition matrix of an absorbing chain, and write the matrix in canonical form.

Solution State 2 is an absorbing state, so condition (*a*) of the definition of an absorbing Markov chain is satisfied. Also $p_{32} > 0$ and $p_{12}(2) > 0$, so condition (*b*) is satisfied. Therefore, **P** is the transition matrix of an absorbing Markov chain.

To write the transition matrix in canonical form, we relabel the states so that absorbing states are listed first. In this example the only absorbing state is state 2. We relabel the states so that the original state 2 is relabeled as state 1. The labels assigned to the remaining states are not important. We relabel the states as follows:

Old label	1	2	3
New label	2	1	3

After the states are relabeled, the transition matrix changes to reflect the new labels. For instance, the old (1, 3) entry $p_{13} = \frac{2}{3}$ becomes the new (2, 3) entry; the old (2, 2) entry $p_{22} = 1$ becomes the new (1, 1) entry, and so on. The transition matrix written in canonical form is

$$\begin{array}{c} \\ 2 \\ 1 \\ 3 \end{array} \begin{array}{c} \begin{array}{ccc} 2 & 1 & 3 \end{array} \\ \begin{bmatrix} 1 & 0 & 0 \\ 0 & \frac{1}{3} & \frac{2}{3} \\ \frac{1}{2} & 0 & \frac{1}{2} \end{bmatrix} \end{array}$$

The labels beside the rows and above the columns are the old state labels. Thus the first row corresponds to the relabeled state 1, which is state 2 in terms of the original transition matrix. ■

In Example 8.22 we noted the original state labels along the rows and columns of the canonical form. These labels are useful both in forming the canonical form and in answering questions by using the canonical form. Consequently, we frequently use such labels in the pages that follow.

Example 8.23 Show that a transition matrix may have more than one canonical form by using different relabelings to find two canonical forms for the matrix in Example 8.21.

Solution The transition matrix of Example 8.21 is

$$\mathbf{P} = \begin{bmatrix} \frac{2}{3} & \frac{1}{6} & \frac{1}{6} & 0 \\ \frac{1}{9} & \frac{2}{3} & \frac{1}{9} & \frac{1}{9} \\ 0 & 0 & 1 & 0 \\ 0 & \frac{1}{3} & 0 & \frac{2}{3} \end{bmatrix}$$

The state relabeling

Old label	1	2	3	4
New label	3	2	1	4

results in the canonical form

$$\begin{array}{c} \\ 3 \\ 2 \\ 1 \\ 4 \end{array}\begin{array}{c} \begin{array}{cccc} 3 & 2 & 1 & 4 \end{array} \\ \begin{bmatrix} 1 & 0 & 0 & 0 \\ \frac{1}{9} & \frac{2}{3} & \frac{1}{9} & \frac{1}{9} \\ \frac{1}{6} & \frac{1}{6} & \frac{2}{3} & 0 \\ 0 & \frac{1}{3} & 0 & \frac{2}{3} \end{bmatrix} \end{array}$$

On the other hand, the state relabeling

Old label	1	2	3	4
New label	4	3	1	2

results in the canonical form

$$\begin{array}{c} \\ 3 \\ 4 \\ 2 \\ 1 \end{array}\begin{array}{c} \begin{array}{cccc} 3 & 4 & 2 & 1 \end{array} \\ \begin{bmatrix} 1 & 0 & 0 & 0 \\ 0 & \frac{2}{3} & \frac{1}{3} & 0 \\ \frac{1}{9} & \frac{1}{9} & \frac{2}{3} & \frac{1}{9} \\ \frac{1}{6} & 0 & \frac{1}{6} & \frac{2}{3} \end{bmatrix} \end{array} \tag{8.4}$$

■

We conclude from Example 8.23 that the entries in a canonical form are not necessarily uniquely defined. However, the form of the matrix is unique. In particular, when written in canonical form, the transition matrix of an absorbing chain has a $k \times k$ identity matrix in the upper left-hand corner. Also the submatrix in the upper right-hand corner consists of all zeros. It is conventional to identify the submatrices in **P** as follows:

$$\mathbf{P} = \begin{bmatrix} \mathbf{I} & \mathbf{O} \\ \mathbf{R} & \mathbf{Q} \end{bmatrix} \tag{8.5}$$

Here **I** is a $k \times k$ identity matrix, **O** is a matrix with all zeros, and **R** and **Q** consist of transition probabilities which correspond to transitions which lead directly to absorption, **R**, and transitions which do not lead directly to absorption, **Q**.

In Example 8.23 matrix (8.4) is in the form shown in (8.5), with a 1×1 identity matrix $\mathbf{I} = [1]$ and with

$$\mathbf{R} = \begin{bmatrix} 0 \\ \frac{1}{9} \\ \frac{1}{6} \end{bmatrix} \quad \text{and} \quad \mathbf{Q} = \begin{array}{c} \\ 4 \\ 2 \\ 1 \end{array}\begin{array}{c} \begin{array}{ccc} 4 & 2 & 1 \end{array} \\ \begin{bmatrix} \frac{2}{3} & \frac{1}{3} & 0 \\ \frac{1}{9} & \frac{2}{3} & \frac{1}{9} \\ 0 & \frac{1}{6} & \frac{2}{3} \end{bmatrix} \end{array} \tag{8.6}$$

In this example $N - k = 4 - 1 = 3$.

The matrix $\mathbf{Q}$ is always square, and in fact, for an identity matrix $\mathbf{I}$ the same size as $\mathbf{Q}$, the matrix $\mathbf{I} - \mathbf{Q}$ is invertible.

Fundamental Matrix

Consider an absorbing Markov chain whose transition matrix is written in canonical form as (8.5). The matrix $\mathbf{N} = (\mathbf{I} - \mathbf{Q})^{-1}$ is called the *fundamental matrix* for the absorbing Markov chain. It is understood that since $\mathbf{Q}$ is $(N - k) \times (N - k)$, the identity matrix in the expression $\mathbf{I} - \mathbf{Q}$ must be the same size, and so is $\mathbf{N}$.

The computation of $\mathbf{N}$ is illustrated in the next example.

Example 8.24 An absorbing Markov chain has the transition matrix

$$\begin{bmatrix} 1 & 0 & 0 & 0 \\ 0 & 1 & 0 & 0 \\ \frac{1}{3} & 0 & \frac{2}{3} & 0 \\ 0 & \frac{1}{5} & \frac{2}{5} & \frac{2}{5} \end{bmatrix} \tag{8.7}$$

Problem Find the fundamental matrix.

Solution The transition matrix is already written in canonical form. The matrix $\mathbf{Q}$ is

$$\mathbf{Q} = \begin{bmatrix} \frac{2}{3} & 0 \\ \frac{2}{5} & \frac{2}{5} \end{bmatrix}$$

and therefore

$$\mathbf{I} - \mathbf{Q} = \begin{bmatrix} 1 & 0 \\ 0 & 1 \end{bmatrix} - \begin{bmatrix} \frac{2}{3} & 0 \\ \frac{2}{5} & \frac{2}{5} \end{bmatrix} = \begin{bmatrix} \frac{1}{3} & 0 \\ -\frac{2}{5} & \frac{3}{5} \end{bmatrix}$$

We determine $(\mathbf{I} - \mathbf{Q})^{-1} = \mathbf{N}$ by using the techniques introduced in Chapter 6. We have

$$\mathbf{N} = \begin{bmatrix} 3 & 0 \\ 2 & \frac{5}{3} \end{bmatrix}$$

■

Example 8.25 Find the fundamental matrix for the absorbing Markov chain defined in Example 8.21.

Solution The transition matrix for Example 8.21 is written in canonical form in (8.4). The matrix **Q** was determined in (8.6) to be

$$\mathbf{Q} = \begin{bmatrix} \frac{2}{3} & \frac{1}{3} & 0 \\ \frac{1}{9} & \frac{2}{3} & \frac{1}{9} \\ 0 & \frac{1}{6} & \frac{2}{3} \end{bmatrix}$$

Therefore,

$$\mathbf{I} - \mathbf{Q} = \begin{bmatrix} 1 & 0 & 0 \\ 0 & 1 & 0 \\ 0 & 0 & 1 \end{bmatrix} - \begin{bmatrix} \frac{2}{3} & \frac{1}{3} & 0 \\ \frac{1}{9} & \frac{2}{3} & \frac{1}{9} \\ 0 & \frac{1}{6} & \frac{2}{3} \end{bmatrix} = \begin{bmatrix} \frac{1}{3} & -\frac{1}{3} & 0 \\ -\frac{1}{9} & \frac{1}{3} & -\frac{1}{9} \\ 0 & -\frac{1}{6} & \frac{1}{3} \end{bmatrix}$$

We use the techniques of Chapter 6 to find

$$\mathbf{N} = (\mathbf{I} - \mathbf{Q})^{-1} = \begin{bmatrix} 5 & 6 & 2 \\ 2 & 6 & 2 \\ 1 & 3 & 4 \end{bmatrix}$$

■

Let us now consider the long-run behavior of an absorbing Markov chain. Since there is a positive probability that the system moves from each nonabsorbing state to some absorbing state in some number of transitions, we expect that sooner or later it reaches an absorbing state. Of course, once the system reaches an absorbing state, it does not leave it. However, this is a stochastic process, and we do not know exactly when the system will reach an absorbing state. As usual, in such circumstances we turn to the notion of expected value to help us describe the behavior of the process. It turns out that the fundamental matrix is very useful in this respect. In fact, it fully deserves its name!

Assume for the moment that we are given an absorbing Markov chain whose transition matrix **P** is written in canonical form (8.5), with absorbing states numbered 1 through k and nonabsorbing states numbered $k + 1$ through N. We refer to state $k + 1$ as the first nonabsorbing state, state $k + 2$ as the second nonabsorbing state, etc. The rows and columns of **Q** are identified with nonabsorbing states in order. This identification carries over to the fundamental matrix **N**.

Our results are contained in the following theorem, stated here without proof.

Theorem on the Interpretation of the Fundamental Matrix

The (i, j) entry in the fundamental matrix **N** gives the expected number of times that a system which begins in the ith nonabsorbing state will be in the jth nonabsorbing state before it reaches an absorbing state. The sum of the entries in the ith row of **N** gives the expected number of transitions of a system which begins in the ith nonabsorbing state and continues until it first reaches an absorbing state.

Example 8.26 (Continuation of Example 8.24) An absorbing Markov chain whose transition matrix is (8.7) is initially in state 3, the first nonabsorbing state.

Problem Find the expected number of transitions before the system first reaches an absorbing state.

Solution The fundamental matrix is $\mathbf{N} = \begin{bmatrix} 3 & 0 \\ 2 & \frac{5}{3} \end{bmatrix}$. The expected number of times the system is in state 3 (the first nonabsorbing state) before absorption is 3, and the expected number of times it is in state 4 (the second nonabsorbing state) is 0. Therefore, the expected number of transitions before the system reaches an absorbing state is $3 + 0 = 3$. ■

If we are given an absorbing Markov chain whose transition matrix is not in canonical form, the methods described above can still be applied. It is only necessary to relabel the states, so that the transition matrix is in canonical form, and then to consistently attach original state labels to the appropriate rows and columns of **Q** and **N**. We illustrate this technique in the following examples.

Example 8.27 (Continuation of Example 8.21) An arrangement of connected compartments is described in Example 8.21. A mouse is released in compartment 2.

Psychology

Problem Find the expected number of transitions before the mouse first reaches compartment 3, the compartment with the one-way doors.

Solution We relabel the states as follows:

Old label	1	2	3	4
New label	4	3	1	2

The resulting fundamental matrix (computed in Example 8.25) is

$$\mathbf{N} = \begin{bmatrix} 5 & 6 & 2 \\ 2 & 6 & 2 \\ 1 & 3 & 4 \end{bmatrix}$$

Attaching the original state labels to the appropriate rows and columns of **N** [see labeling in (8.6)], we have

$$\mathbf{N} = \begin{array}{c} \\ 4 \\ 2 \\ 1 \end{array} \begin{array}{c} \begin{array}{ccc} 4 & 2 & 1 \end{array} \\ \begin{bmatrix} 5 & 6 & 2 \\ 2 & 6 & 2 \\ 1 & 3 & 4 \end{bmatrix} \end{array}$$

Since the mouse was released in compartment 2, we look for old state label 2. We see that it is attached to the second row of **N**. Thus, if the system is initially in old state 2, then the expected number of times the mouse occupies the first nonabsorbing state (old state 4) before reaching an absorbing state is 2. The expected numbers of times the mouse occupies the second and third nonabsorbing states (old states 2 and 1) are 6 and 2, respectively. Therefore, the expected total number of transitions before the mouse first reaches an absorbing state is

$$2 + 6 + 2 = 10$$

Thus, in terms of the original situation, if the mouse begins in compartment 2, then the expected number of transitions before the mouse first reaches compartment 3, the one with the one-way doors, is 10. Naturally, in any specific case, the number of transitions may be more or less than 10. This number is simply an average over many repetitions of the experiment. ■

The method of Example 8.27 can also be used to answer this question: If the mouse begins in compartment 2 of the original maze (no one-way doors), what is the expected number of transitions before the mouse *first reaches compartment 3*? We simply imagine the original maze to be altered to one with a one-way door on compartment 3. The resulting system is an absorbing Markov chain (as in Example 8.21), and the expected number of transitions before absorption, given that the mouse began in compartment 2, provides an answer to the question.

The fundamental matrix can also be used to obtain other types of information about the system. For example, if the system has a single absorbing state, then the system will eventually reach that state, but if there is more than one absorbing state, then it may in general be absorbed in any one of them. Given the state in which the system begins, the likelihood that it will be absorbed in various absorbing states can be computed by using the fundamental matrix. A method to carry out the computation is described in Exercise 25.

Exercises for Section 8.4

1. A transition matrix of a Markov chain is given as

$$\begin{bmatrix} 0 & .5 & .5 \\ 0 & 1 & 0 \\ .8 & 0 & .2 \end{bmatrix}$$

(*a*) Determine whether it is the transition matrix of an absorbing Markov chain.
(*b*) If not, which condition of the definition of an absorbing Markov chain is violated?
(*c*) If so, determine for each nonabsorbing state the *minimum* number of transitions necessary to reach some absorbing state.

For Exercises 2 through 6, repeat Exercise 1 with the given transition matrix.

2. $\begin{bmatrix} .5 & 0 & .5 \\ 0 & 1 & 0 \\ .8 & 0 & .2 \end{bmatrix}$

3. $\begin{bmatrix} 1 & 0 & 0 & 0 \\ 0 & .2 & .2 & .6 \\ 0 & 0 & 1 & 0 \\ 0 & .5 & 0 & .5 \end{bmatrix}$

4. $\begin{bmatrix} .5 & 0 & .5 & 0 \\ 1 & 0 & 0 & 0 \\ 0 & 1 & 0 & 0 \\ 0 & 0 & 1 & 0 \end{bmatrix}$

5. $\begin{bmatrix} 1 & 0 & 0 & 0 \\ 0 & 1 & 0 & 0 \\ 0 & 0 & .8 & .2 \\ 0 & 0 & 1 & 0 \end{bmatrix}$

6. $\begin{bmatrix} 1 & 0 & 0 & 0 \\ \frac{1}{3} & 0 & \frac{2}{3} & 0 \\ 0 & 1 & 0 & 0 \\ 0 & 0 & 1 & 0 \end{bmatrix}$

7. For the transition matrix of Example 8.21, find the canonical form which corresponds to this relabeling:

Old label	1	2	3	4
New label	2	3	1	4

8. For the transition matrix of Example 8.21, find the canonical form which corresponds to this relabeling:

Old label	1	2	3	4
New label	2	4	1	3

9. Find the canonical form for the transition matrix (8.3) which corresponds to this relabeling:

Old label	1	2	3	4	5
New label	2	1	3	4	5

10. Repeat Exercise 9 for this relabeling:

Old label	1	2	3	4	5
New label	5	1	2	3	4

11. Write the transition matrix

$$\begin{bmatrix} 0 & \frac{1}{2} & \frac{1}{2} \\ 0 & 1 & 0 \\ \frac{1}{3} & \frac{1}{3} & \frac{1}{3} \end{bmatrix}$$

in canonical form, and find the associated fundamental matrix.

For Exercises 12 through 16, repeat Exercise 11 for the given transition matrix.

12. $\begin{bmatrix} 1 & 0 & 0 \\ 0 & 1 & 0 \\ \frac{1}{3} & 0 & \frac{2}{3} \end{bmatrix}$

13. $\begin{bmatrix} 0 & \frac{1}{2} & 0 & \frac{1}{2} \\ 0 & 1 & 0 & 0 \\ \frac{1}{4} & 0 & \frac{1}{2} & \frac{1}{4} \\ 0 & 0 & 0 & 1 \end{bmatrix}$

14. $\begin{bmatrix} \frac{1}{3} & \frac{2}{3} & 0 & 0 \\ 0 & \frac{1}{2} & \frac{1}{2} & 0 \\ 0 & 0 & 1 & 0 \\ 1 & 0 & 0 & 0 \end{bmatrix}$

15. $\begin{bmatrix} \frac{1}{2} & 0 & \frac{1}{4} & \frac{1}{4} \\ 1 & 0 & 0 & 0 \\ 0 & 1 & 0 & 0 \\ 0 & 0 & 0 & 1 \end{bmatrix}$

16. $\begin{bmatrix} \frac{1}{2} & \frac{1}{2} & 0 & 0 \\ 0 & 0 & 1 & 0 \\ 0 & 0 & 1 & 0 \\ 1 & 0 & 0 & 0 \end{bmatrix}$

17. Suppose that the absorbing Markov chain whose transition matrix is (8.7) is initially in state 4. Find the expected number of transitions before it reaches an absorbing state. (*Hint*: Use Example 8.24.)

Psychology

18. In the situation described in Example 8.21, suppose the mouse begins in compartment 1.
   (*a*) Find the expected number of times the mouse is in compartment 2 before it first reaches compartment 3.
   (*b*) Find the expected total number of transitions before the mouse first reaches compartment 3.

Psychology

19. Consider the arrangement of compartments introduced in Example 8.10. If the mouse begins in the nest, find the expected number of transitions before the mouse first reaches the compartment with the exercise device.

Employment

20. Consider the employment classification situation described in Exercise 21, Section 8.1. If a person is self-employed now, what is the expected number of years before she is first employed in small business?

Consumer choices

21. The automobile purchasing patterns of a computer scientist are described in Exercise 24, Section 8.2. If she purchases a Cadillac this year, in how many years will she first purchase a Mercedes?

Games of chance

22. A game is played on the board pictured in Figure 8.14. On each play a marker either remains in place or moves to an adjacent space on the board. On each play the marker is twice as likely to move as to remain in place, and if it moves, it is twice as likely to move clockwise as counterclockwise. If the marker is initially in the red space, find the expected number of plays before the marker first reaches the green space.

*23. An absorbing Markov chain has the transition matrix shown below. If the system is initially in state 3, find the expected number of transitions before it first reaches an absorbing state.

$$\begin{bmatrix} .5 & 0 & 0 & 0 & .5 & 0 \\ 0 & 1 & 0 & 0 & 0 & 0 \\ .2 & 0 & .2 & .2 & .2 & .2 \\ 0 & .4 & .2 & .2 & 0 & .2 \\ 0 & 0 & 0 & 0 & 1 & 0 \\ 1 & 0 & 0 & 0 & 0 & 0 \end{bmatrix}$$

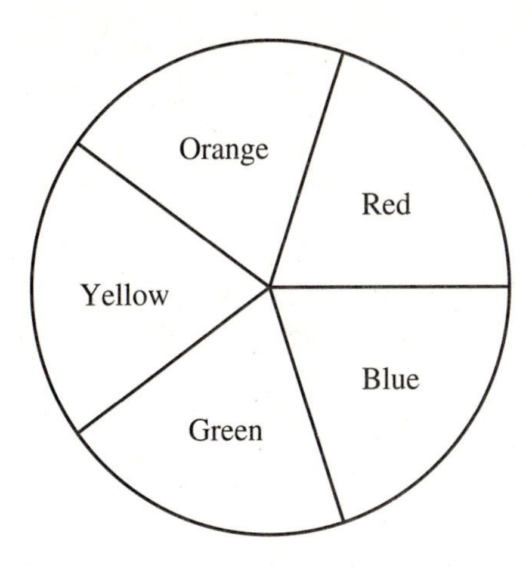

FIGURE 8.14

24. A Markov chain has the transition matrix shown below. If it begins in state 5, find the expected number of transitions before it first reaches state 3.

$$\begin{bmatrix} 0 & 1 & 0 & 0 & 0 & 0 \\ 0 & 0 & 1 & 0 & 0 & 0 \\ 0 & 0 & 0 & 1 & 0 & 0 \\ 0 & 0 & 0 & 0 & 1 & 0 \\ 0 & 0 & 0 & 0 & 0 & 1 \\ .5 & 0 & 0 & 0 & .5 & 0 \end{bmatrix}$$

25. The matrix $\mathbf{R}$ in the canonical form $\begin{bmatrix} \mathbf{I} & \mathbf{O} \\ \mathbf{R} & \mathbf{Q} \end{bmatrix}$ is an $(N - k) \times k$ matrix. Thus, the matrix product $\mathbf{NR}$ is also $(N - k) \times k$. Each row of $\mathbf{NR}$ is associated with a nonabsorbing state, and each column is associated with an absorbing state. Each row of $\mathbf{NR}$ is a probability vector. The (i, j) entry of $\mathbf{NR}$ is the probability that if the system begins in the ith nonabsorbing state, it will be absorbed in the jth absorbing state.
    (*a*) Suppose that the Markov chain with transition matrix (8.7) begins in the first nonabsorbing state. What is the probability that it is absorbed in the first absorbing state?
    (*b*) Suppose that this Markov chain begins in the second nonabsorbing state. What is the probability that it is absorbed in the first absorbing state?
26. Consider the absorbing Markov chain of Exercise 23. If the system is initially in state 3, find the probability that it is absorbed in state 2. (*Hint*: Use Exercise 25.)
27. Let a be a number such that $0 < a < 1$, and consider the transition matrix

$$\mathbf{P} = \begin{bmatrix} 1 & 0 \\ a & 1 - a \end{bmatrix}$$

Show that if the Markov chain with transition matrix $\mathbf{P}$ starts in state 2, then the expected number of times it is in state 2 before it is absorbed is $1/a$.

28. Let a and b be positive numbers such that $0 < a + b < 1$, and consider the transition matrix

$$\mathbf{P} = \begin{bmatrix} 1 & 0 & 0 \\ 0 & 1 & 0 \\ a & b & 1-a-b \end{bmatrix}$$

Suppose the Markov chain with transition matrix $\mathbf{P}$ starts in state 3. Find the expected number of moves before it is absorbed, and find the probability it is absorbed in state 1.

8.5 A MARKOV CHAIN MODEL FOR PLANT SUCCESSION

In many areas in temperate climates there is a natural progression of vegetation as time passes: from open meadow grasslands through brush and shrubs to young forests and eventually to mature forests. Even in the absence of interference by humans, there are events which significantly alter, and sometimes reverse, this progression. Such an event, and a very important one, is fire. Fires arise naturally through lightning strikes and are an important contributor to the perpetuation of grasslands. The natural progression of vegetation, and especially the occurrence of fires, is influenced by random events and therefore can be modeled by using stochastic processes. In addition to fires, other random events influencing vegetation include the introduction of seeds of plants not currently represented, the amount and timing of rainfall, the feeding habits of wildlife, and similar natural events. Under certain circumstances, Markov chains are appropriate as models for such situations.

To construct a Markov chain model for plant succession, we focus on a single area which we suppose is small enough that it can be classified into exactly one of four states. The state is determined by the dominant vegetation form: grassland (G), brush and shrubs (B), young forest (YF), and mature forest (MF). We suppose that the area is observed every decade and that the state or character of the area is noted at each observation. For the moment we consider the progression from state to state in the absence of fire. We assume that the period between observations is such that progression proceeds at most one step between successive observations. That is, if the area is grassland at one observation, then at the next observation it is either grassland or brush and shrubs, and so on for other plant types. We also assume that once an area becomes a mature forest, it remains so throughout the time of observation.

Continuing to consider the situation in the absence of fires, suppose the data support the assumption that one-step (10-year) transition probabilities are as follows:

$$\Pr[G \mid G] = .7 \text{ and } \Pr[B \mid G] = .3$$
$$\Pr[B \mid B] = .8 \text{ and } \Pr[YF \mid B] = .2$$
$$\Pr[YF \mid YF] = .5 \text{ and } \Pr[MF \mid YF] = .5$$
$$\Pr[MF \mid MF] = 1$$

Next, we turn to what happens when there is a fire. We consider only fires which are severe enough to cause an area to revert to grassland. Fires which have no effect on the state of the system or which cause the system to revert to a state other than grassland are not considered in this model. Suppose that in any decade a fire which reverts the area to grassland occurs with probability .1. We note that if the occurrence of such fires can be viewed as a Bernoulli process, then this assumption leads to the conclusion that the expected number of such fires is one per century. It follows that, in each decade, with probability .9 the area does not have a fire which affects the state of the system.

Combining our assumptions about the occurrence of fires and what happens in the two situations (no fire and fire), we find that between two successive observations we have the following:

Probability of transition from grassland to grassland
$$= \Pr[G \mid G \text{ and no fire}] \cdot \Pr[\text{no fire}] + \Pr[G \mid G \text{ and fire}] \cdot \Pr[\text{fire}]$$
$$= .7(.9) + 1(.1) = .73$$

This approach also yields transition probabilities for the other possible transitions. For example,

Probability of transition from grassland to brush
$$= \Pr[B \mid G \text{ and no fire}] \cdot \Pr[\text{no fire}] + \Pr[B \mid G \text{ and fire}] \cdot \Pr[\text{fire}]$$
$$= .3(.9) + 0(.1) = .27$$

Probability of transition from brush to grassland
$$= \Pr[G \mid B \text{ and no fire}] \cdot \Pr[\text{no fire}] + \Pr[G \mid B \text{ and fire}] \cdot \Pr[\text{fire}]$$
$$= 0(.9) + 1(.1) = .1$$

Probability of transition from brush to brush
$$= \Pr[B \mid B \text{ and no fire}] \cdot \Pr[\text{no fire}] + \Pr[B \mid B \text{ and fire}] \cdot \Pr[\text{fire}]$$
$$= .8(.9) + 0(.1) = .72$$

Probability of transition from brush to young forest
$$= \Pr[YF \mid B \text{ and no fire}] \cdot \Pr[\text{no fire}] + \Pr[YF \mid B \text{ and fire}] \cdot \Pr[\text{fire}]$$
$$= .2(.9) + 0(.1) = .18$$

We now have the entries in the first two rows of the transition matrix for this Markov chain. Entries in the last two rows can be determined by using the same approach. We collect all this information in a transition matrix which describes the plant succession, including the possibility of fire. The transition matrix **P**, with states as the types of vegetation which dominate the area, is

$$\mathbf{P} = \begin{array}{c} \\ G \\ B \\ YF \\ MF \end{array} \begin{array}{c} \begin{array}{cccc} G & B & YF & MF \end{array} \\ \begin{bmatrix} .73 & .27 & 0 & 0 \\ .1 & .72 & .18 & 0 \\ .1 & 0 & .45 & .45 \\ .1 & 0 & 0 & .9 \end{bmatrix} \end{array} \tag{8.8}$$

Suppose that the succession of vegetation is described by this model and that the area is observed over many decades. What is the probability that over the long run it is grassland? Since a power of the matrix **P** has all positive entries, this is a regular Markov chain, and we can answer the question by determining the stable vector for the chain. That is, we find the unique probability vector **W** which satisfies the equation **WP** = **W**. The system of equations **WP** = **W** is

$$[w_1 \quad w_2 \quad w_3 \quad w_4] \begin{bmatrix} .73 & .27 & 0 & 0 \\ .1 & .72 & .18 & 0 \\ .1 & 0 & .45 & .45 \\ .1 & 0 & 0 & .9 \end{bmatrix} = [w_1 \quad w_2 \quad w_3 \quad w_4]$$

or

$$\begin{aligned} .27w_1 - .1w_2 - .1w_3 - .1w_4 &= 0 \\ .27w_1 - .28w_2 \qquad\qquad\qquad &= 0 \\ .18w_2 - .55w_3 \qquad\quad &= 0 \\ .45w_3 - .1w_4 &= 0 \end{aligned}$$

The solution of this system which satisfies the additional condition $w_1 + w_2 + w_3 + w_4 = 1$ is

$$\mathbf{W} = [w_1 \quad w_2 \quad w_3 \quad w_4] = [.270 \quad .261 \quad .085 \quad .384]$$

From this we conclude that over the long term we would expect the area to be grassland about 27 percent of the time, brush about 26.1 percent of the time, young forest about 8.5 percent of the time, and mature forest about 38.4 percent of the time.

Next, suppose that through intervention it is possible to control fires if a decision is made to do so. Also suppose it is social policy to control fires in mature forests, and therefore once the area reaches a mature forest, it remains in that state. In this situation the transition matrix becomes

$$\begin{array}{c} \\ G \\ B \\ YF \\ MF \end{array} \begin{array}{c} \begin{array}{cccc} G & B & YF & MF \end{array} \\ \begin{bmatrix} .73 & .27 & 0 & 0 \\ .1 & .72 & .18 & 0 \\ .1 & 0 & .45 & .45 \\ 0 & 0 & 0 & 1 \end{bmatrix} \end{array}$$

This is the transition matrix of an absorbing Markov chain.

Suppose that the area is initially a grassland. How many years before it becomes a mature forest?

To answer the question, we determine the expected number of transitions required for the system to first reach state *MF*, given that it began in state *G*. Labeling nonabsorbing states in the order *G*, *B*, *YF*, the matrices **Q** and **N** (as defined in Section 8.4) are

$$\mathbf{Q} = \begin{bmatrix} .73 & .27 & 0 \\ .1 & .72 & .18 \\ .1 & 0 & .45 \end{bmatrix}$$

$$\mathbf{N} = \begin{bmatrix} 7.04 & 6.79 & 2.22 \\ 3.34 & 6.79 & 2.22 \\ 1.28 & 1.23 & 2.22 \end{bmatrix}$$

From this we conclude that if the system is initially in state 1 (grassland), then the expected number of transitions until the system first reaches state 4 (mature forest) is $7.04 + 6.79 + 2.22 = 16.05$, and consequently the expected number of years is about 160.

Exercises for Section 8.5

1. Suppose the area is now grassland. Using the model developed in this section which includes the possibility of fire, the model with transition matrix **P**, find the probability that it is a young forest in 40 years. Find the probability that it is grassland after 80 years.
2. As in Exercise 1, suppose the area is now grassland. Find the probability that it is a young forest in 40 years and grassland in 80 years. Find the probability that it is a young forest in 40 years and either grassland or a young forest in 80 years.
3. Suppose that a change in rainfall results in a decrease in the probability of a transition from a young forest to a mature forest. In particular, suppose that, in the absence of fire, the transition from young forest to mature forest is .4. If all other conditions remain the same as discussed in the text, find the stable vector for this model. In the long run, what percentage of the time will the area be mature forest?
4. Suppose that an increase in pollution results in a decrease in the probability of a transition from grassland to brush. In this case suppose that, in the absence of fire, the transition from grassland to brush is .2. If all other conditions remain the same as discussed in the text, find the stable vector for this model. In the long run, what percentage of the time will the area be mature forest?
5. Suppose that both the changes in rainfall and pollution described in Exercises 3 and 4 hold, but otherwise the situation is as described in the text. Find the new stable vector. What percentage of the time, in the long run, will the area be mature forest?
6. Show that the Markov chain with transition matrix **P** is a regular Markov chain.
7. Consider a situation similar to that described in this section except that the probability of a fire in any decade is .2. Find the stable probability vector in this case, and interpret each of the entries.

8. In the situation described in this section, suppose the area is initially a young forest. Find the expected number of observations before it first becomes a mature forest.
9. Suppose that there are two types of fires, minor and major. In a minor fire, mature forest remains mature forest, and young forest and brush revert to grassland. Suppose a minor fire occurs each decade with probability .05 and a major fire with probability .1. Construct a Markov chain model for this situation, and find the stable vector. Assume that the probabilities for succession in the absence of fire are as given in this section.
10. Consider the situation described in Exercise 9. If the area is initially grassland, find the expected number of years before it first becomes a mature forest.

IMPORTANT TERMS AND CONCEPTS

You should be able to describe, define, or give examples of and use each of the following:

State
Transition
Transition diagram
Transition matrix
Markov chain
State vector
Regular Markov chain
Long-run behavior
Stable probabilities
Absorbing state
Absorbing Markov chain
Canonical form
Fundamental matrix
Unit vector

REVIEW EXERCISES

1. A Markov chain has the transition matrix

$$\begin{bmatrix} .5 & .5 \\ .8 & .2 \end{bmatrix}$$

   (*a*) Find the transition diagram of this Markov chain.
   (*b*) If the system is initially in state 2, find the probability that it is in state 2 after two transitions.

2. A Markov chain has the transition matrix shown below.

$$\begin{bmatrix} 0 & 1 & 0 \\ .4 & .4 & .2 \\ .5 & 0 & .5 \end{bmatrix}$$

   (*a*) Find the transition diagram of this Markov chain.
   (*b*) If the system is initially in state 1, find the probability that it is in state 2 after two transitions. After four transitions.

3. A Markov chain has the transition diagram shown in Figure 8.15.
   (*a*) Find the transition matrix for this Markov chain.
   (*b*) If the system is initially in state 1, find the probability that it is in state 2 after two transitions.

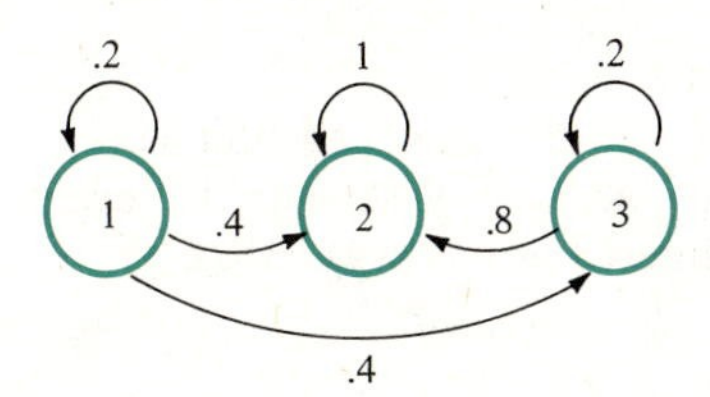

FIGURE 8.15

4. A Markov chain has the transition matrix

$$\begin{bmatrix} .8 & .2 \\ .4 & .6 \end{bmatrix}$$

   (*a*) If the initial state vector is $[.3 \quad .7]$, in what state is the system most likely to be after three transitions?
   (*b*) Find the vector of stable probabilities.

5. A Markov chain has the transition matrix

$$\mathbf{P} = \begin{bmatrix} 0 & 0 & 1 \\ 1 & 0 & 0 \\ 0 & .4 & .6 \end{bmatrix}$$

   (*a*) Find the matrix $\mathbf{P}(2)$ of two-step transition probabilities.
   (*b*) If the initial state vector is $[.3 \quad 0 \quad .7]$, find the state vector after two transitions.

6. A Markov chain has the transition matrix

$$\begin{bmatrix} \frac{1}{4} & \frac{1}{4} & \frac{1}{2} \\ \frac{1}{4} & \frac{3}{8} & \frac{3}{8} \\ \frac{1}{2} & 0 & \frac{1}{2} \end{bmatrix}$$

   (*a*) Find the transition diagram for this Markov chain.
   (*b*) Find the matrix of two-step transition probabilities.
   (*c*) If the initial state vector is $[\frac{3}{5} \quad \frac{1}{5} \quad \frac{1}{5}]$, what is the state vector after one transition?

7. A Markov chain which has the transition matrix

$$\begin{bmatrix} .3 & .2 & .5 \\ .1 & .6 & .3 \\ .4 & .3 & .3 \end{bmatrix}$$

   is initially in state 1. In what state is it most likely to be after two transitions?

8. A Markov chain has the transition matrix

$$\mathbf{P} = \begin{bmatrix} .5 & .5 \\ .2 & .8 \end{bmatrix}$$

(*a*) Find $\mathbf{P}(2)$ and $\mathbf{P}(4)$.
(*b*) Find the vector $\mathbf{W}$ of stable probabilities.
(*c*) How close are the rows of $\mathbf{P}(4)$ to the vector obtained in part (*b*)? That is, what is the largest entry in $\mathbf{P}(4) - \mathbf{WI}$?

9. A Markov chain has the transition matrix

$$\mathbf{P} = \begin{bmatrix} 0 & 0 & 1 \\ .6 & 0 & .4 \\ 0 & 1 & 0 \end{bmatrix}$$

(*a*) Determine whether this Markov chain is regular.
(*b*) If it is regular, find the vector of stable probabilities.

10. A Markov chain has the transition matrix

$$\begin{bmatrix} .5 & 0 & 0 & .5 \\ 1 & 0 & 0 & 0 \\ 0 & .4 & .4 & .2 \\ 0 & 0 & .5 & .5 \end{bmatrix}$$

(*a*) Determine whether this Markov chain is regular.
(*b*) If it is regular, find the vector of stable probabilities.
(*c*) If it is regular, in which state is the system most likely to be in the long run?

11. A Markov chain has the transition matrix

$$\begin{bmatrix} 0 & 1 & 0 & 0 \\ 0 & 0 & 0 & 1 \\ .5 & .5 & 0 & 0 \\ 0 & 0 & .5 & .5 \end{bmatrix}$$

(*a*) Determine whether this Markov chain is regular.
(*b*) If it is regular, find the vector of stable probabilities.
(*c*) If it is regular, in which state is the system most likely to be in the long run?

12. A Markov chain has the transition matrix

$$\begin{bmatrix} 0 & .5 & .5 & 0 \\ 1 & 0 & 0 & 0 \\ 0 & 0 & 0 & 1 \\ 0 & .5 & .5 & 0 \end{bmatrix}$$

(*a*) Determine whether this Markov chain is regular.
(*b*) If it is regular, find the vector of stable probabilities.
(*c*) If it is regular, in which state is the system most likely to be in the long run?

Ecology 13. A simplified model for forest succession is formulated as a Markov chain with the transition matrix given below.

$$\begin{array}{l} \\ \text{Blackgum} \\ \text{Red maple} \\ \text{Beech} \end{array} \begin{array}{c} \begin{array}{ccc} \text{Blackgum} & \text{Red maple} & \text{Beech} \end{array} \\ \begin{bmatrix} .55 & .25 & .20 \\ .15 & .55 & .30 \\ 0 & .05 & .95 \end{bmatrix} \end{array}$$

Suppose that a transition represents the evolution of the forest over 50 years. The forest now consists of 50 percent blackgum and 50 percent red maple.

(*a*) Find a state vector which describes the current composition of the forest.
(*b*) Find the composition of the forest in 100 years.
(*c*) Find the long-run composition of the forest.

14. The transition matrix of a Markov chain is

$$\begin{bmatrix} 1 & 0 & 0 & 0 \\ \frac{1}{2} & \frac{1}{2} & 0 & 0 \\ 0 & 0 & 1 & 0 \\ 0 & 0 & \frac{2}{3} & \frac{1}{3} \end{bmatrix}$$

(*a*) Verify that it is the transition matrix of an absorbing Markov chain.
(*b*) Determine the expected number of transitions to absorption for each initial state.

15. A Markov chain has the transition matrix

$$\begin{bmatrix} \frac{1}{4} & 0 & \frac{1}{4} & \frac{1}{2} \\ 0 & 1 & 0 & 0 \\ \frac{1}{3} & \frac{1}{3} & \frac{1}{3} & 0 \\ 0 & 0 & 0 & 1 \end{bmatrix}$$

(*a*) Verify that this is the transition matrix of an absorbing Markov chain.
(*b*) If the system is initially in state 3, determine the expected number of transitions before it reaches an absorbing state.

Psychology 16. Consider the arrangement of compartments introduced in Example 8.10. If the mouse begins in the nest, find the expected number of times the mouse is in the compartment with the exercise device before it first reaches the compartment with the water.

Games of chance 17. A game is played on the board pictured in Figure 8.14. On each play a marker either remains in place or moves to an adjacent space on the board. On each play the marker is twice as likely to move as to remain in place, and if it moves, it is equally likely to move clockwise and counterclockwise. If the marker is initially in the red space, find the expected number of times it is in the yellow space before it first reaches the green space.

18. Suppose that the Markov chain with the transition matrix given in Exercise 10 is initially in state 3. Find the expected number of transitions before it first reaches state 1.

19. An absorbing Markov chain has the transition diagram shown in Figure 8.16.
    (*a*) If the system begins in state 2, find the expected number of transitions before it reaches an absorbing state.
    (*b*) If the system begins in state 3, find the expected number of visits to state 2 before it reaches an absorbing state.

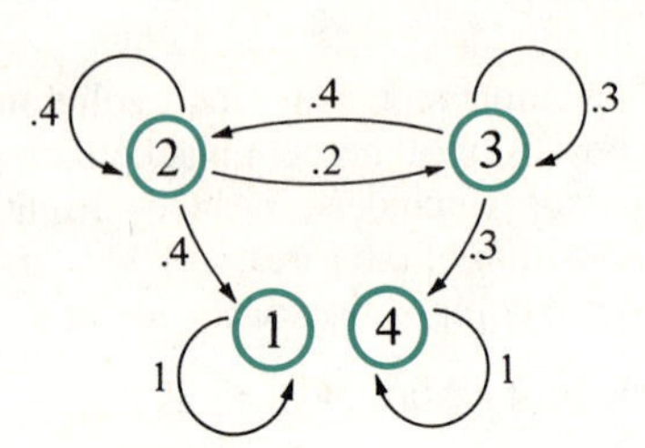

FIGURE 8.16

Employment

20. The employment shifts of people living in a small town are described in Exercise 19, Section 8.1. Using a Markov chain model, find the expected number of years before a person who is now self-employed is first employed in industry.

Consumer choices

21. The Sunday evening eating habits of a student are described in Exercise 23, Section 8.1. If she eats this week in a Chinese restaurant, use a Markov chain model to find the expected number of weeks before she first eats in an Italian restaurant.

Consumer choices

22. A consumer organization surveyed 1000 individuals immediately after each had purchased a headache remedy. Four brands were considered, A, B, C, and X, and purchase probabilities were determined. In particular, part of the data obtained in the survey can be summarized as follows:

PURCHASE PROBABILITIES

Previous Purchase	This Purchase			
	A	B	C	X
A	.2	.3	.2	.3
B	0	.5	.2	.3
C	.1	0	.8	.1
X	.2	0	0	.8

Assume that this situation can be modeled as a Markov chain.
   (*a*) Determine the probability that an individual who buys brand A this time will buy brand X on the second subsequent purchase.
   (*b*) Which users (A, B, or C) will become brand X users in the smallest expected number of purchases?
   (*c*) Determine the anticipated long-run distribution of users.

Consumer choices

23. In the situation described in Exercise 22, a person is currently using brand X. Find the expected number of purchases before brand A is first used.

Psychology

24. A mouse moves at random in the maze shown in Figure 8.17. In particular, suppose that the mouse moves in such a way that
    (1) It is twice as likely to change compartments as to remain where it is.
    (2) If it changes compartments, then the mouse is equally likely to make any of the possible moves.

    Observations are made every 5 minutes and every time the mouse changes compartments.
    (*a*) Model this as a Markov chain, and find the transition matrix.
    (*b*) Find the stable probability vector for this situation.

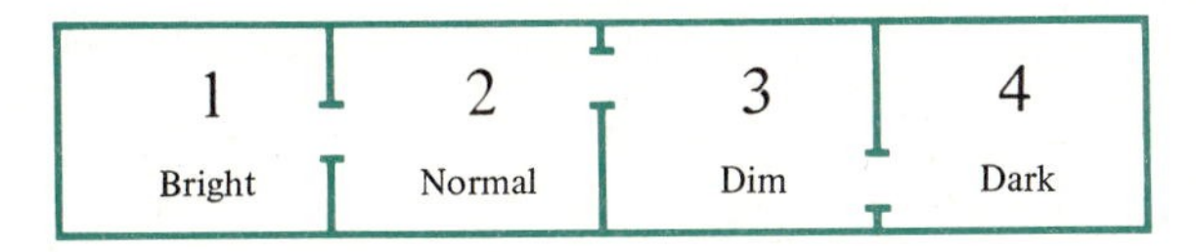

FIGURE 8.17

Psychology

25. In the situation described in Exercise 24, suppose that the mouse begins in the dark compartment. Find the expected number of observations before the mouse first reaches the bright compartment.

Psychology

26. Consider the situation described in Exercise 24. Suppose that (1) remains valid and that (2) is replaced by the following: If the mouse changes compartments and has a choice, then the mouse is twice as likely to move toward a darker compartment as a lighter one. Answer (*a*) and (*b*) in this situation.

27. An absorbing Markov chain has the transition matrix

$$\begin{bmatrix} .2 & .2 & .2 & .2 & .2 \\ 0 & 1 & 0 & 0 & 0 \\ .5 & 0 & .5 & 0 & 0 \\ 0 & 0 & 0 & 1 & 0 \\ 0 & 0 & 0 & 0 & 1 \end{bmatrix}$$

    (*a*) Find the fundamental matrix for this absorbing chain.
    (*b*) If the system begins in state 3, find the expected number of transitions before it reaches an absorbing state.

28. An absorbing Markov chain has the transition matrix

$$\begin{bmatrix} 0 & .5 & 0 & .5 & 0 \\ 0 & 1 & 0 & 0 & 0 \\ 0 & 0 & 1 & 0 & 0 \\ .2 & 0 & .4 & 0 & .4 \\ 0 & .3 & 0 & .3 & .4 \end{bmatrix}$$

    (*a*) Find the fundamental matrix for this absorbing chain.
    (*b*) If the system begins in state 1, find the expected number of visits to state 5 before it reaches an absorbing state.

29. Suppose a is a number such that $0 < a < 1$. Show that the Markov chain with the transition matrix $\mathbf{P}$ is regular, where

$$\mathbf{P} = \begin{bmatrix} 0 & 1 & 0 \\ 0 & 0 & 1 \\ a & 0 & 1-a \end{bmatrix}$$

30. Find the vector of stable probabilities for matrix $\mathbf{P}$ of Exercise 29.

HAPTER

9

Interest, Mortgages, and Financial Decision Making

THE SETTING AND OVERVIEW

Making decisions about the use of money is almost unavoidable. Can I afford to buy or lease that new car? What is the best way for me to finance the purchase of a house or condominium? How should I invest my inheritance from Aunt Matilda? To give useful answers to such questions, we need to understand and be able to compare various ways of measuring and expressing interest, time payment plans, mortgages, and investment options. Such questions are the topic of this chapter. We begin with a discussion of savings accounts and continue with annuities, time payment plans, stocks, bonds, and similar investments.

Many of the examples and exercises in this chapter require more computation than has been common in previous chapters, and a calculator or computer should be used. In the computations in the examples (and in the answers to the exercises, Appendix C), 10 digits were used in the calculations. In most cases, a calculator which uses eight or even six digits will be adequate.

INTEREST

If you borrow money, you should expect to pay a fee to do so, and if you lend money, you should expect to receive a fee. *Interest* describes the fee paid (or received) for the use of money. The *interest rate* is the percentage obtained by expressing the amount of interest as a percentage of the amount of money used. Interest is paid for the use of money for a certain period of time, and consequently an interest rate is expressed as a percentage per unit time. For example, if the interest is \$80 when \$1000 is borrowed for 1 year, then the interest rate is 80/1000 = .08 = 8 *percent per year.* As another example, if the interest is \$50 when \$1000 is used for 6 months, then the interest rate is 50/1000 = .05 = 5 *percent per half-year.*

Because interest rates are compared so frequently, it is customary, and in some cases required, for them to be expressed using a year as the time period. This can be done in more than one way. Thus, in the last example we could assume that the fee for using \$1000 for the second 6 months of the year is the same as for the first 6 months, namely, \$50. Therefore, with this assumption the fee for the year is \$50 + \$50 = \$100, and the interest rate is 100/1000 = .10, or 10 percent per year. Other common assumptions would yield slightly different rates. We return to this topic later in the section with a more complete discussion, including definitions of commonly used terms.

In any transaction involving borrowed money, the amount of money borrowed is called the *principal* of the loan. The basic formula

$$I = Prt \tag{9.1}$$

relates the principal P, the interest rate per unit time r, the number of time periods t, and the amount of interest I. Given any three of the four quantities I, P, r, and t, the fourth can be determined by using Equation (9.1).

Example 9.1 Find the interest on a loan of \$5000, given that the interest rate is 9 percent per year and the term of the loan is 6 months.

Solution Using formula (9.1) with P and I measured in dollars and t measured in years, we have $P = 5000$, $r = .09$, and $t = .5$. Thus, the amount of interest (in dollars) is

$$I = 5000(.09)(.5) = 225$$

■

Example 9.2 *Agriculture*

A farmer has an opportunity to lease an additional field to plant a larger soybean crop. The lease on the land will cost \$6000, which must be paid in advance. Seed, fertilizer, herbicide, and other costs associated with planting the additional field will amount to \$9000. The farmer estimates that he can sell the crop for \$17,800.

The farmer has no excess cash, and he will need to borrow for 8 months the \$15,000 necessary to produce the crop. What interest rate per year can he afford to pay and still make a profit of at least \$1000?

Solution The operating profit is \$17,800 − \$15,000 = \$2800, and the farmer will make a profit of at least \$1000 as long as his interest expense is less than \$1800.

We apply Equation (9.1) with $P = \$15{,}000$, $I = \$1800$, and $t = \frac{8}{12}$: $1800 = 15{,}000r(\frac{8}{12})$. We have

$$r = 1800\left(\frac{12}{8}\right)\left(\frac{1}{15{,}000}\right) = .18$$

We conclude that if the farmer can borrow money at an interest rate of less than 18 percent per year, then he can make a profit of at least \$1000 by leasing the field. ■

If a person deposits money in a savings account at a bank, the bank has, in effect, borrowed that person's money, and accordingly, the bank pays the depositor interest. The bank may compute and pay this interest in different ways with different rates, depending upon the type of account and the prevailing interest rates in the economy. If interest is paid on only the original amount in the account and new deposits (but *not* on accumulated interest), then the account is said to pay *simple interest*. A more common type of account is one in which interest is added to the account periodically and in effect becomes a deposit into the account. In this case the depositor earns "interest on interest." Interest computed in this way is said to be *compounded*, and the account is said to pay *compound interest*.

Example 9.3

Savings

The Fourth Federal Bank of Mulberry offers a savings account in which the interest is paid at a rate of 4 percent per year, compounded quarterly (every 3 months) and automatically credited to the account. At the end of each quarter, the interest is computed by using as the principal the minimum amount in the account during the quarter. Assuming no deposits or withdrawals, the amount on deposit at the beginning of each quarter consists of the original amount deposited into the account plus all interest credited to the account.

Jessica opens such an account on July 1 into which she initially deposits \$1000. How much will be in the account after 5 years, assuming there are no deposits or withdrawals other than the interest being credited to the account by the bank?

Solution At the start of the first quarter (3-month period), there is \$1000 in Jessica's account. At the end of the quarter, interest is computed by using formula (9.1), and the result is

$$I = Prt = \$1000(.04)\left(\tfrac{1}{4}\right) = \$10.00$$

This amount is now added to the account, and we assume that the total, \$1010, is left in the account for the second quarter. At the end of the second quarter, interest is computed on a principal of \$1010 and is added to the account. By using formula (9.1) again the interest is $I = Prt = \$1010(.04)(\frac{1}{4}) = \$10.10$, and this is added to the amount in the account, giving a new balance of \$1020.10. Repeating this argument a total of 20 times (5 years is 20 quarters) gives the final amount. The columns headed "Compound Interest" in Table 9.1 give the amounts deposited in her account each quarter and the balance at the end of each quarter. The answer to Jessica's question is the entry in color in Table 9.1. ■

Table 9.1 gives a comparison of the amounts in the account at the end of each quarter for 5 years when simple interest and compound interest are paid.[1] In this instance there is only one deposit, and consequently simple interest payments are always the same. This contrasts with the compound interest payments which increase during the 5-year period. More precisely, the entries in Table 9.1 under the headings "Quarterly Payment" are all obtained from the basic interest formula $I = Prt$. In the case of simple interest, P, r, and t are the same every quarter, and hence I is always \$10 [$= \$1000(.04)(\frac{1}{4})$]. In the case of compound interest, the quantities r and t are always the same, but the principal P increases each quarter by the amount of interest paid the previous quarter. If we denote the original principal by P, the principal at the end of quarter 1 by P_1, and the principal at the end of quarter k by P_k, then the interest paid at the end of quarter $k + 1$ is $P_k rt$. We denote this interest by I_{k+1}, that is $I_{k+1} = P_k rt$. Thus

$$P_{k+1} = P_k + I_{k+1} = P_k + P_k rt = P_k(1 + rt)$$

We can use this last result to obtain a formula which will enable us to compute P_n for any n without first computing all P_k and I_k for $k < n$. Such a formula is very useful in problems such as that posed in Example 9.3. In that example we were interested in the last row of Table 9.1, but we had to compute all the intermediate rows to get there. The general formula can be understood by looking at some particular cases from Example 9.3. Since $P = \$1000$, we have (in dollars)

$$\begin{aligned} P_1 &= P(1 + rt) = 1000[1 + (.04)(\tfrac{1}{4})] = 1000(1 + .01) \\ P_2 &= P_1(1 + rt) = 1000(1.01)[1 + (.04)(\tfrac{1}{4})] = 1000(1.01)^2 \\ P_3 &= P_2(1 + rt) = 1000(1.01)^2[1 + (.04)(\tfrac{1}{4})] = 1000(1.01)^3 \end{aligned}$$

In fact, for any number n of periods, we have

$$P_n = 1000(1.01)^n \tag{9.2}$$

[1]In actual practice, a savings institution may obtain slightly different values for the entries in Table 9.1 depending on the number of digits used in the computations and on its policy of rounding numbers.

TABLE 9.1

Quarter	Simple Interest		Compound Interest	
	Quarterly Payment	Total in Account at End of Quarter	Quarterly Payment	Total in Account at End of Quarter
1	$10	$1010	$10.00	$1010.00
2	10	1020	10.10	1020.10
3	10	1030	10.20	1030.30
4	10	1040	10.30	1040.60
5	10	1050	10.41	1051.01
6	10	1060	10.51	1061.52
7	10	1070	10.62	1072.14
8	10	1080	10.72	1082.86
9	10	1090	10.83	1093.69
10	10	1100	10.94	1104.62
11	10	1110	11.05	1115.67
12	10	1120	11.16	1126.83
13	10	1130	11.27	1138.09
14	10	1140	11.38	1149.47
15	10	1150	11.49	1160.97
16	10	1160	11.61	1172.58
17	10	1170	11.73	1184.30
18	10	1180	11.84	1196.15
19	10	1190	11.96	1208.11
20	10	1200	12.08	1220.19

In Equation (9.2) the quantity .01 is the *interest per quarter*, $(.04)(\frac{1}{4}) = .01$, and n is the number of quarters in the period of concern. The general formula for compound interest can be derived in a similar manner.

Formula for Compound Interest

If a principal P is invested for a period consisting of n time units, and if interest is compounded and added at rate k per unit of time, then the total investment at the end of the period is

$$P_n = P(1 + k)^n \tag{9.3}$$

Note In formula (9.3), k is the annual rate of interest *only* if the unit of time is 1 year; otherwise, it is the rate for the unit of time considered. In the situation of Example 9.3, the annual rate of interest was 4 percent, and the unit of time was $\frac{1}{4}$ year. Thus in that case $k = (.04)(\frac{1}{4}) = .01$. In general, if r is the annual interest rate and T is the number of units of time per year, then $k = r/T$. Just above, the unit of time was a quarter, so $T = 4$.

Example 9.4

Savings

An amount of $1000 is invested for 10 years, and interest is compounded and added every 6 months at the rate of 5 percent per year. Find the total amount of interest added to the investment by the end of the tenth year.

Solution We use formula (9.3) to compute the total of the original investment (the principal) and interest. Here the unit of time is 6 months, or $\frac{1}{2}$ year, and there are 20 units of time in the period considered; that is, $n = 20$. Since the interest rate is 5 percent *per year*, it is 2.5 percent per *unit time*; that is, $k = .05(\frac{1}{2}) = .025$. Consequently, applying Equation (9.3), we conclude that the total investment at the end of 10 years is

$$\begin{aligned} P_{20} &= \$1000(1 + .025)^{20} \\ &= \$1000(1.025)^{20} \\ &= \$1000(1.63862) = \$1638.62 \end{aligned}$$

Therefore, the amount of interest added is

$$\$1638.62 - \$1000 = \$638.62$$

■

Frequently, especially in their advertising, savings institutions will extoll the virtues of their own savings accounts, and in particular they will call attention to the frequency with which they compute and pay compound interest. Certainly the frequency of payment makes a difference when interest is compounded, and it is interesting (no pun intended) to consider and evaluate this difference. Suppose that four different institutions pay the same basic rate, say, 6 percent per year, but one compounds and pays interest yearly, one semiannually (every 6 months), one quarterly, and one monthly. What is the difference in the amounts in the various accounts after 10 years, given the same initial investment of $1000 and no deposits or withdrawals?

We can answer the question just posed by using formula (9.3) for each of the four accounts. The results are shown in Table 9.2.

It is clear from Table 9.2 that a saver profits by having interest compounded and paid frequently. However, the gain for the entire 10 years in going from two periods per year to four periods ($7.91) is not as great as the gain in going from one to two periods per year ($15.26). Moreover, the gain in going from four periods per year to 12 is smaller yet ($5.38). Thus, at least for this example, it appears that there would be relatively little gain in compounding and paying interest even more frequently, say, weekly or daily. In fact, the amount in the account after 10 years assuming daily compounding is $1822.03 (Exercise 12), only $2.63 more than the amount resulting from monthly compounding, slightly more than 25 cents a year.

TABLE 9.2

Frequency of Interest Payments	Initial Investment	Formula for P_n	Amount in Account after 10 Years
Annually	\$1000	$1000(1+.06)^{10}$	\$1790.85
Semiannually	1000	$1000(1+.03)^{20}$	1806.11
Quarterly	1000	$1000(1+.015)^{40}$	1814.02
Monthly	1000	$1000(1+.005)^{120}$	1819.40

The comparisons illustrated in this example hold in general; that is, unless the amounts involved are quite large, the differences between quarterly, monthly, and daily compounding are relatively small. It is important to remember throughout this discussion that the rate of interest is assumed to be constant. Many money market mutual funds compound interest daily, but normally the *rate* also changes daily. The amount of the investment in such funds can be computed by using Equation (9.1) repeatedly with changing values of r as well as changing values of P.

We conclude this section by returning to the description of interest rates in ways that make comparisons relatively straightforward. The basic objective of many rules and regulations governing advertising by banks, loan companies, and other businesses dealing with money matters is that information given to customers make it easy to compare one rate with another. Most such rules require that rates be given on an annual basis. The term "annual percentage yield (APY)" is commonly used by financial institutions in describing rates for certificates of deposit and savings accounts. It provides an answer to the following question: If interest is compounded and paid at time intervals other than once a year, what rate of interest compounded and paid once a year gives the same return? A precise definition is as follows.

Annual Percentage Yield

Suppose a principal P increases to an amount A in a period of t years. Then the *annual percentage yield* (APY) is

$$\text{APY} = 100\left[\left(\frac{A}{P}\right)^{1/t} - 1\right] \tag{9.4}$$

measured in percent per year.

When interest is *paid* by the consumer rather than received, the corresponding term is *Annual Percentage Rate*, APR.

Example 9.5 Find the annual percentage yield for an account in which the interest is compounded quarterly at the rate of 6 percent per year.

Savings

Solution We use Equation (9.3) to determine the value of $1 on deposit in such an account for 1 year. We have $P = 1$, $k = (.06)(\frac{1}{4}) = .015$, and $n = 4$. The amount A after 1 year is

$$A = (1 + .015)^4 = 1.06136$$

Now, applying Equation (9.4) with $A = 1.06136$, $P = 1$, and $t = 1$, we find the annual percentage yield to be

$$\text{APY} = 100\left[\left(\frac{1.06136}{1}\right)^{1/1} - 1\right] = 100(1.06136 - 1) = 6.136$$

That is, the annual percentage yield is 6.136 percent. ■

Example 9.6 A $10,000 six-month certificate of deposit pays interest at the rate of 4.8 percent per year. What is the annual percentage yield of such a certificate?

Savings

Solution We begin by using Equation (9.3) to determine the value of the certificate after 6 months, or $\frac{1}{2}$ year. The interest paid at the end of 6 months is

$$\begin{aligned} I &= \$10{,}000(.048)(\tfrac{1}{2}) \\ &= \$240 \end{aligned}$$

Thus, the value of the certificate after 6 months is $10,240. We have $P = \$10{,}000$ and $t = .5$. Therefore, applying Equation (9.4), we have

$$\begin{aligned} \text{APY} &= 100\left[\left(\frac{\$10{,}240}{\$10{,}000}\right)^{1/.5} - 1\right] \\ &= 100[(1.024)^2 - 1] \\ &= 4.86 \end{aligned}$$

and the annual percentage yield is 4.86 percent. ■

Example 9.7 Steve has an account in a money market mutual fund that pays interest daily at a rate which usually changes from day to day. In a quarter during which he makes no deposits or withdrawals, his account increases from $6754.85 to $6842.00. Find the annual percentage yield for this period.

Solution We use Equation (9.4) with $A = 6842.00$, $P = 6754.85$, and $t = .25$. Then $1/t = 4$ and

$$\begin{aligned} \text{APY} &= 100\left[\left(\frac{6842.00}{6754.85}\right)^4 - 1\right] \\ &= 100[(1.0129)^4 - 1] \\ &= 5.26 \end{aligned}$$

Thus, his annual percentage yield for the quarter was 5.26 percent. ■

Exercises for Section 9.1

1. In each of the following situations, find the amount of interest earned on an investment.
   (*a*) Principal of \$2000, interest rate of 5 percent per year compounded annually, period of 1 year
   (*b*) Principal of \$4000, interest rate of 6 percent per year compounded semiannually, period of 6 months
2. In each of the following situations, find the amount of interest earned on an investment.
   (*a*) Principal of \$2000, interest rate of 5 percent per year compounded semiannually, period of 1 year
   (*b*) Principal of \$5000, interest rate of 6 percent per year compounded quarterly, period of 1 year
3. In each of the following situations, find the amount of interest earned on an investment.
   (*a*) Principal of \$200, interest rate of 4 percent per year compounded quarterly, period of 2 years
   (*b*) Principal of \$10,000, interest rate of 5 percent per year compounded quarterly, period of 5 years
4. In each of the following situations, find the amount of interest earned on an investment.
   (*a*) Principal of \$500, interest rate of 4.5 percent per year compounded semiannually, period of 2 years
   (*b*) Principal of \$1000, interest rate of 6.5 percent per year compounded quarterly, period of 4 years
5. Prepare a table similar to Table 9.2 for an interest rate of 4 percent per year and a period of 10 years.
6. Prepare a table similar to Table 9.2 for an interest rate of 7.5 percent per year and a period of 8 years.

Real estate

7. A land speculator has an opportunity to buy 20 acres of vacant land near the Metroburg Industrial Park for \$100,000. She feels certain that she can resell the land within 5 years for \$175,000. The Fourth Federal Bank of Metroburg is willing to lend her \$100,000 for 5 years at an annual interest rate of 10 percent compounded yearly with principal and interest due in 5 years. If she borrows the money and holds the land for the full 5 years and then sells it for \$175,000, will she make a profit or incur a loss? How much?

Real estate 8. In the situation of Exercise 7, what is the maximum annual interest rate she can pay and still make a profit of at least $10,000?

Real estate 9. In the situation of Exercise 7, the speculator believes that if she holds the land for 10 years, she will be able to sell it for $250,000. In this case, the Fourth Federal Bank of Metroburg is willing to lend her $100,000 for 10 years at an annual interest rate of 11 percent compounded yearly with principal and interest due in 10 years. If she borrows the money and holds the land for the full 10 years and then sells it for $250,000, will she make a profit or incur a loss? How much?

Savings 10. Sam has $1000 to invest for a period of 5 years. Metroburg Bank offers an account with interest at the annual rate of 4.35 percent compounded annually, and Fourth National Bank offers an account with interest paid at the rate of 4.2 percent per year compounded quarterly. Which account will have the greater value at the end of 5 years?

Savings 11. An amount of $10,000 is to be invested for 5 years. Is it more profitable to invest it in an account which pays 6 percent per year compounded daily or in an account which pays 6.25 percent per year compounded quarterly?

Savings 12. Show that if interest is 6 percent per year and is compounded daily, then in 10 years a deposit of $1000 grows to $1822.03.

Savings 13. Not all financial institutions have the same conventions regarding daily compounding of interest. Some use a 360-day year, and others use the actual number of days (365 or 366). Suppose that $100,000 is deposited in an account which pays 8 percent interest compounded daily. Find the interest for 1 year, using (*a*) a 360-day year and (*b*) a 365-day year. Note that the difference is only 2 cents on a deposit of $100,000.

Savings 14. Show that a deposit drawing interest at 8 percent compounded quarterly will double in value in less than 9 years.

Savings 15. The Safety First Savings and Loan Association increases the rate paid on savings accounts from 5 to 5.5 percent, each compounded quarterly. What are the old and new annual percentage yields?

Economics 16. If inflation is at an annual rate of 4.5 percent, what interest rate must the Safety First Savings Bank offer its customers so that their savings grow at least as fast as inflation? Safety First compounds interest quarterly.

Real estate 17. An investor purchased a tract of unimproved land in 1980 for $80,000 and sold it in 1995. After all expenses were paid, he had a net profit (before taxes) of $60,000. What is the effective annual yield on his investment of $80,000? That is, at what interest rate r compounded yearly would $80,000 grow to $140,000 in 15 years?

Real estate 18. In the situation of Exercise 17, the investor believes that if he holds the land until the year 2000, then he will be able to sell it for $170,000. If he does so, what would be his annual percentage yield on his original investment of $80,000?

Savings 19. On January 1, 1994, Jane deposited $2000 in an account which pays 4.4 percent interest compounded quarterly. On July 1 she withdrew $500, and on January 1, 1995, she withdrew another $500. How much was in the account on July 1, 1995?

Savings 20. On January 1, 1992, Sam deposited $5000 in an account which pays 4 percent interest compounded quarterly. On January 1, 1993, he withdrew $500, and on July 1, 1993, he withdrew $1000. How much was in the account on July 1, 1995?

Financial planning 21. Henry has won a \$100,000 prize in the lottery, and he plans to set aside a portion of his winnings for his daughter's college education. She will begin college in 12 years. He sets aside \$20,000 and deposits it in an account which pays interest at a rate of 4.5 percent compounded quarterly.

(*a*) How much will be available in 12 years?

(*b*) If he withdraws \$5000 each year for 4 years, starting at the end of year 12, how much remains in the account after the last withdrawal?

Financial planning 22. Sarah inherits \$100,000 from Aunt Matilda, and Sarah plans to set aside a portion of her inheritance for her son's college education. She would like to have a fund of \$30,000 when he begins college in 14 years. If she can earn 5 percent per year compounded quarterly, how much does she need to deposit now to achieve her goal?

Financial planning 23. In the situation of Exercise 22, suppose Sarah would like to have the \$30,000 for her son's education in 14 years and an additional \$1000 for him as a graduation present in 18 years. How much does she need to deposit now to achieve this goal?

Financial planning 24. In the situation of Exercise 22, Sarah would like to have \$7000 in 14 years, \$7300 in 15 years, \$7700 in 16 years, and \$8000 in 17 years. How much does she need to deposit now to achieve this goal?

Investments 25. It is common for interest rates available on investments to fluctuate from one period to another. Suppose that over 5 consecutive years the interest rates available on 1-year certificates of deposit are 5, 4, 5, 5, and 6 percent. Find the annual percentage yield on \$1000 deposited at the beginning of this period and reinvested together with accumulated interest for a total of 5 years. (*Hint*: One way to solve the problem is to begin with an estimate of the annual percentage yield, compare the results of compounding at this rate with the results obtained with the given rates, and then refine your estimate to obtain the annual percentage yield correct to 0.1 percent. A reasonable initial estimate would be the average of the five yearly rates.)

Investments 26. An investor believes that he will be able to earn interest over the next 10 years at the following rates: 5 percent per year for 4 years, 6 percent per year for the next 4 years, and 7 percent per year for the next 2 years. All rates are compounded quarterly. What is the annual percentage yield of a 10-year investment of \$1000 under these circumstances? (*Hint*: See the hint in Exercise 25.)

Investments 27. An investment increases by 10 percent in one year and decreases by 10 percent in the next year. Find the annual percentage yield for the 2-year period.

Economics 28. The consumer price index increases from 275 to 310 in a 5-year period. Find the yearly rate of increase which, when compounded, leads to this result. [*Hint*: One way to solve the problem is to begin with an estimate of the yearly rate, compare the effect of compounding at this rate with the given increase, and then refine your estimate to obtain a rate correct to 0.1 percent. A reasonable initial estimate of the yearly rate would be the total percent increase divided by 5 (the number of years).]

Investments 29. An investor purchases 100 shares of Universal Widget Corporation stock on October 1, 1994, for \$2000. She receives dividends of \$100 on December 1, 1994, March 1, 1995, and June 1, 1995; and she immediately deposits the funds in a savings account, where they earn interest at the rate of 4 percent compounded daily. On July 1, 1995, she sells the stock for \$2200. Find the annual percentage yield on this investment for the period from October 1, 1994, to July 1, 1995.

Investments 30. In the situation described in Exercise 29, suppose the stock is sold on July 1, 1995, for \$1800. All other aspects of the situation remain unchanged. In this case, find the annual percentage yield on the investment for the period from October 1, 1994, to July 1, 1995.

9.2 THE PRESENT VALUE OF FUTURE PAYMENTS

How much should you be willing to pay for the right to receive a specified amount of money at a specified time in the future? To answer this question intelligently, you need more information. In particular, you must know exactly when the money is to be paid and what interest rate you could obtain on money invested between now and the time of payment. Normally interest rates fluctuate, but you must assume that there is one interest rate which holds for the entire period. In addition to the interest rate, you must also know how often interest will be compounded and paid.

The term "present value" is usually used to indicate the amount you should be willing to pay to receive a certain payment at a specified time in the future.

Present Value

The *present value* of an amount A which is to be paid after M periods of time with an interest rate of k per period is the principal P which, when invested at rate k for M periods, increases to the value A.

In Section 9.1 we showed that a principal P invested for M periods of time with interest compounded at a rate of k per period will increase in value to $P(1 + k)^M$. In the present situation we know A, k, and M, and we seek P such that $A = P(1 + k)^M$. Thus

$$P = \frac{A}{(1 + k)^M} \tag{9.5}$$

is the present value of A.

Example 9.8 Find the present value of \$100 to be paid in 5 years if the interest rate is 6 percent per year compounded semiannually.

Solution The semiannual interest rate is 3 percent, and in 5 years there are 10 semiannual periods. From Equation (9.5), the present value P of the \$100 is

$$P = \frac{\$100}{(1 + .03)^{10}} = \frac{\$100}{1.3439} = \$74.41$$

■

In some consumer loans, the interest is deducted from the funds provided to the borrower. Such loans are called *discount loans*, although they are not necessarily bargains in the usual sense. For example, a person who borrows \$3000 for 1 year may be given only \$2500 and yet be expected to pay back \$3000 a year later. The annual interest on such a loan is

$$\frac{\$3000 - \$2500}{\$2500} = \frac{\$500}{\$2500} = .20, \text{ or 20 percent per year}$$

A related problem, again involving present value, is the following.

Example 9.9 Suppose a person borrows money for 6 months at an annual rate of 10 percent. If he must pay back \$3000 in 6 months, and if this amount includes both principal and interest, how much does he receive at the beginning of the 6-month period?

Solution We seek a principal P which will grow to \$3000 in 6 months at an annual interest rate of 10 percent. In other words, we seek the present value of \$3000 to be paid in 6 months at a 10 percent annual interest rate. Using Equation (9.5) with a 6-month rate of 5 percent, we have

$$P = \frac{\$3000}{(1 + .05)^1} = \$2857.14$$

■

Example 9.10 The Homestake Construction Company owes the Fourth Federal Bank \$10,000 payable in 4 years and \$20,000 payable in 6 years. The company asks the bank to consolidate the loans into one loan due in 5 years. If interest is 12 percent compounded semiannually, what should be the amount due after 5 years?

Corporate finance

Solution 1 One method of solving this problem is to compute the combined present values of the two loans and then to consider what this amount will be in 5 years if it is invested at 12 percent compounded semiannually.

The present value of the \$10,000 loan is [using (9.5), with $k = .06$ and $M = 8$]

$$P_1 = \frac{\$10,000}{(1.06)^8} = \frac{\$10,000}{1.593848} = \$6274.12$$

and the present value of the \$20,000 loan is [using (9.5), with $k = .06$ and $M = 12$]

$$P_2 = \frac{\$20,000}{(1.06)^{12}} = \frac{\$20,000}{2.012196} = \$9939.39$$

Thus the present value of both loans is

$$P_3 = P_2 + P_1 = \$16{,}213.51$$

In 5 years a principal with value P_3, invested at 12 percent compounded semiannually, will increase to an amount

$$A_1 = P_3(1.06)^{10} = (\$16{,}213.51)(1.7908477) = \$29{,}035.93$$

In other words, the company should replace the two loans with one loan of \$29,035.93 payable in 5 years.

Solution 2 A second method of converting these two loans to a single 5-year loan is to consider the value of the \$10,000 loan after 5 years and to add to this the value of the \$20,000 loan after 5 years. The amount \$10,000 payable after 4 years will grow during the fifth year to an amount

$$A_2 = (\$10{,}000)(1.06)^2 = \$11{,}236.00$$

Also after 5 years the \$20,000 loan payable after 6 years will have the present value

$$P_4 = \frac{\$20{,}000}{(1.06)^2} = \$17{,}799.93$$

Thus a single 5-year loan should be for the amount

$$A_2 + P_4 = \$29{,}035.93$$

■

Example 9.11 Suppose that an amount P is to be invested in an account which pays 4 percent interest compounded quarterly. After 2 years \$1000 is to be withdrawn, and after 4 years another \$1000 is to be withdrawn. After the second withdrawal, the account should have value zero. What is the correct value of P?

Solution After 2 years the principal P will grow to the amount

$$A = P(1.01)^8$$

A withdrawal of \$1000 is to be made from the amount A and the remainder, $(A - 1000)$, is to be invested for another 2 years. After these last 2 years the value is $(A - 1000)(1.01)^8$. This new value is to be exactly \$1000, so that a second withdrawal of \$1000 can be made to close the account. We have the equation (in dollars)

$$(A - 1000)(1.01)^8 = 1000$$

$$A = \frac{1000}{(1.01)^8} + 1000$$

$$= 923.48 + 1000 = 1923.48$$

Finally, $P = A/(1.01)^8$, and we obtain

$$P = \frac{1923.48}{(1.01)^8} = 1776.30$$

Thus, the initial deposit should be $1776.30. ■

Bonds and notes are promises to pay specified sums (called *face* or *redemption* values) at specified times in the future. They are issued by corporations and governmental units to raise funds for construction and other uses. The bonds and notes usually pay interest semiannually at a specified rate on the face value of the instrument. However, some corporations issue *zero coupon notes* on which no interest is paid but which are sold at a substantial discount from face value. The difference between the purchase price and the face value is essentially the interest. In contrast to a note that pays interest periodically, the interest on a zero coupon note is available only at maturity.

Example 9.12 *Corporate finance*

On November 1, 1994, GKB Corporation offers zero coupon notes due in 8 years at $56.00. That is, for $56.00 you can buy the right to receive $100 on November 1, 2002.

Problem Find the implicit annual rate of interest in this note. That is, find the annual rate of interest k for which $56.00 will increase to $100 in 8 years. (Note that although corporate bonds usually pay interest semiannually, we are interested in an *annual* rate, and hence we assume yearly compounding.)

Solution We again use the equation $A = P(1 + k)^M$, but this time we know A, P, and M and we are to find k. We have $A = \$100$, $P = \$56$, and $M = 8$. Consequently,

$$\$100 = \$56(1 + k)^8$$

from which we conclude

$$1 + k = \left(\frac{\$100}{\$56}\right)^{1/8} = (1.7857)^{.125} = 1.0752$$

Thus, $k = .0752$, and the implicit annual interest rate is 7.52 percent. ■

Example 9.13

Financial planning

Steve and Susan Saver hold a winning ticket in a state lottery. This ticket will bring them payments of \$1000 per month for 10 years. The Savers plan to use the money to establish a fund to provide for their children's college educations. They agree to deposit \$1000 on the last day of each month in an account at the Safety First Savings and Loan which earns interest at the rate of 6 percent per year compounded monthly. They made their decision on January 1 and their first payment on January 31. What is the value of the account on December 31 two years later?

Solution We assume that all payments (deposits) are made and that interest is credited at the end of each month. The first payment of \$1000 is made at the end of the first month, and of course no interest is paid at that time. However, the first payment will accumulate interest for the remaining 23 months of the 2 years. Since the interest rate is 0.5 percent per month, at the end of 2 years the first payment will have a value (in dollars) of $1000(1 + .005)^{23}$. The second payment will accumulate interest for 22 months, and at the end of 2 years it will have the value $1000(1.005)^{22}$. The value of the remaining payments can be determined similarly, and at the end of 2 years the sum of all payments and accumulated interest is

$$\begin{aligned} S &= 1000(1.005)^{23} + 1000(1.005)^{22} + \cdots + 1000(1.005) + 1000 \\ &= 1000[1 + (1.005) + \cdots + (1.005)^{22} + (1.005)^{23}] \end{aligned}$$

In this formula for S, the sum of the powers of the term 1.005 forms a geometric series, and there is a simple formula for the value of such a series:

Formula for the Sum of a Geometric Series

If a is any number, $a \neq 1$, then the sum $1 + a + a^2 + \cdots + a^n$ has the value

$$\frac{a^{n+1} - 1}{a - 1}$$

Using this formula in the expression for S (for $a = 1.005$ and $n = 23$), we have

$$S = \$1000 \frac{(1.005)^{24} - 1}{1.005 - 1} = \frac{\$1000}{.005} [(1.005)^{24} - 1] = \$25{,}431.96$$ ■

The situation described in Example 9.13 is an instance of a financial arrangement known as an *annuity*.

Annuity

An *annuity* is a sequence of a specific number of equal payments made or received by an individual at equally spaced times. To determine the amount of each payment, an interest rate must be specified. The times at which payments are made (or received) are called *payment dates,* and the interval between successive payment dates is the *payment period.* In an *ordinary annuity,* interest is credited on payment dates. The *amount of an annuity* is the sum of all payments plus all interest accumulated at the specified interest rate.

In Example 9.13 the payment dates are the last day in each month, and the payment period is 1 month. In this example, "payments" are actually deposits.

The technique used in Example 9.13 to find S can be used to find the amount of any annuity. If Y represents the payment made on each payment date, and if the interest rate is k per period, then after n periods the value of the first payment (which is made at the end of the first period) is $Y(1 + k)^{n-1}$. After n periods the value of the second payment is $Y(1 + k)^{n-2}, \ldots,$ the value of the last payment (made at the end of the last period) is Y. The amount of the annuity is

$$S = Y(1 + k)^{n-1} + Y(1 + k)^{n-2} + \cdots + Y(1 + k) + Y$$

Using the formula for the sum of a geometric series, we have

$$S = Y\frac{(1 + k)^n - 1}{(1 + k) - 1} = \frac{Y}{k}[(1 + k)^n - 1]$$

To summarize:

The *amount S of an annuity* with payments Y, n payment dates, and an interest rate of k per period is

$$S = \frac{Y}{k}[(1 + k)^n - 1] \qquad (9.6)$$

Annuities are bought and sold (in Example 9.13, the Savers could sell their winning ticket), and to determine a fair price for an annuity, it is important to know its present value. Since an annuity is a sequence of payments made on specific dates together with an interest rate, it is reasonable to make the definition:

The *present value of an annuity* is the sum of the present values of all payments, using the interest rate of the annuity.

Example 9.14 An ordinary annuity consists of payments of \$1000 per quarter for 4 years at 8 percent per year.

Problem Find the present value of this annuity.

Solution The first payment is made at the end of the first quarter. We use formula (9.5) with $k = .02$, and we conclude that the first payment has a present value (in dollars) of $1000(1 + .02)^{-1}$. Likewise, we conclude that the ith payment has a present value of $\$1000(1 + .02)^{-i}$. From the formula for the sum of a geometric series, the present value of the annuity is (in dollars)

$$\begin{aligned} V &= 1000(1.02)^{-1} + 1000(1.02)^{-2} + \cdots + 1000(1.02)^{-15} + 1000(1.02)^{-16} \\ &= \frac{1000}{(1.02)^{16}}[(1.02)^{15} + (1.02)^{14} + \cdots + 1.02 + 1] \\ &= \frac{1000}{(1.02)^{16}}\left[\frac{(1.02)^{16} - 1}{(1.02) - 1}\right] \\ &= 728.4458\left(\frac{.3727857}{.02}\right) \\ &= 13{,}577.71 \end{aligned}$$

■

Again, the method introduced in this example is a general one which can be used to compute the present value of any ordinary annuity. If the annuity consists of n payments of amount P, with interest rate k per period, then the present value of the annuity is

$$\begin{aligned} V &= P(1 + k)^{-1} + P(1 + k)^{-2} + \cdots + P(1 + k)^{-n+1} + P(1 + k)^{-n} \\ &= \frac{P}{(1 + k)^n}\left[\frac{(1 + k)^n - 1}{(1 + k) - 1}\right] \\ &= \frac{P}{k}\left[1 - \left(\frac{1}{1 + k}\right)^n\right] \end{aligned}$$

To summarize:

The present value V of an annuity with payments Y, n payment dates, and an interest rate of k per period is

$$V = \frac{Y}{k}\left[1 - \left(\frac{1}{1 + k}\right)^n\right] \qquad (9.7)$$

Remark If you already know the amount of an annuity, then formula (9.7) for finding the present value of an annuity can be simplified. Recall that the amount of an annuity with n payments of size Y and interest rate k per payment period is $S = (Y/k)[(1 + k)^n - 1]$. Using this and (9.7), we have

$$V = \frac{Y}{k}\left[1 - \left(\frac{1}{1+k}\right)^n\right] = \frac{1}{(1+k)^n}\left\{\frac{Y}{k}[(1+k)^n - 1]\right\} = \frac{S}{(1+k)^n}$$

This shows that *the present value of an annuity is simply the present value of the amount of the annuity.*

It is common for annuities to arise as periodic payments from a sum deposited with a pension fund or an insurance company. In such cases it is usual for the amounts of the payments to be determined on an actuarial basis. That is, the amounts are determined by the expected life of the owner. However, the idea of annuity is a very general one, and annuities occur in settings which are unrelated to pension plans or retirement options.

Example 9.15

Financial planning

Steve and Susan Saver plan to establish a savings account to partially pay for their daughter's education. They propose to make a single deposit in an amount sufficient to provide \$1000 quarterly for 4 years with no balance remaining after that time. They intend to open an account at Safety First Savings and Loan which pays interest at a rate of 6 percent per year compounded quarterly.

Problem They plan to make the deposit on July 1, 1995, and to begin withdrawals on September 30 of that year and continue quarterly through June 30, 1999. How large does their initial deposit need to be?

Solution Withdrawals of \$1000 are to be made for 16 consecutive quarters, after which no balance is to remain. The entire amount deposited, call it D, earns interest for one quarter. Thus the amount in the account on September 30 (after interest is added) is $D(1.015)$. After the first withdrawal of \$1000 is made, there remains $D(1.015) - \$1000$, and this amount earns interest for the second quarter. The amount in the account at the end of the second quarter, after interest is added but before any withdrawal is made, is (in dollars)

$$[D(1.015) - 1000](1.015)$$

After the withdrawal is made at the beginning of the third quarter, the amount in the account is

$$[D(1.015) - 1000](1.015) - 1000 = D(1.015)^2 - 1000(1.015) - 1000$$

Likewise, the amount in the account at the beginning of the fourth quarter is

$$D(1.015)^3 - [1000 + 1000(1.015) + 1000(1.015)^2]$$

Continuing, the amount in the account after the last withdrawal (at the end of the sixteenth quarter) is

$$D(1.015)^{16} - [1000 + 1000(1.015) + \cdots + 1000(1.105)^{15}]$$

But this amount is to be zero, so we have

$$D(1.015)^{16} = 1000[1 + 1.015 + \cdots + (1.015)^{15}]$$

or

$$D = \frac{1000}{(1.015)^{16}}[1 + 1.015 + \cdots + (1.015)^{15}]$$

In this formula for D, the sum of the powers of the term 1.015 forms a geometric series, and we can use our formula to determine the value of such a series:

$$D = \frac{1000}{(1.015)^{16}}\left[\frac{(1.015)^{16} - 1}{1.015 - 1}\right] = \frac{1000}{0.15}\left[1 - \frac{1}{(1.015)^{16}}\right]$$
$$= 14,131.26$$

The initial deposit should be \$14,131.26. If we view this sequence of withdrawals as an annuity, then the amount of this annuity is (\$1000/.015)[$(1.015)^{16} - 1$], and the initial deposit required is simply the present value of this amount. ■

Exercises for Section 9.2

1. Find the present value of \$10,000 payable in 10 years if the interest rate is 5 percent per year compounded yearly.
2. Find the present value of \$10,000 payable in 10 years if the interest rate is 5 percent per year compounded quarterly.
3. Find the present value of \$1000 payable in 5 years if the interest rate is 6 percent per year compounded quarterly.
4. Find the present value of \$10,000 payable in 10 years if the interest rate is 8 percent per year compounded semiannually.
5. On a 9-month discount loan, a borrower must pay back \$1000 at the end of the loan period. If the interest rate is 7.5 percent per year and if the \$1000 payment includes both principal and interest, how much did the borrower receive at the beginning of the loan period?
6. A borrower must pay a lender \$4000, including both principal and interest, at the end of 18 months. With an interest rate of 8 percent per year compounded quarterly, how much did the borrower receive from the lender at the beginning of the loan period?
7. Sam has a choice of two alternatives to raise some cash. In the first, he must repay \$1000, including both principal and interest at a rate of 9 percent per year compounded quarterly, after 1 year. In the second, he must repay \$1200, including both principal and interest at a rate of 10 percent per year compounded quarterly, after 2 years. In which case will he receive more cash at the beginning of the loan period?

8. Mike applies for a discount loan at the Safety First Savings Bank, and he is offered terms under which he repays \$5000, including both principal and interest at the rate of 7.5 percent per year compounded yearly, in 2 years. He is told that interest rates are declining, and it is likely that in a month the same arrangement would involve interest at the rate of 7 percent per year compounded yearly. How much more would Mike receive at the beginning of the loan period if he waited and interest rates declined, as the banker believes?

9. Find the amounts of the following annuities:
   (*a*) \$1200 paid semiannually for 4 years at 8 percent per year interest
   (*b*) \$600 paid quarterly for 4 years at 8 percent per year interest

10. Find the amounts of the following annuities:
    (*a*) \$12,000 paid annually for 5 years at 6 percent per year interest
    (*b*) \$1000 paid monthly for 5 years at 6 percent per year interest

11. Find the present value of each of the annuities in Exercise 9.

12. Find the present value of each of the annuities in Exercise 10.

13. Find the amount and present value of an annuity for which one pays \$10 per week for 128 weeks with an interest rate of .1 percent per week.

Car financing

14. Jennifer makes a down payment of \$4000 on a new car. She then agrees to pay the balance of the car's price plus interest by paying \$200 per month for 48 months. If interest is charged at the rate of 8 percent per year and there are no other charges added to the price of the car, what is the cash price of the car?

Car financing

15. A car is advertised in a newspaper as follows: \$49 down payment, payments of \$166.36 per month for 48 months, interest at 8.25 percent per year. Find the cash price of the car.

Car financing

16. A car is advertised in a newspaper as follows: \$500 down payment, payments of \$286.68 per month for 60 months, interest at 6.95 percent per year. Find the cash price of the car.

Personal finance

17. Jack owes Jill \$5000 payable in 3 years and \$3000 payable in 9 years. The two loans are to be consolidated into one loan payable in 5 years. What is the amount payable in 5 years if the interest rate is 9 percent per year and interest is compounded semiannually?

Corporate finance

18. The Ampsunwatts Electric Automobile Company is restructuring, and it plans to defer its debt payments. It now has three loans outstanding. The first is for \$100,000 payable in 1 year, the second is for \$200,000 payable in 2 years, and the third is for \$300,000 payable in 3 years. The company seeks to defer payment of these loans and to consolidate them into one loan payable in 5 years. The interest rate is 10 percent per year compounded annually. Assuming the lenders accept this proposal, what amount will have to be paid at the end of 5 years?

Car financing

19. George agrees to purchase a used car from Fred. Fred says that the purchase price is \$4800. He agrees to accept a down payment of \$1000 and a payment of \$4600 24 months later. What is the annual percentage rate of interest implicit in this proposal?

Personal finance

20. A houseboat is advertised in a newspaper for an initial payment of \$10,000 plus a second \$10,000 payment 3 years later. If the stated price is \$16,695, what is the annual percentage rate of interest implicit in this advertisement?

Corporate finance 21. The Interstate Bran Muffin Company (whose employees call it the "other IBM") decides to raise funds for expansion by issuing zero coupon notes payable in 5 years. The offering price is \$800 for a \$1000 note. If such a note is purchased and held to maturity, what is the annual percentage yield to the investor?

Corporate finance 22. In the situation described in Exercise 21, suppose that the notes are due in 4 years rather than 5. Find the annual percentage yield to an investor under these circumstances.

Corporate finance 23. In the situation described in Exercise 21, suppose that the company decides that it must offer investors the equivalent of 6 percent interest compounded annually. Find the offering price which should be assigned to the notes to provide this annual percentage yield.

Corporate finance 24. In the situation described in Exercise 21, suppose Paul pays \$8000 for notes when they are issued. One year later he sells them for \$8500. At the time of the sale, what interest rate are investors willing to accept for funds to be paid 4 years in the future?

Corporate finance 25. A zero coupon note payable in 8 years is offered by Brown Brothers Box Company. The offering price of the note was set by the company to provide a rate of return equal to 8 percent per year compounded annually.

(*a*) Find the offering price of the note as a percentage of the redemption value.

(*b*) After 1 year investors would be willing to buy notes with a 7-year maturity which provided a rate of return equal to 7 percent per year compounded annually. If Audrey bought \$10,000 of the Brown Brothers notes at the original price and sold them 1 year later, find her profit.

(*c*) Find the *rate* of return to Audrey in part *b*.

Corporate finance 26. In the situation described in Exercise 25, suppose that after 1 year interest rates have increased, and investors buying zero coupon notes with a 7-year maturity ask for return equivalent to 8.5 percent per year compounded annually. If Audrey bought \$10,000 of the notes at the original price and sold them 1 year later, does she have a profit or a loss? How much?

Financial planning 27. The Safety First Savings and Loan pays interest at a rate of 6 percent per year compounded quarterly. Nick Saver wants to establish an annuity for his father which will pay \$2000 quarterly for a period of 10 years. How much should Nick deposit so that the annuity can be paid and there will be no remaining balance after 10 years?

Financial planning 28. Suppose that in the situation described in Exercise 27 the interest rate is 8 percent instead of 6 percent. How much should Nick deposit with the higher interest rate to provide the same annuity?

Corporate finance 29. A company is considering buying a new computer which costs \$1,200,000. As an alternative, it could lease the computer for \$15,000 per month. The life of the computer is estimated to be 10 years, after which the computer will have a salvage value of \$100,000. If the company buys the computer, it must pay all maintenance costs, and a 10-year maintenance agreement will cost \$150,000. If current interest rates are 8 percent compounded quarterly, should the company buy or lease the computer? (*Hint*: Find the present value of the amount paid to lease the computer and compare it with the present value of the money needed to buy the computer.)

Corporate finance

30. A company has the option of buying or leasing a new phone system. The purchase option involves an initial cost of $1,000,000 and monthly maintenance of $2000. After 5 years the system will have a salvage value of $500,000. The lease option requires monthly payments of $10,000. Current interest rates are 10 percent per year compounded yearly. Should the company purchase or lease the phone system to obtain phone service at least cost?

9.3 TIME PAYMENTS, AMORTIZATION, AND MORTGAGES

A consumer who makes a purchase but does not have funds to pay cash for it must, in one way or another, take out a loan from the seller or from a third party. It is common for both the principal and the interest of such a loan to be repaid by making a sequence of equal payments at regular intervals (usually monthly). In such a case, the loan is said to be *amortized*. Naturally the amount of each payment will depend on the number of payments and the interest rate charged for the loan. The payments are determined so that an annuity with these payments for the specified period and interest rate has a present value equal to the amount of the loan.

Example 9.16

Car financing

The cash price for a 2-year-old car is $8000. If the total amount is to be paid by monthly payments and if interest is .75 percent per month, what monthly payments will pay off the loan in 36 months? In 48 months?

Solution Let the monthly payment be denoted by Y. If there are 36 payments and interest is .75 percent per payment period, then the present value V of the total amount of the payments is [from (9.7)]

$$V = \frac{Y}{.0075}[1 - (1 + .0075)^{-36}]$$

Since Y is to be selected so that the present value is $8000, we can determine the payments by solving for Y (in dollars) in the equation

$$8000 = \frac{Y}{.0075}[1 - (1 + .0075)^{-36}]$$

This gives

$$Y = \frac{.0075\,(8000)}{1 - (1.0075)^{-36}} = 254.40$$

If the loan is amortized over 48 months, then the payments are

$$Y = \frac{.0075\,(8000)}{1 - (1.0075)^{-48}} = 199.08$$

We see that the payments are about \$55 per month lower when the loan is repaid over 48 months than when it is repaid over 36 months. However, we also see that the amount paid over 36 months is \$9158.40, while the amount paid over 48 months is \$9555.84. ■

The method used in Example 9.16 can be used in general.

Amortization Payments

If a loan of L dollars at interest rate k per payment period is amortized with n payments of Y dollars each, then each payment must be (in dollars)

$$Y = \frac{kL}{1 - (1 + k)^{-n}} \tag{9.8}$$

Example 9.17 *House financing*

The Young family plans to purchase a new house. One house in which they are interested would require them to take a \$60,000 mortgage. They can obtain a mortgage for 25 years at an interest rate of 8.4 percent.

Problem Find the monthly payment.

Solution We use Equation (9.8) with $k = .084/12 = .007, n = 25 \times 12 = 300$, and $L = 60{,}000$. We have

$$Y = \frac{.007(60{,}000)}{1 - (1.007)^{-300}} = \frac{420}{1 - .12336} = 479.10$$

Each monthly payment is \$479.10. ■

Example 9.18 *House financing*

In the situation described in Example 9.17, the Youngs consider postponing their purchase for a year because they believe that within that time, interest rates will fall below 8 percent. However, because of inflation the price of the house will increase, and they will need a loan of \$65,000.

Problem Find the monthly payments required to amortize a \$65,000 loan over a period of 25 years with interest rates of 6.5, 7, 7.5, and 8 percent per year.

Solution Again we use Equation (9.8), this time with $L = 65{,}000$, $n = 300$, and four different values of k: $k = .065/12$, $.07/12$, $.075/12$, and $.08/12$. The results are shown in the table.

Interest Rate, %	Monthly Payment, $
6.5	438.88
7.0	459.41
7.5	480.34
8.0	501.68

Thus, if interest rates decline to only about 7.5 percent, then the Youngs' payments will be no lower than if they purchase the house now. On the other hand, if interest rates decline to 7 percent or below, then they will have lower payments if they wait. ■

Example 9.19 *Condominium financing*

Alice plans to buy a new condominium, and she has $18,000 for a down payment. Current interest rates are 8.4 percent for loans up to 80 percent of the value of the condominium and 9 percent for loans between 80 and 90 percent of the value. Loans are for 20 years. Alice can afford payments of at most $1000 per month.

Problem Find the value of the most expensive condominium that Alice can afford.

Solution Alice's $18,000 would serve as a 20 percent down payment on a condominium with value up to $90,000. If she chose a condominium with the largest value, she would need a loan of $72,000. To determine the monthly payments on such a loan, we use Equation (9.8) and take $k = .084/12 = .007$, $n = 240$, and $L = 72{,}000$. The payments are

$$Y = \frac{.007(72{,}000)}{1 - (1.007)^{-240}} = \$620.28$$

clearly within her range of less than $1000.

Next, if she takes out a loan of more than $72,000, she will have to pay 9 percent interest. We solve for the amount L of a loan which results in payments of $1000 per month. We take $k = .09/12 = .0075$, and

$$\$1000 = \frac{.0075L}{1 - (1.0075)^{-240}} = .008997L$$

Consequently, $L = \$111{,}144.95$. Using a loan of this amount together with her \$18,000, she could purchase a condominium with a value of \$129,144.95. Since her \$18,000 provides a down payment of more than 10 percent (actually almost 14 percent) of the purchase price, this solves the problem. ■

The remainder of this section is concerned with *sinking funds*. Frequently a company or an individual recognizes in advance that certain expenses will occur in the future. If the magnitude and timing of the expenses are known, or can be estimated, then provision can be made to cover them. A common approach is to make regular deposits into an account in such a way that the deposits together with the interest earned will be enough to cover the expenses. Such an account is called a *sinking fund*. Sinking funds are frequently used by companies to finance the replacement of old equipment.

Example 9.20

Corporate finance

The treasurer of the International Bran Muffin Company is told that a set of ovens at the company's Mulberry plant will have to be replaced in 3 years at an estimated cost of \$400,000. The treasurer proposes to make semiannual payments into an account which pays 5 percent interest compounded semiannually.

Problem Find the amount of each semiannual deposit so that after 3 years there will be \$400,000 in the account.

Solution Let each semiannual deposit be denoted by Y, and let the amount in the account after 3 years be denoted by S. Then by Equation (9.6), S and Y are related by

$$S = \frac{Y}{.025}[(1.025)^6 - 1]$$

Since our goal is to have $S = 400{,}000$ (in dollars), Y must be taken equal to

$$Y = \frac{400{,}000(.025)}{(1.025)^6 - 1} = 62{,}619.99$$

That is, semiannual deposits of \$62,619.99 each, for a period of 3 years with interest of 5 percent compounded semiannually, will result in a fund of \$400,000 to be used to replace the ovens. ■

This technique can be used to solve similar problems, and the following formula can frequently be applied.

Sinking Fund Payments

If a sinking fund of S dollars is to be accumulated through n payments of Y dollars each with interest rate k per period, then each payment must be (in dollars)

$$Y = \frac{kS}{(1 + k)^n - 1} \tag{9.9}$$

Example 9.21 *Corporate finance*

The Electric Company has $1,000,000 in bonds payable in 5 years. The corporation plans to make quarterly payments into a sinking fund at 7.5 percent interest compounded quarterly.

Problem How much should be deposited each quarter so that the fund will contain $1,000,000 after 5 years?

Solution We use Equation (9.9) with $S = 1{,}000{,}000$, $n = 20$, and $k = .075/4 = .01875$. We have

$$Y = \frac{.01875(1{,}000{,}000)}{(1.01875)^{20} - 1}$$
$$= 41{,}671.48$$

■

Exercises for Section 9.3

1. Find the monthly payments that are necessary to amortize a loan of $5000 in 1 year with interest at 9 percent per year.
2. Find the monthly payments that are necessary to amortize a loan of $10,000 in 3 years with interest at 7.5 percent per year.
3. Find the quarterly payments that are necessary to amortize a loan of $50,000 in 5 years with interest at 8.5 percent per year.
4. Find the quarterly payments that are necessary to amortize a loan of $150,000 in 6 years with interest at 7 percent per year.
5. Find the yearly payments that are necessary to amortize a loan of $20,000 in 10 years with interest at 6.5 percent per year.
6. Find the yearly payments that are necessary to amortize a loan of $60,000 in 15 years with interest at 8.5 percent per year.
7. Find the quarterly payments that are necessary to amortize a loan of $25,000 in 2 years with interest at 8 percent per year.
8. In the situation of Exercise 7, how does the *total amount paid* change if payments are made semiannually instead of quarterly? Yearly instead of quarterly? All other conditions of the problem remain the same.

Corporate finance

9. The Wonder Widget Corporation can afford payments of $2000 per month to amortize a loan. The rate of interest for companies with credit ratings similar to that of Wonder Widget is 9 percent per year. How large a loan can the company afford if the loan is amortized over 5 years? Over 10 years?

10. Find the monthly payment if a $100,000 loan is amortized over 10 years at an interest rate of 7.5 percent per year. What fraction of the first monthly payment is interest? (*Hint*: Find the interest due on a loan of 1 month with a principal of $100,000 at a rate of 7.5 percent per year.) Repeat the exercise if the loan is amortized over 15 years.

Car financing

11. Sarah is shopping for a car, and she has $2000 in cash. The interest rate for financing the purchase of a car is 7.2 percent per year for a 4-year loan. She finds a car she likes for $9500. How much will her monthly payments be if she makes a down payment of $2000 and finances the remainder for 4 years? Find the total of her payments (principal and interest) over the term of the loan.

Car financing

12. Repeat Exercise 11 if she uses only $500 for a down payment. All other conditions remain the same.

Car financing

13. Sarah can finance her car over a period of 5 years at an interest rate of 7.8 percent. Find her monthly payments if she makes a down payment of $2000 and finances the remainder for 5 years.

14. Which has the larger monthly amortization payments, a loan of $10,000 for 10 years or a loan of $20,000 for 25 years? In both cases the rate of interest is 6 percent per year.

House financing

15. The Fontanez family plans to purchase a house for $95,000. They have $20,000 to use as a down payment, and they plan to finance the remainder over 25 years. If the current rate is 6.8 percent per year for such loans, find the monthly payment necessary to amortize the loan.

House financing

16. Carl plans to buy a new house, and he will need to finance $80,000. He has several options:
    6.0 percent per year for 15 years
    6.6 percent per year for 20 years
    7.2 percent per year for 25 years
    In each case interest is compounded monthly.
    (*a*) Which option results in the lowest monthly payments?
    (*b*) What is the total amount of interest paid in each case?

Personal finance

17. Tamara plans to borrow $10,000, and she can do so at a rate of 8.7 percent per year. She can afford payments of $180 per month. What is the shortest period in which she can repay the loan?

Condominium financing

18. In the situation of Example 9.19, suppose that Alice has $12,000 to be used as a down payment. If all other conditions remain the same, find the most expensive condominium she can afford.

House financing

19. Karita and Carl would like to buy a $95,000 house, but they are unable to arrange for a mortgage. They have $25,000 to be used as a down payment. The owner of the house is willing to provide short-term financing for a period of 5 years at a rate of 10 percent per year.
    (*a*) Determine their monthly payment if they pay only the interest for 5 years.
    (*b*) Determine their monthly payment if they amortize $20,000 of the loan over 5 years and pay only interest on the remaining $50,000. In this case they will have to refinance $50,000 at the end of the 5-year period.

20. Small differences in interest rates result in rather large differences in monthly payments and in the total amount paid when a loan is amortized over a fairly long period. Assume a loan of $50,000 is to be amortized over 30 years. Find the monthly payments and the total amount paid at interest rates of 6, 7, and 8 percent per year.

Consumer finance

21. The Carson family takes out a consumer loan at Fred's Friendly Finance Company. The loan is for $5000 for 2 years at an interest rate of 18 percent per year.
    (*a*) Find the monthly payment necessary to amortize the loan.
    (*b*) Find the total amount paid (principal and interest).
    (*c*) Suppose the loan is amortized over 1 year instead of 2 years at the same interest rate. Find the total amount paid in this situation.

Corporate finance

22. The Fast Haul Trucking Company plans to buy a new $80,000 truck in 5 years. If money is to be deposited into an account which pays 6.5 percent per year compounded semiannually, how much should be deposited each 6 months so that at the end of 5 years there will be enough cash to buy a new truck?

Corporate finance

23. Shortshocks Electronics, Inc., arranges a loan of $500,000 to expand its semiconductor division. The interest rate on the loan is 9 percent compounded yearly, and the loan together with accumulated interest is to be repaid in 3 years. A sinking fund is established in an account which pays 6 percent per year compounded quarterly. What quarterly payments into the sinking fund will give the amount required to pay off the loan plus accumulated interest in 3 years?

Corporate finance

24. In the situation of Exercise 23, suppose the corporation prefers to make semiannual sinking fund payments. If all other conditions remain the same, what semiannual payments into the account will give the amount required to pay off the loan plus accumulated interest in 3 years?

Financial planning

25. David and Deborah Baker wish to set up a college fund for their children, 10-year-old Diane and 6-year-old David, Jr. They plan to invest quarterly, and their goal is to have a fund containing $30,000 when Diane begins college. How much should the Bakers invest each quarter if the investments are to stop when Diane begins college at age 18? Interest is 8 percent compounded quarterly.

Financial planning

26. In the situation described in Exercise 25, suppose that the Bakers' goal is to provide $3000 per year for 4 years for each child. Payments to the children will be made at the beginning of each year, and each child will begin college at age 18 and go for exactly 4 years. How much should the Bakers invest each quarter if the last deposit is to be made when Diane begins college (exactly 8 years from now)?

Corporate finance

27. The Fast Haul Trucking Company can arrange to make sinking fund payments of $6000 semiannually. The best interest rate it can receive on its money is 10 percent per year compounded semiannually. How many periods are necessary for Fast Haul to accumulate $100,000?

Corporate finance

28. Everest Securities, Inc., sells $1,000,000 of 8 percent (interest rate per year) sinking fund debentures due in 10 years with interest payable semiannually. The company can earn 9 percent per year compounded semiannually on money it deposits in a sinking fund. What amount does the company need every 6 months to meet interest *and* sinking fund payments?

University finance

29. Gigantic State University plans to replace its mainframe computer in 5 years, and a new machine will cost $2,000,000. If the university can obtain 8 percent (interest rate per year) on its money, how much must be deposited in a sinking fund each year to finance the replacement of the computer?

Corporate finance

30. XTZ Corporation plans to add an office in Mulberry, and it will cost \$800,000 to do so. Beginning in 5 years, that office will generate a net profit of \$200,000 per year. These profits can be invested to yield 6 percent per year. What interest rate can the corporation pay to finance the expansion if the goal is to have the total income from the new office generate enough to pay off the debt (principal and interest) in exactly 10 years?

9.4 EVALUATING INVESTMENT OPTIONS AND FINANCIAL DECISION MAKING

As an alternative to investing in savings accounts at savings institutions, one can invest in the stock issued by corporations and in notes, bonds, and commercial paper issued by corporations and various governmental units. Such investments can be made either in individual issues or in collections of issues through mutual funds or money market funds. In such cases there is usually an element of uncertainty about the return on the investment which does not exist for investments in savings accounts. On the other hand, there is also the potential for much greater gain. Since notes and bonds correspond more closely to savings accounts, we begin our study with them.

A *bond* is an agreement to pay a fixed sum, the *redemption value*, at a certain date in the future, the *redemption date*, together with fixed interest payments at prescribed intervals up to the redemption date. The annual interest is a percentage of the redemption value, and this rate is called the *stated* interest rate. *Notes* are very similar to bonds (the difference is in the degree of security of the investment), and for our purposes the two can be studied together. Bonds are sold on the open market, and the selling price may differ substantially from the redemption value. A person who purchases a bond obtains the right to all *future* interest payments as well as the right to the redemption value. Normally the next interest payment is prorated between the current holder and the purchaser. At the time of purchase the purchaser must pay a fraction of the next interest payment to the current holder. The fraction is that proportion of the current period which has passed. In computing the return on a bond, both the redemption value and future interest payments must be considered. If the purchase price of a bond is less than the redemption value, then the bond is said to have been purchased at a *discount*. If the purchase price is more than the redemption value, then the bond is purchased at a *premium*. Finally, if the bond is purchased for the same price as the redemption value, then the bond is purchased *at par*. The stated interest rate is fixed at the time the bond is issued, and it does not vary with the price of a bond on the resale market.

Example 9.22 Describe the payments received by a purchaser of a bond with a \$1000 redemption value, a \$960 purchase price, and semiannual interest payments at a stated interest rate of 7 percent per year. The bond is purchased on January 1, 1995, the

Bond investing

redemption date is January 1, 1998, and interest is paid every January 1 and July 1 up to and including the redemption date.

Solution Since interest is based on the redemption value and is 3.5 percent per half-year, each interest payment is .035($1000) = $35. There are six future interest payments made to the purchaser. They are made on July 1, 1995; January 1 and July 1, 1996; January 1 and July 1, 1997; and January 1, 1998. Thus, the total interest paid is $210. Moreover, there is a payment of $1000 (the redemption value) on January 1, 1998. Thus, the total payments to the purchaser amount to $1210. Since the purchase price is $960, the purchaser of the bond makes a net profit of $250 over the period of ownership of the bond (January 1, 1995, to January 1, 1998). ■

Since the decision to purchase a bond is normally based on a comparison of the returns on this and other investments, the rate of return is a significant item. One measure of the rate of return on a bond is its *current yield*, the yearly interest divided by the price at which the bond can be purchased. A more refined measure is the *yield to maturity*.

The *yield to maturity* for a specific purchase price of a bond is the *interest rate* for which the sum of the present values of the interest payments yet to be made and the present value of the redemption value is equal to the purchase price.

For very long-term bonds, those due in 20 years or more, the current yield will usually be close to the yield to maturity. Unlike the stated interest rate, both the current yield and the yield to maturity of a bond normally change as time passes.

Example 9.23 *Bond investing*

A $1000 bond with 7 percent stated interest rate is purchased on January 1, 1995, for $960. The bond is due on January 1, 1998. (This is the bond of Example 9.22).

Problem Find the yield to maturity of this bond on January 1, 1995.

Solution The quantity we seek, the yield to maturity, is an interest rate. Since interest is paid semiannually, it is natural to work with the (unknown) semiannual interest rate. We denote it by k. We determine k by taking the sum S of the present values of interest payments yet to be made and the present value of the redemption value, and then we set S equal to the purchase price, $960. Each of the present values involves the unknown interest rate k.

The present value of each of the interest payments is shown in Table 9.3. The present value of the \$1000 due on redemption is $\$1000/(1 + k)^6$. Using these values, we have an expression for S in terms of k:

$$S = \frac{\$1000}{(1+k)^6} + \frac{\$35}{1+k}\left[1 + \frac{1}{1+k} + \frac{1}{(1+k)^2} + \frac{1}{(1+k)^3} + \frac{1}{(1+k)^4} + \frac{1}{(1+k)^5}\right]$$

The expression on the right-hand side can be simplified with the help of the formula for the sum of a geometric series. We have

$$S = \frac{\$1000}{(1+k)^6} + \frac{\$35}{1+k}\left[\frac{1 - 1/(1+k)^6}{1 - 1/(1+k)}\right]$$

or

$$S = \frac{\$1000}{(1+k)^6} + \frac{\$35}{k}\left[1 - \frac{1}{(1+k)^6}\right] \tag{9.10}$$

The problem has been reduced to one of finding k such that S, as given by (9.10), has the value \$960, the purchase price of the bond. When phrased in this way, the problem is essentially one of finding a root of a sixth-degree polynomial. Although there are computational algorithms for solving such problems, we use a systematic trial-and-error approach. In proceeding by trial and error—perhaps better described as proceeding by successive approximations—it is use

TABLE 9.3

Date	Payment, \$	Present Value, \$
July 1, 1995	35	$\frac{35}{(1+k)}$
January 1, 1996	35	$\frac{35}{(1+k)^2}$
July 1, 1996	35	$\frac{35}{(1+k)^3}$
January 1, 1997	35	$\frac{35}{(1+k)^4}$
July 1, 1997	35	$\frac{35}{(1+k)^5}$
January 1, 1998	35	$\frac{35}{(1+k)^6}$
January 1, 1998	1000	$\frac{1000}{(1+k)^6}$

ful to keep in mind that in formula (9.10) an increase in k results in a decrease in the present value S. For $k = 0.4$, 0.45, 0.5, and 0.55, the present value S in (9.10) has the following values (in dollars):

$$
\begin{array}{ll}
k = .040 & S = 790.31 + 183.47 = 973.49 \\
k = .045 & S = 767.90 + 180.53 = 948.42 \\
k = .050 & S = 746.22 + 177.65 = 923.86 \\
k = .055 & S = 725.25 + 174.84 = 900.09
\end{array}
$$

We see from these choices of k that $k = 0.045$ gives a value of S that is too small, and $k = .040$ gives a value of S that is too large. At the next stage we try $k = .041$, .042, .043, and .044, and we obtain

$$
\begin{array}{ll}
k = .041 & S = 785.77 + 182.88 = 968.65 \\
k = .042 & S = 781.26 + 182.29 = 963.54 \\
k = .043 & S = 776.66 + 181.70 = 958.47 \\
k = .044 & S = 772.32 + 181.11 = 953.43
\end{array}
$$

Thus, $k = .043$ gives a value of S that is a little too small, and $k = .042$ gives a value of S that is a little too large. Since the value of S for $k = .043$ is a little closer to 960 than the value of S for $k = .042$, we next try $k = .0426$, .0427, and .0428. We obtain $S = 960.49$ for $k = .0426$, $S = 959.99$ for $k = .0427$, and $S = 959.48$ for $k = .0428$. The value of S for $k = .0427$ is quite close to 960, and since k is now accurate to at least three decimal places, we consider the problem to be solved. We take $k = .0427$ as the proper interest rate for a half-year. Thus the yield to maturity, based on the interest rate per year, is .0854, or equivalently 8.54 percent. ■

The price of a bond on the resale market is related to many factors, including the fiscal state of the corporation or municipality issuing the bond, the stated interest rate, and general economic conditions. In particular, the price of a bond on the resale market is strongly influenced by the relation between the stated interest rate and the prevailing interest rate in the economy. If the prevailing interest rate is higher than the stated interest rate, then the price of the bond tends to be below par so that the bond can compete as an investment. Conversely, if the prevailing interest rate is lower than the stated interest rate, then the price of a bond tends to be above par. It is for this reason that some bonds sell for prices above their redemption values. Suppose that interest rates in the economy are higher than the stated interest rate on a bond and you buy it below par. If interest rates in the economy decline, then the market price of your bond may increase. Thus, in addition to buying a bond to obtain the interest payments, you can also purchase a bond as a speculative investment, hoping to buy the bond at one price and sell it at some future date at a higher price. Of course, not all such plans work, and it is always possible that at the time you wish to sell the bond its market price will be the same as or even less than the amount you paid for it. In such a case, at least you have the consolation of having received the interest payments while you held the bond.

Another investment option is the purchase of stock issued by a corporation. The stockholders (owners of stock) of a corporation are the owners of that corporation, whereas bondholders are its creditors. Each stockholder is a part owner. For instance, if a company issues 3,000,000 shares of stock, then an individual who owns 3000 shares owns $\frac{1}{1000}$ of the company. All company decisions are, in theory, made by the stockholders (persons owning the company's stock) and by individuals hired by the stockholders to run the day-to-day affairs of the company (the management). In practice, most stockholders of a company do not buy the stock to enable them to share in the process of owning and managing the company. Instead, they buy the stock as an investment in the hope of obtaining a return which exceeds returns on other investments such as bonds and savings accounts. A corporation may issue several types or classes of stock. The two general classifications are *common* and *preferred*. Holders of preferred stock are given precedence (in ways defined precisely in the articles of incorporation of the company) over the holders of common stock. Here we consider only a single type of stock, and we have common stock in mind.

One who invests in stocks can profit in two ways. First, the company may issue dividends to all stockholders. A *dividend* is a payment made by the company that issued the stock to the owners of the stock. Dividends are often paid on a regular basis (such as quarterly), and for certain companies, such as utilities, the amounts rarely decrease, so that a stockholder can depend on these payments as a source of income. However, dividends must come from earnings of the company, and they are not guaranteed. If the company suffers losses, the payment of dividends may have to cease, at least temporarily. Also, even if a company has the money to pay dividends, it is not usually required to do so. Instead, the management may choose to use the money to modernize the plant or equipment or for other purposes (e.g., to buy another company).

The second method by which a shareholder can gain from owning shares is by having the shares increase in value. It is said that the shares have *appreciated* when they increase in value. A shareholder with shares that have appreciated can sell the shares and realize a gain in this way. Shares of stock of most major companies are traded (bought and sold) on stock exchanges each weekday, and the system handles trades almost constantly. The value of a company's stock varies with the public's perception of the company's future, with the general state of the economy, and with the actual financial state of the company. Thus the price of a stock often (in fact usually) varies in the course of a single day and can vary greatly in periods such as a week or month. This provides the opportunity for a rapid gain (or loss) on an investment. If the stock appreciates substantially in a short time, then the *return on investment* can be very large. Here, by "return on investment" we mean the increase in value of the investment per unit time as a percentage of the investment. It is common to give the return on investment as a percent per year.

Example 9.24

Stock investing

Gus buys 1000 shares of MDI (Multi-Decibel, Inc.) at \$4.00 per share. One month after his purchase, the company announces the development of a robot that recognizes human speech, and the company stock appreciates quickly to \$5.50 a share. Gus sells his shares immediately.

Problem Find his return on investment.

Solution Since Gus sells his shares for \$1.50 more than he paid for them, his profit is 1000(\$1.50) = \$1500. His return on investment is

$$\frac{\$1500}{\$4000} = .375 = 37.5 \text{ percent}$$

for 1 month. On a yearly basis this is 450 percent. ■

Remark In Example 9.24 and in the example which follows we ignore the commissions and fees which investors have to pay to a stockbroker to buy and sell stocks. These fees vary depending upon the broker used for the transaction, the price per share of the stock, and the number of shares bought or sold.

Values of shares of stock may decrease as well as increase and the behavior of stock prices is very unpredictable. One method of attempting to deal with the uncertainty of the future values of a stock is to estimate the probabilities of various actions of the stock market and then to compute the expected value of the return.

Example 9.25

Stock investing

Karina is considering buying 100 shares of Fem Sport, a new company which specializes in women's sports equipment. Fem Sport stock now sells for \$16 a share, and the company has not yet made a substantial inroad into the market. However, a new line of tennis and golf equipment is planned, and if it sells well, the company will profit greatly. Karina estimates the probability that the new line will sell well at .2, she estimates the probability of it selling moderately well at .5, and she estimates the probability of it being a complete failure at .3. If the line does well, she believes the stock price will double in a year to \$32. If the line does moderately well, she believes that the stock will increase in value to \$20; and if the line fails to sell, she believes the stock will fall in value to \$4 per share. Assuming these three possibilities are the only ones which can occur, what is the expected return on an investment in these shares?

Solution The expected value of the price of the stock in 1 year is (see Chapter 4 for the computation of expected values)

$$E = \$32(.2) + \$20(.5) + \$4(.3) = \$17.60$$

The expected return on investment is

$$\frac{\$17.60 - \$16.00}{\$16.00} = \frac{\$1.60}{\$16.00} = .1 = 10 \text{ percent}$$

per year. ■

Remarks As the reader is no doubt aware, the analysis in Example 9.25 is quite naive. In an actual case the future price of the stock of Fem Sport will depend on many factors other than just the success of its new line of equipment. In particular, the state of the entire economy will be an important factor in determining the price of the stock. It is possible that a company does well in the sense of sales and profits, and yet the price of its stock falls because of factors other than the success of the company. To be realistic in an example such as 9.25, many other factors must be considered. This requires the estimation of probabilities for events which are very complex and interrelated. However, the method of computing expected return in these more realistic settings is the same as that used in Example 9.25.

Exercises for Section 9.4

Bond investing

1. Describe the payments received by a purchaser of a bond with a $10,000 redemption value, a $9200 purchase price, and a stated interest rate of 5.5 percent per year paid semiannually. The bond is purchased on July 1, 1995, and the redemption date is January 1, 1998. Interest is paid every January 1 and July 1, up to and including the redemption date.

Bond investing

2. Describe the payments received by a purchaser of a bond with a $20,000 redemption value, a $22,500 purchase price, and a stated interest rate of 7.5 percent per year paid semiannually. The bond is purchased on January 1, 1995, and the redemption date is July 1, 1999. Interest is paid every January 1 and July 1, up to and including the redemption date.

Bond investing

3. A bond with a $5000 redemption value, a purchase price of $5200 plus accrued interest, and a stated interest rate of 8 percent per year paid annually is purchased on October 1, 1995. The redemption date is January 1, 1999. Interest is paid every January 1, up to and including the redemption date.
   (*a*) What is the amount of accrued interest due at the time of purchase?
   (*b*) Describe the payments received by a purchaser up to and including the redemption date.

   Note Accrued interest is paid by the purchaser of a bond to the seller and is that fraction of the next interest payment due the seller as a result of the time the seller held the bond.

Bond investing

4. A bond with a \$10,000 redemption value, a purchase price of \$9800 plus accrued interest, and a stated interest rate of 6 percent per year paid semiannually is purchased on April 1, 1995. The redemption date is July 1, 1999. Interest is paid every January 1 and July 1, up to and including the redemption date.
   (*a*) What is the amount of accrued interest due at the time of purchase?
   (*b*) Describe the payments received by a purchaser up to and including the redemption date.
5. Find the yield to maturity for the purchaser of the bond in Exercise 1.
6. Find the yield to maturity for the purchaser of the bond in Exercise 2.
7. Find the yield to maturity for the purchaser of the bond in Exercise 3.
8. Find the yield to maturity for the purchaser of the bond in Exercise 4.

Stock investing

9. What is the return on investment of a share of stock which is purchased for \$55.25, held 2 years, and then sold for \$65.25?

Stock investing

10. What is the return on investment of a share of stock which is purchased for \$35.50, held 4 years, and then sold for \$45.00?

Stock investing

11. What is the return on investment of a share of stock which is purchased for \$65.25 and sold 6 months later for \$66.75?

Stock investing

12. What is the return on investment of a share of stock which is purchased for \$33.75, held 9 months, and then sold for \$48.75?

Stock investing

13. In a stock split, the number of shares is increased and the price per share is decreased by an appropriate amount to keep the total value of all shares the same. Thus, in a 2-for-1 split, the number of shares is doubled and the price per share is cut in half. Suppose Margaret bought shares in MDI for \$12 per share. Thirty months later MDI has a 3-for-1 split, and shares which were selling for \$27 before the split sell for \$9 immediately after the split. Margaret sells her shares immediately after the split. What is her return on investment?

Stock investing

14. In the situation described in Exercise 13, Margaret holds her shares for an additional year after the split and then sells them for \$10 per share. What is her return on investment in this situation?

Stock investing

15. Wen buys 200 shares in PDQ Corporation at \$20 per share, and 18 months later there is a 5-for-4 stock split. Immediately after the split Wen sells his stock at \$25 per share. Find his return on investment.

Stock investing

16. In the situation described in Exercise 15, Wen holds his shares for an additional 2 years after the split and then sells them for \$40 per share. What is his return on investment in this situation?

Stock investing

17. In the situation described in Exercise 15, Wen sells half his shares immediately after the split and invests the proceeds in a 2-year Certificate of Deposit (CD) with 5 percent interest compounded semiannually. He sells the remaining shares 2 years after the split for \$40 per share. What is his return on the original \$4000 investment?

Stock investing

18. Angela buys stock in TAKY Corporation for \$40 per share. Two years later there is a 3-for-1 stock split, and immediately after the split the stock sells for \$15 per share. Two years later the stock sells for \$12 per share.
    (*a*) Find Angela's return on investment if she sells her stock immediately after the stock split.
    (*b*) Find her return on investment if she sells her stock 2 years after the stock split.

Stock investing

19. Find the expected return on investment in Example 9.25 if the possible future stock prices (1 year from now) and their probabilities are as follows:

Possible Value, $	Probability
40	.1
32	.1
20	.4
16	.3
2	.1

Stock investing

20. Ming is considering the purchase of stock in a new bioengineering firm. A share of stock can be purchased today for $15 per share. She estimates that

    In 2 years the firm will go out of business, and the stock will be worthless, with probability .4.

    The firm will be moderately successful, and the stock will double from its present price, with probability .5.

    The firm will be very successful, and the stock price will increase to $50, with probability .1.

    Find her expected return on investment in 2 years.

Stock investing

21. In the situation in Exercise 20, suppose that the probabilities and associated outcomes for failure and moderate success are as described. In the event that the company is very successful, how high must the stock price rise for Ming to have a return on investment of 10 percent per year?

Bond investing

22. A $1000 bond which pays interest semiannually at the rate of 4.5 percent per year is discounted so that its current yield (the yield based on its current market price) is 6 percent.
    (*a*) What is the current market price of the bond?
    (*b*) If the redemption date of the bond is 5 years in the future, what is its yield to maturity based on the current market price?

Bond investing

23. A $1000 bond which pays interest semiannually at the rate of 7.5 percent per year sells at a premium so that its current yield (the yield based on its current market price) is 6 percent.
    (*a*) What is the current market price of the bond?
    (*b*) If the redemption date of the bond is 4 years in the future, what is its yield to maturity based on the current market price?

Stock investing

24. An engineer has invented a low-cost small computer, and she plans to raise money to market her machine by issuing 1 million shares of stock at $5 per share. An investor estimates the possible future values of the stock (1 year from now) and their probabilities as follows:

Possible Value, $	Probability
100	.01
10	.30
5	.20
0.01	.49

What is the expected return on investment for someone who buys 1000 shares of this new company and who plans to sell them 1 year later?

Financial planning

25. Compare the following two investments and decide which has a better rate of return for the investor for the 30-month period beginning January 1, 1995. Explain your method of comparison.
    (*a*) A 30-month certificate of deposit which pays interest at a rate of 4.5 percent per year, compounded quarterly and paid at maturity
    (*b*) A bond purchased at par on January 1, 1995, and due on July 1, 1997, which pays interest semiannually on January 1 and July 1, at a rate of 5.2 percent per year. Assume that the interest can be reinvested at a rate of 3.6 percent per year, compounded quarterly.

Financial planning

26. A taxpayer in the "40 percent bracket" has an income for which $1 of additional income results in $0.40 of additional income taxes. Although the interest paid on corporate bonds is usually subject to income tax, municipal bonds (bonds issued by state and local governments) are generally *tax-exempt*; i.e., their interest is not subject to income tax.
    (*a*) What interest rate would a corporate bond have to carry to provide the same after-tax income for a taxpayer in the 40 percent bracket as a tax-exempt bond which pays 8 percent?
    (*b*) Repeat (*a*) for a taxpayer in the 50 percent bracket.
    (*c*) A tax-exempt bond pays 6 percent, and a corporate bond pays 9 percent interest. Find the highest income tax bracket for which the corporate bond yields greater after-tax income to a taxpayer in that bracket.

Financial planning

27. A $1000 bond pays interest annually at a stated rate of 8 percent and is selling at par. If there is inflation, then interest rates will increase to 10 percent; if there is a recession, then interest rates will decline to 6 percent; and if there is neither, then interest rates will remain unchanged. In any case, the price of the bond will change so that the current yield on the bond matches the prevailing interest rate. An investor estimates that 1 year from now there will be inflation with probability .3, a recession with probability .2, and neither with probability .5. Find the expected annual rate of return on the bond for the next year.

Financial planning

28. In the situation described in Exercise 27, suppose that the probability of recession is .5, the probability of inflation is .1, and the probability of neither is .4. All other conditions of the problem remain the same. Find the expected annual rate of return on the bond for the next year.

Financial planning

29. An investor can purchase for $950 a bond which pays interest semiannually at a stated rate of 5 percent and which will pay $1000 in 1 year. The same investor has studied the common stock of ZYX Corporation and reached the following conclusions:

    In 1 year the stock will sell for $80 per share with probability .2.
    In 1 year the stock will sell for $60 per share with probability .5.
    In 1 year the stock will sell for $50 per share with probability .3.
    The dividend is $2 per share per year, and it is unlikely that the dividend will be changed in the next year.

    The stock of ZYX Corporation now sells for $58 per share. Ignoring any tax considerations, which investment is the most attractive? Be sure to tell exactly what you are comparing in the two cases.

Financial planning 30. In Exercise 29, find the selling price of the stock of ZYX Corporation for which the return on investment in the stock is exactly equal to the yield to maturity on the bond. All other information in the problem remains unchanged.

IMPORTANT TERMS AND CONCEPTS

You should be able to describe, define, or give examples of and use each of the following:

Interest
Simple interest
Compound interest
Annual percentage rate
Present value
Current yield
Purchase at a premium, at a discount, or at par
Stated interest rate
Annuity
Present value of an annuity
Amortization
Bond
Redemption value or face value
Redemption date
Yield to maturity
Return on investment
Common stock
Stock split
Sinking fund

REVIEW EXERCISES

Savings 1. A certificate of deposit is available with interest compounded quarterly at the rate of 4 percent per year for 4 years. Find the annual percentage yield for this certificate.

Personal finance 2. Pekka now has a balance of $1802.92 in his savings account. He has made no deposits or withdrawals for the last 3 years, and interest has been compounded quarterly at the rate of 4 percent per year. Find the amount in his savings account 3 years ago.

Personal finance 3. In the situation described in Exercise 2, suppose Pekka made a deposit of $1000 in his account 2 years ago, but all the other conditions in Exercise 2 remain the same. Find the amount in his savings account 3 years ago.

Personal finance 4. In the situation described in Exercise 2, suppose that the interest rate paid on his savings for the last 2 years has been 4 percent per year, but the rate was 5 percent per year for the first year of the 3-year period. In both cases interest is compounded quarterly. Find the amount in Pekka's savings account 3 years ago.

Corporate finance 5. The Fast Haul Trucking Company will need $500,000 in 6 years to purchase new equipment. The company plans to invest a fixed amount now in an account paying 5 percent interest compounded yearly. How much must be invested to have $500,000 in 6 years?

Corporate finance 6. In Exercise 5, the company plans to make equal payments into an account which pays 5 percent interest compounded quarterly. How much must be deposited at the start of each quarter in order to have $500,000 in 6 years?

House financing 7. Find the monthly payments to amortize a mortgage of $60,000 over a period of 20 years at an interest rate of 7.2 percent per year.

Financial planning

8. Suppose 20-year home mortgages are currently available at a rate of 7.8 percent per year. Jason and Joline can afford mortgage payments of at most \$850 per month. What is the largest mortgage they can afford? If they take out the largest loan possible, how much interest will they pay over the period of the loan?

Financial planning

9. Suppose 25-year mortgages are currently available at a rate of 8.4 percent per year. Jason and Joline can afford mortgage payments of at most \$850 per month. What is the largest mortgage they can afford? If they take out the largest loan possible, how much interest will they pay over the period of the loan? What is the amount of interest paid per dollar of loan?

10. Find the quarterly payments necessary to amortize a loan of \$10,000 over a period of 4 years with interest at 6 percent per year compounded quarterly.

11. Find the semiannual payments necessary to amortize a loan of \$20,000 over a period of 8 years with interest at 7 percent per year compounded semiannually.

12. Find the monthly payments necessary to amortize a \$50,000 loan in 30 years with interest rates of 6, 7, 8, 9, and 10 percent per year. Graph these points on a coordinate system in which the horizontal axis is the interest rate and the vertical axis is the monthly payment.

Folklore

13. There is a story that in 1650 Indians sold Manhattan Island for \$1. Assume that the average interest rate for the period from 1650 to 1990 was 5 percent per year, and assume that the original \$1 had been deposited in a savings account bearing interest at this rate compounded yearly. Also assume that there were no deposits or withdrawals other than the original deposit of \$1. Find the value of the account in 1990.

    This exercise can be used to illustrate the dramatic changes in total accumulations which result from small differences in interest rates when the period of compounding is long. Indeed, suppose that instead of 5 percent, the average interest rate from 1650 to 1990 was 6 percent, a difference of 20 percent. Show that the accumulated total increases by more than 2200 percent.

Consumer prices

14. A major financial newspaper carried the following statement: "Consumer prices climbed 0.5 percent in March led by a surge in clothing prices. The rise, equal to a 6.4 percent annual rate, was the largest in over a year and is another sign that inflation is heating up." Show that this statement overstates the annual rate based on the assumption of 0.5 percent monthly increases compounded for a year.

Compensation

15. Harold has negotiated a new contract with MDI, Inc., and he has the option of a deferred compensation package. For next year he has a choice of a salary of \$3500 per month or a salary of \$3400 per month and a year-end bonus of \$1250. If Harold can invest in a money market fund at a rate of 6 percent per year compounded monthly, which compensation package will give him the larger yearly income?

Compensation

16. As a part of its compensation package, a corporation plans to purchase an annuity for an executive. The plan is to make monthly payments of \$1000 for 5 years. The annuity has an interest rate of 6 percent per year. Find the amount and present value of this annuity.

Corporate finance

17. The owner of a small business takes out a loan for \$8000 payable in 5 years with accumulated interest. The interest rate is 7 percent per year compounded semiannually. What is the total amount due at the end of 5 years?

Financial planning

18. The *Metroburg Monitor*, the daily newspaper in Metroburg, published a table of monthly payments of principal and interest necessary to amortize a 30-year loan. Verify the entries shown below.

	Amount of Loan, $	
Interest Rate, %	$60,000	$120,000
6.0	359.73	719.46
6.5	379.24	758.48
7.0	399.18	798.36
7.5	419.53	839.06

Financial planning

19. Steven Smith, IV, who is now 5 years old, will begin college when he is 18. Steven Smith, III, estimates that college expenses will be $12,000, $13,000, $14,000, and $16,000 for the first, second, third, and fourth years, respectively, and he plans to open a savings account to accumulate funds for his son's education. He can obtain an interest rate of 5 percent per year compounded quarterly. Assuming that the interest rate remains constant and that he deposits equal amounts once each year, find the amount he should deposit each year to have enough money in the account when his son begins college so that no further deposits will be needed. Assume that he will make a deposit today and 13 additional deposits, ending when Steven Smith, IV, is 18 years old. Withdrawals are to be made on the 13th, 14th, 15th, and 16th anniversaries of the opening of the account.

Stock investing

20. Ivan buys 3500 shares of stock at $2.25 per share. Two years later there is a 2-for-1 stock split, after which he sells his stock for $2.75 per share. Find his return on investment per year.

Bond investing

21. Joan purchases a bond for $4800. It has a redemption value of $5000 and a stated interest rate of 5 percent per year paid semiannually, on January 1 and July 1. If she purchases the bond on January 1, 1995, and the redemption date is July 1, 1998, what is her total gain (total income minus purchase price) over the period between purchase and redemption?

Bond investing

22. What is the yield to maturity of the bond of Exercise 21?

Financial planning

23. Andrew is considering investing $100,000 in a small business. He estimates that the business will be successful with probability .8 and that it will fail with probability .2. If the business is successful, then he expects to be able to sell at a profit of $10,000 in 2 years or a profit of $30,000 in 5 years. If the business fails, he will lose half of his investment. Find his expected return on investment in 2 years. In 5 years.

Financial planning

24. In the situation described in Exercise 23, Andrew considers the situation more carefully, and he decides that his earlier analysis was valid only if the economy was strong, which he expects to happen with probability .7. If the economy does not remain strong, then he believes the probability of the business failing goes up to .4. If his other estimates remain as in Exercise 23, find his expected return on investment in 2 years.

Corporate finance 25. A company is planning to purchase a new computer 3 years from now at an estimated cost of \$4,000,000. One year later the company plans to buy an auxiliary memory unit at an estimated cost of \$1,500,000. The company has established a sinking fund so it will have the cash available at the time of each purchase. The company will make payments into the fund semiannually, and interest will be paid in the fund at a rate of 8 percent per year compounded semiannually. How much should be paid into the fund semiannually?

Corporate finance 26. In January 1995, the officers of MDI, Inc., prepare a corporate plan. They estimate that they can set aside 20 percent of their yearly earnings toward a new corporate headquarters building this year and a similar percentage for each of the next 5 years. They estimate that earnings will grow at a rate of 25 percent per year for the next 5 years. Also, they can invest money at a rate of 10 percent per year compounded yearly. If earnings are \$1,000,000 in 1990, how much will be available for the new corporate headquarters in 5 years?

CHAPTER

10 Linear Programming: The Simplex Method

10.0 THE SETTING AND OVERVIEW

The usefulness of linear programming models for optimization problems arising in business and other fields is due in large part to the dependable and efficient algorithms which have been developed for solving the resulting problems. This chapter is devoted to an introduction to the best-known such algorithm: the simplex method.

The simplex algorithm is, in the time frame of mathematics, a fairly recent result. In fact, it is one of the most recent mathematical results presented in this text. It was invented in the late 1940s by George Dantzig, and it rapidly gained widespread use. The invention of the algorithm was a part of a remarkable coincidence of events in the late 1940s and early 1950s that had a profound impact on our ability to understand, model, and solve many important problems in resource allocation, scheduling, etc. The first event was the recognition that such problems arise in a variety of settings, that many such problems have common mathematical characteristics, and that their solutions may have significant economic consequences. This recognition was accelerated by major transportation and allocation problems arising in World War II. The second event was the invention of the simplex algorithm, an algorithm which makes the solution of linear programming problems relatively straightforward. Finally, the invention of digital computers made it possible to use the simplex algorithm to solve problems of the size that actually occur in practice. Without computers the simplex algorithm would be little more than an interesting mathematical result. Its usefulness rests on the fact that when implemented on a computer, it can be used to quickly solve problems with very large numbers of variables. Versions of the algorithm are regularly used to solve linear programming problems with thousands and, in special cases, even millions of variables.

The basic idea behind the simplex algorithm is the same as that used in Chapter 7 to solve linear programming problems with a graphical technique. That is, if the problem has a solution, then we know that the maximum (or minimum) value of the objective function is attained at a corner point of the feasible set, and we need a method of, first, finding corner points and, second, evaluating the objective function at the corner points. It is clear that in the general case we cannot hope to use a method based on geometry. There is a chance that we could draw figures for simple problems involving three variables, but the situation is hopeless for more than three variables. Also, it is desirable to avoid evaluating the objective function at all the corner points of the feasible set if we can do so. Indeed, a problem with n variables and n constraints can have a feasible set with as many as 2^n corner points, a very large number even for relatively small values of n: if $n = 20$, then $2^n = 1{,}048{,}576$. A problem with 20 variables and 20 constraints can have a feasible set with over 1 million corner points!

The simplex method evaluates the objective function at one corner point after another in a way that gives better and better values of the objective function. At each corner point, the algorithm tells us exactly which corner point to select next and the value of the objective function at that corner point. Although it has this simple geometric interpretation, it is purely an algebraic method.

There are several versions of the simplex algorithm, each version designed to be especially effective on problems of a certain type. These special versions are quite effective, but because linear programming problems are so important, there is great interest in finding methods which are even better than the simplex method. The term "better" usually means faster. It is common to measure the speed of an algorithm by the number of basic arithmetic operations needed to solve a problem with specified numbers of variables and constraints. Improvements which seem relatively unimpressive, say an improvement of 10 percent, can result in enormous dollar savings to a business which solves thousands of such problems each month. Consequently, there is a great deal of interest in finding methods that are significantly better than the simplex algorithm.

This search for new methods to solve linear programming problems has resulted in the creation of fundamentally new approaches to the problems. One such approach is an algorithm invented by N. Karmarkar at Bell Laboratories in 1984. Karmarkar's algorithm finds approximate solutions by evaluating the objective function at carefully selected points *inside* the feasible set. The method enables you to find an approximate solution as close as you wish to an actual solution of the problem. The increased speed of Karmarkar's method has generated a great deal of both theoretical and practical interest. The method seems to be especially successful on problems with large numbers of constraints.

10.1 STANDARD MAXIMUM PROBLEMS, SLACK VARIABLES, AND BASIC SOLUTIONS

The simplex method can be applied to any linear programming problem, no matter how the problem is formulated. However, the method is easier to explain and to understand when the problem is formulated in a special way. In this section we describe the special form, and then we discuss the initial step needed to begin the simplex method on a problem in this form. To illustrate our techniques, we use one of the examples formulated in Chapter 7, namely Example 7.5. That problem can be formulated as follows:

Example 10.1

Task scheduling

An office manager must allocate the time of the office staff among the activities of auditing, business accounting, and tax accounting. We let

x = the number of hours per week of auditing
y = the number of hours per week of business accounting
z = the number of hours per week of tax accounting

Then the linear programming problem is to

$$\begin{array}{ll} \text{Maximize} & \text{Profit} = p = 4x + 10y + 6z \\ \text{subject to} & x \geq 0 \quad y \geq 0 \quad z \geq 0 \\ & 15x + 20y + 30z \leq 4{,}800 \\ & 30x + 60y + 45z \leq 10{,}800 \\ & \qquad 6y + 3z \leq 1{,}800 \end{array} \tag{10.1}$$

■

The inequalities (10.1) are satisfied by $x = y = z = 0$, in which case the profit is 0, and by $x = y = z = 60$, in which case the profit is 1200. The problem is to find values for x, y, and z that satisfy all inequalities in (10.1) and also maximize the objective function, the profit in this case. We use the term "feasible vector" to describe any triple of numbers x, y, and z which satisfies (10.1). If the problem under consideration has three variables, then we write the vectors as $[x \quad y \quad z]$, and in cases where there are n variables, $n > 3$, we write the vectors as $[x_1 \quad x_2 \quad \cdots \quad x_n]$. We adopt this notation to be consistent with our definition of a vector as a special matrix (with either one row or one column). For instance, in Example 10.1 the vectors $[0 \quad 0 \quad 0]$ and $[60 \quad 60 \quad 60]$ are feasible. The triple $[x \quad y \quad z]$ also represents a point in three dimensions, and it would be possible to develop a geometric approach to solve linear programming problems with three variables similar to the method developed in Chapter 7 for two variables. However, our goal here is to develop a technique which can be used for any number of variables and which does not require a graph of the feasible set (the set of all feasible vectors). Thus, we do not exploit the fact that there is a geometric representation for the feasible set; instead we make use of matrix techniques. Although we concentrate on examples with four or fewer variables, the methods of the chapter can be used for larger numbers of variables. In the next section we indicate how the two techniques (graphing points and solving algebraic systems) are related.

Example 10.1 is typical of the type of problem considered in this chapter. Each problem involves a system of inequalities [such as (10.1)] and an objective function (such as a profit function). The goal is to find a feasible vector that maximizes the value of the objective function for all vectors in the feasible set. In all problems considered in this chapter, the inequality constraints will include the condition that the coordinates of the variable must be nonnegative.

Notice that in the constraints (10.1) and in the objective function p of Example 10.1 there are expressions of the form $ax + by + cz$, where a, b, and c are numbers. Such expressions (which were introduced in Chapter 5) will occur frequently in this chapter. A formal mathematical definition follows:

If $a_1, a_2, \ldots, a_n$ are numbers, then

$$a_1x_1 + a_2x_2 + \cdots + a_nx_n$$

is a *linear function* of the variables $x_1, x_2, \ldots, x_n$.

The problem in Example 10.1 is an example of a type of problem called a *standard maximum problem* (SMP). The precise definition of this class of problem is as follows:

Standard Maximum Problem

A linear programming problem is written as a standard maximum problem (SMP) if:

1. The variables are constrained to be nonnegative.
2. The constraints of the problem (other than the nonnegativity constraints given in condition 1) are all of the form

$$\text{Linear function of the variables} \leq \text{positive number}$$

3. The objective function of the problem is a linear function that is to be maximized.

Other treatments of the simplex method may use the term "standard maximum problem" in a different way. This definition is a convenient one for our purposes.

Example 10.2 In each case, decide whether the given problem can be written as an SMP, and do so when possible.

(*a*) Maximize $p = x - y + 3z$

subject to

$$\begin{aligned} x \geq 0 \quad y \geq 0 \quad z \geq 0 \\ 3x - y + 4z \leq 0 \\ -x + 5y - 3z \geq 8 \end{aligned}$$

(*b*) Maximize $p = x_1 - 4x_2 + 5x_3 - x_4$

subject to

$$\begin{aligned} x_1 \geq 0 \quad x_2 \geq 0 \\ x_1 + 5x_3 - x_4 \leq 5 \\ x_1 + 5x_2 - x_3 \leq 12 \end{aligned}$$

(*c*) Maximize $p = 8x - 25y$

subject to

$$\begin{aligned} x \geq 0 \quad y \geq 0 \\ 3x - 4y \geq -6 \\ 5x + y \geq -2 \end{aligned}$$

Solution (*a*) Conditions 1 and 3 are clearly satisfied. The constraint $-x + 5y - 3z \geq 8$ can be written as an inequality with a $\leq$ sign by multiplying both sides by -1. The result is $x - 5y + 3z \leq -8$. However, the right-hand side of this inequality is not a positive constant, and (*a*) is *not* an SMP.

(*b*) Condition 2 holds, but condition 1 does not hold since only the first two of the four variables are required to be nonnegative. This is not an SMP.

(*c*) All conditions are satisfied, and this is an SMP. To show that condition 2 holds, it is necessary to multiply each of the constraints by -1. The problem written as an SMP is

Maximize $p = 8x - 25y$

subject to

$$\begin{aligned} x \geq 0 \quad y \geq 0 \\ -3x + 4y \leq 6 \\ -5x - y \leq 2 \end{aligned}$$

■

Example 10.3 (See also Example 7.4.) Decide whether the following problem is an SMP.

Maximize Profit $= p = 3x + 8y + 6z$

subject to

$$\begin{aligned} x \geq 0 \quad y \geq 0 \quad z \geq 0 \\ 20x + 4y + 4z \leq 6{,}000 \\ 8x + 8y + 4z \leq 10{,}000 \\ 8x + 4y + 2z \leq 4{,}000 \end{aligned} \tag{10.2}$$

Solution (*a*) This problem has three variables (x, y, and z), and each is constrained to be nonnegative. Thus, condition 1 is satisfied.

(*b*) The constraints of the problem, other than the nonnegativity constraints, i.e., (10.2), are all of the form $ax + by + cz \leq$ positive number. Hence, condition 2 is satisfied.

(*c*) The objective function ($p = 3x + 8y + 6z$) is of the form $p = ax + by + cz$ and is to be maximized, so condition 3 is satisfied.

Therefore, this problem is an SMP. ■

The linear programming problem of Example 10.3, shown to be an SMP, arose as a resource allocation problem. In fact, most of the usual resource allocation problems are naturally formulated as SMPs.

The constraints in a standard maximum problem (SMP) are inequalities which state that linear functions of the variables are to be less than or equal to positive numbers. In Example 10.3 the variables are x, y, and z; these variables must be nonnegative, and they must satisfy the constraints (10.2). We recall from Example 7.4 that the constraints (10.2) represent limitations on the production of fertilizer due to the availability of the resources: nitrate (6000 pounds available), phosphate (10,000 pounds available), and potash (4000 pounds available).

To satisfy the first of these inequalities, we must find nonnegative numbers x, y, and z such that $20x + 4y + 4z$ is either less than or equal to 6000. If $20x + 4y + 4z < 6000$, then there is a positive number u such that

$$20x + 4y + 4z + u = 6000$$

If we consider u to be a variable also, then to satisfy the original constraint, we must find nonnegative numbers x, y, z, and u that satisfy this new equation. If $u = 0$, then the original constraint is satisfied in the form $20x + 4y + 4z = 6000$; and if $u > 0$, the original constraint is satisfied in the form $20x + 4y + 4z < 6000$. The variable u is called a *slack variable* for the first constraint. In terms of the problem of Example 7.4, since the constraint concerns the use of nitrate, u measures the amount (in pounds) of nitrate which is available and which is *not* used in the production of x hundreds of pounds of garden fertilizer, y hundreds of pounds of lawn fertilizer, and z hundreds of pounds of general-purpose fertilizer. Thus it is the unused or "slack" amount of nitrate. Similarly, we can introduce slack variables v and w for the second and third inequalities. Thus an equivalent way of representing the constraints of Example 10.3 is as follows:

Find nonnegative numbers x, y, z and nonnegative values of the slack variables u, v, w such that

$$\begin{aligned} 20x + 4y + 4z + u &= 6{,}000 \\ 8x + 8y + 4z + v &= 10{,}000 \\ 8x + 4y + 2z + w &= 4{,}000 \end{aligned} \qquad (10.3)$$

To distinguish easily between the variables of the original problem and the slack variables, we refer to the former as *decision variables*.

The system (10.3) is a system of equations (instead of inequalities), and as such it can be solved for the unknowns, the decision variables x, y, z, and the slack variables u, v, w, by using the methods of Section 5.3. However, in Example 10.3 and in all SMPs, it is not enough to just solve the system of equations (10.3). Instead, it is required to find solutions which are nonnegative, i.e., each variable must be zero or positive. Moreover, our goal is to find solutions that maximize the objective function. In Example 10.3 the objective function is $p = 3x + 8y + 6z$. Note that the slack variables do not enter directly into the objective function. As we observed above, the slack variables in this example measure the amount of unused material, and since profit is determined by the amount of fertilizer produced—consequently by the material actually used—the slack variables are indirectly, not directly, related to profit. If the slack variables take on large values, then x, y, and z are necessarily small because much of the material is unused, and hence profit is small.

Our approach to the solution of linear programming problems in two variables in Chapter 7 depended on the fact that the maximum of the objective function always occurred at a corner point of the feasible set (provided, of course, that a maximum existed). If such a problem has a solution, then with two variables the solution can be found by using graphical methods: Graph the feasible set, find the corner points, and determine the corner point at which the objective has its maximum (or minimum) value. For more than two variables the situation is essentially the same. However, it is difficult to determine the corner points of the set by graphing, and a technique based on algebra is preferable. The algebra and the geometry are connected in the concept of a basic solution.

Basic Solutions

We introduce the idea of a basic solution in the context of Example 10.3. Later we give a general definition. We also note that the simplex method can be used without understanding the nature of a basic solution and without understanding why the method works. Our discussion here is simply to introduce one of the key concepts of the method.

The system of equations (10.3) was obtained from the inequalities of the SMP of Example 10.3 by adding slack variables. Each solution of this system of equations which has nonnegative values for all variables provides a feasible vector for the original SMP. This feasible vector consists of the values of the decision variables for the solution of the system of equations. The situation is similar for all SMPs. By adding slack variables to the inequalities of an SMP, one forms a system of equations; and each solution of this system of equations which has nonnegative values for the variables provides a feasible vector for the original SMP. As we noted in Chapter 5, every system of equations has either no solutions, exactly one solution, or infinitely many solutions. However, for systems obtained by adding slack variables to the inequalities of an SMP, there is always *at least one* solution which has nonnegative values for all variables. Such a solution is the one obtained by setting all decision variables equal to zero and each slack variable equal to the right-hand side of the equation containing that variable. For example, in system (10.3), shown again here,

$$\begin{aligned} 20x + 4y + 4z + u &= 6{,}000 \\ 8x + 8y + 4z + v &= 10{,}000 \\ 8x + 4y + 2z + w &= 4{,}000 \end{aligned}$$

we have the solution

$$x = 0 \qquad y = 0 \qquad z = 0 \qquad u = 6000 \qquad v = 10{,}000 \qquad w = 4000$$

Thus, the associated feasible vector for the original SMP is $[0 \quad 0 \quad 0]$, that is, $x = 0$, $y = 0$, $z = 0$. This feasible vector is the origin of the space containing the set of feasible vectors; and since the feasible vectors always have nonnegative coordinates, this vector is always a corner point of the feasible set. Also the objective function has value 0 at this point.

The basic idea behind the simplex method is the following: Using the method, we derive special feasible vectors for the problem, and we evaluate the objective function at each such special feasible vector. The special feasible vectors will be constructed in such a way that at each step the objective function takes a larger value than at the preceding step. The method stops when we find the largest value of the objective function or when we conclude that there is no largest value. Much of the utility of the method is based on the fact that the special feasible vectors and the evaluations of the objective function are carried out automatically. To appreciate what is going on when the method is used, it helps to know a little about the related geometry. However, we emphasize again, since the method carries out certain steps automatically, it is not essential to understand the geometry to apply the method.

If an SMP has only one feasible vector, then it is the vector noted above, a vector obtained by setting all decision variables equal to zero. In this case the feasible set consists of a single vector, the vector corresponding to the origin in the space containing vectors of decision variables. Moreover, the objective function takes its maximum value at this point.

If the SMP has more than one feasible vector (the typical case), then it has infinitely many feasible vectors and the associated system of equations has infinitely many solutions. Using the results of Chapter 5, we know that all solutions of the system of equations can be obtained by converting the augmented matrix of the system to reduced form. If there are infinitely many solutions of the system, then certain variables of the system can be given arbitrary values and other variables will be expressed in terms of these arbitrary variables. By using the methods of Chapter 5, exactly which variables can be given arbitrary values is determined by the order in which the variables are listed in the augmented matrix and by the reduced form of the augmented matrix. Once the order of the variables is specified and the reduced form of the matrix (and arbitrary variables) is determined, then each choice of values for the arbitrary variables determines a solution of the system of equations. Some of these choices, those that yield nonnegative values for all variables, determine feasible vectors for the original SMP. One special choice of the arbitrary variables is to set them all to zero. For this choice, if all other variables are nonnegative, then the resulting feasible vector is a special solution of the system, called a *basic solution*. It is these basic solutions that the simplex method uses to solve the SMP. The simplex method evaluates the objective function at basic solutions. The important geometric facts are that *each basic solution corresponds to a feasible vector which is a corner point of the feasible set, and if the SMP has a solution, then there is a solution at a corner point*. To summarize:

Basic Solutions

Consider an SMP with n decision variables and m inequality constraints. Introduce m slack variables to form a system of m equations in $m + n$ variables. The number of slack variables is always equal to the number of inequality constraints. Next, convert this system to reduced form by using elementary row operations. A *basic solution* is any nonnegative solution of this system with each variable not corresponding to a leading 1 in the reduced form given the value 0. The remaining variables are then determined by the system of equations.

Note that we could have put the slack variables before, intermixed with, or following the decision variables. Different placements of the slack variables lead to different reduced forms and therefore to different basic solutions. An extremely valuable feature of the simplex method is that the algorithm automatically places the slack variables in the appropriate positions. Before discussing the algorithm, we study an example which illustrates the way basic solutions, reduced forms, and corner points are related.

Example 10.4 An SMP is given by

$$\begin{array}{ll} \text{Maximize} & p = 2x + y \\ \text{subject to} & x \geq 0 \qquad y \geq 0 \\ & x \leq 40 \\ & 3x + 4y \leq 240 \\ & 2x + y \leq 100 \end{array}$$

Problem Introduce slack variables and find three basic solutions. Graph the feasible set and find all corner points.

Solution Since there are three inequality constraints, we introduce three slack variables: u, v, and w. The resulting system of equations is

$$\begin{aligned} x \qquad\quad + u &= 40 \\ 3x + 4y + v &= 240 \\ 2x + y + w &= 100 \end{aligned}$$

To find our first basic solution, we rewrite the system with the slack variables preceding the decision variables:

$$\begin{aligned} u \qquad\quad + x \qquad &= 40 \\ v \quad + 3x + 4y &= 240 \\ w + 2x + y &= 100 \end{aligned}$$

The augmented matrix for this system is

$$\left[\begin{array}{ccccc|c} 1 & 0 & 0 & 1 & 0 & 40 \\ 0 & 1 & 0 & 3 & 4 & 240 \\ 0 & 0 & 1 & 2 & 1 & 100 \end{array}\right]$$

which is clearly already in reduced form. It follows from our work in Chapter 5 that variables x and y can be given arbitrary values. According to the definition of a basic solution, we set $x = 0$ and $y = 0$, and we obtain $u = 40$, $v = 240$, and $w = 100$. Thus our first basic solution is

$$x = 0 \qquad y = 0 \qquad u = 40 \qquad v = 240 \qquad w = 100$$

(Note that a basic solution includes values for all decision variables and all slack variables. Also, as we noted above, it is always the case that an SMP has a basic solution in which all decision variables have the value 0.)

To determine our next basic solution, we write the system in the form

$$\begin{aligned} x \qquad\qquad + u \qquad\quad &= 40 \\ 3x + v \qquad\qquad + 4y &= 240 \\ 2x \qquad + w \qquad + y &= 100 \end{aligned} \tag{10.4}$$

Using row reduction operations, we transform this to the form

$$\begin{aligned} x \qquad + u \qquad &= 40 \\ v \quad - 3u + 4y &= 120 \\ w - 2u + y &= 20 \end{aligned} \tag{10.5}$$

The augmented matrix for this system is in reduced form. Variables u and y can be given arbitrary values, so we set $u = 0$ and $y = 0$. It follows that $x = 40$, $v = 120$, and $w = 20$. Consequently, another basic solution is

$$x = 40 \qquad y = 0 \qquad u = 0 \qquad v = 120 \qquad w = 20$$

To find our next basic solution, we write the original system in the form

$$\begin{array}{rcl} x \qquad\qquad + u & & = 40 \\ 3x + 4y + v & & = 240 \\ 2x + y \qquad\qquad\quad & + w & = 100 \end{array} \tag{10.6}$$

Using row reduction operations, we transform this to

$$\begin{array}{rcl} x \qquad + u \qquad & & = 40 \\ y \quad - 2u + w & & = 20 \\ v + 5u - 4w & & = 40 \end{array} \tag{10.7}$$

We set $u = 0$ and $w = 0$, and it follows that $x = 40$, $y = 20$, and $v = 40$. Consequently, another basic solution is

$$x = 40 \qquad y = 20 \qquad u = 0 \qquad v = 40 \qquad w = 0$$

The feasible set for this problem is shown as the shaded region $\mathcal{F}$ in Figure 10.1. With the notation of Chapter 5 (i.e., ordered pairs of numbers are points in the plane), the corner points of $\mathcal{F}$ are $P = (0, 0)$, $Q = (40, 0)$, $R = (40, 20)$, $S = (32, 36)$, and $T = (0, 60)$. Notice that the x and y coordinates of P are the same as the x and y coordinates of the first basic solution we determined above, the x and y coordinates of Q are the same as the x and y coordinates of the sec-

FIGURE 10.1

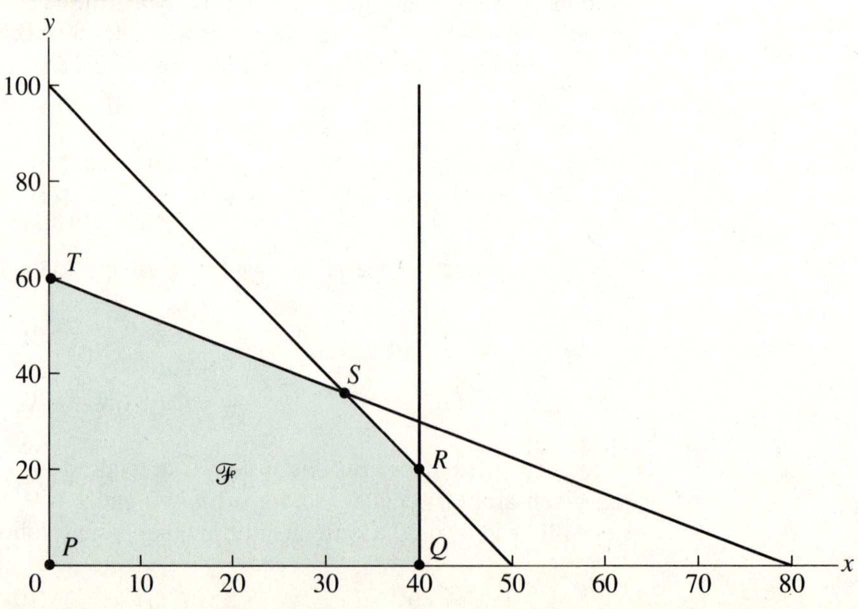

ond basic solution, and the x and y coordinates of R are the same as the x and y coordinates of the third basic solution. In fact, each of the corner points of the feasible set corresponds to a basic solution in this way. ■

Example 10.4 illustrates relations between basic solutions, corner points, and reduced forms. The relations illustrated in this example hold in general. Namely, as we noted earlier: Each basic solution of the system of equations corresponds to a corner point of the feasible set. Each corner point of the feasible set corresponds to a basic solution of the system of equations.

Therefore, a maximum of the objective function (if it exists) will occur at a basic solution of the system of equations. This important fact provides the connection between the geometric techniques of Chapter 7 and the algebraic techniques of this chapter.

In any standard maximum problem the basic solutions of the system of equations obtained by adding slack variables correspond to corner points of the feasible set. Moreover, if there is a maximum for the objective function, it will occur at a basic solution.

It follows that to solve an SMP, we can proceed by finding basic solutions of the related system of equations and evaluating the objective function at these basic solutions. These basic solutions can be found by using a systematic step-by-step procedure called the *simplex method*, which involves operations similar to those used in Section 5.3. This method not only identifies basic solutions but also constructs one basic solution after another in such a way that the value of the objective function steadily increases until its maximum value is reached or until it is clear that there is no maximum value. A description of this method is the topic of the next two sections.

Exercises for Section 10.1

1. Show that the vectors $[5 \quad 5 \quad 5]$ and $[1 \quad 1 \quad 10]$ are feasible for the SMP

Maximize $\quad p = x + 2y + 3z$

subject to

$$\begin{aligned} x \geq 0 \qquad y \geq 0 \qquad z \geq 0 \\ 3x + 4y + 8z \leq 100 \\ x + 8y + 2z \leq 80 \\ 5x + y + 4z \leq 120 \end{aligned}$$

2. If the vector $[x \quad 2 \quad 2]$ is feasible for the SMP of Exercise 1, what is the largest value that x can have?
3. Which of the following vectors are feasible for the SMP of Exercise 1?
(*a*) $[7 \quad 7 \quad 7]$ (*b*) $[3 \quad 4 \quad -2]$
(*c*) $[10 \quad 8 \quad 4]$ (*d*) $[2 \quad 1 \quad 10]$

4. Show that the vectors [5 5 10] and [8 0 19] are feasible for the SMP

$$\begin{array}{ll} \text{Maximize} & p = 2x + 5y + 3z \\ \text{subject to} & x \geq 0 \quad y \geq 0 \quad z \geq 0 \\ & 12x + 10y + z \leq 120 \\ & 5x + 21y + 4z \leq 180 \\ & x + 15y + 8z \leq 160 \end{array}$$

5. Which of the following vectors are feasible for the SMP of Exercise 4?
   (*a*) [1 7 7] (*b*) [6 4 8] (*c*) [7 2 15]
6. Show that the vectors [60 60 60], [0 180 0], and [300 15 0] are feasible for the problem posed in Example 10.1.
7. If a vector [0 120 z] is feasible for the problem posed in Example 10.1, what is the largest value z can have?
8. Which of the following vectors are feasible for the problem formulated in Example 10.1?
   (*a*) [100 60 20] (*b*) [100 20 60]
   (*c*) [20 150 60] (*d*) [60 120 60]

Resource allocation

9. The Kidsports Toy Company makes plastic balls, bats, and rackets for children. The manufacturing of these toys is a three-step process: (1) the plastic is molded, (2) the toy is painted, and (3) special labels are put on the toy. Each ball requires 4 ounces of plastic, 2 minutes of paint time, and 1 minute to apply labels. Each bat requires 9 ounces of plastic, 3 minutes of paint time, and 1 minute to apply labels. Each racket requires 6 ounces of plastic, 2 minutes of paint time, and 3 minutes to apply labels. During each hour that the company's plant is in operation, there are 200 ounces of plastic available, 55 minutes can be used for painting, and 45 minutes can be used for labeling. Also, the company makes \$0.75 per ball, \$1.50 per bat, and \$1.25 per racket.

   Formulate a linear programming problem whose solution gives the number of balls, bats, and rackets to be manufactured each hour to maximize the company's profit from these toys. Is the problem an SMP?

Decide which of the linear programming problems in Exercises 10 through 13 are SMPs. If a problem is not an SMP, state why not.

10. Find the maximum of $p = 3x + 2y - 3z$ subject to

$$\begin{array}{c} x \geq 0 \quad y \geq 0 \\ 3x + 12z \leq 28 \\ 4y + 8z \leq 35 \end{array}$$

11. Find the maximum of $p = 5x - y + 8z$ subject to

$$\begin{array}{c} x \geq 0 \quad y \geq 0 \quad z \geq 0 \\ 30x - 45y + 100z \leq 500 \\ 50x + 80y - 120z \leq 850 \\ 25x - 15y - 250z \geq 100 \end{array}$$

12. Find the minimum of $p = x + y$ subject to

$$\begin{aligned} x \geq 0 \qquad y &\leq 0 \\ 2x + \ y &\leq 7 \\ x + 3y &\leq 6 \end{aligned}$$

13. Find the maximum of $p = x + y + 5z$ subject to

$$\begin{aligned} x \geq 0 \qquad y \geq 0 \qquad z &\geq 0 \\ 10x - 5y + 10z &\leq 150 \\ 5x + 8y - 12z &\leq 165 \\ 4x + 5y + 18z &\leq 120 \end{aligned}$$

14. (*a*) Show that the set shown in Figure 10.2*a* could be the feasible set for an SMP.

(*b*) Show that the set shown in Figure 10.2*b* could not be the feasible set for an SMP.

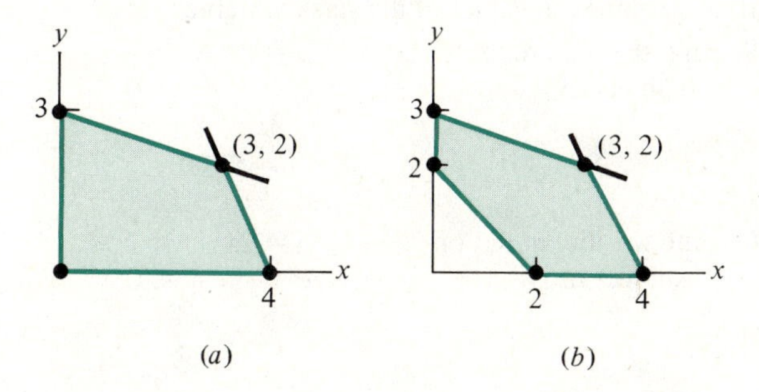

FIGURE 10.2

Convert each of the linear programming problems in Exercises 15 through 18 to a problem with slack variables.

15. Find the maximum of subject to

$$\begin{aligned} p &= 5x + 4y \\ x \geq 0 \qquad y &\geq 0 \\ 3x + 12y &\leq 42 \\ 8x + \ 5y &\leq 60 \end{aligned}$$

16. Find the maximum of subject to

$$\begin{aligned} p &= 15x + 27y \\ x \geq 0 \qquad y &\geq 0 \\ 3x + 12y &\leq 42 \\ 8x + \ 5y &\leq 60 \\ 15x + \ 8y &\leq 120 \end{aligned}$$

17. Find the maximum of $p = x + 4y + 5z$
    subject to
$$x \geq 0 \qquad y \geq 0 \qquad z \geq 0$$
$$x + 2y + z \leq 12$$
$$2x + 5y + 3z \leq 45$$

18. Find the maximum of $p = x + 10y + 20z$
    subject to
$$x \geq 0 \qquad y \geq 0 \qquad z \geq 0$$
$$x + 15y + 10z \leq 200$$
$$10x + 5y + z \leq 120$$
$$15x + y + 20z \leq 300$$
$$x + y + z \leq 100$$

In Exercises 19 through 25, convert the problem to one with equations by adding slack variables to the given inequalities. Graph the feasible set in the plane, and find the corner points. Finally, find the basic solution which corresponds to each corner point. (*Hint*: Use the relation between basic solutions and corner points, *not* the definition of a basic solution. Begin with the values of x and y for the corner point, and then determine the values of the slack variables.)

19. Find the maximum of $p = 5x + 4y$
    subject to
$$x \geq 0 \qquad y \geq 0$$
$$3x \leq 12$$
$$5y \leq 60$$

20. Find the maximum of $p = 2x + 3y$
    subject to
$$x \geq 0 \qquad y \geq 0$$
$$x + y \leq 8$$
$$3y \leq 8$$

21. Find the maximum of $p = 3x + 8y$
    subject to
$$x \geq 0 \qquad y \geq 0$$
$$x + y \leq 5$$
$$3x + 6y \leq 18$$

22. Find the maximum of $p = 3x + 2y$
    subject to
$$x \geq 0 \qquad y \geq 0$$
$$3x + 10y \leq 10$$
$$2x + 2y \leq 3$$
$$10x + y \leq 10$$

23. Find the maximum of $p = 5x + y$
    subject to
$$x \geq 0 \qquad y \geq 0$$
$$4x + 4y \leq 20$$
$$3x + 12y \leq 24$$
$$10x + 3y \leq 30$$

24. Find the maximum of $p = 6x + 4y$
    subject to
$$x \geq 0 \qquad y \geq 0$$
$$2x + y \leq 16$$
$$x + y \leq 10$$
$$x \leq 7$$

25. Find the maximum of $p = 4x + 5y$
    subject to
$$x \geq 0 \qquad y \geq 0$$
$$x \geq y - 1$$
$$x + y \leq 20$$

26. Give the steps which transform system (10.4) to system (10.5).
27. Give the steps which transform system (10.6) to system (10.7).
28. Use the ideas of Example 10.4 to find the basic solutions which correspond to the corner points S and T of Figure 10.1.
29. Convert the following problem to one with slack variables, and use the technique of Example 10.4 to find three basic solutions of the resulting system of equations.

    Find the maximum of $p = 5x + y$
    subject to
$$x \geq 0 \qquad y \geq 0$$
$$2x + y \leq 8$$
$$x + 2y \leq 8$$

30. Convert the following problem to one with slack variables, and use the technique of Example 10.4 to find three basic solutions of the resulting system of equations.

    Find the maximum of $p = 2x + 3y + z$
    subject to
$$x \geq 0 \qquad y \geq 0 \qquad z \geq 0$$
$$x + y \leq 2$$
$$y + z \leq 2$$
$$x + z \leq 2$$

31. The following problem contains slack variables. State an SMP which could be converted to this problem by adding slack variables.

    Find the maximum of $p = 3x + 2y + z$
    subject to
$$x \geq 0 \qquad y \geq 0 \qquad z \geq 0 \qquad u \geq 0 \qquad v \geq 0$$
$$3x - y + 2z + u = 100$$
$$2x + y - z + v = 150$$

32. Repeat Exercise 31 for the following problem:

    Find the maximum of $p = 2x - y + z$
    subject to
$$x \geq 0 \qquad y \geq 0 \qquad z \geq 0 \qquad u \geq 0 \qquad v \geq 0 \qquad w \geq 0$$
$$x + u = 200$$
$$3x + y + v = 700$$
$$y + z + w = 300$$

Maintenance

33. The supervisor of the Metroburg Street Department makes a choice among a "quick fix," a satisfactory but temporary repair job, and a long-lasting repair job on each city street. A quick fix requires 200 cubic yards of asphalt and 200 hours of labor for each mile of street. A satisfactory but temporary job requires 500 cubic yards of asphalt and 100 hours of labor for each mile, and a long-lasting job requires 1000 cubic yards of asphalt and 200 hours of labor. There are 10,000 cubic yards of asphalt and 1000 hours of labor available. The supervisor estimates that a satisfactory but temporary job has twice the value of a quick fix and a long-lasting job has 4 times the value of a quick fix.
    (*a*) Formulate a linear programming problem whose solution gives the number of miles of street to be repaired by each method to give maximum value.
    (*b*) Introduce slack variables and write the inequality constraints as equalities.
    (*c*) Give an interpretation of each of the slack variables.

Farming

34. A farmer has 1000 acres and water rights to 600 acre-feet of water for next season. (An acre-foot of water is the amount of water which covers 1 acre at a depth of 1 foot.) Crop A yields 120 bushels per acre and requires 6 inches of water per acre, crop B yields 80 bushels per acre and requires 5 inches of water per acre, and crop C yields 50 bushels per acre and requires 4 inches of water per acre. The farmer expects crop A to yield a profit of \$1.00 per bushel, crop B a profit of \$1.20 per bushel, and crop C a profit of \$2.00 per bushel. The farmer wants to allocate her land to grow crops which produce maximum profit.
    (*a*) Formulate a linear programming problem for this situation.
    (*b*) Introduce slack variables and write the inequality constraints as equalities.
    (*c*) Give an interpretation of each of the slack variables.

Resource allocation

35. The management of Brown Brothers Box Company is preparing to bid on some new contracts, and the analysis of the resource requirements and profits for boxes for computers, printers, and paint are as follows. A computer box requires 12 square feet of heavy-duty cardboard, 18 square feet of regular cardboard, and 15 square feet of white facing paper. A printer box requires 6 square feet of heavy-duty cardboard, 12 square feet of regular cardboard, and 8 square feet of white facing paper. A box for paint requires 10 square feet of regular cardboard. Current arrangements with suppliers will enable the management of Brown Brothers Box to allocate each week to new production: 1500 square feet of heavy-duty cardboard, 2500 square feet of regular cardboard, and 2000 square feet of facing paper. The profit is \$1.50 for a computer box, \$.80 for a printer box, and \$.25 for a paint box.
    (*a*) Formulate a linear programming problem whose solution gives the weekly allocation of resources which yields maximum profit.
    (*b*) Introduce slack variables and write the inequality constraints as equalities.
    (*c*) Give an interpretation of each slack variable.

10.2 TABLEAUS AND THE PIVOT OPERATION

The simplex method is an algebraic process that solves an SMP by systematically evaluating the objective function at certain corner points of the feasible set. The process continues either until a corner point is found at which the objective function assumes its maximum value or until it is known that the objective function does not assume a maximum value on the feasible set. The steps of the algorithm are carried out on an array of numbers called a *tableau* and the step which constructs one tableau from another is called a *pivot operation*. This section describes how tableaus are formed and how the pivot operation is carried out. The next section describes the complete simplex method.

Formation of the Initial Tableau

Example 10.5 Consider the SMP

$$\begin{array}{ll} \text{Maximize} & p = 10x + 6y + 2z \\ \text{subject to} & x \geq 0 \quad y \geq 0 \quad z \geq 0 \\ & 2x + 2y + 3z \leq 160 \\ & 5x + y + 10z \leq 100 \end{array}$$

Problem Form an initial tableau for this SMP.

Solution First we convert the constraints to a system of equations by introducing slack variables. Since there are two inequality constraints (we do not consider the nonnegativity constraints $x \geq 0$, $y \geq 0$, and $z \geq 0$ in forming the initial tableau), we need two slack variables. We call these slack variables u and v, and we have the new problem:

$$\begin{array}{ll} \text{Maximize} & p = 10x + 6y + 2z \\ \text{subject to} & x \geq 0 \quad y \geq 0 \quad z \geq 0 \\ & u \geq 0 \quad v \geq 0 \\ & 2x + 2y + 3z + u = 160 \\ & 5x + y + 10z + v = 100 \end{array}$$

Setting aside for the moment the nonnegativity conditions, this is a system of two equations in five variables, and the system can be solved by using the methods introduced in Chapter 5. The augmented matrix for the system (see Section 5.3) is

$$\left[\begin{array}{ccccc|c} 2 & 2 & 3 & 1 & 0 & 160 \\ 5 & 1 & 10 & 0 & 1 & 100 \end{array}\right]$$

Our initial tableau consists of the augmented matrix together with additional information listed above, to the left, and below. Above we list the variables in order at the top of the respective columns, and we add the phrase "Basic Solution" above the column to the right of the vertical line. To the left we list the slack variables next to each row. The tableau at this stage, shown as Tableau

10.1, is not yet an initial tableau for the simplex method. We continue by adding a third row to the tableau, and we label this row on the left by p to indicate that the entries in the row are obtained from the coefficients of the objective function p. The entry below the column labeled x is the negative of the coefficient of x in p, or -10 in this example. Similarly, the entries below y, z, u, and v are the negatives of the coefficients of those variables in p. They are -6, -2, 0, and 0, respectively, since 6, 2, 0, and 0 are the coefficients of y, z, u, and v in p. Finally, the entry below the column labeled "Basic Solution" is the value of p when the variables are given the values shown in this column. For example, in Tableau 10.1 variable u has the value 160, v has the value 100, and each of x, y, and z has the value 0. Therefore, since $p = 10x + 6y + 2z$, for these values of the variables, the objective function p has the value 0, and the entry in the row labeled p in the basic solution column is 0. The row labeled p is called the *check row* of the tableau for reasons that will become clear as we develop the algorithm. The tableau so constructed is the initial tableau of the problem. The initial tableau of Example 10.5 is shown in Tableau 10.2. ■

Remarks It is important to keep in mind that the entries in the last row of the initial tableau are the *negatives* of the coefficients in p. Thus if one of these coefficients is a negative number, then the corresponding entry in the last row of the initial tableau is a positive number.

It is also important to note that we are using the term "Basic Solution" on the tableau in the same way as it was defined in Section 10.1. The entries in the column labeled "Basic Solution" (with the convention that variables not listed at the left have value 0) correspond to a basic solution of the system of equations obtained by adding slack variables. Therefore, these values of the variables give us a corner point of the feasible set. This will be true for each tableau obtained by using the simplex method.

TABLEAU 10.1

	x	y	z	u	v	**Basic Solution**
u	2	2	3	1	0	160
v	5	1	10	0	1	100

TABLEAU 10.2

	x	y	z	u	v	**Basic Solution**
u	2	2	3	1	0	160
v	5	1	10	0	1	100
p	−10	−6	−2	0	0	0

Example 10.6 Consider the SMP

$$\begin{array}{ll} \text{Maximize} & p = 3x + 8y + 6z \\ \text{subject to} & x \geq 0 \quad y \geq 0 \quad z \geq 0 \\ & 20x + 4y + 4z \leq 6{,}000 \\ & 8x + 8y + 4z \leq 10{,}000 \\ & 8x + 4y + 2z \leq 4{,}000 \end{array}$$

Problem Form the initial tableau for this SMP.

Solution We convert the three inequalities to equations by introducing slack variables u, v, and w. This gives the new problem:

$$\begin{array}{ll} \text{Maximize} & p = 3x + 8y + 6z \\ \text{subject to} & x \geq 0 \quad y \geq 0 \quad z \geq 0 \\ & u \geq 0 \quad v \geq 0 \quad w \geq 0 \\ & 20x + 4y + 4z + u = 6{,}000 \\ & 8x + 8y + 4z + v = 10{,}000 \\ & 8x + 4y + 2z + w = 4{,}000 \end{array}$$

The initial tableau for this problem is formed in two steps. First, we form the augmented matrix for the system of equations and label the columns at the top; we label the rows with the slack variables at the left; and we label the right-hand column "Basic Solution." This gives Tableau 10.3.

The second and final step in forming the initial tableau (in practice, the two steps are carried out together) is to form the check row. This row is obtained by writing in each column the negative of the coefficient of that variable in the objective function p. In this example the entries are, in order, -3, -8, -6, 0, 0, 0. This follows from the fact that here $p = 3x + 8y + 6z$. The entries in the u, v, and w columns are 0 since these variable do not appear (they have zero coefficients in p). With this row added we obtain Tableau 10.4. The entry in the basic solution column of the p row is the value of p for the basic solution $x = 0$, $y = 0$, $z = 0$, $u = 6000$, $v = 10{,}000$, and $w = 4000$. This value is $p = 3x + 8y + 6z = 3(0) + 8(0) + 6(0) = 0$. ■

TABLEAU 10.3

	x	y	z	u	v	w	Basic Solution
u	20	4	4	1	0	0	6,000
v	8	8	4	0	1	0	10,000
w	8	4	2	0	0	1	4,000

TABLEAU 10.4

	x	y	z	u	v	w	Basic Solution
u	20	4	4	1	0	0	6,000
v	8	8	4	0	1	0	10,000
w	8	4	2	0	0	1	4,000
p	−3	−8	−6	0	0	0	0

In Example 10.6 we noted that the entries in the check row of the initial tableau in the columns of the slack variables u, v, w are all zero. In fact, the slack variables never appear in the objective function; consequently, in the initial tableau the entries in the check row under the slack variables are always zero. Also, the entry in the check row and the basic solution column of the initial tableau is always zero.

The Pivot Operation

The initial tableau provides a basic solution which serves as a starting point for the algorithm. The core of the simplex method is an algebraic technique that produces a new basic solution from a given one. This algebraic technique is called the *pivot operation*, or simply *pivoting*. Given a tableau, we use the pivot operation to produce a new tableau.

The operation of pivoting in a tableau is essentially the operation we used in solving a system of linear equations when we replaced an equation by another one in which certain variables had zero coefficients. In particular, it consists of row operations on the augmented matrix which convert one entry in the matrix to a 1 and all other entries in the column of that entry to 0s. The entry which is converted to a 1 is called the *pivot element*. If the entry which is converted to a 1 is in the ith row and the jth column, then the new tableau has the variable of the jth column listed to the left of the ith row. For example, in Tableau 10.2 (reproduced as Tableau 10.5), suppose that the entry in row 2, column 1 is the pivot element. This entry is circled in Tableau 10.5. In this case the pivot element is

TABLEAU 10.5

	x	y	z	u	v	Basic Solution
u	2	2	3	1	0	160
v	⑤	1	10	0	1	100
p	−10	−6	−2	0	0	0

located in the row of the tableau labeled v and the column labeled x. The variable v is called the *departing variable*, and the variable x is called the *entering variable*. These names reflect the fact that in the basic solution corresponding to the original tableau, variable v had a nonzero value (in general) and x had value 0, while in the new tableau v has value 0 and x has a nonzero value (in general). The new tableau is obtained by multiplying row 2 by $\frac{1}{5}$ (to make the pivot element 1) and then adding a suitable multiple of the new row 2 to each remaining row, including the row labeled p, to convert the other entries in the first column to 0s. For example, after row 2 is multiplied by $\frac{1}{5}$, then -2 times the new row 2 is added to row 1. The result is that the new entry in row 1, column 1 is 0. Similarly, if 10 times the new second row is added to the row labeled p, then the new entry in row 3, column 1 is 0. The tableau which results from this pivot operation is shown in Tableau 10.6.

Note that in Tableau 10.6 the variables used as row labels are u and x. This means that we have a new basic solution in which u and x have the values given in the basic solution column and all other variables have the value 0. The basic solution given by Tableau 10.6 is $x = 20$, $y = 0$, $z = 0$, $u = 120$, and $v = 0$. Moreover, for this basic solution the objective function $p = 10x + 6y + 2z$ has the value $10(20) + 6(0) + 2(0) = 200$, the value in the lower right-hand corner of the tableau. Note that the value of the objective function is larger at this basic solution (corner point) than at the basic solution of the initial tableau: It has the value 200 at this basic solution, whereas originally it had the value 0.

To summarize, in this example the result of the pivot operation is that one variable (here it is x) replaces another (here v) in the basic solution. This is always the case with pivot operations.

As a second example of the pivot operation, we return to Example 10.6 and its initial tableau, Tableau 10.4. Suppose the entry in row 3, column 2 of Tableau 10.4 is used as a pivot element. This entry is circled in Tableau 10.7. The selection of the pivot element identifies the entering and departing variables: The entering variable is y, and the departing variable is w. To carry out a pivot operation with this pivot element, first we multiply the third row by $\frac{1}{4}$ to convert the pivot element to 1. Next, we add a suitable multiple of the new third row to each of the remaining rows to convert all other entries in the column of the pivot element to 0s. The result is Tableau 10.8.

TABLEAU 10.6

	x	y	z	u	v	Basic Solution
u	0	$\frac{8}{5}$	-1	1	$-\frac{2}{5}$	120
x	1	$\frac{1}{5}$	2	0	$\frac{1}{5}$	20
p	0	-4	18	0	2	200

TABLEAU 10.7

	x	y	z	u	v	w	**Basic Solution**
u	20	4	4	1	0	0	6,000
v	8	8	4	0	1	0	10,000
w	8	④	2	0	0	1	4,000
p	−3	−8	−6	0	0	0	0

TABLEAU 10.8

	x	y	z	u	v	w	**Basic Solution**
u	12	0	2	1	0	−1	2000
v	−8	0	0	0	1	−2	2000
y	2	1	$\frac{1}{2}$	0	0	$\frac{1}{4}$	1000
p	13	0	−2	0	0	2	8000

The new basic solution given in Tableau 10.8 is $x = 0$, $y = 1000$, $z = 0$, $u = 2000$, and $v = 2000$. The value of the objective function p for this basic solution is

$$p = 3x + 8y + 6z = 3(0) + 8(1000) + 6(0) = 8000$$

as shown in the lower right-hand corner of the tableau. Since the problem (see Example 10.6) is to maximize p, we see that the basic solution of Tableau 10.8 is better than the basic solution of the initial tableau, Tableau 10.7.

The pivot operation is the crucial part of the simplex method. Pivots are carried out until either a maximum is obtained or it is established that no maximum exists. Criteria for distinguishing these two cases and for determining which entry in the tableau should be used as a pivot element are given in the next section.

Exercises for Section 10.2

Form the initial tableaus for the SMPs in Exercises 1 to 12.

1. Maximize $p = 6x + 4y$

   subject to
   $$x \geq 0 \qquad y \geq 0$$
   $$3x + 5y \leq 20$$
   $$x + 3y \leq 9$$

2. Maximize $p = -x + 5y$

subject to
$$\begin{aligned} x \geq 0 \qquad y &\geq 0 \\ 3x + 2y &\leq 12 \\ -x + y &\leq 1 \end{aligned}$$

3. Maximize $p = 3x + 7y$

subject to
$$\begin{aligned} x \geq 0 \qquad y &\geq 0 \\ 2x &\leq 15 \\ 3y &\leq 8 \end{aligned}$$

4. Maximize $p = 3x$

subject to
$$\begin{aligned} x \geq 0 \qquad y &\geq 0 \\ 5x + 4y &\leq 20 \\ x - 3y &\leq 3 \end{aligned}$$

5. Maximize $p = 2y$

subject to
$$\begin{aligned} x \geq 0 \qquad y &\geq 0 \\ 2x + 3y &\leq 12 \\ 6x + y &\leq 9 \\ x &\leq 4 \end{aligned}$$

6. Maximize $p = -3x + 6y$

subject to
$$\begin{aligned} x \geq 0 \qquad y &\geq 0 \\ 4x + 8y &\leq 16 \\ -x + y &\leq 1 \\ x - y &\leq 1 \end{aligned}$$

7. Maximize $p = 2x - y$

subject to
$$\begin{aligned} x \geq 0 \qquad y &\geq 0 \\ 3y &\leq 24 \\ 2x &\leq 24 \\ 2x + 3y &\leq 36 \end{aligned}$$

8. Maximize $p = 2x - y + z$

subject to
$$\begin{aligned} x \geq 0 \qquad y \geq 0 \qquad z &\geq 0 \\ 5x - y + z &\leq 20 \\ 2x + y + 2z &\leq 30 \end{aligned}$$

9. Maximize $p = 2x + 3y + z$

subject to
$$\begin{aligned} x \geq 0 \qquad y \geq 0 \qquad z &\geq 0 \\ 5x + z &\leq 100 \\ y + 3z &\leq 300 \\ 2x - y + z &\leq 900 \end{aligned}$$

10. Maximize $p = 3x + y + 5z$

subject to
$$x \geq 0 \quad y \geq 0 \quad z \geq 0$$
$$x + 2y + z \leq 20$$
$$x \leq 10$$
$$x + y \leq 12$$

11. Maximize $p = 2x_1 - x_2 + 3x_3 - x_4$

subject to
$$x_1 \geq 0 \quad x_2 \geq 0 \quad x_3 \geq 0 \quad x_4 \geq 0$$
$$x_1 + 3x_2 - x_3 + x_4 \leq 20$$
$$3x_1 + 6x_2 + 3x_3 - x_4 \leq 30$$
$$x_1 - 3x_2 + x_3 + x_4 \leq 15$$

12. Maximize $p = x_1 + 2x_2 - 3x_3 - 4x_4 + 5x_5$

subject to
$$x_1 \geq 0 \quad x_2 \geq 0 \quad x_3 \geq 0 \quad x_4 \geq 0 \quad x_5 \geq 0$$
$$x_1 + x_2 \leq 10$$
$$x_1 + x_3 \leq 20$$
$$x_1 - x_4 \leq 15$$
$$x_3 + x_4 \leq 5$$
$$x_2 - x_5 \leq 30$$

In Exercises 13 through 24, carry out a pivot operation with the circled element as the pivot element. Find the new basic solution.

13.

	x	*y*	*u*	*v*	**Basic Solution**
u	2	③	1	0	12
v	6	1	0	1	9
p	−1	−3	0	0	0

14.

	x	*y*	*u*	*v*	**Basic Solution**
u	2	3	1	0	12
v	⑥	1	0	1	9
p	−1	−3	0	0	0

15.

	x	*y*	*u*	*v*	**Basic Solution**
u	④	4	1	0	8
v	−1	1	0	1	1
p	−2	1	0	0	0

16.

	x	y	z	u	v	**Basic Solution**
u	3	2	8	1	0	24
v	(1)	6	0	0	1	6
p	−8	−5	−1	0	0	0

17.

	x	y	z	u	v	**Basic Solution**
u	10	5	8	1	0	80
v	3	(2)	12	0	1	30
p	−1	−3	2	0	0	0

18.

	x	y	z	u	v	**Basic Solution**
u	(5)	−1	1	1	0	20
v	2	1	2	0	1	30
p	−2	1	−1	0	0	0

19.

	x	y	u	v	w	**Basic Solution**
u	4	8	1	0	0	16
v	−1	(1)	0	1	0	1
w	1	−1	0	0	1	1
p	3	−6	0	0	0	0

20.

	x	y	u	v	w	**Basic Solution**
u	2	10	1	0	0	80
v	3	1	0	1	0	9
w	0	(2)	0	0	1	3
p	−2	−3	0	0	0	0

21.

	x	y	u_1	u_2	u_3	u_4	Basic Solution
u_1	5	5	1	0	0	0	20
u_2	5	3	0	1	0	0	15
u_3	(1)	0	0	0	1	0	2
u_4	0	2	0	0	0	1	3
p	−4	1	0	0	0	0	0

22.

	x	y	z	u	v	w	Basic Solution
u	5	0	1	1	0	0	100
v	0	(1)	3	0	1	0	300
w	2	−1	1	0	0	1	900
p	−2	−3	−1	0	0	0	0

23.

	x	y	z	u	v	w	Basic Solution
u	5	(5)	15	1	0	0	100
v	6	3	9	0	1	0	90
w	2	1	4	0	0	1	20
p	−1	−5	−3	0	0	0	0

24.

	x_1	x_2	x_3	x_4	u_1	u_2	u_3	Basic Solution
u_1	1	3	−1	1	1	0	0	20
u_2	3	6	(3)	−1	0	1	0	30
u_3	1	−3	1	1	0	0	1	15
p	−2	1	−3	1	0	0	0	0

Find the SMP, i.e., the constraints and the objective function, for the initial tableau in each of the following exercises:

25. Exercise 14
26. Exercise 17
27. Exercise 21
28. Exercise 24

In Exercises 29 through 33, carry out a pivot on the tableau shown, using the circled entry as the pivot element. Describe the basic solution given by the original tableau and the basic solution given by the new tableau (the one obtained by pivoting), and decide which is better in terms of the original SMP.

29.

	x	y	z	u	v	Basic Solution
x	1	$\frac{2}{3}$	2	$\frac{1}{3}$	0	7
v	0	(3)	2	-1	1	9
p	0	-1	10	2	1	42

30.

	x	y	z	u	v	Basic Solution
u	1	0	(5)	1	0	50
y	0	1	$\frac{1}{10}$	0	$\frac{1}{10}$	3
p	-1	0	$-\frac{3}{2}$	0	$\frac{1}{2}$	15

31.

	x	y	z	u	v	w	Basic Solution
u	12	0	(2)	1	0	-1	2000
v	-8	0	0	0	1	-2	2000
y	2	1	$\frac{1}{2}$	0	0	$\frac{1}{4}$	1000
p	13	0	-2	0	0	2	8000

32.

	x	y	z	u	v	Basic Solution
z	($\frac{1}{5}$)	0	1	$\frac{1}{5}$	0	10
y	$-\frac{1}{50}$	1	0	$-\frac{1}{50}$	$\frac{1}{10}$	2
p	$-\frac{7}{10}$	0	0	$\frac{3}{10}$	$\frac{1}{2}$	30

33.

	x	y	u	v	w	Basic Solution
u	(12)	0	1	-8	0	8
y	-1	1	0	1	0	1
w	0	0	0	1	1	2
p	-3	0	0	6	0	6

In Exercises 34 and 35, use each of the circled entries as a pivot element and compare the results of the respective pivots in terms of the original SMP.

34.

	x	y	z	u	v	w	Basic Solution
u	1	3	−1	1	0	0	20
v	③	6	③	0	1	0	30
w	1	−3	1	0	0	1	15
p	−2	1	−3	0	0	0	0

35.

	x	y	z	u	v	Basic Solution
u	①	0	⑤	1	0	50
y	0	1	$\frac{1}{10}$	0	$\frac{1}{10}$	3
p	−1	0	$-\frac{3}{2}$	0	$\frac{1}{2}$	15

10.3 OPTIMAL VECTORS VIA THE SIMPLEX METHOD

In Section 10.2 we studied how to obtain an initial tableau for the simplex method and how to pivot to obtain new tableaus. However, we have not yet answered two important questions:

Which element should be the pivot element?

When should the process of pivoting stop?

The answer to the second question depends on the answer to the first. Indeed, the process of pivoting should be stopped as soon as there is no proper pivot element.

The technique of finding the proper pivot element is as follows:

1. Examine the bottom row of the tableau, and find the most negative entry (ignore the entry in the column labeled "Basic Solution"). The variable that labels the column with this most negative entry is the *entering variable*. The situation in which the bottom row of the tableau contains no negative entries will be discussed below.
2. Divide each positive entry in the column of the entering variable into the corresponding entry (same row) in the basic solution column. The variable in the row that gives the smallest quotient is the *departing variable*. The situation in which there are no positive entries in the column is discussed below.

In Section 10.0, we described the simplex method as an algorithm which selects one corner point after another in such a way that the values of the objective function become better and better. With this in mind, we can describe the rationale for step 1 of the technique for finding the proper pivot element as follows: By selecting the entering variable as described, when the value of the entering variable increases, the value of the objective function increases more rapidly than (or at least as rapidly as) for any other choice of entering variable. For example, if the objective function is $p = x + 2y + 3z$, then the check row of the initial tableau has nonzero entries $-1, -2, -3$ in the columns labeled x, y, and z, respectively. Using step 1, we select z as the entering variable. We see that when z increases, say by 1 unit, then the objective function increases more rapidly, by 3 units, than it increases when either x or y increases. After the initial pivot the entries in the check row are not as directly related to the coefficients in the objective function, but the same reasoning holds.

The rationale for step 2 is that making the selection described guarantees that the resulting values of the variables yield a corner point of the feasible set for the problem. In fact, if you choose as departing variable a variable for which the quotient is not the smallest, then the resulting values of the variables may yield a point which is *not* in the feasible set. There are other satisfactory ways to select a pivot element, but we consider only the algorithm described by steps 1 and 2.

To illustrate rules 1 and 2 for finding a pivot element, we examine Tableau 10.9. We begin by examining the bottom row, the check row in the tableau. There are some negative entries, and using rule 1, we select the most negative entry. Here that entry is -10, and it is in the first column. We label this column with a vertical arrow, the arrow shown above the x in the first column in Tableau 10.9. Variable x is the entering variable.

TABLEAU 10.9

	↓ x	y	z	u	v	**Basic Solution**
u	2	2	3	1	0	160
→ v	⑤	1	10	0	1	100
p	−10	−6	−2	0	0	0

Next, we use rule 2 to determine the departing variable. We divide each of the two entries in the basic solution column by the corresponding entry in the first column. The quotient for the first row is $160/2 = 80$, and the quotient for the second row is $100/5 = 20$. Since $20 < 80$, the variable associated with the second row, v in this case, is the departing variable. We denote the departing variable with a small horizontal arrow, the arrow shown next to v in Tableau 10.9. The pivot element is the entry determined by the entering variable (a column) and the departing variable (a row). The pivot element is identified by a circle in Tableau 10.9. We pivot using this pivot element to construct a new tableau corresponding to a new basic solution.

A pivot was carried out earlier in Tableau 10.6 with this pivot element. The result is reproduced in Tableau 10.10.

TABLEAU 10.10

	x	$\downarrow$ y	z	u	v	Basic Solution
$\rightarrow u$	0	($\frac{8}{5}$)	-1	1	$-\frac{2}{5}$	120
x	1	$\frac{1}{5}$	2	0	$\frac{1}{5}$	20
p	0	-4	18	0	2	200

If we again apply rules 1 and 2 for determining a pivot element, we see that y is the entering variable (-4 is the only negative entry in the bottom row) and u is the departing variable ($120 \div \frac{8}{5} = 75$ is the smallest quotient). Thus, the next pivot element is $\frac{8}{5}$, which is circled in Tableau 10.10. Pivoting at this entry gives Tableau 10.11.

TABLEAU 10.11

	x	y	z	u	v	Basic Solution
y	0	1	$-\frac{5}{8}$	$\frac{5}{8}$	$-\frac{1}{4}$	75
x	1	0	$\frac{17}{8}$	$-\frac{1}{8}$	$\frac{1}{4}$	5
p	0	0	$\frac{31}{2}$	$\frac{5}{2}$	1	500

If we attempt to apply rules 1 and 2 to Tableau 10.11, we encounter a problem. *No entry* in the bottom row is negative. In such a case, how should we interpret the rule "find the most negative entry"? The answer is that no further pivoting is necessary:

> If the bottom row of a tableau contains no negative entries, then the basic solution given by the tableau provides a solution of the linear programming problem; i.e., it gives values of the decision variables which maximize the objective function.

Thus, in Tableau 10.11 the basic solution ($x = 5, y = 75, z = 0, u = 0, v = 0$) provides a solution for the original SMP, Example 10.5: $x = 5$, $y = 75$, $z = 0$. Moreover, as we noted earlier, the last entry in the basic solution column is the maximum value of the objective function. Here the objective function is $p = 10x + 6y + 2z$, and the maximum value of p is $p = 10(5) + 6(75) + 2(0) = 50 + 450 = 500$.

As a second illustration of the rules for finding a pivot element, we study another example.

Example 10.7 An SMP is given as follows:

Maximize $p = 3x + 8y + 6z$

subject to

$$\begin{aligned} x \geq 0 \quad y \geq 0 \quad z &\geq 0 \\ 20x + 4y + 4z &\leq 6{,}000 \\ 8x + 8y + 4z &\leq 10{,}000 \\ 8x + 4y + 2z &\leq 4{,}000 \end{aligned}$$

Problem Use the simplex method to solve this SMP.

Solution In Example 10.6 we obtained the initial tableau for this problem (Tableau 10.4). It is reproduced in Tableau 10.12.

Using rule 1 for finding a pivot element, we examine the bottom row of the tableau and find the most negative entry. This is the entry -8. Since it is in the column labeled y, variable y is the entering variable. Next, using rule 2, we divide each positive entry in the y column into the corresponding entry in the basic solution column. The quotients are $6000/4 = 1500$ in the row labeled u, $10{,}000/8 = 1250$ in the row labeled v, and $4000/4 = 1000$ in the row labeled w. The smallest quotient is in the row labeled w. Therefore w is the departing variable. The pivot element is the entry in the column of the entering variable and the row of the departing variable. The entering and departing variables are shown by arrows, and the pivot element is circled in Tableau 10.12.

A pivot was carried out earlier with this pivot element. The result (Tableau 10.8) is reproduced in Tableau 10.13.

If we again apply rules 1 and 2 for obtaining a pivot element, we see that z is the entering variable (-2 is the only negative entry in the bottom row) and u is the departing variable ($2000/2 = 1000$ is the smallest quotient). Thus the next pivot element is the row 1, column 3 entry circled in Tableau 10.13. Pivoting at this element results in Tableau 10.14.

TABLEAU 10.12

	x	↓ y	z	u	v	w	Basic Solution
u	20	4	4	1	0	0	6,000
v	8	8	4	0	1	0	10,000
→ w	8	(4)	2	0	0	1	4,000
p	−3	−8	−6	0	0	0	0

TABLEAU 10.13

	x	y	↓ z	u	v	w	Basic Solution
→ u	12	0	②	1	0	−1	2000
v	−8	0	0	0	1	−2	2000
y	2	1	$\frac{1}{2}$	0	0	$\frac{1}{4}$	1000
p	13	0	−2	0	0	2	8000

TABLEAU 10.14

	x	y	z	u	v	w	Basic Solution
z	6	0	1	$\frac{1}{2}$	0	$-\frac{1}{2}$	1,000
v	−8	0	0	0	1	−2	2,000
y	−1	1	0	$-\frac{1}{4}$	0	$\frac{1}{2}$	500
p	25	0	0	1	0	1	10,000

At this point there are no negative entries in the bottom row, and the original SMP (formulated in Example 7.4) has been solved. The basic solution given by Tableau 10.14 ($x = 0$, $y = 500$, $z = 1000$, $u = 0$, $v = 2000$, $w = 0$) provides values of the decision variables which maximize the objective function. These values are $x = 0$, $y = 500$, and $z = 1000$. ■

The problem solved in Example 10.7 arose in the situation described in Example 7.4, and it is important to interpret this solution in terms of the original situation. The variables x, y, and z represent the amounts (in hundreds of pounds) of 20-8-8, 4-8-4, and 4-4-2 fertilizer, respectively, to be produced by the company. Thus to obtain maximum profit, the company should produce 500 units of 4-8-4 fertilizer and 1000 units of 4-4-2 fertilizer. Since $x = 0$, no 20-8-8 fertilizer should be produced. Translated to pounds, the optimal production schedule is

No 20-8-8

50,000 pounds of 4-8-4

100,000 pounds of 4-4-2

The values of u, v, and w also provide information about the problem. Since these variables measure slack or unused amounts of nitrate, phosphate, and potash, respectively, we see that at optimal production the company will have 2000 pounds of phosphate left over ($v = 2000$). All nitrate and potash will be utilized in the production of fertilizer ($u = 0$ and $w = 0$).

We have now provided a partial answer to the second question posed at the beginning of the section: When should pivoting be stopped? The partial answer is that pivoting should stop whenever the bottom row of the tableau has no negative entries. In such a case the basic solution given in the tableau provides a solution to the problem. There is one other case in which pivoting must also stop. This is when the bottom row contains a negative entry (and therefore an entering variable can be chosen), but the column of the entering variable *contains no positive elements* (and therefore no departing variable can be chosen). In such a case, *the linear programming problem has no solution*. This possibility is illustrated in the next example.

Example 10.8 An SMP is given as follows:

Maximize $p = x + 2y + z$

subject to $x \geq 0 \quad y \geq 0 \quad z \geq 0$

$$x - y + z \leq 5$$
$$2x - z \leq 10$$

Problem Apply the simplex method to show that this problem has no solution.

Solution We form the initial tableau by introducing slack variables u and v to form the system of equations

$$x - y + z + u = 5$$
$$2x - z + v = 10$$

The initial tableau is Tableau 10.15.

TABLEAU 10.15

		x	↓ y	z	u	v	Basic Solution
?	u	1	−1	1	1	0	5
	v	2	0	−1	0	1	10
	p	−1	−2	−1	0	0	0

Applying the rules for finding a pivot element, we see that y is the entering variable. However, since no entry in the column labeled y is positive, no departing variable can be chosen. Therefore, this SMP has *no solution*. The quantity p can be made arbitrarily large and has no maximum value on the feasible set. In particular, if we examine the constraints carefully, we see the difficulty. The first constraint can be written as

$$x + z - 5 \leq y$$

and this will be satisfied for arbitrarily large values of y. The second constraint, $2x - z \leq 10$, does not involve y. We conclude that there are arbitrarily large values of y which satisfy the constraints. This means that the objective function $p = x + 2y + z$ assumes arbitrarily large values. ■

There is one final item to consider in our study of the rules for finding pivot elements. We must decide how to proceed if the rules result in a tie, i.e., if either two or more entries are "most negative" in the bottom row or two or more quotients are "smallest." In such cases we have a choice in selecting the entering variable (if there are two equal most-negative entries) or the departing variable (if there are two equal smallest quotients). Such situations pose no problem; any of the possible choices is acceptable. The optimal value of the objective function will be the same, although the optimal vectors may be different. Thus if two variables are legitimate choices for the entering variable, either can be selected and the simplex method can be continued. A similar comment holds for the departing variable.

Summary of the Simplex Method

1. Check whether the problem is formulated as an SMP. If it is, continue to step 2. If it is not, stop.
2. Introduce slack variables.
3. Form the initial tableau.
4. Carry out pivot operations until a new pivot element cannot be determined.
5. If a pivot element cannot be determined because there is an entering variable but no departing variable, the problem has no solution. If a pivot element cannot be determined because an entering variable cannot be selected, the problem is solved. The basic solution in the final tableau provides values of the decision variables which solve the problem. The entry in the bottom row and in the column labeled "Basic Solution" gives the maximum value of the objective function.

To conclude this section, we carry out one more example of the simplex method. In this case, along with the algebra of the method (the tableaus), we show the related geometry.

Example 10.9 Use the simplex method to solve the following SMP:

$$\text{Maximize} \qquad p = 10x + 7y$$

$$\text{subject to} \qquad x \geq 0 \qquad y \geq 0$$
$$x \leq 200$$
$$2x + 3y \leq 1200$$
$$25x + 15y \leq 7500$$

Also graph the feasible set in the xy plane, and show the values of the objective function $p = 10x + 7y$ as a third dimension above the xy plane. Finally, identify the basic solutions obtained by the simplex method on the graph.

Solution We let u, v, and w be the slack variables, so that the initial tableau is Tableau 10.16. The initial basic solution is noted next to this tableau.

In Figure 10.3 we have graphed the feasible set for the SMP in the xy plane, and we show the values of the objective function p as a third dimension. The ini-

FIGURE 10.3

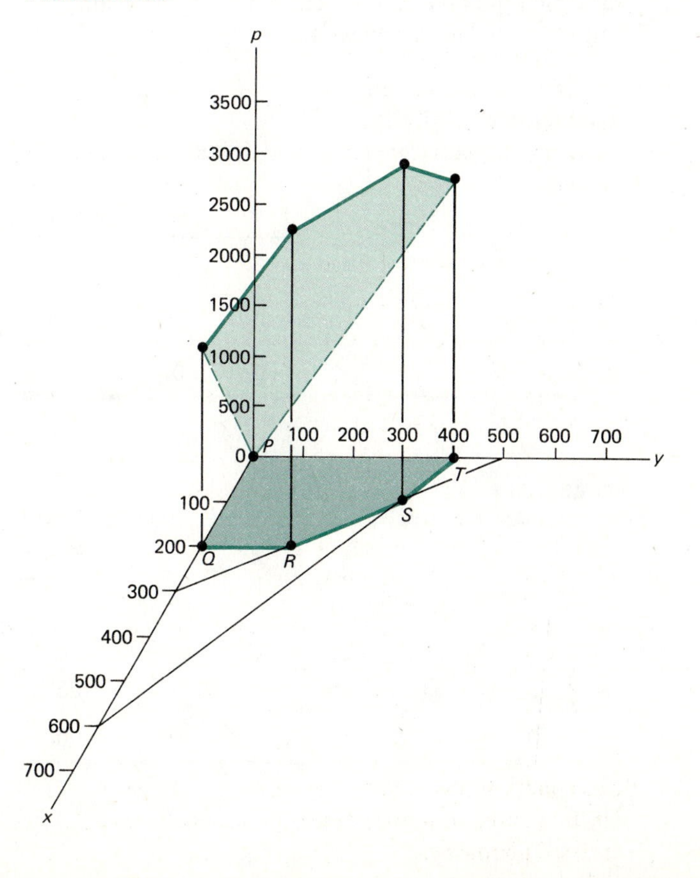

tial basic solution is at the origin of the three-dimensional space since $x = 0$, $y = 0$, and $p = 0$. We label this as point P on the graph.

The pivot element for Tableau 10.16 is circled, and after one pivot we obtain Tableau 10.17. The new basic solution is noted next to the tableau, and the corresponding point on the graph is labeled Q in Figure 10.3. Note that Q is a corner point of the feasible set, and the value of the objective function (p) is larger at Q ($p = 2000$) than it was at P ($p = 0$).

In Tableau 10.17, the pivot element is the entry 15; and after pivoting, the new tableau is Tableau 10.18. The new basic solution (point R on Figure 10.3) has $p = 9500/3$.

In Tableau 10.18, the pivot element is 3; after pivoting, we obtain the final tableau (Tableau 10.19). The new basic solution (which is optimal) is the point S in Figure 10.3. The optimal value of p is 10,000/3.

In Figure 10.3, the feasible set is shown as the dark-shaded area in the xy plane, and the values of the objective function $p = 10x + 7y$ are shown as the light-shaded plane above the feasible set.

Note that at each stage of the simplex method (i.e., at each pivot), the basic solution moves from one corner point of the feasible set to another corner point at which the objective function has a larger value. The method stops when the maximum value is achieved. ■

TABLEAU 10.16

	x	y	u	v	w	Basic Solution
u	(1)	0	1	0	0	200
v	2	3	0	1	0	1200
w	25	15	0	0	1	7500
p	−10	−7	0	0	0	0

Initial Basic Solution

$x = 0, y = 0,$
$u = 200, v = 1200,$
$w = 7500, p = 0$
Shown as point P in Figure 10.3

TABLEAU 10.17

	x	y	u	v	w	Basic Solution
x	1	0	1	0	0	200
v	0	3	−2	1	0	800
w	0	(15)	−25	0	1	2500
p	0	−7	10	0	0	2000

New Basic Solution

$x = 200, y = 0,$
$u = 0, v = 800,$
$w = 2500,$
$p = 2000$
Shown as point Q in Figure 10.3

TABLEAU 10.18

	x	y	u	v	w	Basic Solution
x	1	0	1	0	0	200
v	0	0	③	1	$-\frac{1}{5}$	300
y	0	1	$-\frac{5}{3}$	0	$\frac{1}{15}$	$\frac{500}{3}$
p	0	1	$-\frac{5}{3}$	0	$\frac{7}{15}$	$\frac{9500}{3}$

New Basic Solution

$x = 200$, $y = \frac{500}{3}$, $u = 0$, $v = 300$, $w = 0$, $p = \frac{9500}{3}$

Shown as point R in Figure 10.3

TABLEAU 10.19

	x	y	u	v	w	Basic Solution
x	1	0	0	$-\frac{1}{3}$	$\frac{1}{15}$	100
u	0	0	1	$\frac{1}{3}$	$-\frac{1}{15}$	100
y	0	1	0	$\frac{5}{9}$	$-\frac{2}{45}$	$\frac{1000}{3}$
p	0	0	0	$\frac{5}{9}$	$\frac{16}{45}$	$\frac{10,000}{3}$

Optimal Basic Solution

$x = 100$, $y = \frac{1000}{3}$, $u = 100$, $v = 0$, $w = 0$, $p = \frac{10,000}{3}$

Shown as point S in Figure 10.3

Exercises for Section 10.3

For each of the tableaus in Exercises 1 through 10, find the entering variable, departing variable, and pivot element.

1.

	x	y	u	v	Basic Solution
u	9	4	1	0	36
v	3	6	0	1	9
p	−3	−2	0	0	0

2.

	x	y	u	v	Basic Solution
u	−8	16	1	0	40
v	8	8	0	1	80
p	−6	−4	0	0	0

3.

	x	y	u	v	Basic Solution
y	$\frac{1}{2}$	1	$\frac{1}{2}$	0	4
v	4	0	-1	1	4
p	-1	0	2	0	16

4.

	x	y	u	v	Basic Solution
x	1	$\frac{1}{4}$	$\frac{1}{4}$	0	15
v	0	$\frac{5}{4}$	$-\frac{3}{4}$	1	45
p	0	$-\frac{11}{4}$	$\frac{5}{4}$	0	75

5.

	x	y	z	u	v	Basic Solution
u	6	2	2	1	0	24
v	3	6	0	0	1	18
p	15	20	5	0	0	0

6.

	x	y	z	u	v	Basic Solution
z	$\frac{3}{4}$	$\frac{1}{2}$	1	$\frac{1}{4}$	0	15
v	$-\frac{3}{2}$	1	0	$-\frac{1}{2}$	1	10
p	$-\frac{1}{2}$	-3	0	0	0	150

7.

	x	y	z	u	v	Basic Solution
u	5	0	2	1	$-\frac{1}{3}$	5
y	$\frac{1}{2}$	1	0	0	$\frac{1}{6}$	1
p	-2	0	-1	0	1	6

8.

	x	y	z	u	v	**Basic Solution**
u	2	3	4	1	0	12
v	0	0	5	0	1	10
p	−8	−6	−10	0	0	0

9.

	x	y	u	v	w	**Basic Solution**
u	4	2	1	0	0	20
v	3	−3	0	1	0	9
w	6	6	0	0	1	24
p	−1	−3	0	0	0	0

10.

	x	y	z	u	v	w	**Basic Solution**
u	5	2	0	1	0	0	30
v	0	4	4	0	1	0	160
w	3	0	6	0	0	1	120
p	−4	−8	−10	0	0	0	0

For Exercises 11 through 16, beginning with the given tableau, use the simplex method to complete the solution of the SMP.

11. The tableau of Exercise 1
12. The tableau of Exercise 2
13. The tableau of Exercise 7
14. The tableau of Exercise 8
15. The tableau of Exercise 9
16. The tableau of Exercise 10

Solve each of the SMPs in Exercises 17 through 27 using the simplex method.

17. Maximize $p = 2x + 3y$
    subject to $x \geq 0 \quad y \geq 0$
    $3x + 8y \leq 24$
    $6x + 4y \leq 30$

18. Maximize subject to
$$p = 2x + 3y$$
$$x \geq 0 \quad y \geq 0$$
$$x \leq 5$$
$$y \leq 4$$
$$x + y \leq 8$$

19. Maximize subject to
$$p = x$$
$$x \geq 0 \quad y \geq 0$$
$$x - y \leq 4$$
$$-x + 3y \leq 4$$

20. Maximize subject to
$$p = x - y$$
$$x \geq 0 \quad y \geq 0$$
$$5x - 5y \leq 20$$
$$2x - 10y \leq 18$$

21. Maximize subject to
$$p = 7x + 6y$$
$$x \geq 0 \quad y \geq 0$$
$$2x + 3y \leq 30$$
$$-x + 3y \leq 9$$
$$y \leq 6$$

22. Maximize subject to
$$p = x$$
$$x \geq 0 \quad y \geq 0$$
$$-x + y \leq 2$$
$$y \leq 5$$

23. Maximize subject to
$$p = x + 10y$$
$$x \geq 0 \quad y \geq 0$$
$$x \leq 8$$
$$8x + y \leq 40$$
$$y \leq 4$$
$$x + 2y \leq 10$$

24. Maximize subject to
$$p = 5x - 4y + 3z$$
$$x \geq 0 \quad y \geq 0 \quad z \geq 0$$
$$5x + 5z \leq 100$$
$$5y - 5z \leq 50$$
$$5x - 5y \leq 50$$

25. Maximize subject to
$$p = 3x + 4y + 5z$$
$$x \geq 0 \quad y \geq 0 \quad z \geq 0$$
$$3x + 8y + 6z \leq 24$$
$$6x + 2y + 8z \leq 24$$

26. Maximize subject to

$$p = 6x + y + 3z$$
$$x \geq 0 \qquad y \geq 0 \qquad z \geq 0$$
$$3x + y \leq 15$$
$$2x + 2y + z \leq 20$$

27. Maximize subject to

$$p = 6x + 10y + 2z$$
$$x \geq 0 \qquad y \geq 0 \qquad z \geq 0$$
$$x \leq 5$$
$$5x + 5y \leq 60$$
$$x + 2y + 3z \leq 24$$

28. The following tableau was obtained as the final tableau in the use of the simplex method in an attempt to solve an SMP. Determine whether the problem has a solution; if it does, find it and the maximum value of the objective function.

	x	y	u	v	w	Basic Solution
y	0	1	−2	0	2	4
x	1	0	−1	0	2	6
v	0	0	0	1	−1	3
p	0	0	−3	0	1	10

29. Repeat Exercise 28 for the following tableau.

	x	y	u	v	w	Basic Solution
u	0	1	−1	0	4	10
x	1	0	2	0	5	25
v	0	0	3	1	−3	4
p	0	0	3	0	5	70

Resource allocation

30. Use the simplex method to solve the following: A firm produces two kinds of turkey stuffing mix during the holiday season, regular and special sage. Each box of regular mix requires 12 ounces of bread, 4 ounces of cornbread-sage mix, and $\frac{1}{2}$ ounce of spices. Each box of special sage mix requires 5 ounces of bread, 3 ounces of cornbread-sage mix, and 1 ounce of spices. The profit on one box of regular mix is 45 cents, and the profit on one box of special sage is 30 cents. Each day the firm has 210 pounds of bread, 60 pounds of cornbread-sage mix, and 10 pounds of spices. Determine the amounts of each kind of stuffing mix that should be produced each day to yield maximum profit.

Resource allocation 31. Ralph's Pretty Good Grocery makes two varieties of Raw Bits (the breakfast cereal for above-average people) from oat hulls and pine nuts. A box of Basic Bits uses 6 ounces of oat hulls and 8 ounces of pine nuts, and a box of Lite Bits uses 12 ounces of oat hulls and 4 ounces of pine nuts. Ralph has 2880 ounces of oat hulls and 1248 ounces of pine nuts. He makes a profit of 50 cents on each box of Basic Bits and 30 cents on each box of Lite Bits. How many boxes of each should he make to maximize his profit?

Resource allocation 32. In the situation described in Exercise 31, suppose that the price of a package of Basic Bits can be increased so that the profit per package increases to 70 cents. In this situation, how many packages of each type of cereal should be produced to yield maximum profit?

Manufacturing 33. Use the simplex method to solve the following problem. Larry's Logs manufactures two kinds of pet logs: a standard model and a deluxe model. Each standard model requires 10 minutes of rough sawing and 10 minutes of sanding; each deluxe model requires 5 minutes of rough sawing and 20 minutes of sanding. Larry owns two saws and three sanders, and he hires helpers to run each piece of equipment 5 hours per week. Each standard log brings a profit of \$5 and each deluxe log a profit of \$7. If every log made can be sold, how many of each should be made to maximize profit?

Manufacturing 34. Repeat Exercise 33 with the modification that profit on a standard log is \$3 and profit on a deluxe log is \$7.

Resource allocation 35. Mountain High Expeditions prepares granola breakfast food containing seeds and nuts and, as an optional ingredient, raisins. A large packet of breakfast food including raisins contains 50 grams of seeds, 15 grams of nuts, and 15 grams of raisins. A small packet including raisins contains 20 grams of seeds, 10 grams of nuts, and 10 grams of raisins. A large packet of breakfast food without raisins contains 50 grams of seeds and 20 grams of nuts, and a small packet without raisins contains 30 grams of seeds and 10 grams of nuts. There are 1500 grams of seeds, 900 grams of nuts, and 500 grams of raisins available to be used in the breakfast food. The profit per packet is 3 cents for a large packet containing raisins and 2.5 cents for a large packet without raisins, and 2 cents for a small packet with raisins and 1.5 cents for a small packet without raisins. How many packets of each type should be prepared to yield maximum profit?

10.4 DUAL PROGRAMMING PROBLEMS AND MINIMUM PROBLEMS

Linear programming problems occur either as maximum problems or as minimum problems. In Chapter 7 both types were solved by using geometric methods. In this chapter we have restricted our attention to maximum problems (actually to SMPs); however, our methods are more general than they might at first appear. The generality arises from the fact that linear programming problems, or the situations that give rise to such problems, do not occur in isolation but arise in natural pairs. If one member of the pair is a maximization problem, to maximize profit, for instance, then the second member of the pair is a minimization problem, to minimize cost for instance. These two problems are intimately related in that they are phrased in terms of the same data. Indeed, a complete study of either problem requires consideration of both problems in the pair. In this section we study the second member of the pair when the first is an SMP. In particular, we shall see that the simplex method solves both problems with the same set of tableaus. Also we shall see through an example, that the minimizing problem is often important economically as well as mathematically.

In discussing the way linear programming problems arise in pairs, it is useful to have a shorthand notation for writing such problems. The matrix notation developed in Chapter 6 in connection with systems of linear equations is equally appropriate for systems of linear inequalities.

The column vector **X** with n coordinates was defined in Section 6.1 as

$$\mathbf{X} = \begin{bmatrix} x_1 \\ x_2 \\ \cdot \\ \cdot \\ \cdot \\ x_n \end{bmatrix}$$

If n is 2 or 3, then we usually write $\mathbf{X} = \begin{bmatrix} x \\ y \end{bmatrix}$ or $\mathbf{X} = \begin{bmatrix} x \\ y \\ z \end{bmatrix}$ to avoid subscripts.

Also in Section 6.1, equality of vectors was defined in terms of equality of coordinates. We use the same approach in defining inequality for vectors.

Vector Inequalities

Let **X** and **Y** be column n-vectors.

$$\mathbf{X} = \begin{bmatrix} x_1 \\ x_2 \\ \cdot \\ \cdot \\ \cdot \\ x_n \end{bmatrix} \qquad \mathbf{Y} = \begin{bmatrix} y_1 \\ y_2 \\ \cdot \\ \cdot \\ \cdot \\ y_n \end{bmatrix}$$

The inequality $\mathbf{X} \leq \mathbf{Y}$ means that each coordinate of **X** is less than or equal to the corresponding coordinate of **Y**. Thus $\mathbf{X} \leq \mathbf{Y}$ is equivalent to the n inequalities $x_1 \leq y_1, x_2 \leq y_2, \ldots, x_n \leq y_n$.

With this definition and the matrix-vector product defined in Section 6.1, we can easily write systems of inequalities in matrix notation. For instance, consider the system

$$\begin{aligned} x &\leq 80 \\ x + y &\leq 120 \\ 2x + 5y &\leq 500 \end{aligned}$$

If we define matrix **A** and vectors **X** and **B** as

$$\mathbf{A} = \begin{bmatrix} 1 & 0 \\ 1 & 1 \\ 2 & 5 \end{bmatrix} \qquad \mathbf{X} = \begin{bmatrix} x \\ y \end{bmatrix} \qquad \mathbf{B} = \begin{bmatrix} 80 \\ 120 \\ 500 \end{bmatrix}$$

the system can be written as $\mathbf{AX} \le \mathbf{B}$.

Conceptually a vector is an ordered set of numbers, and it does not matter whether we write these numbers in a vertical column or a horizontal row. However, in multiplying matrices and vectors together, it is important that the sizes be such that the products are defined. Although it is rare for confusion to arise, any potential for misunderstanding can be eliminated by distinguishing notationally between vectors written as a column and as a row. If the symbol **X** represents a column vector, we modify the symbol to denote a row vector with the same coordinates as **X**. We also modify (in a similar way) the symbol for a row vector to denote a column vector with the same coordinates.

Vector Transpose

$$\text{If } \mathbf{X} = \begin{bmatrix} x_1 \\ x_2 \\ \cdot \\ \cdot \\ \cdot \\ x_n \end{bmatrix}, \text{ then } \mathbf{X}^T = [x_1 \quad x_2 \quad \cdots \quad x_n].$$

$$\text{If } \mathbf{Y} = [y_1 \quad y_2 \quad \cdots \quad y_n], \text{ then } \mathbf{Y}^T = \begin{bmatrix} y_1 \\ y_2 \\ \cdot \\ \cdot \\ \cdot \\ y_n \end{bmatrix}.$$

The symbols $\mathbf{X}^T$ and $\mathbf{Y}^T$ are read "**X** transpose" and "**Y** transpose," respectively.

If **X** and **W** are two column n-vectors, then $\mathbf{W}^T$ is a row n-vector and $\mathbf{W}^T\mathbf{X}$ is defined. For example, if

$$\mathbf{X} = \begin{bmatrix} x_1 \\ x_2 \\ x_3 \\ x_4 \end{bmatrix} \quad \text{and} \quad \mathbf{W} = \begin{bmatrix} 1 \\ 2 \\ 4 \\ -1 \end{bmatrix}$$

then [from Equation (6.1)]

$$\mathbf{W}^T\mathbf{X} = x_1 + 2x_2 + 4x_3 - x_4$$

We are now able to write linear programming problems in matrix form.

Example 10.10 An SMP is given as follows:

$$\text{Maximize} \qquad p = 3x + y + 2z$$

$$\text{subject to} \qquad \begin{aligned} x \geq 0 \qquad y \geq 0 \qquad z \geq 0 \\ 2x + y \leq 20 \\ x + 3y + 5z \leq 40 \\ 3x + 5y + z \leq 60 \end{aligned}$$

Problem Write this problem in matrix form.

Solution We define

$$\mathbf{A} = \begin{bmatrix} 2 & 1 & 0 \\ 1 & 3 & 5 \\ 3 & 5 & 1 \end{bmatrix} \qquad \mathbf{B} = \begin{bmatrix} 20 \\ 40 \\ 60 \end{bmatrix} \qquad \mathbf{C} = \begin{bmatrix} 3 \\ 1 \\ 2 \end{bmatrix} \qquad \mathbf{X} = \begin{bmatrix} x \\ y \\ z \end{bmatrix}$$

The constraints (other than $x \geq 0$, $y \geq 0$, $z \geq 0$) can be written as $\mathbf{AX} \leq \mathbf{B}$. We write the constraints $x \geq 0$, $y \geq 0$, $z \geq 0$ as $\mathbf{X} \geq \mathbf{0}$, where $\mathbf{0}$ denotes the vector all of whose coordinates are zero. The objective function p can be written as $\mathbf{C}^T\mathbf{X}$. Consequently, the SMP can be written in matrix form as

$$\text{Find the maximum of } \mathbf{C}^T\mathbf{X} \text{ for } \mathbf{X} \geq \mathbf{0} \text{ and } \mathbf{AX} \leq \mathbf{B}$$ ■

For any $m \times n$ matrix **A**, nonnegative column m-vector **B**, and column n-vector **C**, the SMP defined by **A**, **B**, **C** is

$$\text{Find the maximum of } \mathbf{C}^T\mathbf{X} \text{ for } \mathbf{X} \geq \mathbf{0} \text{ and } \mathbf{AX} \leq \mathbf{B}.$$

This SMP is denoted by SMP[**A**, **B**, **C**].

Of course, if an SMP is given in matrix form, it can also be expressed in terms of linear inequalities (see Exercises 11 to 18).

Feasible and Optimal Vectors

A vector **X** is *feasible* for the SMP[**A**, **B**, **C**] if $\mathbf{X} \geq \mathbf{0}$ and $\mathbf{AX} \leq \mathbf{B}$. A vector $\mathbf{X}^*$ is *optimal* for the SMP[**A**, **B**, **C**] if $\mathbf{X}^*$ is feasible and $\mathbf{C}^T\mathbf{X}^* \geq \mathbf{C}^T\mathbf{X}$ for all feasible vectors **X**.

The matrix formulation of linear programming problems is useful in showing how these problems occur in pairs.

Dual Linear Programming Problems

Let **A** be an $m \times n$ matrix, **B** be a column m-vector, and **C** a column n-vector. The SMP[**A**, **B**, **C**] is the problem:

$$\text{Maximize } \mathbf{C}^T\mathbf{X} \text{ subject to } \mathbf{X} \geq \mathbf{0} \text{ and } \mathbf{AX} \leq \mathbf{B}. \tag{10.8}$$

Using the same matrix **A** and vectors **B** and **C**, we can also form a minimization problem:

$$\text{Minimize } \mathbf{Z}^T\mathbf{B} \text{ subject to } \mathbf{Z} \geq \mathbf{0} \text{ and } \mathbf{Z}^T\mathbf{A} \geq \mathbf{C}. \tag{10.9}$$

Problems (10.8) and (10.9) are said to be *dual linear programming problems*. Each is said to be the *dual* of the other.

Example 10.11 Let

$$\mathbf{A} = \begin{bmatrix} 1 & 3 \\ 6 & 2 \end{bmatrix} \qquad \mathbf{B} = \begin{bmatrix} 6 \\ 12 \end{bmatrix} \qquad \mathbf{C} = \begin{bmatrix} 18 \\ 18 \end{bmatrix}$$

Problem Solve the SMP[**A**, **B**, **C**] and its dual.

Solution The SMP[**A**, **B**, **C**] is the problem

$$\begin{array}{ll} \text{Maximize} & p = 18x + 18y \\ \text{subject to} & x \geq 0 \qquad y \geq 0 \\ & x + 3y \leq 6 \\ & 6x + 2y \leq 12 \end{array}$$

We use the simplex method with Tableau 10.20 as the initial tableau. Tableaus 10.21 and 10.22 are obtained by pivoting.

Since Tableau 10.22 has no negative entries in the last row, we see that a solution of the SMP is given by $x = \frac{3}{2}$, $y = \frac{3}{2}$. The maximum value of $p = 18x + 18y$ is 54.

Next, we seek a solution of the dual of SMP[**A**, **B**, **C**]. Following the definition of the dual problem given above, the dual is

$$\text{Minimize } \mathbf{Z}^T\mathbf{B} \text{ subject to } \mathbf{Z} \geq \mathbf{0} \text{ and } \mathbf{Z}^T\mathbf{A} \geq \mathbf{C}.$$

If we set $\mathbf{Z} = \begin{bmatrix} r \\ s \end{bmatrix}$, then this problem is

$$\begin{array}{ll} \text{Minimize} & 6r + 12s \\ \text{subject to} & r \geq 0 \quad s \geq 0 \\ & r + 6s \geq 18 \\ & 3r + 2s \geq 18 \end{array}$$

TABLEAU 10.20

	↓ x	y	u	v	Basic Solution
u	1	3	1	0	6
→v	⑥	2	0	1	12
p	−18	−18	0	0	0

TABLEAU 10.21

	x	↓ y	u	v	Basic Solution
→u	0	($\frac{8}{3}$)	1	$-\frac{1}{6}$	4
x	1	$\frac{1}{3}$	0	$\frac{1}{6}$	2
p	0	−12	0	3	36

TABLEAU 10.22

	x	y	u	v	Basic Solution
y	0	1	$\frac{3}{8}$	$-\frac{1}{16}$	$\frac{3}{2}$
x	1	0	$-\frac{1}{8}$	$\frac{3}{16}$	$\frac{3}{2}$
p	0	0	$\frac{9}{2}$	$\frac{9}{4}$	54

At the moment, our only technique to solve this minimization problem is the graphical method of Chapter 7. Using this method, we graph the feasible set determined by the constraints, and we evaluate the objective function at the corner points of the set. The feasible set is shown in Figure 10.4. The corner points are (0, 9), (18, 0), and $(\frac{9}{2}, \frac{9}{4})$. Evaluating the objective function $6r + 12s$ at the corner points, we find that the minimum occurs at $(r, s) = (\frac{9}{2}, \frac{9}{4})$. This minimum value is 54. Since the objective function never assumes negative values, it follows (from the theorem of Section 7.3) that 54 is the smallest value of the objective function on the feasible set. We have solved the dual of the SMP[**A**, **B**, **C**]. ■

In Example 10.11 we used a geometric technique which is difficult to apply to problems with more than two variables. We need another method for such problems. To this end, we make the following observations about the solution of the SMP[**A**, **B**, **C**] and the solution of the dual problem in Example 10.11.

1. The maximum value of the objective function in the original problem (SMP[**A**, **B**, **C**]) is the same as the minimum value of the objective function in the dual problem. Both objective functions have an optimal value of 54.
2. The solution of the dual problem (both the coordinates of the optimal vector $[\frac{9}{2} \;\; \frac{9}{4}]$ and the value of the objective function, 54) is displayed in the last row of the tableau which gives the solution of the original SMP (Tableau 10.22).

These observations give evidence that we did not actually need to solve the dual problem by using graphical methods. At least for this example, the solution of the dual problem is contained in the same tableau as the solution of the original problem. The example is not special in this regard, and dual problems always have the following two properties.

FIGURE 10.4

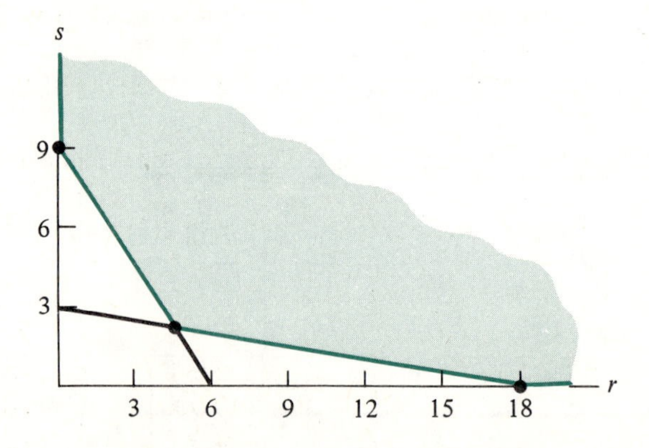

Properties of Dual Programming Problems

1. If one of a pair of dual linear programming problems has an optimal vector, the other also has an optimal vector.
2. Suppose that the simplex method is used to solve an SMP and that the method produces a tableau which yields an optimal vector for the SMP. Then the same tableau yields an optimal vector for the dual problem. An optimal vector for the dual problem is displayed in the last row of the tableau in the columns corresponding to the slack variables for the SMP. The value of the first variable in the dual problem is in the column of the first slack variable, the value of the second variable in the dual problem is in the column of the second slack variable, and so forth, for all dual variables. The optimal value of the objective function for the dual problem is the same as the optimal value of the objective function for the original SMP.

To illustrate the significance of the dual problem in a practical setting, we first consider the SMP posed in Example 7.4 and solved in Example 10.7. Then we provide a detailed interpretation of the result.

Example 10.12 As formulated in Example 7.4, a problem of interest to the Plant Power Fertilizer Company yields the following SMP:

Resource allocation

$$\begin{array}{ll} \text{Maximize} & p = 3x + 8y + 6z \\ \text{subject to} & x \geq 0 \quad y \geq 0 \quad z \geq 0 \\ & 20x + 4y + 4z \leq 6{,}000 \\ & 8x + 8y + 4z \leq 10{,}000 \\ & 8x + 4y + 2z \leq 4{,}000 \end{array} \tag{10.10}$$

We used the simplex method to solve this problem, and we obtained Tableau 10.14 as the final tableau. This tableau is reproduced as Tableau 10.23.

Problem Solve the dual of the SMP posed in (10.10).

Solution Using (10.8) and (10.9), we see that the dual of (10.10) is

$$\text{Minimize } \mathbf{Z}^T\mathbf{B} \text{ subject to } \mathbf{Z} \geq \mathbf{0} \text{ and } \mathbf{Z}^T\mathbf{A} \geq \mathbf{C}.$$

Here

$$\mathbf{A} = \begin{bmatrix} 20 & 4 & 4 \\ 8 & 8 & 4 \\ 8 & 4 & 2 \end{bmatrix} \qquad \mathbf{B} = \begin{bmatrix} 6{,}000 \\ 10{,}000 \\ 4{,}000 \end{bmatrix} \qquad \mathbf{C} = \begin{bmatrix} 3 \\ 8 \\ 6 \end{bmatrix} \qquad \mathbf{Z} = \begin{bmatrix} r \\ s \\ t \end{bmatrix}$$

TABLEAU 10.23

	x	y	z	u	v	w	**Basic Solution**
z	6	0	1	$\frac{1}{2}$	0	$-\frac{1}{2}$	1,000
v	−8	0	0	0	1	−2	2,000
y	−1	1	0	$-\frac{1}{4}$	0	$\frac{1}{2}$	500
p	25	0	0	1	0	1	10,000

We now use the properties of dual linear programming problems noted above. We see by property 1 that this dual problem has an optimal vector. Also by property 2, the minimum value of the objective function $\mathbf{Z}^T\mathbf{B}$ is 10,000. In addition, by property 2 the coordinates of an optimal vector are given by the last row of Tableau 10.23. They are $r = 1$, $s = 0$, and $t = 1$. ■

The values of r, s, and t, and the dual problem itself have an important interpretation in the context of the original SMP. We turn next to this topic.

An Interpretation of the Dual Problem

In the SMP resulting from the problem of the Plant Power Fertilizer Company, variables x, y, and z represent the amounts (in hundred of pounds) of 20-8-8, 4-8-4, and 4-4-2 fertilizer produced. The optimal vector given in Tableau 10.23 has $x = 0$, $y = 500$, and $z = 1000$. In this context the dual problem, with the variables r, s, and t, has the following interpretation and significance.

The optimal value of the variable r (namely, 1) is the value which would be added to the company's product (the fertilizer) if 1 additional pound of nitrate were available. Note that nitrate is the material associated with the first constraint, and the optimal value of the first dual variable appears in the column of the slack variable associated with the nitrate constraint.

Similarly, the optimal values of s and t (0 and 1) are the values which would be added to the company's product if 1 more pound of phosphate and potash, respectively, were available. To indicate how these values affect the problem, suppose 100 additional pounds of nitrate are purchased and used in the production of the same three types of fertilizer. The new constraints are

$$\begin{aligned} 20x + 4y + 4z &\le 6{,}100 \\ 8x + 8y + 4z &\le 10{,}000 \\ 8x + 4y + 2z &\le 4{,}000 \end{aligned} \qquad (10.11)$$

The final tableau which results when the problem is solved by the simplex method is Tableau 10.24. It follows that the solution of this new problem is $x = 0$, $y = 475$, $x = 1050$, and the optimal value of the objective function is 10,100. Thus, by increasing the amount of nitrate by 100 pounds, the profit to the company increases by \$100 (from \$10,000 to \$10,100). If the company can buy the additional nitrate for less than \$1 per pound (recall that the optimal value of r is 1), then it will benefit by buying the additional nitrate. If the cost of additional nitrate is more than \$1 per pound, then the company should not buy it, because its cost would exceed the profit from the additional fertilizer that could be produced with it. Similarly, since $s = 0$ and $t = 1$, if the company has 1 additional pound of phosphate or potash, then the profits will increase by \$0 and \$1, respectively. The result concerning phosphate is reasonable since the company already has 2000 pounds of unused phosphate ($v = 2000$). Buying more phosphate would not help increase profits at all.

Warning

It might appear from the discussion above that the company should buy as much nitrate as possible as long as it can be purchased for less than \$1 per pound. *This is not true. Additional nitrate should be purchased at the price determined by the dual problem only if the solution to the dual problem does not change as a result of the addition.*

Note that the solution of the dual problem for the system (10.11), which is given in Tableau 10.24, is the same as the solution of the dual of the original problem: $r = 1$, $s = 0$, and $t = 1$. This does not continue to be the case if large additional amounts of nitrate are purchased. For instance, if an additional 4000 pounds of nitrate are purchased for \$4000, the company actually reduces its net profit. The increase in profit is less than the cost of the additional nitrate (see Exercise 34).

As a final item concerning dual problems and the simplex method, we return to the topic mentioned at the beginning of the chapter. Namely, although the simplex method was introduced to apply to SMP problems, in many cases we can use it to solve minimum problems. In particular, if a minimum problem is the dual of an SMP, then we can use the simplex method to solve the SMP and simultaneously solve the minimum problem.

TABLEAU 10.24

	x	y	z	u	v	w	Basic Solution
z	6	0	1	$\frac{1}{2}$	$\frac{1}{2}$	$-\frac{1}{2}$	1,050
v	-8	0	0	1	1	-2	2,000
y	-1	1	0	$-\frac{1}{4}$	0	$\frac{1}{2}$	475
p	25	0	0	1	0	1	10,100

Example 10.13 A minimum problem is as follows:

$$\text{Minimize} \quad 50r + 10s + 20t$$

$$\text{subject to} \quad r \geq 0 \quad s \geq 0 \quad t \geq 0$$
$$5r - 5s + 5t \geq 5$$
$$10r + 5s + 5t \geq 10$$

Problem Solve this minimum problem by using the simplex method.

Solution Using the relation between (10.8) and (10.9), we see that this minimum problem is the dual of the following SMP:

$$\text{Maximize} \quad p = 5x + 10y$$

$$\text{subject to} \quad x \geq 0 \quad y \geq 0$$
$$5x + 10y \leq 50$$
$$-5x + 5y \leq 10$$
$$5x + 5y \leq 20$$

We apply the simplex method with slack variables u, v, and w. The initial tableau is Tableau 10.25, and after two pivots the final tableau, Tableau 10.26, is obtained.

TABLEAU 10.25

	x	y	u	v	w	Basic Solution
u	5	10	1	0	0	50
v	−5	5	0	1	0	10
w	5	5	0	0	1	20
p	−5	−10	0	0	0	0

TABLEAU 10.26

	x	y	u	v	w	Basic Solution
u	0	0	1	$-\frac{1}{2}$	$-\frac{3}{2}$	15
y	0	1	0	$\frac{1}{10}$	$\frac{1}{10}$	3
x	1	0	0	$-\frac{1}{10}$	$\frac{1}{10}$	1
p	0	0	0	$\frac{1}{2}$	$\frac{3}{2}$	35

TABLE 10.1

		Milligrams of Nutrient per 100 Grams Food		
Fruit	**Calories per 100 Grams**	**Calcium**	**Phosphorus**	**Iron**
Apple	60	10	10	.3
Orange	50	40	20	.2
Banana	90	60	30	.6

From Tableau 10.26, we see that the basic solution of the SMP (the maximum problem) is $x = 1$, $y = 3$, $u = 15$, $v = 0$, and $w = 0$, and the value of the objective function is 35. Using the properties of the dual problem, we conclude that the solution of the original minimum problem is $r = 0$, $s = \frac{1}{2}$, and $t = \frac{3}{2}$, and the minimum value of $50r + 10s + 20t$ is 35. ■

Example 10.14

Nutrition

Clyde would like to meet more of his nutritional needs from fresh fruits. In particular, he would like to get 250 milligrams of calcium [25 percent of the *recommended daily allowance* (RDA)], 500 milligrams of phosphorous (50 percent of RDA), and 9 milligrams of iron (50 percent of RDA) by eating fresh fruits, specifically apples, oranges, and bananas. The nutritional content of these foods is given in Table 10.1.

Since Clyde is very conscious of the number of calories he consumes, he would like to meet his nutritional goals while consuming as few calories as possible. Find a combination of fruits which satisfies his requirements.

Solution We begin by formulating a mathematical problem. Let r, s, and t denote the number of hundreds of grams of apples, oranges, and bananas in his diet, respectively. Then his nutritional requirements are

$$\begin{array}{ll} \text{Calcium:} & 10r + 40s + 60t \geq 250 \\ \text{Phosphorus:} & 10r + 20s + 30t \geq 500 \\ \text{Iron:} & .3r + .2s + .6t \geq 9 \end{array} \qquad (10.12)$$

The numbers of calories associated with r hundred grams of apples, s hundred grams of oranges, and t hundred grams of bananas are, respectively, $60r$, $50s$, and $90t$. Therefore, the total number of calories is

$$60r + 50s + 90t$$

The mathematical problem to be solved is

$$\begin{array}{ll} \text{Minimize} & 60r + 50s + 90t \\ \text{subject to} & r \geq 0 \quad s \geq 0 \quad t \geq 0 \\ & \text{and the constraints in (10.12)} \end{array} \tag{10.13}$$

This is a minimum problem, and we cannot apply the simplex method directly. Therefore, we first determine the dual of the problem (10.13). It is

$$\begin{array}{ll} \text{Maximize} & 250x + 500y + 9z \\ \text{subject to} & x \geq 0 \quad y \geq 0 \quad z \geq 0 \\ & 10x + 10y + .3z \leq 60 \\ & 40x + 20y + .2z \leq 50 \\ & 60x + 30y + .6z \leq 90 \end{array}$$

This is a standard maximum problem, and we can use the simplex method to solve it. The initial tableau is Tableau 10.27. By using the simplex method, the final tableau is determined to be Tableau 10.28 (Exercise 35).

From Tableau 10.28 we see that Clyde can meet his nutritional goals by eating no apples, 500 grams of oranges, and $\frac{4000}{3} = 1333.3$ grams of bananas. These foods provide 1450 calories, and no combination of the three fruits will meet his goals with fewer calories. Incidentally, since a diet containing 1333 grams of bananas is more suited to a monkey than a person, Clyde may wish to rethink his goals. ■

TABLEAU 10.27

	x	y	z	u	v	w	Basic Solution
u	10	10	.3	1	0	0	60
v	40	20	.2	0	1	0	50
w	60	30	.6	0	0	1	90
p	−250	−500	−9	0	0	0	0

TABLEAU 10.28

	x	y	z	u	v	w	Basic Solution
u	−10	0	0	1	$\frac{1}{2}$	$-\frac{2}{3}$	25
y	2	1	0	0	$\frac{1}{10}$	$-\frac{1}{30}$	2
z	0	0	1	0	−5	$\frac{10}{3}$	50
p	750	0	0	0	5	$\frac{40}{3}$	1450

Exercises for Section 10.4

Express each of the SMPs in Exercises 1 through 10 by using matrix notation.

1. Maximize $p = x + 2y$
   subject to
   $$x \geq 0 \qquad y \geq 0$$
   $$3x + 6y \leq 60$$
   $$6x + y \leq 18$$

2. Maximize $p = 8x + 2y$
   subject to
   $$x \geq 0 \qquad y \geq 0$$
   $$4x + 6y \leq 44$$
   $$x - y \leq 1$$

3. Maximize $p = x$
   subject to
   $$x \geq 0 \qquad y \geq 0$$
   $$y \leq 10$$
   $$x + y \leq 15$$
   $$-x + 4y \leq 12$$

4. Maximize $p = 3x + 4y$
   subject to
   $$x \geq 0 \qquad y \geq 0$$
   $$2x + 8y \leq 16$$
   $$5x + 4y \leq 20$$
   $$7x + 3y \leq 12$$

5. Maximize $p = x + 3y - z$
   subject to
   $$x \geq 0 \qquad y \geq 0 \qquad z \geq 0$$
   $$2x + 3y - 4z \leq 20$$
   $$-x + y + z \leq 5$$

6. Maximize $p = 3x + y + 5z$
   subject to
   $$x \geq 0 \qquad y \geq 0 \qquad z \geq 0$$
   $$x + 2y + z \leq 20$$
   $$x \leq 10$$
   $$x + y \leq 12$$

7. Maximize $p = x + 10y + 20z$
   subject to
   $$x \geq 0 \qquad y \geq 0 \qquad z \geq 0$$
   $$x + 15y + 10z \leq 200$$
   $$10x + 5y + z \leq 120$$
   $$15x + y + 20z \leq 300$$
   $$x + y + z \leq 100$$

8. Maximize $p = 5x + 10z$
   subject to
   $$x \geq 0 \qquad y \geq 0 \qquad z \geq 0$$
   $$5x + 5y + z \leq 100$$
   $$x + 5z \leq 100$$
   $$x + y \leq 15$$

9. Maximize $p = x_1 + 2x_2 + 3x_3 + 4x_4$
   subject to
   $$x_1 \geq 0 \qquad x_2 \geq 0 \qquad x_3 \geq 0 \qquad x_4 \geq 0$$
   $$x_1 - 2x_2 - 3x_3 + 4x_4 \leq 20$$
   $$2x_1 + 4x_2 + x_3 - x_4 \leq 10$$

10. Maximize $p = 3x_1 + 4x_2 - x_3 + 4x_4$
subject to $x_1 \geq 0 \quad x_2 \geq 0 \quad x_3 \geq 0 \quad x_4 \geq 0$

$$\begin{aligned} x_1 + 2x_2 + 3x_3 + 4x_4 &\leq 20 \\ 5x_1 + 6x_2 + 7x_3 + 8x_4 &\leq 30 \\ 10x_1 - x_2 + x_3 + 10x_4 &\leq 40 \\ -x_1 + x_2 + 3x_3 + 7x_4 &\leq 30 \end{aligned}$$

Using the given information concerning **A**, **B**, and **C**, express each SMP[**A**, **B**, **C**] in Exercises 11 through 18 in terms of linear inequalities.

11. $\mathbf{A} = \begin{bmatrix} 2 & 2 \\ 3 & 6 \end{bmatrix} \quad \mathbf{B} = \begin{bmatrix} 16 \\ 18 \end{bmatrix} \quad \mathbf{C} = \begin{bmatrix} 3 \\ 8 \end{bmatrix}$

12. $\mathbf{A} = \begin{bmatrix} 4 & 7 \\ 8 & 3 \end{bmatrix} \quad \mathbf{B} = \begin{bmatrix} 28 \\ 24 \end{bmatrix} \quad \mathbf{C} = \begin{bmatrix} 0 \\ 3 \end{bmatrix}$

13. $\mathbf{A} = \begin{bmatrix} 1 & -2 & 3 \\ -4 & 0 & 6 \end{bmatrix} \quad \mathbf{B} = \begin{bmatrix} 20 \\ 30 \end{bmatrix} \quad \mathbf{C} = \begin{bmatrix} 4 \\ -1 \\ 3 \end{bmatrix}$

14. $\mathbf{A} = \begin{bmatrix} 1 & 4 \\ 2 & 5 \\ 3 & 6 \end{bmatrix} \quad \mathbf{B} = \begin{bmatrix} 10 \\ 20 \\ 15 \end{bmatrix} \quad \mathbf{C} = \begin{bmatrix} 5 \\ 8 \end{bmatrix}$

15. $\mathbf{A} = \begin{bmatrix} 1 & 0 & 1 \\ 0 & 1 & 0 \\ 1 & 1 & 1 \end{bmatrix} \quad \mathbf{B} = \begin{bmatrix} 100 \\ 50 \\ 200 \end{bmatrix} \quad \mathbf{C} = \begin{bmatrix} 10 \\ 15 \\ 20 \end{bmatrix}$

16. $\mathbf{A} = \begin{bmatrix} 1 & -2 & 4 \\ 3 & 6 & -2 \\ -4 & 9 & 1 \end{bmatrix} \quad \mathbf{B} = \begin{bmatrix} 7 \\ 5 \\ 10 \end{bmatrix} \quad \mathbf{C} = \begin{bmatrix} 5 \\ -2 \\ 3 \end{bmatrix}$

17. $\mathbf{A} = \begin{bmatrix} 1 & 5 & 1 \\ 0 & 1 & 2 \\ 4 & 2 & 0 \\ 2 & 1 & 2 \end{bmatrix} \quad \mathbf{B} = \begin{bmatrix} 10 \\ 40 \\ 60 \\ 20 \end{bmatrix} \quad \mathbf{C} = \begin{bmatrix} 2 \\ 6 \\ 4 \end{bmatrix}$

18. $\mathbf{A} = \begin{bmatrix} 1 & 0 & -1 & 5 \\ -1 & 2 & 7 & 4 \\ 0 & 3 & -2 & 0 \\ 0 & 3 & 0 & 0 \\ 0 & -3 & 0 & 6 \end{bmatrix} \quad \mathbf{B} = \begin{bmatrix} 25 \\ 50 \\ 20 \\ 24 \\ 30 \end{bmatrix} \quad \mathbf{C} = \begin{bmatrix} 2 \\ -2 \\ 3 \\ 7 \end{bmatrix}$

State the dual for the problem posed in each of Exercises 19 through 26.

19. Maximize $2x + 3y$
subject to $x \geq 0 \quad y \geq 0$

$$\begin{aligned} 3x + 4y &\leq 12 \\ 6x + 3y &\leq 18 \end{aligned}$$

20. Maximize $2x + y$
subject to
$$\begin{aligned} x \geq 0 \quad & y \geq 0 \\ x &\leq 200 \\ 2x + y &\leq 1200 \\ 25x + 15y &\leq 7500 \end{aligned}$$

21. Minimize $3x + 4y$
subject to
$$\begin{aligned} x \geq 0 \quad & y \geq 0 \\ 8x + 3y &\geq 24 \\ 4x + 6y &\geq 12 \end{aligned}$$

22. Minimize $8x + 10y$
subject to
$$\begin{aligned} x \geq 0 \quad & y \geq 0 \\ x + y &\geq 400 \\ y &\geq 200 \\ 2y - x &\geq 0 \end{aligned}$$

23. Maximize $x + y + z$
subject to
$$\begin{aligned} x \geq 0 \quad y \geq 0 \quad & z \geq 0 \\ x + 4y + z &\leq 1 \\ 4x + 8y + z &\leq 1 \end{aligned}$$

24. Minimize $3x + 4y + 2z$
subject to
$$\begin{aligned} x \geq 0 \quad y \geq 0 \quad & z \geq 0 \\ 2x - y + 3z &\geq 24 \\ x + 3y &\geq 6 \\ 2y - 3z &\geq 6 \end{aligned}$$

25. The SMP[**A**, **B**, **C**] with **A**, **B**, and **C** as in Exercise 13.
26. The SMP[**A**, **B**, **C**] with **A**, **B**, and **C** as in Exercise 15.
27. Solve the SMP and its dual given in Exercise 1.
28. Solve the SMP and its dual given in Exercise 2.
29. Solve the SMP and its dual given in Exercise 3.
30. Solve the SMP and its dual given in Exercise 4.
31. Solve the SMP and its dual given in Exercise 8.
32. Solve the SMP and its dual given in Exercise 9.
33. It is an important fact about linear programming problems that if either an SMP or its dual has a solution, then both have solutions. A consequence of this fact is that if an SMP has no solution, then neither does its dual. For the SMP given in Exercise 13:
    (*a*) Use the simplex method to show that this problem has no solution.
    (*b*) Formulate the dual problem and use the geometric method of Chapter 7 to show that it has no solution.
34. Solve the SMP of Example 10.12 and its dual when the amount of nitrate is increased from 6000 to 10,000 pounds. Interpret the results of the dual problem.
35. Use the simplex method to convert Tableau 10.27 to Tableau 10.28.

IMPORTANT TERMS AND CONCEPTS

You should be able to describe, define, or give examples of and use each of the following:

Feasible vector
Linear function
Decision variables
Slack variables
Standard maximum problem (SMP)
Basic solution
Initial tableau
Pivot element
Pivot operation
Simplex method
Entering variable
Departing variable
Dual of an SMP

REVIEW EXERCISES

In Exercises 1 to 3, decide whether the problem stated is an SMP. If it is not, state the condition which is not met.

1. Maximize $p = 3x - y$
   subject to
$$4x + 5y \leq 240$$
$$-x + 2y \leq 80$$

2. Maximize $p = x + 3y + 5z$
   subject to
$$x \geq 0 \quad y \geq 0 \quad z \geq 0$$
$$10x + 5y + 15z \leq 200$$
$$x + 5y + 20z = 250$$

3. Maximize $p = 5x + y + 5z$
   subject to
$$x \geq 0 \quad y \geq 0 \quad z \geq 0$$
$$25x + 15y + 5z \leq 300$$
$$x \leq 10$$

Resource allocation

4. Formulate the following situation as a linear programming problem. The California Dry Fruit Company sells gift packages called Deluxe, Special, and Supreme. The amounts of fruit per package and the profit per package are as follows:

Type	Apricots, pounds	Apple Slices, pounds	Raisins, pounds	Profit, $
Deluxe	1	2	1	2.50
Special	2	1	1	2.00
Supreme	1.5	1.5	2	2.25

Suppose that the company can sell as many packages as it can make, and it has on hand 300 pounds of dried apricots, 350 pounds of dried apple slices, and 400 pounds of raisins. How many packages of each type should be made to maximize profit?

Resource allocation

5. Suppose that in Exercise 4 there are only enough boxes to make 50 Deluxe packages and 60 Supreme packages. How does the formulation of the problem change (assume enough boxes for Special packages)?

Production

6. Formulate the following situation as a linear programming problem. The Trimetal Mining Company operates three mines from which gold, silver, and copper are mined. The Alpha and Beta mines cost \$1000 per day to operate, and each yields 30 ounces of gold, 200 ounces of silver, and 400 pounds of copper each day. The Omega mine costs \$1200 per day to operate, and it yields 40 ounces of gold, 400 ounces of silver, and 300 pounds of copper each day. The company has a contract to supply at least 500 ounces of gold, 8000 ounces of silver, and 4000 pounds of copper. How many days should each mine be operated so that the contract can be filled at minimum cost?

Resource allocation

7. Formulate the following situation as a linear programming problem. Creative Crate Company makes crates for mirrors, bulletin boards, and cabinets from lumber, plywood, and cardboard. A crate for bulletin boards requires 60 square feet of cardboard and 30 board feet of lumber. A crate for mirrors requires 17 square feet of plywood, 20 square feet of cardboard, and 30 board feet of lumber. A crate for a cabinet requires 5 square feet of plywood, 50 square feet of cardboard, and 30 board feet of lumber. Each week the company has 2700 square feet of cardboard, 1800 board feet of lumber, and 500 square feet of plywood. The firm makes a profit of \$2, \$5, and \$10 for each crate for bulletin boards, mirrors, and cabinets, respectively, and it can sell as many as it makes. How many crates of each type should be made to maximize profit?

8. For Exercise 4:
   (*a*) Convert the problem to one with slack variables.
   (*b*) Find the initial tableau.
9. Repeat Exercise 8 for the problem of Exercise 5.
10. Repeat Exercise 8 for the problem of Exercise 7.
11. Carry out one pivot on the initial tableau for the problem of Exercise 4.
12. Repeat Exercise 11 for the problem of Exercise 7.
13. Convert the following problem to one with slack variables. Then graph the feasible set in the plane, find the corner points, and find the basic solutions which correspond to the corner points.

$$\begin{array}{ll} \text{Maximize} & p = 2x + 3y \\ \text{subject to} & x \geq 0 \quad y \geq 0 \\ & 3x + 5y \leq 30 \\ & 6x + 2y \leq 18 \end{array}$$

14. Repeat Exercise 13 with this problem:

$$\begin{array}{ll} \text{Maximize} & p = 2x - 3y \\ \text{subject to} & x \geq 0 \quad y \geq 0 \\ & y - x \leq 2 \\ & x - y \leq 2 \\ & y \leq 4 \end{array}$$

15. Find the initial tableau for the SMP of Exercise 13, and carry out one pivot operation. What is the basic solution that results from this pivot operation?

16. Repeat Exercise 15 with the SMP of Exercise 14.

Resource allocation

17. Find the initial tableau for the following linear programming problem. The Plant Power Fertilizer Company produces three fertilizer brands: Standard, Special, and Super. One sack (100 pounds) of each contains the nutrients nitrogen, phosphorus, and potash in the amounts shown below. The profit per sack is shown at the right, and the available material is shown at the bottom. How many sacks of each brand should be produced to maximize profit?

	Amount per Sack, Pounds			
Brand	**Nitrogen**	**Phosphorus**	**Potash**	**Profit, $**
Standard	20	5	5	10
Special	10	10	10	5
Super	5	15	10	15
Available	2000	1500	1500	

18. Carry out a pivot on the following tableau. Use the element circled as the pivot element, and describe the new basic solution.

	x	y	u	v	Basic Solution
u	10	5	1	0	70
v	(15)	6	0	1	90
p	−5	−3	0	0	0

19. Carry out a pivot on the following tableau. Use the element circled as the pivot element, and describe the new basic solution.

	x	y	u	v	w	Basic Solution
u	4	5	1	0	0	40
v	2	(6)	0	1	0	30
w	5	12	0	0	1	60
p	−3	−8	0	0	0	0

20. Carry out a pivot on the following tableau. Use the element circled as the pivot element, and describe the new basic solution. Note that the initial tableau for this problem is the same as that of Exercise 19; however, in this case after the first pivot, no further pivoting is possible.

	x	y	u	v	w	Basic Solution
u	4	5	1	0	0	40
v	2	6	0	1	0	30
w	5	(12)	0	0	1	60
p	−3	−8	0	0	0	0

21. Carry out a pivot on the following tableau. Use the element circled as the pivot element, and describe the new basic solution.

	x	y	z	u	v	w	Basic Solution
u	1	2	1	1	0	0	300
v	(2)	1	1	0	1	0	350
w	1	1	2	0	0	1	400
p	$-\frac{5}{2}$	−2	$-\frac{9}{4}$	0	0	0	0

22. Carry out a pivot on the following tableau. Use the element circled as the pivot element, and describe the new basic solution.

	x	y	z	u	v	w	Basic Solution
u	25	10	10	1	0	0	1000
v	10	15	(25)	0	1	0	250
w	50	5	5	0	0	1	300
p	−100	−75	−125	0	0	0	0

23. (*a*) Find the pivot element and carry out one pivot on the following tableau.
(*b*) Identify the new solution obtained by pivoting in (*a*). Is this solution optimal?

	x	y	z	u	v	w	Basic Solution
u	16	8	4	1	0	0	16
v	32	24	8	0	1	0	40
w	18	6	−16	0	0	1	36
p	2	−2	−4	0	0	0	0

24. Find the pivot element in the following tableau and carry out one pivot.

	x	y	z	u	v	w	**Basic Solution**
u	50	100	25	1	0	0	100
v	100	25	−25	0	1	0	50
w	150	200	50	0	0	1	250
p	−50	25	−100	0	0	0	0

25. Use the simplex method to solve the following linear programming problem:

Maximize $p = 5x - 2y$

subject to $x \geq 0 \quad y \geq 0$

$$x + y \leq 20$$
$$x + 4y \leq 40$$
$$x \leq 15$$

26. Graph the feasible set for the problem of Exercise 25, and find all corner points. Find the basic solutions for the corner points on the line $x = 15$.

27. (*a*) Express the SMP of Exercise 25 in matrix form.
(*b*) Formulate the dual of the SMP in Exercise 25.

28. Use the simplex method to solve the following linear programming problem:

Maximize $p = 2x + y + 4z$

subject to $x \geq 0 \quad y \geq 0 \quad z \geq 0$

$$x + 2y + z \leq 12$$
$$2x + y + 2z \leq 24$$

29. (*a*) Express the SMP of Exercise 28 in matrix form.
(*b*) Formulate the dual of the SMP in Exercise 28.

30. Formulate the dual of the SMP for which the initial tableau is that shown in Exercise 19.

31. Use the simplex method to solve the following linear programming problem:

Maximize $p = x - y$

subject to $x \geq 0 \quad y \geq 0$

$$5x - 10y \leq 20$$
$$x + y \leq 5$$

32. Formulate the dual of the problem in Exercise 31, and use the final tableau of that exercise to solve the dual problem.

33. (*a*) Use the initial tableau of Exercise 19 to solve the problem formulated in Exercise 30.
(*b*) Use the initial tableau of Exercise 20 to solve the problem formulated in Exercise 30.

Note that (*a*) and (*b*) give two different solutions for the problem formulated in Exercise 30, but the value of the objective function is the same for both solutions.

34. Solve the following linear programming problem:

$$\begin{array}{ll} \text{Minimize} & 16r + 3s + 2t \\ \text{subject to} & r \geq 0 \quad s \geq 0 \quad t \geq 0 \\ & 2r + s \geq 2 \\ & 4r + t \geq 4 \end{array}$$

35. Solve the following linear programming problem:

$$\begin{array}{ll} \text{Minimize} & 36r + 24s + 12t \\ \text{subject to} & r \geq 0 \quad s \geq 0 \quad t \geq 0 \\ & 3r + 8s + 12t \geq 48 \\ & 9r + 3s + 6t \geq 24 \\ & 2r + 4s \geq 12 \end{array}$$

Resource allocation 36. Backwoods Co-op makes trail mix from raisins and dried apricots. A small package uses 4 ounces of raisins and 4 ounces of dried apricots, and a large package uses 8 ounces of raisins and 4 ounces of apricots. The profit per package is 10 cents and 15 cents for small and large packages, respectively. If there are 240 ounces of raisins and 160 ounces of apricots, how many packages of each size should be prepared to yield maximum profit?

Resource allocation 37. In the situation described in Exercise 36, suppose that the price of large packages of trail mix can be increased so that the profit of a large package increases to 30 cents. In this situation, how many packages of each size should be prepared to yield maximum profit?

Resource allocation 38. Clara's Bakery makes a nut delight cake and a sugar plum cake for sale during the holidays. Each nut delight cake requires 40 grams of nuts and 600 grams of cake mix, and each sugar plum cake requires 40 grams of raisins, 32 grams of nuts, and 800 grams of cake mix. The bakery has a daily supply of 1 kilogram of raisins, 800 grams of nuts, and 24.4 kilograms of cake mix (1 kilogram is equal to 1000 grams). Clara makes a profit of \$1.80 on each nut delight cake and \$1.50 on each sugar plum cake. How many cakes of each type should she produce to maximize her profit?

Recycling 39. The World Aluminum Company has three methods for obtaining high-grade aluminum from recycled beverage cans. With method 1, each ton of cans requires 1 hour of mechanical separation, 2 hours of smelting, and 0.5 hour of chemical treatment, and it yields 1500 pounds of high-grade aluminum. With method 2, each ton of cans requires 2 hours of mechanical separation, 3 hours of smelting, and 0.5 hour of chemical treatment, and it yields 1800 pounds of high-grade aluminum. Finally, with method 3, each ton of cans requires 2 hours of mechanical separation, 1 hour of smelting, and 1 hour of chemical treatment, and it yields 1200 pounds of high-grade aluminum. The mechanical separator is available 60 hours each week, the smelter can be used 45 hours each week, and the chemical treatment facility can operate for 15 hours each week. How many tons of cans should be processed by each method to maximize the weekly output of high-grade aluminum?

CHAPTER 11

Two-Person Zero-Sum Games

11.0 THE SETTING AND OVERVIEW

We have all encountered situations involving uncertainty in which we must make decisions. In such situations often the consequences of our decisions are determined not only by our own choices, but also by chance and by the choices of others. For example, when we select a checkout lane at the grocery store, our checkout time depends on chance (will the scanner fail to read the price on one of our items?) and on the choices of others (will the customer ahead of us ask a question or write a check?). As another example, suppose Sam plays poker. Sam's gain (or loss) depends on chance (the cards dealt), on choice (the bets made), and on competition (the actions of other players designed to enhance their chances of winning and to diminish Sam's). In this chapter we study mathematical techniques which can be used to help analyze situations involving chance, choice, and competition. The word "game" is frequently associated with such situations, and we use that word as a general term to describe any such situation. Our study will be restricted to games involving exactly two players and for which the gain of one player always comes at the expense of the other player. We begin by considering two examples; then we turn to the study of methods to solve these games.

11.1 TWO SIMPLE GAMES

The examples which follow are typical of the types of games we will study in this chapter. The first has an especially simple structure which permits a straightforward analysis and solution. The second example considers a game which is more complex than the first. Because of its complexity, the second exhibits more of the features of general games than the first.

Example 11.1 **Carol's complaint:** Rhoda and Carol play the following simple game. Each holds two cards, one labeled 1 and one labeled 2. To play the game, each selects one of her cards, and they simultaneously show the card selected to the other player. Then money is exchanged between the players, the amount depending on the cards selected. Specifically, if Rhoda (R) selects card 1 and Carol (C) selects card 1, then the amount exchanged is \$2, and it is paid by C to R. If R selects 1 and C selects 2, then the amount is \$4 and it is paid by C to R. On the other hand, if R selects 2 and C selects either 1 or 2, then an amount of \$2 is paid by R to C. These amounts are shown in the array (11.1). In this array the numbers are payments from C to R, and hence the negatives of these numbers represent amounts paid by R to C.

		C: 1	C: 2
R	1	\$2	\$4
	2	−\$2	−\$2

(11.1)

The numbers to the left and the top of the array represent the possible selections by R and C, respectively.

Problem Assuming that Rhoda and Carol are each interested in maximizing the amount of money they receive, how should they play the game? That is, what card should each select?

Solution By considering Rhoda's choices (1 and 2) and examining the amounts in array (11.1), it is clear that Rhoda should *always* select card 1. She always gains (either \$2 or \$4) by selecting card 1, and she always loses (\$2) by playing card 2. Carol, on the other hand, is in the sad position of always losing when Rhoda plays 1. Her only choice is between losing \$2 and losing \$4 per play of the game. In the hope that the game will end before she goes broke, she should always play 1. ■

Example 11.1 is such a simple game that it poses no challenge to the two players. They always do the same thing, and the play of the game is quite dull (not to mention expensive from Carol's point of view). The game in the next example provides more of a challenge to the players. Before going ahead, however, you should first ask yourself how Rhoda and Carol should play if the amounts exchanged are as shown in the array (11.2) instead of (11.1). The answer to this question involves some of the basic ideas of the chapter.

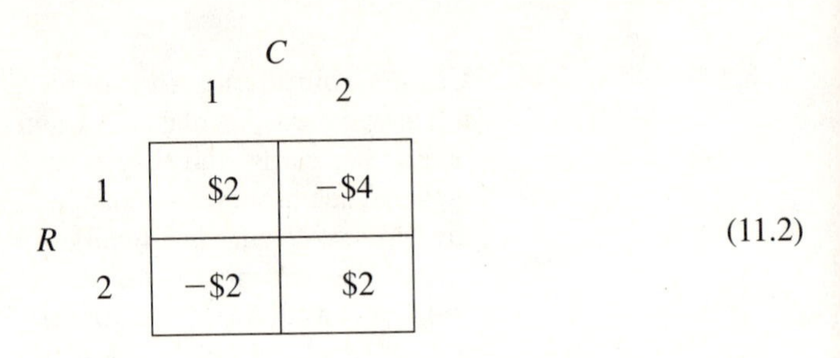

		C 1	*C* 2
R	1	\$2	−\$4
R	2	−\$2	\$2

(11.2)

Example 11.2 **A word-matching game:** Rolf is resting under a chestnut tree when his friend Chester joins him and suggests that they match coins for fun and (perhaps) profit. Rolf expresses a dislike for strenuous exercise, and he suggests that they simultaneously state the word "heads" or "tails" and make the payments accordingly. In Rolf's opinion, the payoffs should be as follows: If they match at heads, Chester pays him 40 cents; if they match at tails, Chester pays him 25 cents; and if they fail to match, Chester pays him 10 cents. Chester is not overly impressed with this proposal, but after some thought he agrees to play the game provided Rolf pays him 20 cents before the start of each play.

Problem How should Rolf and Chester play the game? That is, how should each player choose between heads and tails? Is this game fair to each of them? ■

To study the questions posed in this example (the questions are answered in Section 11.3), it is useful to introduce some notation and terminology. We will use these concepts throughout the chapter. We refer to the participants as *players* and the amounts which the players gain (or lose) as a result of playing the game once as *payoffs*. In games of the type we study, it is customary to represent the choices and payoffs in a matrix called the *payoff matrix*. Such matrices are shown in arrays (11.1) to (11.3). The rows of each of these payoff matrices correspond to the options available to one of the two players (in these examples Rhoda or Rolf), and the columns correspond to the options available to the other player (Carol or Chester). The entries in the matrices are the payoffs to be made *by the column player* (the player associated with the columns) *to the row player* (the player associated with the rows). The payoff matrix for the game between Rolf and Chester is shown in array (11.3). These payoffs include the 20 cents that Rolf pays Chester before each play of the game.

		Chester Heads	Chester Tails
Rolf	Heads	+20¢	−10¢
	Tails	−10¢	+5¢

(11.3)

Several features of the word-matching game should be noted since they can be used to characterize an important class of games. First, there are two players. Second, each player must choose one of two options at each play of the game, and the player must choose without knowledge of the choice of the opponent. Also, after the players make their choices, there is an exchange between them to effect the payoff determined by their choices. Thus, the gain of one player is always equal to the loss of the other player. We shall see in the next section that these characteristics—the number of players, method of play, and payoff system—are used to classify games. We shall also see that this game between Rolf and Chester is an unusually simple one, and it is possible to analyze it completely by using only elementary mathematical ideas.

Since Rolf and Chester both know the payoffs resulting from different choices by each of them, it stands to reason that they will select one of their options (words) in such a way as to achieve their aims. But what are their aims? Here we must make an assumption about the objectives of Rolf and Chester in playing the game. *We assume that each is motivated by a desire to maximize his overall gain.* On a certain play of the game Rolf may win (show a net profit), and on another play of the game Chester may win. Thus on any specific play of the game, each player could show a profit or a loss. Of course, each player would like to win on every play, but in this game (assuming both players are reasonably intelligent) there is no method of selecting words that will ensure a win on every play to either player. In such circumstances, the natural quantity for each player to consider and to attempt to maximize is his *average net gain per play of the game*. Thus, although each may sometimes win and sometimes lose, he would like to play in such a way that over many plays of the game his net gain will be as large as possible.

An initial consideration of array (11.3) might indicate that Rolf should always choose the word "heads" (i.e., say "heads") since this gives him a chance to achieve his greatest net gain. This gain occurs when Chester plays (says) "heads," and it amounts to 20 cents since Rolf receives 40 cents from Chester and he has already paid 20 cents to Chester. However, if Rolf always plays heads, then Chester will always play tails and Rolf will consistently show a net loss of 10 cents. On the other hand, if Rolf always plays tails, then Chester will always play heads and once again Rolf will show a loss of 10 cents. Of course, the answer to this dilemma for Rolf is obvious. He should mix up his choices, sometimes playing heads and sometimes playing tails. But exactly how should they be mixed? This is the central question for Rolf, and also for Chester since the reasoning is the same for him. To properly discuss the problem of finding the best method of mixing choices, we need a more formal mathematical setting than that of the present discussion. We provide this setting in the next section.

Exercises for Section 11.1

1. Suppose that in Example 11.1, the payoff when Rhoda selects card 2 and Carol selects either card 1 or 2 is \$1 paid by Carol to Rhoda, and all other payoffs are as in Example 11.1. How should Rhoda and Carol play in this situation?
2. Suppose that in Example 11.1, the payoff when Rhoda plays card 1 and Carol plays card 1 is \$2 from Rhoda to Carol, and the payoff when Rhoda plays card 1 and Carol plays card 2 is \$4 from Rhoda to Carol. All other payoffs are as in Example 11.1. How should Rhoda and Carol play in this situation?
3. Suppose that in Example 11.1, the payoff matrix is as shown in array (11.4). How should Rhoda and Carol play in this case?

		C 1	*C* 2
R	1	4	−8
R	2	2	−2

(11.4)

4. Repeat Exercise 3, but change the payoff \$2 in array (11.4) to −\$2. Does the answer change?
5. Repeat Exercise 3, but replace the payoff to Rhoda when Rhoda plays card 2 and Carol plays card 1 by the letter A, where A represents a number of dollars, and all other payoffs are as in array (11.4). For which values of A is the answer the same as in Exercise 3?
6. Repeat Exercise 1, but replace the payoff to Rhoda when Rhoda plays card 2 and Carol plays either card 1 or 2 by the letter A, where A represents a number of dollars, and all other payoffs are as in Example 11.1. For which values of A is the answer the same as in Example 11.1?
7. Consider a situation similar to that described in Example 11.1 with the change that now Rhoda has three cards, labeled 1, 2, and 3, and Carol has cards labeled 1 and 2. The payoffs are shown in array (11.5). How should Rhoda and Carol play in this case?

		C 1	C 2
	1	-1	3
R	2	0	1
	3	-2	0

(11.5)

8. Repeat Exercise 7, but change the payoff -1 in array (11.5) to 1.
9. Suppose that in the Rolf-Chester word-matching game the payoffs are changed so that when the players match words, Rolf pays Chester 10 cents and when they do not match, Chester pays Rolf 20 cents. Also, suppose that Chester pays Rolf 5 cents at the start of each play of the game. Form a matrix similar to that of array (11.3) for this new game.
10. Form a matrix as in Exercise 9, but now suppose that Rolf pays Chester 30 cents when they match at heads and 15 cents when they match at tails, and Chester pays Rolf 15 cents when they do not match. No other payments are made.
11. In the Rolf-Chester game with payoff matrix (11.3), suppose each player says heads one-half the time and tails one-half the time (in a random way). What is the expected value of the game to Rolf per play of the game? (*Hint*: Decide how often each payoff will occur, and compute the expected payoff as defined in Chapter 4.)
12. Answer the question of Exercise 11, but suppose that each player says heads with probability $\frac{2}{3}$ and tails with probability $\frac{1}{3}$.
13. Answer the question of Exercise 11, but suppose that Rolf says heads with probability $\frac{3}{4}$ and tails with probability $\frac{1}{4}$, and Chester says heads with probability $\frac{1}{3}$ and tails with probability $\frac{2}{3}$.
14. Answer the question of Exercise 11 for the payoff matrix given in array (11.2). Here Rhoda and Carol each play 1 and 2 equally often (at random), and we are interested in the expected gain to Rhoda.
15. Suppose that in the game with payoff matrix (11.2) Rhoda plays 1 one-third of the time and 2 two-thirds of the time. Carol plays in the same way. What is the expected gain to Rhoda?
16. Answer the question of Exercise 11 for the payoff matrix of Exercise 9 and for both players saying heads with probability $\frac{3}{4}$ and tails with probability $\frac{1}{4}$.
17. In each of the following areas, describe a situation which has the characteristics of a game. In each case list the options open to each player, and note the payoffs to each player for the different outcomes. It is not necessary that the sum of the payoffs for a particular outcome be zero as it is in the Rolf-Chester word-matching game.
    (*a*) Labor-management negotiations for a contract
    (*b*) Political parties campaigning in an election

18. As in Exercise 17, describe game situations for:
    (*a*) Disarmament talks between two superpowers
    (*b*) Divorce negotiations between a husband and wife
19. Ray and Carlos play a card game with the following rules. Ray has cards labeled 3, 4, and 5 and Carlos has cards labeled +1 and −1. Each player selects a card (without the other player knowing which one), and the two cards are displayed simultaneously. If the sum of the numbers on the cards is 3 or 4, then Carlos pays Ray $1, and if the sum is 2, 5, or 6, then Ray pays Carlos $1. By using an array similar to (11.1), the amounts paid by Carlos (*C*) to Ray (*R*) can be displayed as

		C −1	*C* +1
	3	−1	1
R	4	1	−1
	5	1	−1

    Find the expected gain to Ray if he selects his card at random (i.e., each with probability $\frac{1}{3}$) and Carlos selects his card at random (i.e., each with probability $\frac{1}{2}$).
20. In the situation described in Exercise 19, find the expected gain to Ray if Carlos selects the card labeled +1 with probability .7 and the card labeled −1 with probability .3, and Ray selects the cards labeled 3, 4, and 5 with probabilities .2, .3, and .5, respectively.
21. Ray and Carlos play a card game with the following rules. Ray has cards labeled 3, 4, and 5, and Carlos has cards labeled +1 and −1. Each player selects a card (without the other player knowing which one), and the two cards are displayed simultaneously. If the sum of the numbers on the cards is 3 or 4, then Carlos pays Ray that number of dollars, and if the sum is 2, 5, or 6, then Ray pays Carlos that number of dollars. By using an array similar to (11.1), the amounts paid by Carlos (*C*) to Ray (*R*) can be displayed as

		C −1	*C* +1
	3	−2	4
R	4	3	−5
	5	4	−6

Find the expected gain to Ray if he selects his card at random (i.e., each with probability $\frac{1}{3}$) and Carlos selects his card at random (i.e., each with probability $\frac{1}{2}$).

22. In the situation described in Exercise 21, find the expected gain to Ray if Carlos selects his card at random and Ray selects his card at random from the cards labeled 3 and 5. What if Carlos selects his card at random and Ray selects his card at random from the cards labeled 3 and 4?

11.2 SADDLE POINTS AND DOMINANCE

Each of the examples in Section 11.1 involves two players; each player has a fixed finite number of options (two in these cases) available on each play; and on each play of the game the gain of one player is the loss of the other player. Such a situation is an example of a *two-person zero-sum game.* In general, "zero sum" is used to describe a game with the property that the sum of the payoffs to the players on each play of the game is zero. In the examples of Section 11.1 there are two payoffs, and since one is the negative of the other, the sum is always zero. A precise definition is the following:

Two-Person Zero-Sum Game

A *two-person zero-sum game* consists of two (finite) sets $R = \{r_1, r_2, \ldots, r_m\}$ and $C = \{c_1, c_2, \ldots, c_n\}$, together with an $m \times n$ payoff matrix $\mathbf{P} = (p_{ij})$ which associates a number p_{ij} with each pair (r_i, c_j). The elements of the set R are called the *pure strategies for the row player*, and the elements of the set C are called the *pure strategies for the column player.* The number p_{ij} is the *payoff to the row player* associated with the pair of strategies (r_i, c_j), and $-p_{ij}$ is the corresponding *payoff to the column player* associated with this pair of strategies.

Two-person zero-sum games with finite sets R and C (as defined above) are also called *matrix games*, since the payoffs are all given by a matrix.

The word-matching game played by Rolf and Chester is a convenient one to use to illustrate our definition of a two-person zero-sum game. The sets R and C consist of the options or pure strategies which are available to Rolf and Chester, respectively. In this case both players have the same two options [they must play heads (H) or tails (T) on each play of the game]. Thus the sets R and C are $R = C = \{H, T\}$. The entries in the matrix $\mathbf{P}$ are determined from the description of the game given in Example 11.2, and they were noted in array (11.3). Thus the payoff matrix is

$$\mathbf{P} = \begin{array}{c} \\ H \\ T \end{array} \begin{array}{c} \begin{array}{cc} H & \quad T \end{array} \\ \begin{bmatrix} 20 & -10 \\ -10 & 5 \end{bmatrix} \end{array}$$

The process of representing two-person zero-sum games by matrices is a reversible process. Any matrix with real numbers as entries may be viewed as the payoff matrix for a two-person zero-sum game. The entry in the ith row and the jth column of the matrix is the payoff to the row player when she or he uses the ith strategy and the opponent (the column player) uses his or her jth strategy. The negative of this entry is the return or the payoff to the column player. The exact nature of the strategies for each player is not known, nor is it important. It is possible to analyze, and in one sense solve, a two-person zero-sum game based only on the knowledge of the payoff matrix.

We have used the word "solve" in regard to games, and we should be more precise about this word and about what we mean by the phrase "a solution to the game." First we note that we are considering only two-person zero-sum matrix games. For such games a solution consists of three things: *methods of play for both players*, so that they will have the largest possible expected return that they can guarantee themselves, and the *value* of this expected return to the row player. By methods of play we mean prescriptions for each player which tell the players whether they should play one pure strategy all the time or mix pure strategies, and if they should mix their strategies, the method tells them which ones to mix and how often to play each of them.

As we noted in our discussion of the word-matching game, it is natural for both Rolf and Chester to play by mixing their choice of words and not by saying the same word every time. This situation often but not always occurs in two-person zero-sum games. In some games the "best" strategy is to select the same option or pure strategy every time. This is the case with the game of Example 11.1, and it is also the case with the next example.

Example 11.3 The matrix below is the payoff matrix of a two-person zero-sum game.

$$\mathbf{P} = \begin{array}{c} \\ r_1 \\ r_2 \\ r_3 \\ r_4 \end{array} \begin{array}{c} \begin{array}{ccc} c_1 & c_2 & c_3 \end{array} \\ \begin{bmatrix} 1 & -1 & 2 \\ 0 & 1 & 1 \\ 0 & -2 & 1 \\ 2 & 1 & 3 \end{bmatrix} \end{array}$$

Problem Decide how the row and column players should play the game.

Solution Recall that the entries in this matrix represent payoffs *to* the row player *from* the column player. Note that there are four options or pure strategies for the row player and three pure strategies for the column player. We proceed by inspecting the payoff matrix with the goal of determining which pure strategies are most attractive to each player. Since we are concerned only with pure strategies in this example, indeed, we have not yet defined any other type of strategy, let us use the term "strategy" here to refer to a pure strategy. We note that if the row player uses strategy (row) 4, then for any choice of strategy by the column player the payoff to the row player is at least as high as if he or she used any other strategy, and if the column player uses strategy (column) 3, then the payoff to the row player is actually higher with strategy 4 than with any other strategy. Thus the row player can be assured of a maximum return against any column strategy by simply playing r_4 every time. On the other hand, the column player can also see that the row player should always play r_4, and thus, the column player can minimize losses by always playing c_2. Hence, in this simple (and somewhat dull) game, each player should always use the same strategy, and the payoff will always be $+1$ to the row player. In a sense we have solved this game by determining a method of play for each player which in the long run is as good as or better than any other method of play. Also we know the average gain (or loss) for each player if both players use these best methods of play. ■

It is interesting that in Example 11.3 knowledge of a player's intentions cannot be used to reduce her or his payoff. Thus if the row player announces in advance that he will use strategy r_4, then the column player cannot use this information to reduce the payoff to the row player. Similarly, if the column player announces in advance that she plans to use strategy c_2, then the row player cannot use this information to increase his own payoff, i.e., reduce the payoff to the column player. We will contrast this state of affairs with another very different one in the next section.

The situation which occurs in Example 11.3 is an example of a game which is *strictly determined*. It has a *saddle point*. To be specific:

Saddle Point

A two-person zero-sum game has a *saddle point* at row i and column j if the (i, j) entry in the payoff matrix is both the minimum of the ith row and the maximum of the jth column. Any game with a saddle point is *strictly determined.*

Let us consider Example 11.3 from a somewhat different point of view. In addition to providing a perspective for our work, this viewpoint will give a method of identifying saddle points when they exist. We begin by asking, What is the worst payoff that the row player can receive when he plays strategy 1? An inspection of the payoff matrix shows that the minimum payoff the row player can receive when he plays strategy 1 is -1. Likewise, 0 is the minimum payoff the row player can receive if he plays strategy 2; and -2 and 1 are the minimum payoffs for strategies 3 and 4, respectively. Write the row minimums to the right of the payoff matrix as shown below.

$$\begin{bmatrix} 1 & -1 & 2 \\ 0 & 1 & 1 \\ 0 & -2 & 1 \\ 2 & 1 & 3 \end{bmatrix} \qquad \begin{matrix} \text{Row} \\ \text{minimums} \\ -1 \\ 0 \\ -2 \\ 1 \end{matrix}$$

In a similar manner we can analyze the options available to the column player, remembering that higher payoffs are undesirable from the column player's point of view (since they represent what she must pay out). The column maximums—the worst that can happen to the column player—are listed below the payoff matrix.

$$
\begin{array}{rcc}
 & & \text{Row minimums} \\
 & \begin{bmatrix} 1 & -1 & 2 \\ 0 & 1 & 1 \\ 0 & -2 & 1 \\ 2 & 1 & 3 \end{bmatrix} & \begin{array}{c} -1 \\ 0 \\ -2 \\ \boxed{1}\ \text{Maximum} \end{array} \\
\text{Column maximums} & \begin{array}{ccc} 2 & \boxed{1} & 3 \\ & \text{Minimum} & \end{array} &
\end{array}
$$

Next, the row player observes that if he selects strategy r_4, then he is guaranteed a payoff of at least 1 unit, and this is the best he can guarantee himself. Also, the column player observes that if she plays strategy c_2, then she is guaranteed a loss of not more than 1 unit, and this is the best she can guarantee herself. In this example these two numbers are equal, and the (4, 2) entry in the matrix is by definition a saddle point. The game is strictly determined since each player should always play the same (pure) strategy. There is no choice to be made at each play of the game. Thus if the row player seeks to maximize his average winning per game, then his best strategy is to play r_4 every time. Likewise, the column player's best strategy is to play c_2 every time. It is in this sense that the game is strictly determined.

The procedure illustrated above is a general one which can always be used to find saddle points. The method is as follows.

Method of Finding a Saddle Point

1. Identify the row minimums (list them to the right of the payoff matrix for easy reference) and the column maximums (list them below the payoff matrix).
2. Check whether the maximum of the row minimums is equal to the minimum of the column maximums.
   a. If so, then there is at least one saddle point.
   b. If not, then there are no saddle points.
3. If there is a saddle point, then there is one in a row corresponding to a maximum row minimum and in a column corresponding to a minimum column maximum.

This method is illustrated in the next example.

Example 11.4 Payoff matrices $\mathbf{P}_1$ and $\mathbf{P}_2$ for two games are

$$\mathbf{P}_1 = \begin{bmatrix} 1 & -1 & 2 \\ 1 & 1 & 3 \\ -1 & 1 & -2 \\ 3 & -2 & 3 \end{bmatrix} \qquad \mathbf{P}_2 = \begin{bmatrix} 1 & -1 & 2 \\ 1 & 1 & 3 \\ -1 & 1 & -2 \\ 3 & 2 & -3 \end{bmatrix}$$

Problem Decide whether each game is strictly determined. That is, find all saddle points (if there are any) for each game.

Solution The row minimums and column maximums for $\mathbf{P}_1$ are

$$\begin{array}{rc|l} & \begin{bmatrix} 1 & -1 & 2 \\ 1 & \textcircled{1} & 3 \\ -1 & 1 & -2 \\ 3 & -2 & 3 \end{bmatrix} & \begin{array}{l} \text{Row minimums} \\ -1 \\ \boxed{1} \text{ Maximum} \\ -2 \\ -2 \end{array} \\ \text{Column maximums} & 3 \quad \boxed{1} \quad 3 & \\ & \text{Minimum} & \end{array}$$

The maximum of the row minimums and the minimum of the column maximums, shown in boxes, are both 1. Thus there is a saddle point, the circled entry in row 2, column 2. In this matrix there is only one saddle point. Note that other entries have the value 1, but in this matrix they are not saddle points. It is possible, however, to have more than one saddle point in a payoff matrix (see Exercise 19). The game with payoff matrix $\mathbf{P}_1$ is strictly determined. The best strategy for the row player is to always play r_2, and the best strategy for the column player is to always play c_2.

The row minimums, column maximums, the maximum row minimum, and the minimum column maximum for matrix $\mathbf{P}_2$ are

$$\begin{array}{rc|l} & \begin{bmatrix} 1 & -1 & 2 \\ 1 & 1 & 3 \\ -1 & 1 & -2 \\ 3 & 2 & -3 \end{bmatrix} & \begin{array}{l} \text{Row minimums} \\ -1 \\ \boxed{1} \text{ Maximum} \\ -2 \\ -3 \end{array} \\ \text{Column maximums} & 3 \quad \boxed{2} \quad 3 & \\ & \text{Minimum} & \end{array}$$

In this matrix the maximum of the row minimums (1) is not equal to the minimum of the column maximums (2), and there is no saddle point. ■

Even though matrix $\mathbf{P}_2$ of Example 11.4 does not have a saddle point, it is possible to deduce some information regarding the best way to play the game by simply inspecting the matrix. For example, from the row player's point of view, pure strategies r_1 and r_3 should never be played because pure strategy r_2 is always at least as good as r_1 and r_3 and sometimes it is better (in terms of the payoff to the row player). Thus in deciding how to play this game, the row player can ignore strategies r_1 and r_3 and simply consider the matrix

$$\begin{array}{c} \\ r_2 \\ r_4 \end{array} \begin{array}{c} \begin{array}{ccc} c_1 & c_2 & c_3 \end{array} \\ \begin{bmatrix} 1 & 1 & 3 \\ 3 & 2 & -3 \end{bmatrix} \end{array}$$

At this stage, however, the column player can also simplify the game (we assume the players are equally intelligent so that the column player knows the row player is not going to play r_1 or r_3). The column player observes that pure strategy c_2 is always a better play than c_1 since it never results in a larger payoff to the row player and it sometimes results in a smaller payoff. Thus the column player should never use c_1, and it can be eliminated from further consideration. Taking these observations into account, we see that the matrix to be considered has the form

$$\begin{array}{c} \\ r_2 \\ r_4 \end{array} \begin{array}{c} \begin{array}{cc} c_2 & c_3 \end{array} \\ \begin{bmatrix} 1 & 3 \\ 2 & -3 \end{bmatrix} \end{array} \tag{11.6}$$

At this stage no further simplifications are possible, and the game is similar to the word-matching game in that the best results cannot be achieved by using pure strategies. It is necessary to mix pure strategies in an appropriate way. The problem of determining the proper mix will be discussed in the next section.

The simplifications which were used to convert matrix $\mathbf{P}_2$ of Example 11.4 to matrix (11.6) are called *simplifications due to dominance*. In making this notion precise, it is convenient to use vector notation. Each row and each column of a payoff matrix can be considered to be a vector. In fact, the rows are row vectors and the columns are column vectors. We denote the vector which is the ith row of the payoff matrix by $\mathbf{R}_i$ and the vector which is the jth column of the matrix by $\mathbf{C}_j$. Vector inequalities were introduced in Section 10.4, and they provide a useful means of defining dominance.

Row i of a payoff matrix *dominates* row j if $\mathbf{R}_i \geq \mathbf{R}_j$. Column i of a payoff matrix dominates column j if $\mathbf{C}_i \leq \mathbf{C}_j$.

Recall that our convention is that the entries in the payoff matrix represent amounts paid by the column player to the row player. Hence, small numbers are better for the column player than large numbers.

In matrix $\mathbf{P}_2$ of Example 11.4, row 2 ($\mathbf{R}_2$) dominates both rows 1 and 3 ($\mathbf{R}_1$ and $\mathbf{R}_3$). Any row which is dominated by another row of the payoff matrix need not be considered as an option for the row player since results which are at least as good can be achieved by using other rows. In effect, that row (and hence that pure strategy) can be eliminated from the payoff matrix. Similarly, if column i dominates column j, then column j (pure strategy c_j) need not be considered by the column player. That column can effectively be eliminated from the payoff matrix.

As a final comment concerning reducing the size of a payoff matrix by using dominance, we note that the reductions may be made sequentially. For example, first the row player may eliminate a row by dominance; then, in the new reduced matrix, the column player may eliminate a column by dominance; then, in the twice-reduced matrix, the row player may eliminate another row by dominance; and so on, until the matrix contains no further dominance. In the case of a matrix reduced by dominance to size 1×1, the matrix has a saddle point, namely, the single remaining entry.

Example 11.5 Reduce the following payoff matrix by using dominance.

$$\begin{bmatrix} 3 & 1 & 0 & 1 & 1 & 2 \\ 2 & 1 & -1 & -1 & 0 & 1 \\ 1 & -1 & 1 & 0 & 1 & -1 \\ 2 & -1 & -1 & 2 & -1 & 0 \\ 4 & -1 & -1 & 2 & -2 & -2 \end{bmatrix}$$

Solution Row 1 dominates row 2, and column 6 dominates column 1. After row 2 (strategy r_2) and column 1 (strategy c_1) have been removed, the payoff matrix has the form (we include the strategies adjacent to the matrix to help us to keep track of the strategies still available to use)

$$\begin{array}{c} \\ r_1 \\ r_3 \\ r_4 \\ r_5 \end{array} \begin{array}{c} \begin{array}{ccccc} c_2 & c_3 & c_4 & c_5 & c_6 \end{array} \\ \begin{bmatrix} 1 & 0 & 1 & 1 & 2 \\ -1 & 1 & 0 & 1 & -1 \\ -1 & -1 & 2 & -1 & 0 \\ -1 & -1 & 2 & -2 & -2 \end{bmatrix} \end{array}$$

In this matrix $\mathbf{R}_4$ dominates $\mathbf{R}_5$ (note that this is not true in the original matrix). Also, in this new matrix $\mathbf{C}_2$ dominates $\mathbf{C}_4$. Removing $\mathbf{R}_5$ and $\mathbf{C}_4$, we have the payoff matrix

$$\begin{array}{c} \\ r_1 \\ r_3 \\ r_4 \end{array} \begin{array}{c} \begin{array}{cccc} c_2 & c_3 & c_5 & c_6 \end{array} \\ \begin{bmatrix} 1 & 0 & 1 & 2 \\ -1 & 1 & 1 & -1 \\ -1 & -1 & -1 & 0 \end{bmatrix} \end{array}$$

We continue the process. Now $\mathbf{R}_1$ dominates $\mathbf{R}_4$, and $\mathbf{C}_2$ dominates $\mathbf{C}_5$ and $\mathbf{C}_6$. Therefore, we have

$$\begin{array}{c} \\ r_1 \\ r_3 \end{array} \begin{array}{c} \begin{array}{cc} c_2 & c_3 \end{array} \\ \begin{bmatrix} 1 & 0 \\ -1 & 1 \end{bmatrix} \end{array}$$

At this stage we must stop since there is no further dominance. However, a substantial reduction of the problem has been achieved. The original payoff matrix was 5×6, and by dominance this was reduced to 2×2. As we will see in Section 11.3, a game with a matrix of this size is fairly easy to solve. ■

Exercises for Section 11.2

Decide whether each of the matrices in Exercises 1 to 12 has a saddle point. Find *all* such saddle points.

1. $\begin{bmatrix} 1 & 4 \\ -1 & 3 \end{bmatrix}$

2. $\begin{bmatrix} 1 & 0 \\ -1 & 3 \end{bmatrix}$

3. $\begin{bmatrix} 1 & 2 & 3 \\ 2 & 2 & 3 \\ 2 & 2 & 2 \end{bmatrix}$

4. $\begin{bmatrix} 1 & -2 & 3 \\ 0 & 2 & -1 \\ 1 & -1 & 3 \end{bmatrix}$

5. $\begin{bmatrix} 1 & -2 & 3 \\ 0 & -1 & -1 \\ 1 & -1 & 3 \end{bmatrix}$

6. $\begin{bmatrix} 0 & 5 & 2 \\ 6 & 0 & 1 \\ 3 & 4 & 3 \end{bmatrix}$

7. $\begin{bmatrix} 0 & -1 & -3 & 3 \\ 1 & 0 & 0 & 3 \\ 1 & 0 & 0 & 1 \\ 0 & -2 & -1 & 1 \end{bmatrix}$

8. $\begin{bmatrix} 1 & -1 & 2 & 3 \\ 0 & 3 & -3 & 1 \\ 1 & 0 & 2 & 4 \\ -1 & 2 & -4 & 0 \end{bmatrix}$

9. $\begin{bmatrix} 3 & 0 & 1 & 2 \\ 1 & -1 & 0 & 1 \\ 2 & 0 & 1 & -1 \\ 3 & 1 & 1 & 0 \end{bmatrix}$

10. $\begin{bmatrix} 3 & 0 & 1 & 2 \\ 1 & -1 & 0 & 1 \\ 2 & 0 & 1 & -1 \\ 3 & -1 & 1 & 0 \end{bmatrix}$

11. $\begin{bmatrix} 1 & -1 & 2 & -2 \\ -3 & 2 & -1 & 4 \\ 10 & -3 & 6 & -8 \end{bmatrix}$

12. $\begin{bmatrix} -2 & 1 & -3 & 4 & -1 \\ 2 & 1 & -2 & 6 & 0 \\ -3 & 4 & 1 & -6 & 3 \end{bmatrix}$

13. Use dominance to reduce the size of the matrix in Exercise 7.
14. Use dominance to reduce the size of the matrix in Exercise 8.
15. Use dominance to reduce the size of the matrix in Exercise 12.

16. Use dominance to reduce the size of the following matrix:

$$\begin{bmatrix} 5 & -1 & 1 & 5 \\ -5 & 1 & 1 & 9 \\ 5 & 1 & -3 & 1 \\ 9 & 1 & 1 & 7 \end{bmatrix}$$

17. Use dominance to reduce the size of the following matrix.

$$\begin{bmatrix} 5 & -1 & 1 & 5 & 3 \\ -2 & 1 & 2 & 6 & 1 \\ 3 & 1 & -2 & 2 & 2 \\ 6 & 2 & 1 & 4 & 2 \end{bmatrix}$$

18. Use dominance to reduce the size of the following matrix.

$$\begin{bmatrix} 4 & -2 & 3 & 5 & -2 \\ 1 & 1 & 0 & 4 & 1 \\ 1 & 3 & 1 & 1 & 2 \\ -1 & 2 & 2 & 0 & -1 \end{bmatrix}$$

19. Find an example of a 3×3 two-person zero-sum game which has exactly two saddle points.
20. Find an example of a 3×5 two-person zero-sum game which has exactly four saddle points.
21. Rolf and Chester decide to match numbers instead of coins. The choices for each player are the numbers 1, 2, and 3. The payoffs to Rolf (from Chester) are as follows: If Rolf plays number r and Chester plays number c, the payoff to Rolf is

$$r^2 - 2rc + c^2 \qquad \text{if } r \geq c$$

and

$$2rc - r^2 - c^2 \qquad \text{if } r < c$$

Form the payoff matrix for this 3×3 game, and check for saddle points and dominance.
22. Suppose that the payoff matrix for a two-person zero-sum game is

$$\begin{bmatrix} a & 2 \\ -1 & b \end{bmatrix}$$

where $-1 < a < b < 2$. Decide if this game has a saddle point; if it does, find it.
23. Suppose that the payoff matrix for a two-person zero-sum game is

$$\begin{bmatrix} a & b \\ c & a \end{bmatrix}$$

where $b < a < c$. Decide if this game has a saddle point; if it does, find it.

24. Use the payoff matrix of Exercise 23, but assume $a < b < c$. Decide if the game has a saddle point; if it does, find it.

11.3 SOLVING 2 × *n* AND *m* × 2 GAMES

If a game has a saddle point, there is a pure strategy r_i for the row player and a pure strategy c_j for the column player such that the row player does best by always playing r_i and the column player does best by always playing c_j. On the other hand, if the game does not have a saddle point, then it is not advantageous for either player to always use the same pure strategy. The goal of *maximizing the average return per game* is achieved only by mixing pure strategies in some way, i.e., by playing one pure strategy part of the time and other pure strategies at other times. The fractions of the time that various strategies are to be played must be determined for each game. For a person not to be taken advantage of by an opponent, the method of play should not follow a pattern. Thus, in Example 11.2, if Rolf decides to play heads and tails each one-half the time, he should not simply alternate between heads and tails. Instead he should use some chance device (such as flipping a coin) and let this chance device determine his play. We can represent Rolf's mixed strategy of playing each pure strategy one-half the time by using the vector $[\frac{1}{2} \quad \frac{1}{2}]$. Other mixed strategies for Rolf are $[\frac{1}{3} \quad \frac{2}{3}]$, $[\frac{5}{6} \quad \frac{1}{6}]$, and in general $[r \quad 1-r]$ for any number r such that $0 \leq r \leq 1$. In this vector notation the first coordinate, r, is the probability that Rolf plays row 1 on any specific play of the game, and the second coordinate, $1 - r$, is the probability that he plays row 2. The problem for Rolf is to decide which choice of r gives him the "best" mixed strategy. In general, to *solve* a game is to find the best mixed strategies for both players and to find their expected return per game if both use these strategies.

The goals of this section are to illustrate the concept of an optimal mixed strategy and to show how to obtain such a strategy in some special situations. If one of the two players has only two pure strategies (or if by using dominance the game can be reduced to one in which one of the players has two pure strategies), then a solution for the game can be found by a relatively simple technique which involves the graphs of lines. We initially describe this method for a game where the row player has only two pure strategies. Later we will indicate how the method can be adapted to a game in which the column player has two pure strategies. If a game has a payoff matrix with m rows and n columns, we say the game is $m \times n$. Thus, in this section we solve $2 \times n$ and $m \times 2$ games.

A Technique for Solving 2 × *n* Games

We study an example first and then the general method.

Example 11.6 Find a solution for the two-person zero-sum game with the following payoff matrix.

$$\mathbf{P} = \begin{bmatrix} 1 & -1 & 2 \\ -1 & 2 & -2 \end{bmatrix}$$

Solution We first note that this game does not have a saddle point and that it cannot be reduced by dominance. Next we let $\mathbf{S}(r) = [r \quad 1-r]$ be an arbitrary mixed strategy for the row player. Thus, $0 \leq r \leq 1$. We consider the average payoff to the row player when he uses $\mathbf{S}(r)$ and the column player uses pure strategy c_i (column i) on every play of the game. This payoff depends on i and on r, and we denote it by $p_i(r)$. Using the definition of expected value (Section 4.2), the expected return to the row player when he uses $\mathbf{S}(r)$ and the column player uses c_1 is

$$p_1(r) = 1(r) + (-1)(1 - r) = 2r - 1$$

This formula for $p_1(r)$ is obtained by noting that the row player plays row 1 with probability r and hence receives payoff $+1$ with probability r (see column 1 of the payoff matrix **P**). Similarly, the row player plays row 2 with probability $1 - r$ and thus obtains payoff -1 with this probability.

In the same way as the formula for $p_1(r)$ was obtained, the formulas for $p_2(r)$ and $p_3(r)$ are

$$p_2(r) = -1(r) + 2(1 - r) = -3r + 2$$
$$p_3(r) = 2(r) + (-2)(1 - r) = 4r - 2$$

where $p_2(r)$ and $p_3(r)$ represent, of course, the expected payoff to the row player when he uses mixed strategy $\mathbf{S}(r)$ and the column player uses pure strategies c_2 and c_3, respectively.

The functions p_1, p_2, and p_3 are linear in r, and their graphs are shown in Figure 11.1. The horizontal axis is the r axis, and the vertical axis represents the expected payoffs to the row player. If we take the point of view of the row player, then for each choice of r, the values $p_1(r)$, $p_2(r)$, and $p_3(r)$ represent, respectively, the expected payoffs when the strategy $\mathbf{S}(r) = [r \quad 1-r]$ is played against each of the three pure strategies of the column player. For example, if $r = \frac{1}{4}$, then $p_1(r) = -\frac{1}{2}$, $p_2(r) = \frac{5}{4}$, and $p_3(r) = -1$. Thus, if the row player uses strategy $\mathbf{S}(\frac{1}{4}) = [\frac{1}{4} \quad \frac{3}{4}]$, then the column player can (by using strategy c_3 every time) hold the expected payoff to the row player to -1. Of course, the row player would like to make this expected payoff higher, and hence the row player will choose r in such a way as to achieve the maximum possible expected payoff, considering at all times that the row player has no control over the column player's choice.

From Figure 11.1, it is clear that in one sense the best choice of r is the value of r such that $p_1(r) = p_2(r)$. To be more precise, for this choice of r the minimum expected payoff to the row player is as large as possible. We claim that this choice of r is such that $\mathbf{S}(r) = [r \quad 1-r]$ is an optimal mixed strategy for the row player. That is, using it, the row player maximizes his expected gain not only

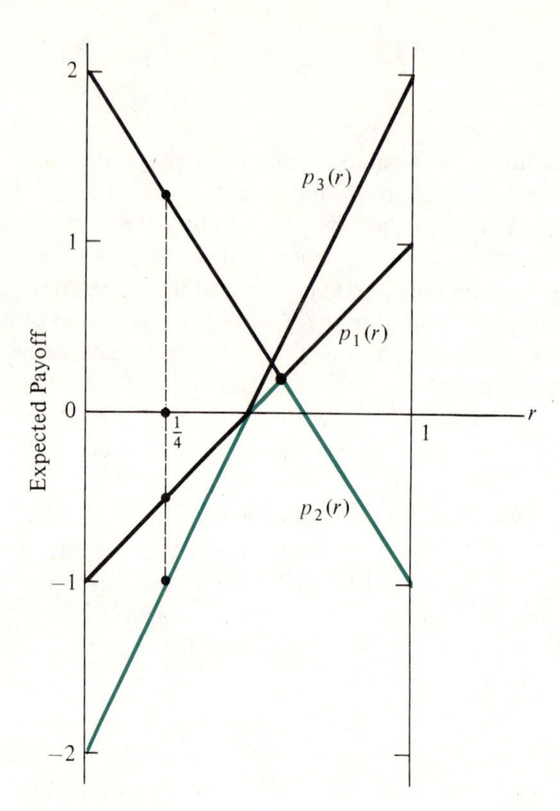

FIGURE 11.1

against the pure strategies of the column player but also against any mixed strategy of the column player. If the column player uses a pure strategy, then the expected payoff to the row player is exactly $p_i(r)$ for some i. If the column player uses a mixed strategy, then the payoff to the row player will still be between the highest and the lowest of the payoffs associated with pure strategies. In any case the payoff is at least equal to the smallest payoff associated with a pure strategy. A similar argument holds for any fixed value of r. This method of finding an optimal strategy is called *max-min* reasoning since the row player chooses r in such a way that the worst thing that can happen to him (i.e., the minimum expected payoff he can receive) is the best possible (i.e., a maximum). The row player *max*imizes his *min*imum expected payoff. The graph of the function which gives the minimum expected payoff for each value of r is shown by the colored line in Figure 11.1. It is clear that in this example the maximum value of this minimum payoff is attained when r is selected so that $p_1(r) = p_2(r)$, that is, with $r = \frac{3}{5}$. Thus we claim that $\mathbf{S}(\frac{3}{5}) = [\frac{3}{5} \quad \frac{2}{5}]$ is an optimal strategy for the row player.

To find an optimal strategy for the column player, we consider only the pure strategies c_1 and c_2. Indeed, if the row player is using the strategy $\mathbf{S}(\frac{3}{5}) = [\frac{3}{5} \quad \frac{2}{5}]$, then the payoff to the row player will be smaller if the column player uses c_1 and c_2

than if the column player uses c_3 [in Figure 11.1, $p_3(\frac{3}{5})$ is greater than $p_1(\frac{3}{5}) = p_2(\frac{3}{5})$]. Thus the reduced game has the payoff matrix

$$\begin{array}{cc} & \begin{array}{cc} c_1 & c_2 \end{array} \\ \begin{array}{c} r_1 \\ r_2 \end{array} & \begin{bmatrix} 1 & -1 \\ -1 & 2 \end{bmatrix} \end{array}$$

To obtain the optimal mixed strategy for the column player, we consider an arbitrary strategy $\mathbf{T}(c) = [c \quad 1-c]$, $0 \le c \le 1$, and we consider the expected payoff when this strategy is played against each row strategy. As before, this payoff is a function of c. We denote the expected payoff when strategy $\mathbf{T}(c)$ is played against pure strategy r_i by $q_i(c)$. These payoffs are

$$q_1(c) = 1(c) + (-1)(1-c) = 2c - 1$$
$$q_2(c) = -1(c) + 2(1-c) = -3c + 2$$

Finally, to obtain the value of c which gives an optimal mixed strategy for the column player, we solve for that value of c such that $q_1(c) = q_2(c)$. The desired value of c is $c = \frac{3}{5}$. Thus an optimal strategy for the column player is $[\frac{3}{5} \quad \frac{2}{5} \quad 0]$. The 0 in the third coordinate reflects the fact that pure strategy c_3 is never used.

The reasoning behind the use of the value of c determined by setting $q_1(c) = q_2(c)$ is analogous to the reasoning behind the selection of the value of r for an optimal strategy for the row player. Namely, one considers the graphs of functions q_1 and q_2 (lines) as shown in Figure 11.2. In this figure the interval $0 \le c \le 1$ on the horizontal axis represents the possible values of c which correspond to the mixed strategies available to the column player, and the vertical axis represents the expected payoffs to the row player. Thus for $c = \frac{1}{4}$, if the row player uses row 1 every time, the payoff to the row player is $-\frac{1}{2} = q_1(\frac{1}{4})$. On the other hand, if the row player uses row 2 every time, then the payoff to the row player is $\frac{5}{4} = q_2(\frac{1}{4})$. Since the column player makes these payoffs to the row

FIGURE 11.2

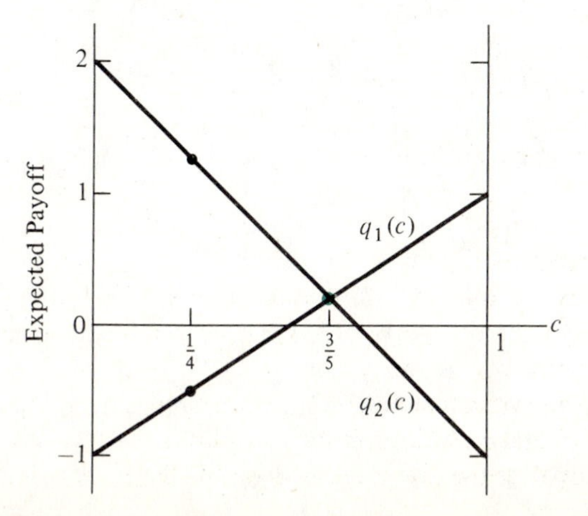

player (the game is zero-sum) and since the column player does not know which strategy the row player will use, the column player does not consider $c = \frac{1}{4}$ to be the best possible choice for minimizing the average gain to the row player. Indeed, to hold the average gain of the row player to a minimum and thus to maximize the long-term gain to the column player, the column player must choose c such that $q_1(c) = q_2(c)$. For any other choice of c, the column player runs the risk of being exploited by the row player and having to pay more than would be the case with $c = \frac{3}{5}$. ■

The discussion in Example 11.6 supports the claim that the optimal strategies for the row and column players are $\mathbf{S}^* = [\frac{3}{5} \quad \frac{2}{5}]$ and $\mathbf{T}^* = [\frac{3}{5} \quad \frac{2}{5} \quad 0]$, respectively. A rigorous proof that this is the case can be given.

The average payoff to the row player in this game, called the *value* of the game, is the number $v = p_1(\frac{3}{5}) = p_2(\frac{3}{5}) = q_1(\frac{3}{5}) = q_2(\frac{3}{5}) = \frac{1}{5}$.

The method described in Example 11.6 works for any two-person zero-sum game in which one player has only two pure strategies. Using Example 11.6 as a guide, we can outline a method of solving games with payoff matrices which are $2 \times n$. Before using this method, you should first check whether the game has a saddle point. If it does, then the solution can be obtained by using the methods of Section 11.2.

Method for Solving 2 × *n* Games by Using Lines

1. Denote an arbitrary strategy for the row player by $\mathbf{S}(r) = [r \quad 1-r]$, and denote the expected payoff when $\mathbf{S}(r)$ is played against column i by $p_i(r)$, $i = 1, 2, \ldots, n$. Graph the functions $p_i(r)$, $i = 1, 2, \ldots, n$, for $0 \le r \le 1$.
2. Find the maximum point on the polygonal line which bounds the graphs of $p_1, p_2, \ldots, p_n$ from below. Let k and l be defined by the property that the graphs of $p_k(r)$ and $p_l(r)$ lie on this polygon and intersect at the highest point on the polygon, $k < l$. If the payoff matrix does not have a saddle point, k and l are certain to exist.
3. Solve for r^* such that $p_k(r^*) = p_l(r^*)$. *An optimal strategy for the row player* is given by $\mathbf{S}^* = [r^* \quad 1-r^*]$.
4. Let $\mathbf{T}(c)$ be the mixed strategy for the column player which involves playing column k with probability c and column l with probability $1 - c$, $0 \le c \le 1$. Also let $q_i(c)$ be the payoff when $\mathbf{T}(c)$ is played against row i, $i = 1, 2$.
5. Solve for the value of c^* such that $q_1(c^*) = q_2(c^*)$. Since the payoff matrix does not have a saddle point, c^* will lie in the interval $0 \le c \le 1$. An optimal strategy for the column player is

$$\mathbf{T}^* = [0 \quad 0 \quad \cdots \quad 0 \quad \underbrace{c^* \quad 0}_{k\text{th coordinate}} \quad \cdots \quad 0 \quad \underbrace{1 - c^* \quad 0}_{l\text{th coordinate}} \quad \cdots \quad 0]$$

6. The value of the game (the expected return to the row player) is $v = p_k(r^*) = p_l(r^*) = q_1(c^*) = q_2(c^*)$.

A Technique for Solving $m \times 2$ Games

The method outlined above is applicable to a game with a payoff matrix which is $2 \times n$. To solve a game which has an $m \times 2$ payoff matrix, one simply interchanges the rows and columns of the payoff matrix, multiplies each entry by -1, and solves the new game by using the method given above (the new game is $2 \times m$). If $\tilde{\mathbf{S}}$ and $\tilde{\mathbf{T}}$ are optimal for the new game and if the value is $\tilde{v}$, then $\mathbf{T}^* = \tilde{\mathbf{S}}$, $\mathbf{S}^* = \tilde{\mathbf{T}}$, and $v = -\tilde{v}$ give a solution of the original game. The next example illustrates this method.

Example 11.7 Find a solution for the two-person zero-sum game with the following payoff matrix.

$$\mathbf{P} = \begin{bmatrix} -1 & 2 \\ 3 & -2 \\ 2 & -1 \end{bmatrix}$$

Solution We begin by multiplying each entry of the payoff matrix by -1 and interchanging rows and columns. This gives the payoff matrix $\tilde{\mathbf{P}}$ for a new game. The new game is 2×3, and

$$\tilde{\mathbf{P}} = \begin{bmatrix} 1 & -3 & -2 \\ -2 & 2 & 1 \end{bmatrix}$$

Using the method outlined above, we have $p_1(r) = 3r - 2$, $p_2(r) = -5r + 2$, and $p_3(r) = -3r + 1$. The graphs of these functions are shown in Figure 11.3.

The lines in Figure 11.3 are coincident at the point $(\frac{1}{2}, -\frac{1}{2})$ for which $r = \frac{1}{2}$. Hence an optimal strategy for the row player (in this new game) is $\tilde{\mathbf{S}} = [\frac{1}{2} \ \ \frac{1}{2}]$. The lines which bound the graph in Figure 11.3 from below correspond to strategies 1 and 2 for the column player. By using columns 1 and 2 and the method outlined above, an optimal strategy for the column player is given by $\tilde{\mathbf{T}} = [\frac{5}{8} \ \ \frac{3}{8} \ \ 0]$. The value of the new game is $\tilde{v} = -\frac{1}{2}$. Thus a solution of the original 3×2 game is given by $\mathbf{S}^* = [\frac{5}{8} \ \ \frac{3}{8} \ \ 0]$, $\mathbf{T}^* = [\frac{1}{2} \ \ \frac{1}{2}]$, and $v = \frac{1}{2}$. Other solutions can also be obtained (see Exercise 14 for examples). ■

The concept of a best mixed strategy can be made precise in a formal mathematical setting, but for our purposes such a precise formulation is unnecessary. Instead, as we noted above, it is sufficient to consider a mixed strategy to be best or optimal if the player using it is guaranteed an average return which is as large as or larger than the average return guaranteed by any other mixed strategy. A game is said to be *solved* when optimal strategies have been found for both players and when the average guaranteed return to both players has been determined.

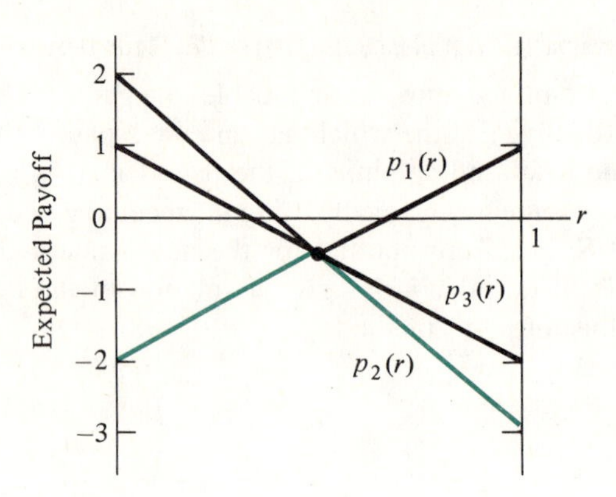

FIGURE 11.3

Optimal strategies need not be unique. In Example 11.7 the column player has several optimal strategies (actually infinitely many). But the value of a game is unique. Thus the payoff to the row player for all pairs of optimal strategies, one for each player, is the same number v, the value of the game.

Exercises for Section 11.3

Using the methods described in this section, solve the games with the payoff matrices shown in Exercises 1 to 12.

1. $\begin{bmatrix} 2 & -2 \\ -1 & 5 \end{bmatrix}$

2. $\begin{bmatrix} -8 & 10 \\ 4 & 0 \end{bmatrix}$

3. $\begin{bmatrix} 1 & -4 & 1 \\ -2 & 3 & -1 \end{bmatrix}$

4. $\begin{bmatrix} -4 & 5 & -1 \\ 2 & 0 & 1 \end{bmatrix}$

5. $\begin{bmatrix} 5 & 4 & 3 \\ -2 & 1 & 5 \end{bmatrix}$

6. $\begin{bmatrix} -1 & 4 \\ \frac{5}{2} & 1 \\ 2 & 2 \end{bmatrix}$

7. $\begin{bmatrix} -1 & 3 \\ 2 & -1 \\ 7 & -10 \end{bmatrix}$

8. $\begin{bmatrix} 3 & -1 \\ -2 & 2 \\ 1 & -3 \end{bmatrix}$

9. $\begin{bmatrix} 2 & 4 & 1 & 8 \\ 4 & 1 & 8 & 0 \end{bmatrix}$

10. $\begin{bmatrix} -3 & 2 \\ -1 & -1 \\ 3 & -4 \\ 0 & -2 \end{bmatrix}$

11. $\begin{bmatrix} -1 & 2 & -3 \\ 0 & 3 & -1 \\ -1 & 5 & -4 \end{bmatrix}$

12. $\begin{bmatrix} -2 & 1 & -2 & 3 \\ -1 & 2 & -2 & 4 \\ 6 & -3 & 4 & -1 \\ 2 & 1 & 0 & 2 \end{bmatrix}$

13. Suppose that the graphs of two functions $p_l(r)$ and $p_m(r)$ do not intersect for $0 \le r \le 1$. What can you conclude about the lth and mth columns of the payoff matrix?

14. In Example 11.7 find an optimal strategy for the row player which uses only rows 1 and 3.

Solve the two-person zero-sum games with the payoff matrices of Exercises 15 to 25.

15. $\begin{bmatrix} 0 & 1 & -1 & -2 \\ 2 & 1 & -1 & 2 \\ 2 & -3 & -2 & 3 \\ -4 & 4 & -3 & 5 \end{bmatrix}$

16. $\begin{bmatrix} -1 & -3 & 0 & 2 \\ 3 & 2 & -1 & 0 \\ 0 & -2 & 1 & 3 \\ 2 & 1 & 0 & 1 \end{bmatrix}$

17. $\begin{bmatrix} -2 & 2 & -3 \\ 6 & -4 & 5 \\ 6 & -5 & 4 \end{bmatrix}$

18. $\begin{bmatrix} 1 & -2 & 2 \\ 0 & 3 & 0 \\ 1 & -2 & 1 \end{bmatrix}$

19. $\begin{bmatrix} 5 & -1 & 1 & 5 & 3 \\ -2 & 1 & 2 & 6 & 1 \\ 3 & 1 & -2 & 2 & 2 \\ 6 & 2 & 1 & 4 & 2 \end{bmatrix}$

(This is the payoff matrix of Exercise 17, Section 11.2.)

20. $\begin{bmatrix} 4 & -2 & 3 & 5 & -2 \\ 1 & 1 & 0 & 4 & 1 \\ 1 & 3 & 4 & 1 & 2 \\ -1 & 2 & 2 & 0 & -1 \end{bmatrix}$

21. $\begin{bmatrix} 1 & 3 & 2 \\ 3 & 0 & 4 \\ 2 & 1 & 2 \\ 0 & 1 & 2 \\ 1 & 4 & 2 \\ 2 & 0 & 2 \end{bmatrix}$

22. $\begin{bmatrix} a & b \\ b & a \end{bmatrix}$ where $a < b$

23. $\begin{bmatrix} a & b \\ c & a \end{bmatrix}$ where $a < b < c$

24. $\begin{bmatrix} a & 2 & -1 \\ -2 & a & 2 \end{bmatrix}$ where $0 < a < 2$

25. $\begin{bmatrix} -a & b & -b \\ b & -a & a \end{bmatrix}$ where $0 < a < b < 2$

11.4 SOLVING $m \times n$ GAMES BY USING THE SIMPLEX METHOD

Every two-person zero-sum game has a solution in the sense described in Section 11.3. We discussed how a solution can be obtained in the special case of a game in which at least one of the players has only two pure strategies. In this section we will describe a method which can be used to solve an $m \times n$ game. The basic idea is simple. First, a two-person zero-sum game is converted into a linear programming problem (an SMP in the terminology of Chapter 10), and then the SMP and its dual are solved by using the simplex method (again, see Chapter 10). The solution of the SMP and its dual provide the optimal strategies for the column and row players, respectively. The method also yields the value of the game. We begin by illustrating the method with an example. Following the example we provide a summary of the steps for use in general situations.

Example 11.8 Solve the two-person zero-sum game with the following payoff matrix:

$$\mathbf{P} = \begin{bmatrix} 1 & 0 & 0 \\ -1 & 1 & -1 \\ 0 & -1 & 1 \end{bmatrix}$$

Solution We first observe that the payoff matrix of this game does not have a saddle point and that it cannot be reduced in size by dominance. Thus the technique of Section 11.3 cannot be used.

To solve this game by the simplex method, first we convert the payoff matrix to a matrix with all positive entries by adding the same suitable constant to every entry. In this case if we add 2 to every entry in **P**, we obtain a new matrix which has all positive entries. We denote this new matrix by $\mathbf{P}_2$ (the subscript 2 indicates that we added 2 to each entry in **P**):

$$\mathbf{P}_2 = \begin{bmatrix} 3 & 2 & 2 \\ 1 & 3 & 1 \\ 2 & 1 & 3 \end{bmatrix}$$

The second step in solving this game by the simplex method is to consider the SMP[$\mathbf{P}_2$, **B**, **C**], where **B** and **C** are column vectors with all coordinates equal to 1. The number of coordinates of **B** is the same as the number of rows of $\mathbf{P}_2$, and the number of coordinates of **C** is the same as the number of columns of $\mathbf{P}_2$. Since $\mathbf{P}_2$ is a 3×3 matrix, we have

$$\mathbf{B} = \begin{bmatrix} 1 \\ 1 \\ 1 \end{bmatrix} \quad \text{and} \quad \mathbf{C} = \begin{bmatrix} 1 \\ 1 \\ 1 \end{bmatrix}$$

Next we solve the SMP[$\mathbf{P}_2$, **B**, **C**] by using the simplex method. The tableaus used in the solution are shown in Tableaus 11.1 to 11.4.

TABLEAU 11.1

	↓ x	y	z	u	v	w	Basic Solution
→ u	③	2	2	1	0	0	1
v	1	3	1	0	1	0	1
w	2	1	3	0	0	1	1
p	−1	−1	−1	0	0	0	0

TABLEAU 11.2

	x	$\downarrow$ y	z	u	v	w	Basic Solution
x	1	$\frac{2}{3}$	$\frac{2}{3}$	$\frac{1}{3}$	0	0	$\frac{1}{3}$
$\rightarrow v$	0	$\textcircled{\frac{7}{3}}$	$\frac{1}{3}$	$-\frac{1}{3}$	1	0	$\frac{2}{3}$
w	0	$-\frac{1}{3}$	$\frac{5}{3}$	$-\frac{2}{3}$	0	1	$\frac{1}{3}$
p	0	$-\frac{1}{3}$	$-\frac{1}{3}$	$\frac{1}{3}$	0	0	$\frac{1}{3}$

TABLEAU 11.3

	x	y	$\downarrow$ z	u	v	w	Basic Solution
x	1	0	$\frac{4}{7}$	$\frac{9}{21}$	$-\frac{2}{7}$	0	$\frac{1}{7}$
y	0	1	$\frac{1}{7}$	$-\frac{1}{7}$	$\frac{3}{7}$	0	$\frac{2}{7}$
$\rightarrow w$	0	0	$\textcircled{\frac{12}{7}}$	$-\frac{5}{7}$	$\frac{1}{7}$	1	$\frac{3}{7}$
p	0	0	$-\frac{2}{7}$	$\frac{2}{7}$	$\frac{1}{7}$	0	$\frac{3}{7}$

TABLEAU 11.4

	x	y	z	u	v	w	Basic Solution
x	1	0	0	$\frac{8}{7}$	$-\frac{1}{3}$	$-\frac{1}{3}$	0
y	0	1	0	$-\frac{1}{12}$	$\frac{5}{15}$	$-\frac{1}{12}$	$\frac{1}{4}$
z	0	0	1	$-\frac{5}{12}$	$\frac{1}{12}$	$\frac{7}{12}$	$\frac{1}{4}$
p	0	0	0	$\frac{1}{6}$	$\frac{1}{6}$	$\frac{1}{6}$	$\frac{1}{2}$

Since the last row of Tableau 11.4 has no negative entries, the SMP[$\mathbf{P}_2$, **B**, **C**] and its dual have been solved. These two problems are the following:

SMP[$\mathbf{P}_2$, **B, C**]	**Dual**
Find the maximum of $x + y + z$ for $x \geq 0$, $y \geq 0$, $z \geq 0$, and $3x + 2y + 2z \leq 1$ $x + 3y + z \leq 1$ $2x + y + 3z \leq 1$	Find the minimum of $r + s + t$ for $r \geq 0$, $s \geq 0$, $t \geq 0$, and $3r + s + 2t \geq 1$ $2r + 3s + t \geq 1$ $2r + s + 3t \geq 1$

According to Tableau 11.4, an optimal vector for the SMP[$\mathbf{P}_2$, **B**, **C**] is $[x \quad y \quad z] = [0 \quad \frac{1}{4} \quad \frac{1}{4}]$, and an optimal vector for the dual problem is $[r \quad s \quad t] = [\frac{1}{6} \quad \frac{1}{6} \quad \frac{1}{6}]$. The relationship between these optimal vectors and optimal strategies for the players in the game with the payoff matrix **P** is as follows:

1. Let M be the sum of the coordinates of the optimal vector for the SMP[$\mathbf{P}_2$, **B**, **C**], $M = 0 + \frac{1}{4} + \frac{1}{4} = \frac{1}{2}$. An optimal strategy for the column player is

$$\mathbf{T} = \begin{bmatrix} \frac{0}{M} & \frac{\frac{1}{4}}{M} & \frac{\frac{1}{4}}{M} \end{bmatrix} = \begin{bmatrix} 0 & \frac{1}{2} & \frac{1}{2} \end{bmatrix}$$

and an optimal strategy for the row player is

$$= \begin{bmatrix} \frac{\frac{1}{6}}{M} & \frac{\frac{1}{6}}{M} & \frac{\frac{1}{6}}{M} \end{bmatrix} = \begin{bmatrix} \frac{1}{3} & \frac{1}{3} & \frac{1}{3} \end{bmatrix}$$

2. The value of the game with the payoff matrix **P** is $v = (1/M) - k$, where k is the amount added to the entries of **P** to make each entry positive. Here, $k = 2$ and $M = \frac{1}{2}$, so $v = 2 - 2 = 0$.

Thus, according to statements 1 and 2, the game with payoff matrix **P** has the following solution: The optimal strategy for the row player is the mixed strategy $[\frac{1}{3} \quad \frac{1}{3} \quad \frac{1}{3}]$, the optimal strategy for the column player is the mixed strategy $[0 \quad \frac{1}{2} \quad \frac{1}{2}]$, and the value of the game is 0. ■

The approach described in Example 11.8 can be used to solve any two-person zero-sum game. The method can be summarized as follows:

The Solution of Games by Using the Simplex Method

1. Convert the payoff matrix **P** for the game to a payoff matrix with all positive entries by adding the same positive constant k to each entry. Denote the new payoff matrix by $\mathbf{P}_k$.
2. Use the simplex method to solve the SMP[$\mathbf{P}_k$, **B**, **C**], where **B** and **C** are vectors with each coordinate having value 1. The number of coordinates of **B** is the same as the number of rows of $\mathbf{P}_k$, and the number of coordinates of **C** is the same as the number of columns of $\mathbf{P}_k$.
3. Let the optimal vector obtained for the SMP[$\mathbf{P}_k$, **B**, **C**] be $\mathbf{X} = [x_1 \quad x_2 \quad \cdots \quad x_n]$, and let the optimal vector for the dual be $\mathbf{Z} = [z_1 \quad z_2 \quad \cdots \quad z_m]$. Also let $M = x_1 + x_2 + \cdots + x_n$. Since the SMP[$\mathbf{P}_k$, **B**, **C**] maximizes $x_1 + \cdots + x_n$ and the dual problem minimizes $z_1 + \cdots + z_m$, we also have $M = z_1 + z_2 + \cdots + z_m$ (see Section 10.4).
4. An optimal strategy for the row player in the game with payoff matrix **P** is $[z_1/M \quad z_2/M \quad \cdots \quad z_m/M]$. An optimal strategy for the column player is $[x_1/M \quad x_2/M \quad \cdots \quad x_n/M]$. The value of the game is $v = (1/M) - k$.

This technique of solving a game by using the simplex method can be applied to any two-person zero-sum game. Thus a game in which at least one of the players has exactly two pure strategies can be solved either by using the method of Section 11.3 or by using the simplex method.

Example 11.9 Solve the two-person zero-sum game with the following payoff matrix:

$$\mathbf{P} = \begin{bmatrix} -1 & 0 & -2 & 1 \\ 0 & -1 & 0 & 2 \\ 2 & 1 & -1 & 2 \end{bmatrix}$$

Solution Although **P** does not have a saddle point, there is dominance in the matrix. In particular, row 3 dominates row 1, and column 3 dominates columns 1 and 4. Consequently, we need only consider the game with the reduced payoff matrix

$$\tilde{\mathbf{P}} = \begin{matrix} & c_2 & c_3 \\ r_2 & -1 & 0 \\ r_3 & 1 & -1 \end{matrix}$$

Since $\tilde{\mathbf{P}}$ has two rows, we could use the method of Section 11.3. Instead, we use the simplex method in the way outlined above.

Beginning with step 1, we seek a constant k such that when k is added to each entry in the payoff matrix, the result is a matrix with all positive entries. In this case $k = 2$ will do. Adding 2 to each entry in $\tilde{\mathbf{P}}$, we obtain $\tilde{\mathbf{P}}_2$:

$$\tilde{\mathbf{P}}_2 = \begin{bmatrix} 1 & 2 \\ 3 & 1 \end{bmatrix}$$

Proceeding to step 2, we use the simplex method to solve SMP[$\tilde{\mathbf{P}}_2$, **B**, **C**], where **B** and **C** are column 2-vectors with both coordinates having value 1. The tableaus which arise in applying the simplex method are Tableaus 11.5 to 11.7.

TABLEAU 11.5

	↓ x	y	u	v	Basic Solution
u	1	2	1	0	1
→ v	③	1	0	1	1
p	−1	−1	0	0	0

TABLEAU 11.6

	x	$\downarrow$ y	u	v	Basic Solution
$\rightarrow u$	0	$\textcircled{\frac{5}{3}}$	1	$-\frac{1}{3}$	$\frac{2}{3}$
x	1	$\frac{1}{3}$	0	$\frac{1}{3}$	$\frac{1}{3}$
p	0	$-\frac{2}{3}$	0	$\frac{1}{3}$	$\frac{1}{3}$

TABLEAU 11.7

	x	y	u	v	Basic Solution
y	0	1	$\frac{3}{5}$	$-\frac{1}{5}$	$\frac{2}{5}$
x	1	0	$-\frac{1}{5}$	$\frac{2}{5}$	$\frac{1}{5}$
p	0	0	$\frac{2}{5}$	$\frac{1}{5}$	$\frac{3}{5}$

According to steps 3 and 4 of our solution process, $M = \frac{3}{5}$, and optimal strategies for the row and column player are, respectively,

$$\mathbf{S} = \begin{bmatrix} \frac{\frac{2}{5}}{\frac{3}{5}} & \frac{\frac{1}{5}}{\frac{3}{5}} \end{bmatrix} = \begin{bmatrix} \frac{2}{3} & \frac{1}{3} \end{bmatrix}$$

$$\mathbf{T} = \begin{bmatrix} \frac{\frac{1}{5}}{\frac{3}{5}} & \frac{\frac{2}{5}}{\frac{3}{5}} \end{bmatrix} = \begin{bmatrix} \frac{1}{3} & \frac{2}{3} \end{bmatrix}$$

The value of the game is (recall that $k = 2$)

$$v = \frac{1}{M} - k = \frac{5}{3} - 2 = -\frac{1}{3}$$

However, remember that these optimal strategies are for the game with payoff matrix $\tilde{\mathbf{P}}_2$. Since $\tilde{\mathbf{P}}_2$ was obtained from $\mathbf{P}$ by using dominance, optimal strategies for the row and column players of the original game are, respectively,

$$\mathbf{S}^* = [0 \quad \tfrac{2}{3} \quad \tfrac{1}{3}] \qquad \mathbf{T}^* = [0 \quad \tfrac{1}{3} \quad \tfrac{2}{3} \quad 0]$$

■

Exercises for Section 11.4

Exercises 1 through 8 give payoff matrices for two-person zero-sum games. Solve each game by using both the simplex method and the graphical method of Section 11.3.

1. $\begin{bmatrix} 2 & 5 \\ 4 & 2 \end{bmatrix}$

2. $\begin{bmatrix} 4 & -2 \\ 0 & 8 \end{bmatrix}$

3. $\begin{bmatrix} -3 & 3 \\ 1 & -3 \end{bmatrix}$

4. $\begin{bmatrix} 4 & -2 \\ 2 & 8 \end{bmatrix}$

5. $\begin{bmatrix} 6 & 8 & -2 \\ 4 & -4 & 6 \end{bmatrix}$

6. $\begin{bmatrix} 1 & -3 \\ -5 & 1 \\ -1 & 3 \end{bmatrix}$

7. $\begin{bmatrix} 1 & -1 & 3 \\ 0 & 2 & -2 \\ -1 & -1 & 1 \end{bmatrix}$

8. $\begin{bmatrix} 1 & 3 & -4 \\ -2 & 0 & 4 \\ -1 & 1 & 2 \end{bmatrix}$

Exercises 9 through 14 give payoff matrices for two-person zero-sum games. Solve each game by using any method you choose.

9. $\begin{bmatrix} 1 & -1 & 2 \\ 2 & 1 & -1 \\ -1 & 0 & 1 \end{bmatrix}$

10. $\begin{bmatrix} 1 & -1 & 2 \\ 2 & -2 & 2 \\ -1 & 0 & 1 \end{bmatrix}$

11. $\begin{bmatrix} 2 & 1 & 2 & -5 \\ -1 & 4 & -2 & -1 \\ -3 & -2 & 7 & 1 \end{bmatrix}$

12. $\begin{bmatrix} 4 & -1 & 2 \\ 1 & 2 & 2 \\ 1 & 3 & 1 \end{bmatrix}$

13. $\begin{bmatrix} 3 & -1 & -4 \\ 2 & 0 & 1 \\ -4 & -1 & 3 \end{bmatrix}$

14. $\begin{bmatrix} 1 & -1 & 1 & -1 \\ -1 & 0 & 1 & 0 \\ 1 & 1 & 0 & -1 \end{bmatrix}$

15. Solve the two games with payoff matrices **P** and **Q** shown below. Use your results to propose a relationship for the effect on the solution of a game of multiplying the payoff matrix by a positive number.

$$\mathbf{P} = \begin{bmatrix} 2 & -3 \\ -1 & 3 \end{bmatrix} \qquad \mathbf{Q} = \begin{bmatrix} 20 & -30 \\ -10 & 30 \end{bmatrix}$$

16. Solve the two games with payoff matrices **P** and **Q** shown below.

$$\mathbf{P} = \begin{bmatrix} \frac{1}{2} & -\frac{1}{4} \\ -\frac{1}{2} & 1 \end{bmatrix} \qquad \mathbf{Q} = \begin{bmatrix} 2 & -1 \\ -2 & 4 \end{bmatrix}$$

How does the value of a game change if the payoff matrix is multiplied by a positive number?

17. Solve the game with payoff matrix **P** given below. (*Hint*: You may wish to use the information obtained in Exercise 16.)

$$\mathbf{P} = \begin{bmatrix} .1 & -.1 & .2 \\ -.1 & 0 & .1 \\ .2 & .1 & -.2 \end{bmatrix}$$

Investments

18. Two investment advisory services are competing for the opportunity to manage a part of a large pension fund. The managers of the pension fund have decided that they will assign funds to the two advisory services in the following way. At the beginning of the competition, each advisory service will make investment decisions for 50 percent of the assets of the pension fund. On the first day of each investment quarter, the two services must deliver their advice for the upcoming quarter. Their advice must be one of the following: buy stocks or buy bonds. If in a given quarter both services give the same advice, both give equally correct or incorrect advice, or the prices of stocks and bonds remain essentially the same, then no change will be made in the percentage of assets allocated to each service. However, if the services disagree and one is right (makes a good recommendation) and the other is wrong, then the service which gives the better recommendation increases its share of the assets by 10 percent of the total and the other loses the same amount.

    Question: Suppose that both services know that if a market change occurs in the present economic climate, then stocks will go up and bond prices will go down with probability .6, while the reverse situation will occur with probability .4. How should each service choose its advice? That is, how often should each advise to buy stocks? How often to buy bonds? (*Hint*: Use expected values as payoffs.)

Investments

19. Consider the setting described in Exercise 18, with the following changes. One of the advisory services, call it A, has an established relationship with the pension fund, and if both advisory services give equally correct advice, then service A is rewarded and the other service, call it Z, is penalized. If both services give incorrect advice, then neither is rewarded. As above, if one service gives correct advice and the other gives incorrect advice, then the firm giving correct advice is rewarded.

    Question: Suppose that both advisory services know that if a market change occurs in the present economic climate, then the probabilities for stock and bond price changes are as follows:

    Stocks up and bonds up with probability .50
    Stocks up and bonds down with probability .20
    Stocks down and bonds up with probability .15
    Stocks down and bonds down with probability .15

    How should each service choose its advice?

20. Solve the game with the payoff matrix:

$$\begin{bmatrix} 0 & 2 & 1 & 1 \\ 1 & 0 & 2 & 0 \\ 2 & 1 & 0 & 2 \end{bmatrix}$$

21. Solve the game with the payoff matrix

$$\begin{bmatrix} 1 & 4 & 3 \\ 2 & 1 & 4 \\ -2 & 1 & 5 \end{bmatrix}$$

22. Solve the game with the payoff matrix

$$\begin{bmatrix} 1 & 1 & 2 & 0 & 3 \\ 0 & 1 & 3 & 1 & 0 \\ 1 & 0 & 0 & 1 & 2 \end{bmatrix}$$

23. Solve the game with the payoff matrix

$$\begin{bmatrix} 0 & -2 & 2 & -1 \\ 2 & 0 & 0 & -2 \\ -2 & 0 & 0 & 2 \\ 1 & 2 & -2 & 0 \end{bmatrix}$$

24. Solve the game with the payoff matrix

$$\begin{bmatrix} 2 & -1 & 1 \\ 1 & 0 & 1 \\ -1 & 3 & 1 \\ 0 & 1 & 1 \end{bmatrix}$$

IMPORTANT TERMS AND CONCEPTS

You should be able to describe, define, or give examples of and use each of the following:

Two-person zero-sum game
Payoff
Payoff matrix
Pure strategy
Mixed strategy
Saddle point
Strictly determined game
Dominance
Optimal mixed strategy
Value of a game
Solution of a game
Solution by graphing
Solution by the simplex method

REVIEW EXERCISES

Find all saddle points for the games with the payoff matrices of Exercises 1 to 5.

1. $\begin{bmatrix} 3 & 1 & 2 \\ -1 & -2 & 1 \\ 2 & 0 & 4 \end{bmatrix}$

2. $\begin{bmatrix} 1 & 0 & 3 \\ 0 & -1 & 2 \\ -2 & 0 & 3 \end{bmatrix}$

3. $\begin{bmatrix} 7 & -1 & -2 \\ 4 & -3 & 6 \\ 2 & -1 & 3 \\ 6 & -3 & -4 \\ -5 & -1 & 3 \end{bmatrix}$

4. $\begin{bmatrix} 1 & 2 & 3 & 1 \\ -1 & 5 & -3 & 0 \\ 0 & -2 & 6 & -2 \\ 1 & 4 & 1 & 1 \end{bmatrix}$

5. $\begin{bmatrix} 1 & -1 & 2 & -3 & 5 \\ 3 & 0 & 1 & 0 & 4 \end{bmatrix}$

Use the technique based on graphing lines to solve the games with the payoff matrices of Exercises 6 to 10.

6. $\begin{bmatrix} 2 & -3 \\ -1 & 6 \end{bmatrix}$

7. $\begin{bmatrix} -1 & 3 & -2 & 4 \\ 2 & -1 & 3 & -2 \end{bmatrix}$

8. $\begin{bmatrix} 2 & -2 \\ -3 & 3 \\ 4 & -4 \\ -5 & 5 \\ 6 & -3 \end{bmatrix}$

9. $\begin{bmatrix} 1 & -1 & 2 & -3 \\ 3 & 0 & -2 & 0 \end{bmatrix}$

10. $\begin{bmatrix} 0 & 4 \\ 2 & -3 \\ -1 & -1 \\ 1 & 2 \\ -1 & -3 \end{bmatrix}$

Use the simplex method to solve the games with each of the payoff matrices of Exercises 11 to 15.

11. $\begin{bmatrix} 2 & -3 \\ -1 & 6 \end{bmatrix}$

12. $\begin{bmatrix} 2 & 1 & -4 \\ 3 & 2 & -4 \\ 5 & -2 & 5 \end{bmatrix}$

13. $\begin{bmatrix} 1 & -1 & 2 \\ -2 & 2 & -4 \\ 3 & -2 & 6 \end{bmatrix}$

14. $\begin{bmatrix} 1 & -1 & 2 & -3 \\ 3 & 0 & -2 & 0 \end{bmatrix}$

15. $\begin{bmatrix} 4 & -1 & 0 \\ -1 & -1 & 1 \\ 0 & 1 & -2 \end{bmatrix}$

Solve (using any method) the games with the payoff matrices of Exercises 16 to 24.

16. $\begin{bmatrix} 3 & -1 \\ -2 & 1 \end{bmatrix}$

17. $\begin{bmatrix} 3 & -2 & 1 \\ -1 & -1 & 0 \\ 3 & -1 & 1 \end{bmatrix}$

18. $\begin{bmatrix} 3 & -2 & 1 \\ -1 & 2 & 0 \\ 3 & -1 & 1 \end{bmatrix}$

19. $\begin{bmatrix} 3 & 3 & -1 & 5 \\ 1 & -1 & 5 & 2 \\ -2 & -5 & -3 & 5 \end{bmatrix}$

20. $\begin{bmatrix} 3 & -3 & -1 & 5 \\ 1 & -1 & 5 & 2 \\ -2 & -5 & -3 & 5 \end{bmatrix}$

21. $\begin{bmatrix} 1 & 2 & 0 & 3 \\ 3 & 1 & 1 & 1 \\ 2 & -1 & 2 & -1 \end{bmatrix}$

22. $\begin{bmatrix} 1 & -2 & 1 \\ 0 & 0 & -1 \\ -1 & 2 & 0 \end{bmatrix}$

23. $\begin{bmatrix} 3 & 0 & -1 \\ 1 & 1 & 1 \\ -1 & 2 & 1 \\ 2 & 1 & -1 \end{bmatrix}$

24. $\begin{bmatrix} 1 & 2 & 0 & 3 \\ 0 & 1 & 2 & 1 \\ 1 & 0 & 1 & 0 \end{bmatrix}$

CHAPTER 12

Logic for Finite Mathematics

12.0 THE SETTING AND OVERVIEW

In the process of using mathematics to analyze a situation arising in another field, one of the first steps is to formulate a specific problem. Usually this formulation is in words, normally the words used in describing the situation. The next step is to convert this problem expressed in words to one expressed in mathematical symbols and terms. Since this conversion is crucial to the conclusions drawn from the analysis, care must be taken that the mathematical problem represents the original problem as closely as possible. Since mathematical symbols and expressions have precise meanings, the mathematical problem can be stated very precisely, whereas the original problem may contain ambiguities resulting from the imprecision of some of the words used in stating it. In some situations it is possible to form several mathematical problems, each corresponding to a different interpretation of the words of the original problem. The analysis of some of these mathematical problems may result in worthwhile predictions while the analysis of others may not.

For instance, suppose that your automobile mechanic tells you, "Your heater needs a new switch and control or motor." Does he mean that you need either a new switch and control, on the one hand, or a new motor on the other; or does he mean that you need either a new switch and a new control, on the one hand, or a new switch and a new motor on the other? The ambiguity of the verbal statement makes it difficult to proceed to a decision.

In this chapter, we describe the formal method used in mathematics to translate verbal expressions into mathematical expressions. We also describe the method used in mathematics to assign a truth value (true or false) to a compound statement, based on the truth values of the simple statements in the compound statement. Finally, we discuss the important concepts of *implication* and *proof*.

The topics which we study in this chapter form only a very small part of the subject matter of the field of logic. However, these basic notions are fundamental for fields such as mathematics and law which require clear statements.

12.1 STATEMENTS, CONNECTIVES, AND NEGATION

The basic building blocks in our treatment of logic are simple statements. For us, a simple statement is a declarative sentence (such as "Eileen likes soccer.") which is either a true statement or a false statement. We do not allow statements which can be, in some sense, partly true or partly false. Of course, simple statements are not generally sufficient to express all ideas. Many statements of interest in mathematics are compound statements, i.e., statements which contain two or more simple statements. Thus, it is important for us to consider how simple statements can be combined to convey complicated ideas. In considering ways of combining simple statements to form compound statements, it is essential that we adopt strict conventions regarding the meanings of the connecting words. This is necessary to avoid problems, such as those in the example of switches, controls, and motors, which can arise from the semantics of a sentence.

Simple statements can be combined to form more complex or compound statements by using connecting words such as "and" and "or." The meaning of a compound statement depends on both the meanings of the simple statements used in it and the connectives used in forming it. One of our goals is to develop a method to determine whether a compound statement is true or false from a knowledge of the truth (or falsity) of each of the simple statements used in the compound statement. We do not attempt to solve the important and complex problem of deciding whether a simple statement is true or false. Instead, we will be satisfied with the assumption that each of the simple statements considered is known to be either a true statement or a false statement about some aspect of the real world. We then combine simple statements, using a few special connectives, and we determine the truth value of the resulting statement (i.e., decide if it is true or false) by agreeing to standard rules.

In the discussion of the details of the method, we adopt the notation that lowercase letters p, q, and r represent statements with known truth values.

The three logical concepts which we discuss here are the symbolic equivalents of the words "and," "or," and "not." These concepts will be denoted, respectively, by $\wedge$, $\vee$, and $\sim$, and they are defined in the following way:

1. $p \wedge q$ (read "p and q") is the compound statement which is true when both p and q are true and which is false otherwise.
2. $p \vee q$ (read "p or q") is the compound statement which is false when both p and q are false and which is true otherwise.
3. $\sim p$ (read "not p") is the statement which is false when p is true and true when p is false.

To illustrate the use of these three connectives, we consider examples of the process of converting statements in verbal form to symbolic form and vice versa.

Example 12.1 Consider the following three sentences:

(a) The blind man has a red hat or the blind man has a white hat.
(b) Mr. Robinson is the brakeman and Ms. Smith is the firefighter.
(c) It is not the case that the blind man has a red hat.

To write these equations in symbolic form, we let p, q, r, and s be defined as

$$\begin{aligned} p &= \text{The blind man has a red hat.} \\ q &= \text{The blind man has a white hat.} \\ r &= \text{Mr. Robinson is the brakeman.} \\ s &= \text{Ms. Smith is the firefighter.} \end{aligned}$$

Using the connectives $\wedge$ and $\vee$ and negation $\sim$, we see that the sentences given above can be represented in symbolic form as follows:

(a) $p \vee q$
(b) $r \wedge s$
(c) $\sim p$ ■

Our second example carries out the conversion process in reverse order, i.e., from symbolic to verbal.

Example 12.2 Let p and q be statements which are defined as follows:

p is the statement "The batch of vaccine has been tested."
q is the statement "The batch of vaccine contains no live viruses."

Then the following translations hold:

1. $p \wedge q$ is the statement "The batch of vaccine has been tested and it contains no live viruses." This compound statement is true only if both p and q are true statements.
2. $p \vee q$ is the statement that at least one of the two statements is true, namely, "Either the vaccine has been tested or the vaccine contains no live viruses." It is, of course, possible that both p and q are true, and this is compatible with $p \vee q$ being true.
3. $\sim p$ is the statement "The batch of vaccine has not been tested."
4. $p \wedge (\sim q)$ is the statement "The batch of vaccine has been tested, and it does contain live viruses." ■

When the connective $\wedge$ is used between two statements p and q, the result is a new statement $p \wedge q$. This new statement can be used together with the connective $\vee$ and the statement r to form a new statement $(p \wedge q) \vee r$. The parentheses about the statement $p \wedge q$ indicate that $p \wedge q$ is to be considered as a single statement. It is necessary to use these parentheses or some other means of indicating grouping because the statement $p \wedge q \vee r$ has no well-defined meaning. Indeed, this was illustrated in the example of switches, controls, and motors of Section 12.0. In general, the statements $(p \wedge q) \vee r$ and $p \wedge (q \vee r)$ are different statements, and it is necessary to distinguish between them. An easy rule of thumb to aid you in determining whether an expression is meaningful is that there must be a single statement, or a bracketed expression representing a single statement, on either side of $\wedge$ and $\vee$, and there must be a single statement, or the equivalent in bracketed form, to the right of the symbol $\sim$.

Example 12.3 Using the definitions of statements p, q, r, and s given in Example 12.1, convert the following verbal statements to symbolic form:

(*a*) Ms. Smith is the firefighter and Mr. Robinson is the brakeman, but the blind man does not have a red hat.

(*b*) Ms. Smith is not the firefighter and either the blind man does not have a red hat or he does not have a white hat.

Solution (*a*) "Ms. Smith is the firefighter and Mr. Robinson is the brakeman" is expressed symbolically as $s \wedge r$, while "the blind man does not have a red hat" is expressed as $\sim p$. Thus, the combined statement is expressed as $(s \wedge r) \wedge (\sim p)$.

(*b*) "Ms. Smith is not the firefighter" is expressed as $\sim s$, "the blind man does not have a red hat" is $\sim p$, and "the blind man does not have a white hat" is $\sim q$. Thus, the combined statement is expressed as $(\sim s) \wedge [(\sim p) \vee (\sim q)]$. ■

Exercises for Section 12.1

1. Using the definitions of statements p, q, and r given below, translate each of the following verbal expressions into symbolic form.

p = Your auto heater needs a new switch.
q = Your auto heater needs a new control.
r = Your auto heater needs a new motor.

(*a*) Your auto heater needs a new switch and a new control.
(*b*) Your auto heater needs a new switch and a new motor.
(*c*) Your auto heater needs a new switch and a new motor and a new control.

2. Using the same notation as Exercise 1, translate the following into symbolic form:
   (*a*) Your auto heater needs a new switch or a new control or a new motor.
   (*b*) Your auto heater needs a new switch, and either a new control or a new motor or both.

3. Using the definitions of statements p, q, and r given below, translate each of the following symbolic expressions into a verbal statement.

p = Peter is the president.
q = Mary is a mathematician.
r = Roger is a teacher.

   (*a*) $(p \vee q) \wedge r$
   (*b*) $p \vee (q \wedge r)$
   (*c*) $(\sim p) \vee (q \vee r)$

4. Using the definitions of p, q, and r given in Exercise 3, translate each of the following into a verbal expression.
   (*a*) $p \wedge (\sim q)$
   (*b*) $p \vee [(\sim q) \wedge r]$
   (*c*) $(p \wedge q) \vee (q \wedge r)$

5. Using the definitions of statements p, q, and r given below, translate each of the following statements into verbal form.

p = The stock market average went up.
q = The bond market average went up.
r = The inflation index went up.

   (*a*) $[(\sim p) \vee (\sim q)] \wedge r$ (*b*) $(p \wedge q) \wedge (\sim r)$
   (*c*) $[(\sim p) \wedge (\sim q)] \vee (\sim r)$

6. Using the definitions of p, q, and r given in Exercise 5, express each of the following verbal statements in symbolic form.
   (*a*) The stock market and bond market averages both went down, but the inflation index went up.
   (*b*) The stock market average went up or both the bond market average and the inflation index went down.

7. Let A, B, and C be subsets of a universal set U. Also let x be an element of U, and let p, q, and r be statements defined as follows:

p: x is an element of A
q: x is an element of B
r: x is an element of C

   Translate the following symbolic statements into verbal statements.
   (*a*) $(p \wedge q) \wedge r$ (*b*) $(p \vee q) \vee r$
   (*c*) $(p \wedge q) \wedge (\sim r)$ (*d*) $(p \vee q) \vee (\sim r)$

8. Using the setting of Exercise 7, translate the following verbal statements into symbolic form.
   (*a*) The element x is in the intersection of A and B, but it is not in C.
   (*b*) The element x is in the complement of the union of the sets A, B, and C.
9. Let A and B be subsets of a universal set U, and let x be an element of U. Also let p, q, and r be statements defined as follows:

   p: x is an element of A.
   q: x is not an element of B.
   r: x is not an element of A and x is not an element of B.

   Translate the following verbal statements into symbolic form, using the connectives $\wedge$, $\vee$, and $\sim$.
   (*a*) The element x is in the intersection of A and B.
   (*b*) The element x is in exactly one of the sets A and B.
10. Using the setting of Exercise 9, translate the following symbolic statements into verbal form.
   (*a*) $\sim q$ (*b*) $\sim r$
   (*c*) $(\sim p) \vee r$ (*d*) $p \wedge (\sim q)$

12.2 TRUTH TABLES

All but the very simplest problems which arise in the social and life sciences and in business involve several statements and connectives. Sorting out the precise information contained in a complex statement can be a challenging task, and we now turn to an organizational scheme which aids in analyzing complex statements. To begin, let us consider in detail the compound statement

$$[(p \wedge q) \vee (r \wedge q)] \wedge p$$

which utilizes the simple statements p, q, and r and the connectives $\wedge$ and $\vee$. According to our convention, this compound statement is false unless both statements p and $[(p \wedge q) \vee (r \wedge q)]$ are true. To determine the cases in which an expression such as $[(p \wedge q) \vee (r \wedge q)]$ is true in terms of the truth values of p, q, and r, there is a handy notational device called a *truth table*.

If we represent the truth values "true" and "false" by the letters T and F, then the relationship between the truth values of statements p and q and the truth value of the compound statement $p \wedge q$ can be represented as follows:

p	q	$p \wedge q$
T	T	T
T	F	F
F	T	F
F	F	F

In this table, each row displays a possible set of truth values for p and q and (on the right) the associated truth value of $p \wedge q$. For example, from row 2 we see that when p is true and q is false, then $p \wedge q$ is false.

Truth tables are a convenient and useful way of analyzing logical expressions to see when they are true and when they are false. Using the definitions of $\vee$ and $\sim$ given above, the associated truth tables are

p	q	$p \vee q$
T	T	T
T	F	T
F	T	T
F	F	F

p	$\sim p$
T	F
F	T

In constructing truth tables, it is important to keep in mind that all possible arrangements of truth values of the constituent statements must be considered. The truth table has one row for each arrangement of truth values. For example, the truth table for the expression $(p \wedge q) \vee r$ has eight rows and has the form

p	q	r	$p \wedge q$	$(p \wedge q) \vee r$
T	T	T	T	T
T	T	F	T	T
T	F	T	F	T
T	F	F	F	F
F	T	T	F	T
F	T	F	F	F
F	F	T	F	T
F	F	F	F	F

Thus, for the eight possible arrangements of truth values for p, q, and r, five result in the statement $(p \wedge q) \vee r$ being true and three result in it being false.

We now consider another example in which we represent compound statements in symbolic form.

Example 12.4 Let statements p, q, and r be as follows:

p = Patti likes hard rock.
q = Sam does not like motorcycles.
r = Rhonda likes loud noises.

Several compound statements formed from p, q, and r are given below. Immediately following each one is its symbolic form expressed with the letters p, q, and r and the symbols $\wedge$, $\vee$, and $\sim$.
(a) Patti likes hard rock and Sam likes motorcycles.

$$p \wedge (\sim q)$$

(b) Sam does not like motorcycles and Rhonda likes loud noises, but Patti does not like hard rock.

$$(q \wedge r) \wedge (\sim p)$$

(c) Either Patti likes hard rock or Rhonda likes loud noises, but not both.

$$(p \vee r) \wedge [\sim(p \wedge r)]$$

■

The symbol $\vee$ is called the *inclusive or*. In Example 12.4c we have shown how to represent the *exclusive or*, i.e., the symbol $\veebar$, which is defined so that $p \veebar q$ is true when p is true and q is false and when p is false and q is true. If p and q are both true or both false, then $p \veebar q$ is false. In this text the connective "or" will always be used in the inclusive sense.

Example 12.5 Let p, q, and r be statements. Construct a compound statement from p, q, and r which will be true when exactly two of the statements p, q, and r are true and one of these statements is false.

Solution The statement $u = (p \wedge q) \wedge (\sim r)$ is true when both p and q are true and r is false. Likewise, the statement $v = p \wedge (\sim q) \wedge r$ is true when q is false and both p and r are true, and the statement $w = (\sim p) \wedge (q \wedge r)$ is true when q and r are true and p is false. Since we are looking for a statement which satisfies one of these three conditions, we can obtain such a statement by taking the statement $(u \vee v) \vee w$. It is important to note that no two of the statements u, v, and w can be true at the same time, because each pair contains a contradiction, such as p and $\sim p$. The truth table for these statements is shown below.

p	q	r	$u = (p \wedge q) \wedge (\sim r)$	$v = p \wedge (\sim q) \wedge r$	$w = (\sim p) \wedge (q \wedge r)$	$(u \vee v) \vee w$
T	T	T	F	F	F	F
T	T	F	T	F	F	T
T	F	T	F	T	F	T
T	F	F	F	F	F	F
F	T	T	F	F	T	T
F	T	F	F	F	F	F
F	F	T	F	F	F	F
F	F	F	F	F	F	F

■

Connections with Set Theory

The basic operations of set theory (defined and discussed in Chapter 1) are intersection ($\cap$), union ($\cup$), and complement ($'$). These set operations are directly related to the local connectives. We have

$$A \cap B = \{x: (x \in A) \wedge (x \in B)\}$$
$$A \cup B = \{x: (x \in A) \vee (x \in B)\}$$
$$A' = \{x: \sim(x \in A)\}$$

Recall that in the definition of A', we assume that A is a subset of a universal set U, and we are considering only elements x of U.

Example 12.6 Let A, B, and C be subsets of a universal set U, and let x be an element of the set U. Use a truth table to show that $(A \cup B)'$ is equal to $A' \cap B'$.

Solution To show that $(A \cup B)'$ is equal to $A' \cap B'$, we need to show that these two sets contain exactly the same elements. In other words, the statements $x \in (A \cup B)'$ and $x \in A' \cap B'$ should be true at the same time and false at the same time. The truth table shown below verifies that this is the case.

$x \in A$	$x \in B$	$x \in (A \cup B)$	$x \in (A \cup B)'$	$x \in A'$	$x \in B'$	$x \in (A' \cap B')$
T	T	T	F	F	F	F
T	F	T	F	F	T	F
F	T	T	F	T	F	F
F	F	F	T	T	T	T

■

Exercises for Section 12.2

In Exercises 1 through 10, use a truth table to determine the truth values, if any, of p, q, r, and s for which the given statement is true.

1. (*a*) $(p \wedge q) \vee (\sim q)$
   (*b*) $(p \vee q) \wedge (\sim q)$
2. (*a*) $(p \wedge q) \vee [(\sim p) \wedge (\sim q)]$
   (*b*) $(p \vee q) \wedge [(\sim p) \vee (\sim q)]$
3. (*a*) $(p \vee q) \wedge r$
   (*b*) $(p \wedge q) \vee r$
4. (*a*) $(p \wedge q) \vee (p \wedge r)$
   (*b*) $(p \vee q) \wedge (p \vee r)$
5. (*a*) $(p \wedge q) \vee (r \wedge s)$
   (*b*) $(p \vee q) \wedge (r \vee s)$

6. (*a*) $[(p \wedge q) \wedge r] \vee s$
   (*b*) $[(p \vee q) \vee r] \wedge s$
7. (*a*) $[(\sim p) \wedge (\sim q)] \vee [(\sim r) \wedge (\sim s)]$
   (*b*) $\sim[(p \vee q) \wedge (r \vee s)]$
8. (*a*) $[p \wedge (\sim q)] \vee [r \wedge (\sim s)]$
   (*b*) $[q \vee (\sim p)] \wedge [s \vee (\sim r)]$
9. (*a*) $(p \wedge q) \vee (p \wedge r) \vee (p \wedge s)$
   (*b*) $p \wedge (q \vee r \vee s)$
10. (*a*) $(\sim p) \vee (\sim q) \vee (\sim r) \vee (\sim s)$
    (*b*) $\sim(p \vee q \vee r \vee s)$

For Exercises 11 through 16, use the notation of logical connectives to define the sets given.

11. (*a*) $A \cap (B \cap C)$ (*b*) $(A \cup B) \cup C$
12. (*a*) $A \cup (B \cap C)$ (*b*) $(A \cap B) \cup C$
13. (*a*) $A' \cap B$ (*b*) $A \cup B'$
14. (*a*) $A \cup (B \cap C)'$ (*b*) $(A \cap B)' \cup C$
15. (*a*) $(A' \cap B) \cap (C \cup A')$ (*b*) $(A' \cup B' \cup C')'$
16. (*a*) $A \cap (B' \cup C')$ (*b*) $A' \cup (B \cap C)$

For Exercises 17 through 20, use a truth table to show that the two sets given are the same set:

17. $A \cap (B \cup C)$ and $(A \cap B) \cup (A \cap C)$
18. $(A \cup B')'$ and $A' \cap B$
19. $A \cap (B' \cup C')'$ and $A \cap B \cap C$
20. $(A' \cup B')' \cup (A' \cup C')'$ and $A \cap (B \cup C)$

12.3 EQUIVALENCE, IMPLICATION, AND DEDUCTION

New mathematical results are established by careful reasoning and by logical deduction. The concepts from logic which are important for this process are logical equivalence, logical implication, and logical deduction.

Logical Equivalence

Two symbolic logical expressions are said to be *logically equivalent* or simply *equivalent* if they are true for the same truth values of their component statements and false for the same truth values of their component statements. For example, the expressions $(\sim p) \wedge (\sim q)$ and $\sim(p \vee q)$ are false in all cases except when p is false and q is false. Thus these expressions are logically equivalent. Logically equivalent statements may be interchanged for all matters which involve truth values.

Example 12.7 Show that the statements $(p \wedge q) \vee (\sim q)$ and $\sim[(\sim p) \wedge q]$ are logically equivalent.

Solution The truth tables for the two statements are as shown below:

p	q	$p \wedge q$	$\sim q$	$(p \wedge q) \vee (\sim q)$
T	T	T	F	T
T	F	F	T	T
F	T	F	F	F
F	F	F	T	T

p	q	$\sim p$	$(\sim p) \wedge q$	$\sim[(\sim p) \wedge q]$
T	T	F	F	T
T	F	F	F	T
F	T	T	T	F
F	F	T	F	T

Since for each pair of truth values for p and q, the first two columns of each table, the truth values of the two statements are the same, i.e., the last column of each table, we conclude that the two statements are logically equivalent.

Implication

Mathematics can be considered to be the study of conditional statements about mathematical concepts and symbols. For example, in euclidean geometry one considers the conditional statement "If a triangle T has an interior angle of 90 degrees, then the sum of the two remaining interior angles of T is 90 degrees."

Statements which have the form "if p, then q," where p and q are statements, are called *conditional statements*. Such statements are represented in symbolic terms by writing $p \rightarrow q$. Just as the symbols $\wedge$ and $\vee$ are used to connect statements to form new compound statements, so is the symbol $\rightarrow$ used as a connective. The compound statement $p \rightarrow q$ is true or false depending upon the truth of p and q. The truth table for $p \rightarrow q$ follows:

p	q	$p \rightarrow q$
T	T	T
T	F	F
F	T	T
F	F	T

Logical Deductions

The conditional symbol $\rightarrow$ is very important in the symbolic representation of verbal statements which form a deduction or argument. For example, consider the deduction:

If the drug company releases a vaccine with live viruses, then many people will become ill from the company's vaccine.

If many people become ill from the company's vaccine, then the drug company will lose its license.

Therefore, if the drug company releases a vaccine with live viruses, then the drug company will lose its license.

Symbolically, this argument has the form

$$\begin{array}{c} p \rightarrow q \\ q \rightarrow r \\ \hline \therefore p \rightarrow r \end{array}$$

where p, q, and r are appropriately defined statements, and $\therefore$ means "therefore."

An argument such as this one is said to be *valid* if whenever the hypotheses are true (here $p \rightarrow q$ and $q \rightarrow r$), then the conclusion is true (here $p \rightarrow r$). Equivalently, the argument is valid if the compound statement $[(p \rightarrow q) \wedge (q \rightarrow r)] \rightarrow (p \rightarrow r)$ is true for all truth values of p, q, and r.

Example 12.8 Construct a truth table to show that the argument "if $p \rightarrow q$ and $q \rightarrow r$, then $p \rightarrow r$" is a valid argument.

Solution The truth table shown below verifies that the argument is valid by showing that for all truth values of the simple statements p, q, r, the compound statement $[(p \rightarrow q) \wedge (q \rightarrow r)] \rightarrow (p \rightarrow r)$ is true.

p	q	r	$p \rightarrow q$	$q \rightarrow r$	$p \rightarrow r$	$(p \rightarrow q) \wedge (q \rightarrow r)$	$[(p \rightarrow q) \wedge (q \rightarrow r)] \rightarrow (p \rightarrow r)$
T	T	T	T	T	T	T	T
T	T	F	T	F	F	F	T
T	F	T	F	T	T	F	T
T	F	F	F	T	F	F	T
F	T	T	T	T	T	T	T
F	T	F	T	F	T	F	T
F	F	T	T	T	T	T	T
F	F	F	T	T	T	T	T

■

In addition to showing that certain arguments are valid, truth tables are important in showing that some arguments are *not* valid.

Example 12.9 Two arguments are given below. Use truth tables to show that one is valid and the other is not.

Argument 1	Argument 2
$p \to q$	$p \to q$
p	q
$\therefore q$	$\therefore p$

Solution The truth tables for the two arguments are shown below. Note that for argument 1, all entries in the column of the compound statement $[(p \to q) \wedge p] \to q$ are true, and consequently argument 1 is valid. In the truth table for argument 2, however, there is one false entry in the column of the compound statement $[(p \to q) \wedge q] \to p$, and thus argument 2 is not valid.

ARGUMENT 1

p	q	$p \to q$	$(p \to q) \wedge p$	$[(p \to q) \wedge p] \to q$
T	T	T	T	T
T	F	F	F	T
F	T	T	F	T
F	F	T	F	T

ARGUMENT 2

p	q	$p \to q$	$(p \to q) \wedge q$	$[(p \to q) \wedge q] \to p$
T	T	T	T	T
T	F	F	F	T
F	T	T	T	F
F	F	T	F	T

■

Example 12.10 Express the following verbal argument in symbolic form, and then decide if the argument is valid:

If the temperature is below zero, then Manuel's car will not start and he has to walk to work. Manuel had to walk to work today; therefore, it is below zero today.

Solution Let p, q, and r be defined as follows:

p = The temperature is below zero.
q = Manuel's car will not start.
r = Manuel has to walk to work.

Given these definitions, the verbal argument has the following symbolic form: $p \to (q \wedge r)$ and r, therefore p.

The truth table for this argument is shown below, and since there is a false entry in the column under $[[p \to (q \wedge r)] \wedge r] \to p$, we see that the argument is not valid.

p	q	r	$q \wedge r$	$p \to (q \wedge r)$	$[p \to (q \wedge r)] \wedge r$	$[[p \to (q \wedge r)] \wedge r] \to p)$
T	T	T	T	T	T	T
T	T	F	F	F	F	T
T	F	T	F	F	F	T
T	F	F	F	F	F	T
F	T	T	T	T	T	F
F	T	F	F	T	F	T
F	F	T	F	T	F	T
F	F	F	F	T	F	T

■

Exercises for Section 12.3

For Exercises 1 through 6 show that the two statements given are logically equivalent:

1. $(\sim p) \vee q$ and $p \to q$
2. $p \wedge (\sim q)$ and $\sim(p \to q)$
3. $\sim(p \wedge q)$ and $(\sim p) \vee (\sim q)$
4. $[p \wedge (\sim q)] \vee q$ and $p \vee q$
5. $[p \vee (\sim q)] \wedge q$ and $p \wedge q$
6. $p \to q$ and $(\sim q) \to (\sim p)$

For Exercises 7 through 12, decide whether the argument given is a valid argument.

7. If $p \wedge q$, then $p \vee q$.
8. If $p \vee q$, then $p \wedge q$.
9. If $p \vee q$ and $q \vee r$, then $(p \wedge r) \vee q$.
10. If $p \wedge q$ or $q \wedge r$, then $(p \vee r) \wedge q$.

11. If $p \to (q \to r)$, then $(p \vee q) \to r$.
12. If $p \to (q \to r)$, then $(p \wedge q) \to r$.

For Exercises 13, 14, and 15, express the argument in symbolic form and decide whether the argument is valid.

13. If Sam studies in his mathematics course, he will get at least a grade of B; and if he studies in his physics course, he will get at least a C. Sam received a C in both mathematics and physics. Therefore, Sam did not study in both courses.
14. If Sam studies in his mathematics course, he will get at least a B; and if he studies in his physics course, he will get at least a C. Sam received a C in both mathematics and physics. Therefore, Sam studied in physics, but not in mathematics.
15. If Sam studies in his mathematics course, he will get at least a B; and if he studies in his physics course, he will get at least a C. Sam received a B in both mathematics and physics. Therefore, Sam studied in both courses.

IMPORTANT TERMS AND CONCEPTS

You should be able to describe, define, or give examples of and use each of the following:

Logical connective "and"
Logical connective "or"
Logical connective "not"
Truth table
Equivalence
Implication
Valid argument

REVIEW EXERCISES

1. Using the definitions of statements p, q, and r given below, translate each of the following verbal statements into symbolic form.

p = Paul likes jazz.
q = Sally likes rock.
r = Rob likes rap.

(*a*) Paul likes jazz and Sally likes rock, but Rob does not like rap.
(*b*) Either Paul likes jazz or Rob likes rap.
(*c*) Paul does not like jazz, but either Sally likes rock or Rob likes rap.

2. Using the definitions of Exercise 1, translate the following symbolic statements into verbal form.
(*a*) $p \vee (q \wedge r)$
(*b*) $(p \vee q) \wedge r$
(*c*) $p \to (q \wedge r)$

3. Using the definitions of statements p, q, and r given below, translate each of the following verbal statements into symbolic form.

p = Amy is a mathematics teacher.
q = Beth is a basketball player.
r = Carol is a clarinet player.

(*a*) Amy is not a mathematics teacher and Beth is not a basketball player, but Carol is a clarinet player.
(*b*) Either Amy is a mathematics teacher or both Beth is a basketball player and Carol is a clarinet player.
(*c*) At least one of the following three statements is not true: "Amy is a mathematics teacher," "Beth is a basketball player," and "Carol is a clarinet player."

4. Using the definitions of Exercise 1, translate the following symbolic statements into verbal form.
(*a*) $p \wedge (q \rightarrow r)$
(*b*) $(p \wedge q) \rightarrow r$
(*c*) $(\sim p) \vee (\sim q) \vee (\sim r)$

5. Which truth values of p, q, and r in Exercise 1 make the statement in (*a*) of that exercise a true statement? Statement (*b*)? Statement (*c*)? Which truth values of p, q, and r make both (*a*) true and (*b*) false?

6. Which truth values of p, q, and r in Exercise 2 make both statements (*a*) and (*b*) of that exercise true? Which values make both (*a*) false and (*b*) true?

7. Which truth values of p, q, and r of Exercise 3 make the statement (*a*) of that exercise a true statement? Statement (*b*)? Statement (*c*)? Which values make both (*a*) false and (*b*) true?

8. Which truth values of p, q, and r make the following expressions true statements?
(*a*) $(\sim p) \wedge q$
(*b*) $p \vee (\sim q)$
(*c*) $(p \wedge q) \vee r$

9. Which truth values of p, q, and r make the following expressions true statements?
(*a*) $p \vee [(\sim q) \wedge (\sim r)]$
(*b*) $[p \vee (\sim q)] \wedge [p \vee (\sim r)]$

10. Define the letters p, q, and r to be statements used in the following sentences; then express these sentences in symbolic form, using p, q, r, and $\wedge$, $\vee$, $\sim$.
(*a*) Pauline likes piano music, but Quinn does not.
(*b*) Quinn and Rex like piano music, but Pauline does not.

11. Repeat Exercise 10 for these sentences.
(*a*) Neither Pauline nor Quinn nor Rex likes piano music.
(*b*) Either Quinn or Rex likes piano music, but both do not, and Pauline does not.

12. Construct truth tables for the symbolic expressions representing parts (*a*) and (*b*) of Exercise 10.

13. Construct truth tables for the symbolic expressions representing parts (*a*) and (*b*) of Exercise 11.

14. Use the notation of logical connectives to define the following sets.
(*a*) $(A \cup B) \cap C$
(*b*) $(A \cap B) \cup C'$
(*c*) $(A' \cap B) \cup C$

15. Show that the following pairs of logical expressions are equivalent.
    (*a*) $\sim(p \wedge q)$ and $(\sim p) \vee (\sim q)$
    (*b*) $p \wedge (q \vee r)$ and $(p \wedge q) \vee (p \wedge r)$
16. Show that the following pairs of logical expressions are equivalent.
    (*a*) $(p \wedge q) \vee r$ and $(p \vee r) \wedge (q \vee r)$
    (*b*) $\{\sim[(\sim p) \wedge (\sim q)]\} \vee r$ and $p \vee q \vee r$
17. Decide which, if any, of the following logical expressions are equivalent.
    (*a*) $\sim[(\sim p) \vee q]$
    (*b*) $[p \vee (\sim q)] \wedge (p \vee r)$
    (*c*) $p \vee [(\sim q) \wedge r)]$
    (*d*) $p \wedge (\sim q)$
18. Using the definition of $\rightarrow$, construct truth tables for the following compound statements.
    (*a*) $p \rightarrow (\sim q)$
    (*b*) $(\sim p) \rightarrow q$
19. Repeat Exercise 18 for these statements:
    (*a*) $(p \vee q) \rightarrow p$
    (*b*) $(p \wedge q) \rightarrow p$
    (*c*) $[\sim(p \vee q)] \rightarrow [(\sim p) \vee (\sim q)]$
20. Repeat Exercise 18 for these statements:
    (*a*) $p \rightarrow p \vee q$
    (*b*) $\{p \wedge [p \rightarrow (\sim q)]\} \rightarrow (\sim q)$
21. Let p, q, and r be the statements

p = She is an accountant.
q = She does not like to work with numbers.
r = She is unemployed.

Translate each of the following symbolic statements into a verbal expression:
    (*a*) $(p \wedge q) \rightarrow r$
    (*b*) $p \rightarrow (\sim q)$
    (*c*) $(p \vee r) \rightarrow q$
22. Use a truth table to show that the argument given below is valid.

$$\begin{array}{c} p \rightarrow q \\ q \rightarrow r \\ \underline{r \rightarrow s} \\ \therefore p \rightarrow s \end{array}$$

23. Use truth tables to decide which, if any, of the following arguments are valid.

(*a*) $$\begin{array}{c} p \rightarrow q \\ \underline{q \rightarrow r} \\ \therefore r \rightarrow q \end{array}$$

(*b*) $$\begin{array}{c} \underline{p \rightarrow q} \\ \therefore (\sim q) \rightarrow (\sim p) \end{array}$$

(*c*) $$\begin{array}{c} p \rightarrow q \\ \underline{\sim p} \\ \therefore \sim q \end{array}$$

GLOSSARY

Absorbing Markov chain A Markov chain with at least one absorbing state and in which it is possible to reach some absorbing state from any state.

Absorbing state in a Markov chain A state for which the probability of a transition back to itself is 1.

Adjacency matrix of a graph A matrix whose rows and columns are identified with the vertices of a graph and whose entries tell whether there is an edge between two vertices.

Amortization Method of repaying a debt with equal payments over a specified period of time.

Annual percentage rate (APR) The rate of interest, compounded and paid once a year, which gives the same return as a given rate compounded and paid more than once a year.

Annuity A sequence of equal payments made or received at equally spaced times.

Assignment of probabilities Assignment of nonnegative numbers to the outcomes of an experiment such that the sum of the numbers is 1.

Augmented matrix A matrix obtained from a system of equations of the form $\mathbf{AX} = \mathbf{B}$ by adjoining the column vector $\mathbf{B}$ to the coefficient matrix $\mathbf{A}$, $[\mathbf{A}|\mathbf{B}]$.

Basic solution A solution of a system of linear equations (resulting from the addition of slack variables in a linear programming problem) produced by the simplex method.

Bayes probabilities Conditional probabilities which can be computed by partitioning a sample space in a specific way based on the data of the problem.

Bernoulli process A sequence of a specific number of Bernoulli trials with a given success probability.

Bernoulli trials Independent repetitions of an experiment with exactly 2 outcomes

Binomial coefficients Numbers which arise as the number of ways a subset of r elements can be selected from a set of n elements.

Binomial random variable A random variable which assigns to a Bernoulli process the number of successes.

Bond A promise to pay a specific amount of money at a specified date in the future together with periodic interest.

Bonds sold at a discount Bonds whose selling price is less than face value.

Bonds sold at par Bonds whose selling price is the same as face value.

Bonds sold at a premium Bonds whose selling price is higher than face value.

Bounded set of points in the plane A set whose elements all lie inside some circle in the plane.

Canonical form A form of the transition matrix of an absorbing Markov chain which results from listing all absorbing states first.

Cartesian product of two sets A set whose elements are ordered pairs of elements from the given sets.

Coefficient matrix of a system of equations An array (matrix) in which the entries are the coefficients of the variables in a system of equations.

Column n vector An ordered set of n numbers written one above the other, equivalent to an $n \times 1$ matrix.

Combination principle A method of obtaining the number of ways of selecting a subset of r elements from a set of n elements: $C(n, r) = n!/r!(n - r)!$.

Common stock An investment representing part ownership of a corporation.

Communication matrix A matrix whose rows and columns are identified with the units of an organization and whose elements indicate which units communicate.

Complement of a set The set of elements in the universal set but not in the given set.

Compound interest Interest added to the principal at each payment date, thus increasing the principal on which interest is paid at the next payment date.

Conditional probability The probability of an event with respect to a specified subset of a sample space rather than the entire sample space.

Connected graph A graph $(V, E]$ for which, given any two vertices $v, w \in V$, there is a path from v to w using edges in E.

Consistent system A system of equations which has at least one solution.

Constraints Conditions which must be satisfied by the variables in a linear programming problem.

Coordinates of a vector The numbers which appear as entries in the vector.

Corner point An element in a convex set which has the property that if it belongs to a line completely contained in the set, then it is an endpoint of that line.

Current yield of a bond The amount of interest paid per year divided by the current price of the bond.

Cycle A path which starts and ends at the same vertex.

Decision variables in a linear programming problem The variables whose values are sought as a solution to the original linear programming function.

Density function of a random variable The function which assigns to each set on which the random variable takes a specific value the probability of that event.

Departing variable The variable which departs from the set of row labels in the simplex tableau during a pivot operation.

Digraph A set V (of vertices) together with a set E (of edges) in $V \times V$.

Disjoint sets Two sets whose intersection is empty.

Distance matrix of a digraph A matrix whose rows and columns are identified with the vertices of a digraph and whose entries are the distances between pairs of vertices.

Dominance One strategy in a game dominates another if the first yields a payoff at least as large as the second when played against every pure strategy of the opponent.

Dual of an SMP [A, B, C] A linear programming problem defined as: Minimize $\mathbf{X}^T\mathbf{B}$ subject to $\mathbf{X} \geq \mathbf{0}$ and $\mathbf{A}^T\mathbf{X} \geq \mathbf{C}$.

Edge in a digraph An ordered pair (v, w) which is an element of $E \subset V \times V$, where V is the set of vertices of the digraph.

Effective annual rate of interest See *Annual percentage rate.*

Effective annual yield See *Annual percentage rate.*

Element Any member of a set.

Empty set The set with no elements.

Entering variable The variable which enters the set of row labels in the simplex tableau during a pivot operation.

Equal sets Sets that have the same elements.

Equality of matrices Two matrices of the same size are equal if the corresponding entries in the two matrices are the same.

Equally likely outcomes A set of outcomes of an experiment that each have the same probability.

Equiprobable measure The probability measure in which each outcome is assigned the same probability.

Event A subset of a sample space.

Expected value of a random variable The sum of the products obtained by multiplying each value of the random variable by its probability.

Experiment A set of operations or activities which results in a well-defined set of outcomes called the sample space.

External demand vector The vector of demands other than those resulting from production in a Leontief model.

Face value of a bond The amount to be paid on maturity of the bond.

Feasible set for a linear programming problem The set of vectors satisfying all the constraints of the problem.

Feasible vector An element of the feasible set of a linear programming problem, i.e., a set of values for the variables which satisfies all constraints of the problem.

Fundamental matrix A matrix which arises as the inverse of a portion of the canonical form of the transition matrix of an absorbing Markov chain.

General equation of a line An equation of the form

$$Ax + By = C$$

Geodesic A path of least cost in a digraph.

Graph A pair $[V, E]$, where V is a nonempty set of elements called vertices, and E is a subset of $V \times V$ of elements called edges, such that E does not contain pairs of the form (v, v) and if $(v, w) \in E$, then $(w, v) \in E$ $(v \neq w)$.

Independent events Events A and B for which

$$\Pr[A]\Pr[B] = \Pr[A \cap B]$$

Initial tableau The first tableau formed with the simplex method is used to solve a linear programming problem.

Intercepts of a line Points on a line with one coordinate equal to zero.

Interest Money paid for the use of money.

Intersection of two sets The elements in both sets.

Length of a path in a digraph The number of edges in the path.

Leontief model A linear model for an economic system in which goods are produced and consumed.

Line A set of points (x, y) which can satisfy an equation of the form $Ax + By = C$.

Linear function A function which assigns to a vector $[x_1\, x_2 \ldots x_n]$ the number $a_1x_1 + a_2x_2 + \ldots + a_nx_n$.

Linear programming problem The problem of finding the maximum (or minimum) value of a linear function over a set described by linear constraints.

Long-range behavior The behavior of a Markov chain after a large number of transitions.

Markov chain A stochastic process in which a system makes transitions from one state to another with probabilities which depend only on the two states involved.

Matrix A rectangular array whose entries are either numbers or variables.

Matrix addition An operation which assigns to two matrices of the same size a matrix whose entries are the sum of the corresponding entries of the original two matrices.

Matrix multiplication An operation which assigns to two matrices a matrix whose (i, j) entry is the "row by column" product of the ith row of the first matrix and the jth column of the second.

Mean of a random variable The expected value of that random variable.

Mixed strategy A strategy obtained by playing each of the pure strategies with a specified probability.

Multiplication principle A method to obtain the number of outcomes of certain multistage experiments.

Network A digraph in which each edge is assigned a cost or value.

Nominal interest rate of a bond The interest rate specified on the bond.

Normal approximation to a binomial random variable A method of determining probabilities for a binomial random variable based on tabulated values for the standard normal random variable.

Normal random variable A random variable whose density function is related in a simple way to the density function of a standard normal random variable.

Objective function The function whose value is to be maximized (or minimized) in a linear programming problem.

Optimal mixed strategy A mixed strategy which gives the best average payoff to a player in a game.

Parallel lines Lines with the same slope. If the lines have one point in common, then they are identical.

Partition A collection of nonempty subsets of a set which are pairwise disjoint and whose union is the given set.

Path in a digraph A set of connected edges in the digraph.

Path in a tree diagram A connected set of branches in a tree diagram from the beginning point to the right-hand edge of the diagram.

Payoff The amount gained or lost by a player on one play of a game.

Payoff matrix A matrix whose rows and columns are labeled with the strategies of two players in a game and whose entries are the payoffs to one of the players.

Permutation principle A method of obtaining the number of ways of selecting and ordering r elements from a set of n elements: $P(n, r) = n!(n - r)!$

Pivot element Entry in the simplex tableau identified with the entering and departing variables and which is used in the computation of a new basic solution (see *Pivot operation*).

Pivot operation An operation on a simplex tableau in which a new basic solution is obtained by using the pivot element to replace all rows other than the row of the departing variable by new rows with a zero in the column of the pivot element. The row of the pivot element is replaced by a row with a 1 in the column of the pivot element.

Plane For any numbers A, B, C, and D, with A, B, and C, not all zero, the sets of points $\{(x, y, z): Ax + By + Cz = D\}$.

Present value The principal which will grow to a specified amount in a given period of time with a given interest rate.

Present value of an annuity The present value of the amount of an annuity.

Probability of an event The number assigned to a subset of a sample space by the probability measure.

Probability measure A function which assigns a number to each subset of a sample space and which satisfies certain conditions.

Production schedule The levels at which the productive processes in an economic system are operated.

Pure strategy One of the options available to a player on a specific play of a game.

Random selection A selection method in which the outcomes are all equally likely.

Random variable A function which assigns a number to each outcome in a sample space.

Redemption date of a bond The date on which the face value of the bond will be paid to the owner.

Redemption value of a bond The face value of a bond, the amount paid to the owner on the redemption date.

Reduced form of a matrix A specific form of the augmented matrix for a system of equations from which the solutions of the system can be easily obtained.

Regular Markov chain A Markov chain for which some power of the transition matrix contains only positive entries.

Return on investment Profit per unit time (usually a year) divided by cost of an investment.

Row vector A matrix with only one row.

Saddle point An entry in a payoff matrix which is simultaneously the minimum of its row and the maximum of its column.

Sample space The set of outcomes of an experiment.

Scalar multiple of a matrix The result of multiplying each entry in a matrix by the same number (a scalar).

Set A collection of objects for which it can be determined which objects are in the collection and which are not.

Simple interest Interest paid only on the original principal at each payment date.

Simplex method A method of solving linear programming problems which involves forming a tableau and then obtaining successive tableaus by pivot operations.

Slack variables Variables introduced into the constraints of a linear programming problem to convert inequalities to equalities.

Slope-intercept form of a line The equation $y = mx + b$, where m is the slope and b is the y intercept.

Slope of a line The ratio of the difference of the y coordinates of two distinct points on the line to the difference of the x coordinates, provided the latter is not zero.

Solution of a game Optimal strategies for each of the players and the value of the game.

Solution by graphing A method of solving a two-person zero-sum game by graphing lines.

Solution of a system of equations Values for each of the variables for which the system of equations is satisfied.

Spanning tree In a connected graph $[V, E]$, a pair $[V, E']$ where E' is a subset of E and $[V, E']$ is a tree.

Stable probabilities for a regular Markov chain A set of probabilities $\{p_i\}$ such that in the long run the system will be in state i with probability p_i, independent of the state in which the system began.

Standard deviation of a random variable A measure of the dispersion of the values of the random variable about its mean; standard deviation $= \sqrt{\text{variance}}$.

Standard maximum problem Maximize $\mathbf{C}^T\mathbf{X}$ subject to $\mathbf{X} \geq \mathbf{0}$ and $\mathbf{AX} \leq \mathbf{B}$, where $\mathbf{B}$ has strictly positive coordinates.

Standard normal curve The familiar "bell-shaped curve" of statistics given by the equation $y = (1/\sqrt{2\pi})e^{-z^2/2}$.

Standard normal random variable A random variable whose density function has the standard normal curve as its graph.

State of a Markov chain One of the several possible results of observing a Markov chain.

State vector for a Markov chain A vector whose coordinates give the probabilities that the Markov chain is in each of the possible states.

Stated interest rate of a bond The interest rate specified on the bond.

Status A measure of the position of an individual (or unit) in an organization.

Stochastic process A probabilistic process which takes place in steps.

Stock split The result of replacing each share of common stock by more than one share while retaining the same fractional ownership for each shareholder.

Strictly determined game A game which has a saddle point.

Subset of a set Any set A whose elements are also in a set B is a subset of B.

System of linear equations A set of linear equations which are to be solved simultaneously.

Technology matrix A matrix which describes the production process in a Leontief model.

Transition The result of two successive observations of a Markov chain.

Transition diagram A diagram in which the states are denoted by points or circles, and transition probabilities are noted on lines connecting the points.

Transition matrix of a Markov chain A matrix whose rows and columns are identified with state labels and whose entries are the transition probabilities between respective states.

Tree (as a special graph) A connected graph with no cycles.

Tree diagram (as a tool for counting and probability problems) A pictorial method of representing the outcomes of a multistage experiment.

Two-person zero-sum game A game with two players in which the gain of one is the loss of the other.

Unbounded set A set such that no circle in the plane contains all of its elements (a set which is not bounded).

Union of two sets The set of elements which are in either or both of the two sets.

Universal set A set determined by a particular problem which contains all other sets involved in the problem as subsets.

Value of a game The greatest payoff that the row player can be guaranteed and the smallest loss that the column player can be guaranteed.

Value of a path in a network The sum of the values of the edges in the path.

Variance of a random variable A measure of the dispersion of the values of the random variable about its mean; $\sqrt{\text{variance}} =$ standard deviation.

Venn diagram A pictorial representation of a set called the universal set and certain subsets.

Vertex of a digraph An element of the set V used to define the digraph.

Yield to maturity of a bond The interest rate for which the sum of the present values of the interest payments yet to be made and the present value of the redemption value is equal to the current price.

APPENDIX

Areas Under The Standard Normal Curve

A.0 INTRODUCTION

Many of the problems in Chapter 4 involving normal random variables require that we determine $\Pr[a \le Z \le b]$ where Z is the standard normal random variable. To determine $\Pr[a \le Z \le b]$, it is useful to recall that

$$\Pr[a \le Z \le b] = \Pr[0 \le Z \le |a|] + \Pr[0 \le Z \le b] \qquad \text{if } a < 0 < b$$

and $$\Pr[a \le Z \le b] = \Pr[0 \le Z \le b] - \Pr[0 \le Z \le a] \qquad \text{if } 0 < a < b$$

Consequently, our task reduces to computing $\Pr[0 \le Z \le z]$ where z is any number. This probability is the area under the standard normal curve between 0 and z shown in Figure B.1.

FIGURE A.1

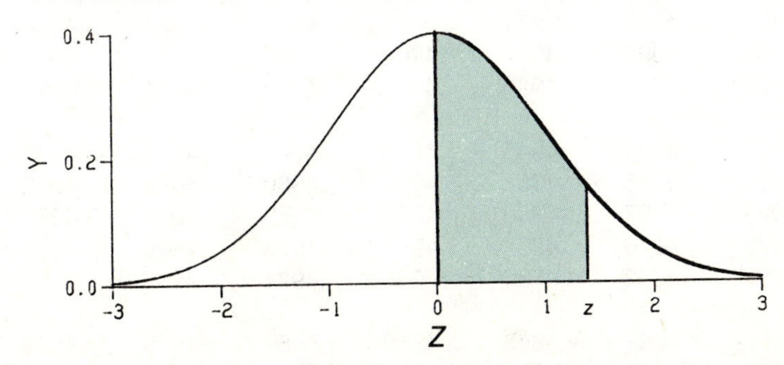

A.1 USE OF THE TABLE

The entries in Table B.1 are the areas under the standard normal curve from 0 to z where the units and tenths digits of z are given in the left-hand column and the hundredths digit is given in the top line. For example, if $z = 1.32$, then we look under z for the row labeled "1.3" and follow that row to the column labeled ".02." The entry there tells us that the area under the standard normal curve from 0 to 1.32 is .4066. That is, $\Pr[0 \leq Z \leq 1.32] = .4066$.

TABLE A.1 Areas under the Standard Normal Curve

z	.00	.01	.02	.03	.04	.05	.06	.07	.08	.09
0.0	.0000	.0040	.0080	.0120	.0160	.0199	.0239	.0279	.0319	.0359
0.1	.0398	.0438	.0478	.0517	.0557	.0596	.0636	.0675	.0714	.0753
0.2	.0793	.0832	.0871	.0910	.0948	.0987	.1026	.1064	.1103	.1141
0.3	.1179	.1217	.1255	.1293	.1331	.1368	.1406	.1443	.1480	.1517
0.4	.1554	.1591	.1628	.1664	.1700	.1736	.1772	.1808	.1844	.1879
0.5	.1915	.1950	.1985	.2019	.2054	.2088	.2123	.2157	.2190	.2224
0.6	.2257	.2291	.2324	.2357	.2389	.2422	.2454	.2486	.2517	.2549
0.7	.2580	.2611	.2642	.2673	.2704	.2734	.2764	.2794	.2823	.2852
0.8	.2881	.2910	.2939	.2967	.2995	.3023	.3051	.3078	.3106	.3133
0.9	.3159	.3186	.3212	.3238	.3264	.3289	.3315	.3340	.3365	.3389
1.0	.3413	.3438	.3461	.3485	.3508	.3531	.3554	.3577	.3599	.3621
1.1	.3643	.3665	.3686	.3708	.3729	.3749	.3770	.3790	.3810	.3830
1.2	.3849	.3869	.3888	.3907	.3925	.3944	.3962	.3980	.3997	.4015
1.3	.4032	.4049	.4066	.4082	.4099	.4115	.4131	.4147	.4162	.4177
1.4	.4192	.4207	.4222	.4236	.4251	.4265	.4279	.4292	.4306	.4319
1.5	.4332	.4345	.4357	.4370	.4382	.4394	.4406	.4418	.4429	.4441
1.6	.4452	.4463	.4474	.4484	.4495	.4505	.4515	.4525	.4535	.4545
1.7	.4554	.4564	.4573	.4582	.4591	.4599	.4608	.4616	.4625	.4633
1.8	.4641	.4649	.4656	.4664	.4671	.4678	.4686	.4693	.4699	.4706
1.9	.4713	.4719	.4726	.4732	.4738	.4744	.4750	.4756	.4761	.4767
2.0	.4772	.4778	.4783	.4788	.4793	.4798	.4803	.4808	.4812	.4817
2.1	.4821	.4826	.4830	.4834	.4838	.4842	.4846	.4850	.4854	.4857
2.2	.4861	.4864	.4868	.4871	.4875	.4878	.4881	.4884	.4887	.4890
2.3	.4893	.4896	.4898	.4901	.4904	.4906	.4909	.4911	.4913	.4916
2.4	.4918	.4920	.4922	.4925	.4927	.4929	.4931	.4932	.4934	.4936
2.5	.4938	.4940	.4941	.4943	.4945	.4946	.4948	.4949	.4951	.4952
2.6	.4953	.4955	.4956	.4957	.4959	.4960	.4961	.4962	.4963	.4964
2.7	.4965	.4966	.4967	.4968	.4969	.4970	.4971	.4972	.4973	.4974
2.8	.4974	.4975	.4976	.4977	.4977	.4978	.4979	.4979	.4980	.4981
2.9	.4981	.4982	.4982	.4983	.4984	.4984	.4985	.4985	.4986	.4986
3.0	.4987	.4987	.4987	.4988	.4988	.4989	.4989	.4989	.4990	.4990

A.2 INTERPOLATION IN THE TABLE

In some applications it is necessary to determine areas defined by values of z which are not listed in Table B.1. For example, to determine $\Pr[0 \le Z \le 1.363]$ we cannot use the method described in Section 1. For our purposes it is adequate to use linear interpolation between the values given in the table. To obtain the value (area) corresponding to 1.363, we interpolate between the values for 1.36 and 1.37. Since 1.363 is three-tenths of the way from 1.36 to 1.37, we obtain an approximation to the value for 1.363 by finding the value which is three-tenths of the way between the values for 1.36 and 1.37. The value for 1.36 is .4131 and the value for 1.37 is .4147. Therefore, an approximate value for 1.363 is

$$\begin{aligned} .4131 + .3(.4147 - .4131) &= .4131 + .3(.0016) \\ &= .4131 + .0005 \\ &= .4136 \end{aligned}$$

We therefore have $\Pr[0 \le Z \le 1.363] = .4136$.

In general, if we wish to obtain $\Pr[0 \le Z \le z_2]$ where $z_1 \le z_2 \le z_1 + 0.01$ and the value for z_1 is given in the table, we use the formula

$$\begin{aligned} &\Pr[0 \le Z \le z_2] \\ &= \Pr[0 \le Z \le z_1] + 100(z_2 - z_1)(\Pr[0 \le Z \le z_1 + .01] - \Pr[0 \le Z \le z_1]) \end{aligned}$$

APPENDIX

B Success Probabilities For Bernoulli Processes

The probability of obtaining exactly r successes in a Bernoulli process consisting of n trials with success probability p is $C(n, r)p^r(1 - p)^{n-r}$. These probabilities are given for $n = 4$ through 10 and selected values of p in Table C.1. The probabilities for $n = 1$, 2, and 3 can easily be obtained by direct calculation. Success probabilities for $p = .55, .60, \ldots, .95$ can be obtained from Table C.1 by using the following fact:

Probability of r successes in n trials with success probability p
= probability of $n - r$ successes in n trials with success probability $1 - p$

TABLE B.1 Success Probabilities for Bernoulli Processes

n	r	.05	.10	.15	.20	.25	.30	.35	.40	.45	.50
4	0	.8145	.6561	.5220	.4096	.3164	.2401	.1785	.1296	.0915	.0625
	1	.1715	.2916	.3685	.4096	.4219	.4116	.3845	.3456	.2995	.2500
	2	.0135	.0486	.0975	.1536	.2109	.2646	.3105	.3456	.3675	.3750
	3	.0005	.0036	.0115	.0256	.0469	.0756	.1115	.1536	.2005	.2500
	4	.0000	.0001	.0005	.0016	.0039	.0081	.0150	.0256	.0410	.0625
5	0	.7783	.5905	.4437	.3277	.2373	.1681	.1160	.0778	.0503	.0313
	1	.2036	.3280	.3915	.4096	.3955	.3601	.3124	.2592	.2059	.1563
	2	.0214	.0729	.1382	.2048	.2637	.3087	.3364	.3456	.3369	.3125
	3	.0011	.0081	.0244	.0512	.0879	.1323	.1811	.2304	.2757	.3125
	4	.0000	.0005	.0022	.0064	.0146	.0284	.0488	.0768	.1128	.1563
	5	.0000	.0000	.0001	.0003	.0010	.0024	.0053	.0102	.0185	.0313

TABLE B.1 (continued)

n	r	.05	.10	.15	.20	.25	.30	.35	.40	.45	.50
						p					
6	0	.7351	.5314	.3771	.2621	.1780	.1176	.0754	.0467	.0277	.0156
	1	.2321	.3543	.3993	.3932	.3560	.3025	.2437	.1866	.1359	.0938
	2	.0305	.0984	.1762	.2458	.2966	.3241	.3280	.3110	.2780	.2344
	3	.0021	.0146	.0415	.0819	.1318	.1852	.2355	.2765	.3032	.3125
	4	.0001	.0012	.0055	.0154	.0330	.0595	.0951	.1382	.1861	.2344
	5	.0000	.0001	.0004	.0015	.0044	.0102	.0205	.0369	.0609	.0938
	6	.0000	.0000	.0000	.0001	.0002	.0007	.0018	.0041	.0083	.0156
7	0	.6983	.4783	.3206	.2097	.1335	.0824	.0490	.0280	.0152	.0078
	1	.2573	.3720	.3960	.3670	.3115	.2471	.1848	.1306	.0872	.0547
	2	.0406	.1240	.2097	.2753	.3115	.3177	.2985	.2613	.2140	.1641
	3	.0036	.0230	.0617	.1147	.1730	.2269	.2679	.2903	.2918	.2734
	4	.0002	.0026	.0109	.0287	.0577	.0972	.1442	.1935	.2388	.2734
	5	.0000	.0002	.0012	.0043	.0115	.0250	.0466	.0774	.1172	.1641
	6	.0000	.0000	.0001	.0004	.0013	.0036	.0084	.0172	.0320	.0547
	7	.0000	.0000	.0000	.0000	.0001	.0002	.0006	.0016	.0037	.0078
8	0	.6634	.4305	.2725	.1678	.1001	.0576	.0319	.0168	.0084	.0039
	1	.2793	.3826	.3847	.3355	.2670	.1977	.1373	.0896	.0548	.0313
	2	.0515	.1488	.2376	.2936	.3115	.2965	.2587	.2090	.1569	.1094
	3	.0054	.0331	.0839	.1468	.2076	.2541	.2786	.2787	.2568	.2188
	4	.0004	.0046	.0185	.0459	.0865	.1361	.1875	.2322	.2627	.2734
	5	.0000	.0004	.0026	.0092	.0231	.0467	.0808	.1239	.1719	.2188
	6	.0000	.0000	.0002	.0011	.0038	.0100	.0217	.0413	.0703	.1094
	7	.0000	.0000	.0000	.0001	.0004	.0012	.0033	.0079	.0164	.0313
	8	.0000	.0000	.0000	.0000	.0000	.0001	.0002	.0007	.0017	.0039
9	0	.6302	.3874	.2316	.1342	.0751	.0404	.0207	.0101	.0046	.0020
	1	.2985	.3874	.3679	.3020	.2253	.1556	.1004	.0605	.0339	.0176
	2	.0629	.1722	.2597	.3020	.3003	.2668	.2162	.1612	.1110	.0703
	3	.0077	.0446	.1069	.1762	.2336	.2668	.2716	.2508	.2119	.1641
	4	.0006	.0074	.0283	.0661	.1168	.1715	.2194	.2508	.2600	.2461
	5	.0000	.0008	.0050	.0165	.0389	.0735	.1181	.1672	.2128	.2461
	6	.0000	.0001	.0006	.0028	.0087	.0210	.0424	.0743	.1160	.1641
	7	.0000	.0000	.0000	.0003	.0012	.0039	.0098	.0212	.0407	.0703
	8	.0000	.0000	.0000	.0000	.0001	.0004	.0013	.0035	.0083	.0176
	9	.0000	.0000	.0000	.0000	.0000	.0000	.0001	.0003	.0008	.0020
10	0	.5987	.3487	.1969	.1074	.0563	.0282	.0135	.0060	.0025	.0010
	1	.3151	.3874	.3474	.2684	.1877	.1211	.0725	.0403	.0207	.0098
	2	.0746	.1937	.2759	.3020	.2816	.2335	.1757	.1209	.0763	.0439
	3	.0105	.0574	.1298	.2013	.2503	.2668	.2522	.2150	.1665	.1172
	4	.0010	.0112	.0401	.0881	.1460	.2001	.2377	.2508	.2384	.2051
	5	.0001	.0015	.0085	.0264	.0584	.1029	.1536	.2007	.2340	.2461
	6	.0000	.0001	.0012	.0055	.0162	.0368	.0689	.1115	.1596	.2051
	7	.0000	.0000	.0001	.0008	.0031	.0090	.0212	.0425	.0746	.1172
	8	.0000	.0000	.0000	.0001	.0004	.0014	.0043	.0106	.0229	.0439
	9	.0000	.0000	.0000	.0000	.0000	.0001	.0005	.0016	.0042	.0098
	10	.0000	.0000	.0000	.0000	.0000	.0000	.0000	.0001	.0003	.0010

APPENDIX C

Answers to Odd-Numbered Exercises

CHAPTER 1

Section 1.1

1. (*a*) False (*b*) false (*c*) true
3. (*a*) False (*b*) true
5. $B = \{p\}, C = \{q, r\}; B = \{q\}, C = \{p, r\}; B = \{r\}, C = \{p, q\}$; also $B = \{q, r\}, C = \{p\}; B = \{p, r\}, C = \{q\}; B = \{p, q\}, C = \{r\}$
7. (*a*) True (*b*) false (*c*) true (*d*) false (*e*) false (*f*) true
9. (*a*) $X' = \{b, 4\}$ (*b*) $Y' = \{c, d, 4, 6\}$ (*c*) $X \cap Y' = \{c, d, 6\}$
11. (*a*) $A \cup B = \{a, b, c, 2, 3\}$ (*b*) $B \cap C = \{2, 3\}$ (*c*) $(A \cup B) \cap (B \cup C) = \{a, 2, 3\}$ (*d*) $A' = \{1, 2, 3\}$ (*e*) $A \cap B' = \{b, c\}$ (*f*) $A \cup C' = \{a, b, c\}$
13. (*a*) $A \cap B = \{x$: x owns a GM car *and* x works for GM) (*b*) $B \cap A' = \{x$: x works for GM and x does not own a GM car$\}$ (*c*) $(A \cup B) \cap D = \{x$: x owns stock in GM and also x owns a GM car or x works for GM$\}$ (*d*) $(C \cap A) = \{x$: x owns a GM car and x is the president of GM$\}$
15. (*a*) No (*b*) yes (*c*) yes (*d*) no
17. (*a*) Yes (*b*) no
19. $A = X \cap Y, B = X \cap Z', C = Y \cup Z$
21. (*a*) $\varnothing, \{x\}$ (*b*) $\varnothing, \{x\}, \{y\}, \{x, y\}$ (*c*) $\varnothing, \{x\}, \{y\}, \{z\}, \{x, y\}, \{y, z\}, \{x, z\}, \{x, y, z\}$
23. $Z = \{b, 4, 6\}$ $(Z' = \{a, c, 2\})$, $Y = \{b, 2, 4\}$ $(Y' = \{a, c, 6\})$, $X = \{b, c, 2, 4, 6\}$ or $Z = \{b, 4, 6\}$ $(Z' = \{a, c, z\})$, $Y = \{b, 2, 4, 6\}$ $(Y' = \{a, c\})$, $X = \{b, c, 2, 4\}$
25. (*a*) $A \times B = \{(a, a), (a, b), (a, d), (b, a), (b, b), (b, d), (c, a), (c, b), (c, d)\}$ (*b*) $(A \times B) \cap (B \times A) = \{(a, a), (a, b), (b, a), (b, b)\}$
27. $A = \{a, b\}, B = \{1, 2, 3\}$
29. (*a*) Yes (*b*) yes (*c*) no (*d*) no

Section 1.2

1. (Figure C1.1)
3. (*a*) v, x, y, z (*b*) y (*c*) x, y, z, v, w (*d*) y, v
5. (*a*) True (*b*) true (*c*) false (*d*) true
7. (*a*) $(B \cap C) \cap A'$ (*b*) $[B \cap (A \cup C)'] \cup [C \cap (A \cup B)']$ (*c*) $(A \cup B \cup C)' \cup (B \cap C)$
9. (*a*) No (*b*) no (*c*) yes (*d*) no
11. 5
13. 60
15. (*b*), (*d*)
17. $A \times B = \{(1, v), (1, w), (2, v), (2, w), (3, v), (3, w)\}$, $A \times \{v\} = \{(1, v), (2, v), (3, v)\}$, and $A \times \{w\} = \{(1, w), (2, w), (3, w)\}$; clearly, $A \times \{v\} \cup A \times \{w\} = A \times B$ and $A \times \{v\} \cap A \times \{w\} = \varnothing$. Hence, these sets form a partition of $A \times B$.
19. (*a*) $n(A \times B) = 5 \cdot 2 = 10$ (*b*) $n(B \times B \times B) = 2 \cdot 2 \cdot 2 = 8$
21. 15
23. If $n(B) = 4$, there are 4 ways because B must contain $A[n(A) = 3]$ so you must choose one element from the four left in $C \cap A'$; If $n(B) = 5$, there are 6 ways:
    If $C = \{1, 2, 3, 4, 5, 6, 7\}$ and $A = \{1, 2, 3\}$, then you may choose:
    $B = \{1, 2, 3, 4, 5\}$
    $B = \{1, 2, 3, 4, 6\}$
    $B = \{1, 2, 3, 4, 7\}$
    $B = \{1, 2, 3, 5, 6\}$
    $B = \{1, 2, 3, 5, 7\}$
    $B = \{1, 2, 3, 6, 7\}$
25. 34
27. $n(X_1) = 10, n(X_2) = 5, n(X_3) = 25$
29. $n(X_1) = 16$

FIGURE C1.1

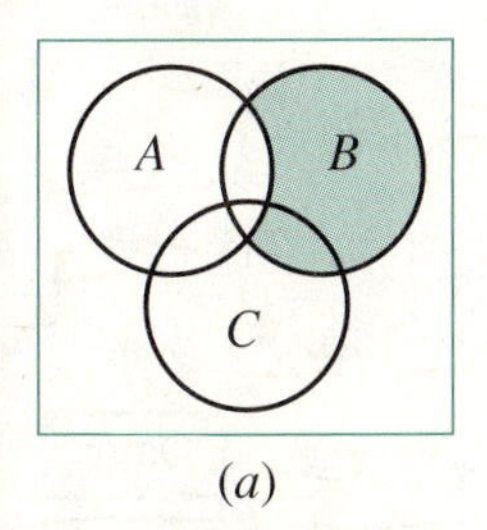

(*a*)

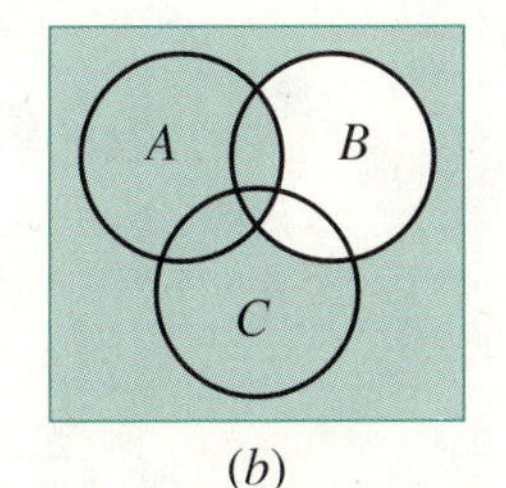

(*b*)

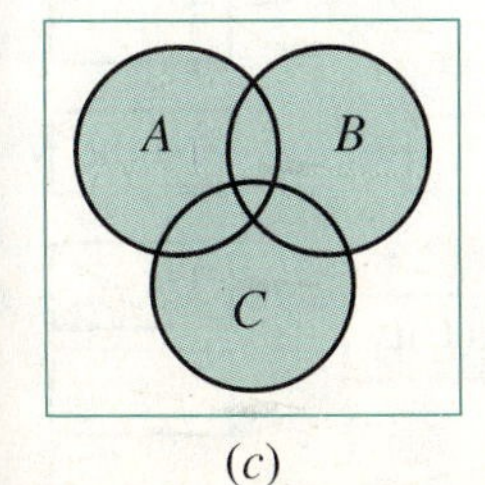

(*c*)

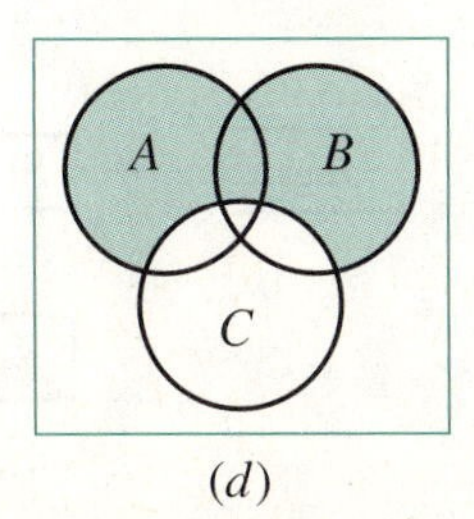

(*d*)

Section 1.3

1. 32
3. 55
5. 110
7. 23
9. 7
11. 36
13. 7
15. 37
17. 4
19. 52
21. 80
23. 15
25. 8
27. (*a*) 50 (*b*) 0 (*c*) 30
29. There is not enough information to determine the exact number of students who like canoeing and exactly one of hiking or camping; however, it is known that this number is at most 30.
31. 9

Section 1.4

1. $\{TTTT, TTTF, TTFT, TTFF, TFTT, TFTF, TFFT, TFFF, FTTT, FTTF, FTFT, FTFF, FFTT, FFTF, FFFT, FFFF\}$
3. Each outcome consists of a pair of suits. The sample space is $S = \{\text{heart, spade, club, diamond}\} \times \{\text{heart, spade, club, diamond}\}$. Thus

$$S = \{HH, HS, HC, HD, SH, SS, SC, SD, CH, CS, CC, CD, DH, DS, DC, DD\}$$

5. $S = \{2, 3, 4, 5, 6, 7, 8, 9, 10, 11, 12\}$
7. Each outcome is an ordered pair taken from $A = \{0, 2, 4, 6\}$. $S = A \times A$, $n(S) = 16$
9. 84
11. 36
13. 243
15. U = untrained, T = trained, $S = \{UTU, UUT, UUU, TUU\}$ (see Figure C1.2)
17. (*a*) $5^4 = 625$ (*b*) $4^4 = 256$ (*c*) $4^4 = 256$
19. (*a*) (Figure C1.3) (*b*) (Figure C1.4)
21. (Figure C1.5) There are 16 possible logs.
23. 75
25. $2 \times 2 + 3 \times 2 = 10$
27. 19
29. $1 \times 4 + 1 \times 2 = 6$

Review Exercises

1. (*a*) $A \cap B = \{2, x\}$ (*b*) $A \cup B = \{1, 2, 3, 4, u, v, x, y, z\}$ (*c*) $(A \cap B) \cup C = \{2, 3, x, z\}$

FIGURE C1.2

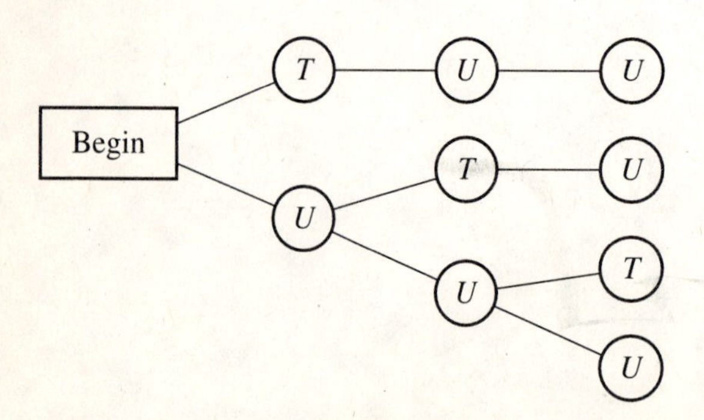

FIGURE C1.3

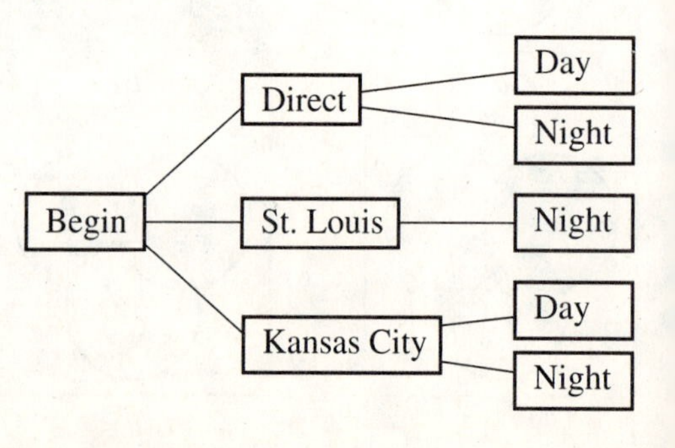

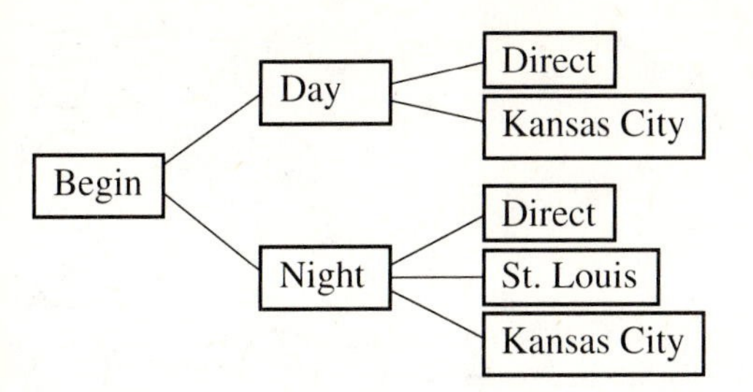

FIGURE C1.4

3. (*a*) $A \cap B = \{6, 12\}$ (*b*) $(A \cup B) \cap C = \{4, 6, 8, 18\}$ (*c*) $A \cup (B \cap C) = \{4, 6, 8, 12, 18, 24\}$ (*d*) $C' = \{2, 10, 12, 14, 16, 20, 22, 24, 26\}$ (*e*) $A \cap B' = \{18, 24\}$ (*f*) $B \cap (A \cup C)' = \{2, 10, 14\}$
5. $(A \times B) \cap (B \times A) = \{(1, 1), (1, y), (y, 1), (y, y)\}$
7. (*a*) S = {(Montana, Maine), (Montana, Minnesota), (Montana, Michigan), (Minnesota, Maine), (Minnesota, Michigan), (Michigan, Maine)}; (*b*) S' = {(Montana, Montana), (Minnesota, Montana), (Minnesota, Minnesota), (Michigan, Montana), (Michigan, Minnesota), (Michigan, Michigan), (Maine, Montana), (Maine, Minnesota), (Maine, Michigan), (Maine, Maine)}
9. (*a*) Cannot be determined (*b*) can be determined, 8
11. $4 \times 5 \times 6 = 120$
13. $65 - 10 = 55$
15. (*a*) 38 (*b*) 20

FIGURE C1.5

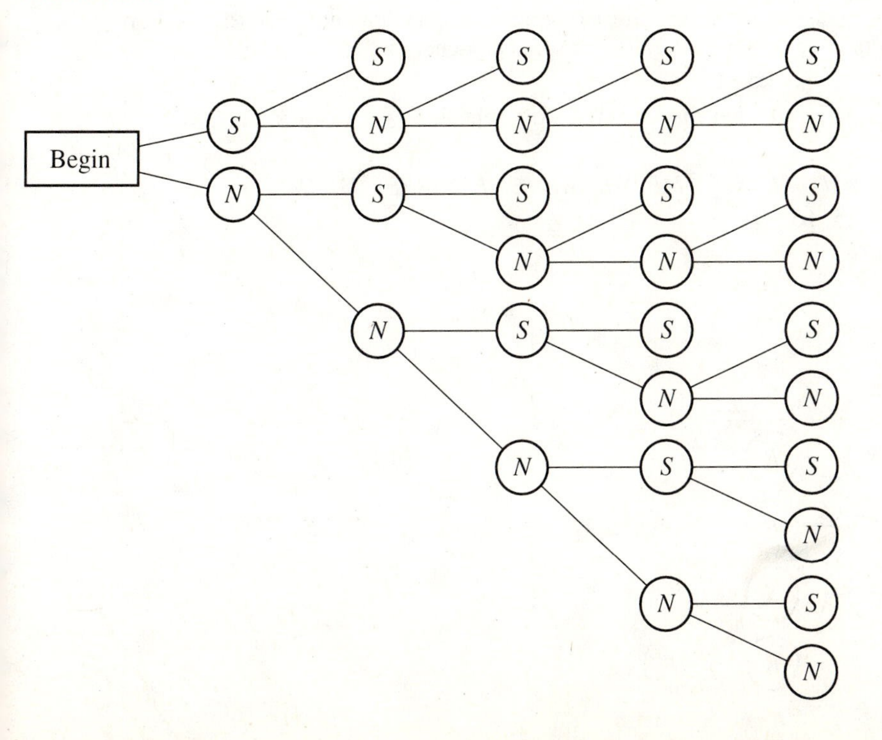

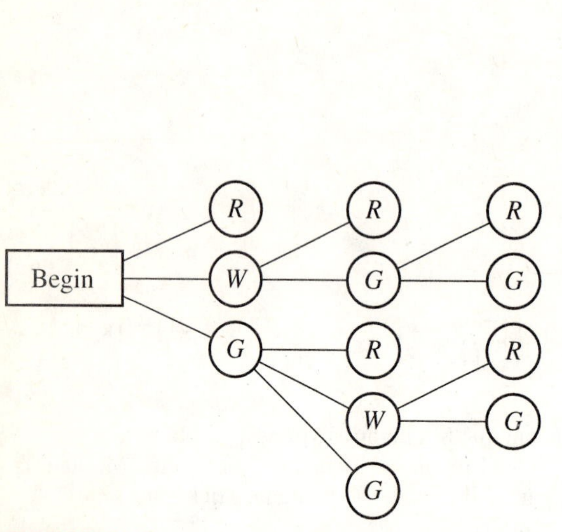

FIGURE C1.6

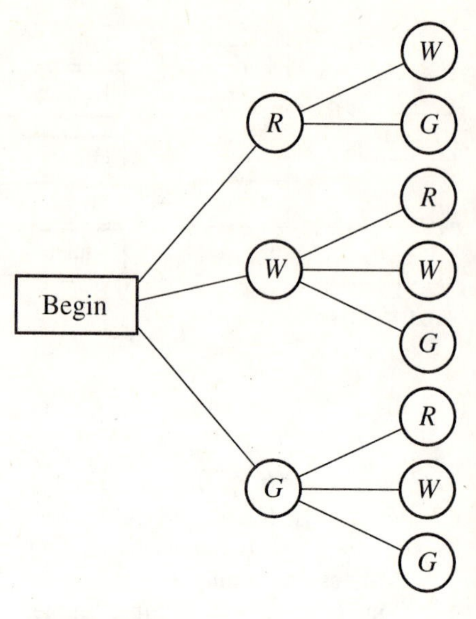

FIGURE C1.7

17. Figure C1.6 $S = \{R, WR, WGR, WGG, GR, GWR, GWG, GG\}$
19. Figure C1.7 $S = \{RW, RG, WR, WW, WG, GR, GW, GG\}$
21. Five possible surveys: (phone, short questionnaire), (phone, long questionnaire), (in person, short questionnaire), (in person, long questionnaire), (in person, open-ended)
23. 56
25. 15
27. 24
29. Figure C1.8 R = report, S = market survey, set of work schedules = $\{SS, SRR, RSR, RRS, RRRR\}$
31. 8
33. Figure C1.9 $S = \{EEMM, EMEM, EMME, MEEM, MEME, MMEE, MMMMM\}$
35. $X = \{2, 3, 4, 8, 9, 10, 14, 15, 16, 20\}$

FIGURE C1.8

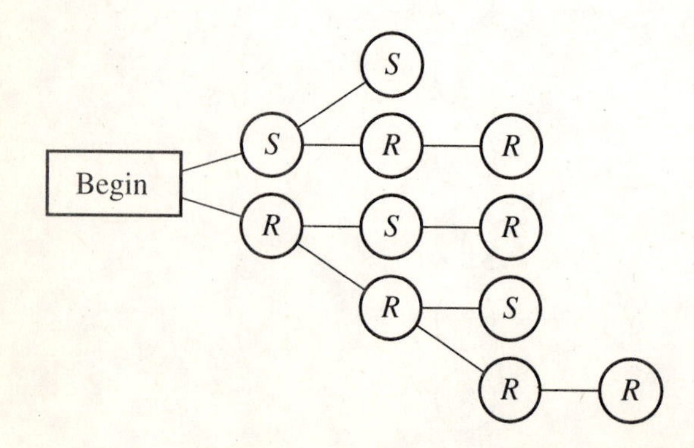

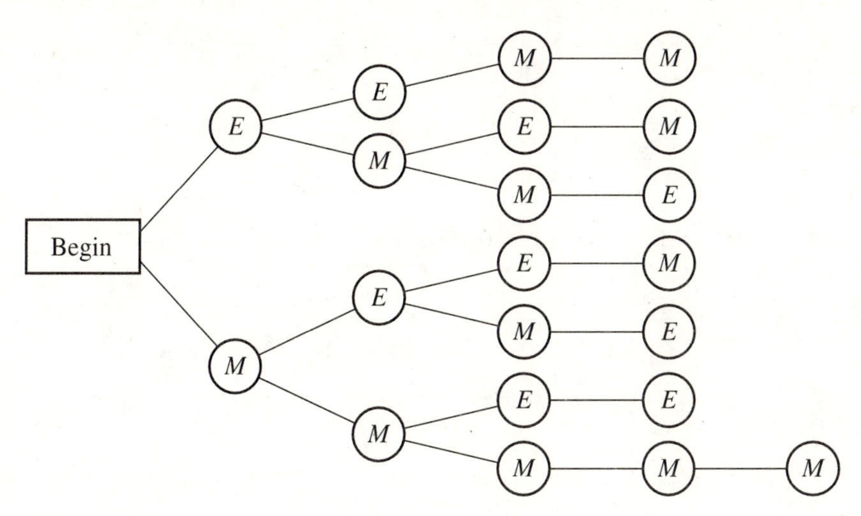

FIGURE C1.9

CHAPTER 2

Section 2.1

1. Let S denote sale; N, no sale.

$$S = \{SS, SNS, SNN, NSS, NSN, NN\}$$

3. $E = \{SSSS, SSSN, SSNS, SNSS, NSSS\}$
   $F = \{SSSS, SSSN, SSNS, SSNN, NSSS, NSSN, NSN\}$
5. $E = \{(3, 1), (3, 2), (3, 4), (3, 5), (3, 6), (1, 3), (2, 3), (4, 3), (5, 3), (6, 3)\}$
   $F = E \cup \{(3, 3)\}$
   $G = \{(1, 2), (1, 4), (1, 6), (2, 1), (2, 3), (2, 5), (3, 2), (3, 4), (3, 6), (4, 1), (4, 3), (4, 5), (5, 2), (5, 4), (5, 6), (6, 1), (6, 3), (6, 5)\}$
   $H = \{(1, 1), (1, 2), (1, 3), (1, 4), (1, 5), (1, 6), (2, 1), (2, 2), (2, 3), (2, 4), (2, 5), (3, 1), (3, 2), (3, 2), (3, 4), (4, 1), (4, 2), (4, 3), (5, 1), (5, 2), (6, 1)\}$
7. (a) $S = \{WL, WM, WH, SL, SM, SH, EL, EM, EH\}$
   (b) $\{SL, SM, SH\}$
   (c) $\{SL, SM, SH, WL, EL\}$
   (d) $\{SL\}$
9. (a) $S = \{UTI, UIT, TUI, TIU, ITU, IUT\}$
   (b) $\{ITU, IUT\}$
   (c) $\{UTI, TUI, TIU\}$
11. (a) $\{S, WS, NS, WWS, WNS, NWS, NNS\}$
    (b) $\{WS, WNS, WNN, WNW, WWS, WWN, WWW, NWS, NWN, NWW, NNW\}$
    (c) $\{WS, WNS, WNN, NWS, NWN, NNW\}$

13. .15
15. $w_1 = w_2 = \frac{3}{10}$, $w_3 = w_4 = w_5 = w_6 = \frac{1}{10}$
17. $a = 6$
19. Pr[heads] $= \frac{3}{4}$, Pr[tails] $= \frac{1}{4}$
21. $\Pr[E] = \frac{3}{10}$, $\Pr[F'] = 1 - \Pr[F] = \frac{7}{10}$
23. Pass = grade of A, B, C, D.
    Pr[pass] = .15 + .2 + .35 + .15 = .85
    Pr[withdraw or fail] = .15
25. Pr[win 10] = .01, Pr[win 1000] = .0002, Pr[win 10,000] = .00001, Pr[lose] = .98979
27. (*a*) $4 \times 3 \times 3 = 36$ (*b*) $\frac{1}{36}$ (*c*) $\frac{1}{4}$
29. (*a*) 20 (*b*) $\frac{1}{20}$ (*c*) $\frac{9}{10}$
31. 8, $\frac{1}{8}$
33. 6
35. (*a*) 25 (*b*) $\frac{1}{25}$ (*c*) $\frac{21}{25}$

Section 2.2

1. (*a*) 720 (*b*) 6 (*c*) 70 (*d*) 270,725
3. (*a*) 20 (*b*) 7 (*c*) 720 (*d*) 6840
5. $P(6, 6) = 720$
7. $22 \times 22 = 484$
9. 60 numbers can be formed, 24 of which are less than 500.
11. $P(8, 2) = 56$
13. $P(10, 8) = 1{,}814{,}400$
15. $P(9, 3) = 504$
17. $P(7, 4) = 840$
19. (*a*) $P(6, 4) = 360$ (*b*) $P(5, 3) = 60$
21. $P(5, 3) = 60$
23. (*a*) $P(10, 3) = 720$ (*b*) $3 \cdot P(9, 2) = 216$ (*c*) $P(9, 2) = 72$
25. 96
27. (*a*) $P(7, 4) = 840$ (*b*) 816
29. 3240
31. 180
33. 420
35. $n = 5$ to $n = 9$ works.

Section 2.3

1. $C(7, 5) = 35$, $C(8, 3) = 56$, $C(9, 3) = 84$
3. (*a*) 840 (*b*) 35 (*c*) 210 (*d*) 1 (*e*) 8
5. $C(5, 3) + C(5, 4) + C(5, 5) = 16$
7. (*a*) $C(8, 2) = 28$ (*b*) $C(8, 2) - C(3, 2) = 25$
9. (*a*) $C(5, 2) = 10$ (*b*) $C(3, 2) = 3$ (*c*) $3 \cdot 2 = 6$ (*d*) $C(5, 2) - C(2, 2) = 9$
11. 17
13. (*a*) $C(9, 3) = 84$ (*b*) $C(9, 3) - C(3, 3) - C(4, 3) = 79$
15. (*a*) $C(3, 2) \cdot C(5, 2) = 30$ (*b*) $C(3, 2) + C(5, 2) = 13$ (*c*) $C(13, 2) = 78$
17. $C(5, 2) \cdot C(4, 2) = 60$
19. (*a*) $C(6, 3) \cdot C(4, 2) = 120$ (*b*) $120 + C(6, 4) \cdot C(4, 1) = 180$
21. $3 \cdot C(6, 4) = 45$
23. (*a*) $C(14, 5) = 2002$ (*b*) $C(14, 5) - C(6, 5) - C(8, 5) = 1940$
25. $C(10, 3) - C(6, 3) - C(4, 3) = 96$
27. (*a*) $C(13, 3) \cdot C(39, 2) = 211{,}926$ (*b*) $4 \cdot 211{,}926 = 847{,}704$ (*c*) 685,464
29. $P(6, 3) - P(4, 3) - P(3, 3) = 90$
31. 35

33. $$C(n,2)+C(n,3)=\frac{n!}{(n-2)!2!}+\frac{n!}{(n-3)!3!}$$
$$=\frac{n(n-1)}{2}+\frac{n(n-1)(n-2)}{3\cdot 2}=\frac{n(n-1)}{2}\left(\frac{3+n-2}{3}\right)$$
$$=\frac{n(n-1)}{2}\frac{n+1}{3}=\frac{(n+1)(n)(n-1)}{3\cdot 2}=\frac{(n+1)!}{(n-2)!3!}=C(n+1,3)$$

35. $n = 1, 2, 3, 4, 5$

Section 2.4

1. $\frac{C(3,2)}{C(5,2)}=\frac{3}{10}$

3. $\frac{C(13,3)}{C(52,3)}=\frac{286}{22{,}100}=\frac{143}{11{,}050}=\frac{11}{850}$

5. $\frac{C(4,3)}{C(7,3)}=\frac{4}{35}$

7. $\frac{1}{216}$

9. Pr[both hired] $=\frac{3}{10}$, Pr[one, but not both] $=\frac{3}{5}$

11. $\frac{1}{C(8,2)}=\frac{1}{28}$

13. (*a*) $\frac{2}{6}=\frac{1}{3}$ (*b*) $\frac{1}{2}$

15. $\frac{C(5,2)C(5,2)}{C(10,4)}=\frac{100}{210}=\frac{10}{21}$

17. $\frac{3!+2}{4!}=\frac{1}{3}$

19. $\frac{C(4,2)C(5,3)+C(4,3)C(5,2)+C(4,4)C(5,1)}{C(9,5)}=\frac{105}{126}$

21. (*a*) $\frac{C(9,2)}{C(10,3)}=\frac{3}{10}$ (*b*) $\frac{P(9,2)}{P(10,3)}=\frac{1}{10}$

23. $\frac{5}{8}$

25. $\frac{7}{729}$

27. $\frac{8}{32}=\frac{1}{4}$

29. (*a*) $\frac{C(4,1)\cdot C(5,2)}{C(9,3)}=\frac{10}{21}$ (*b*) $\frac{C(8,2)}{C(9,3)}=\frac{1}{3}$

31. $\frac{C(14,7)-C(8,7)-C(8,6)\cdot C(6,1)-C(8,1)\cdot C(6,6)}{C(14,7)}=\frac{3248}{3432}$

33. $n > 6$

35. $1 < n \leq 9$

Review Exercises

1. $S = \{(R,R,R), (R,R,B), (R,B,R), (R,B,B), (B,R,R), (B,R,B), (B,B,R)\}$. The event that at least one blue ball is drawn is $\{(R,R,B), (R,B,R), (R,B,B), (B,R,R), (B,R,B), (B,B,R)\}$.

3. $S = \{MMM, MMF, MFM, MFF, FMM, FMF, FFM, FFF\}$
$E = \{MMF, MFM, FMM, MMM\}$

5. $\Pr[1] = \Pr[2] = \Pr[5] = \Pr[6] = \frac{1}{10}$
$\Pr[3] = \Pr[4] = \frac{3}{10}$

7. $4 \times 12 = 48$

9. (*a*) $3 \times 14 \times 14 = 588$
(*b*) $3 \times 14 \times 13 = 546$

11. (*a*) 24 (*b*) $C(4, 2) = 6$

13. 24

15. (*a*) $7! = 5040$ (*b*) $\frac{1}{7}$ (*c*) $C(7,3) - C(4,3) - C(3,3) = 30$

17. $\frac{14}{36} = \frac{7}{18}$

19. $P(5, 2) \cdot P(6, 2) = 600$

21. $\frac{1}{3}$

23. (*a*) $\frac{1}{4}$ (*b*) $\frac{1}{12}$

25. (*a*) $C(24, 3) = 2024$ (*b*) $C(6, 3) = 20$ (*c*) $C(18, 3) = 816$ (*d*) $816 + C(18, 2) \cdot 6 = 1734$

27. (*a*) $\dfrac{48}{C(52, 5)}$ (*b*) $\dfrac{13 \cdot 48}{C(52, 5)}$ (*c*) $\dfrac{4 \cdot 10}{C(52, 5)}$

29. 30 ways

31. (*a*) $\dfrac{C(3, 2)}{C(6, 2)} = \dfrac{1}{5}$ (*b*) $\dfrac{C(6, 2) - C(3, 2)}{C(6, 2)} = \dfrac{4}{5}$ (*c*) $\frac{4}{15}$

33. 420

35. (*a*) $P(6, 2) \cdot 3 \cdot 2 = 180$ (*b*) 90 (*c*) $\frac{1}{6}$

37. $18 \le n \le 32$

Chapter 3

Section 3.1

1. (*a*) .2 (*b*) .5 (*c*) .05

3. (*a*) .2 (*b*) .35 (*c*) .3

5. (*a*) .55 (*b*) .25 (*c*) .4

7. (*a*) .9 (*b*) .2 (*c*) .3

9. (*a*) $\frac{1}{8}$ (*b*) $\frac{1}{4}$ (*c*) $\frac{7}{8}$

11. (*a*) .35 (*b*) 1 (*c*) .8 (*d*) .9 (*e*) .25 (*f*) 0

13. .2

15. .3

17. (*a*) $\frac{4}{7}$ (*b*) $\frac{6}{7}$

19. .86

21. $.6 + .7 - .5 = .8$

23. $.65 + .25 - .85 = .05$

25. (*a*) and (*d*)

27. .1

29. Note that $E \cap F'$, $E \cap F$, and $E' \cap F$ form a partition of $E \cup F$.
By (iii): $\Pr[E \cup F] = \Pr[(E \cap F') \cup (E \cap F) \cup (E' \cap F)]$
$= \Pr[E \cap F'] + \Pr[E \cap F] + \Pr[E' \cap F]$
Now $E \cap F'$ and $E \cap F$ partition E, and $E \cap F$ and $E' \cap F$ partition F.
Thus $\Pr[E \cup F] = \Pr[E \cap F'] + \Pr[E \cap F]$
$+ \Pr[E' \cap F] + \Pr[E \cap F] - \Pr[E \cap F]$
$= \Pr[E] + \Pr[F] - \Pr[E \cap F]$

31. Yes. Using a computer we find that K is approximately 0.1928

Section 3.2

1. $\Pr[A|B] = \frac{2}{3}$, $\Pr[B|A] = \frac{4}{7}$, $\Pr[A \cup B] = .7 + .6 - .4 = .9$

3. $\Pr[F|E] = \Pr[E|F] = 0$

5. $\Pr[B|A] = \frac{3}{8}$, $\Pr[B|A'] = \frac{1}{8}$

7. $\frac{1}{2} = \Pr[A|B] = \Pr[A \cap B]/\Pr[B]$ so that $\Pr[B] = \frac{2}{5}$
$\frac{1}{3} = \Pr[B|A] = \dfrac{\Pr[A \cap B]}{\Pr[A]}$ so that $\Pr[A] = \frac{3}{5}$
Consequently, $\Pr[A] \times \Pr[B] = \frac{6}{25} \ne \frac{1}{5} = \Pr[A \cap B]$; so A and B are *not* independent.

9. (*a*) .5 (*b*) .2

11. $\Pr[A] = \frac{8}{9}$, $\Pr[B] = \frac{4}{9}$, $\Pr[A \cap B] = \frac{1}{3}$

13. Not independent

15. If A and B are independent and if $A \cap B = \varnothing$, then $\Pr[A \cap B] = \Pr[\varnothing] = 0 = \Pr[A] \cdot \Pr[B]$. But the product of two nonzero numbers is always nonzero, so either $\Pr[A] = 0$ or $\Pr[B] = 0$.

17. (*a*) $\frac{2}{5}$ (*b*) $\frac{1}{5}$ (*c*) $\frac{2}{11}$

19. $\frac{2}{5}$

21. $\frac{2}{3}$

23. $\frac{2}{7}$

25. $\frac{2}{11}$

27. (*a*) $\frac{\frac{1}{66}}{\frac{10}{66}} = \frac{1}{10}$ (*b*) $\frac{\frac{2(3)}{66}}{\frac{10}{66}} = \frac{6}{10}$

29. $\Pr[2 \text{ gray} \mid 2 \text{ males}] = \frac{\frac{3}{66}}{\frac{21}{66}} = \frac{3}{21}$

$\Pr[2 \text{ white} \mid 2 \text{ males}] = \frac{\frac{6}{66}}{\frac{21}{66}} = \frac{6}{21}$

$\Pr[1 \text{ gray and } 1 \text{ white} \mid 2 \text{ males}] = \frac{\frac{3(4)}{66}}{\frac{21}{66}} = \frac{12}{21}$

Therefore, one of each color is the most likely outcome.

31. (Figure C3.1)

33. Three cards in a run

Section 3.3

1. (Figure C3.2)
3. (*a*) (Figure C3.3) (*b*) $\frac{1}{3} + \frac{6}{30} = \frac{8}{15}$
5. (Figure C3.4) *Let* S = small, NS = not small, O = one, NO = not one
7. (*a*) $\frac{5}{9}$ (*b*) $\frac{5}{6}$

9. .20

11. (*a*) .67 (*b*) .845

FIGURE C3.1

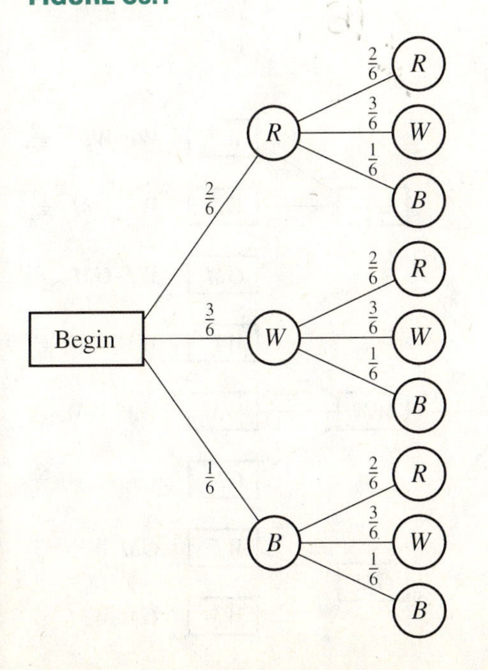

FIGURE C3.2

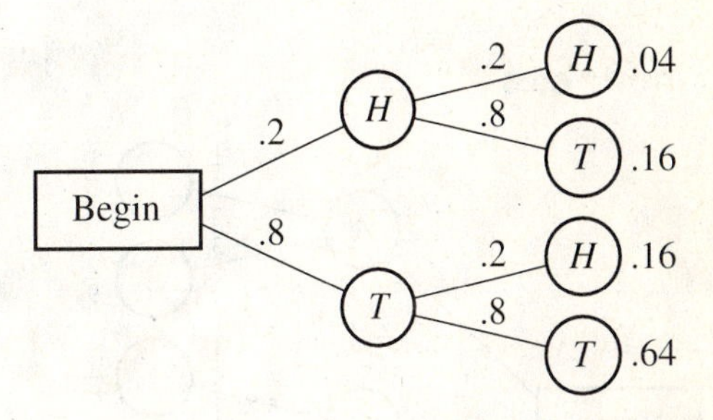

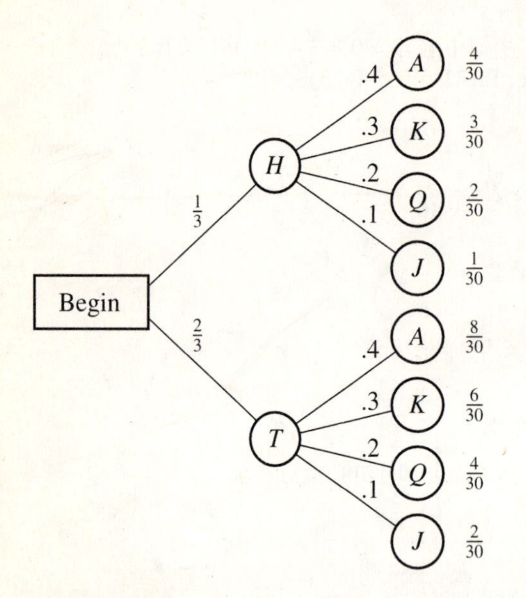

FIGURE C3.3

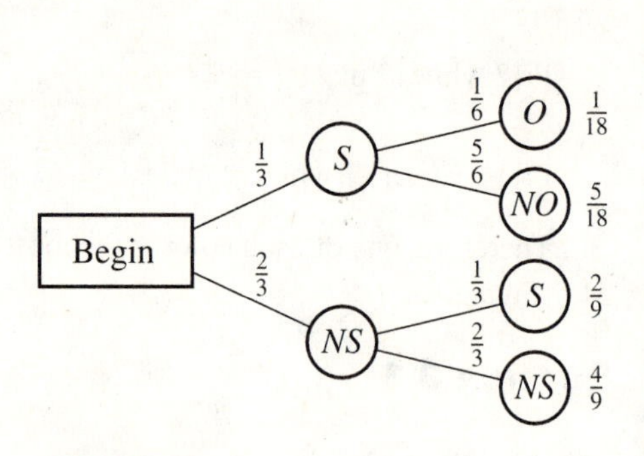

FIGURE C3.4

13. (*a*) $\frac{4}{7}$ (*b*) $\frac{3}{5}$
15. $\frac{5}{18}$
17. (Figure C3.5)
19. (*a*) (Figure C3.6) (*b*) $\frac{6}{56} + \frac{3}{56} + \frac{3}{56} = \frac{12}{56}$
21. (Figure C3.7) $\Pr[\text{2 red balls}] = \frac{1}{3}$, $\Pr[\text{2 red} \mid \text{first is red}] = \frac{1}{2}$

FIGURE C3.5

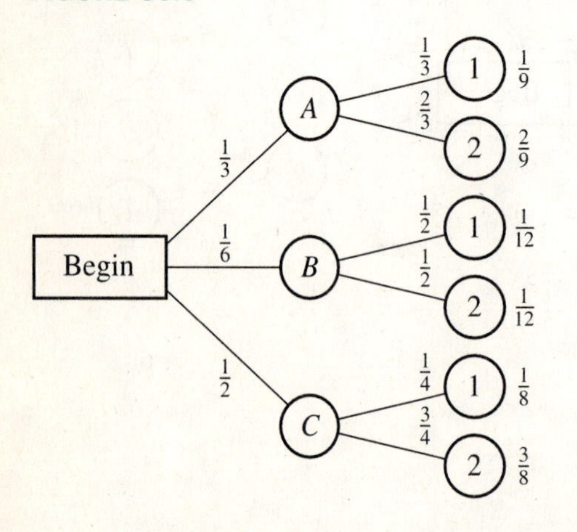

FIGURE C3.6

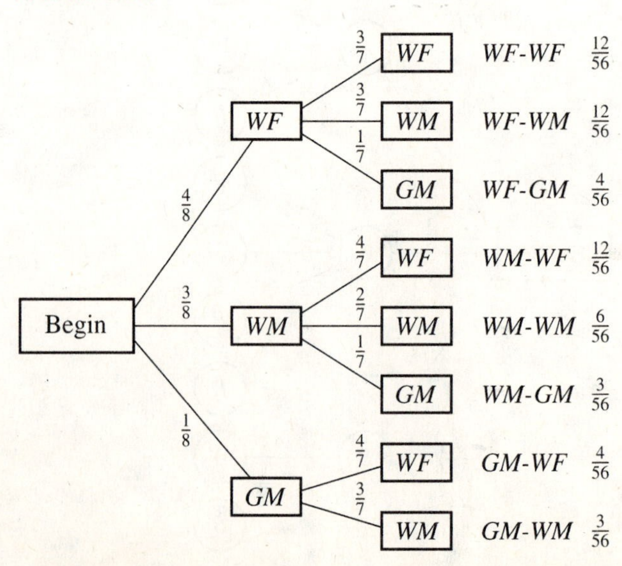

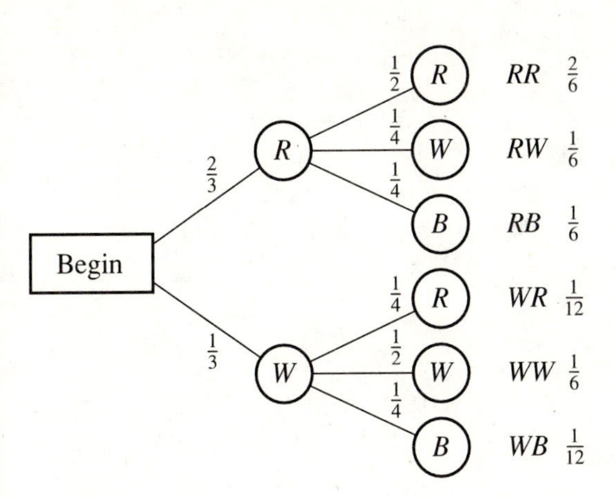

FIGURE C3.7

23. $\frac{21}{64}$

25. Pr[exactly 2 profitable years in first 3 years] = (.2)(.6)(.4) + (.2)(.4)(.4) + (.8)(.4)(.6) = .272

27. Pr[makes exactly 2 of next 3] $= \frac{2}{18} + \frac{2}{18} + \frac{1}{18} = \frac{5}{18}$

29. $\{0BB, 0BY, 0YB, 0YY, 1BB, 1BY, 1YB, 2BB\}$
$\Pr[0BB] = \frac{1}{7}$, $\Pr[0BY] = \frac{1}{7}$, $\Pr[0YB] = \frac{1}{7}$, $\Pr[0YY] = \frac{1}{21}$,
$\Pr[1BB] = \frac{2}{7}$, $\Pr[1BY] = \frac{2}{21}$, $\Pr[1YB] = \frac{2}{21}$, $\Pr[2BB] = \frac{1}{21}$

31. .007969

Section 3.4

1. (*a*) $\frac{2}{3}$ (*b*) $\frac{31}{60}$ (*c*) $\frac{2}{5}$ (*d*) $\frac{16}{31}$

3. $\frac{665}{935}$

5. (*a*) $\frac{1}{5}$ (*b*) $\frac{1}{5}$

7. $\Pr[X \mid B] = \frac{1}{2}$, $\Pr[Y \mid B \text{ or } W] = \frac{1}{4}$

9. $\Pr[\text{first B} \mid \text{second R}] = \dfrac{\frac{3}{6}\cdot\frac{2}{5}}{\frac{2}{6}\cdot\frac{1}{5}+\frac{1}{6}\cdot\frac{2}{5}+\frac{3}{6}\cdot\frac{2}{5}} = \dfrac{3}{5}$

11. $\frac{25}{61}$

13. (*a*) .29 (*b*) $\frac{1}{29}$

15. $\dfrac{.6(.5)}{.6(.5) + .2(.75) + .2(.6)} = \dfrac{.30}{.57} = \dfrac{10}{19}$

17. $\dfrac{.5(.4)}{.5(.4) + .25(.7) + .25(.5)} = \dfrac{.20}{.50} = .4$

19. $\frac{67}{94}$

21. (*a*) .22 (*b*) $\frac{10}{22}$

23. $\frac{245}{690}$

25. $\dfrac{.6(.3)}{.6(.3) + .3(.4) + .1(.7)} = \dfrac{.18}{.37} = \dfrac{18}{37}$

27. $\dfrac{\frac{2}{6}(\frac{2}{6})}{\frac{1}{6}(\frac{2}{5}) + \frac{2}{6}(\frac{2}{6}) + \frac{3}{6}(\frac{2}{5})} = \dfrac{20}{68} = \dfrac{5}{17}$

29. (*a*) $\frac{2}{3}$ (*b*) $\frac{4}{9}$

31. Pr[Best Air] $= \frac{10}{58}$, Pr[East − West] $= \frac{28}{58}$, Pr[Bi-Coastal] $= \frac{20}{58}$

33. Undergraduate degree

Section 3.5

1. (*a*) $C(3, 1)(.2)^1(.8)^2 = .384$ (*b*) $C(3, 2)(.4)^2(.6)^1 = .288$ (*c*) $C(3, 3)(\frac{2}{3})^3 = \frac{8}{27}$
3. (*a*) $C(6, 4)(.6)^4(.4)^2 = .31104$ (*b*) $C(5, 3)(.4)^3(.6)^2 = .2304$
5. (*a*) $1 - C(6, 0)(.8)^6 - C(6, 1)(.2)^1(.8)^5 - C(6, 2)(.2)^2(.8)^4 = .09888$
   (*b*) $C(5, 2)(.8)^2(.2)^3 + C(5, 1)(.8)^1(.2)^4 + C(5, 0)(.2)^5 = .05792$
7. $1 - C(5, 5)(\frac{2}{5})^5 - C(5, 0)(\frac{3}{5})^5 = .912$
9. $C(5, 2)(\frac{3}{4})^2(\frac{1}{4})^3 + C(5, 3)(\frac{3}{4})^3(\frac{1}{4})^2 = \frac{360}{1024}$
11. $1 - (\frac{2}{3})^4 - (\frac{1}{3})^4$
13. (*a*) $15(.4)^2(.6)^4 = .3110$
    (*b*) $C(6, 0)(.4)^0(.6)^6 + C(6, 1)(.4)^1(.6)^5 + C(6, 2)(.4)^2(.6)^4 = (.6)^6 + 6(.4)(.6)^5 + 15(.4)^2(.6)^4$
    $= .0467 + .1866 + .3110 = .5443$
15. $C(20, 18)(\frac{1}{2})^{20}$
17. $C(8, 2)(\frac{1}{4})^2(\frac{3}{4})^6$
19. $1 - C(3, 0)(\frac{1}{10})^0(\frac{9}{10})^3 = .271$
21.

Number of Lines	Probability of Catching at Least 4 Fish
4	$C(4, 4)(.7)^4 = (.7)^4 = .2401$
5	$C(5, 4)(.6)^4(.4) + C(5, 5)(.6)^5 = 5(.6)^4(.4) + 1(.6)^5 = .3370$
6	$C(6, 4)(.5)^4(.5)^2 + C(6, 5)(.5)^5(.5) + C(6, 6)(.5)^6 = 15(.5)^6 + 6(.5)^6 + (.5)^6 = .3438$

Therefore, the fisherman's chances of catching at least 4 fish are best with 6 lines.

23. 3
25. 29
27. $\frac{11}{64}$
29. $\frac{47}{128}$
31. $[210(\frac{1}{5})^4(\frac{4}{5})^6]^2 = .00776$
33. .2500
35. .6080

Review Exercises

1. (*a*) .54 (*b*) 1.0 (*c*) .23 (*d*) .27
3. (*a*) 0 (*b*) .35 (*c*) .5
5. (*a*) $\frac{1}{7}$ (*b*) 0 (*c*) $\frac{1}{10}$
7. (*a*) .6 (*b*) .5 (*c*) .6
9. $\frac{1}{2}$
11. (*a*) Independent (*b*) not independent (*c*) independent
13. $\frac{2}{11}$
15. 1.0 (if exactly one mouse is gray, the other must be white, therefore male)
17. Pr[exactly one evaluated favorably] = .03 + .16 + .6 = .25
19. $C(10, 3)(\frac{1}{3})^3(\frac{2}{3})^7 + C(10, 7)(\frac{1}{3})^7(\frac{2}{3})^3$
21. $C(10, 4)p^4(1 - p)^6 = C(10, 5)p^5(1 - p)^5$ implies $p = \frac{5}{11}$.
23. (Figure C3.8)
25. $\frac{3}{8}$
27. (*a*) $\frac{3}{5}$ (*b*) $\frac{3}{5}$ (*c*) $\frac{7}{12}$ (*d*) $\frac{7}{12}$
29. (*a*) $\frac{29}{40}$ (*b*) $\frac{11}{15}$ (*c*) $\frac{11}{29}$ (*d*) $\frac{1}{3}$
31. $$\frac{C(10, 6)(.6)^6(.4)^4 + C(10, 7)(.6)^7(.4)^3 + C(10, 8)(.6)^8(.4)^2 + C(10, 9)(.6)^9(.4)^1}{1 - C(10, 0)(.6)^0(.4)^{10} - C(10, 10)(.6)^{10}(.4)^0}$$
33. (*a*) .00376 (*b*) .0004
35. .9412

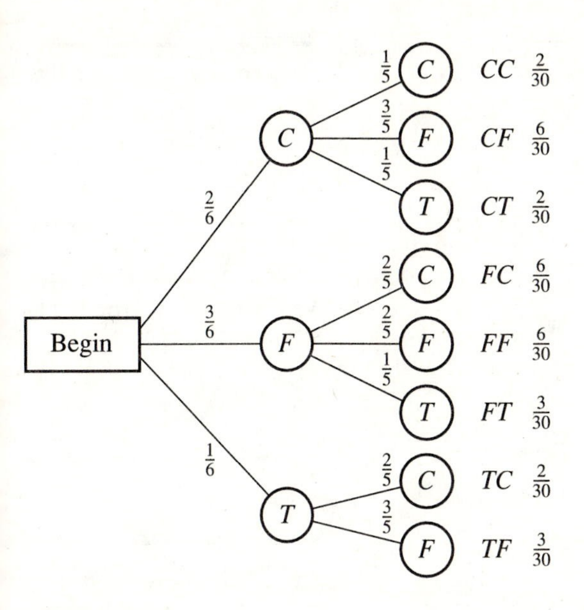

FIGURE C3.8

CHAPTER 4

Section 4.1

1.

Value of X	Probability
−2	$\frac{1}{4}$
0	$\frac{1}{2}$
2	$\frac{1}{4}$

3.

Value of X	Probability
0	$\frac{1}{2}$
1	$\frac{1}{6}$
3	$\frac{1}{6}$
5	$\frac{1}{6}$

5.

Value of X	Probability
0	$\frac{1}{15}$
1	$\frac{8}{15}$
2	$\frac{6}{15}$

7.

Value of X	Probability
−1	$\frac{3}{4}$
2	$\frac{1}{4}$

9.

Value of X	Probability
0	.008
1	.096
2	.384
3	.512

11.

Value of X	Probability
0	$\frac{188}{221}$
1	$\frac{32}{221}$
2	$\frac{1}{221}$

13.

Value of X	Probability
0	$C(36, 3)/C(52, 3) = \frac{7140}{22,100}$
2	$C(36, 2)C(12, 1)/C(52, 3) = \frac{7560}{22,100}$
4	$C(36, 1)C(12, 2)/C(52, 3) = \frac{2376}{22,100}$
5	$C(36, 2)C(4, 1)/C(52, 3) = \frac{2520}{22,100}$
6	$C(12, 3)/C(52, 3) = \frac{220}{22,100}$
7	$C(36, 1)C(12, 1)C(4, 1)/C(52, 3) = \frac{1728}{22,100}$
9	$C(12, 2)C(4, 1)/C(52, 3) = \frac{264}{22,100}$
10	$C(36, 1)C(4, 2)/C(52, 3) = \frac{216}{22,100}$
12	$C(12, 1)C(4, 2)/C(52, 3) = \frac{72}{22,100}$
15	$C(4, 3)/C(52, 3) = \frac{4}{22,100}$

15.

Value of X	Probability
10	$\frac{1}{10}$
15	$\frac{6}{10}$
20	$\frac{3}{10}$

17.

Value of X	Probability
0	$\frac{1}{10}$
1	$\frac{6}{10}$
2	$\frac{3}{10}$

19.

Value of X	Probability
0	$C(48, 5)/C(52, 5)$
1	$C(48, 4)C(4, 1)/C(52, 5)$
2	$C(48, 3)C(4, 2)/C(52, 5)$
3	$C(48, 2)C(4, 3)/C(52, 5)$
4	$C(48, 1)C(4, 4)/C(52, 5)$

21.

Value of X	Probability
0	$\frac{20}{56}$
1	$\frac{30}{56}$
2	$\frac{6}{56}$

23.

Value of X	Probability
0	.4096
1	.4096
2	.1536
3	.0256
4	.0016

25.

Value of X	Probability
1	$\frac{18}{36} = .50000$
2	$\frac{15}{36} = .41667$
3	$\frac{17}{216} = .07870$
4	$\frac{1}{216} = .00463$

27.

Value of X	Probability
3	$\frac{9}{27}$
4	$\frac{10}{27}$
5	$\frac{8}{27}$

29.

Value of X	Probability
0	$\frac{13}{72}$
1	$\frac{34}{72}$
2	$\frac{25}{72}$

31.

Value of X	Probability
−1	.9999
500	.0001

Section 4.2

1. $\frac{5}{3}$
3. $\frac{91}{21}$
5. .9
7.

Value of X	Probability	Product
−1	.3	−.3
−2	.2	−.4
1.5	.2	.3
2	.3	.6

$E[X] = .2$

9. \$0.15
11. 2.97
13. 4
15. $\frac{7}{6}$
17. $E[X] = 2(\frac{1}{4}) + 3(\frac{1}{8}) + 4(\frac{5}{8}) = \frac{27}{8}$
19. −\$9.30—a loss of \$9.30
21. 12.25
23. $E[X] = 0(\frac{2}{9}) + 1(\frac{3}{9}) + 2(\frac{4}{9}) = \frac{11}{9}$
25. 1.4
27. 84.788
29. −2
31. $\sigma[x] = 1.4907$

Section 4.3

1. (*a*) .3413 (*b*) .3413
3. (*a*) .2580 (*b*) .1179
5. $a = 0, b = 3$
7. (*a*) .4332 (*b*) .2417
9. .3023
11. (*a*)

Value of X	Probability
0	0.0039
1	0.0312
2	0.1094
3	0.2188
4	0.2734
5	0.2188
6	0.1094
7	0.0312
8	0.0039

(*b*) (Figure C4.1)

FIGURE C4.1

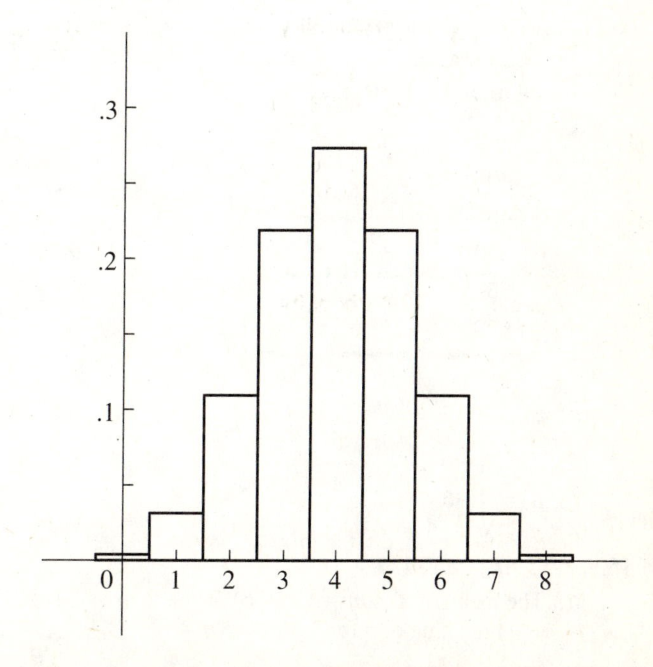

13. (*a*) No (*b*) yes (*c*) yes (*d*) yes

15. .6846

17. $\Pr[X \le 1] > \Pr[X \ge 3.5]$

19. $\sigma = .0593$

21. .3811

23. About 91 senior males are expected to be 6 feet 2 inches or taller, and about 23 are expected to be 5 feet 4 inches or shorter.

25. 1

27. .0088

29. .9666

Review Exercises

1.

Value of X	Probability
-1	$\frac{1}{6}$
2	$\frac{1}{3}$
5	$\frac{1}{2}$

3.

Value of X	Probability
0	$\frac{1}{10}$
1	$\frac{6}{10}$
2	$\frac{3}{10}$

5. (*a*) 2, 3, 4, 5

(*b*)

Value of X	Probability
2	$\frac{1}{4}$
3	$\frac{1}{8}$
4	$\frac{1}{8}$
5	$\frac{1}{2}$

7.

Value of X	Probability
$-.4$	$\frac{10}{56}$
$-.1$	$\frac{30}{56}$
.2	$\frac{15}{56}$
.5	$\frac{1}{56}$

9. $E[X] = 5(\frac{4}{9}) + 10(\frac{3}{9}) + 25(\frac{2}{9}) = \frac{100}{9}$

11. (*a*)

Value of X	Probability
350	$\frac{2}{35}$
400	$\frac{5}{35}$
450	$\frac{12}{35}$
500	$\frac{9}{35}$
550	$\frac{6}{35}$
600	$\frac{1}{35}$

(*b*) $E[Y] = \dfrac{3300}{7}$

(*c*) $\sigma = 58.90$

13. (*a*) $\mu = .4(365) = 146$

(*b*) $\sigma = \sqrt{87.6} = 9.36$

15. $\mu = 5$; $\sigma = 2.179$

17. (*a*) The mean of X is $np = 100(.6)$. The mean of Y is $150p$. If the two means are equal, then $150p = 60$ and the value of p for Y must be $p = \frac{60}{150} = .4$.

(*b*) The variance for X is $100(.6)(.4) = 24$, and the variance for Y is $150(.4)(.6) = 36$. The variance for Y is larger.

19. Y takes values between -3 and 3, inclusive.

21. (*a*) .7327 (*b*) .3446 (*c*) .8531

23. (*a*) $\mu = 54$, $\sigma = 6$ (*b*) $np = 54 > 5$, $n(1-p) = 108 > 5$, and normal approximation is valid (*c*) .9247

25. .0466

27. (*a*) 0 (*b*) .0044

29. Expected income from bonuses = \$76.56; Standard deviation of income from bonuses = \$80.51

31.

Value of $X + Y$	Probability	Product
0	$\frac{6}{60}$	0
1	$\frac{24}{60}$	.4
2	$\frac{6}{60}$	.2
6	$\frac{12}{60}$	1.2
7	$\frac{12}{60}$	1.4

$E[X + Y] = 3.2$

33.

Value of X	Probability
0	.352
5	.432
20	.216

$E[X] = 6.48$, $\sigma = 7.43$

35. Expected gain to the state is \$764,012.91

CHAPTER 5

Section 5.1

1. (*a*) (Figure C5.1) (*b*) (Figure C5.2)

3. Let x = Canadian price per liter, and let y = USA price per gallon; $y = 3.027x$; 65 cents/liter ≈ 1.97 dollars/gallon

5. (Figure C5.3)

7. (Figure C5.4) (*a*) x intercept $(2, 0)$; y intercept $(0, -2)$ (Figure C5.5) (*b*) no x intercept; y intercept $(0, 3)$; (Figure C5.6); (*c*) x intercept $(-9, 0)$; y intercept $(0, \frac{9}{2})$ (Figure C5.7); (*d*) x intercept $(0, 0)$, using our definition, there is no y intercept

9. (*a*) $\frac{3}{2}$ (*b*) 1

11. (Figure C5.8); (*a*) $y = (-\frac{2}{3})x + \frac{11}{3}$ (*b*) (Figure C5.9); $y = x + 2$

13. Slope $\frac{1}{3}$; x intercept $(-6, 0)$; y intercept $(0, 2)$

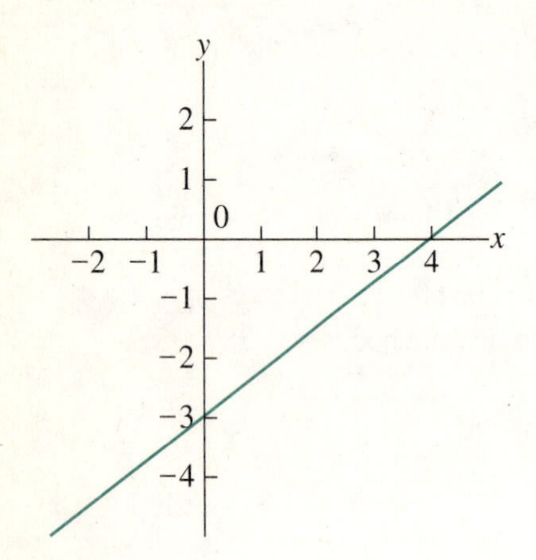

FIGURE C5.1

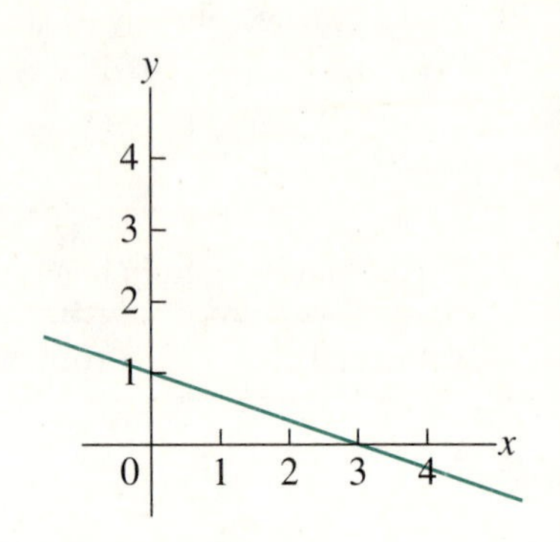

FIGURE C5.2

15. (Figure C5.10); $y = -x - 3$
17. 18 defects
19. (Figure C5.11); $y = (\frac{5}{9})(x - 32)$; slope $= \frac{5}{9}$
    x-axis is degrees fahrenheit
    y-axis is degrees celsius
21. (*a*) $y = .3x + 40$ (*b*) $y = .5x + 30$; (*b*) is less expensive if $x = 30$
23. 3000 miles

FIGURE C5.3

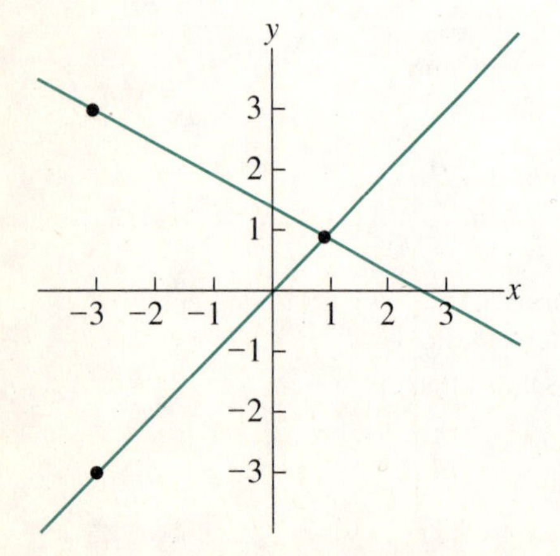

FIGURE C5.4

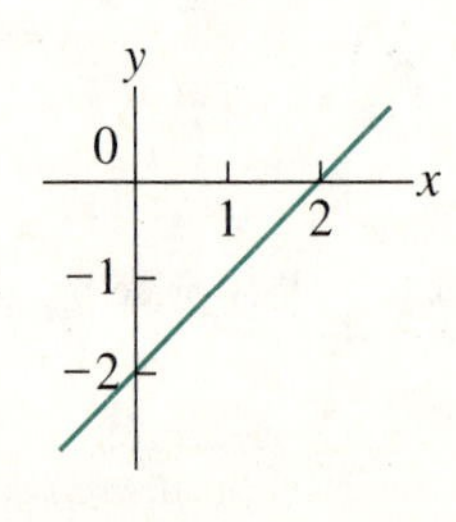

FIGURE C5.5

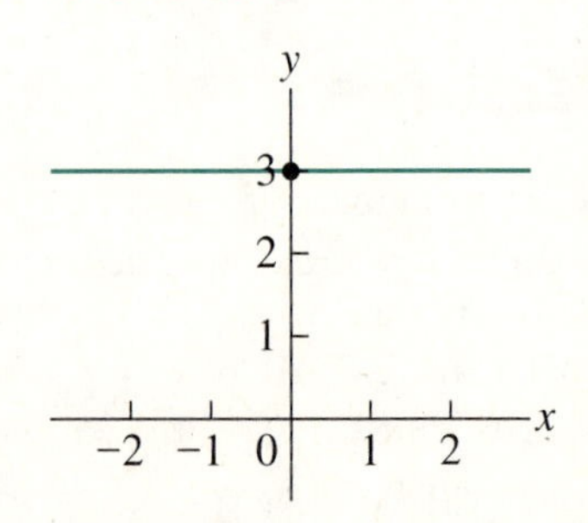

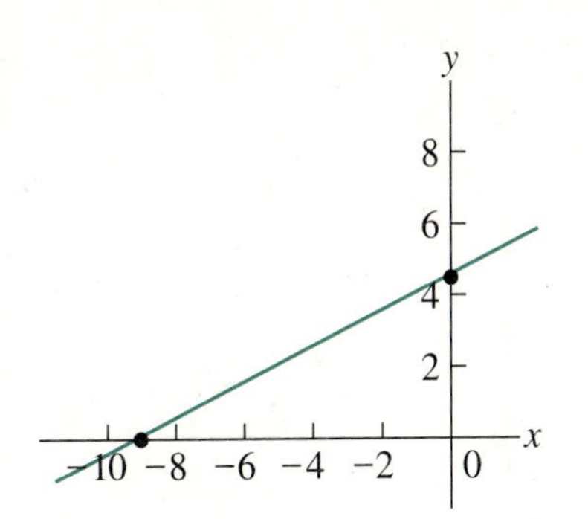

FIGURE C5.6

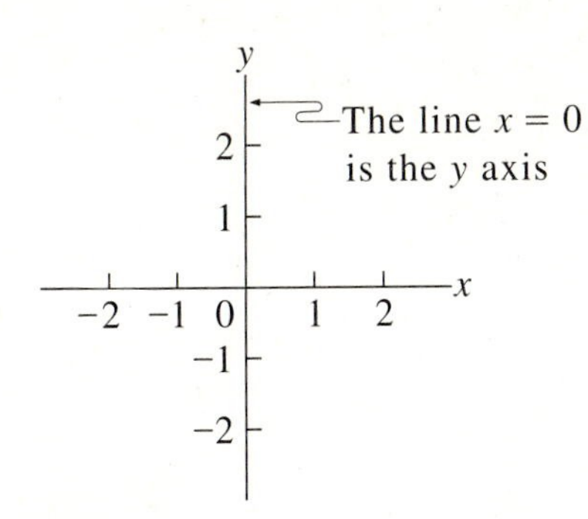

FIGURE C5.7

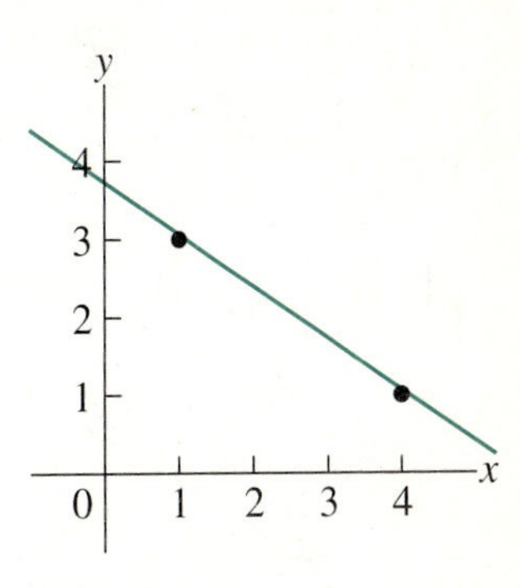

FIGURE C5.8

25. $Ax_1 + By_1 = C$ and $Ax_2 + By_2 = C$
Subtract to obtain:
$Ax_1 + By_1 - (Ax_2 + By_2) = 0$
$A(x_1 - x_2) + B(y_1 - y_2) = 0$
Proof by contradiction:
Suppose $x_1 = x_2$; then $x_1 - x_2 = 0$.
By the above equation: $B(y_1 - y_2) = 0$, and since $B \neq 0$ we conclude $y_1 - y_2 = 0$ or $y_1 = y_2$.
So $x_1 = x_2$ and $y_1 = y_2$ and therefore the points are not distinct.

FIGURE C5.9

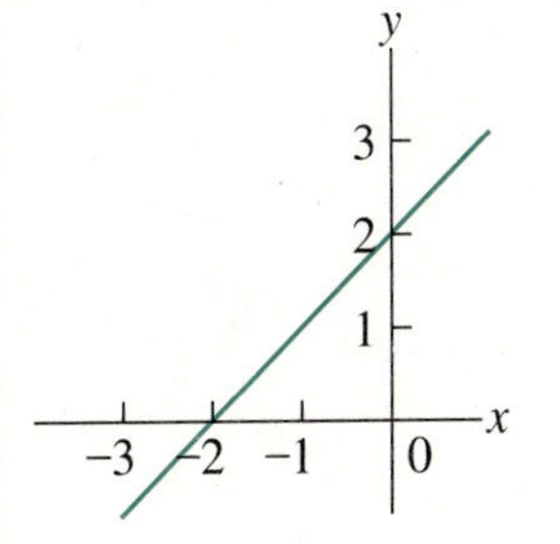

FIGURE C5.10

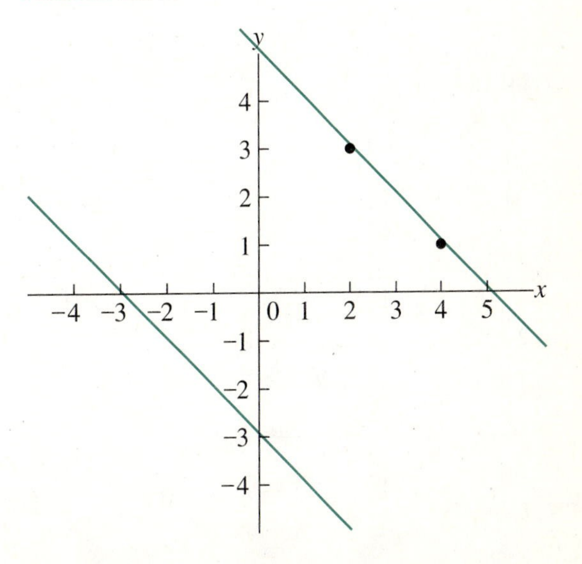

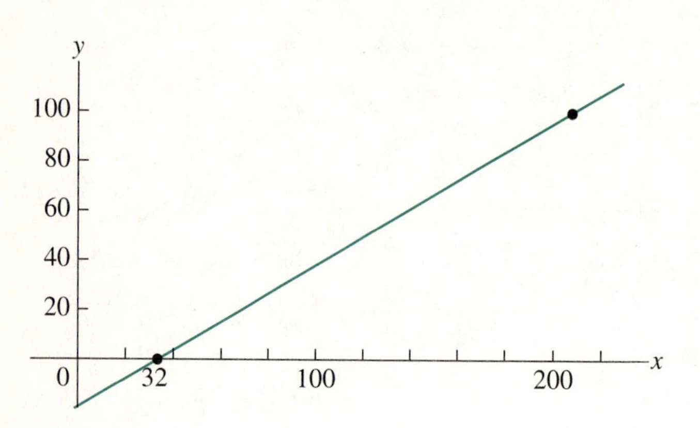

FIGURE C5.11

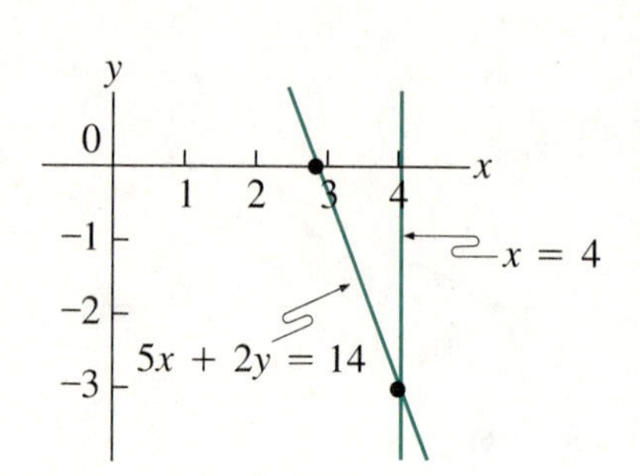

FIGURE C5.12

Section 5.2

1. l = length; w = width; $l = 1.3w$; $l + w = 115$
3. $4x + y = 300$
   $2x + y = 60$
   $x + (\frac{1}{3})y = 75$
5. (*a*) (Figure C5.12); $x = 4$; $y = -3$ (*b*) (Figure C5.13); $x = 5$; $y = 3$
7. (*a*) (Figure C5.14); No solution—lines are parallel; (*b*) (Figure C5.15); $x = -12$; $y = -10$
9. (*a*) Lines are the same, solution set is the whole line; (*b*) (Figure C5.16); $x = -1$; $y = 6$
11. Width = 50; length = 65

FIGURE C5.13

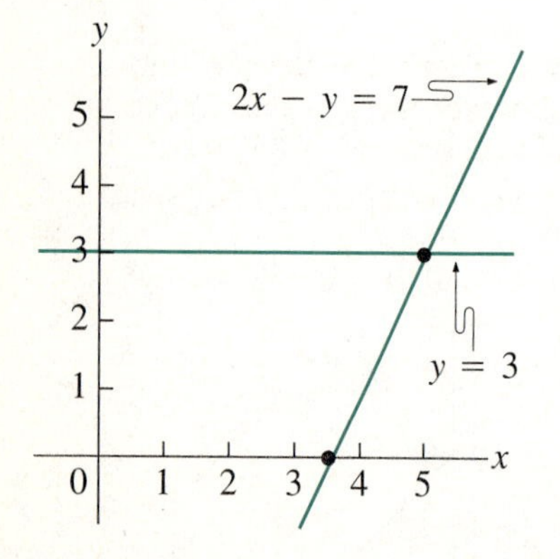

FIGURE C5.14

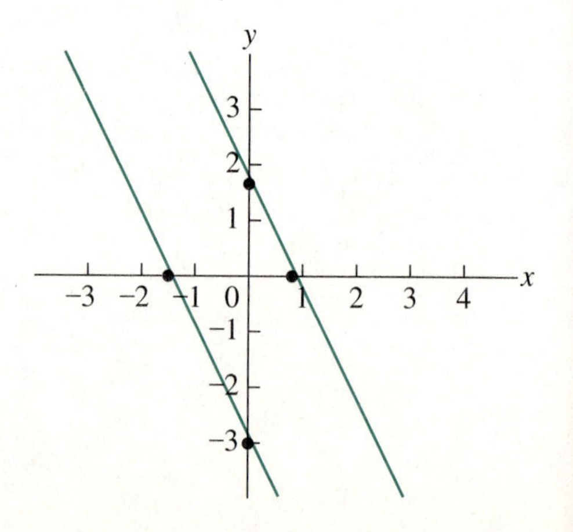

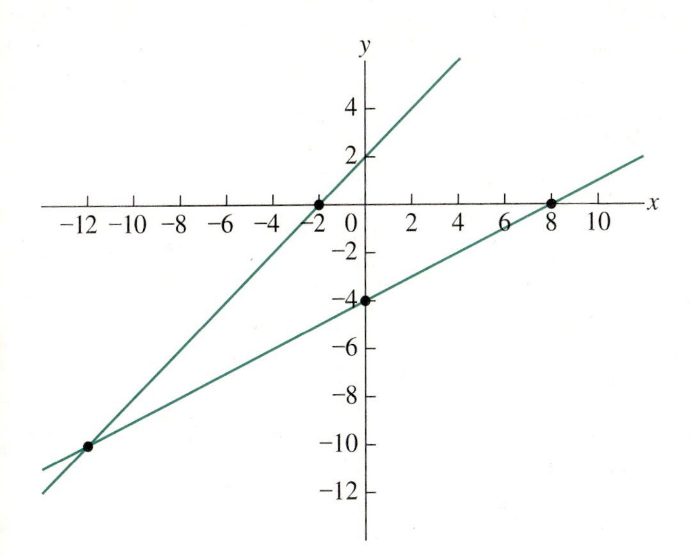

FIGURE C5.15

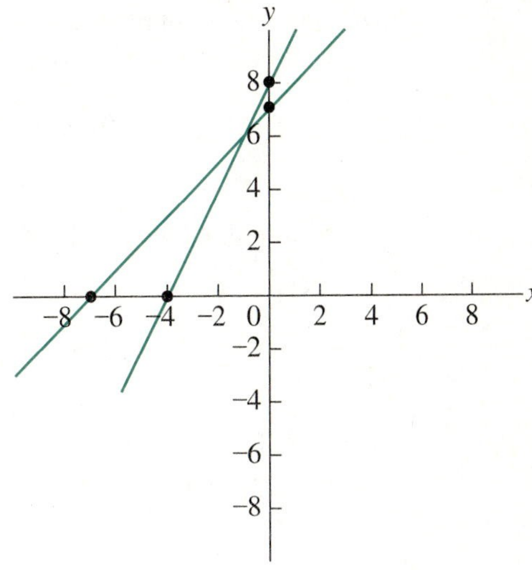

FIGURE C5.16

13. No solution. Using the solution to the first two equations, the third equation gives: $x + (\frac{1}{3})y = 70 + (\frac{1}{3})20 = 77.67 \neq 75$
15. $50{,}000 + 7500x = 1{,}000{,}000 - 125{,}000x$ gives $x = 7.17$ years.
17. \$400,000 in utility bonds; \$600,000 in savings
19. -40 degrees celsius $= -40$ degrees fahrenheit
21. $(.1)x + (.2)(100{,}000 - x) = 17{,}000$ gives $x = 30{,}000$. Thus, Fred invested \$30,000 in the first stock and \$70,000 in the second stock.
23. \$2,500,000 in bonds; \$5,000,000 in stocks; \$2,500,000 in real estate
25. $a = \frac{1}{2}$; $b = \frac{5}{2}$
27. Tom is lying. His figures are not compatible with either Dick's or Harry's. In each case, the resulting system of equations does not have a solution with nonnegative prices. However, the claims by Dick and Harry are compatible. They are satisfied if each apple costs 35 cents and each orange costs 30 cents.
29. No, with only 25 ounces of raisins, all dough and apples cannot be used. He should make 50 large and no small muffins to use all raisins and apples.

Section 5.3

1. x = number of airplanes; y = number of boats; z = number of cars

$$\begin{aligned} 100x + 50y + 50z &= 10{,}500 \\ 10x + 100y &= 1{,}500 \\ 200x + 50y + 150z &= 25{,}500 \end{aligned}$$

3. x = return on stocks; y = return on bonds; z = return on money market funds

$$\begin{aligned} 50{,}000x + 25{,}000y + 25{,}000z &= 9{,}000 \\ (100{,}000/3)x + (100{,}000/3)y + (100{,}000/3)z &= 6000 \\ 50{,}000x + 50{,}000y &= 8{,}000 \end{aligned}$$

5. (*a*) Not reduced: does not satisfy A, B, or C (*b*) reduced (*c*) not reduced: does not satisfy B.

7. (*a*) $\left[\begin{array}{ccc|c} 2 & -1 & -1 & 2 \\ 1 & 1 & 2 & 4 \end{array}\right]$ (*b*) $\left[\begin{array}{cc|c} 3 & -2 & 4 \\ 1 & 3 & -1 \end{array}\right]$

9. $3(1) + 2(2) - (-1) = 8$
   $1(1) + 1(2) + (-1) = 2$
   $2(1) + 1(2) - (-1) = 5$

11. $x = 3, y = 1, z = -2$
13. $x = 2 - (\frac{3}{2})z,\ y = -\frac{3}{4} - (\frac{1}{2})z$, z arbitrary
15. $x = 3 - z,\ y = -(\frac{1}{2})z$, z arbitrary
17. $x = 2, y = -2, z = -1$
19. $x = 0, y = 4, z = 0$
21. 80 feathers
23. 50 airplanes, 10 boats, 100 cars
25. The system of equations in Exercise 3 does not have a solution with all nonnegative values, so the goals cannot be achieved with nonnegative returns. If we allow negative returns, then the goals can be achieved with returns of 18% on stocks, −2% on bonds, and 2% on money market funds.
27. (*a*) x = time for old clients, y = time for new clients, z = time for office man., w = time for long range planning

$$\begin{aligned} 2(x + y) &= x + y + z + w \\ x &= 2y \\ y &= 2w \\ x + y + z + w &= 40 \end{aligned} \qquad \text{or} \qquad \begin{aligned} x + y &= 20 \\ x - 2y &= 0 \\ y - 2w &= 0 \\ x + y + z + w &= 40 \end{aligned}$$

   (*b*) $\frac{40}{3}$ hours for old clients; $\frac{20}{3}$ hours for new clients; $\frac{50}{3}$ hours for office management; $\frac{10}{3}$ hours for long range planning

29. (*a*) $x_1 = \frac{7}{4} - (\frac{7}{8})x_4,\ x_2 = (\frac{1}{2})x_3 - (\frac{5}{16})x_4 - \frac{3}{8}$, x_3 arbitrary, x_4 arbitrary
   (*b*) Setting $x_1 = 0$, we obtain $x_2 = -1 + (\frac{1}{2})x_3$, x_3 arbitrary, $x_4 = 2$
31. No solution, this is an inconsistent system.
33. (*a*) $x = 1 + (\frac{1}{5})z,\ y = 3 - (\frac{3}{5})z$, z arbitrary (*b*) The intersection with $y = 0$ is the point P: $x = 2$, $y = 0$, $z = 5$; The intersection with $z = 0$ is the point Q: $x = 1, y = 3, z = 0$ (*c*) (Figure C5.17)

Review Exercises

1. 2.5 years
3. (*a*) Slope $= \frac{1}{2}$, x intercept $= 6$, y intercept $= -3$ (*b*) slope $= \frac{3}{5}$, x intercept $= \frac{7}{3}$, y intercept $= -\frac{7}{5}$
5. $x = -\frac{5}{3}, y = 2$
7. $x = 2, y = -\frac{5}{2}$
9. Stephanie catches Tina in 22.5 minutes. At this time Rachael has run 11.25 laps = 4500 meters.
11. 9 years
13. $1(3) + 1(1) = 5 - 1$, $3(3) - 3(1) = 2(3)$
    $4 = 4$ $\qquad$ $6 = 6$

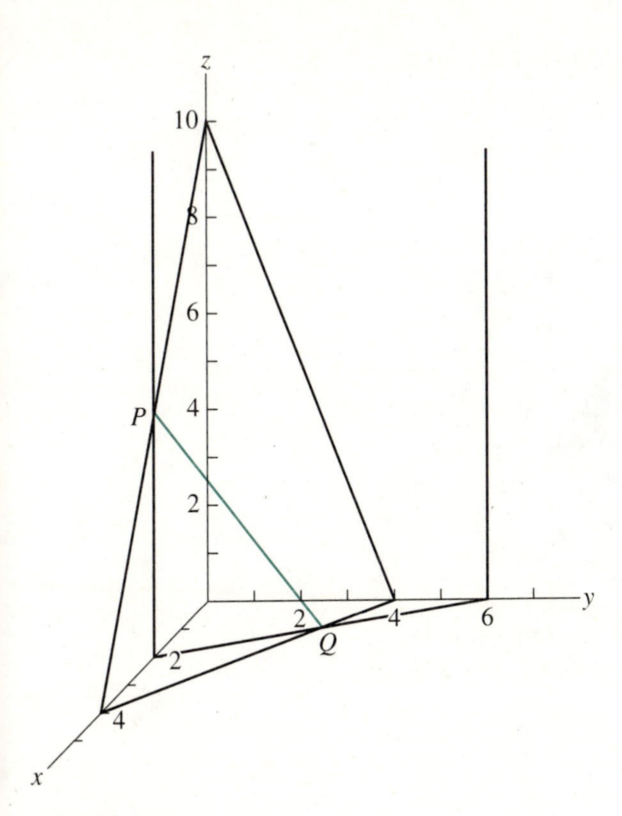

FIGURE C5.17

15. $x_1 = \frac{8}{5}, x_2 = \frac{6}{5}$

17. $x = 5 - 5z, y = 2 - 2z, z$ arbitrary

19. $x_1 = 0, x_2 = -1, x_3 = -2$

21. (*a*) $x_1 = \frac{3}{2} - (\frac{1}{2})x_4, x_2 = 3 - x_4, x_3 = \frac{5}{2} - \frac{3}{2}x_4, x_4$ arbitrary
    (*b*) $x_1 = 0, x_2 = 0, x_2 = -2, x_4 = 3$
    (*c*) $x_1 = 1, x_2 = 2, x_3 = 1, x_4 = 1$

23. x = age of truck in years; y = miles driven in units of 1,000; z = cost of maintenance for one year; $z = 50 + 70x + 20y$ (Figure C5.18) This plane is shown in Figure C5.18, and the point of intersection is (1, 10, 320).

25. *b*

27. *a*

29. *c*

31. (Figure C5.19)

CHAPTER 6

Section 6.1

1. (*a*) 3×3 (*b*) 3×2 (*c*) 3×1, column vector

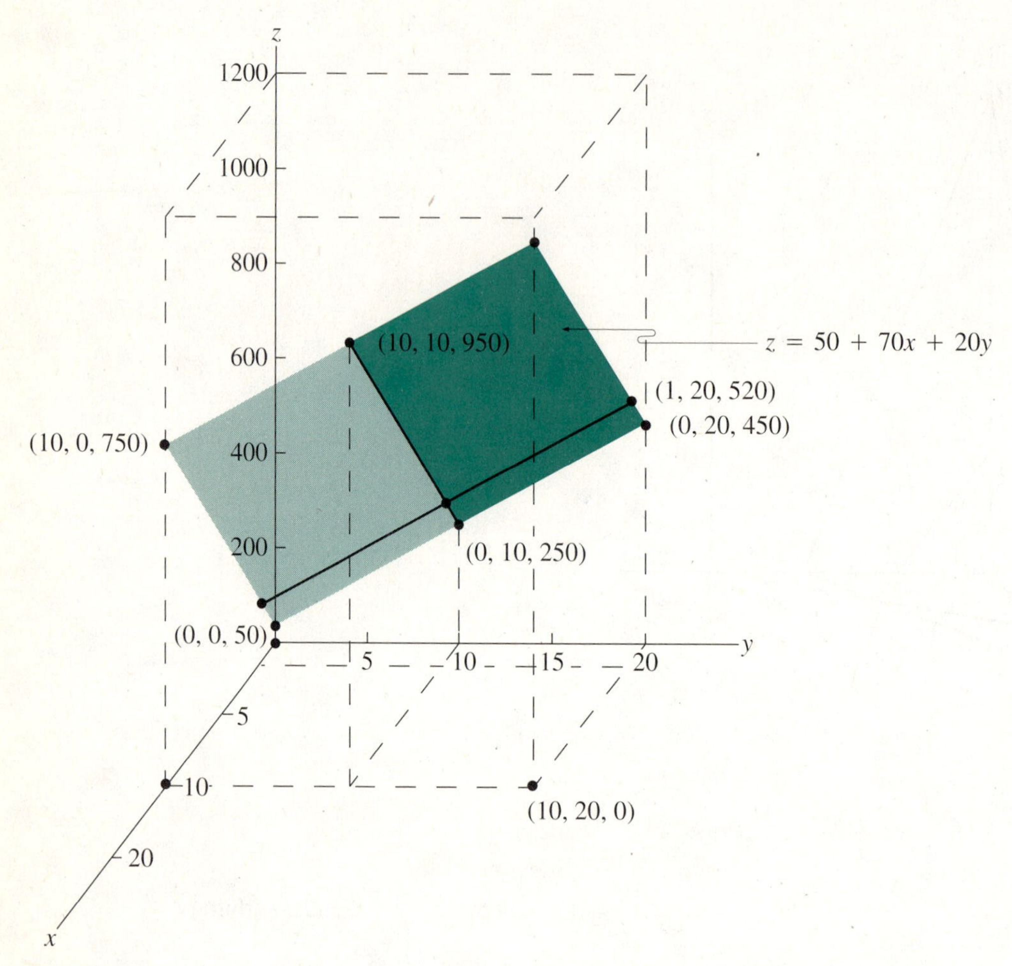

FIGURE C5.18

3. $A + B = \begin{bmatrix} 6 & 6 \\ 3 & 3 \end{bmatrix} \quad A - B = \begin{bmatrix} 4 & -2 \\ -1 & -3 \end{bmatrix}$

5. $3A + 2B = \begin{bmatrix} 7 & 1 & 7 \\ 4 & 4 & 3 \\ 12 & -14 & 0 \end{bmatrix}$

$$2A - B = \begin{bmatrix} 7 & -4 & 7 \\ -2 & 5 & 2 \\ 8 & 0 & -7 \end{bmatrix}$$

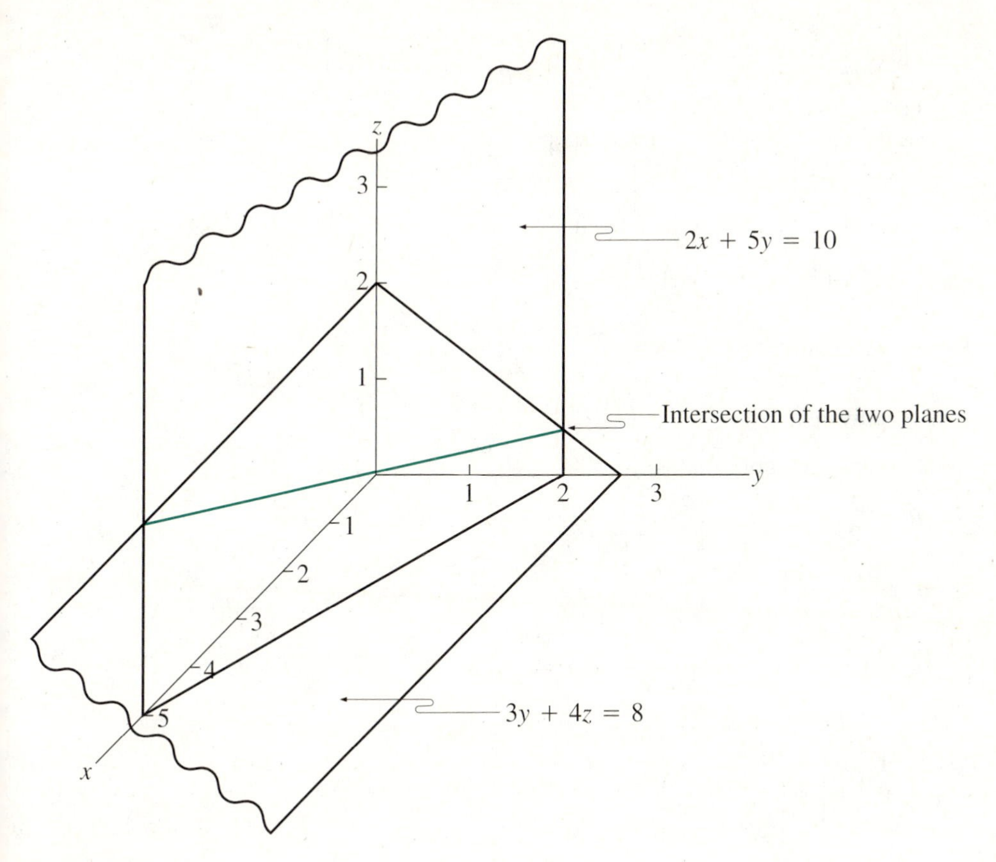

FIGURE C5.19

7. (*a*) Not defined

(*b*) $AB = \begin{bmatrix} 0 & -5 & -2 \\ 9 & -4 & -13 \end{bmatrix}$

(*c*) Not defined

(*d*) $(A + C)B = \begin{bmatrix} 1 & -1 & -1 \\ 6 & -4 & 3 \end{bmatrix} \quad B = \begin{bmatrix} -4 & -5 & 2 \\ 7 & -10 & -15 \end{bmatrix}$

9. $2A - D = C$ implies $2A - C = D$

$D = \begin{bmatrix} 5 & -2 & 1 \\ 6 & -2 & 6 \end{bmatrix}$

11.

	Alice	Barbara	Charles	David	Ellen
Life	16	22	16	15	13
Auto	13	8	16	24	26
Home	19	22	13	6	20

13. (*a*) $2B = \begin{bmatrix} 6 & 2 \\ -2 & 0 \\ 8 & 4 \end{bmatrix}$ (*b*) $BA = \begin{bmatrix} 9 & 7 \\ -2 & -2 \\ 14 & 10 \end{bmatrix}$ (*c*) $CBA = \begin{bmatrix} 50 & 38 \\ 39 & 29 \\ 39 & 29 \end{bmatrix}$

15. (*a*) $3A - B = \begin{bmatrix} 7 & -2 & -9 \\ 2 & 8 & 15 \end{bmatrix}$ (*b*) $CA = \begin{bmatrix} 7 & 4 & -3 \\ 4 & 0 & -6 \end{bmatrix}$

    (*c*) $(A - 2B)C$ is undefined (*d*) ABC is undefined

17. $AB = \begin{bmatrix} 1 & c-3 \\ 0 & c-2 \end{bmatrix}$ so if $AB = I$, then $c = 3$

19. (*a*) $2A + C = B$ is equivalent to $C = B - 2A = \begin{bmatrix} -6 & 5 \\ -9 & 2 \end{bmatrix}$

    (*b*) $A + B + D = I$ is equivalent to $D = I - A - B = \begin{bmatrix} 1 & -2 \\ -3 & -1 \end{bmatrix}$

21. There are many possible answers. For example,

    Set $A = \begin{bmatrix} 1 & -1 \\ 1 & -1 \end{bmatrix}, B = \begin{bmatrix} 2 & -2 \\ 2 & -2 \end{bmatrix}, C = \begin{bmatrix} 2 & 3 \\ 2 & 3 \end{bmatrix}$

23. $A^2 = \begin{bmatrix} 5 & 2 \\ 2 & 1 \end{bmatrix}$ $A^3 = \begin{bmatrix} 12 & 5 \\ 5 & 2 \end{bmatrix}$

25. $[2 \quad 1]\begin{bmatrix} \frac{1}{2} & \frac{1}{2} \\ 1 & 0 \end{bmatrix} = [2 \quad 1]$, so

    $XP^2 = (XP)P = XP = X$

27. $$P^2 = \begin{bmatrix} 0 & 1 & 0 \\ 0 & 0 & 1 \\ \frac{2}{3} & 0 & \frac{1}{3} \end{bmatrix}\begin{bmatrix} 0 & 1 & 0 \\ 0 & 0 & 1 \\ \frac{2}{3} & 0 & \frac{1}{3} \end{bmatrix} = \begin{bmatrix} 0 & 0 & 1 \\ \frac{2}{3} & 0 & \frac{1}{3} \\ \frac{2}{9} & \frac{2}{3} & \frac{1}{9} \end{bmatrix}$$

    $$P^4 = \begin{bmatrix} 0 & 0 & 1 \\ \frac{2}{3} & 0 & \frac{1}{3} \\ \frac{2}{9} & \frac{2}{3} & \frac{1}{9} \end{bmatrix}\begin{bmatrix} 0 & 0 & 1 \\ \frac{2}{3} & 0 & \frac{1}{3} \\ \frac{2}{9} & \frac{2}{3} & \frac{1}{9} \end{bmatrix} = \begin{bmatrix} \frac{2}{9} & \frac{2}{3} & \frac{1}{9} \\ \frac{2}{27} & \frac{2}{9} & \frac{19}{27} \\ \frac{38}{81} & \frac{2}{27} & \frac{37}{81} \end{bmatrix}$$

29. (*a*) If $XP = X$, then $k(XP) = kX$ or $(kX)P = kX$; that is, if $Y = kX$, then $YP = Y$;
    (*b*) let $Y = (\frac{2}{7})[1 \quad 1 \quad \frac{3}{2}] = [\frac{2}{7} \quad \frac{2}{7} \quad \frac{3}{7}]$

31. $X = [1 \quad 2 \quad 4]$ is one such vector and another is $[\frac{1}{4} \quad \frac{1}{2} \quad 1]$.

33. $$AX = \begin{bmatrix} 1 & 2 & 1 \\ 2 & 3 & 1 \\ -2 & 4 & 6 \end{bmatrix}\begin{bmatrix} 2+t \\ 1-t \\ t \end{bmatrix} = \begin{bmatrix} 1(2+t) + 2(1-t) + t \\ 2(2+t) + 3(1-t) + t \\ -2(2+t) + 4(1-t) + 6t \end{bmatrix} = \begin{bmatrix} 4 \\ 7 \\ 0 \end{bmatrix}$$

35. Solve $\begin{bmatrix} 1 & -2 \\ -3 & 4 \end{bmatrix}\begin{bmatrix} x_{11} & x_{12} \\ x_{21} & x_{22} \end{bmatrix} = \begin{bmatrix} x_{11} - 2x_{21} & x_{12} - 2x_{22} \\ -3x_{11} + 4x_{21} & 3x_{12} + 4x_{22} \end{bmatrix} = \begin{bmatrix} 10 & -7 \\ -22 & 15 \end{bmatrix}$

    to obtain $x_{11} = 2, x_{12} = -1, x_{21} = -4, x_{22} = 3$. Therefore, $B = \begin{bmatrix} 2 & -1 \\ -4 & 3 \end{bmatrix}$

Section 6.2

1. Yes, $B = A^{-1}$

3. No, $B \neq A^{-1}$

5. $AA^{-1} = \begin{bmatrix} 3 & 2 \\ 4 & 3 \end{bmatrix}\begin{bmatrix} 3 & -2 \\ -4 & 3 \end{bmatrix} = \begin{bmatrix} 1 & 0 \\ 0 & 1 \end{bmatrix}$

7. $C^{-1} = \begin{bmatrix} \frac{3}{5} & \frac{1}{5} \\ -\frac{2}{5} & \frac{1}{5} \end{bmatrix}$ $D^{-1} = \begin{bmatrix} 20 & -10 \\ -\frac{25}{2} & \frac{15}{2} \end{bmatrix}$

9. $A^{-1} = \begin{bmatrix} \frac{1}{2} & 0 \\ -\frac{3}{2} & 1 \end{bmatrix}$

If $AX = \begin{bmatrix} 1 \\ 0 \end{bmatrix}$, then $X = \begin{bmatrix} \frac{1}{2} \\ -\frac{3}{2} \end{bmatrix}$

If $AX = \begin{bmatrix} 1 \\ -1 \end{bmatrix}$, then $X = \begin{bmatrix} \frac{1}{2} \\ -\frac{5}{2} \end{bmatrix}$

11. No inverse

13. $A^{-1} = \begin{bmatrix} 0 & \frac{1}{2} & -\frac{1}{2} \\ 1 & 0 & -1 \\ 0 & 0 & 1 \end{bmatrix}$

15. (*a*) $\begin{bmatrix} 2 & 4 \\ 1 & 3 \end{bmatrix} \begin{bmatrix} x_1 \\ x_2 \end{bmatrix} = \begin{bmatrix} -2 \\ 4 \end{bmatrix}$

(*b*) $A^{-1} = \begin{bmatrix} \frac{3}{2} & -2 \\ -\frac{1}{2} & 1 \end{bmatrix}$ $X = A^{-1}\begin{bmatrix} -2 \\ 4 \end{bmatrix} = \begin{bmatrix} -11 \\ 5 \end{bmatrix}$

17. $A^{-1} = \begin{bmatrix} 1 & -2 \\ -2 & 5 \end{bmatrix}$, $(A^{-1})^{-1} = \begin{bmatrix} 5 & 2 \\ 2 & 1 \end{bmatrix}$

19. $A^{-1}B^{-1} = \begin{bmatrix} 1 & -2 \\ -2 & 5 \end{bmatrix} \begin{bmatrix} -5 & 3 \\ 2 & -1 \end{bmatrix} = \begin{bmatrix} -9 & 5 \\ 20 & -11 \end{bmatrix} = (BA)^{-1}$

$B^{-1}A^{-1} = \begin{bmatrix} -5 & 3 \\ 2 & -1 \end{bmatrix} \begin{bmatrix} 1 & -2 \\ -2 & 5 \end{bmatrix} = \begin{bmatrix} -11 & 25 \\ 4 & -9 \end{bmatrix} = (AB)^{-1}$

21. (*a*) $A^{-1} = \begin{bmatrix} -10 & 0 & -15 \\ -8 & 2 & -8 \\ -5 & 0 & -5 \end{bmatrix}$ (*b*) $AA^{-1}A = A$

23. $A^{-1} = \begin{bmatrix} \frac{15}{14} & \frac{4}{14} & \frac{1}{14} \\ \frac{4}{14} & \frac{16}{14} & \frac{4}{14} \\ \frac{1}{14} & \frac{4}{14} & \frac{15}{14} \end{bmatrix}$

25. $A^{-1} = \begin{bmatrix} \frac{2}{5} & -\frac{1}{5} & \frac{1}{5} \\ \frac{1}{5} & \frac{2}{5} & -\frac{2}{5} \\ \frac{1}{5} & \frac{2}{5} & \frac{3}{5} \end{bmatrix}$ $AX = \begin{bmatrix} 1 \\ 3 \\ 1 \end{bmatrix}$ has solution $X = A^{-1}\begin{bmatrix} 1 \\ 3 \\ 1 \end{bmatrix} = \begin{bmatrix} 0 \\ 1 \\ 2 \end{bmatrix}$.

$AX = \begin{bmatrix} 2 \\ 1 \\ 0 \end{bmatrix}$ has solution $X = A^{-1}\begin{bmatrix} 2 \\ 1 \\ 0 \end{bmatrix} = \begin{bmatrix} \frac{3}{5} \\ \frac{4}{5} \\ \frac{4}{5} \end{bmatrix}$

27. Each of the systems has the same coefficient matrix, call it A. Then A^{-1} exists and

$$A^{-1} = \begin{bmatrix} -1 & 1 & 2 \\ 1 & 0 & -1 \\ 1 & -1 & -1 \end{bmatrix}$$

(a) The system has solution $X = A^{-1}\begin{bmatrix} 15 \\ 1 \\ 10 \end{bmatrix} = \begin{bmatrix} 6 \\ 5 \\ 4 \end{bmatrix}$, therefore $x = 6, y = 5, z = 4$

(b) $x = 3, y = 3, z = 3$; (c) $x = 1, y = 0, z = -1$

29. $A^2 = \begin{bmatrix} 4 & 18 \\ 6 & 28 \end{bmatrix}, A^{-1} = \begin{bmatrix} \frac{5}{2} & -\frac{3}{2} \\ -\frac{1}{2} & \frac{1}{2} \end{bmatrix}, (A^{-1})(A^{-1}) = \begin{vmatrix} 7 & -\frac{9}{2} \\ -\frac{3}{2} & 1 \end{vmatrix},$

$(AA)^{-1} = (A^2)^{-1} = \begin{bmatrix} 7 & -\frac{9}{2} \\ -\frac{3}{2} & 1 \end{bmatrix}$. Yes, $(A^{-1})(A^{-1}) = (AA)^{-1}$.

31. $A^{-1} = \begin{bmatrix} 1 & -b \\ 0 & 1 \end{bmatrix}$

33. $A^{-1} = \begin{bmatrix} \dfrac{d}{ad - bc} & -\dfrac{b}{ad - bc} \\ -\dfrac{c}{ad - bc} & \dfrac{a}{ad - bc} \end{bmatrix}, ad \neq bc$

35. $A^{-1} = \begin{bmatrix} 1 & -a & ac-b \\ 0 & 1 & -c \\ 0 & 0 & 1 \end{bmatrix}$

Section 6.3

1. $A = \begin{bmatrix} .1 & .3 \\ 2 & 0 \end{bmatrix}$ $\quad (I - A)^{-1} = \begin{bmatrix} \frac{10}{3} & 1 \\ \frac{20}{3} & 3 \end{bmatrix}$

(a) $\begin{bmatrix} \frac{10}{3} & 1 \\ \frac{20}{3} & 3 \end{bmatrix}\begin{bmatrix} 15 \\ 5 \end{bmatrix} = \begin{bmatrix} 55 \\ 115 \end{bmatrix}$

(b) $\begin{bmatrix} \frac{10}{3} & 1 \\ \frac{20}{3} & 3 \end{bmatrix}\begin{bmatrix} 1000 \\ 125 \end{bmatrix} = \begin{bmatrix} \frac{10{,}375}{3} \\ \frac{21{,}125}{3} \end{bmatrix}$

3. (a) $X = \begin{bmatrix} 7.6 & 4 & 2 \\ 16 & 15 & 5 \\ 14 & 10 & 5 \end{bmatrix}\begin{bmatrix} 20 \\ 20 \\ 8 \end{bmatrix} = \begin{bmatrix} 248 \\ 660 \\ 520 \end{bmatrix}$

(b) $\begin{bmatrix} 7.6 & 4 & 2 \\ 16 & 15 & 5 \\ 14 & 10 & 5 \end{bmatrix}\begin{bmatrix} 100 \\ 120 \\ 50 \end{bmatrix} = \begin{bmatrix} 1340 \\ 3650 \\ 2850 \end{bmatrix}$

5. $X = \begin{bmatrix} 7.6 & 4 & 2 \\ 16 & 15 & 5 \\ 14 & 10 & 5 \end{bmatrix}\begin{bmatrix} 100 \\ 60 \\ 0 \end{bmatrix} = \begin{bmatrix} 1000 \\ 2500 \\ 2000 \end{bmatrix}$

7. $(I - A)^{-1} = \begin{bmatrix} \frac{35}{18} & \frac{20}{18} \\ \frac{25}{18} & \frac{40}{18} \end{bmatrix}$ (*a*) Yes (*b*) yes

9. Yes, $(I - A)^{-1} = \begin{bmatrix} \frac{8}{3} & \frac{2}{3} \\ 3 & 2 \end{bmatrix}$

11. $(I - A)^{-1} = \begin{bmatrix} \frac{79}{60} & \frac{32}{60} & \frac{4}{60} \\ \frac{2}{15} & \frac{16}{15} & \frac{2}{15} \\ \frac{1}{60} & \frac{8}{60} & \frac{76}{60} \end{bmatrix}$

13. (*a*) $\begin{bmatrix} 74 \\ 102 \end{bmatrix}$ (*b*) $\begin{bmatrix} 158 \\ 204 \end{bmatrix}$ (*c*) If D is doubled, then the production schedule X must be doubled.

15. $(I - A)^{-1} = \begin{bmatrix} 8 & 4 \\ 7.5 & 5 \end{bmatrix}$ and $X = \begin{bmatrix} 20 \\ 20 \end{bmatrix}$

17. $(I - A)^{-1} = \begin{bmatrix} 20 & 10 \\ \frac{50}{3} & 10 \end{bmatrix}$ and $X = \begin{bmatrix} 170 \\ 150 \end{bmatrix}$

19. $(I - A)^{-1} = \begin{bmatrix} 2 & 3 & 3 \\ 0 & 15 & 10 \\ 0 & 5 & 5 \end{bmatrix}$ and $X = \begin{bmatrix} 15 \\ 35 \\ 15 \end{bmatrix}$

21. $(I - A)^{-1} = \begin{bmatrix} \frac{23}{9} & \frac{8}{9} & \frac{2}{9} \\ \frac{1}{9} & \frac{16}{9} & \frac{4}{9} \\ \frac{2}{3} & \frac{2}{3} & \frac{8}{3} \end{bmatrix}$ and $X = \begin{bmatrix} 174 \\ 78 \\ 288 \end{bmatrix}$

23. x_1 = amount of steel and x_2 = amount of lumber produced.
$A = \begin{bmatrix} .2 & .6 \\ .5 & .6 \end{bmatrix}$ $(I - A)^{-1} = \begin{bmatrix} 20 & 30 \\ 25 & 40 \end{bmatrix}$ and $X = (I - A)^{-1}\begin{bmatrix} 20 \\ 10 \end{bmatrix}$
Required production is 700 units of steel and 900 units of lumber.

25. Yes, the production schedule is 525 units of grain, 1000 units of steel, and 1140 units of coal.

27. Let x_1 = number of units of coal
x_2 = number of units of lumber
x_3 = number of units of steel
Then $A = \begin{bmatrix} .4 & 0 & .2 \\ .5 & .2 & .5 \\ .3 & .6 & .2 \end{bmatrix}$ $(I - A)^{-1} = \begin{bmatrix} 3.5417 & 1.250 & 1.667 \\ 5.7292 & 4.375 & 4.167 \\ 5.625 & 3.750 & 5.000 \end{bmatrix}$
and the required production scheduled is $x_1 = 230$, $x_2 = 455$, and $x_3 = 450$.

29. $A = \begin{bmatrix} .2 & c \\ .6 & .4 \end{bmatrix}$ and $(I - A)^{-1} = \begin{bmatrix} \frac{5}{4-5c} & \frac{25c}{3(4-5c)} \\ \frac{5}{4-5c} & \frac{20}{3(4-5c)} \end{bmatrix}$
Thus, both A and $(I - A)^{-1}$ have nonnegative entries for c satisfying $0 \le c < .8$.

31. $0 < a < \frac{2}{3}$

33. $0 < a < .4$

35. The matrix $(1 - A)^{-1}$ is
$$\begin{bmatrix} 1.25/(1-k) & 2.5k/(1-k) \\ 1/(1-k) & 2/(1-k) \end{bmatrix}$$
Thus, for $0 < k < 1$, the assumptions of the Leontief model are satisfied.
(*a*) If $k = .1$ and $D = \begin{bmatrix} 20 \\ 10 \end{bmatrix}$, then $X = \begin{bmatrix} 275/9 \\ 400/9 \end{bmatrix}$.

(*b*)

k	Amount of good 2
.1	44.44
.2	50.00
.3	57.14
.4	66.67
.5	80.00
.6	100.00
.7	133.33
.8	200.00
.9	400.00

(*c*) (Figure C6.1)

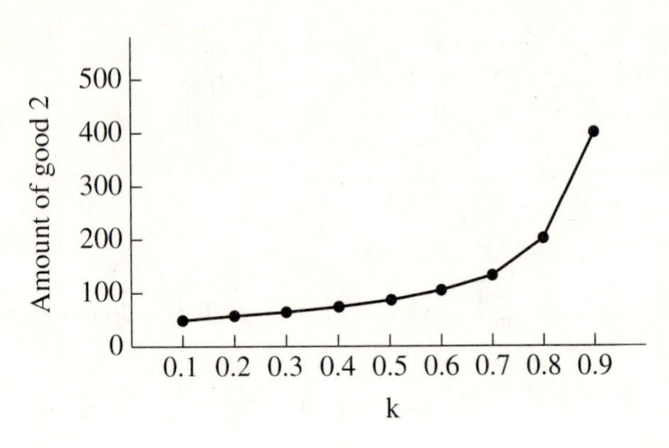

FIGURE C6.1

Review Exercises

1. (*a*) 4×2 (*b*) 2×3 (*c*) 1×4 (*d*) 3×1

3. $5A + B = \begin{bmatrix} 0 & 34 \\ 12 & 83 \\ 65 & -38 \end{bmatrix}$ $A - 2B = \begin{bmatrix} -11 & 9 \\ -24 & -1 \\ 2 & -12 \end{bmatrix}$

5. $Y = \begin{bmatrix} 3 \\ 10 \\ 16 \end{bmatrix}$ $Z = \begin{bmatrix} 7 \\ 15 \\ 42 \end{bmatrix}$

7. $p = -1, q = -1$

9. (*a*) $\begin{bmatrix} 6 & 12 \\ 2 & 10 \end{bmatrix}$ (*b*) Not defined (*c*) $\begin{bmatrix} 8 & 7 \\ 3 & 5 \\ 0 & 6 \end{bmatrix}$

(*d*) $\begin{bmatrix} 33 & 75 \\ 5 & 25 \\ -2 & 8 \end{bmatrix}$ (*e*) Not defined (*f*) $\begin{bmatrix} 36 \\ 10 \\ 2 \end{bmatrix}$

11. (*a*) $\begin{bmatrix} 4 & 0 \\ 6 & -2 \\ -4 & 0 \end{bmatrix}$ (*b*) $\begin{bmatrix} -2 & 4 & -3 \\ -2 & -3 & 1 \end{bmatrix}$ (*c*) Not defined

(*d*) $\begin{bmatrix} 1 & 0 & 3 \\ -2 & -7 & 8 \end{bmatrix}$ (*e*) Not defined (*f*) $\begin{bmatrix} 10 & -2 \\ -7 & 1 \end{bmatrix}$

13. $A^2 = \begin{bmatrix} -9 & -8 \\ 20 & -1 \end{bmatrix}$ $AB = \begin{bmatrix} -9 & -5 & 3 \\ 7 & 1 & 15 \end{bmatrix}$

$A(AB) = \begin{bmatrix} -23 & -7 & -27 \\ -24 & -22 & 60 \end{bmatrix}$

$A^2B = \begin{bmatrix} -23 & -7 & -27 \\ -24 & -22 & 60 \end{bmatrix}$

15. $AC = \begin{bmatrix} 5 & 10 \\ 4 & 8 \end{bmatrix}$ $\quad BC = \begin{bmatrix} 5 & 10 \\ 4 & 8 \end{bmatrix}$

One conclusion is that there are matrices A, B, and C such that $AC = BC$ but $A \neq B$. That is, the fact "for numbers a, b, and $c \neq 0$, if $ac = bc$, then $a = b$" does not carry over to matrices.

17. $A^{-1} = \begin{bmatrix} 3 & -4 \\ -5 & 7 \end{bmatrix}$

19. $A^{-1} = \begin{bmatrix} 6 & -1 & -1 \\ -2 & 1 & 0 \\ -3 & 0 & 1 \end{bmatrix}$

21. (*a*) $\begin{bmatrix} -8 & 11 \\ 11 & -\frac{25}{2} \end{bmatrix}$ (*b*) $\begin{bmatrix} -24 & 14 \\ -\frac{9}{2} & \frac{7}{2} \end{bmatrix}$

23. $k = -6$

25. $X = \begin{bmatrix} -11 \\ 4 \\ 6 \end{bmatrix}$, $\quad X = \begin{bmatrix} -7 \\ 2 \\ 4 \end{bmatrix}$, $\quad X = \begin{bmatrix} 1 \\ 0 \\ 0 \end{bmatrix}$

27. $0 < a < .8$

29. $\begin{bmatrix} 15 \\ 35 \end{bmatrix}$

31. $\begin{bmatrix} 152 \\ 75 \\ 130 \end{bmatrix}$

33. $(I - A)^{-1}$ exists, but has negative entries.

35. $\begin{bmatrix} 3205.6 \\ 3375.0 \\ 2990.2 \\ 4206.6 \end{bmatrix}$

CHAPTER 7

Section 7.1

1. Let x = number of regular sandwiches
y = number of large sandwiches

Maximize: $.8x + 1.2y$

Subject to:
$$\begin{aligned} x \geq 0, y &\geq 0 \\ 6x + 10y &\leq 1320 \quad \text{(bread)} \\ 2x + 4y &\leq 480 \quad \text{(meat)} \end{aligned}$$

3. Let x = number of sacks of 25-10-5
y = number of sacks of 8-10-10

Maximize: $7x + 5y$

Subject to:
$$\begin{aligned} x \geq 0, y &\geq 0 \\ 25x + 8y &\leq 12000 \quad \text{(nitrate)} \\ 10x + 10y &\leq 10000 \quad \text{(phosphate)} \\ 5x + 10y &\leq 7000 \quad \text{(potash)} \end{aligned}$$

5. Let x = material from subsidiary, pounds
   y = material from independent supplier, pounds

   Minimize: $.8x + y$

   Subject to:
   $$x \geq 0, y \geq 0$$
   $$x + y \geq 45000$$
   $$x \leq 35000$$

7. As number 5, with the additional constraint:
   $$y \leq 20000$$

9. Let x = gallons of regular ice cream
   y = gallons of low-calory ice cream

   Maximize: $x + 1.2y$

   Subject to:
   $$x \geq 0, y \geq 0$$
   $$.6x + .7y \leq 800 \quad \text{(skim milk)}$$
   $$x + .3y \leq 400 \quad \text{(sugar)}$$
   $$.4x + .4y \leq 400 \quad \text{(cream)}$$

11. Let x = number of subject hours
    y = number of consultant minutes

    Maximize: $x + .04y$

    Subject to:
    $$x \geq 0, y \geq 0$$
    $$x \leq 15$$
    $$x \geq 6$$
    $$y \leq 200$$
    $$y \geq 30x$$
    $$y \leq 50x$$

13. Let x = number of full teams
    y = number of half teams

    Maximize: $180x + 100y$

    Subject to:
    $$x \geq 0, y \geq 0$$
    $$x + y \leq 200 \quad \text{(doctors)}$$
    $$3x + 2y \leq 450 \quad \text{(nurses)}$$

15. Let x = number of standard boxes produced each week
    y = number of heavy duty boxes produced each week

    Maximize: $3x + 4y$

    Subject to:
    $$x \geq , y \geq 0$$
    $$x + 5y \leq 5000 \quad \text{(100-pound test)}$$
    $$3x + y \leq 4500 \quad \text{(liner)}$$
    $$x \geq 500$$

17. Let x = number of Scary Harry
y = number of Horrible Harriet
z = number of The Glob

Maximize: $x + 1.25y + 1.5z$

Subject to:
$x \geq 0, y \geq 0, z \geq 0$
$4x + 3y + 9z \leq 160$ (plastic)
$3x + 4y + z \leq 50$ (clothes)
$2x + 4y + 3z \leq 50$ (features)

19. [5 5 10] is not feasible because the time constraint for features is violated; [10 3 2] is not feasible because the time constraint for features is violated.

21. Let x = number of type A
y = number of type B
z = number of type C

Maximize: $x + .75y + 1.25z$

Subject to:
$x \geq 0, y \geq 0, z \geq 0$
$25x + 10y + 50z \leq 1000$ (worms)
$10x + 15y + 5z \leq 250$ (minnows)
$10x + 25y + 5z \leq 300$ (grasshoppers)

23. Let x = number of hundreds of small widgets without locks
y = number of hundreds of small widgets with locks
z = number of hundreds of medium widgets with locks
w = number of hundreds of large widgets with locks

Maximize: $2x + 10y + 11z + 20w$

Subject to:
$x \geq 0, y \geq 0, z \geq 0, w \geq 0$
$x + 2y + 3z + 6w \leq 8$ (assembly)
$x + 5y + 4z + 8w \leq 9$ (painting)
$3y + z + 4w \leq 2$ (installation)

25. Let x = number of Deluxe Packs
y = number of Special Packs
z = number of Standard Packs

Maximize: $3x + 2y + 1.5z$

Subject to:
$x \geq 0, y \geq 0, z \geq 0$
$16x + 20y + 16z \leq 1200$ (dates)
$24x + 12y + 8z \leq 900$ (apricots)
$12x + 3y \leq 360$ (candied fruit)

27. Let x = number of type 1 layouts,
y = number of type 2 layouts,
z = number of type 3 layouts,

Maximize: $50x + 30y + 60z$

Subject to:
$$x \geq 0, y \geq 0, z \geq 0$$
$$30x + 10y + 20z \leq 1000 \quad \text{(tulips)}$$
$$20x + 40y + 50z \leq 800 \quad \text{(daffodils)}$$
$$4x + 3y + 2z \leq 100 \quad \text{(flowering shrubs)}$$

29. Let x = units of furniture timber
y = units of plywood timber
z = units of pulpwood timber

Maximize: $500x + 400y + 200z$

Subject to:
$$x \geq 0, y \geq 0, z \geq 0$$
$$100x + 80y + 50z \leq 1000 \quad \text{(labor)}$$
$$20x + 30y + 30z \leq 500 \quad \text{(machine time)}$$

Section 7.2

1. (Figure C7.1)
3. (Figure C7.2)
5. (Figure C7.3)
7. (Figure C7.4)
9. (Figure C7.5)
11. $(4, 8), (4, 3), (-\frac{8}{3}, \frac{4}{3})$
13. $(\frac{5}{2}, \frac{7}{2}), (\frac{3}{5}, \frac{8}{5}), (12, -6)$
15. $(\frac{20}{3}, \frac{10}{3}), (\frac{15}{4}, \frac{25}{4})$
17. Corner point $(12, 2)$; (Figure C7.6)
19. (Figure C7.7); Corner points: $(-8, 6), (-2, 0), (-1, \frac{5}{2})$
21. (Figure C7.8); Corner points: $(0, 0), (0, -2), (\frac{3}{2}, \frac{3}{2}), (3, 1)$

FIGURE C7.1

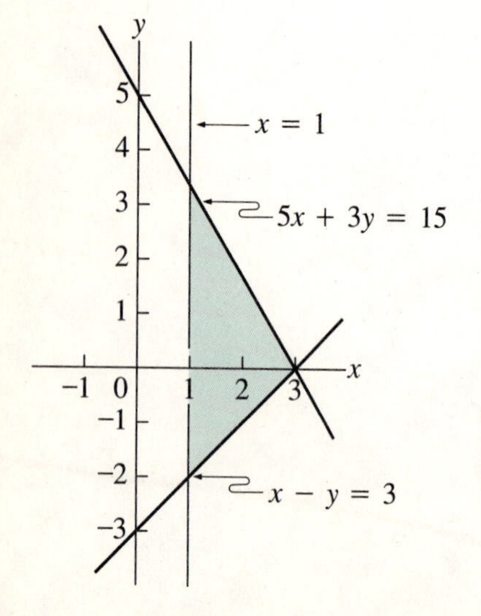

FIGURE C7.2

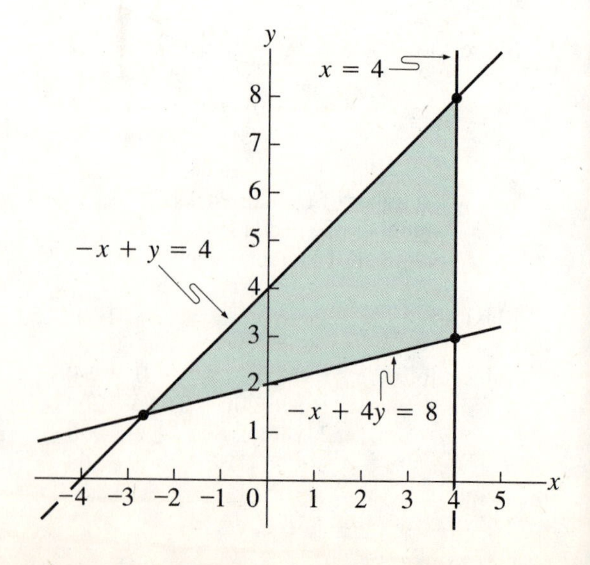

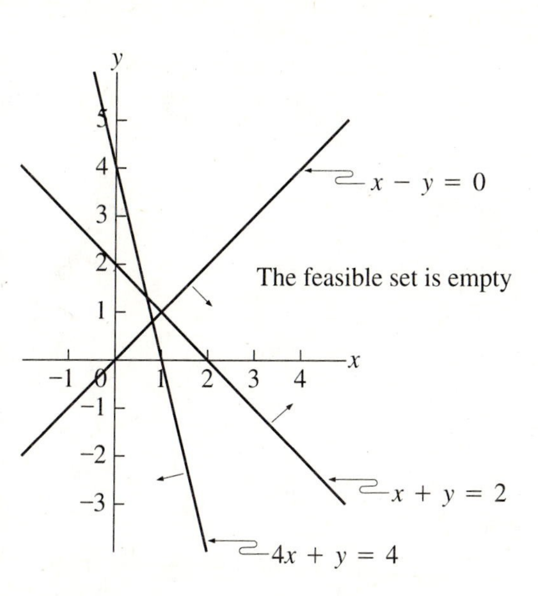

FIGURE C7.3

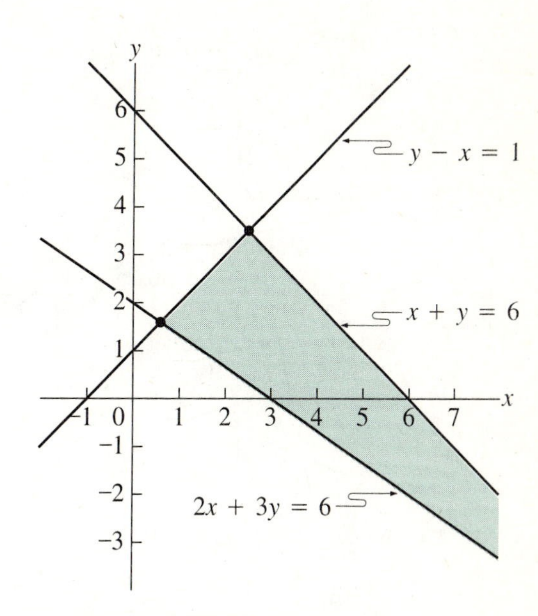

FIGURE C7.4

23. (Figure C7.9)
 Let x = number of standard boxes, hundreds
 y = number of heavy duty boxes, hundreds

$$\begin{aligned} x \geq 0 \; y &\geq 0 \\ 200x + 400y &\leq 100{,}000 \\ 8x + \quad 3y &\leq \quad 2400 \\ y &\geq \quad 100 \end{aligned}$$

FIGURE C7.5

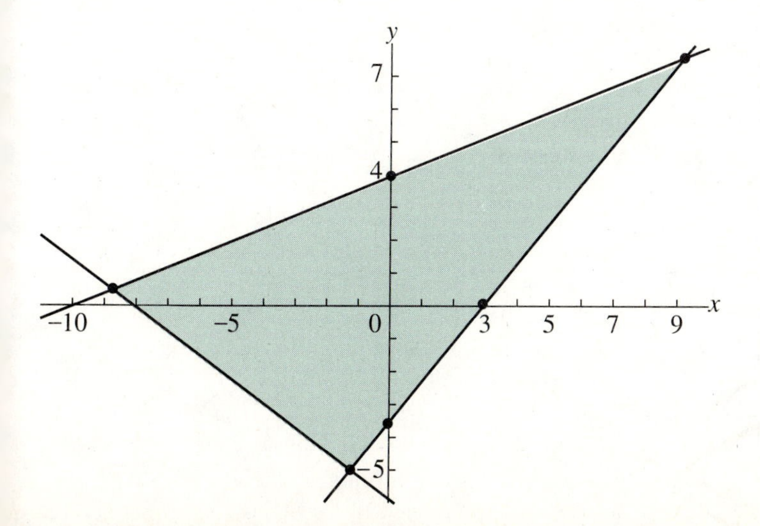

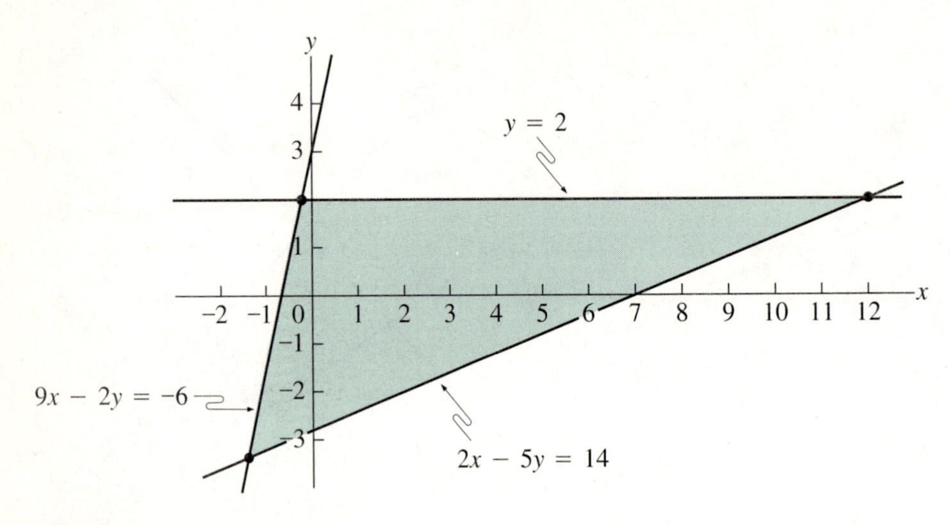

FIGURE C7.6

25. (Figure C7.10)
 Let x = acres for first experiment
 y = acres for second experiment

$$x \geq 0 \quad y \geq 0$$
$$x + y \leq 3$$
$$2x + 8y \leq 16$$
$$12x + 4y \leq 24$$

FIGURE C7.7

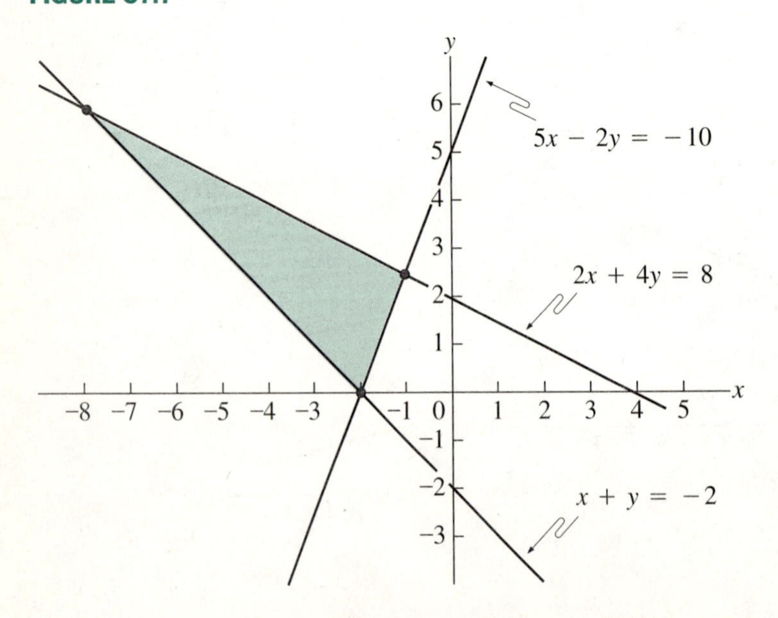

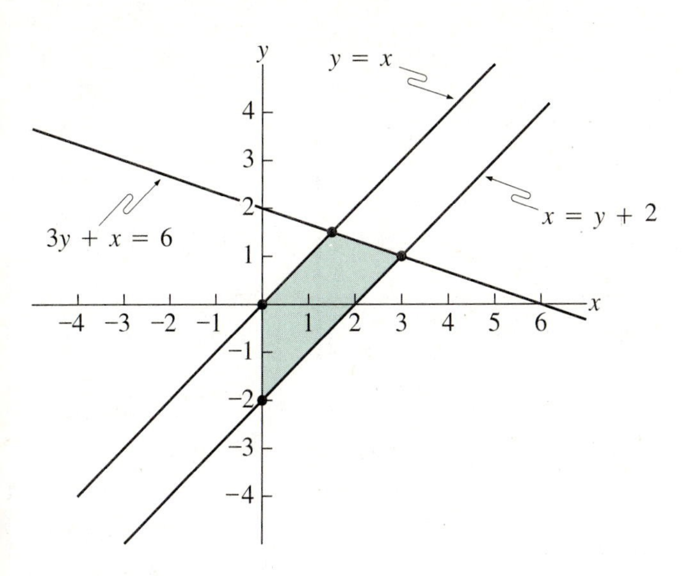

FIGURE C7.8

27. (Figure C7.11)
29. (Figure C7.12)
    The feasible set lies on the z axis
31. (Figure C7.13)
    At (0, 4), $p = 28$
    At (6, 0), $p = 24$
    At (3, 1), $p = 19$

FIGURE C7.9

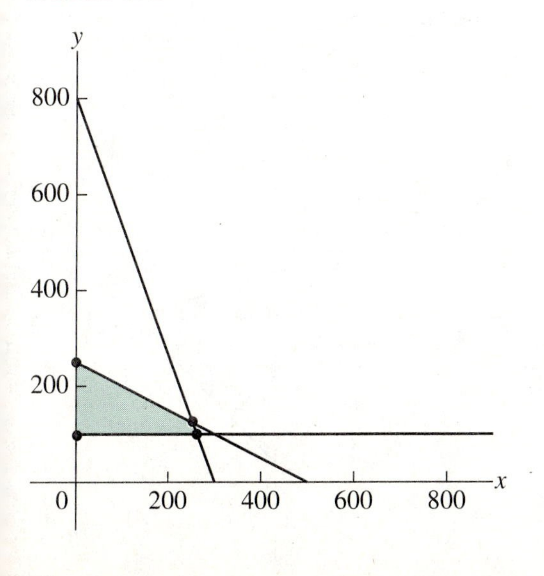

FIGURE C7.10

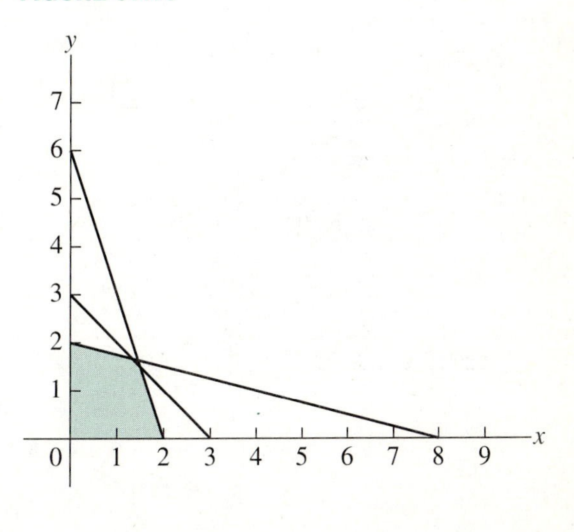

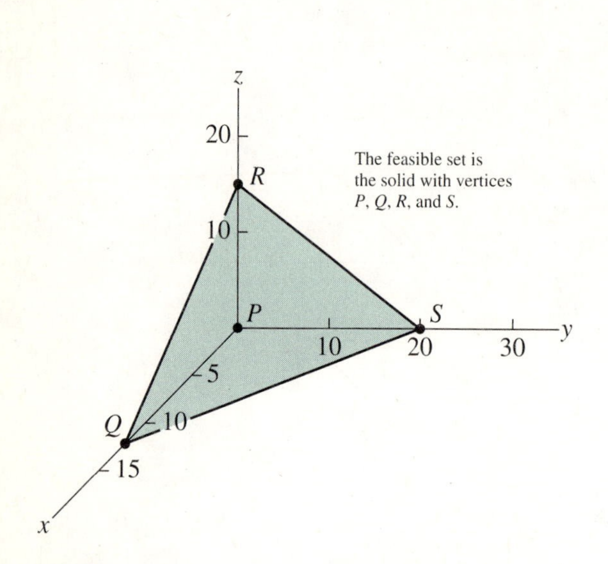

FIGURE C7.11

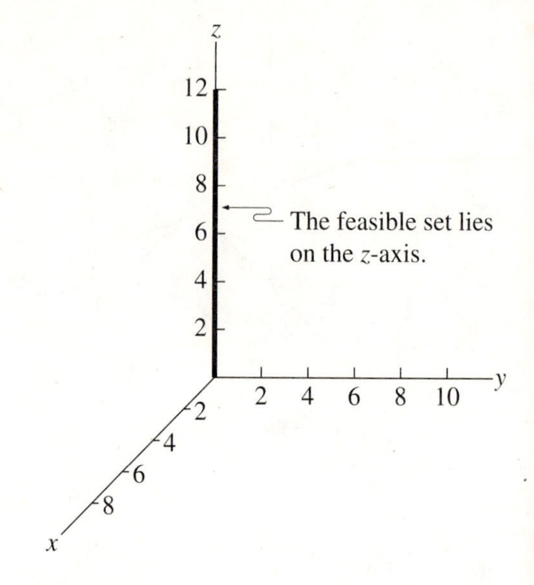

FIGURE C7.12

Section 7.3

1. (Figure C7.14); Bounded — Corner points: $(0, 3), (3, 0), (-\frac{3}{2}, \frac{3}{2})$
3. (Figure C7.15); Bounded — Corner points: $(1, 0), (0, 0), (0, 3), (2, 1)$

FIGURE C7.13

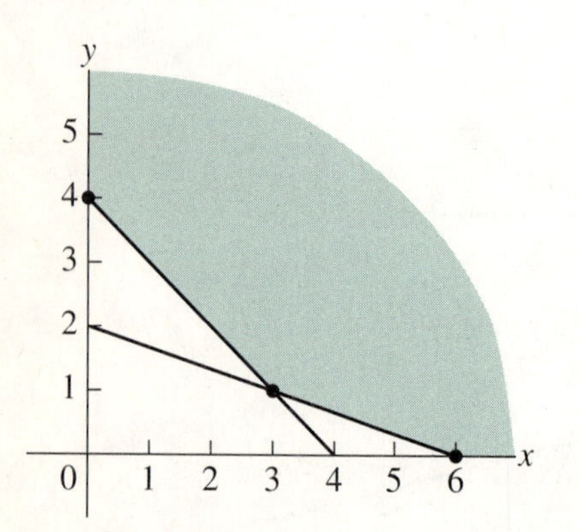

FIGURE C7.14

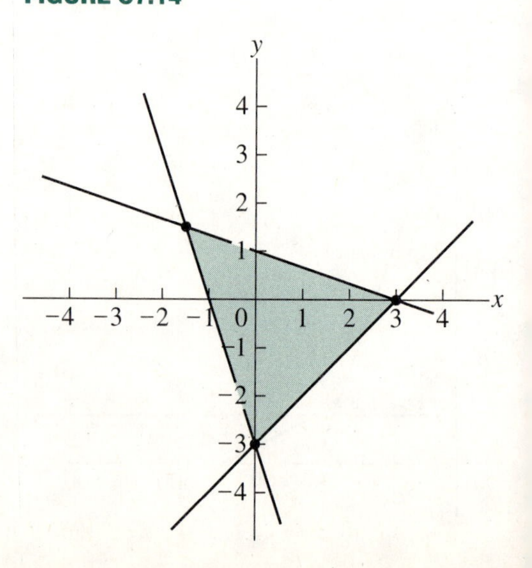

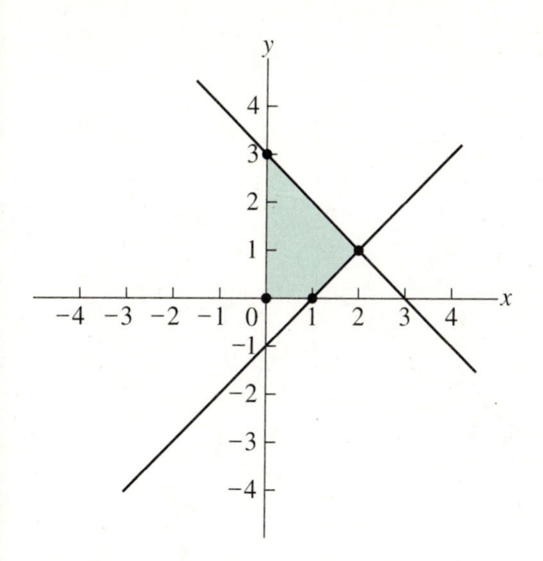

FIGURE C7.15

Empty feasible set

FIGURE C7.16

5. (Figure C7.16)
   Empty feasible set
7. Maximum of $2x + y$ is 11 at (5, 1); minimum of $x - 2y$ is -9 at (1, 5)
9. No minimum values; no maximum values
11. No minimum values; no maximum values
13. see Figure C7.17

Corner point	Value of $3x - 2y$
(0, 0)	0
$(-\frac{5}{6}, \frac{5}{6})$	$-\frac{25}{6}$
$(\frac{5}{2}, \frac{5}{2})$	$\frac{5}{2}$ ← maximum

15. see Figure C7.18

Corner point	Value of $x - 3y$
(0, 0)	0
(0, 1)	−3
$(\frac{36}{5}, \frac{14}{5})$	$-\frac{6}{5}$
(3, 0)	3 ← maximum

17. Maximum of \$176 for 220 regular sandwiches and no large sandwiches
19. Maximum of $\$\frac{34.100}{7}$ for $\frac{6400}{21}$ sacks of 25-10-5 and 11, 500/21 sacks of 8-10-10
21. Maximum of \$10,140 for 264 desks and 324 file cabinets.
23. Maximum of $\frac{44}{3}$ for $\frac{20}{3}$ subject hours and 200 consultant minutes.
25. Maximum of 2700 at $x = 150$ and $y = 0$.
27. Maximum of 675 at $x = 1250$ and $y = 750$.
29. (*a*) 3 poppy seed and 14 German chocolate for a profit of \$62; (*b*) no; (*c*) yes, 400 grams
31. (*a*) Maximum of $\frac{8}{3}$ at $(\frac{22}{3}, \frac{7}{3})$; minimum does not exist
    (*b*) Maximum does not exist; minimum of 12 at $(\frac{22}{3}, \frac{7}{3})$. Minimum is not unique
    (*c*) Maximum does not exist; minimum of $\frac{64}{3}$ at $(\frac{22}{3}, \frac{7}{3})$; minimum is unique
    (*d*) Maximum of $\frac{8}{3}$ at $(\frac{7}{3}, \frac{22}{3})$; maximum is unique; minimum does not exist

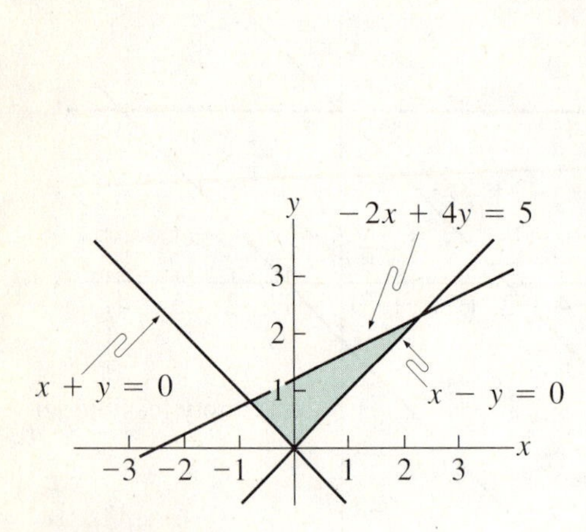

FIGURE C7.17

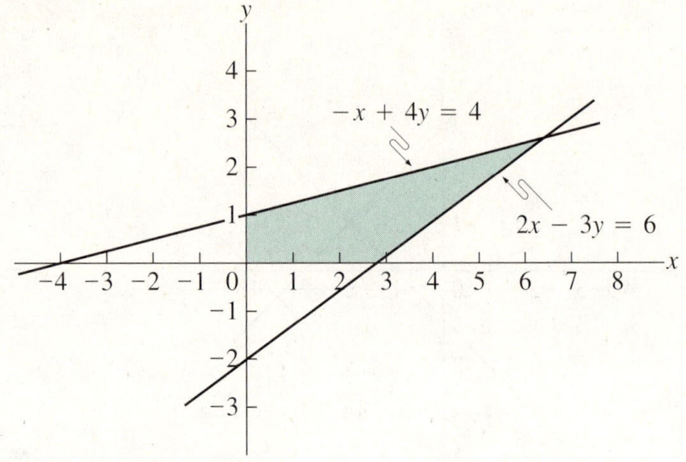

FIGURE C7.18

33. 48 boxes of regular mix and 32 boxes of deluxe mix for a profit of \$176
35. (*a*) Maximum of $-\frac{8}{5}$ at $(\frac{16}{5}, \frac{4}{5})$
    (*b*) No maximum
    (*c*) Maximum of 0 attained at $(\frac{16}{5}, \frac{4}{5})$ and at other points too
37. Minimum does not exist

Review Exercises

1. Let x = number of passenger planes, y = number of cargo planes

Maximize: $3x + 4y$

subject to:

$$\begin{aligned} x \geq 0,\ y &\geq 0 \\ 4x + 5y &\leq 320 \quad \text{(steel)} \\ x + 3y &\leq 240 \quad \text{(plastic)} \\ 8x + 12y &\leq 240 \quad \text{(wood)} \end{aligned}$$

3. Let x = number of acres of crop A, y = number of acres of crop B

Maximize: $170x + 190y$

subject to:

$$\begin{aligned} x \geq 0,\ y &\geq 0 \\ x + 2y &\leq 3000 \quad \text{(labor)} \\ 90x + 60y &\leq 150{,}000 \quad \text{(capital)} \\ x + y &\leq 2000 \quad \text{(land)} \end{aligned}$$

5. Let x = number of large bran muffins, y = number of small bran muffins

Maximize: $.25x + .10y$

subject to:

$$\begin{aligned} x \geq 0,\ y &\geq 0 \\ 4x + y &\leq 300 \quad \text{(dough)} \\ 2x + y &\leq 155 \quad \text{(bran)} \end{aligned}$$

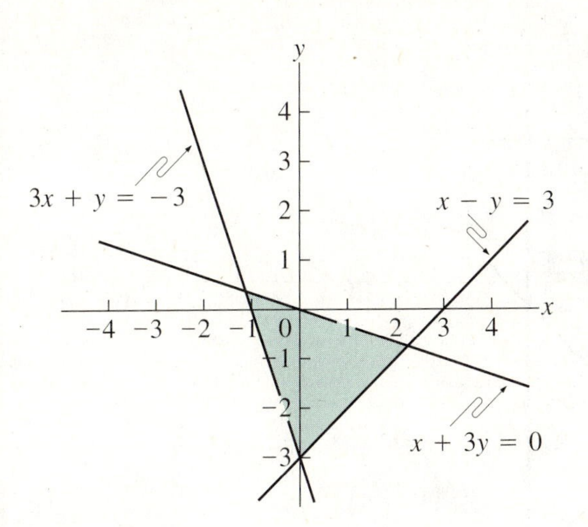

FIGURE C7.19

7. Let x = number of sacks of Standard, y = number of sacks of Special, z = number of sacks of Super

Maximize: $10x + 5y + 15z$

Subject to: $x \geq 0, y \geq 0, z \geq 0$

$20x + 10y + 5z \leq 2000$ (nitrogen)

$5x + 10y + 15z \leq 1500$ (phosphorus)

$5x + 10y + 10z \leq 1500$ (potash)

9. Maximum of 90 at (30, 0). Thus, make 30 passenger planes and no cargo planes. All resources are not consumed, only the wood strips are used up.
11. Maximum of 360,000 at (1000, 1000). Yes, all resources are consumed.
13. Maximum of 19.125 at $(\frac{145}{2}, 10)$. Yes, all resources are consumed.
15. 50 gross soft chalk, no hard chalk, cost is $75.00
17. She should produce only right-hand widgets. She should make 32.5 sets of 100, i.e., 3250 of them.
19. (Figure C7.19); corner points at $(0, -3)$, $(\frac{9}{4}, -\frac{3}{4})$, $(-\frac{9}{8}, \frac{3}{8})$
21. (*a*) (Figure C7.20) (*b*) $(6, 2)$, $(6, 6)$, $(-9, 6)$, $(\frac{12}{5}, -\frac{8}{5})$ (*c*) $\frac{64}{5}$
23. Maximum does not exist; minimum does not exist
25. Maximum does not exist; minimum of 9 at (0, 3)
27. Maximum of $\frac{27}{2}$ at $(\frac{3}{2}, \frac{7}{2})$; minimum of -14 at $(-4, -2)$
29. (*a*) Maximum of 3 at $(1, -1)$ (*b*) There is no minimum

CHAPTER 8

Section 8.1

1. (Figure C8.1)
3. State 2

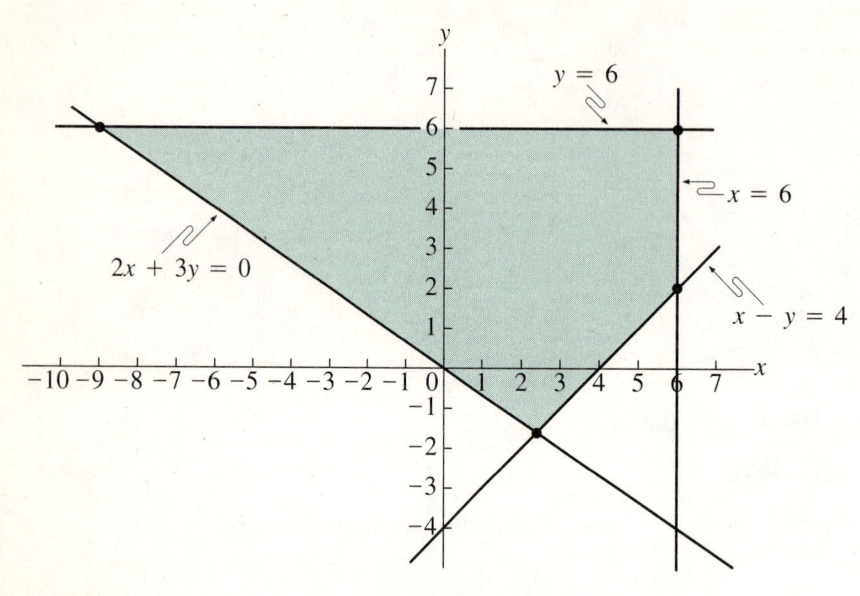

FIGURE C7.20

5. (Figure C8.2)

7. $\begin{bmatrix} 0 & 1 & 0 \\ 0 & .5 & .5 \\ .8 & 0 & .2 \end{bmatrix}$

9. (Figure C8.3)

11. (*a*) .3 (*b*) State 3

13. State 2

15. The states are high volume and low volume. If state 1 is high volume and state 2 is low volume, then the transition matrix is

$$\begin{bmatrix} .7 & .3 \\ .5 & .5 \end{bmatrix}$$

17. (*a*) System is in state 1 if the small animal is in the meadow and in state 2 if it is in the woods.
(*b*) (Figure C8.4)
(*c*) $\begin{bmatrix} \frac{2}{3} & \frac{1}{3} \\ \frac{1}{3} & \frac{2}{3} \end{bmatrix}$

FIGURE C8.1

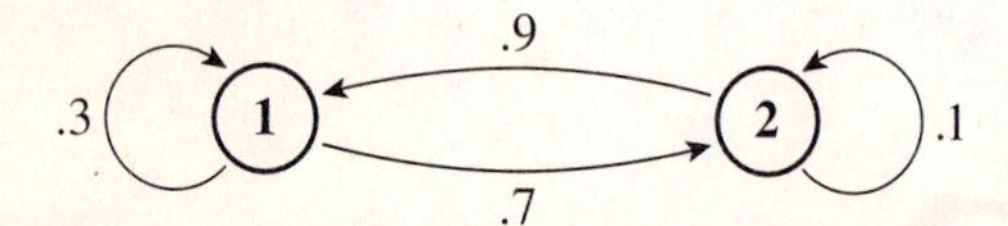

FIGURE C8.2

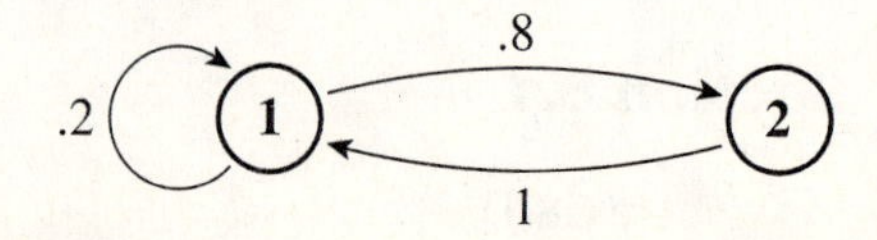

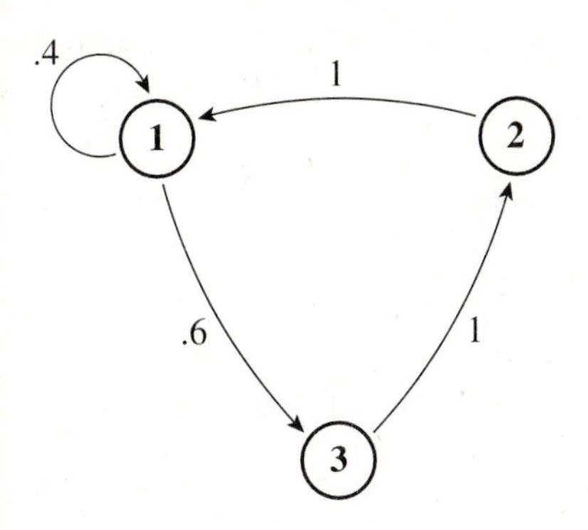

FIGURE C8.3

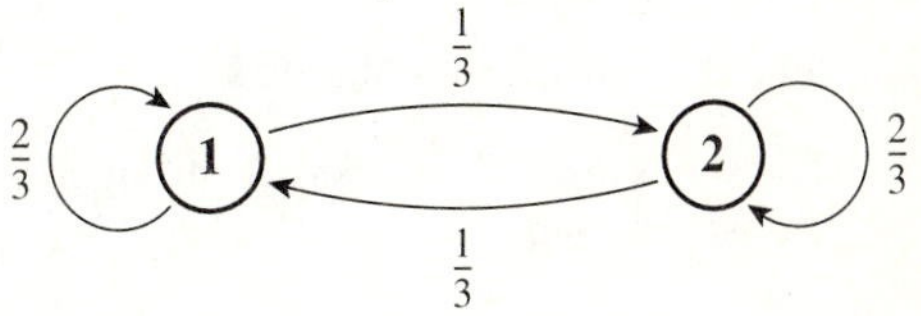

$\frac{1}{3}$
$\frac{2}{3}$ 1 2 $\frac{2}{3}$
$\frac{1}{3}$

FIGURE C8.4

19. (*a*) The states are the three employment groups: industry, small business, and self-employed. Identify these states as 1, 2, and 3, respectively.

(*b*) (Figure C8.5)

(*c*) $\begin{bmatrix} .70 & .20 & .10 \\ .30 & .50 & .20 \\ .30 & .30 & .40 \end{bmatrix}$

21.
$$\begin{array}{c} \\ R \\ G \\ Y \\ B \end{array} \begin{array}{c} \begin{array}{cccc} R & G & Y & B \end{array} \\ \begin{bmatrix} 0 & \frac{1}{2} & 0 & \frac{1}{2} \\ 0 & \frac{1}{2} & \frac{1}{3} & \frac{1}{6} \\ \frac{1}{6} & 0 & \frac{1}{2} & \frac{1}{3} \\ \frac{1}{3} & \frac{1}{6} & 0 & \frac{1}{2} \end{bmatrix} \end{array}$$

FIGURE C8.5

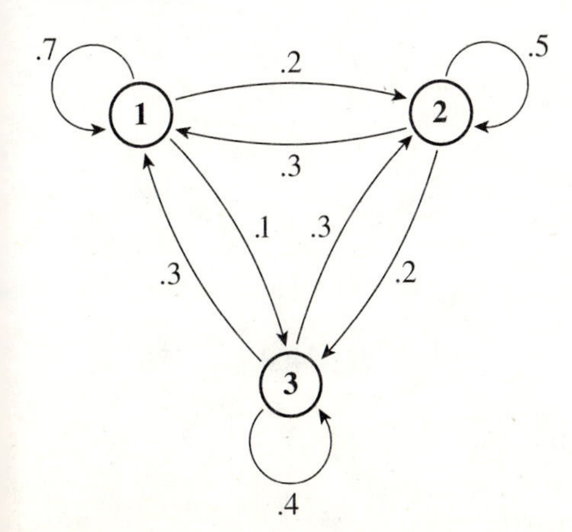

23. The states are: 1, she eats Chinese food; 2, she eats Greek food; and 3, she eats Italian food. The transition matrix is

$$\begin{bmatrix} 0 & \frac{2}{3} & \frac{1}{3} \\ \frac{1}{2} & 0 & \frac{1}{2} \\ \frac{5}{6} & \frac{1}{6} & 0 \end{bmatrix}$$

25. State 1 is make a free throw, state 2 is miss. The transition matrix is $\begin{bmatrix} .8 & .2 \\ .4 & .6 \end{bmatrix}$.

27. The states are: 1, if stocks are stronger; 2, if bonds are stronger; and 3, if both are equally strong. Transition matrix is

$$\begin{bmatrix} .6 & .3 & .1 \\ .3 & .5 & .2 \\ .4 & .4 & .2 \end{bmatrix}$$

29. Let X be a random variable which associates with each attendance schedule for the first 4 weeks the number of times the student attends class.

Values of X	Probabilities
1	$\frac{1}{4}$
2	$\frac{3}{2}$

$E[X] = 1(\frac{1}{4}) + 2(\frac{3}{4}) = \frac{7}{4}$

Section 8.2

1. (*a*) $\mathbf{P}(2) = \begin{bmatrix} .75 & .25 \\ .5 & .5 \end{bmatrix}$ (*b*) $p_{12}(2) = .25, \quad p_{21}(2) = .5$

3. $\mathbf{P}(3) = \begin{bmatrix} .625 & .375 \\ .75 & .25 \end{bmatrix}$ $\mathbf{P}(4) = \begin{bmatrix} .6875 & .3125 \\ .625 & .375 \end{bmatrix}$

5. $\mathbf{P}(3) = \begin{bmatrix} .752 & .248 \\ .744 & .256 \end{bmatrix}$

7. (*a*) $\mathbf{P}(2) = \begin{bmatrix} 0 & 0 & 1 \\ 1 & 0 & 0 \\ 0 & 1 & 0 \end{bmatrix}$ (*b*) $\mathbf{P}(3) = \begin{bmatrix} 1 & 0 & 0 \\ 0 & 1 & 0 \\ 0 & 0 & 1 \end{bmatrix}$

9. (*a*) $\mathbf{P}(2) = \begin{bmatrix} .1 & .8 & .1 \\ 0 & 1 & 0 \\ .25 & .40 & .35 \end{bmatrix}$ $\mathbf{P}(3) = \begin{bmatrix} .05 & .88 & .07 \\ 0 & 1 & 0 \\ .175 & .600 & .225 \end{bmatrix}$ $\mathbf{P}(4) = \begin{bmatrix} .035 & .920 & .045 \\ 0 & 1 & 0 \\ .1125 & .7400 & .1475 \end{bmatrix}$

   (*b*) $p_{22}(2) = 1, \quad p_{22}(3) = 1, \quad p_{22}(4) = 1$ (*c*) $p_{22}(k) = 1$ for all k

11. (*a*) Figure C8.6 (*b*) [.5 0 .5] (*c*) $\mathbf{P}(2) = \begin{bmatrix} 0 & 0 & 1 \\ .5 & 0 & .5 \\ .25 & .5 & .25 \end{bmatrix}$

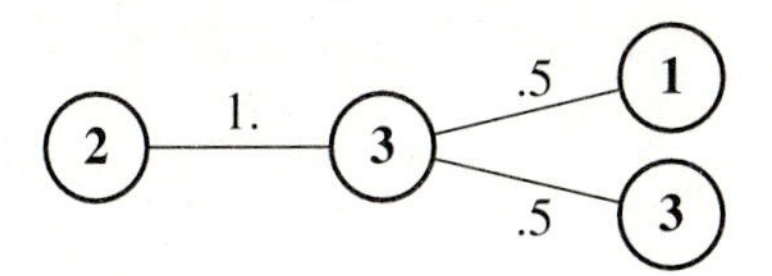

FIGURE C8.6

13. $\mathbf{P}(2) = \begin{bmatrix} .42 & .08 & .50 \\ .36 & .12 & .52 \\ .35 & .10 & .55 \end{bmatrix}$

$\mathbf{P}(3) = \begin{bmatrix} .366 & .100 & .534 \\ .380 & .096 & .524 \\ .385 & .090 & .525 \end{bmatrix}$

For $\mathbf{P}(2)$: $.42 + .08 + .50 = 1$
$.36 + .12 + .52 = 1$
$.35 + .10 + .55 = 1$

For $\mathbf{P}(3)$: $.366 + .100 + .534 = 1$
$.380 + .096 + .524 = 1$
$.385 + .090 + .525 = 1$

15. (*a*) .72 (*b*) .56

17. $\mathbf{P}(2) = \begin{bmatrix} .40 & .36 & .24 \\ .33 & .34 & .33 \\ .27 & .30 & .43 \end{bmatrix}$, $p_{33}(3) = .295$

19. .18, central

21. (*a*)

$$\mathbf{P}(2) = \begin{array}{c} \\ I \\ SB \\ SE \end{array} \begin{array}{c} \begin{array}{ccc} I & SB & SE \end{array} \\ \begin{bmatrix} .82 & .10 & .08 \\ .18 & .675 & .145 \\ .18 & .425 & .395 \end{bmatrix} \end{array}$$

(*b*) .10

23. Self-employed (*SE*), .395

25. Cadillac (with probability $\frac{683}{2048}$), although probability of each is very close to $\frac{1}{3}$.

27. State 4

29. (*a*) $\mathbf{P}(2) = \begin{bmatrix} .40 & .48 & .12 & 0 \\ .2 & .40 & .08 & .2 \\ 0 & 0 & .5 & .5 \\ 0 & 0 & .25 & .75 \end{bmatrix}$

$\mathbf{P}(4) = \begin{bmatrix} .3136 & .3840 & .1464 & .156 \\ .2560 & .3136 & .1604 & .270 \\ 0 & 0 & .375 & .625 \\ 0 & 0 & .3125 & .6875 \end{bmatrix}$

$\mathbf{P}(8) = \begin{bmatrix} .1966 & .2408 & .2112 & .3514 \\ .1606 & .1966 & .2323 & .4105 \\ 0 & 0 & .3359 & .6641 \\ 0 & 0 & .3320 & .6680 \end{bmatrix}$

(*b*) No

(*c*) No, from state 3 it is possible to reach only states 3 and 4.

Section 8.3

1. (*a*) Yes (*b*) no (*c*) no (*d*) yes
3. $[\frac{1}{4} \;\; \frac{1}{4} \;\; \frac{2}{4}]$
5. (*a*) $[\frac{2}{3} \;\; \frac{1}{3}]$ (*b*) $\frac{188}{300}$ (*c*) $\frac{469}{750}$
7. (*a*) If the meadow is state 1, then the state vector is $[\frac{3}{4} \;\; \frac{1}{4}]$ (*b*) $\frac{19}{36}$
9. $[\frac{5}{8} \;\; \frac{3}{8}]$
11. $[\frac{4}{7} \;\; \frac{3}{7}]$
13. Regular, stable vector is $[\frac{1}{4} \;\; \frac{3}{4}]$.
15. Not regular
17. Regular, stable vector is $[\frac{5}{23} \;\; \frac{4}{23} \;\; \frac{10}{23} \;\; \frac{4}{23}]$.
19. She makes 80 percent of her free throws.
21. The student attends class on Friday afternoons with probability $\frac{7}{13}$.
23. (*a*) $\mathbf{P}(4) = \begin{bmatrix} .3411 & .3207 & .2786 & .0595 \\ .2138 & .4180 & .2138 & .1543 \\ .2786 & .3207 & .3411 & .0595 \\ .1190 & .4630 & .1190 & .2990 \end{bmatrix}$

$$\mathbf{P}(8) = \begin{bmatrix} .2697 & .3604 & .2658 & .1042 \\ .2403 & .3834 & .2403 & .1361 \\ .2658 & .3604 & .2697 & .1042 \\ .2083 & .4083 & .2083 & .1750 \end{bmatrix}$$

(*b*) .2221 in $\mathbf{P}^4$, .0614 in $\mathbf{P}^8$
(*c*) $[\frac{1}{4} \;\; \frac{3}{8} \;\; \frac{1}{4} \;\; \frac{1}{8}]$
(*d*) State 2, food

25. (*a*) (Figure C8.7)
(*b*) $k = 10$ for P_1; $k = 6$ for P_2

FIGURE C8.7

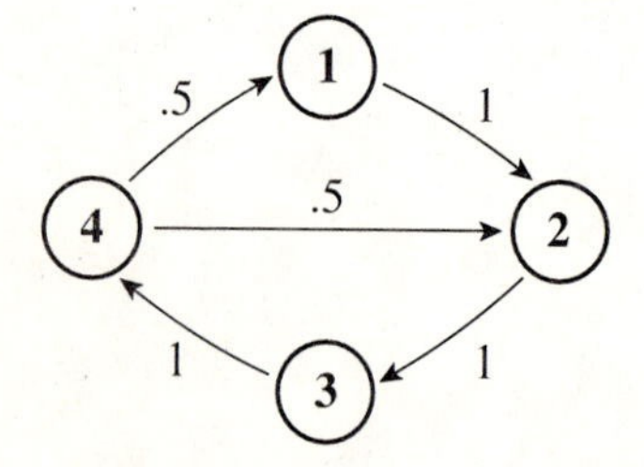

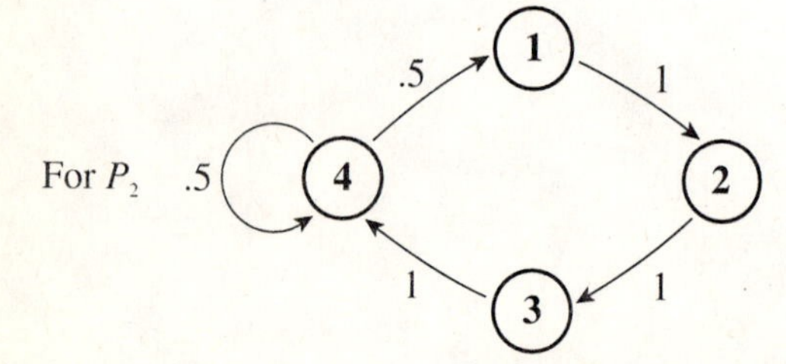

27. The eighth power of the matrix **P** has a zero in the (1, 1) entry for any values of x, y, and z.

29. $$\begin{bmatrix} \dfrac{b}{1+b-a} & \dfrac{1-a}{1+b-a} \end{bmatrix} \begin{bmatrix} a & 1-a \\ b & 1-b \end{bmatrix} = \begin{bmatrix} \dfrac{b}{1+b-a} & \dfrac{1-a}{1+b-a} \end{bmatrix}$$

31. Stable vector is $\begin{bmatrix} \dfrac{b}{1+b} & \dfrac{1}{1+b} \end{bmatrix}$.

Section 8.4

1. (*a*) Yes (*b*), (*c*) From state 1, one transition; state 2 is absorbing; from state 3, two transitions.

3. (*a*) Yes (*b*), (*c*) State 1 is absorbing; from state 2, one transition; state 3 is absorbing; from state 4, two transitions

5. (*a*) No (*b*) impossible to reach an absorbing state from either state 3 or state 4

7. $$\begin{array}{c} \\ 3 \\ 1 \\ 2 \\ 4 \end{array} \begin{array}{c} \begin{array}{cccc} 3 & 1 & 2 & 4 \end{array} \\ \begin{bmatrix} 1 & 0 & 0 & 0 \\ \frac{1}{6} & \frac{2}{3} & \frac{1}{6} & 0 \\ \frac{1}{9} & \frac{1}{9} & \frac{2}{3} & \frac{1}{9} \\ 0 & 0 & \frac{1}{3} & \frac{2}{3} \end{bmatrix} \end{array}$$

9. $$\begin{array}{c} \\ 2 \\ 1 \\ 3 \\ 4 \\ 5 \end{array} \begin{array}{c} \begin{array}{ccccc} 2 & 1 & 3 & 4 & 5 \end{array} \\ \begin{bmatrix} 1 & 0 & 0 & 0 & 0 \\ 0 & \frac{1}{3} & \frac{1}{3} & 0 & \frac{1}{3} \\ 0 & \frac{1}{2} & \frac{1}{2} & 0 & 0 \\ \frac{2}{10} & \frac{1}{10} & \frac{3}{10} & \frac{4}{10} & 0 \\ 0 & 0 & 0 & 1 & 0 \end{bmatrix} \end{array}$$

11. $$\begin{array}{c} \\ 2 \\ 1 \\ 3 \end{array} \begin{array}{c} \begin{array}{ccc} 2 & 1 & 3 \end{array} \\ \begin{bmatrix} 1 & 0 & 0 \\ \frac{1}{2} & 0 & \frac{1}{2} \\ \frac{1}{3} & \frac{1}{3} & \frac{1}{3} \end{bmatrix} \end{array} \qquad \mathbf{N} = \begin{bmatrix} \frac{4}{3} & 1 \\ \frac{2}{3} & 2 \end{bmatrix}$$

13. $$\begin{array}{c} \\ 2 \\ 4 \\ 1 \\ 3 \end{array} \begin{array}{c} \begin{array}{cccc} 2 & 4 & 1 & 3 \end{array} \\ \begin{bmatrix} 1 & 0 & 0 & 0 \\ 0 & 1 & 0 & 0 \\ \frac{1}{2} & \frac{1}{2} & 0 & 0 \\ 0 & \frac{1}{4} & \frac{1}{4} & \frac{1}{2} \end{bmatrix} \end{array} \qquad \mathbf{N} = \begin{bmatrix} 1 & 0 \\ \frac{1}{2} & 2 \end{bmatrix}$$

15. $$\begin{array}{c} \\ 4 \\ 2 \\ 3 \\ 1 \end{array} \begin{array}{c} \begin{array}{cccc} 4 & 2 & 3 & 1 \end{array} \\ \begin{bmatrix} 1 & 0 & 0 & 0 \\ 0 & 0 & 0 & 1 \\ 0 & 1 & 0 & 0 \\ \frac{1}{4} & 0 & \frac{1}{4} & \frac{1}{2} \end{bmatrix} \end{array} \qquad \mathbf{N} = \begin{bmatrix} 2 & 1 & 4 \\ 2 & 2 & 4 \\ 1 & 1 & 4 \end{bmatrix}$$

17. $2 + \frac{5}{3} = \frac{11}{3}$
19. 27
21. $\frac{8}{3}$ years
23. 3.2
25. (*a*) 1 (*b*) $\frac{2}{3}$
27. The matrix **P** is in canonical form with $\mathbf{Q} = [1 - a]$. Therefore, $\mathbf{I} - \mathbf{Q} = [1] - [1 - a] = [a]$ and $\mathbf{N} = \left[\frac{1}{a}\right]$.

Section 8.5

1. .1195, .2884
3. [.2703 .2606 .1020 .3671], about 37 percent of the time
5. [.3571 .2296 .0898 .3234], about 32 percent of the time
7. The stable vector is [.4545 .3030 .0808 .1616].
9. With states defined as in this section, the transition matrix is

$$\mathbf{P} = \begin{array}{c} \\ G \\ B \\ YF \\ MF \end{array}\begin{array}{c} \begin{array}{cccc} G & B & YF & MF \end{array} \\ \begin{bmatrix} .745 & .255 & 0 & 0 \\ .150 & .680 & .170 & 0 \\ .150 & 0 & .425 & .425 \\ .100 & 0 & 0 & .900 \end{bmatrix} \end{array}$$

and the stable vector is [.3296 .2627 .0777 .3301].

Review Exercises

1. (*a*) (Figure C8.8) (*b*) (.2)(.2) + (.8)(.5) = .44
3. (*a*) $\begin{bmatrix} .2 & .4 & .4 \\ 0 & 1 & 0 \\ 0 & .8 & .2 \end{bmatrix}$ (*b*) (.2)(.4) + (.4)(1) + (.4)(.8) = .80
5. (*a*) $\mathbf{P}(2) = \begin{bmatrix} 0 & .4 & .6 \\ 0 & 0 & 1 \\ .4 & .24 & .36 \end{bmatrix}$ (*b*) [.280 .288 .432]
7. State 3, probability .36
9. (*a*) Regular (*b*) $[\frac{6}{26} \quad \frac{10}{26} \quad \frac{10}{26}]$
11. (*a*) Regular (*b*) $[\frac{1}{9} \quad \frac{2}{9} \quad \frac{2}{9} \quad \frac{4}{9}]$ (*c*) State 4
13. (*a*) [.5 .5 0] (*b*) 25.25 percent blackgum, 32 percent red maple, and 42.75 percent beech (*c*) 3.85 percent blackgum, 11.54 percent red maple, and 84.61 percent beech
15. $\frac{13}{5}$
17. 1.2

FIGURE C8.8

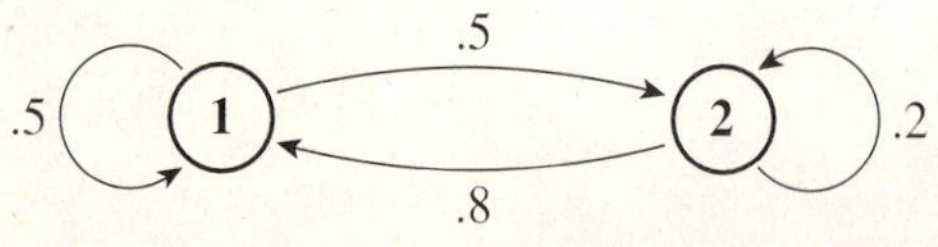

19. (*a*) $\frac{45}{17}$ (*b*) $\frac{20}{17}$
21. 2.5 weeks
23. 5
25. 13.5 (including the initial observation)
27. (*a*) $\mathbf{N} = \begin{bmatrix} \frac{5}{3} & \frac{2}{3} \\ \frac{5}{3} & \frac{8}{3} \end{bmatrix}$ (*b*) $\frac{13}{3}$
29. **P**(4) has all positive entries.

CHAPTER 9

Section 9.1

1. (*a*) $100 (*b*) $120
3. (*a*) $16.57 (*b*) $2820.37
5.

Frequency of Interest Payments	Initial Investment, $	Formula for P_n	Amount in Account after 10 years, $
Annually	1000	$1000(1+.04)^{10}$	1480.24
Semiannually	1000	$1000(1+.20)^{20}$	1485.95
Quarterly	1000	$1000(1+.01)^{40}$	1488.86
Monthly	1000	$1000[1+(\frac{.04}{12})]^{120}$	1490.83

7. Profit of $13,949
9. Loss of $33,942.10
11. More profitable to invest at 6.25 percent compounded quarterly
13. (*a*) Using 360 days/year: 8327.74
    (*b*) Using 365 days/year: 8327.76
15. 5.095 and 5.614 percent
17. 3.80 percent per year
19. $1102.26
21. (*a*) $34,216.82 (*b*) $17,717.58
23. $15,371.18 (assuming she withdraws $30,000 at the beginning of the 15th year)
25. 5.00 percent per year
27. −.50 percent per year for the 2 years
29. 33.7 percent per year

Section 9.2

1. $6139.13
3. $742.47
5. $946.75
7. Repay $1200 in 2 years, he receives $984.90.
9. (*a*) $11,057.07 (*b*) $11,183.57
11. (*a*) $8079.29 (*b*) $8146.63
13. Amount = $1364.80, present value = $1200.90
15. $6781.78 + $49 = $6830.78
17. $8072.15
19. 13.04 percent per year
21. 4.564 percent per year
23. $747.26
25. (*a*) 54.0 percent; (*b*) $824.81; (*c*) 15.27 percent for 1 year
27. $59,831.69
29. The company should lease the computer.

Section 9.3

1. \$437.26
3. \$3094.85
5. \$2782.09
7. \$3412.74
9. For 5 years \$96,346.75; for 10 years \$157,883.39
11. Monthly payment = \$180.29, total payment = \$8653.92, principal = \$7500, interest = \$1153.92
13. Monthly payment = \$151.36
15. \$520.55
17. 6 years
19. (*a*) \$583.33 (*b*) \$841.61
21. (*a*) \$249.62; (*b*) \$5990.88; (*c*) \$5500.80
23. \$49,651.41
25. \$678.32
27. 13 payments: After 12 semiannual payments the account has \$95,502.76, and after 13 payments it has \$106,277.90.
29. \$340,912.91

Section 9.4

1. January 1, 1996 \$275.00 interest
   July 1, 1996 \$275.00 interest
   January 1, 1997 \$275.00 interest
   July 1, 1997 \$275.00 interest
   January 1, 1998 \$275.00 interest, \$10,000 principal
3. (*a*) \$300.00
   (*b*) January 1, 1996 \$400.00 interest
   January 1, 1997 \$400.00 interest
   January 1, 1998 \$400.00 interest
   January 1, 1999 \$400.00 interest, \$5000 principal
5. 9.152 percent per year
7. 6.56 percent per year
9. 9.05 percent per year
11. 4.60 percent per year
13. 50 percent per year
15. 37.5 percent per year
17. After 4 years Paul has \$8449.42, a return of 111.24 percent on his original investment of \$4000.
19. Expected return = 26.25 percent
21. \$31.50 per share, assuming that the 10 percent per year is compounded.
23. (*a*) \$1250 (*b*) 1.095 percent per year
25. The bond is a better investment (has a better rate of return).
27. Expected rate of return is 8.67 percent.
29. Compare the expected gain on the stock investment with the yield to maturity on the bond. The bond has the higher rate of return.

Review Exercises

1. 4.06 percent
3. \$639.02
5. \$373,107.71
7. \$472.41
9. Largest possible mortgage is \$106,449.68. Amount of interest is \$148,550.32. Interest per \$1 of loan is \$1.40.
11. \$1653.70

13. Value in 1990 with 5 percent interest is \$16,008,907.08. Value in 1990 with 6 percent interest is \$401,785,435.74. Percentage increase is (401,785,435.74 − 16,008,907.08)/16,008,907.08 = .241, or an increase of 24.1 percent.
15. The salary of \$3400 per month plus bonus.
17. \$11,284.79
19. \$2576.56
21. \$1075
23. Expected loss of 2 percent in 2 years, expected gain of 14 percent in 5 years
25. Six payments of \$765,839.36 each followed by two payments of \$162,791.75 each

CHAPTER 10

Section 10.1

1. [5 5 5] has all coordinates ≥ 0,
$$\begin{aligned} 3(5) + 4(5) + 8(5) &= 75 \leq 100 \\ 1(5) + 8(5) + 2(5) &= 55 \leq 80 \\ 5(5) + 1(5) + 4(5) &= 50 \leq 120 \end{aligned}$$
therefore [5 5 5] is feasible for the given SMP
[1 1 10] has all coordinates ≥ 0,
$$\begin{aligned} 3(1) + 4(1) + 8(10) &= 87 \leq 100 \\ 1(1) + 8(1) + 2(10) &= 29 \leq 80 \\ 5(1) + 1(1) + 4(10) &= 46 \leq 120 \end{aligned}$$
therefore [1 1 10] is feasible for the given SMP.
3. (*a*) No (*b*) no (*c*) no (*d*) yes
5. (*a*) No (*b*) yes (*c*) yes
7. The constraints are satisfied for $z = 80$ but for no larger value of z.
9. Let x = number of balls produced each hour
y = number of bats produced each hour
z = number of rackets produced each hour

Maximize: $.75x + 1.50y + 1.25z$

subject to:
$$\begin{aligned} &x \geq 0, y \geq 0, z \geq 0 \\ 4x + 9y + 6z &\leq 200 \quad \text{(plastic)} \\ 2x + 3y + 2z &\leq 55 \quad \text{(paint time)} \\ x + y + 3z &\leq 45 \quad \text{(label time)} \end{aligned}$$
The problem is an SMP
11. Not an SMP; last constraint not in correct form.
13. The problem is an SMP.
15. Maximize: $p = 5x + 4y$

subject to:
$$\begin{aligned} &x \geq 0, y \geq 0, u \geq 0, v \geq 0 \\ 3x + 12y + u &= 42 \\ 8x + 5y + v &= 60 \end{aligned}$$

17. Maximize: $$p = x + 4y + 5z$$

subject to:
$$\begin{aligned} x \geq 0, y \geq 0, z \geq 0, u \geq 0, v \geq 0 \\ x + 2y + z + u = 12 \\ 2x + 5y + 3z + v = 45 \end{aligned}$$

19. Maximize: $$p = 5x + 4y$$

subject to:
$$\begin{aligned} x \geq 0, y \geq 0, u \geq 0, v \geq 0 \\ 3x + u = 12 \\ 5y + v = 60 \end{aligned}$$

(See Figure C10.1)

Corner point	Basic solution
P	$x = 0, y = 0, u = 12, v = 60$
Q	$x = 0, y = 12, u = 12, v = 0$
R	$x = 4, y = 0, u = 0, v = 60$
S	$x = 4, y = 12, u = 0, v = 0$

21. Maximize: $$p = 3x + 8y$$

subject to:
$$\begin{aligned} x \geq 0, y \geq 0, u \geq 0, v \geq 0 \\ x + y + u = 5 \\ 3x + 6y + v = 18 \end{aligned}$$

FIGURE C10.1

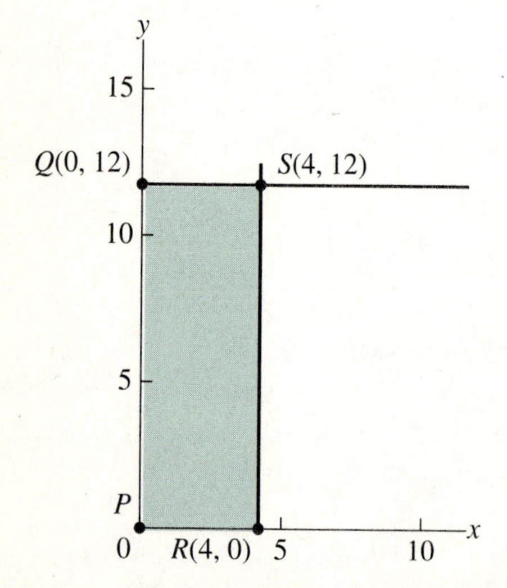

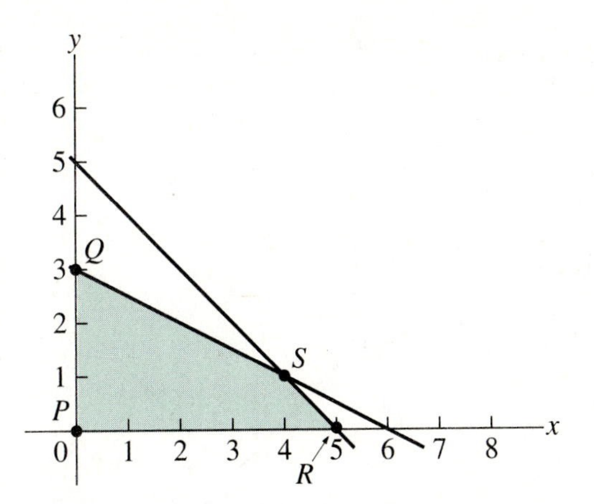

FIGURE C10.2

(See Figure C10.2)

Corner point	Basic solution
P	$x = 0, y = 0, u = 5, v = 18$
Q	$x = 0, y = 3, u = 2, v = 0$
R	$x = 5, y = 0, u = 0, v = 3$
S	$x = 4, y = 1, u = 0, v = 0$

23. Maximize: $p = 5x + y$

subject to:

$$x \geq 0, y \geq 0, u \geq 0, v \geq 0, w \geq 0$$
$$4x + 4y + u = 20$$
$$3x + 12y + v = 24$$
$$10x + 3y + w = 30$$

(See Figure C10.3)

Corner point	Basic solution
P	$x = 0, y = 0, u = 20, v = 24, w = 30$
Q	$x = 0, y = 2, u = 12, v = 0, w = 24$
R	$x = 3, y = 0, u = 8, v = 15, w = 0$
S	$x = \frac{96}{37}, y = \frac{50}{37}, u = \frac{156}{37}, v = 0, w = 0$

25. Maximize: $p = 4x + 5y$

subject to:

$$x \geq 0, y \geq 0, u \geq 0, v \geq 0$$
$$-x + y + u = 1$$
$$x + y + v = 20$$

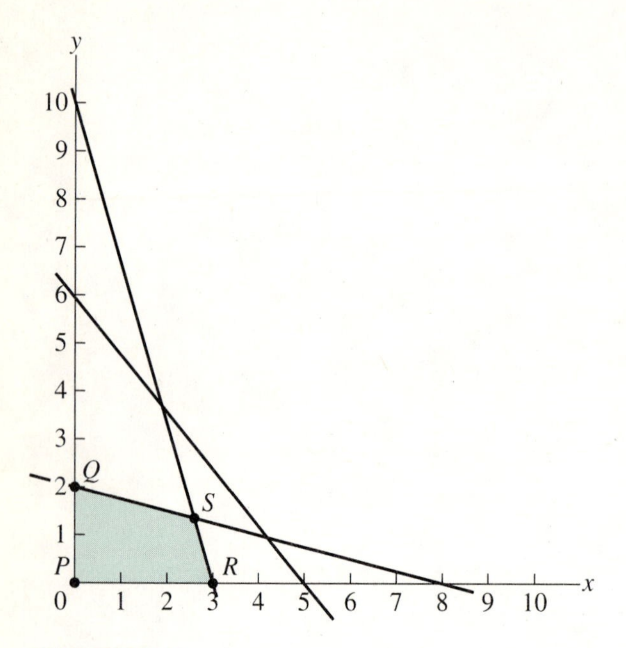

FIGURE C10.3

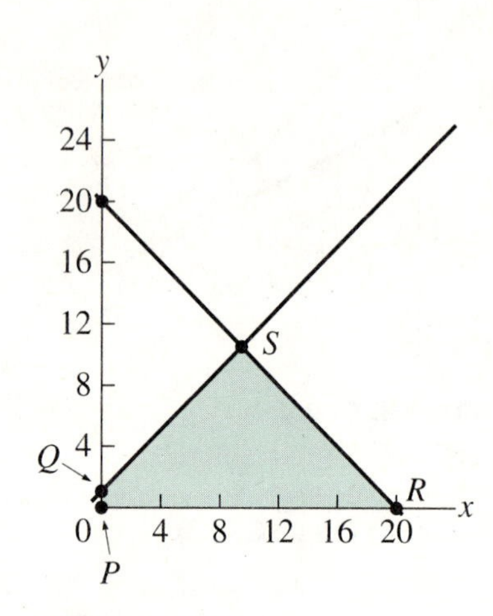

FIGURE C10.4

(See Figure C10.4)

Corner point	Basic solution
P	$x = 0, y = 0, u = 1, v = 20$
Q	$x = 0, y = 1, u = 0, v = 19$
R	$x = 20, y = 0, u = 21, v = 0$
S	$x = \frac{19}{2}, y = \frac{21}{2}, u = 0, v = 0$

27. Add -3 times the first equation to the second equation; add -2 times the first equation to the third equation, then switch the second and third equations. Add -4 times the second equation to the third equation.

29. Maximize: $p = 5x + y$

subject to:
$$x \geq 0, y \geq 0, u \geq 0, v \geq 0$$
$$2x + y + u = 8$$
$$x + 2y + v = 8$$

Basic solutions:

(*a*) $x = 0, y = 0, u = 8, v = 8$; (*b*) $x = 0, y = 4, u = 4, v = 0$; (*c*) $x = 4, y = 0, u = 0, v = 4$; (*d*) $x = \frac{8}{3}, y = \frac{8}{3}, u = 0, v = 0$

31. Maximize: $p = 3x + 2y + z$

subject to:
$$x \geq 0, y \geq 0, z \geq 0$$
$$3x - y + 2z \leq 100$$
$$2x + y - z \leq 150$$

33. (*a*) Let x = number of miles repaired by "quick fix"
y = number of miles repaired by satisfactory but temporary method
z = number of miles repaired with long lasting repairs

Maximize: $p = x + 2y + 4z$

subject to: $x \geq 0, y \geq 0, z \geq 0$
$200x + 500y + 1000z \leq 10{,}000$
$200x + 100y + 200z \leq 1000$

(*b*) $200x + 500y + 1000z + u = 10{,}000$
$200x + 100y + 200z + v = 1000$

(*c*) u = number of cubic yards of asphalt not used
v = number of hours of labor not used

35. (*a*) Let x = number of computer boxes
y = number of printer boxes
z = number of paint boxes

Maximize: $p = 1.5x + .8y + .25z$

subject to: $x \geq 0, y \geq 0, z \geq 0$
$12x + 6y \leq 1500$ (heavy cardboard)
$18x + 12y + 10z \leq 2500$ (regular cardboard)
$15x + 8y \leq 2000$ (facing paper)

(*b*) $12x + 6y + u = 1500$
$18x + 12y + 10z + v = 2500$
$15x + 8y + w = 2000$

(*c*) u = number of square feet of unused heavy cardboard
v = number of square feet of unused regular cardboard
w = number of square feet of unused facing paper

Section 10.2

1.

	x	y	u	v	Basic Solution
u	3	5	1	0	20
v	1	3	0	1	9
p	−6	−4	0	0	0

3.

	x	y	u	v	Basic Solution
u	2	0	1	0	15
v	0	3	0	1	8
p	−3	−7	0	0	0

5.

	x	y	u	v	w	Basic Solution
u	2	3	1	0	0	12
v	6	1	0	1	0	9
w	1	0	0	0	1	4
p	0	−2	0	0	0	0

7.

	x	y	u	v	w	Basic Solution
u	0	3	1	0	0	24
v	2	0	0	1	0	24
w	2	3	0	0	1	36
p	−2	1	0	0	0	0

9.

	x	y	z	u	v	w	Basic Solution
u	5	0	1	1	0	0	100
v	0	1	3	0	1	0	300
w	2	−1	1	0	0	1	900
p	−2	−3	−1	0	0	0	0

11.

	x_1	x_2	x_3	x_4	u	v	w	Basic Solution
u	1	3	−1	1	1	0	0	20
v	3	6	3	−1	0	1	0	30
w	1	−3	1	1	0	0	1	15
p	−2	1	−3	1	0	0	0	0

13.

	x	y	u	v	Basic Solution
y	$\frac{2}{3}$	1	$\frac{1}{3}$	0	4
v	$\frac{16}{3}$	0	$-\frac{1}{3}$	1	5
p	1	0	1	0	12

New basic solution: $x = 0, y = 4, u = 0, v = 5, p = 12$

15.

	x	y	u	v	Basic Solution
x	1	1	$\frac{1}{4}$	0	2
v	0	2	$\frac{1}{4}$	1	3
p	0	3	$\frac{1}{2}$	0	4

New basic solution: $x = 2, y = 0, u = 0, v = 3, p = 4$

17.

	x	y	z	u	v	Basic Solution
u	$\frac{5}{2}$	0	-22	1	$-\frac{5}{2}$	5
y	$\frac{3}{2}$	1	6	0	$\frac{1}{2}$	15
p	$\frac{7}{2}$	0	20	0	$\frac{3}{2}$	45

New basic solution: $x = 0, y = 15, z = 0, u = 5, v = 0, p = 45$

19.

	x	y	u	v	w	Basic Solution
u	12	0	1	-8	0	8
y	-1	1	0	1	0	1
w	0	0	0	1	1	2
p	-3	0	0	6	0	6

New basic solution: $x = 0, y = 1, u = 8, v = 0, w = 2, p = 6$

21.

	x	y	u_1	u_2	u_3	u_4	Basic Solution
u_1	0	5	1	0	-5	0	10
u_2	0	3	0	1	-5	0	5
x	1	0	0	0	1	0	2
u_4	0	2	0	0	0	1	3
p	0	1	0	0	4	0	8

New basic solution: $x = 2, y = 0, u_1 = 10, u_2 = 5, u_3 = 0, u_4 = 3, p = 8$

23.

	x	y	z	u	v	w	Basic Solution
y	1	1	3	$\frac{1}{5}$	0	0	20
v	3	0	0	$-\frac{3}{5}$	1	0	30
w	1	0	1	$-\frac{1}{5}$	0	1	0
p	4	0	12	1	0	0	100

New basic solution: $x = 0, y = 20, z = 0, u = 0, v = 30, w = 0, p = 100$

25. Maximize: $x + 3y$

subject to:

$$x \geq 0, y \geq 0$$
$$2x + 3y \leq 12$$
$$6x + y \leq 9$$

27. Maximize: $4x - y$

subject to:

$$x \geq 0, y \geq 0$$
$$5x + 5y \leq 20$$
$$5x + 3y \leq 15$$
$$x \leq 2$$
$$2y \leq 3$$

29. Original basic solution: $x = 7, y = 0, z = 0, u = 0, v = 9, p = 42$
New tableau:

	x	y	z	u	v	Basic Solution
x	1	0	$\frac{14}{9}$	$\frac{5}{9}$	$-\frac{2}{9}$	5
y	0	1	$\frac{2}{3}$	$-\frac{1}{3}$	$\frac{1}{3}$	3
p	0	0	$\frac{32}{3}$	$\frac{5}{3}$	$\frac{4}{3}$	45

New basic solution: $x = 5, y = 3, z = 0, u = 0, v = 0, p = 45$

31. Original basic solution: $x = 0, y = 1000, z = 0, u = 2000, v = 2000, w = 0, p = 8000$
New tableau:

	x	y	z	u	v	w	Basic Solution
z	6	0	1	$\frac{1}{2}$	0	$-\frac{1}{2}$	1000
v	−8	0	0	0	1	−2	2000
y	−1	1	0	$-\frac{1}{4}$	0	$\frac{1}{2}$	500
p	25	0	0	1	0	1	10000

New basic solution: $x = 0, y = 500, z = 1000, u = 0, v = 2000, w = 0, p = 10000$

33. Original basic solution: $x = 0, y = 1, u = 8, v = 0, w = 2, p = 6$
New tableau:

	x	y	u	v	w	Basic Solution
x	1	0	$\frac{1}{12}$	$-\frac{2}{3}$	0	$\frac{2}{3}$
y	0	1	$\frac{1}{12}$	$\frac{1}{3}$	0	$\frac{5}{3}$
w	0	0	0	1	1	2
p	0	0	$\frac{1}{4}$	4	0	8

New basic solution: $x = \frac{2}{3}, y = \frac{5}{3}, u = 0, v = 0, w = 2, p = 8$

35. Using the 1 in the x column as a pivot element:

	x	y	z	u	v	Basic Solution
x	1	0	5	1	0	50
y	0	1	$\frac{1}{10}$	0	$\frac{1}{10}$	3
p	0	0	$\frac{7}{2}$	1	$\frac{1}{2}$	65

Using the 5 in the z column as a pivot element:

	x	y	z	u	v	Basic Solution
z	$\frac{1}{5}$	0	1	$\frac{1}{5}$	0	10
y	$-\frac{1}{50}$	1	0	$-\frac{1}{50}$	$\frac{1}{10}$	2
p	$-\frac{7}{10}$	0	0	$\frac{3}{10}$	$\frac{1}{2}$	30

Using the (1, 1) entry as a pivot element yields a new basic solution for which the objective function has the value 65. This is larger than the value of the objective function for the basic solution which results from using the (1, 3) entry as a pivot element.

Section 10.3

For Exercises 1, 3, 5, 7, and 9, the pivot element is specified by its position (row, column)

Exercise	Entering variable	Departing variable	Pivot element
1.	x	v	(2, 1) entry
3.	x	v	(2, 1) entry
5.	None		
7.	x	u	(1, 1) entry
9.	y	w	(3, 2) entry

11. $x = 3, y = 0, p = 9$
13. $x = 0, y = 1, z = \frac{5}{2}, p = \frac{17}{2}$
15. $x = 0, y = 4, p = 12$
17. $x = 4, y = \frac{3}{2}, p = \frac{25}{2}$
19. $x = 8, y = 4, p = 8$
21. $x = 15, y = 0, p = 105$
23. $x = 2, y = 4, p = 42$
25. $x = 0, y = \frac{12}{13}, z = \frac{36}{13}, p = \frac{228}{13}$
27. $x = 0, y = 12, z = 0, p = 120$
29. There is a solution: $x = 25$, $y = 10$, maximum value $= 70$
31. 48 boxes of Basic Bits, 216 boxes of Lite Bits, maximum profit = \$88.80.
33. 50 standard logs and 20 deluxe logs. Maximum profit is \$390
35. Maximum profit is \$1.25 and is achieved by preparing 50 small packets with raisins and $\frac{50}{3}$ small packets without raisins. No large packets should be prepared.

Section 10.4

The matrices **A** and vectors **B**, **C**, and **X** for Exercises 1, 3, 5, 7, and 9 are as shown:

	A	**B**	**C**	**X**
1.	$\begin{bmatrix} 3 & 6 \\ 6 & 1 \end{bmatrix}$	$\begin{bmatrix} 60 \\ 18 \end{bmatrix}$	$\begin{bmatrix} 1 \\ 2 \end{bmatrix}$	$\begin{bmatrix} x \\ y \end{bmatrix}$
3.	$\begin{bmatrix} 0 & 1 \\ 1 & 1 \\ -1 & 4 \end{bmatrix}$	$\begin{bmatrix} 10 \\ 15 \\ 12 \end{bmatrix}$	$\begin{bmatrix} 1 \\ 0 \end{bmatrix}$	$\begin{bmatrix} x \\ y \end{bmatrix}$
5.	$\begin{bmatrix} 2 & 3 & -4 \\ -1 & 1 & 1 \end{bmatrix}$	$\begin{bmatrix} 20 \\ 5 \end{bmatrix}$	$\begin{bmatrix} 1 \\ 3 \\ -1 \end{bmatrix}$	$\begin{bmatrix} x \\ y \\ z \end{bmatrix}$
7.	$\begin{bmatrix} 1 & 15 & 10 \\ 10 & 5 & 1 \\ 15 & 1 & 20 \\ 1 & 1 & 1 \end{bmatrix}$	$\begin{bmatrix} 200 \\ 120 \\ 300 \\ 100 \end{bmatrix}$	$\begin{bmatrix} 1 \\ 10 \\ 20 \end{bmatrix}$	$\begin{bmatrix} x \\ y \\ z \end{bmatrix}$
9.	$\begin{bmatrix} 1 & -2 & -3 & 4 \\ 2 & 4 & 1 & -1 \end{bmatrix}$	$\begin{bmatrix} 20 \\ 10 \end{bmatrix}$	$\begin{bmatrix} 1 \\ 2 \\ 3 \\ 4 \end{bmatrix}$	$\begin{bmatrix} x_1 \\ x_2 \\ x_3 \\ x_4 \end{bmatrix}$

11. Maximize: $3x + 8y$

subject to:
$$\begin{aligned} x \ge 0, y &\ge 0 \\ 2x + 2y &\le 16 \\ 3x + 6y &\le 18 \end{aligned}$$

13. Maximize: $4x - y + 3z$

subject to:
$$\begin{aligned} x \ge 0, y \ge 0, z &\ge 0 \\ x - 2y + 3z &\le 20 \\ -4x \qquad + 6z &\le 30 \end{aligned}$$

15. Maximize: $10x + 15y + 20z$

subject to:
$$\begin{aligned} x \ge 0, y \ge 0, z &\ge 0 \\ x \qquad + z &\le 100 \\ y \qquad &\le 50 \\ x + y + z &\le 200 \end{aligned}$$

17. Maximize: $2x + 6y + 4z$

subject to:
$$\begin{aligned} x \ge 0, y \ge 0, z &\ge 0 \\ x + 5y + z &\le 10 \\ y + 2z &\le 40 \\ 4x + 2y \qquad &\le 60 \\ 2x + y + 2z &\le 20 \end{aligned}$$

19. Minimize: $12r + 18s$

subject to:
$$\begin{aligned} r \ge 0, s &\ge 0 \\ 3r + 6s &\ge 2 \\ 4r + 3s &\ge 3 \end{aligned}$$

21. Maximize: $24r + 12s$

subject to:
$$\begin{aligned} r \ge 0, s &\ge 0 \\ 8r + 4s &\le 3 \\ 3r + 6s &\le 4 \end{aligned}$$

23. Minimize: $r + s$

subject to:
$$\begin{aligned} r \ge 0, s &\ge 0 \\ r + 4s &\ge 1 \\ 4r + 8s &\ge 1 \\ r + s &\ge 1 \end{aligned}$$

25. Maximize: $20r + 30s$

subject to:
$$\begin{aligned} r \ge 0, s &\ge 0 \\ r - 4s &\ge 4 \\ -2r \qquad &\ge -1 \\ 3r + 6s &\ge 3 \end{aligned}$$

27. Solution of SMP: $x = \frac{16}{11}, y = \frac{102}{11}, p = 20$
    Solution of dual: $r = \frac{1}{3}, s = 0, p = 20$
29. Solution of SMP: $x = 15, y = 0, p = 15$
    Solution of dual: $r = 0, s = 1, t = 0, p = 15$
31. Solution of SMP: $x = 15, y = 0, z = 17, p = 245$
    Solution of dual: $r = 0, s = 2, t = 3, p = 245$
33. (*a*).

	x	y	z	u	v	Basic Solution
u	1	−2	3	1	0	20
v	−4	0	6	0	1	30
p	−4	1	−3	0	0	0

	x	y	z	u	v	Basic Solution
x	1	−2	3	1	0	20
v	0	−8	18	4	1	110
p	0	−7	9	4	0	80

Since there is an entering variable, y, but no departing variable, the problem has no solution.

(*b*) The dual problem is Minimize: $20r + 30s$

subject to:

$$\begin{aligned} r \geq 0, s &\geq 0 \\ r - 4s &\geq 4 \\ -2r &\geq -1 \\ 3r + 6s &\geq 3 \end{aligned}$$

The feasible set for this problem is empty.

35.

	x	y	z	u	v	w	Basic Solution
u	10	10	.3	1	0	0	60
v	40	20	.2	0	1	0	50
w	60	30	.6	0	0	1	90
p	−250	−500	−9	0	0	0	0

	x	y	z	u	v	w	Basic Solution
u	−10	0	.2	1	−.5	0	35
y	2	1	.01	0	.05	0	2.5
w	0	0	.3	0	−1.5	1	15
p	750	0	−4	0	25	0	1250

	x	y	z	u	v	w	Basic Solution
u	−10	0	0	1	.5	$-\frac{2}{3}$	25
y	2	1	0	0	.1	$-\frac{1}{30}$	2
z	0	0	1	0	−5	$\frac{10}{3}$	50
p	750	0	0	0	5	$\frac{40}{3}$	1450

Review Exercises

1. Not an SMP: missing constraints $x \geq 0$, $y \geq 0$
3. This problem is an SMP.
5. The constraints $x \leq 50$, $z \leq 60$ must be added
7. Set x = number of bulletin board crates
   y = number of mirror crates
   z = number of cabinet crates

Maximize: $2x + 5y + 10z$

subject to:
$$\begin{aligned} x \geq 0,\ y \geq 0,\ z \geq 0 \\ 60x + 20y + 50z \leq 2700 \\ 30x + 30y + 30z \leq 1800 \\ 17y + 5z \leq 500 \end{aligned}$$

Solution: $x = 0$, $y = 10$, $z = 50$

9. (*a*) Maximize: $2.5x + 2y + 2.25z$

subject to: $x \geq 0, y \geq 0, z \geq 0, u_1 \geq 0, u_2 \geq 0, u_3 \geq 0, u_4 \geq 0, u_5 \geq 0$
$$\begin{aligned} x + 2y + 1.5z + u_1 &= 300 \\ 2x + y + 1.5z + u_2 &= 350 \\ x + y + 2z + u_3 &= 400 \\ x + u_4 &= 50 \\ z + u_5 &= 60 \end{aligned}$$

(*b*)

	x	y	z	u_1	u_2	u_3	u_4	u_5	Basic Solution
u_1	1	2	1.5	1	0	0	0	0	300
u_2	2	1	1.5	0	1	0	0	0	350
u_3	1	1	2	0	0	1	0	0	400
u_4	1	0	0	0	0	0	1	0	50
u_5	0	0	1	0	0	0	0	1	60
p	−2.5	−2	−2.25	0	0	0	0	0	0

11.

	x	y	z	u	v	w	Basic Solution
u	0	1.5	.75	1	−.5	0	125
x	1	.5	.75	0	.5	0	175
w	0	.5	1.25	0	−.5	1	225
p	0	−.75	−.375	0	1.25	0	$\frac{875}{2}$

13. Maximize:

$$p = 2x + 3y$$

subject to:

$$x \geq 0, y \geq 0, u \geq 0, v \geq 0$$
$$3x + 5y + u = 30$$
$$6x + 2y + v = 18$$

(See Figure C10.5)

Corner Point	Basic Solution
P	$x = 0, y = 0, u = 30, v = 18$
Q	$x = 0, y = 6, u = 0, v = 6$
R	$x = \frac{5}{4}, y = \frac{21}{4}, u = 0, v = 0$
S	$x = 3, y = 0, u = 21, v = 0$

FIGURE C10.5

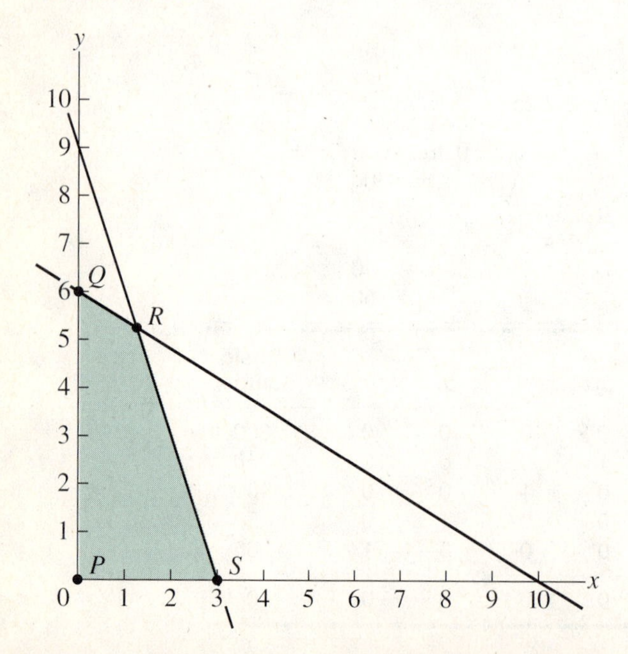

15. Initial tableau:

	x	y	u	v	Basic Solution
u	3	5	1	0	30
v	6	2	0	1	18
p	−2	−3	0	0	0

	x	y	u	v	Basic Solution
y	$\frac{3}{5}$	1	$\frac{1}{5}$	0	6
v	$\frac{24}{5}$	0	$-\frac{2}{5}$	1	6
p	$-\frac{1}{5}$	0	$\frac{3}{5}$	0	18

Basic solution: $x = 0, y = 6, u = 0, v = 6, p = 18$

17.

	x	y	z	u	v	w	Basic Solution
u	20	10	5	1	0	0	2000
v	5	10	15	0	1	0	1500
w	5	10	10	0	0	1	1500
p	−10	−5	−15	0	0	0	0

19.

	x	y	u	v	w	Basic Solution
u	$\frac{7}{3}$	0	1	$-\frac{5}{6}$	0	15
y	$\frac{1}{3}$	1	0	$\frac{1}{6}$	0	5
w	1	0	0	−2	1	0
p	$-\frac{1}{3}$	0	0	$\frac{4}{3}$	0	40

Basic solution: $x = 0, y = 5, u = 15, v = 0, w = 0, p = 40$

21.

	x	y	z	u	v	w	Basic Solution
u	0	$\frac{3}{2}$	$\frac{1}{2}$	1	$-\frac{1}{2}$	0	125
x	1	$\frac{1}{2}$	$\frac{1}{2}$	0	$\frac{1}{2}$	0	175
w	0	$\frac{1}{2}$	$\frac{3}{2}$	0	$-\frac{1}{2}$	1	225
p	0	$-\frac{3}{4}$	−1	0	$\frac{5}{4}$	0	$\frac{875}{2}$

Basic solution: $x = 175, y = 0, z = 0, u = 125, v = 0, w = 225, p = \frac{875}{2}$

23. (*a*) The pivot element is the (1, 3) entry

	x	*y*	*z*	*u*	*v*	*w*	Basic Solution
z	4	2	1	$\frac{1}{4}$	0	0	4
v	0	8	0	−2	1	0	8
w	82	38	0	4	0	1	100
p	18	6	0	1	0	0	16

(*b*) The new basic solution is $x = 0, y = 0, z = 4, u = 0, v = 8, w = 100, p = 16$. It is optimal since there is no choice for an entering variable.

25.

	x	*y*	*u*	*v*	*w*	Basic Solution
u	1	1	1	0	0	20
v	1	4	0	1	0	40
w	1	0	0	0	1	15
p	−5	2	0	0	0	0

	x	*y*	*u*	*v*	*w*	Basic Solution
u	0	1	1	0	−1	5
v	0	4	0	1	−1	25
x	1	0	0	0	1	15
p	0	2	0	0	5	75

Solution: $x = 15, y = 0, p = 75$

27. (*a*) Maximize: $\boldsymbol{C}^T\boldsymbol{X}$ subject to $\boldsymbol{X} \geq \boldsymbol{0}, \boldsymbol{AX} \leq \boldsymbol{B}$

$$\boldsymbol{A} = \begin{bmatrix} 1 & 1 \\ 1 & 4 \\ 1 & 0 \end{bmatrix} \quad \boldsymbol{B} = \begin{bmatrix} 20 \\ 40 \\ 15 \end{bmatrix} \quad \boldsymbol{C} = \begin{bmatrix} 5 \\ -2 \end{bmatrix} \quad \boldsymbol{X} = \begin{bmatrix} x \\ y \end{bmatrix}$$

(*b*) Minimize $\boldsymbol{B}^T\mathbf{Z}$ subject to $\mathbf{Z} \geq \mathbf{0}, \mathbf{Z}^T\mathbf{A} \geq \mathbf{C}$ with **A**, **B**, **C** as above and

$$\mathbf{Z} = \begin{bmatrix} r \\ s \\ t \end{bmatrix}$$

29. (*a*) Maximize $\mathbf{C}^T X$ subject to $\mathbf{X} \geq \mathbf{0}$ and $\mathbf{AX} \leq \mathbf{B}$

$$\boldsymbol{A} = \begin{bmatrix} 1 & 2 & 1 \\ 2 & 1 & 2 \end{bmatrix} \quad \boldsymbol{B} = \begin{bmatrix} 12 \\ 24 \end{bmatrix} \quad \boldsymbol{C} = \begin{bmatrix} 2 \\ 1 \\ 4 \end{bmatrix} \quad \boldsymbol{X} = \begin{bmatrix} x \\ y \\ z \end{bmatrix}$$

(*b*) Minimize $\mathbf{B}^T\mathbf{Z}$ subject to $\mathbf{Z} \geq \mathbf{0}$ and $\mathbf{Z}^T\mathbf{A} \geq \mathbf{C}$ with $\mathbf{A}$, $\mathbf{B}$, $\mathbf{C}$ as above and

$$\mathbf{Z} = \begin{bmatrix} r \\ s \end{bmatrix}$$

31. $x = \frac{14}{3}, y = \frac{1}{3}, p = \frac{13}{3}$
33. (*a*) $r = 0, s = \frac{2}{3}, t = \frac{1}{3}, p = 40$; (*b*) $r = 0, s = 0, t = \frac{2}{3}, p = 40$
35. $r = 0, s = 3, t = \frac{5}{2}, p = 102$
37. 0 small packages and 30 large packages
39. 20 tons by method 1
    0 tons by method 2
    5 tons by method 3

CHAPTER 11

Section 11.1

1. Rhoda should always select card 1, and Carol should always select card 1.
3. Carol should always select card 2, and Rhoda should always select card 2.
5. For $A \geq -2$ the answer is the same.
7. Rhoda should always select card 2, and Carol should always select card 1.
9.

		Chester	
		H	T
Rolf	H	−5	25
	T	25	−5

11. 1.25 cents
13. 0 cents
15. $-\$\frac{2}{9}$
17. (*a*) Management can meet labor's demands, compromise, or refuse to negotiate. Labor has a choice of striking or not. A payoff matrix where the payoff is a gain by labor (including the costs of a strike, if any) might be as follows:

		Management		
		Meet Demands	Compromise	Refuse to Negotiate
Labor	Strike	1.5	1.0	.5
	No Strike	2	.05	0

(*b*) Each of the two political parties, *A* and *B*, has a choice of campaigning on domestic issues, foreign policy questions, or the statements made by the opposition. A payoff matrix where the payoff is the gain in votes by party *A* might be as follows:

		B		
		Domestic Issues	Foreign Policy	Opposition
	Domestic Issues	0	−100	50
A	Foreign Policy	−50	100	50
	Opposition	100	50	−100

19. \$0

21. $-\$\frac{1}{3}$

Section 11.2

1. Yes, (1, 1) entry
3. Yes, there are four saddle points: (2, 1), (2, 2), (3, 1), (3, 2).
5. Yes, there are two saddle points: (2, 2) and (3, 2).
7. Yes, there are four saddle points: (2, 2), (2, 3), (3, 2), (3, 3).
9. No
11. No
13. Row 2 dominates rows 1, 3, and 4; column 2 dominates columns 1 and 4.
15. Column 3 dominates columns 2 and 5; row 2 dominates row 1.
17. Column 2 dominates columns 4 and 5; then row 4 dominates rows 1 and 3.
19. One example is $\begin{bmatrix} 1 & -1 & -2 \\ 2 & 0 & 0 \\ 2 & -2 & -1 \end{bmatrix}$
21. $\begin{bmatrix} 0 & -1 & -4 \\ 1 & 0 & -1 \\ 4 & 1 & 0 \end{bmatrix}$ Row 3 dominates rows 1 and 2; column 3 dominates columns 1 and 2; there is a saddle point at the (3, 3) entry.
23. The (2, 2) entry is a saddle point, value is *a*.

Section 11.3

1. $S^* = [\frac{3}{5} \quad \frac{2}{5}]$, $T^* = [\frac{7}{10} \quad \frac{3}{10}]$, $V = \frac{4}{5}$
3. $S^* = [\frac{1}{2} \quad \frac{1}{2}]$, $T^* = [\frac{7}{10} \quad \frac{3}{10} \quad 0]$, $V = -\frac{1}{2}$
5. $S^* = [\frac{7}{9} \quad \frac{2}{9}]$, $T^* = [\frac{2}{9} \quad 0 \quad \frac{7}{9}]$, $V = \frac{31}{9}$
7. $S^* = [\frac{3}{7} \quad \frac{4}{7} \quad 0]$, $T^* = [\frac{4}{7} \quad \frac{3}{7}]$, $V = \frac{5}{7}$
9. $S^* = [\frac{3}{5} \quad \frac{2}{5}]$, $T^* = [\frac{3}{5} \quad \frac{2}{5} \quad 0 \quad 0]$, $Y = \frac{14}{5}$
11. There is a saddle point at (2, 3): $S^* = [0 \quad 1 \quad 0]$, $T^* = [0 \quad 0 \quad 1]$, $V = -1$.

13. One dominates the other. In particular, if $p_l(r)$ lies above $p_m(r)$ for $0 \le r \le 1$, then C_m dominates C_l.
15. $S^* = [0 \quad 1 \quad 0 \quad 0]$, $T^* = [0 \quad 0 \quad 1 \quad 0]$, $V = -1$ [saddle point at the (2, 3) entry]
17. $S^* = [\frac{9}{14} \quad \frac{5}{14} \quad 0]$, $T^* = [0 \quad \frac{4}{7} \quad \frac{3}{7}]$, $V = -\frac{1}{7}$
19. $S = [0 \quad \frac{5}{9} \quad 0 \quad \frac{4}{9}]$, $T^* = [\frac{1}{9} \quad 0 \quad \frac{8}{9} \quad 0 \quad 0]$, $V = \frac{14}{9}$
21. $S^* = [0 \quad \frac{1}{2} \quad 0 \quad 0 \quad \frac{1}{2} \quad 0]$, $T^* = [\frac{2}{3} \quad \frac{1}{3} \quad 0]$, $V = 2$
23. $S^* = \left[\frac{c-a}{b+c-2a} \quad \frac{b-a}{b+c-2a}\right]$, $T^* = \left[\frac{b-a}{b+c-2a} \quad \frac{c-a}{b+c-2a}\right]$, $V = \frac{bc-a^2}{b+c-2a}$
25. $S^* = \left[\frac{a}{a+b} \quad \frac{b}{a+b}\right]$, $T^* = \left[0 \quad \frac{1}{2} \quad \frac{1}{2}\right]$, $V = 0$

Section 11.4

1. $S^* = [\frac{2}{5} \quad \frac{3}{5}]$, $T^* = [\frac{3}{5} \quad \frac{2}{5}]$, $V = \frac{16}{5}$
3. $S^* = [\frac{2}{5} \quad \frac{3}{5}]$, $T = [\frac{3}{5} \quad \frac{2}{5}]$, $V = -\frac{3}{5}$
5. $S^* = [\frac{1}{2} \quad \frac{1}{2}]$, $T^* = [0 \quad \frac{2}{5} \quad \frac{3}{5}]$, $V = 2$
7. There are infinitely many solutions. Two are:
   $S^* = [\frac{1}{2} \quad \frac{1}{2} \quad 0]$, $T^* = [\frac{3}{4} \quad \frac{1}{4} \quad 0]$, $V = \frac{1}{2}$
   $S^* = [\frac{1}{2} \quad \frac{1}{2} \quad 0]$, $T^* = [0 \quad \frac{5}{8} \quad \frac{3}{8}]$, $V = \frac{1}{2}$
9. $S^* = [\frac{1}{13} \quad \frac{5}{13} \quad \frac{7}{13}]$, $T^* = [\frac{1}{13} \quad \frac{7}{13} \quad \frac{5}{13}]$, $V = \frac{4}{13}$
11. $S^* = [\frac{4}{81} \quad \frac{70}{81} \quad \frac{7}{81}]$, $T^* = [\frac{44}{81} \quad 0 \quad \frac{2}{81} \quad \frac{35}{81}]$, $V = -\frac{83}{81}$
13. There is a saddle point at (2, 2): $S^* = [0 \quad 1 \quad 0]$, $T^* = [0 \quad 1 \quad 0]$, $V = 0$.
15. In both cases $S^* = [\frac{4}{9} \quad \frac{5}{9}]$ and $T^* = [\frac{6}{9} \quad \frac{3}{9}]$; for P, $V = \frac{3}{9}$, and for Q, $V = \frac{30}{9}$. If a payoff matrix is multiplied by a positive number k, then the optimal strategies for the row and column players remain the same and the value of the game is multiplied by k.
17. $S^* = [0 \quad \frac{3}{4} \quad \frac{1}{4}]$, $T^* = [0 \quad \frac{3}{4} \quad \frac{1}{4}]$, $V = \frac{1}{40}$
19. Advisory service A should suggest stocks with probability $\frac{4}{7}$ and bonds with probability $\frac{3}{7}$. Advisory service Z should suggest stocks and bonds with probabilities $\frac{2}{7}$ and $\frac{5}{7}$, respectively.
21. $S^* = [\frac{1}{4} \quad \frac{3}{4} \quad 0]$, $T^* = [\frac{3}{4} \quad \frac{1}{4} \quad 0]$, $V = \frac{7}{4}$
23. There are infinitely many optimal strategies. One is $S = [0 \quad \frac{1}{2} \quad \frac{1}{2} \quad 0]$, $T^* = [0 \quad \frac{1}{2} \quad \frac{1}{2} \quad 0]$, $V = 0$. In fact, $S^* = [\alpha, \frac{1}{2} - \frac{5}{4}\alpha, \frac{1}{2} - \frac{3}{4}\alpha, \alpha] = T^*$ is optimal for $0 \le \alpha \le \frac{2}{5}$.

Review Exercises

1. Saddle point at the (1, 2) entry, and the value of the game is 1.
3. Saddle point at the (3, 2) entry, and the value of the game is -1.
5. The (2, 2) and (2, 4) entries are saddle points, and the value of the game is 0.
7. $S^* = [\frac{4}{9} \quad \frac{5}{9}]$, $T^* = [\frac{6}{9} \quad 0 \quad 0 \quad \frac{3}{9}]$, $V = \frac{6}{9}$
9. $S^* = [\frac{2}{7} \quad \frac{5}{7}]$, $T^* = [0 \quad 0 \quad \frac{3}{7} \quad \frac{4}{7}]$, $V = -\frac{6}{7}$
11. $S^* = [\frac{7}{12} \quad \frac{5}{12}]$, $T^* = [\frac{9}{12} \quad \frac{3}{12}]$, $V = \frac{9}{12}$
13. $S^* = [0 \quad \frac{5}{9} \quad \frac{4}{9}]$, $T^* = [\frac{4}{9} \quad \frac{5}{9} \quad 0]$, $V = \frac{2}{9}$
15. $S^* = [\frac{1}{13} \quad \frac{7}{13} \quad \frac{5}{13}]$, $T^* = [\frac{1}{13} \quad \frac{7}{13} \quad \frac{5}{13}]$, $V = -\frac{3}{13}$
17. $T^* = [0 \quad 1 \quad 0]$, $V = -1$. There are infinitely many choices for S^*; two are $[0 \quad 0 \quad 1]$ and $[0 \quad 1 \quad 0]$. In general, $[0 \quad r \quad 1-r]$ is a solution for $0 \le r \le 1$.
19. $S^* = [\frac{3}{5} \quad \frac{2}{5} \quad 0]$, $T^* = [0 \quad \frac{3}{5} \quad \frac{2}{5} \quad 0]$, $V = \frac{7}{5}$

21. There are infinitely many optimal strategies. Two are $S^* = [0 \quad 1 \quad 0]$, $T^* = [0 \quad \frac{1}{3} \quad \frac{2}{3} \quad 0]$, $V = 1$ and $S^* = [0 \quad 1 \quad 0]$, $T^* = [0 \quad \frac{1}{2} \quad \frac{1}{2} \quad 0]$, $V = 1$. In fact, $S^* = [0 \quad 1 \quad 0]$ and $T^* = [0 \quad c \quad 1-c \quad 0]$ are optimal for $\frac{1}{3} \le c \le \frac{1}{2}$.
23. $S^* = [0 \quad 1 \quad 0 \quad 0]$, $T^* = [0 \quad 0 \quad 1]$, $V = 1$ (saddle point)

CHAPTER 12

Section 12.1

1. (*a*) $p \wedge q$
   (*b*) $p \wedge r$
   (*c*) $p \wedge (q \wedge r)$ or $(p \wedge q) \wedge$ r
3. (*a*) Roger is a teacher, and either Peter is president or Mary is a mathematician or both.
   (*b*) Either Peter is president, or both Mary is a mathematician and Roger is a teacher.
   (*c*) At least one of the following is true: Peter is not president or Mary is a mathematician or Roger is a teacher.
5. (*a*) The inflation index went up, and the stock market average did not go up or the bond market average did not go up or neither went up.
   (*b*) The stock market average went up, and the bond market average went up, but the inflation index did not go up.
   (*c*) Both the stock and the bond markets did not go up or the inflation index did not go up.
7. (*a*) x is an element of A and x is an element of B and x is an element of C. (i.e., x is an element of $A \cap B \cap C$).
   (*b*) x is an element of A or x is an element of B, or x is an element of C (i.e., x is an element of $A \cup B \cup C$).
   (*c*) x is an element of A and x is an element of B, but x is not an element of C (that is, x is an element of $A \cap B \cap C'$).
   (*d*) x is an element of A or x is an element of B or x is not an element of C (that is, x is an element of $A \cup B \cup C'$.
9. (*a*) $p \wedge (\sim q)$; (*b*) $(\sim r) \wedge \{\sim[p \wedge (\sim q)]\}$

Section 12.2

1. (*a*) p and q true; q false and p either true or false; (*b*) p true and q false
3. (*a*) r true and either p or q true, or both;
   (*b*) p and q true and r either true or false; r true and p, q either true or false
5. (*a*) p and q true, r and s can be either true or false;
   r and s true, p and q can be either true or false
   (*b*) Either p or q (or both) true and either r or s (or both) true
7. (*a*) p and q false, r and s can be either true or false; r and s false, p and q can be either true or false
   (*b*) Both p and q false or both r and s false

9. (*a*) p and q true, r and s can be either true or false;
p and r true, q and s can be either true or false;
p and s true, q and r can be either true or false
(*b*) Same as in (*a*)

11. (*a*) $\{x: (x \in A) \wedge (x \in B) \wedge (x \in C)\}$
(*b*) $\{x: (x \in A) \vee (x \in B) \vee (x \in C)\}$

13. (*a*) $\{x:[\sim(x \in A)] \wedge (x \in B)\}$ (*b*) $\{x:(x \in A) \vee [\sim(x \in B)]\}$

15. (*a*) $\{x: [\sim(x \in A)] \wedge (x \in B)\} \wedge \{x \in C) \vee [\sim(x \in A)]\}$
(*b*) $\{x: \sim\{[\sim(x \in A)] \vee [\sim(x \in B)] \vee [\sim(x \in C)]\}\}$

17.

$x \in A$	$x \in B$	$x \in C$	$x \in (B \cup C)$	$x \in [A \cap (B \cup C)]$	$x \in (A \cap B)$	$x \in (A \cap C)$	$x \in [(A \cap B) \cup (A \cap C)]$
T	T	T	T	T	T	T	T
T	T	F	T	T	T	F	T
T	F	T	T	T	F	T	T
T	F	F	F	F	F	F	F
F	T	T	T	F	F	F	F
F	T	F	T	F	F	F	F
F	F	T	T	F	F	F	F
F	F	F	F	F	F	F	F

19.

$x \in A$	$x \in B$	$x \in C$	$x \in B'$	$x \in C'$	$x \in (B' \cup C')$	$x \in (B' \cup C')'$	$x \in [A \cap (B' \cup C')']$	$x \in (A \cap B \cap C)$
T	T	T	F	F	F	T	T	T
T	T	F	F	T	T	F	F	F
T	F	T	T	F	T	F	F	F
T	F	F	T	T	T	F	F	F
F	T	T	F	F	F	T	F	F
F	T	F	F	T	T	F	F	F
F	F	T	T	F	T	F	F	F
F	F	F	T	T	T	F	F	F

Section 12.3

1.

p	q	$\sim p$	$(\sim p) \vee q$	$p \rightarrow q$
T	T	F	T	T
T	F	F	F	F
F	T	T	T	T
F	F	T	T	T

Compare the two rightmost columns.

3.

p	q	$p \wedge q$	$\sim(p \wedge q)$	$\sim p$	$\sim q$	$(\sim p) \vee (\sim q)$
T	T	T	F	F	F	F
T	F	F	T	F	T	T
F	T	F	T	T	F	T
F	F	F	T	T	T	T

Compare the fourth and seventh columns.

5.

p	q	$\sim q$	$p \vee (\sim q)$	$[p \vee (\sim q)] \wedge q$	$p \wedge q$
T	T	F	T	T	T
T	F	T	T	F	F
F	T	F	F	F	F
F	F	T	T	F	F

Compare the two rightmost columns.

7. Valid

9. Valid

11. Not valid. Consider p true, q false, and r false.

13. Let

p = Sam studies in the mathematics course.
q = Sam studies in the physics course.
r = Sam receives at least a B in mathematics.
s = Sam receives at least a C in physics.

The argument is $[(p \to r) \wedge (q \to s) \wedge (\sim r) \wedge s] \to [(\sim q) \wedge (\sim p)]$. The argument is not valid. Consider p false, q true, r false, and s true. Then $(p \to r)$ is true, $(q \to s)$ is true, $\sim r$ is true, and s is true. Thus $(p \to r) \wedge (q \to s) \wedge (\sim r) \wedge s$ is true. But $(\sim q) \wedge (\sim p)$ is false, which is inconsistent with the definition of implication.

15. With p, q, r, and s as in Exercise 13, the argument is $[(p \to r) \wedge (q \to s) \wedge r \wedge s] \to (p \wedge q)$. The argument is not valid. Consider p false, q false, r true, and s true.

Review Exercises

1. (*a*) $p \wedge q \wedge (\sim r)$ (*b*) $p \vee r$ (*c*) $(\sim p) \wedge (q \vee r)$

3. (*a*) $(\sim p) \wedge (\sim q) \wedge r$ (*b*) $p \vee (q \wedge r)$ (*c*) $(\sim p) \vee (\sim q) \vee (\sim r)$

5. For part (*a*): p true, q true, r false
For part (*b*): p true, r either true or false; r true, p either true or false
For part (*c*): p false, q true, r true; p false, q true, r false; p false, q false, r true
No truth values of p, q, and r make part (*a*) true and part (*b*) false.

7. For part (*a*): p true, q false, r true
for part (*b*): p true, q and r either true or false;
p false, q true, r true
for part (*c*): at least one of p, q, r must be false
All truth values which make part (*b*) true make part (*a*) false

9. (*a*) p true, q and r either true or false; q false, r false, p either true or false
   (*b*) Same as (*a*)

11. (*a*) $(\sim p) \wedge (\sim q) \wedge (\sim r)$ (*b*) $(\sim p) \wedge ((q \wedge (\sim r) \vee [(\sim q) \wedge r])$

13. (*a*)

p	q	r	$(\sim p) \wedge (\sim q) \wedge (\sim r)$
T	T	T	F
T	T	F	F
T	F	T	F
T	F	F	F
F	T	T	F
F	T	F	F
F	F	T	F
F	F	F	T

(*b*)

p	q	r	$(\sim p) \wedge ((q \wedge (\sim r)) \vee ((\sim q) \wedge r))$
T	T	T	F
T	T	F	F
T	F	T	F
T	F	F	F
F	T	T	F
F	T	F	T
F	F	T	T
F	F	F	F

15. Compare truth tables

(*a*)

p	q	$\sim(p \wedge q)$	$(\sim p) \vee (\sim q)$
T	T	F	F
T	F	T	T
F	T	T	T
F	F	T	T

(*b*)

p	q	r	$p \wedge (q \vee r)$	$(p \wedge q) \vee (p \wedge r)$
T	T	T	T	T
T	T	F	T	T
T	F	T	T	T
T	F	F	F	F
F	T	T	F	F
F	T	F	F	F
F	F	T	F	F
F	F	F	F	F

17. (*a*) is equivalent to (*d*), (*b*) is equivalent to (*c*)

19. (*a*)

p	q	$(p \vee q) \rightarrow p$
T	T	T
T	F	T
F	T	F
F	F	T

(*b*)

p	q	$(p \wedge q) \rightarrow p$
T	T	T
T	F	T
F	T	T
F	F	T

(*c*)

p	q	$[\sim(p \vee q)] \rightarrow [(\sim p) \vee (\sim q)]$
T	T	T
T	F	T
F	T	T
F	F	T

21. (*a*) If she is an accountant and she does not like to work with numbers, then she is unemployed.
(*b*) If she is an accountant, then she likes to work with numbers.
(*c*) If she is an accountant or she is unemployed, then she does not like to work with numbers.

23. (*a*) Not a valid argument

p	q	r	$p \rightarrow q$	$q \rightarrow r$	$(p \rightarrow r) \wedge (q \rightarrow r)$	$r \rightarrow q$	$[(p \rightarrow q) \wedge (q \rightarrow r)] \rightarrow (r \rightarrow q)$
T	T	T	T	T	T	T	T
T	T	F	T	F	F	T	T
T	F	T	F	T	F	F	T
T	F	F	F	T	F	T	T
F	T	T	T	T	T	T	T
F	T	F	T	F	F	T	T
F	F	T	T	T	T	F	F
F	F	F	T	T	T	T	T

(*b*) A valid argument

p	q	$\sim p$	$\sim q$	$p \to q$	$(\sim q) \to (\sim p)$	$(p \to q) \to [(\sim q \to (\sim p)]$
T	T	F	F	T	T	T
T	F	F	T	F	F	T
F	T	T	F	T	T	T
F	F	T	T	T	T	T

(*c*) Not a valid argument

p	q	$\sim p$	$p \to q$	$(p \to q) \wedge (\sim p)$	$\sim q$	$[(p \to q) \wedge (\sim p)] \to (\sim q)$
T	T	F	T	F	F	T
T	F	F	F	F	T	T
F	T	T	T	T	F	F
F	F	T	T	T	T	T

Index